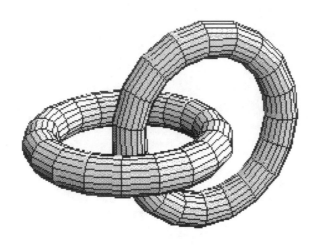

Introduction to
Tensor Calculus
and
Continuum Mechanics

by J.H. Heinbockel

Emeritus Professor of Mathematics

Old Dominion University

 www.trafford.com

North America & international
toll-free: 844-688-6899 (USA & Canada)
fax: 812 355 4082

PREFACE

This is an introductory text which presents fundamental concepts from the subject areas of tensor calculus, differential geometry and continuum mechanics. The material presented is suitable for a two semester course in applied mathematics and is flexible enough to be presented to either upper level undergraduate or beginning graduate students majoring in applied mathematics, engineering or physics. The presentation assumes the students have some knowledge from the areas of matrix theory, linear algebra and advanced calculus. Each section includes many illustrative worked examples. At the end of each section there is a large collection of exercises which range in difficulty. Many new ideas are presented in the exercises and so the students should be encouraged to read all the exercises.

The purpose of preparing these notes is to condense into an introductory text the basic definitions and techniques arising in tensor calculus, differential geometry and continuum mechanics. In particular, the material is presented to (i) develop a physical understanding of the mathematical concepts associated with tensor calculus and (ii) develop the basic equations of tensor calculus, differential geometry and continuum mechanics which arise in engineering applications. From these basic equations one can go on to develop more sophisticated models of applied mathematics. The material is presented in an informal manner and uses mathematics which minimizes excessive formalism.

The material has been divided into two parts. The first part deals with an introduction to tensor calculus and differential geometry which covers such things as the indicial notation, tensor algebra, covariant differentiation, dual tensors, bilinear and multilinear forms, special tensors, the Riemann Christoffel tensor, space curves, surface curves, curvature, fundamental quadratic forms and concludes with an introduction to relativity. The second part is an introduction to basic concepts from continuum mechanics which emphasizes the application of tensor algebra and calculus to a wide variety of applied areas from engineering and physics. The selected applications are from the areas of dynamics, elasticity, fluids and electromagnetic theory. The continuum mechanics portion focuses on an introduction of the basic concepts from linear elasticity and fluids. The second part concludes with a brief introduction to quaternions and Clifford algebra. The Appendix A contains units of measurements from the Système International d'Unitès along with some selected physical constants. The Appendix B contains a listing of Christoffel symbols of the second kind associated with various coordinate systems. The Appendix C is a summary of useful vector identities. The Appendix D contains solutions to selected exercises.

J.H. Heinbockel, 2001

INTRODUCTION TO TENSOR CALCULUS AND CONTINUUM MECHANICS

PART 1: INTRODUCTION TO TENSOR CALCULUS

A scalar field describes a one-to-one correspondence between a single scalar number and a point. This idea is extended to develop n-dimensional vector fields which are described by a one-to-one correspondence between n-numbers and a point. Let us generalize these concepts by assigning n-squared numbers to a single point or n-cubed numbers to a single point. When these numbers obey certain transformation laws they become examples of tensor fields. In general, scalar fields are referred to as tensor fields of rank or order zero whereas vector fields are called tensor fields of rank or order one.

Closely associated with tensor calculus is the indicial or index notation. In section 1 the indicial notation is defined and illustrated. We also define and investigate scalar, vector and tensor fields when they are subjected to various coordinate transformations. It turns out that tensors have certain properties which are independent of the coordinate system used to describe the tensor. Because of these useful properties, we can use tensors to represent various fundamental laws occurring in physics, engineering, science and mathematics. These representations are extremely useful as they are independent of the coordinate systems considered.

§1.1 INDEX NOTATION

Two vectors $\vec{A}$ and $\vec{B}$ can be expressed in the component form

$$\vec{A} = A_1\,\widehat{\mathbf{e}}_1 + A_2\,\widehat{\mathbf{e}}_2 + A_3\,\widehat{\mathbf{e}}_3 \qquad \text{and} \qquad \vec{B} = B_1\,\widehat{\mathbf{e}}_1 + B_2\,\widehat{\mathbf{e}}_2 + B_3\,\widehat{\mathbf{e}}_3,$$

where $\widehat{\mathbf{e}}_1$, $\widehat{\mathbf{e}}_2$ and $\widehat{\mathbf{e}}_3$ are orthogonal unit basis vectors. Often when no confusion arises, the vectors $\vec{A}$ and $\vec{B}$ are expressed for brevity sake as number triples. For example, we can write

$$\vec{A} = (A_1,\ A_2,\ A_3) \qquad \text{and} \qquad \vec{B} = (B_1,\ B_2,\ B_3)$$

where it is understood that only the components of the vectors $\vec{A}$ and $\vec{B}$ are given. The unit vectors would be represented

$$\widehat{\mathbf{e}}_1 = (1,0,0), \qquad \widehat{\mathbf{e}}_2 = (0,1,0), \qquad \widehat{\mathbf{e}}_3 = (0,0,1).$$

A still shorter notation, depicting the vectors $\vec{A}$ and $\vec{B}$ is the index or indicial notation. In the index notation, the quantities

$$A_i, \quad i = 1,2,3 \qquad \text{and} \qquad B_p, \quad p = 1,2,3$$

represent the components of the vectors $\vec{A}$ and $\vec{B}$. This notation focuses attention only on the components of the vectors and employs a dummy subscript whose range over the integers is specified. The symbol A_i refers to all of the components of the vector $\vec{A}$ simultaneously. The dummy subscript i can have any of the integer values $1, 2$ or 3. For $i = 1$ we focus attention on the A_1 component of the vector $\vec{A}$. Setting $i = 2$ focuses attention on the second component A_2 of the vector $\vec{A}$ and similarly when $i = 3$ we can focus attention on the third component of $\vec{A}$. The subscript i is a dummy subscript and may be replaced by another letter, say p, so long as one specifies the integer values that this dummy subscript can have.

It is also convenient at this time to mention that higher dimensional vectors may be defined as ordered n−tuples. For example, the vector

$$\vec{X} = (X_1, X_2, \ldots, X_N)$$

with components X_i, $i = 1, 2, \ldots, N$ is called a N−dimensional vector. Another notation used to represent this vector is

$$\vec{X} = X_1 \,\widehat{\mathbf{e}}_1 + X_2 \,\widehat{\mathbf{e}}_2 + \cdots + X_N \,\widehat{\mathbf{e}}_N$$

where

$$\widehat{\mathbf{e}}_1, \ \widehat{\mathbf{e}}_2, \ldots, \widehat{\mathbf{e}}_N$$

are linearly independent unit base vectors. Note that many of the operations that occur in the use of the index notation apply not only for three dimensional vectors, but also for N−dimensional vectors.

In future sections it is necessary to define quantities which can be represented by a letter with subscripts or superscripts attached. Such quantities are referred to as systems. When these quantities obey certain transformation laws they are referred to as tensor systems. For example, quantities like

$$A_{ij}^k \qquad e^{ijk} \qquad \delta_{ij} \qquad \delta_i^j \qquad A^i \qquad B_j \qquad a_{ij}.$$

The subscripts or superscripts are referred to as indices or suffixes. When such quantities arise, the indices must conform to the following rules:

1. They are lower case Latin or Greek letters.
2. The letters at the end of the alphabet (u, v, w, x, y, z) are never employed as indices.

The number of subscripts and superscripts determines the order of the system. A system with one index is a first order system. A system with two indices is called a second order system. In general, a system with N indices is called a Nth order system. A system with no indices is called a scalar or zeroth order system.

The type of system depends upon the number of subscripts or superscripts occurring in an expression. For example, A_{jk}^i and B_{st}^m, (all indices range 1 to N), are of the same type because they have the same number of subscripts and superscripts. In contrast, the systems A_{jk}^i and C_p^{mn} are not of the same type because one system has two superscripts and the other system has only one superscript. For certain systems the number of subscripts and superscripts is important. In other systems it is not of importance. The meaning and importance attached to sub- and superscripts will be addressed later in this section.

In the use of superscripts one must not confuse "powers "of a quantity with the superscripts. For example, if we replace the independent variables (x, y, z) by the symbols $(x^1, \ x^2, \ x^3)$, then we are letting $y = x^2$ where x^2 is a variable and not x raised to a power. Similarly, the substitution $z = x^3$ is the replacement of z by the variable x^3 and this should not be confused with x raised to a power. In order to write a superscript quantity to a power, use parentheses. For example, $(x^2)^3$ is the variable x^2 cubed. One of the reasons for introducing the superscript variables is that many equations of mathematics and physics can be made to take on a concise and compact form.

There is a range convention associated with the indices. This convention states that whenever there is an expression where the indices occur unrepeated it is to be understood that each of the subscripts or superscripts can take on any of the integer values $1, 2, \ldots, N$ where N is a specified integer. For example, the Kronecker delta symbol δ_{ij}, defined by $\delta_{ij} = 1$ if $i = j$ and $\delta_{ij} = 0$ for $i \neq j$, with i, j ranging over the values 1,2,3, represents the 9 quantities

$$\delta_{11} = 1 \qquad \delta_{12} = 0 \qquad \delta_{13} = 0$$
$$\delta_{21} = 0 \qquad \delta_{22} = 1 \qquad \delta_{23} = 0$$
$$\delta_{31} = 0 \qquad \delta_{32} = 0 \qquad \delta_{33} = 1.$$

The symbol δ_{ij} refers to all of the components of the system simultaneously. As another example, consider the equation

$$\widehat{\mathbf{e}}_m \cdot \widehat{\mathbf{e}}_n = \delta_{mn} \qquad m, n = 1, 2, 3 \tag{1.1.1}$$

the subscripts m, n occur unrepeated on the left side of the equation and hence must also occur on the right hand side of the equation. These indices are called "free "indices and can take on any of the values 1, 2 or 3 as specified by the range. Since there are three choices for the value for m and three choices for a value of n we find that equation (1.1.1) represents nine equations simultaneously. These nine equations are

$$\widehat{\mathbf{e}}_1 \cdot \widehat{\mathbf{e}}_1 = 1 \qquad \widehat{\mathbf{e}}_1 \cdot \widehat{\mathbf{e}}_2 = 0 \qquad \widehat{\mathbf{e}}_1 \cdot \widehat{\mathbf{e}}_3 = 0$$
$$\widehat{\mathbf{e}}_2 \cdot \widehat{\mathbf{e}}_1 = 0 \qquad \widehat{\mathbf{e}}_2 \cdot \widehat{\mathbf{e}}_2 = 1 \qquad \widehat{\mathbf{e}}_2 \cdot \widehat{\mathbf{e}}_3 = 0$$
$$\widehat{\mathbf{e}}_3 \cdot \widehat{\mathbf{e}}_1 = 0 \qquad \widehat{\mathbf{e}}_3 \cdot \widehat{\mathbf{e}}_2 = 0 \qquad \widehat{\mathbf{e}}_3 \cdot \widehat{\mathbf{e}}_3 = 1.$$

Symmetric and Skew-Symmetric Systems

A system defined by subscripts and superscripts ranging over a set of values is said to be symmetric in two of its indices if the components are unchanged when the indices are interchanged. For example, the third order system T_{ijk} is symmetric in the indices i and k if

$$T_{ijk} = T_{kji} \quad \text{for all values of } i, j \text{ and } k.$$

A system defined by subscripts and superscripts is said to be skew-symmetric in two of its indices if the components change sign when the indices are interchanged. For example, the fourth order system T_{ijkl} is skew-symmetric in the indices i and l if

$$T_{ijkl} = -T_{ljki} \quad \text{for all values of } i, j, k \text{ and } l.$$

As another example, consider the third order system a_{prs}, $p, r, s = 1, 2, 3$ which is completely skew-symmetric in all of its indices. We would then have

$$a_{prs} = -a_{psr} = a_{spr} = -a_{srp} = a_{rsp} = -a_{rps}.$$

It is left as an exercise to show this completely skew- symmetric systems has 27 elements, 21 of which are zero. The 6 nonzero elements are all related to one another thru the above equations when $(p, r, s) = (1, 2, 3)$. This is expressed as saying that the above system has only one independent component.

Summation Convention

The summation convention states that whenever there arises an expression where there is an index which occurs twice on the same side of any equation, or term within an equation, it is understood to represent a summation on these repeated indices. The summation being over the integer values specified by the range. A repeated index is called a summation index, while an unrepeated index is called a free index. The summation convention requires that one must never allow a summation index to appear more than twice in any given expression. Because of this rule it is sometimes necessary to replace one dummy summation symbol by some other dummy symbol in order to avoid having three or more indices occurring on the same side of the equation. The index notation is a very powerful notation and can be used to concisely represent many complex equations. For the remainder of this section there is presented additional definitions and examples to illustrated the power of the indicial notation. This notation is then employed to define tensor components and associated operations with tensors.

EXAMPLE 1.1-1 The two equations

$$y_1 = a_{11}x_1 + a_{12}x_2$$

$$y_2 = a_{21}x_1 + a_{22}x_2$$

can be represented as one equation by introducing a dummy index, say k, and expressing the above equations as

$$y_k = a_{k1}x_1 + a_{k2}x_2, \qquad k = 1, 2.$$

The range convention states that k is free to have any one of the values 1 or 2, (k is a free index). This equation can now be written in the form

$$y_k = \sum_{i=1}^{2} a_{ki}x_i = a_{k1}x_1 + a_{k2}x_2$$

where i is the dummy summation index. When the summation sign is removed and the summation convention is adopted we have

$$y_k = a_{ki}x_i \qquad i, k = 1, 2.$$

Since the subscript i repeats itself, the summation convention requires that a summation be performed by letting the summation subscript take on the values specified by the range and then summing the results. The index k which appears only once on the left and only once on the right hand side of the equation is called a free index. It should be noted that both k and i are dummy subscripts and can be replaced by other letters. For example, we can write

$$y_n = a_{nm}x_m \qquad n, m = 1, 2$$

where m is the summation index and n is the free index. Summing on m produces

$$y_n = a_{n1}x_1 + a_{n2}x_2$$

and letting the free index n take on the values of 1 and 2 we produce the original two equations.

■

EXAMPLE 1.1-2 To show that the product $a_{ik}a_{jk} = b_{ij}, i, j, k = 1, 2, 3$ is a representation of the matrix product $AA^T = B$ we sum on k to obtain

$$a_{i1}a_{j1} + a_{i2}a_{j2} + a_{i3}a_{j3} = b_{ij}.$$

Now i, j are free indices which can be assigned any of the values 1, 2 or 3. We have

$$\begin{array}{ll} \text{for } i = 1, j = 1 & a_{11}a_{11} + a_{12}a_{12} + a_{13}a_{13} = b_{11} \\ \text{for } i = 1, j = 2 & a_{11}a_{21} + a_{12}a_{22} + a_{13}a_{23} = b_{12} \\ \text{for } i = 1, j = 3 & a_{11}a_{31} + a_{12}a_{32} + a_{13}a_{33} = b_{13} \\ \text{for } i = 2, j = 1 & a_{21}a_{11} + a_{22}a_{12} + a_{23}a_{13} = b_{21} \\ \text{for } i = 2, j = 2 & a_{21}a_{21} + a_{22}a_{22} + a_{23}a_{23} = b_{22} \\ \text{for } i = 2, j = 3 & a_{21}a_{31} + a_{22}a_{32} + a_{23}a_{33} = b_{23} \\ \text{for } i = 3, j = 1 & a_{31}a_{11} + a_{32}a_{12} + a_{33}a_{13} = b_{31} \\ \text{for } i = 3, j = 2 & a_{31}a_{21} + a_{32}a_{22} + a_{33}a_{23} = b_{32} \\ \text{for } i = 3, j = 3 & a_{31}a_{31} + a_{32}a_{32} + a_{33}a_{33} = b_{33} \end{array}$$

It is readily verified that the above index notation is a shorthand for the matrix notation

$$AA^T = \begin{pmatrix} a_{11} & a_{12} & a_{13} \\ a_{21} & a_{22} & a_{23} \\ a_{31} & a_{32} & a_{33} \end{pmatrix} \begin{pmatrix} a_{11} & a_{21} & a_{31} \\ a_{12} & a_{22} & a_{32} \\ a_{13} & a_{23} & a_{33} \end{pmatrix} = \begin{pmatrix} b_{11} & b_{12} & b_{13} \\ b_{21} & b_{22} & b_{23} \\ b_{31} & b_{32} & b_{33} \end{pmatrix} = B$$

∎

EXAMPLE 1.1-3. For $y_i = a_{ij}x_j$, $i, j = 1, 2, 3$ and $x_i = b_{ij}z_j$, $i, j = 1, 2, 3$ solve for the y variables in terms of the z variables.

Solution: In matrix form the given equations can be expressed:

$$\begin{pmatrix} y_1 \\ y_2 \\ y_3 \end{pmatrix} = \begin{pmatrix} a_{11} & a_{12} & a_{13} \\ a_{21} & a_{22} & a_{23} \\ a_{31} & a_{32} & a_{33} \end{pmatrix} \begin{pmatrix} x_1 \\ x_2 \\ x_3 \end{pmatrix} \quad \text{and} \quad \begin{pmatrix} x_1 \\ x_2 \\ x_3 \end{pmatrix} = \begin{pmatrix} b_{11} & b_{12} & b_{13} \\ b_{21} & b_{22} & b_{23} \\ b_{31} & b_{32} & b_{33} \end{pmatrix} \begin{pmatrix} z_1 \\ z_2 \\ z_3 \end{pmatrix}.$$

Now solve for the y variables in terms of the z variables and obtain

$$\begin{pmatrix} y_1 \\ y_2 \\ y_3 \end{pmatrix} = \begin{pmatrix} a_{11} & a_{12} & a_{13} \\ a_{21} & a_{22} & a_{23} \\ a_{31} & a_{32} & a_{33} \end{pmatrix} \begin{pmatrix} b_{11} & b_{12} & b_{13} \\ b_{21} & b_{22} & b_{23} \\ b_{31} & b_{32} & b_{33} \end{pmatrix} \begin{pmatrix} z_1 \\ z_2 \\ z_3 \end{pmatrix}.$$

The index notation employs indices that are dummy indices and so we can write

$$y_n = a_{nm}x_m, \quad n, m = 1, 2, 3 \quad \text{and} \quad x_m = b_{mj}z_j, \quad m, j = 1, 2, 3.$$

Here we have purposely changed the indices so that when we substitute for x_m, from one equation into the other, a summation index does not repeat itself more than twice. Substituting we find the indicial form of the above matrix equation as

$$y_n = a_{nm}b_{mj}z_j, \quad m, n, j = 1, 2, 3$$

where n is the free index and m, j are the dummy summation indices. It is left as an exercise to expand both the matrix equation and the indicial equation and verify that they are different ways of representing the same thing.

∎

6

EXAMPLE 1.1-4. The dot product of two vectors A_q, $q = 1, 2, 3$ and B_j, $j = 1, 2, 3$ can be represented with the index notation by the product $A_i B_i = AB \cos \theta$ $i = 1, 2, 3$, $A = |\vec{A}|$, $B = |\vec{B}|$. Since the subscript i is repeated it is understood to represent a summation index. Summing on i over the range specified, there results

$$A_1 B_1 + A_2 B_2 + A_3 B_3 = AB \cos \theta.$$

Observe that the index notation employs dummy indices. At times these indices are altered in order to conform to the above summation rules, without attention being brought to the change. As in this example, the indices q and j are dummy indices and can be changed to other letters if one desires. Also, in the future, if the range of the indices is not stated it is assumed that the range is over the integer values $1, 2$ and 3.

∎

To systems containing subscripts and superscripts one can apply certain algebraic operations. We present in an informal way the operations of addition, multiplication and contraction.

Addition, Multiplication and Contraction

The algebraic operation of addition or subtraction applies to systems of the same type and order. That is we can add or subtract like components in systems. For example, the sum of A^i_{jk} and B^i_{jk} is again a system of the same type and is denoted by $C^i_{jk} = A^i_{jk} + B^i_{jk}$, where like components are added.

The product of two systems is obtained by multiplying each component of the first system with each component of the second system. Such a product is called an outer product. The order of the resulting product system is the sum of the orders of the two systems involved in forming the product. For example, if A^i_j is a second order system and B^{mnl} is a third order system, with all indices having the range 1 to N, then the product system is fifth order and is denoted $C^{imnl}_j = A^i_j B^{mnl}$. The product system represents N^5 terms constructed from all possible products of the components from A^i_j with the components from B^{mnl}.

The operation of contraction occurs when a lower index is set equal to an upper index and the summation convention is invoked. For example, if we have a fifth order system C^{imnl}_j and we set $i = j$ and sum, then we form the system

$$C^{mnl} = C^{jmnl}_j = C^{1mnl}_1 + C^{2mnl}_2 + \cdots + C^{Nmnl}_N.$$

Here the symbol C^{mnl} is used to represent the third order system that results when the contraction is performed. Whenever a contraction is performed, the resulting system is always of order 2 less than the original system. Under certain special conditions it is permissible to perform a contraction on two lower case indices. These special conditions will be considered later in the section.

The above operations will be more formally defined after we have explained what tensors are.

The e-permutation symbol and Kronecker delta

Two symbols that are used quite frequently with the indicial notation are the e-permutation symbol and the Kronecker delta. The e-permutation symbol is sometimes referred to as the alternating tensor. The e-permutation symbol, as the name suggests, deals with permutations. A permutation is an arrangement of things. When the order of the arrangement is changed, a new permutation results. A transposition is an interchange of two consecutive terms in an arrangement. As an example, let us change the digits 1 2 3 to 3 2 1 by making a sequence of transpositions. Starting with the digits in the order 1 2 3 we interchange

2 and 3 (first transposition) to obtain 1 3 2. Next, interchange the digits 1 and 3 (second transposition) to obtain 3 1 2. Finally, interchange the digits 1 and 2 (third transposition) to achieve 3 2 1. Here the total number of transpositions of 1 2 3 to 3 2 1 is three, an odd number. Other transpositions of 1 2 3 to 3 2 1 can also be written. However, these are also an odd number of transpositions.

EXAMPLE 1.1-5.   The total number of possible ways of arranging the digits 1 2 3 is six. We have three choices for the first digit. Having chosen the first digit, there are only two choices left for the second digit. Hence the remaining number is for the last digit. The product $(3)(2)(1) = 3! = 6$ is the number of permutations of the digits $1, 2$ and 3. These six permutations are

$$
\begin{aligned}
&1\,2\,3 \quad \text{even permutation}\\
&1\,3\,2 \quad \text{odd permutation}\\
&3\,1\,2 \quad \text{even permutation}\\
&3\,2\,1 \quad \text{odd permutation}\\
&2\,3\,1 \quad \text{even permutation}\\
&2\,1\,3 \quad \text{odd permutation.}
\end{aligned}
$$

Here a permutation of 1 2 3 is called even or odd depending upon whether there is an even or odd number of transpositions of the digits. A mnemonic device to remember the even and odd permutations of 123 is illustrated in the figure 1.1-1. Note that even permutations of 123 are obtained by selecting any three consecutive numbers from the sequence 123123 and the odd permutations result by selecting any three consecutive numbers from the sequence 321321.

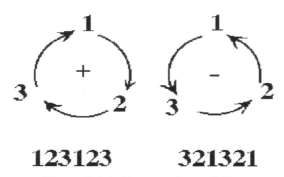

123123   **321321**
Figure 1.1-1. Permutations of 123.

■

In general, the number of permutations of n things taken m at a time is given by the relation

$$P(n,m) = n(n-1)(n-2)\cdots(n-m+1).$$

By selecting a subset of m objects from a collection of n objects, $m \leq n$, without regard to the ordering is called a combination of n objects taken m at a time. For example, combinations of 3 numbers taken from the set $\{1, 2, 3, 4\}$ are $(123), (124), (134), (234)$. Note that ordering of a combination is not considered. That is, the permutations $(123), (132), (231), (213), (312), (321)$ are considered equal. In general, the number of combinations of n objects taken m at a time is given by $C(n,m) = \binom{n}{m} = \dfrac{n!}{m!(n-m)!}$ where $\binom{n}{m}$ are the binomial coefficients which occur in the expansion

$$(a+b)^n = \sum_{m=0}^{n} \binom{n}{m} a^{n-m} b^m.$$

The definition of permutations can be used to define the e-permutation symbol.

Definition: (e-Permutation symbol or alternating tensor)

The e-permutation symbol is defined

$$e^{ijk\ldots l} = e_{ijk\ldots l} = \begin{cases} 1 & \text{if } ijk\ldots l \text{ is an even permutation of the integers } 123\ldots n \\ -1 & \text{if } ijk\ldots l \text{ is an odd permutation of the integers } 123\ldots n \\ 0 & \text{in all other cases} \end{cases}$$

EXAMPLE 1.1-6. Find e_{612453}.

Solution: To determine whether 612453 is an even or odd permutation of 123456 we write down the given numbers and below them we write the integers 1 through 6. Like numbers are then connected by a line and we obtain figure 1.1-2.

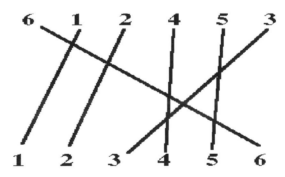

Figure 1.1-2. Permutations of 123456.

In figure 1.1-2, there are seven intersections of the lines connecting like numbers. The number of intersections is an odd number and shows that an odd number of transpositions must be performed. These results imply $e_{612453} = -1$.

■

Another definition used quite frequently in the representation of mathematical and engineering quantities is the Kronecker delta which we now define in terms of both subscripts and superscripts.

Definition: **(Kronecker delta)** The Kronecker delta is defined:

$$\delta_{ij} = \delta_i^j = \begin{cases} 1 & \text{if } i \text{ equals } j \\ 0 & \text{if } i \text{ is different from } j \end{cases}$$

EXAMPLE 1.1-7. Some examples of the $e-$permutation symbol and Kronecker delta are:

$$e_{123} = e^{123} = +1 \qquad \delta_1^1 = 1 \qquad \delta_{12} = 0$$

$$e_{213} = e^{213} = -1 \qquad \delta_2^1 = 0 \qquad \delta_{22} = 1$$

$$e_{112} = e^{112} = 0 \qquad \delta_3^1 = 0 \qquad \delta_{32} = 0.$$

■

EXAMPLE 1.1-8. When an index of the Kronecker delta δ_{ij} is involved in the summation convention, the effect is that of replacing one index with a different index. For example, let a_{ij} denote the elements of an $N \times N$ matrix. Here i and j are allowed to range over the integer values $1, 2, \ldots, N$. Consider the product

$$a_{ij}\delta_{ik}$$

where the range of i, j, k is $1, 2, \ldots, N$. The index i is repeated and therefore it is understood to represent a summation over the range. The index i is called a summation index. The other indices j and k are free indices. They are free to be assigned any values from the range of the indices. They are not involved in any summations and their values, whatever you choose to assign them, are fixed. Let us assign a value of j and k to the values of j and k. The underscore is to remind you that these values for j and k are fixed and not to be summed. When we perform the summation over the summation index i we assign values to i from the range and then sum over these values. Performing the indicated summation we obtain

$$a_{ij}\delta_{ik} = a_{1j}\delta_{1k} + a_{2j}\delta_{2k} + \cdots + a_{kj}\delta_{kk} + \cdots + a_{Nj}\delta_{Nk}.$$

In this summation the Kronecker delta is zero everywhere the subscripts are different and equals one where the subscripts are the same. There is only one term in this summation which is nonzero. It is that term where the summation index i was equal to the fixed value k This gives the result

$$a_{kj}\delta_{kk} = a_{kj}$$

where the underscore is to remind you that the quantities have fixed values and are not to be summed. Dropping the underscores we write

$$a_{ij}\delta_{ik} = a_{kj}$$

Here we have substituted the index i by k and so when the Kronecker delta is used in a summation process it is known as a substitution operator. This substitution property of the Kronecker delta can be used to simplify a variety of expressions involving the index notation. Some examples are:

$$B_{ij}\delta_{js} = B_{is}$$

$$\delta_{jk}\delta_{km} = \delta_{jm}$$

$$e_{ijk}\delta_{im}\delta_{jn}\delta_{kp} = e_{mnp}.$$

Some texts adopt the notation that if indices are capital letters, then no summation is to be performed. For example,

$$a_{KJ}\delta_{KK} = a_{KJ}$$

as δ_{KK} represents a single term because of the capital letters. Another notation which is used to denote no summation of the indices is to put parenthesis about the indices which are not to be summed. For example,

$$a_{(k)j}\delta_{(k)(k)} = a_{kj},$$

since $\delta_{(k)(k)}$ represents a single term and the parentheses indicate that no summation is to be performed. At any time we may employ either the underscore notation, the capital letter notation or the parenthesis notation to denote that no summation of the indices is to be performed. To avoid confusion altogether, one can write out parenthetical expressions such as "(no summation on k)".

∎

EXAMPLE 1.1-9. In the Kronecker delta symbol δ_j^i we set j equal to i and perform a summation. This operation is called a contraction. There results δ_i^i, which is to be summed over the range of the index i. Utilizing the range $1, 2, \ldots, N$ we have

$$\delta_i^i = \delta_1^1 + \delta_2^2 + \cdots + \delta_N^N$$

$$\delta_i^i = 1 + 1 + \cdots + 1$$

$$\delta_i^i = N.$$

In three dimension we have δ_j^i, $i, j = 1, 2, 3$ and

$$\delta_k^k = \delta_1^1 + \delta_2^2 + \delta_3^3 = 3.$$

In certain circumstances the Kronecker delta can be written with only subscripts. For example, δ_{ij}, $i, j = 1, 2, 3$. We shall find that these circumstances allow us to perform a contraction on the lower indices so that $\delta_{ii} = 3$.

∎

EXAMPLE 1.1-10. The determinant of a matrix $A = (a_{ij})$ can be represented in the indicial notation. Employing the e-permutation symbol the determinant of an $N \times N$ matrix is expressed

$$|A| = e_{ij\ldots k} a_{1i} a_{2j} \cdots a_{Nk}$$

where $e_{ij\ldots k}$ is an Nth order system. In the special case of a 2×2 matrix we write

$$|A| = e_{ij} a_{1i} a_{2j}$$

where the summation is over the range 1,2 and the e-permutation symbol is of order 2. In the special case of a 3×3 matrix we have

$$|A| = \begin{vmatrix} a_{11} & a_{12} & a_{13} \\ a_{21} & a_{22} & a_{23} \\ a_{31} & a_{32} & a_{33} \end{vmatrix} = e_{ijk} a_{i1} a_{j2} a_{k3} = e_{ijk} a_{1i} a_{2j} a_{3k}$$

where i, j, k are the summation indices and the summation is over the range 1,2,3. Here e_{ijk} denotes the e-permutation symbol of order 3. Note that by interchanging the rows of the 3×3 matrix we can obtain more general results. Consider (p, q, r) as some permutation of the integers $(1, 2, 3)$, and observe that the determinant can be expressed

$$\Delta = \begin{vmatrix} a_{p1} & a_{p2} & a_{p3} \\ a_{q1} & a_{q2} & a_{q3} \\ a_{r1} & a_{r2} & a_{r3} \end{vmatrix} = e_{ijk} a_{pi} a_{qj} a_{rk}.$$

If (p, q, r) is an even permutation of $(1, 2, 3)$ then $\Delta = |A|$

If (p, q, r) is an odd permutation of $(1, 2, 3)$ then $\Delta = -|A|$

If (p, q, r) is not a permutation of $(1, 2, 3)$ then $\Delta = 0$.

We can then write

$$e_{ijk} a_{pi} a_{qj} a_{rk} = e_{pqr} |A|.$$

Each of the above results can be verified by performing the indicated summations. A more formal proof of the above result is given in EXAMPLE 1.1-26, later in this section.

■

EXAMPLE 1.1-11. The expression $e_{ijk} B_{ij} C_i$ is meaningless since the index i repeats itself more than twice and the summation convention does not allow this. If you really did want to sum over an index which occurs more than twice, then one must use a summation sign. For example the above expression would be written $\sum_{i=1}^{3} e_{ijk} B_{ij} C_i$.

■

EXAMPLE 1.1-12.

The cross product of the unit vectors $\widehat{\mathbf{e}}_1$, $\widehat{\mathbf{e}}_2$, $\widehat{\mathbf{e}}_3$ can be represented in the index notation by

$$\widehat{\mathbf{e}}_i \times \widehat{\mathbf{e}}_j = \begin{cases} \widehat{\mathbf{e}}_k & \text{if } (i, j, k) \text{ is an even permutation of } (1, 2, 3) \\ -\widehat{\mathbf{e}}_k & \text{if } (i, j, k) \text{ is an odd permutation of } (1, 2, 3) \\ 0 & \text{in all other cases} \end{cases}$$

This result can be written in the form $\widehat{\mathbf{e}}_i \times \widehat{\mathbf{e}}_j = e_{kij} \widehat{\mathbf{e}}_k$. This later result can be verified by summing on the index k and writing out all 9 possible combinations for i and j.

■

EXAMPLE 1.1-13. Given the vectors A_p, $p = 1, 2, 3$ and B_p, $p = 1, 2, 3$ the cross product of these two vectors is a vector C_p, $p = 1, 2, 3$ with components

$$C_i = e_{ijk}A_jB_k, \quad i, j, k = 1, 2, 3. \tag{1.1.2}$$

The quantities C_i represent the components of the cross product vector

$$\vec{C} = \vec{A} \times \vec{B} = C_1\,\widehat{\mathbf{e}}_1 + C_2\,\widehat{\mathbf{e}}_2 + C_3\,\widehat{\mathbf{e}}_3.$$

The equation (1.1.2), which defines the components of $\vec{C}$, is to be summed over each of the indices which repeats itself. We have summing on the index k

$$C_i = e_{ij1}A_jB_1 + e_{ij2}A_jB_2 + e_{ij3}A_jB_3. \tag{1.1.3}$$

We next sum on the index j which repeats itself in each term of equation (1.1.3). This gives

$$\begin{aligned}
C_i = {} & e_{i11}A_1B_1 + e_{i21}A_2B_1 + e_{i31}A_3B_1 \\
& + e_{i12}A_1B_2 + e_{i22}A_2B_2 + e_{i32}A_3B_2 \\
& + e_{i13}A_1B_3 + e_{i23}A_2B_3 + e_{i33}A_3B_3.
\end{aligned} \tag{1.1.4}$$

Now we are left with i being a free index which can have any of the values of $1, 2$ or 3. Letting $i = 1$, then letting $i = 2$, and finally letting $i = 3$ produces the cross product components

$$\begin{aligned}
C_1 &= A_2B_3 - A_3B_2 \\
C_2 &= A_3B_1 - A_1B_3 \\
C_3 &= A_1B_2 - A_2B_1.
\end{aligned}$$

The cross product can also be expressed in the form $\vec{A} \times \vec{B} = e_{ijk}A_jB_k\,\widehat{\mathbf{e}}_i$. This result can be verified by summing over the indices i, j and k.

■

EXAMPLE 1.1-14. Show $e_{ijk} = -e_{ikj} = e_{jki}$ for $i, j, k = 1, 2, 3$

Solution: The array $i\ k\ j$ represents an odd number of transpositions of the indices $i\ j\ k$ and to each transposition there is a sign change of the e-permutation symbol. Similarly, $j\ k\ i$ is an even transposition of $i\ j\ k$ and so there is no sign change of the e-permutation symbol. The above holds regardless of the numerical values assigned to the indices i, j, k.

■

13

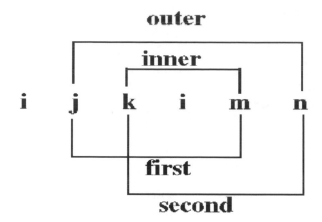

Figure 1.1-3. Mnemonic device for position of subscripts.

The e-δ Identity

An identity relating the e-permutation symbol and the Kronecker delta, which is useful in the simplification of tensor expressions, is the e-δ identity. This identity can be expressed in different forms. The subscript form for this identity is

$$e_{ijk}e_{imn} = \delta_{jm}\delta_{kn} - \delta_{jn}\delta_{km}, \qquad i,j,k,m,n = 1,2,3$$

where i is the summation index and j,k,m,n are free indices. A device used to remember the positions of the subscripts is given in the figure 1.1-3.

The subscripts on the four Kronecker delta's on the right-hand side of the e-δ identity then are read

$$(\text{first})(\text{second})\text{-}(\text{outer})(\text{inner}).$$

This refers to the positions following the summation index. Thus, j,m are the first indices after the summation index and k,n are the second indices after the summation index. The indices j,n are outer indices when compared to the inner indices k,m as the indices are viewed as written on the left-hand side of the identity.

Another form of this identity employs both subscripts and superscripts and has the form

$$e^{ijk}e_{imn} = \delta_m^j\delta_n^k - \delta_n^j\delta_m^k. \tag{1.1.5}$$

One way of proving this identity is to observe the equation (1.1.5) has the free indices j,k,m,n. Each of these indices can have any of the values of $1,2$ or 3. There are 3 choices we can assign to each of j,k,m or n and this gives a total of $3^4 = 81$ possible equations represented by the identity from equation (1.1.5). By writing out all 81 of these equations we can verify that the identity is true for all possible combinations that can be assigned to the free indices.

An alternate proof of the $e-\delta$ identity is to consider the determinant

$$\begin{vmatrix} \delta_1^1 & \delta_2^1 & \delta_3^1 \\ \delta_1^2 & \delta_2^2 & \delta_3^2 \\ \delta_1^3 & \delta_2^3 & \delta_3^3 \end{vmatrix} = \begin{vmatrix} 1 & 0 & 0 \\ 0 & 1 & 0 \\ 0 & 0 & 1 \end{vmatrix} = 1.$$

By performing a permutation of the rows of this matrix we can use the permutation symbol and write

$$\begin{vmatrix} \delta_1^i & \delta_2^i & \delta_3^i \\ \delta_1^j & \delta_2^j & \delta_3^j \\ \delta_1^k & \delta_2^k & \delta_3^k \end{vmatrix} = e^{ijk}.$$

By performing a permutation of the columns, we can write

$$\begin{vmatrix} \delta_r^i & \delta_s^i & \delta_t^i \\ \delta_r^j & \delta_s^j & \delta_t^j \\ \delta_r^k & \delta_s^k & \delta_t^k \end{vmatrix} = e^{ijk} e_{rst}.$$

Now perform a contraction on the indices i and r to obtain

$$\begin{vmatrix} \delta_i^i & \delta_s^i & \delta_t^i \\ \delta_i^j & \delta_s^j & \delta_t^j \\ \delta_i^k & \delta_s^k & \delta_t^k \end{vmatrix} = e^{ijk} e_{ist}.$$

Summing on i we have $\quad \delta_i^i = \delta_1^1 + \delta_2^2 + \delta_3^3 = 3$ and expand the determinant to obtain the desired result

$$\delta_s^j \delta_t^k - \delta_t^j \delta_s^k = e^{ijk} e_{ist}.$$

Generalized Kronecker delta

The generalized Kronecker delta is defined by the $(n \times n)$ determinant

$$\delta_{mn\ldots p}^{ij\ldots k} = \begin{vmatrix} \delta_m^i & \delta_n^i & \cdots & \delta_p^i \\ \delta_m^j & \delta_n^j & \cdots & \delta_p^j \\ \vdots & \vdots & \ddots & \vdots \\ \delta_m^k & \delta_n^k & \cdots & \delta_p^k \end{vmatrix}.$$

For example, in three dimensions we can write

$$\delta_{mnp}^{ijk} = \begin{vmatrix} \delta_m^i & \delta_n^i & \delta_p^i \\ \delta_m^j & \delta_n^j & \delta_p^j \\ \delta_m^k & \delta_n^k & \delta_p^k \end{vmatrix} = e^{ijk} e_{mnp}.$$

Performing a contraction on the indices k and p we obtain the fourth order system

$$\delta_{mn}^{rs} = \delta_{mnp}^{rsp} = e^{rsp} e_{mnp} = e^{prs} e_{pmn} = \delta_m^r \delta_n^s - \delta_n^r \delta_m^s.$$

As an exercise one can verify that the definition of the e-permutation symbol can also be defined in terms of the generalized Kronecker delta as

$$e_{j_1 j_2 j_3 \cdots j_N} = \delta_{j_1 j_2 j_3 \cdots j_N}^{1 \ 2 \ 3 \cdots N}.$$

Additional definitions and results employing the generalized Kronecker delta are found in the exercises. In section 1.3 we shall show that the Kronecker delta and epsilon permutation symbol are numerical tensors which have fixed components in every coordinate system.

Additional Applications of the Indicial Notation

The indicial notation, together with the $e - \delta$ identity, can be used to prove various vector identities.

EXAMPLE 1.1-15.   Show, using the index notation, that $\vec{A} \times \vec{B} = -\vec{B} \times \vec{A}$

Solution: Let

$$\vec{C} = \vec{A} \times \vec{B} = C_1\,\widehat{\mathbf{e}}_1 + C_2\,\widehat{\mathbf{e}}_2 + C_3\,\widehat{\mathbf{e}}_3 = C_i\,\widehat{\mathbf{e}}_i \qquad \text{and let}$$

$$\vec{D} = \vec{B} \times \vec{A} = D_1\,\widehat{\mathbf{e}}_1 + D_2\,\widehat{\mathbf{e}}_2 + D_3\,\widehat{\mathbf{e}}_3 = D_i\,\widehat{\mathbf{e}}_i.$$

We have shown that the components of the cross products can be represented in the index notation by

$$C_i = e_{ijk}A_jB_k \quad \text{and} \quad D_i = e_{ijk}B_jA_k.$$

We desire to show that $D_i = -C_i$ for all values of i. Consider the following manipulations: Let $B_j = B_s\delta_{sj}$ and $A_k = A_m\delta_{mk}$ and write

$$D_i = e_{ijk}B_jA_k = e_{ijk}B_s\delta_{sj}A_m\delta_{mk} \tag{1.1.6}$$

where all indices have the range $1,2,3$. In the expression (1.1.6) note that no summation index appears more than twice because if an index appeared more than twice the summation convention would become meaningless. By rearranging terms in equation (1.1.6) we have

$$D_i = e_{ijk}\delta_{sj}\delta_{mk}B_sA_m = e_{ism}B_sA_m.$$

In this expression the indices s and m are dummy summation indices and can be replaced by any other letters. We replace s by k and m by j to obtain

$$D_i = e_{ikj}A_jB_k = -e_{ijk}A_jB_k = -C_i.$$

Consequently, we find that $\vec{D} = -\vec{C}$ or $\vec{B} \times \vec{A} = -\vec{A} \times \vec{B}$. That is, $\vec{D} = D_i\,\widehat{\mathbf{e}}_i = -C_i\,\widehat{\mathbf{e}}_i = -\vec{C}$.

Note 1. The expressions

$$C_i = e_{ijk}A_jB_k \qquad \text{and} \qquad C_m = e_{mnp}A_nB_p$$

with all indices having the range $1,2,3$, appear to be different because different letters are used as subscripts. It must be remembered that certain indices are summed according to the summation convention and the other indices are free indices and can take on any values from the assigned range. Thus, after summation, when numerical values are substituted for the indices involved, none of the dummy letters used to represent the components appear in the answer.

Note 2. A second important point is that when one is working with expressions involving the index notation, the indices can be changed directly. For example, in the above expression for D_i we could have replaced j by k and k by j simultaneously (so that no index repeats itself more than twice) to obtain

$$D_i = e_{ijk}B_jA_k = e_{ikj}B_kA_j = -e_{ijk}A_jB_k = -C_i.$$

Note 3. Be careful in switching back and forth between the vector notation and index notation. Observe that a vector $\vec{A}$ can be represented $\vec{A} = A_i\,\widehat{\mathbf{e}}_i$ or its components can be represented $\vec{A} \cdot \widehat{\mathbf{e}}_i = A_i, \quad i = 1,2,3$. Do not set a vector equal to a scalar. That is, do not make the mistake of writing $\vec{A} = A_i$ as this is a misuse of the equal sign. It is not possible for a vector to equal a scalar because they are two entirely different quantities. A vector has both magnitude and direction while a scalar has only magnitude.

■

16

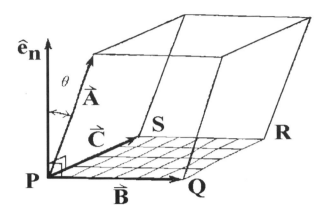

Figure 1.1-4. Triple scalar product and volume

EXAMPLE 1.1-16.   Verify the vector identity

$$\vec{A} \cdot (\vec{B} \times \vec{C}) = \vec{B} \cdot (\vec{C} \times \vec{A})$$

Solution: Let

$$\vec{B} \times \vec{C} = \vec{D} = D_i \,\hat{\mathbf{e}}_i \qquad \text{where} \qquad D_i = e_{ijk}B_jC_k \qquad \text{and let}$$

$$\vec{C} \times \vec{A} = \vec{F} = F_i \,\hat{\mathbf{e}}_i \qquad \text{where} \qquad F_i = e_{ijk}C_jA_k$$

where all indices have the range $1, 2, 3$. To prove the above identity, we have

$$\vec{A} \cdot (\vec{B} \times \vec{C}) = \vec{A} \cdot \vec{D} = A_iD_i = A_ie_{ijk}B_jC_k$$
$$= B_j(e_{ijk}A_iC_k)$$
$$= B_j(e_{jki}C_kA_i)$$

since $e_{ijk} = e_{jki}$. We also observe from the expression $F_i = e_{ijk}C_jA_k$ that we may obtain, by permuting the symbols, the equivalent expression $F_j = e_{jki}C_kA_i$. This allows us to write

$$\vec{A} \cdot (\vec{B} \times \vec{C}) = B_jF_j = \vec{B} \cdot \vec{F} = \vec{B} \cdot (\vec{C} \times \vec{A})$$

which was to be shown.

The quantity $\vec{A} \cdot (\vec{B} \times \vec{C})$ is called a triple scalar product. The above index representation of the triple scalar product implies that it can be represented as a determinant (See example 1.1-10). We can write

$$\vec{A} \cdot (\vec{B} \times \vec{C}) = \begin{vmatrix} A_1 & A_2 & A_3 \\ B_1 & B_2 & B_3 \\ C_1 & C_2 & C_3 \end{vmatrix} = e_{ijk}A_iB_jC_k$$

A physical interpretation that can be assigned to this triple scalar product is that its absolute value represents the volume of the parallelepiped formed by the three noncoplaner vectors $\vec{A}, \vec{B}, \vec{C}$. The absolute value is needed because sometimes the triple scalar product is negative. This physical interpretation can be obtained from an analysis of the figure 1.1-4.

In figure 1.1-4 observe that: (i) $|\vec{B} \times \vec{C}|$ is the area of the parallelogram $PQRS$. (ii) the unit vector

$$\hat{\mathbf{e}}_n = \frac{\vec{B} \times \vec{C}}{|\vec{B} \times \vec{C}|}$$

is normal to the plane containing the vectors $\vec{B}$ and $\vec{C}$. (iii) The dot product

$$\left| \vec{A} \cdot \hat{\mathbf{e}}_n \right| = \left| \vec{A} \cdot \frac{\vec{B} \times \vec{C}}{|\vec{B} \times \vec{C}|} \right| = h$$

equals the projection of $\vec{A}$ on $\hat{\mathbf{e}}_n$ which represents the height of the parallelepiped. These results demonstrate that

$$\left| \vec{A} \cdot (\vec{B} \times \vec{C}) \right| = |\vec{B} \times \vec{C}| \, h = (\text{area of base})(\text{height}) = \text{volume}.$$

■

EXAMPLE 1.1-17. Verify the vector identity

$$(\vec{A} \times \vec{B}) \times (\vec{C} \times \vec{D}) = \vec{C}(\vec{D} \cdot \vec{A} \times \vec{B}) - \vec{D}(\vec{C} \cdot \vec{A} \times \vec{B})$$

Solution: Let $\vec{F} = \vec{A} \times \vec{B} = F_i \hat{\mathbf{e}}_i$ and $\vec{E} = \vec{C} \times \vec{D} = E_i \hat{\mathbf{e}}_i$. These vectors have the components

$$F_i = e_{ijk} A_j B_k \qquad \text{and} \qquad E_m = e_{mnp} C_n D_p$$

where all indices have the range $1, 2, 3$. The vector $\vec{G} = \vec{F} \times \vec{E} = G_i \hat{\mathbf{e}}_i$ has the components

$$G_q = e_{qim} F_i E_m = e_{qim} e_{ijk} e_{mnp} A_j B_k C_n D_p.$$

From the identity $e_{qim} = e_{mqi}$ this can be expressed

$$G_q = (e_{mqi} e_{mnp}) e_{ijk} A_j B_k C_n D_p$$

which is now in a form where we can use the $e - \delta$ identity applied to the term in parentheses to produce

$$G_q = (\delta_{qn} \delta_{ip} - \delta_{qp} \delta_{in}) e_{ijk} A_j B_k C_n D_p.$$

Simplifying this expression we have:

$$\begin{aligned} G_q &= e_{ijk} \left[(D_p \delta_{ip})(C_n \delta_{qn}) A_j B_k - (D_p \delta_{qp})(C_n \delta_{in}) A_j B_k \right] \\ &= e_{ijk} \left[D_i C_q A_j B_k - D_q C_i A_j B_k \right] \\ &= C_q \left[D_i e_{ijk} A_j B_k \right] - D_q \left[C_i e_{ijk} A_j B_k \right] \end{aligned}$$

which are the vector components of the vector

$$\vec{C}(\vec{D} \cdot \vec{A} \times \vec{B}) - \vec{D}(\vec{C} \cdot \vec{A} \times \vec{B}).$$

■

Transformation Equations

Consider two sets of N independent variables which are denoted by the barred and unbarred symbols $\overline{x}^i$ and x^i with $i = 1,\ldots,N$. The independent variables $x^i, i = 1,\ldots,N$ can be thought of as defining the coordinates of a point in a $N-$dimensional space. Similarly, the independent barred variables define a point in some other $N-$dimensional space. These coordinates are assumed to be real quantities and are not complex quantities. Further, we assume that these variables are related by a set of transformation equations.

$$x^i = x^i(\overline{x}^1, \overline{x}^2, \ldots, \overline{x}^N) \qquad i = 1,\ldots,N. \tag{1.1.7}$$

It is assumed that these transformation equations are independent. A necessary and sufficient condition that these transformation equations be independent is that the Jacobian determinant be different from zero, that is

$$J(\frac{x}{\overline{x}}) = \left| \frac{\partial x^i}{\partial \overline{x}^j} \right| = \begin{vmatrix} \frac{\partial x^1}{\partial \overline{x}^1} & \frac{\partial x^1}{\partial \overline{x}^2} & \cdots & \frac{\partial x^1}{\partial \overline{x}^N} \\ \frac{\partial x^2}{\partial \overline{x}^1} & \frac{\partial x^2}{\partial \overline{x}^2} & \cdots & \frac{\partial x^2}{\partial \overline{x}^N} \\ \vdots & \vdots & \ddots & \vdots \\ \frac{\partial x^N}{\partial \overline{x}^1} & \frac{\partial x^N}{\partial \overline{x}^2} & \cdots & \frac{\partial x^N}{\partial \overline{x}^N} \end{vmatrix} \neq 0.$$

This assumption allows us to obtain a set of inverse relations

$$\overline{x}^i = \overline{x}^i(x^1, x^2, \ldots, x^N) \qquad i = 1,\ldots,N, \tag{1.1.8}$$

where the $\overline{x}'s$ are determined in terms of the $x's$. Throughout our discussions it is to be understood that the given transformation equations are real and continuous. Further all derivatives that appear in our discussions are assumed to exist and be continuous in the domain of the variables considered.

EXAMPLE 1.1-18. The following is an example of a set of transformation equations of the form defined by equations (1.1.7) and (1.1.8) in the case $N = 3$. Consider the transformation from cylindrical coordinates (r, α, z) to spherical coordinates (ρ, β, α). From the geometry of the figure 1.1-5 we can find the transformation equations

$$r = \rho \sin \beta$$
$$\alpha = \alpha \qquad 0 < \alpha < 2\pi$$
$$z = \rho \cos \beta \qquad 0 < \beta < \pi$$

with inverse transformation

$$\rho = \sqrt{r^2 + z^2}$$
$$\alpha = \alpha$$
$$\beta = \arctan(\frac{r}{z})$$

Now make the substitutions

$$(x^1, x^2, x^3) = (r, \alpha, z) \qquad \text{and} \qquad (\overline{x}^1, \overline{x}^2, \overline{x}^3) = (\rho, \beta, \alpha).$$

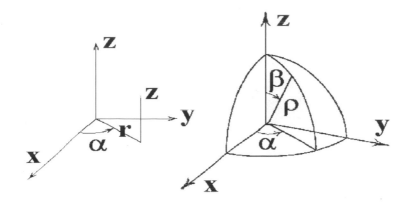

Figure 1.1-5. Cylindrical and Spherical Coordinates

The resulting transformations then have the forms of the equations (1.1.7) and (1.1.8).

■

Calculation of Derivatives

We now consider the chain rule applied to the differentiation of a function of the bar variables. We represent this differentiation in the indicial notation. Let $\Phi = \Phi(\overline{x}^1, \overline{x}^2, \ldots, \overline{x}^n)$ be a scalar function of the variables $\overline{x}^i$, $i = 1, \ldots, N$ and let these variables be related to the set of variables x^i, with $i = 1, \ldots, N$ by the transformation equations (1.1.7) and (1.1.8). The partial derivatives of Φ with respect to the variables x^i can be expressed in the indicial notation as

$$\frac{\partial \Phi}{\partial x^i} = \frac{\partial \Phi}{\partial \overline{x}^j} \frac{\partial \overline{x}^j}{\partial x^i} = \frac{\partial \Phi}{\partial \overline{x}^1} \frac{\partial \overline{x}^1}{\partial x^i} + \frac{\partial \Phi}{\partial \overline{x}^2} \frac{\partial \overline{x}^2}{\partial x^i} + \cdots + \frac{\partial \Phi}{\partial \overline{x}^N} \frac{\partial \overline{x}^N}{\partial x^i} \tag{1.1.9}$$

for any fixed value of i satisfying $1 \leq i \leq N$.

The second partial derivatives of Φ can also be expressed in the index notation. Differentiation of equation (1.1.9) partially with respect to x^m produces

$$\frac{\partial^2 \Phi}{\partial x^i \partial x^m} = \frac{\partial \Phi}{\partial \overline{x}^j} \frac{\partial^2 \overline{x}^j}{\partial x^i \partial x^m} + \frac{\partial}{\partial x^m}\left[\frac{\partial \Phi}{\partial \overline{x}^j}\right] \frac{\partial \overline{x}^j}{\partial x^i}. \tag{1.1.10}$$

This result is nothing more than an application of the general rule for differentiating a product of two quantities. To evaluate the derivative of the bracketed term in equation (1.1.10) it must be remembered that the quantity inside the brackets is a function of the bar variables. Let

$$G = \frac{\partial \Phi}{\partial \overline{x}^j} = G(\overline{x}^1, \overline{x}^2, \ldots, \overline{x}^N)$$

to emphasize this dependence upon the bar variables, then the derivative of G is

$$\frac{\partial G}{\partial x^m} = \frac{\partial G}{\partial \overline{x}^k} \frac{\partial \overline{x}^k}{\partial x^m} = \frac{\partial^2 \Phi}{\partial \overline{x}^j \partial \overline{x}^k} \frac{\partial \overline{x}^k}{\partial x^m}. \tag{1.1.11}$$

This is just an application of the basic rule from equation (1.1.9) with Φ replaced by G. Hence the derivative from equation (1.1.10) can be expressed

$$\frac{\partial^2 \Phi}{\partial x^i \partial x^m} = \frac{\partial \Phi}{\partial \overline{x}^j} \frac{\partial^2 \overline{x}^j}{\partial x^i \partial x^m} + \frac{\partial^2 \Phi}{\partial \overline{x}^j \partial \overline{x}^k} \frac{\partial \overline{x}^j}{\partial x^i} \frac{\partial \overline{x}^k}{\partial x^m} \tag{1.1.12}$$

where i, m are free indices and j, k are dummy summation indices.

EXAMPLE 1.1-19. Let $\Phi = \Phi(r, \theta)$ where r, θ are polar coordinates related to the Cartesian coordinates (x, y) by the transformation equations $x = r\cos\theta \quad y = r\sin\theta$. Find the partial derivatives $\dfrac{\partial\Phi}{\partial x}$ and $\dfrac{\partial^2\Phi}{\partial x^2}$

Solution: The partial derivative of Φ with respect to x is found from the relation (1.1.9) and can be written

$$\frac{\partial\Phi}{\partial x} = \frac{\partial\Phi}{\partial r}\frac{\partial r}{\partial x} + \frac{\partial\Phi}{\partial\theta}\frac{\partial\theta}{\partial x}. \tag{1.1.13}$$

The second partial derivative is obtained by differentiating the first partial derivative. From the product rule for differentiation we can write

$$\frac{\partial^2\Phi}{\partial x^2} = \frac{\partial\Phi}{\partial r}\frac{\partial^2 r}{\partial x^2} + \frac{\partial r}{\partial x}\frac{\partial}{\partial x}\left[\frac{\partial\Phi}{\partial r}\right] + \frac{\partial\Phi}{\partial\theta}\frac{\partial^2\theta}{\partial x^2} + \frac{\partial\theta}{\partial x}\frac{\partial}{\partial x}\left[\frac{\partial\Phi}{\partial\theta}\right]. \tag{1.1.14}$$

To further simplify (1.1.14) it must be remembered that the terms inside the brackets are to be treated as functions of the variables r and θ and that the derivative of these terms can be evaluated by reapplying the basic rule from equation (1.1.13) with Φ replaced by $\frac{\partial\Phi}{\partial r}$ and then Φ replaced by $\frac{\partial\Phi}{\partial\theta}$. This gives

$$\begin{aligned}\frac{\partial^2\Phi}{\partial x^2} &= \frac{\partial\Phi}{\partial r}\frac{\partial^2 r}{\partial x^2} + \frac{\partial r}{\partial x}\left[\frac{\partial^2\Phi}{\partial r^2}\frac{\partial r}{\partial x} + \frac{\partial^2\Phi}{\partial r\partial\theta}\frac{\partial\theta}{\partial x}\right]\\ &+ \frac{\partial\Phi}{\partial\theta}\frac{\partial^2\theta}{\partial x^2} + \frac{\partial\theta}{\partial x}\left[\frac{\partial^2\Phi}{\partial\theta\partial r}\frac{\partial r}{\partial x} + \frac{\partial^2\Phi}{\partial\theta^2}\frac{\partial\theta}{\partial x}\right].\end{aligned} \tag{1.1.15}$$

From the transformation equations we obtain the relations $r^2 = x^2 + y^2$ and $\tan\theta = \dfrac{y}{x}$ and from these relations we can calculate all the necessary derivatives needed for the simplification of the equations (1.1.13) and (1.1.15). These derivatives are:

$$2r\frac{\partial r}{\partial x} = 2x \quad \text{or} \quad \frac{\partial r}{\partial x} = \frac{x}{r} = \cos\theta$$

$$\sec^2\theta\frac{\partial\theta}{\partial x} = -\frac{y}{x^2} \quad \text{or} \quad \frac{\partial\theta}{\partial x} = -\frac{y}{r^2} = -\frac{\sin\theta}{r}$$

$$\frac{\partial^2 r}{\partial x^2} = -\sin\theta\frac{\partial\theta}{\partial x} = \frac{\sin^2\theta}{r} \qquad \frac{\partial^2\theta}{\partial x^2} = \frac{-r\cos\theta\frac{\partial\theta}{\partial x} + \sin\theta\frac{\partial r}{\partial x}}{r^2} = \frac{2\sin\theta\cos\theta}{r^2}.$$

Therefore, the derivatives from equations (1.1.13) and (1.1.15) can be expressed in the form

$$\frac{\partial\Phi}{\partial x} = \frac{\partial\Phi}{\partial r}\cos\theta - \frac{\partial\Phi}{\partial\theta}\frac{\sin\theta}{r}$$

$$\frac{\partial^2\Phi}{\partial x^2} = \frac{\partial\Phi}{\partial r}\frac{\sin^2\theta}{r} + 2\frac{\partial\Phi}{\partial\theta}\frac{\sin\theta\cos\theta}{r^2} + \frac{\partial^2\Phi}{\partial r^2}\cos^2\theta - 2\frac{\partial^2\Phi}{\partial r\partial\theta}\frac{\cos\theta\sin\theta}{r} + \frac{\partial^2\Phi}{\partial\theta^2}\frac{\sin^2\theta}{r^2}.$$

By letting $\overline{x}^1 = r$, $\overline{x}^2 = \theta$, $x^1 = x$, $x^2 = y$ and performing the indicated summations in the equations (1.1.9) and (1.1.12) there is produced the same results as above.

■

Vector Identities in Cartesian Coordinates

Employing the substitutions $x^1 = x$, $x^2 = y$, $x^3 = z$, where superscript variables are employed and denoting the unit vectors in Cartesian coordinates by $\widehat{e}_1$, $\widehat{e}_2$, $\widehat{e}_3$, we illustrated how various vector operations are written by using the index notation.

Gradient. In Cartesian coordinates the gradient of a scalar field is

$$\operatorname{grad}\phi = \frac{\partial \phi}{\partial x}\,\widehat{\mathbf{e}}_1 + \frac{\partial \phi}{\partial y}\,\widehat{\mathbf{e}}_2 + \frac{\partial \phi}{\partial z}\,\widehat{\mathbf{e}}_3.$$

The index notation focuses attention only on the components of the gradient. In Cartesian coordinates these components are represented using a comma subscript to denote the derivative

$$\widehat{\mathbf{e}}_j \cdot \operatorname{grad}\phi = \phi_{,j} = \frac{\partial \phi}{\partial x^j}, \quad j = 1, 2, 3.$$

The comma notation will be discussed in section 4. For now we use it to denote derivatives. For example $\phi_{,j} = \dfrac{\partial \phi}{\partial x^j}$, $\phi_{,jk} = \dfrac{\partial^2 \phi}{\partial x^j \partial x^k}$, etc.

Divergence. In Cartesian coordinates the divergence of a vector field $\vec{A}$ is a scalar field and can be represented

$$\nabla \cdot \vec{A} = div\,\vec{A} = \frac{\partial A_1}{\partial x} + \frac{\partial A_2}{\partial y} + \frac{\partial A_3}{\partial z}.$$

Employing the summation convention and index notation, the divergence in Cartesian coordinates can be represented

$$\nabla \cdot \vec{A} = div\,\vec{A} = A_{i,i} = \frac{\partial A_i}{\partial x^i} = \frac{\partial A_1}{\partial x^1} + \frac{\partial A_2}{\partial x^2} + \frac{\partial A_3}{\partial x^3}$$

where i is the dummy summation index.

Curl. To represent the vector $\vec{B} = \operatorname{curl}\vec{A} = \nabla \times \vec{A}$ in Cartesian coordinates, we note that the index notation focuses attention only on the components of this vector. The components B_i, $i = 1, 2, 3$ of $\vec{B}$ can be represented

$$B_i = \widehat{\mathbf{e}}_i \cdot \operatorname{curl}\vec{A} = e_{ijk}A_{k,j}, \qquad \text{for} \qquad i, j, k = 1, 2, 3$$

where e_{ijk} is the permutation symbol introduced earlier and $A_{k,j} = \frac{\partial A_k}{\partial x^j}$. To verify this representation of the curl $\vec{A}$ we need only perform the summations indicated by the repeated indices. We have summing on j that

$$B_i = e_{i1k}A_{k,1} + e_{i2k}A_{k,2} + e_{i3k}A_{k,3}.$$

Now summing each term on the repeated index k gives us

$$B_i = e_{i12}A_{2,1} + e_{i13}A_{3,1} + e_{i21}A_{1,2} + e_{i23}A_{3,2} + e_{i31}A_{1,3} + e_{i32}A_{2,3}$$

Here i is a free index which can take on any of the values $1, 2$ or 3. Consequently, we have

$$\text{For} \quad i = 1, \quad B_1 = A_{3,2} - A_{2,3} = \frac{\partial A_3}{\partial x^2} - \frac{\partial A_2}{\partial x^3}$$

$$\text{For} \quad i = 2, \quad B_2 = A_{1,3} - A_{3,1} = \frac{\partial A_1}{\partial x^3} - \frac{\partial A_3}{\partial x^1}$$

$$\text{For} \quad i = 3, \quad B_3 = A_{2,1} - A_{1,2} = \frac{\partial A_2}{\partial x^1} - \frac{\partial A_1}{\partial x^2}$$

which verifies the index notation representation of curl $\vec{A}$ in Cartesian coordinates.

22

Other Operations. The following examples illustrate how the index notation can be used to represent additional vector operators in Cartesian coordinates.

1. In index notation the components of the vector $(\vec{B} \cdot \nabla)\vec{A}$ are

$$\{(\vec{B} \cdot \nabla)\vec{A}\} \cdot \widehat{\mathbf{e}}_p = A_{p,q} B_q \qquad p, q = 1, 2, 3$$

This can be verified by performing the indicated summations. We have by summing on the repeated index q

$$A_{p,q} B_q = A_{p,1} B_1 + A_{p,2} B_2 + A_{p,3} B_3.$$

The index p is now a free index which can have any of the values $1, 2$ or 3. We have:

$$
\begin{aligned}
\text{for} \quad p = 1, \quad A_{1,q} B_q &= A_{1,1} B_1 + A_{1,2} B_2 + A_{1,3} B_3 \\
&= \frac{\partial A_1}{\partial x^1} B_1 + \frac{\partial A_1}{\partial x^2} B_2 + \frac{\partial A_1}{\partial x^3} B_3 \\
\text{for} \quad p = 2, \quad A_{2,q} B_q &= A_{2,1} B_1 + A_{2,2} B_2 + A_{2,3} B_3 \\
&= \frac{\partial A_2}{\partial x^1} B_1 + \frac{\partial A_2}{\partial x^2} B_2 + \frac{\partial A_2}{\partial x^3} B_3 \\
\text{for} \quad p = 3, \quad A_{3,q} B_q &= A_{3,1} B_1 + A_{3,2} B_2 + A_{3,3} B_3 \\
&= \frac{\partial A_3}{\partial x^1} B_1 + \frac{\partial A_3}{\partial x^2} B_2 + \frac{\partial A_3}{\partial x^3} B_3
\end{aligned}
$$

2. The scalar $(\vec{B} \cdot \nabla)\phi$ has the following form when expressed in the index notation:

$$
\begin{aligned}
(\vec{B} \cdot \nabla)\phi = B_i \phi_{,i} &= B_1 \phi_{,1} + B_2 \phi_{,2} + B_3 \phi_{,3} \\
&= B_1 \frac{\partial \phi}{\partial x^1} + B_2 \frac{\partial \phi}{\partial x^2} + B_3 \frac{\partial \phi}{\partial x^3}.
\end{aligned}
$$

3. The components of the vector $(\vec{B} \times \nabla)\phi$ is expressed in the index notation by

$$\widehat{\mathbf{e}}_i \cdot \left[(\vec{B} \times \nabla)\phi \right] = e_{ijk} B_j \phi_{,k}.$$

This can be verified by performing the indicated summations and is left as an exercise.

4. The scalar $(\vec{B} \times \nabla) \cdot \vec{A}$ may be expressed in the index notation. It has the form

$$(\vec{B} \times \nabla) \cdot \vec{A} = e_{ijk} B_j A_{i,k}.$$

This can also be verified by performing the indicated summations and is left as an exercise.

5. The vector components of $\nabla^2 \vec{A}$ in the index notation are represented

$$\widehat{\mathbf{e}}_p \cdot \nabla^2 \vec{A} = A_{p,qq}.$$

The proof of this is left as an exercise.

EXAMPLE 1.1-20.   In Cartesian coordinates prove the vector identity

$$\text{curl }(f\vec{A}) = \nabla \times (f\vec{A}) = (\nabla f) \times \vec{A} + f(\nabla \times \vec{A}).$$

Solution: Let $\vec{B} = \text{curl }(f\vec{A})$ and write the components as

$$\begin{aligned} B_i &= e_{ijk}(fA_k)_{,j} \\ &= e_{ijk}[fA_{k,j} + f_{,j}A_k] \\ &= fe_{ijk}A_{k,j} + e_{ijk}f_{,j}A_k. \end{aligned}$$

This index form can now be expressed in the vector form

$$\vec{B} = \text{curl }(f\vec{A}) = f(\nabla \times \vec{A}) + (\nabla f) \times \vec{A}$$

EXAMPLE 1.1-21.   Prove the vector identity $\nabla \cdot (\vec{A} + \vec{B}) = \nabla \cdot \vec{A} + \nabla \cdot \vec{B}$
Solution: Let $\vec{A} + \vec{B} = \vec{C}$ and write this vector equation in the index notation as $A_i + B_i = C_i$. We then have

$$\nabla \cdot \vec{C} = C_{i,i} = (A_i + B_i)_{,i} = A_{i,i} + B_{i,i} = \nabla \cdot \vec{A} + \nabla \cdot \vec{B}.$$

EXAMPLE 1.1-22.   In Cartesian coordinates prove the vector identity $(\vec{A} \cdot \nabla)f = \vec{A} \cdot \nabla f$
Solution: In the index notation we write

$$\begin{aligned} (\vec{A} \cdot \nabla)f = A_i f_{,i} &= A_1 f_{,1} + A_2 f_{,2} + A_3 f_{,3} \\ &= A_1 \frac{\partial f}{\partial x^1} + A_2 \frac{\partial f}{\partial x^2} + A_3 \frac{\partial f}{\partial x^3} = \vec{A} \cdot \nabla f. \end{aligned}$$

EXAMPLE 1.1-23.   In Cartesian coordinates prove the vector identity

$$\nabla \times (\vec{A} \times \vec{B}) = \vec{A}(\nabla \cdot \vec{B}) - \vec{B}(\nabla \cdot \vec{A}) + (\vec{B} \cdot \nabla)\vec{A} - (\vec{A} \cdot \nabla)\vec{B}$$

Solution: The *pth* component of the vector $\nabla \times (\vec{A} \times \vec{B})$ is

$$\begin{aligned} \hat{\mathbf{e}}_p \cdot [\nabla \times (\vec{A} \times \vec{B})] &= e_{pqk}[e_{kji}A_jB_i]_{,q} \\ &= e_{pqk}e_{kji}A_jB_{i,q} + e_{pqk}e_{kji}A_{j,q}B_i \end{aligned}$$

By applying the $e - \delta$ identity, the above expression simplifies to the desired result. That is,

$$\begin{aligned} \hat{\mathbf{e}}_p \cdot [\nabla \times (\vec{A} \times \vec{B})] &= (\delta_{pj}\delta_{qi} - \delta_{pi}\delta_{qj})A_jB_{i,q} + (\delta_{pj}\delta_{qi} - \delta_{pi}\delta_{qj})A_{j,q}B_i \\ &= A_pB_{i,i} - A_qB_{p,q} + A_{p,q}B_q - A_{q,q}B_p \end{aligned}$$

In vector form this is expressed

$$\nabla \times (\vec{A} \times \vec{B}) = \vec{A}(\nabla \cdot \vec{B}) - (\vec{A} \cdot \nabla)\vec{B} + (\vec{B} \cdot \nabla)\vec{A} - \vec{B}(\nabla \cdot \vec{A})$$

24

EXAMPLE 1.1-24. In Cartesian coordinates prove the vector identity $\nabla \times (\nabla \times \vec{A}) = \nabla(\nabla \cdot \vec{A}) - \nabla^2\vec{A}$

Solution: The ith component of $\nabla \times \vec{A}$ is given by $\hat{e}_i \cdot [\nabla \times \vec{A}] = e_{ijk}A_{k,j}$ and consequently the pth component of $\nabla \times (\nabla \times \vec{A})$ is

$$\hat{e}_p \cdot [\nabla \times (\nabla \times \vec{A})] = e_{pqr}[e_{rjk}A_{k,j}]_{,q}$$
$$= e_{pqr}e_{rjk}A_{k,jq}.$$

The $e - \delta$ identity produces

$$\hat{e}_p \cdot [\nabla \times (\nabla \times \vec{A})] = (\delta_{pj}\delta_{qk} - \delta_{pk}\delta_{qj})A_{k,jq}$$
$$= A_{k,pk} - A_{p,qq}.$$

Expressing this result in vector form we have $\nabla \times (\nabla \times \vec{A}) = \nabla(\nabla \cdot \vec{A}) - \nabla^2\vec{A}$.

∎

Indicial Form of Integral Theorems

The divergence theorem, in both vector and indicial notation, can be written

$$\iiint_V \text{div} \cdot \vec{F} \, d\tau = \iint_S \vec{F} \cdot \hat{n} \, d\sigma \qquad \int_V F_{i,i} \, d\tau = \int_S F_i n_i \, d\sigma \qquad i = 1,2,3 \qquad (1.1.16)$$

where n_i are the direction cosines of the unit exterior normal to the surface, $d\tau$ is a volume element and $d\sigma$ is an element of surface area. Note that in using the indicial notation the volume and surface integrals are to be extended over the range specified by the indices. This suggests that the divergence theorem can be applied to vectors in $n-$dimensional spaces.

The vector form and indicial notation for the Stokes theorem are

$$\iint_S (\nabla \times \vec{F}) \cdot \hat{n} \, d\sigma = \int_C \vec{F} \cdot d\vec{r} \qquad \int_S e_{ijk}F_{k,j}n_i \, d\sigma = \int_C F_i \, dx^i \qquad i,j,k = 1,2,3 \qquad (1.1.17)$$

and the Green's theorem in the plane, which is a special case of the Stoke's theorem, can be expressed

$$\iint_S \left(\frac{\partial F_2}{\partial x} - \frac{\partial F_1}{\partial y}\right) dxdy = \int_C F_1 \, dx + F_2 \, dy \qquad \int_S e_{3jk}F_{k,j} \, dS = \int_C F_i \, dx^i \quad i,j,k = 1,2 \quad (1.1.18)$$

Other forms of the above integral theorems are

$$\iiint_V \nabla\phi \, d\tau = \iint_S \phi\hat{n} \, d\sigma$$

obtained from the divergence theorem by letting $\vec{F} = \phi\vec{C}$ where $\vec{C}$ is a constant vector. By replacing $\vec{F}$ by $\vec{F} \times \vec{C}$ in the divergence theorem one can derive

$$\iiint_V \left(\nabla \times \vec{F}\right) d\tau = -\iint_S \vec{F} \times \vec{n} \, d\sigma.$$

In the divergence theorem make the substitution $\vec{F} = \phi\nabla\psi$ to obtain

$$\iiint_V [(\phi\nabla^2\psi + (\nabla\phi) \cdot (\nabla\psi)] \, d\tau = \iint_S (\phi\nabla\psi) \cdot \hat{n} \, d\sigma.$$

The Green's identity

$$\iiint_V \left(\phi \nabla^2 \psi - \psi \nabla^2 \phi\right)\, d\tau = \iint_S (\phi \nabla \psi - \psi \nabla \phi) \cdot \hat{\mathbf{n}}\, d\sigma$$

is obtained by first letting $\vec{F} = \phi \nabla \psi$ in the divergence theorem and then letting $\vec{F} = \psi \nabla \phi$ in the divergence theorem and then subtracting the results.

Determinants, Cofactors

For $A = (a_{ij})$, $i, j = 1, \ldots, n$ an $n \times n$ matrix, the determinant of A can be written as

$$\det A = |A| = e_{i_1 i_2 i_3 \ldots i_n} a_{1 i_1} a_{2 i_2} a_{3 i_3} \ldots a_{n i_n}.$$

This gives a summation of the $n!$ permutations of products formed from the elements of the matrix A. The result is a single number called the determinant of A.

EXAMPLE 1.1-25.   In the case $n = 2$ we have

$$|A| = \begin{vmatrix} a_{11} & a_{12} \\ a_{21} & a_{22} \end{vmatrix} = e_{nm} a_{1n} a_{2m}$$

$$= e_{1m} a_{11} a_{2m} + e_{2m} a_{12} a_{2m}$$

$$= e_{12} a_{11} a_{22} + e_{21} a_{12} a_{21}$$

$$= a_{11} a_{22} - a_{12} a_{21}$$

∎

EXAMPLE 1.1-26.   In the case $n = 3$ we can use either of the notations

$$A = \begin{pmatrix} a_{11} & a_{12} & a_{13} \\ a_{21} & a_{22} & a_{23} \\ a_{31} & a_{32} & a_{33} \end{pmatrix} \qquad \text{or} \qquad A = \begin{pmatrix} a_1^1 & a_2^1 & a_3^1 \\ a_1^2 & a_2^2 & a_3^2 \\ a_1^3 & a_2^3 & a_3^3 \end{pmatrix}$$

and represent the determinant of A in any of the forms

$$\det A = e_{ijk} a_{1i} a_{2j} a_{3k}$$

$$\det A = e_{ijk} a_{i1} a_{j2} a_{k3}$$

$$\det A = e_{ijk} a_1^i a_2^j a_3^k$$

$$\det A = e_{ijk} a_i^1 a_j^2 a_k^3.$$

These represent row and column expansions of the determinant.

An important identity results if we examine the quantity $B_{rst} = e_{ijk} a_r^i a_s^j a_t^k$. It is an easy exercise to change the dummy summation indices and rearrange terms in this expression. For example,

$$B_{rst} = e_{ijk} a_r^i a_s^j a_t^k = e_{kji} a_r^k a_s^j a_t^i = e_{kji} a_t^i a_s^j a_r^k = -e_{ijk} a_t^i a_s^j a_r^k = -B_{tsr},$$

and by considering other permutations of the indices, one can establish that B_{rst} is completely skew-symmetric. In the exercises it is shown that any third order completely skew-symmetric system satisfies $B_{rst} = B_{123} e_{rst}$. But $B_{123} = \det A$ and so we arrive at the identity

$$B_{rst} = e_{ijk} a_r^i a_s^j a_t^k = |A| e_{rst}.$$

Other forms of this identity are

$$e^{ijk}a_i^r a_j^s a_k^t = |A|e^{rst} \quad \text{and} \quad e_{ijk}a_{ir}a_{js}a_{kt} = |A|e_{rst}. \tag{1.1.19}$$

∎

Consider the representation of the determinant

$$|A| = \begin{vmatrix} a_1^1 & a_2^1 & a_3^1 \\ a_1^2 & a_2^2 & a_3^2 \\ a_1^3 & a_2^3 & a_3^3 \end{vmatrix}$$

by use of the indicial notation. By column expansions, this determinant can be represented

$$|A| = e_{rst}a_1^r a_2^s a_3^t \tag{1.1.20}$$

and if one uses row expansions the determinant can be expressed as

$$|A| = e^{ijk}a_i^1 a_j^2 a_k^3. \tag{1.1.21}$$

Define A_m^i as the cofactor of the element a_i^m in the determinant $|A|$. From the equation (1.1.20) the cofactor of a_1^r is obtained by deleting this element and we find

$$A_r^1 = e_{rst}a_2^s a_3^t. \tag{1.1.22}$$

The result (1.1.20) can then be expressed in the form

$$|A| = a_1^r A_r^1 = a_1^1 A_1^1 + a_1^2 A_2^1 + a_1^3 A_3^1. \tag{1.1.23}$$

That is, the determinant $|A|$ is obtained by multiplying each element in the first column by its corresponding cofactor and summing the result. Observe also that from the equation (1.1.20) we find the additional cofactors

$$A_s^2 = e_{rst}a_1^r a_3^t \quad \text{and} \quad A_t^3 = e_{rst}a_1^r a_2^s. \tag{1.1.24}$$

Hence, the equation (1.1.20) can also be expressed in one of the forms

$$|A| = a_2^s A_s^2 = a_2^1 A_1^2 + a_2^2 A_2^2 + a_2^3 A_3^2$$
$$|A| = a_3^t A_t^3 = a_3^1 A_1^3 + a_3^2 A_2^3 + a_3^3 A_3^3$$

The results from equations (1.1.22) and (1.1.24) can be written in a slightly different form with the indicial notation. From the notation for a generalized Kronecker delta defined by

$$e^{ijk}e_{lmn} = \delta_{lmn}^{ijk},$$

the above cofactors can be written in the form

$$A_r^1 = e^{123}e_{rst}a_2^s a_3^t = \frac{1}{2!}e^{1jk}e_{rst}a_j^s a_k^t = \frac{1}{2!}\delta_{rst}^{1jk}a_j^s a_k^t$$
$$A_r^2 = e^{123}e_{srt}a_1^s a_3^t = \frac{1}{2!}e^{2jk}e_{rst}a_j^s a_k^t = \frac{1}{2!}\delta_{rst}^{2jk}a_j^s a_k^t$$
$$A_r^3 = e^{123}e_{tsr}a_1^t a_2^s = \frac{1}{2!}e^{3jk}e_{rst}a_j^s a_k^t = \frac{1}{2!}\delta_{rst}^{3jk}a_j^s a_k^t.$$

These cofactors are then combined into the single equation

$$A_r^i = \frac{1}{2!}\delta_{rst}^{ijk}a_j^s a_k^t \qquad (1.1.25)$$

which represents the cofactor of a_i^r. When the elements from any row (or column) are multiplied by their corresponding cofactors, and the results summed, we obtain the value of the determinant. Whenever the elements from any row (or column) are multiplied by the cofactor elements from a different row (or column), and the results summed, we get zero. This can be illustrated by considering the summation $a_r^m A_m^i = \frac{1}{2!}\delta_{mst}^{ijk}a_j^s a_k^t a_r^m = \frac{1}{2!}e^{ijk}e_{mst}a_r^m a_j^s a_k^t = \frac{1}{2!}e^{ijk}e_{rjk}|A| = \frac{1}{2!}\delta_{rjk}^{ijk}|A| = \delta_r^i|A|$. Here we have used the $e-\delta$ identity to obtain $\delta_{rjk}^{ijk} = e^{ijk}e_{rjk} = e^{jik}e_{jrk} = \delta_r^i\delta_k^k - \delta_k^i\delta_r^k = 3\delta_r^i - \delta_r^i = 2\delta_r^i$ which was used to simplify the above result. As an exercise one can show that an alternate form of the above summation of elements by its cofactors is $a_m^r A_i^m = |A|\delta_i^r$.

EXAMPLE 1.1-27.   In N-dimensions the quantity $\delta_{k_1 k_2 \ldots k_N}^{j_1 j_2 \ldots j_N}$ is called a generalized Kronecker delta. It can be defined in terms of permutation symbols as

$$e^{j_1 j_2 \ldots j_N} e_{k_1 k_2 \ldots k_N} = \delta_{k_1 k_2 \ldots k_N}^{j_1 j_2 \ldots j_N} \qquad (1.1.26)$$

Observe that

$$\delta_{k_1 k_2 \ldots k_N}^{j_1 j_2 \ldots j_N} e^{k_1 k_2 \ldots k_N} = (N!)\, e^{j_1 j_2 \ldots j_N}$$

This follows because $e^{k_1 k_2 \ldots k_N}$ is skew-symmetric in all pairs of its superscripts. The left-hand side denotes a summation of $N!$ terms. The first term in the summation has superscripts $j_1 j_2 \ldots j_N$ and all other terms have superscripts which are some permutation of this ordering with minus signs associated with those terms having an odd permutation. Because $e^{j_1 j_2 \ldots j_N}$ is completely skew-symmetric we find that all terms in the summation have the value $+e^{j_1 j_2 \ldots j_N}$. We thus obtain $N!$ of these terms.

■

EXAMPLE 1.1-28.   The matrices

$$I = \begin{bmatrix} 1 & 0 \\ 0 & 1 \end{bmatrix}, \quad P_1 = \begin{bmatrix} 0 & 1 \\ 1 & 0 \end{bmatrix}, \quad P_2 = \begin{bmatrix} 0 & -i \\ i & 0 \end{bmatrix}, \quad P_3 = \begin{bmatrix} 1 & 0 \\ 0 & -1 \end{bmatrix}$$

where $i^2 = -1$, are called Pauli matrices. It is readily verified that for $m, n = 1, 2, 3$ that

$$P_m P_n = i e_{mnk} P_k + \delta_{mn} I$$
$$P_m P_n = i\,[e_{mn1} P_1 + e_{mn2} P_2 + e_{mn3} P_3] + \delta_{mn} I.$$

Note that i is being used for an imaginary unit and so do not use i as a summation or free index. Multiply this equation on the right by P_r to obtain $P_m P_n P_r = i e_{mnk} P_k P_r + \delta_{mn} P_r$. By interchanging the dummy indices we can write $P_k P_r = i e_{krs} P_s + \delta_{kr} I$ to find that

$$P_m P_n P_r = i e_{mnk}\,[i e_{krs} P_s + \delta_{kr} I] + \delta_{mn} P_r$$
$$P_m P_n P_r = -\,[(\delta_{mr}\delta_{ns} - \delta_{nr}\delta_{ms}) P_s] + i e_{mnk}\delta_{kr} I + \delta_{mn} P_r$$
$$P_m P_n P_r = -\,\delta_{mr} P_n + \delta_{nr} P_m + \delta_{mn} P_r + i e_{mnr} I$$

It is left as an exercise to verify the above matrix multiplication properties of the Pauli matrices.

■

28

EXERCISE 1.1

▶ **1.** Simplify each of the following by employing the summation property of the Kronecker delta. Perform sums on the summation indices only if your are unsure of the result.

$$(a) \quad e_{ijk}\delta_{kn} \qquad\qquad (c) \quad e_{ijk}\delta_{is}\delta_{jm}\delta_{kn} \qquad\qquad (e) \quad \delta_{ij}\delta_{jn}$$

$$(b) \quad e_{ijk}\delta_{is}\delta_{jm} \qquad\qquad (d) \quad a_{ij}\delta_{in} \qquad\qquad\qquad (f) \quad \delta_{ij}\delta_{jn}\delta_{ni}$$

▶ **2.** Simplify and perform the indicated summations over the range $1, 2, 3$

$$(a) \quad \delta_{ii} \qquad\qquad (c) \quad e_{ijk}A_iA_jA_k \qquad\qquad (e) \quad e_{ijk}\delta_{jk}$$

$$(b) \quad \delta_{ij}\delta_{ij} \qquad\qquad (d) \quad e_{ijk}e_{ijk} \qquad\qquad\quad (f) \quad A_iB_j\delta_{ji} - B_mA_n\delta_{mn}$$

▶ **3.** Express each of the following in index notation. Be careful of the notation you use. Note that $\vec{A} = A_i$ is an incorrect notation because a vector can not equal a scalar. The notation $\vec{A} \cdot \hat{\mathbf{e}}_i = A_i$ should be used to express the *ith* component of a vector.

$$(a) \quad \vec{A} \cdot (\vec{B} \times \vec{C}) \qquad\qquad (c) \quad \vec{B}(\vec{A} \cdot \vec{C})$$

$$(b) \quad \vec{A} \times (\vec{B} \times \vec{C}) \qquad\qquad (d) \quad \vec{B}(\vec{A} \cdot \vec{C}) - \vec{C}(\vec{A} \cdot \vec{B})$$

▶ **4.** Show the e permutation symbol satisfies: $(a) \quad e_{ijk} = e_{jki} = e_{kij} \quad (b) \quad e_{ijk} = -e_{jik} = -e_{ikj} = -e_{kji}$

▶ **5.** Use index notation to verify the vector identity $\vec{A} \times (\vec{B} \times \vec{C}) = \vec{B}(\vec{A} \cdot \vec{C}) - \vec{C}(\vec{A} \cdot \vec{B})$

▶ **6.** Let $y_i = a_{ij}x_j$ and $x_m = a_{im}z_i$ where the range of the indices is $1, 2$

   (a) Solve for y_i in terms of z_i using the indicial notation and check your result to be sure that no index repeats itself more than twice.

   (b) Perform the indicated summations and write out expressions for y_1, y_2 in terms of z_1, z_2

   (c) Express the above equations in matrix form. Expand the matrix equations and check the solution obtained in part (b).

▶ **7.** Use the $e - \delta$ identity to simplify $(a) \quad e_{ijk}e_{jik} \qquad (b) \quad e_{ijk}e_{jki}$

▶ **8.** Prove the following vector identities:

$$(a) \quad \vec{A} \cdot (\vec{B} \times \vec{C}) = \vec{B} \cdot (\vec{C} \times \vec{A}) = \vec{C} \cdot (\vec{A} \times \vec{B}) \quad \text{triple scalar product}$$

$$(b) \quad (\vec{A} \times \vec{B}) \times \vec{C} = \vec{B}(\vec{A} \cdot \vec{C}) - \vec{A}(\vec{B} \cdot \vec{C})$$

▶ **9.** Prove the following vector identities:

$$(a) \quad (\vec{A} \times \vec{B}) \cdot (\vec{C} \times \vec{D}) = (\vec{A} \cdot \vec{C})(\vec{B} \cdot \vec{D}) - (\vec{A} \cdot \vec{D})(\vec{B} \cdot \vec{C})$$

$$(b) \quad \vec{A} \times (\vec{B} \times \vec{C}) + \vec{B} \times (\vec{C} \times \vec{A}) + \vec{C} \times (\vec{A} \times \vec{B}) = \vec{0}$$

$$(c) \quad (\vec{A} \times \vec{B}) \times (\vec{C} \times \vec{D}) = \vec{B}(\vec{A} \cdot \vec{C} \times \vec{D}) - \vec{A}(\vec{B} \cdot \vec{C} \times \vec{D})$$

28

► **10.** For $\vec{A} = (1, -1, 0)$ and $\vec{B} = (4, -3, 2)$ find using the index notation,

$$(a) \quad C_i = e_{ijk} A_j B_k, \quad i = 1, 2, 3$$

$$(b) \quad A_i B_i$$

$$(c) \quad \text{What do the results in (a) and (b) represent?}$$

► **11.** Represent the differential equations $\quad \dfrac{dy_1}{dt} = a_{11} y_1 + a_{12} y_2 \quad$ and $\quad \dfrac{dy_2}{dt} = a_{21} y_1 + a_{22} y_2$
using the index notation.

► **12.**

Let $\Phi = \Phi(r, \theta)$ where r, θ are polar coordinates related to Cartesian coordinates (x, y) by the transformation equations $\quad x = r \cos \theta \quad$ and $\quad y = r \sin \theta$.

(a) Find the partial derivatives $\quad \dfrac{\partial \Phi}{\partial y}, \quad$ and $\quad \dfrac{\partial^2 \Phi}{\partial y^2}$

(b) Combine the result in part (a) with the result from EXAMPLE 1.1-19 to calculate the Laplacian

$$\nabla^2 \Phi = \frac{\partial^2 \Phi}{\partial x^2} + \frac{\partial^2 \Phi}{\partial y^2}$$

in polar coordinates.

► **13.** (Index notation) Let $a_{11} = 3, \quad a_{12} = 4, \quad a_{21} = 5, \quad a_{22} = 6$.
Calculate the quantity $C = a_{ij} a_{ij}, \; i, j = 1, 2$.

► **14.** Verify that the moments of inertia I_{ij} defined by

$$I_{11} = \iiint_R (y^2 + z^2) \rho(x, y, z) \, d\tau \qquad\qquad I_{23} = I_{32} = - \iiint_R yz \rho(x, y, z) \, d\tau$$

$$I_{22} = \iiint_R (x^2 + z^2) \rho(x, y, z) \, d\tau \qquad\qquad I_{12} = I_{21} = - \iiint_R xy \rho(x, y, z) \, d\tau$$

$$I_{33} = \iiint_R (x^2 + y^2) \rho(x, y, z) \, d\tau \qquad\qquad I_{13} = I_{31} = - \iiint_R xz \rho(x, y, z) \, d\tau,$$

can be represented in the index notation as $I_{ij} = \iiint_R \left(x^m x^m \delta_{ij} - x^i x^j \right) \rho \, d\tau$, where ρ is the density, $x^1 = x, \; x^2 = y, \; x^3 = z$ and $d\tau = dx\,dy\,dz$ is an element of volume.

► **15.** Determine if the following relation is true or false. Justify your answer.

$$\hat{e}_i \cdot (\hat{e}_j \times \hat{e}_k) = (\hat{e}_i \times \hat{e}_j) \cdot \hat{e}_k = e_{ijk}, \quad i, j, k = 1, 2, 3.$$

Hint: Let $\hat{e}_m = (\delta_{1m}, \delta_{2m}, \delta_{3m})$.

► **16.** Without substituting values for $i, l = 1, 2, 3$ calculate all nine terms of the given quantities

$$(a) \quad B^{il} = (\delta_j^i A_k + \delta_k^i A_j) e^{jkl} \qquad\qquad (b) \quad A_{il} = (\delta_i^m B^k + \delta_i^k B^m) e_{mlk}$$

► **17.** Given that $A_{mn} x^m y^n = 0$ for arbitrary x^i and $y^i, \quad i = 1, 2, 3$. Show that $A_{ij} = 0$ for all values of i, j.

30

▶ **18.**

(a) Given that a_{mn}, $m,n=1,2,3$ is skew-symmetric. Show that $a_{mn}x^m x^n = 0$.

(b) Given that $a_{mn}x^m x^n = 0$, $m,n=1,2,3$ for all values of x^i, $i=1,2,3$. Show that a_{mn} must be skew-symmetric.

▶ **19.** Let A and B denote 3×3 matrices with elements a_{ij} and b_{ij} respectively. Show that if $C=AB$ is a matrix product, then the determinant of C is given by $det(C)=det(A)\cdot det(B)$.

Hint: Use the result from example 1.1-10.

▶ **20.**

(a) Let u^1, u^2, u^3 be functions of the variables s^1, s^2, s^3. Further, assume that s^1, s^2, s^3 are in turn each functions of the variables x^1, x^2, x^3. Let $\left|\dfrac{\partial u^m}{\partial x^n}\right| = \dfrac{\partial(u^1,\ u^2,\ u^3)}{\partial(x^1,\ x^2,\ x^3)}$ denote the Jacobian of the u's with respect to the x's. Show that

$$\left|\frac{\partial u^i}{\partial x^m}\right| = \left|\frac{\partial u^i}{\partial s^j}\frac{\partial s^j}{\partial x^m}\right| = \left|\frac{\partial u^i}{\partial s^j}\right|\cdot\left|\frac{\partial s^j}{\partial x^m}\right|.$$

(b) Given that $\dfrac{\partial x^i}{\partial \bar{x}^j}\dfrac{\partial \bar{x}^j}{\partial x^m} = \dfrac{\partial x^i}{\partial x^m} = \delta^i_m$, show that $J(\frac{x}{\bar{x}})\cdot J(\frac{\bar{x}}{x})=1$, where $J(\frac{x}{\bar{x}})$ is the Jacobian determinant of the transformation (1.1.7).

▶ **21.** A third order system $a_{\ell mn}$ with $\ell,m,n=1,2,3$ is said to be symmetric in two of its subscripts if the components are unaltered when these subscripts are interchanged. When $a_{\ell mn}$ is completely symmetric then $a_{\ell mn}=a_{m\ell n}=a_{\ell nm}=a_{mn\ell}=a_{nm\ell}=a_{n\ell m}$. Whenever this third order system is completely symmetric, then: (i) How many components are there? (ii) How many of these components are distinct?

Hint: Consider the three cases (i) $\ell=m=n$ (ii) $\ell=m\neq n$ (iii) $\ell\neq m\neq n$.

▶ **22.** A third order system $b_{\ell mn}$ with $\ell,m,n=1,2,3$ is said to be skew-symmetric in two of its subscripts if the components change sign when the subscripts are interchanged. A completely skew-symmetric third order system satisfies $b_{\ell mn}=-b_{m\ell n}=b_{mn\ell}=-b_{nm\ell}=b_{n\ell m}=-b_{\ell nm}$. (i) How many components does a completely skew-symmetric system have? (ii) How many of these components are zero? (iii) How many components can be different from zero? (iv) Show that there is one distinct component b_{123} and that $b_{\ell mn}=e_{\ell mn}b_{123}$.

Hint: Consider the three cases (i) $\ell=m=n$ (ii) $\ell=m\neq n$ (iii) $\ell\neq m\neq n$.

▶ **23.** Let $i,j,k=1,2,3$ and assume that $e_{ijk}\sigma_{jk}=0$ for all values of i. What does this equation tell you about the values σ_{ij}, $i,j=1,2,3$?

▶ **24.** Given that A_{mn} and B_{mn} are symmetric for $m,n=1,2,3$. Let $A_{mn}x^m x^n = B_{mn}x^m x^n$ for arbitrary values of x^i, $i=1,2,3$. Show that $A_{ij}=B_{ij}$ for all values of i and j.

▶ **25.** Given that B_{mn} is symmetric and $B_{mn}x^m x^n=0$ for arbitrary values of x^i, $i=1,2,3$. Show that $B_{ij}=0$.

▶ **26.** (<u>Generalized Kronecker delta</u>) Define the generalized Kronecker delta as the $n \times n$ determinant

$$\delta_{mn...p}^{ij...k} = \begin{vmatrix} \delta_m^i & \delta_n^i & \cdots & \delta_p^i \\ \delta_m^j & \delta_n^j & \cdots & \delta_p^j \\ \vdots & \vdots & \ddots & \vdots \\ \delta_m^k & \delta_n^k & \cdots & \delta_p^k \end{vmatrix} \qquad \text{where } \delta_s^r \text{ is the Kronecker delta.}$$

(a) Show $\quad e_{ijk} = \delta_{ijk}^{123}$

(b) Show $\quad e^{ijk} = \delta_{123}^{ijk}$

(c) Show $\quad \delta_{mn}^{ij} = e^{ij} e_{mn}$

(d) Define $\quad \delta_{mn}^{rs} = \delta_{mnp}^{rsp} \quad$ (summation on p)

and show $\quad \delta_{mn}^{rs} = \delta_m^r \delta_n^s - \delta_n^r \delta_m^s$

Note that by combining the above result with the result from part (c)

we obtain the two dimensional form of the $e - \delta$ identity $\quad e^{rs} e_{mn} = \delta_m^r \delta_n^s - \delta_n^r \delta_m^s$.

(e) Define $\delta_m^r = \frac{1}{2} \delta_{mn}^{rn} \quad$ (summation on n) $\quad$ and show $\quad \delta_{pst}^{rst} = 2\delta_p^r$

(f) Show $\quad \delta_{rst}^{rst} = 3!$

▶ **27.** Let A_r^i denote the cofactor of a_i^r in the determinant $\begin{vmatrix} a_1^1 & a_2^1 & a_3^1 \\ a_1^2 & a_2^2 & a_3^2 \\ a_1^3 & a_2^3 & a_3^3 \end{vmatrix}$ as given by equation (1.1.25).

(a) Show $\quad e^{rst} A_r^i = e^{ijk} a_j^s a_k^t$ $\qquad$ (b) Show $\quad e_{rst} A_i^r = e_{ijk} a_s^j a_t^k$

▶ **28.** (a) Show that if $A_{ijk} = A_{jik}, i, j, k = 1, 2, 3$ there is a total of 27 elements, but only 18 are distinct.
(b) Show that for $i, j, k = 1, 2, \ldots, N$ there are N^3 elements, but only $N^2(N+1)/2$ are distinct.

▶ **29.** Given that $a_{ij} = B_i B_j$ for $i, j = 1, 2, 3$ where B_1, B_2, B_3 are arbitrary constants. Calculate $\det(a_{ij}) = |A|$.

▶ **30.**

(a) For $\quad A = (a_{ij}), \; i, j = 1, 2, 3, \quad$ show $\quad |A| = e_{ijk} a_{i1} a_{j2} a_{k3}$.

(b) For $\quad A = (a_j^i), \; i, j = 1, 2, 3, \quad$ show $\quad |A| = e_{ijk} a_1^i a_2^j a_3^k$.

(c) For $\quad A = (a_j^i), \; i, j = 1, 2, 3, \quad$ show $\quad |A| = e^{ijk} a_i^1 a_j^2 a_k^3$.

(d) For $\quad I = (\delta_j^i), \; i, j = 1, 2, 3, \quad$ show $\quad |I| = 1$.

▶ **31.** Given the determinant $|A| = e_{ijk} a_{i1} a_{j2} a_{k3}$ and define A_{im} as the cofactor of a_{im}. Show the determinant can be expressed in any of the forms:

(a) $|A| = A_{i1} a_{i1} \quad$ where $\quad A_{i1} = e_{ijk} a_{j2} a_{k3}$

(b) $|A| = A_{j2} a_{j2} \quad$ where $\quad A_{i2} = e_{jik} a_{j1} a_{k3}$

(c) $|A| = A_{k3} a_{k3} \quad$ where $\quad A_{i3} = e_{jki} a_{j1} a_{k2}$

▶ **32.** Show the results in problem 31 can be written in the forms:

$$A_{i1} = \frac{1}{2!}e_{1st}e_{ijk}a_{js}a_{kt}, \quad A_{i2} = \frac{1}{2!}e_{2st}e_{ijk}a_{js}a_{kt}, \quad A_{i3} = \frac{1}{2!}e_{3st}e_{ijk}a_{js}a_{kt}, \quad \text{or} \quad A_{im} = \frac{1}{2!}e_{mst}e_{ijk}a_{js}a_{kt}$$

▶ **33.** Use the results in problems 31 and 32 to prove that $a_{pm}A_{im} = |A|\delta_{ip}$.

▶ **34.** Let $(a_{ij}) = \begin{pmatrix} 1 & 2 & 1 \\ 1 & 0 & 3 \\ 2 & 3 & 2 \end{pmatrix}$ and calculate $C = a_{ij}a_{ij}$, $i, j = 1, 2, 3$.

▶ **35.** Let

$$a_{111} = -1, \quad a_{112} = 3, \quad a_{121} = 4, \quad a_{122} = 2$$

$$a_{211} = 1, \quad a_{212} = 5, \quad a_{221} = 2, \quad a_{222} = -2$$

and calculate the quantity $C = a_{ijk}a_{ijk}$, $i, j, k = 1, 2$.

▶ **36.** Let

$$a_{1111} = 2, \quad a_{1112} = 1, \quad a_{1121} = 3, \quad a_{1122} = 1$$

$$a_{1211} = 5, \quad a_{1212} = -2, \quad a_{1221} = 4, \quad a_{1222} = -2$$

$$a_{2111} = 1, \quad a_{2112} = 0, \quad a_{2121} = -2, \quad a_{2122} = -1$$

$$a_{2211} = -2, \quad a_{2212} = 1, \quad a_{2221} = 2, \quad a_{2222} = 2$$

and calculate the quantity $C = a_{ijkl}a_{ijkl}$, $i, j, k, l = 1, 2$.

▶ **37.** Simplify the expressions:

(a) $(A_{ijkl} + A_{jkli} + A_{klij} + A_{lijk})x_i x_j x_k x_l$

(b) $(P_{ijk} + P_{jki} + P_{kij})x^i x^j x^k$

(c) $\dfrac{\partial x^i}{\partial x^j}$

(d) $a_{ij}\dfrac{\partial^2 x^i}{\partial \overline{x}^t \partial \overline{x}^s}\dfrac{\partial x^j}{\partial \overline{x}^r} - a_{mi}\dfrac{\partial^2 x^m}{\partial \overline{x}^s \partial \overline{x}^t}\dfrac{\partial x^i}{\partial \overline{x}^r}$

▶ **38.** Let g denote the determinant of the matrix having the components g_{ij}, $i, j = 1, 2, 3$. Show that

(a) $g\,e_{rst} = \begin{vmatrix} g_{1r} & g_{1s} & g_{1t} \\ g_{2r} & g_{2s} & g_{2t} \\ g_{3r} & g_{3s} & g_{3t} \end{vmatrix}$

(b) $g\,e_{rst}e_{ijk} = \begin{vmatrix} g_{ir} & g_{is} & g_{it} \\ g_{jr} & g_{js} & g_{jt} \\ g_{kr} & g_{ks} & g_{kt} \end{vmatrix}$

▶ **39.** Show that $e^{ijk}e_{mnp} = \delta^{ijk}_{mnp} = \begin{vmatrix} \delta^i_m & \delta^i_n & \delta^i_p \\ \delta^j_m & \delta^j_n & \delta^j_p \\ \delta^k_m & \delta^k_n & \delta^k_p \end{vmatrix}$

▶ **40.** Show that $e^{ijk}e_{mnp}A^{mnp} = A^{ijk} - A^{ikj} + A^{kij} - A^{jik} + A^{jki} - A^{kji}$

Hint: Use the results from problem 39.

▶ **41.** Show that

(a) $e^{ij}e_{ij} = 2!$

(b) $e^{ijk}e_{ijk} = 3!$

(c) $e^{ijkl}e_{ijkl} = 4!$

(d) Guess at the result $e^{i_1 i_2 \ldots i_n}e_{i_1 i_2 \ldots i_n}$

▶ **42.** Determine if the following statement is true or false. Justify your answer. $e_{ijk}A_iB_jC_k = e_{ijk}A_jB_kC_i$.

▶ **43.** Let a_{ij}, $i,j = 1,2$ denote the components of a 2×2 matrix A, which are functions of time t.

(a) Expand both $|A| = e_{ij}a_{i1}a_{j2}$ and $|A| = \begin{vmatrix} a_{11} & a_{12} \\ a_{21} & a_{22} \end{vmatrix}$ to verify that these representations are the same.

(b) Verify the equivalence of the derivative relations

$$\frac{d|A|}{dt} = e_{ij}\frac{da_{i1}}{dt}a_{j2} + e_{ij}a_{i1}\frac{da_{j2}}{dt} \quad \text{and} \quad \frac{d|A|}{dt} = \begin{vmatrix} \frac{da_{11}}{dt} & \frac{da_{12}}{dt} \\ a_{21} & a_{22} \end{vmatrix} + \begin{vmatrix} a_{11} & a_{12} \\ \frac{da_{21}}{dt} & \frac{da_{22}}{dt} \end{vmatrix}$$

(c) Let a_{ij}, $i,j = 1,2,3$ denote the components of a 3×3 matrix A, which are functions of time t. Develop appropriate relations, expand them and verify, similar to parts (a) and (b) above, the representation of a determinant and its derivative.

▶ **44.** For $f = f(x^1, x^2, x^3)$ and $\phi = \phi(f)$ differentiable scalar functions, use the indicial notation to find a formula to calculate grad ϕ.

▶ **45.** Use the indicial notation to prove (a) $\quad \nabla \times \nabla \phi = \vec{0}$ $\qquad$ (b) $\quad \nabla \cdot \nabla \times \vec{A} = 0$

▶ **46.** If A_{ij} is symmetric and B_{ij} is skew-symmetric, $i,j = 1,2,3$, then calculate $C = A_{ij}B_{ij}$.

▶ **47.** Assume $\overline{A}_{ij} = \overline{A}_{ij}(\overline{x}^1, \overline{x}^2, \overline{x}^3)$ and $A_{ij} = A_{ij}(x^1, x^2, x^3)$ for $i,j = 1,2,3$ are related by the expression $\overline{A}_{mn} = A_{ij}\frac{\partial x^i}{\partial \overline{x}^m}\frac{\partial x^j}{\partial \overline{x}^n}$. Calculate the derivative $\frac{\partial \overline{A}_{mn}}{\partial \overline{x}^k}$.

▶ **48.** Prove that if any two rows (or two columns) of a matrix are interchanged, then the value of the determinant of the matrix is multiplied by minus one. Construct your proof using 3×3 matrices.

▶ **49.** Prove that if two rows (or columns) of a matrix are proportional, then the value of the determinant of the matrix is zero. Construct your proof using 3×3 matrices.

▶ **50.** Prove that if a row (or column) of a matrix is altered by adding some constant multiple of some other row (or column), then the value of the determinant of the matrix remains unchanged. Construct your proof using 3×3 matrices.

▶ **51.** Simplify the expression $\phi = e_{ijk}e_{\ell mn}A_{i\ell}A_{jm}A_{kn}$.

▶ **52.** Let A_{ijk} denote a third order system where $i,j,k = 1,2$. (a) How many components does this system have? (b) Let A_{ijk} be skew-symmetric in the last pair of indices, how many independent components does the system have?

▶ **53.** Let A_{ijk} denote a third order system where $i,j,k = 1,2,3$. (a) How many components does this system have? (b) In addition let $A_{ijk} = A_{jik}$ and $A_{ikj} = -A_{ijk}$ and determine the number of distinct nonzero components for A_{ijk}.

▶ **54.** Show that every second order system T_{ij} can be expressed as the sum of a symmetric system A_{ij} and skew-symmetric system B_{ij}. Find A_{ij} and B_{ij} in terms of the components of T_{ij}.

▶ **55.** Consider the system A_{ijk}, $\quad i, j, k = 1, 2, 3, 4$.

(a) How many components does this system have?

(b) Assume A_{ijk} is skew-symmetric in the last pair of indices, how many independent components does this system have?

(c) Assume that in addition to being skew-symmetric in the last pair of indices, $A_{ijk} + A_{jki} + A_{kij} = 0$ is satisfied for all values of i, j, and k, then how many independent components does the system have?

▶ **56.** (a) Write the equation of a line $\vec{r} = \vec{r}_0 + t\vec{A}$ in indicial form. (b) Write the equation of the plane $\vec{n} \cdot (\vec{r} - \vec{r}_0) = 0$ in indicial form. (c) Write the equation of a general line in scalar form. (d) Write the equation of a plane in scalar form. (e) Find the equation of the line defined by the intersection of the planes $2x + 3y + 6z = 12$ and $6x + 3y + z = 6$. (f) Find the equation of the plane through the points $(5, 3, 2), (3, 1, 5), (1, 3, 3)$. Find also the normal to this plane.

▶ **57.** The angle $0 \le \theta \le \pi$ between two skew lines in space is defined as the angle between their direction vectors when these vectors are placed at the origin. Show that for two lines with direction numbers a_i and b_i $i = 1, 2, 3$, the cosine of the angle between these lines satisfies

$$\cos\theta = \frac{a_i b_i}{\sqrt{a_i a_i}\sqrt{b_i b_i}}$$

▶ **58.** Let $a_{ij} = -a_{ji}$ for $i, j = 1, 2, \ldots, N$ and prove that for N odd $det(a_{ij}) = 0$.

▶ **59.** Let $\lambda = A_{ij}x_i x_j$ where $A_{ij} = A_{ji}$ and calculate (a) $\dfrac{\partial\lambda}{\partial x_m}$ (b) $\dfrac{\partial^2\lambda}{\partial x_m \partial x_k}$

▶ **60.** Given an arbitrary nonzero vector U_k, $k = 1, 2, 3$, define the matrix elements $a_{ij} = e_{ijk}U_k$, where e_{ijk} is the e-permutation symbol. Determine if a_{ij} is symmetric or skew-symmetric. Suppose U_k is defined by the above equation for arbitrary nonzero a_{ij}, then solve for U_k in terms of the a_{ij}.

▶ **61.** If $A_{ij} = A_i B_j \ne 0$ for all i, j values and $A_{ij} = A_{ji}$ for $i, j = 1, 2, \ldots, N$, show that $A_{ij} = \lambda B_i B_j$ where λ is a constant. State what λ is.

▶ **62.** Assume that A_{ijkm}, with $i, j, k, m = 1, 2, 3$, is completely skew-symmetric. How many independent components does this quantity have?

▶ **63.** Consider $R_{ijkm}, i, j, k, m = 1, 2, 3, 4$. (a) How many components does this quantity have? (b) If $R_{ijkm} = -R_{ijmk} = -R_{jikm}$ then how many independent components does R_{ijkm} have? (c) If in addition $R_{ijkm} = R_{kmij}$ determine the number of independent components.

▶ **64.** Let $x_i = a_{ij}\bar{x}_j$, $\quad i, j = 1, 2, 3$ denote a change of variables from a barred system of coordinates to an unbarred system of coordinates and assume that $\bar{A}_i = a_{ij}A_j$ where a_{ij} are constants, $\bar{A}_i$ is a function of the $\bar{x}_j$ variables and A_j is a function of the x_j variables. Calculate $\dfrac{\partial\bar{A}_i}{\partial\bar{x}_m}$.

§1.2 TENSOR CONCEPTS AND TRANSFORMATIONS

For $\widehat{\mathbf{e}}_1, \widehat{\mathbf{e}}_2, \widehat{\mathbf{e}}_3$ independent orthogonal unit vectors (base vectors), we may write any vector $\vec{A}$ as

$$\vec{A} = A_1\widehat{\mathbf{e}}_1 + A_2\widehat{\mathbf{e}}_2 + A_3\widehat{\mathbf{e}}_3$$

where (A_1, A_2, A_3) are the coordinates of $\vec{A}$ relative to the base vectors chosen. These components are the projection of $\vec{A}$ onto the base vectors and

$$\vec{A} = (\vec{A}\cdot\widehat{\mathbf{e}}_1)\widehat{\mathbf{e}}_1 + (\vec{A}\cdot\widehat{\mathbf{e}}_2)\widehat{\mathbf{e}}_2 + (\vec{A}\cdot\widehat{\mathbf{e}}_3)\widehat{\mathbf{e}}_3.$$

Select any three independent orthogonal vectors, $(\vec{E}_1, \vec{E}_2, \vec{E}_3)$, not necessarily of unit length, we can then write

$$\widehat{\mathbf{e}}_1 = \frac{\vec{E}_1}{|\vec{E}_1|}, \qquad \widehat{\mathbf{e}}_2 = \frac{\vec{E}_2}{|\vec{E}_2|}, \qquad \widehat{\mathbf{e}}_3 = \frac{\vec{E}_3}{|\vec{E}_3|},$$

and consequently, the vector $\vec{A}$ can be expressed as

$$\vec{A} = \left(\frac{\vec{A}\cdot\vec{E}_1}{\vec{E}_1\cdot\vec{E}_1}\right)\vec{E}_1 + \left(\frac{\vec{A}\cdot\vec{E}_2}{\vec{E}_2\cdot\vec{E}_2}\right)\vec{E}_2 + \left(\frac{\vec{A}\cdot\vec{E}_3}{\vec{E}_3\cdot\vec{E}_3}\right)\vec{E}_3.$$

Here we say that

$$\frac{\vec{A}\cdot\vec{E}_{(i)}}{\vec{E}_{(i)}\cdot\vec{E}_{(i)}}, \quad i = 1,2,3$$

are the components of $\vec{A}$ relative to the chosen base vectors $\vec{E}_1, \vec{E}_2, \vec{E}_3$. Recall that the parenthesis about the subscript i denotes that there is no summation on this subscript. It is then treated as a free subscript which can have any of the values $1, 2$ or 3.

Reciprocal Basis

Consider a set of any three independent vectors $(\vec{E}_1, \vec{E}_2, \vec{E}_3)$ which are not necessarily orthogonal, nor of unit length. In order to represent the vector $\vec{A}$ in terms of these vectors we must find components (A^1, A^2, A^3) such that

$$\vec{A} = A^1\vec{E}_1 + A^2\vec{E}_2 + A^3\vec{E}_3.$$

This can be done by taking appropriate projections and obtaining three equations and three unknowns from which the components are determined. A much easier way to find the components (A^1, A^2, A^3) is to construct a reciprocal basis $(\vec{E}^1, \vec{E}^2, \vec{E}^3)$. Recall that two bases $(\vec{E}_1, \vec{E}_2, \vec{E}_3)$ and $(\vec{E}^1, \vec{E}^2, \vec{E}^3)$ are said to be reciprocal if they satisfy the condition

$$\vec{E}_i \cdot \vec{E}^j = \delta_i^j = \begin{cases} 1 & \text{if } i = j \\ 0 & \text{if } i \neq j \end{cases}.$$

Note that $\vec{E}_2 \cdot \vec{E}^1 = \delta_2^1 = 0$ and $\vec{E}_3 \cdot \vec{E}^1 = \delta_3^1 = 0$ so that the vector $\vec{E}^1$ is perpendicular to both the vectors $\vec{E}_2$ and $\vec{E}_3$. (i.e. A vector from one basis is orthogonal to two of the vectors from the other basis.) We can therefore write $\vec{E}^1 = V^{-1}\vec{E}_2 \times \vec{E}_3$ where V is a constant to be determined. By taking the dot product of both sides of this equation with the vector $\vec{E}_1$ we find that $V = \vec{E}_1 \cdot (\vec{E}_2 \times \vec{E}_3)$ is the volume of the parallelepiped formed by the three vectors $\vec{E}_1, \vec{E}_2, \vec{E}_3$ when their origins are made to coincide. In a similar manner it can be demonstrated that for $(\vec{E}_1, \vec{E}_2, \vec{E}_3)$ a given set of basis vectors, then the reciprocal basis vectors are determined from the relations

$$\vec{E}^1 = \frac{1}{V}\vec{E}_2 \times \vec{E}_3, \qquad \vec{E}^2 = \frac{1}{V}\vec{E}_3 \times \vec{E}_1, \qquad \vec{E}^3 = \frac{1}{V}\vec{E}_1 \times \vec{E}_2,$$

where $V = \vec{E}_1 \cdot (\vec{E}_2 \times \vec{E}_3) \neq 0$ is a triple scalar product and represents the volume of the parallelepiped having the basis vectors for its sides.

Let $(\vec{E}_1, \vec{E}_2, \vec{E}_3)$ and $(\vec{E}^1, \vec{E}^2, \vec{E}^3)$ denote a system of reciprocal bases. We can represent any vector $\vec{A}$ with respect to either of these bases. If we select the basis $(\vec{E}_1, \vec{E}_2, \vec{E}_3)$ and represent $\vec{A}$ in the form

$$\vec{A} = A^1\vec{E}_1 + A^2\vec{E}_2 + A^3\vec{E}_3, \tag{1.2.1}$$

then the components (A^1, A^2, A^3) of $\vec{A}$ relative to the basis vectors $(\vec{E}_1, \vec{E}_2, \vec{E}_3)$ are called the contravariant components of $\vec{A}$. These components can be determined from the equations

$$\vec{A} \cdot \vec{E}^1 = A^1, \qquad \vec{A} \cdot \vec{E}^2 = A^2, \qquad \vec{A} \cdot \vec{E}^3 = A^3.$$

Similarly, if we choose the reciprocal basis $(\vec{E}^1, \vec{E}^2, \vec{E}^3)$ and represent $\vec{A}$ in the form

$$\vec{A} = A_1\vec{E}^1 + A_2\vec{E}^2 + A_3\vec{E}^3, \tag{1.2.2}$$

then the components (A_1, A_2, A_3) relative to the basis $(\vec{E}^1, \vec{E}^2, \vec{E}^3)$ are called the covariant components of $\vec{A}$. These components can be determined from the relations

$$\vec{A} \cdot \vec{E}_1 = A_1, \qquad \vec{A} \cdot \vec{E}_2 = A_2, \qquad \vec{A} \cdot \vec{E}_3 = A_3.$$

The contravariant and covariant components are different ways of representing the same vector with respect to a set of reciprocal basis vectors. There is a simple relationship between these components which we now develop. We introduce the notation

$$\vec{E}_i \cdot \vec{E}_j = g_{ij} = g_{ji}, \qquad \text{and} \qquad \vec{E}^i \cdot \vec{E}^j = g^{ij} = g^{ji} \tag{1.2.3}$$

where g_{ij} are called the metric components of the space and g^{ij} are called the conjugate metric components of the space. We can then write

$$\vec{A} \cdot \vec{E}_1 = A_1(\vec{E}^1 \cdot \vec{E}_1) + A_2(\vec{E}^2 \cdot \vec{E}_1) + A_3(\vec{E}^3 \cdot \vec{E}_1) = A_1$$
$$\vec{A} \cdot \vec{E}_1 = A^1(\vec{E}_1 \cdot \vec{E}_1) + A^2(\vec{E}_2 \cdot \vec{E}_1) + A^3(\vec{E}_3 \cdot \vec{E}_1) = A_1$$

or

$$A_1 = A^1 g_{11} + A^2 g_{12} + A^3 g_{13}. \tag{1.2.4}$$

In a similar manner, by considering the dot products $\vec{A} \cdot \vec{E}_2$ and $\vec{A} \cdot \vec{E}_3$ one can establish the results

$$A_2 = A^1 g_{21} + A^2 g_{22} + A^3 g_{23} \qquad A_3 = A^1 g_{31} + A^2 g_{32} + A^3 g_{33}.$$

These results can be expressed with the index notation as

$$A_i = g_{ik}A^k. \tag{1.2.6}$$

Forming the dot products $\vec{A} \cdot \vec{E}^1, \quad \vec{A} \cdot \vec{E}^2, \quad \vec{A} \cdot \vec{E}^3$ it can be verified that

$$A^i = g^{ik}A_k. \tag{1.2.7}$$

The equations (1.2.6) and (1.2.7) are relations which exist between the contravariant and covariant components of the vector $\vec{A}$. Similarly, if for some value j we have $\vec{E}^j = \alpha\vec{E}_1 + \beta\vec{E}_2 + \gamma\vec{E}_3$, then one can show that $\vec{E}^j = g^{ij}\vec{E}_i$. This is left as an exercise.

Coordinate Transformations

Consider a coordinate transformation from a set of coordinates (x, y, z) to (u, v, w) defined by a set of transformation equations

$$x = x(u, v, w) \qquad y = y(u, v, w) \qquad z = z(u, v, w) \tag{1.2.8}$$

It is assumed that these transformations are single valued, continuous and possess the inverse transformation

$$u = u(x, y, z) \qquad v = v(x, y, z) \qquad w = w(x, y, z). \tag{1.2.9}$$

These transformation equations define a set of coordinate surfaces and coordinate curves. The coordinate surfaces are defined by the equations

$$u(x, y, z) = c_1 \qquad v(x, y, z) = c_2 \qquad w(x, y, z) = c_3 \tag{1.2.10}$$

where c_1, c_2, c_3 are constants. These surfaces intersect in the coordinate curves

$$\vec{r}(u, c_2, c_3), \qquad \vec{r}(c_1, v, c_3), \qquad \vec{r}(c_1, c_2, w), \tag{1.2.11}$$

where

$$\vec{r}(u, v, w) = x(u, v, w)\,\widehat{\mathbf{e}}_1 + y(u, v, w)\,\widehat{\mathbf{e}}_2 + z(u, v, w)\,\widehat{\mathbf{e}}_3.$$

The general situation is illustrated in the figure 1.2-1.

Consider the vectors

$$\vec{E}^1 = \operatorname{grad} u = \nabla u, \qquad \vec{E}^2 = \operatorname{grad} v = \nabla v, \qquad \vec{E}^3 = \operatorname{grad} w = \nabla w \tag{1.2.12}$$

evaluated at the common point of intersection (c_1, c_2, c_3) of the coordinate surfaces. The system of vectors $(\vec{E}^1, \vec{E}^2, \vec{E}^3)$ can be selected as a system of basis vectors which are normal to the coordinate surfaces. Similarly, the vectors

$$\vec{E}_1 = \frac{\partial \vec{r}}{\partial u}, \qquad \vec{E}_2 = \frac{\partial \vec{r}}{\partial v}, \qquad \vec{E}_3 = \frac{\partial \vec{r}}{\partial w} \tag{1.2.13}$$

when evaluated at the common point of intersection (c_1, c_2, c_3) forms a system of vectors $(\vec{E}_1, \vec{E}_2, \vec{E}_3)$ which we can select as a basis. This basis is a set of tangent vectors to the coordinate curves. It is now demonstrated that the normal basis $(\vec{E}^1, \vec{E}^2, \vec{E}^3)$ and the tangential basis $(\vec{E}_1, \vec{E}_2, \vec{E}_3)$ are a set of reciprocal bases.

Recall that $\vec{r} = x\,\widehat{\mathbf{e}}_1 + y\,\widehat{\mathbf{e}}_2 + z\,\widehat{\mathbf{e}}_3$ denotes the position vector of a variable point. By substitution for x, y, z from (1.2.8) there results

$$\vec{r} = \vec{r}(u, v, w) = x(u, v, w)\,\widehat{\mathbf{e}}_1 + y(u, v, w)\,\widehat{\mathbf{e}}_2 + z(u, v, w)\,\widehat{\mathbf{e}}_3. \tag{1.2.14}$$

A small change in $\vec{r}$ is denoted

$$d\vec{r} = dx\,\widehat{\mathbf{e}}_1 + dy\,\widehat{\mathbf{e}}_2 + dz\,\widehat{\mathbf{e}}_3 = \frac{\partial \vec{r}}{\partial u}\,du + \frac{\partial \vec{r}}{\partial v}\,dv + \frac{\partial \vec{r}}{\partial w}\,dw \tag{1.2.15}$$

38

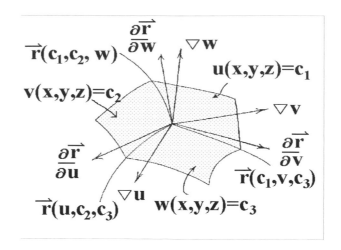

Figure 1.2-1. Coordinate curves and coordinate surfaces.

where

$$\frac{\partial \vec{r}}{\partial u} = \frac{\partial x}{\partial u}\,\widehat{\mathbf{e}}_1 + \frac{\partial y}{\partial u}\,\widehat{\mathbf{e}}_2 + \frac{\partial z}{\partial u}\,\widehat{\mathbf{e}}_3$$

$$\frac{\partial \vec{r}}{\partial v} = \frac{\partial x}{\partial v}\,\widehat{\mathbf{e}}_1 + \frac{\partial y}{\partial v}\,\widehat{\mathbf{e}}_2 + \frac{\partial z}{\partial v}\,\widehat{\mathbf{e}}_3 \tag{1.2.16}$$

$$\frac{\partial \vec{r}}{\partial w} = \frac{\partial x}{\partial w}\,\widehat{\mathbf{e}}_1 + \frac{\partial y}{\partial w}\,\widehat{\mathbf{e}}_2 + \frac{\partial z}{\partial w}\,\widehat{\mathbf{e}}_3.$$

In terms of the u, v, w coordinates, this change can be thought of as moving along the diagonal of a parallelepiped having the vector sides $\dfrac{\partial \vec{r}}{\partial u}\,du$, $\dfrac{\partial \vec{r}}{\partial v}\,dv$, and $\dfrac{\partial \vec{r}}{\partial w}\,dw$.

Assume $u = u(x, y, z)$ is defined by equation (1.2.9) and differentiate this relation to obtain

$$du = \frac{\partial u}{\partial x}\,dx + \frac{\partial u}{\partial y}\,dy + \frac{\partial u}{\partial z}\,dz. \tag{1.2.17}$$

The equation (1.2.15) enables us to represent this differential in the form:

$$du = \operatorname{grad} u \cdot d\vec{r}$$

$$du = \operatorname{grad} u \cdot \left(\frac{\partial \vec{r}}{\partial u}\,du + \frac{\partial \vec{r}}{\partial v}\,dv + \frac{\partial \vec{r}}{\partial w}\,dw \right) \tag{1.2.18}$$

$$du = \left(\operatorname{grad} u \cdot \frac{\partial \vec{r}}{\partial u} \right) du + \left(\operatorname{grad} u \cdot \frac{\partial \vec{r}}{\partial v} \right) dv + \left(\operatorname{grad} u \cdot \frac{\partial \vec{r}}{\partial w} \right) dw.$$

By comparing like terms in this last equation we find that

$$\vec{E}^1 \cdot \vec{E}_1 = 1, \qquad \vec{E}^1 \cdot \vec{E}_2 = 0, \qquad \vec{E}^1 \cdot \vec{E}_3 = 0. \tag{1.2.19}$$

Similarly, from the other equations in equation (1.2.9) which define $v = v(x, y, z)$, and $w = w(x, y, z)$ it can be demonstrated that

$$dv = \left(\operatorname{grad} v \cdot \frac{\partial \vec{r}}{\partial u} \right) du + \left(\operatorname{grad} v \cdot \frac{\partial \vec{r}}{\partial v} \right) dv + \left(\operatorname{grad} v \cdot \frac{\partial \vec{r}}{\partial w} \right) dw \tag{1.2.20}$$

and

$$dw = \left(\operatorname{grad} w \cdot \frac{\partial \vec{r}}{\partial u} \right) du + \left(\operatorname{grad} w \cdot \frac{\partial \vec{r}}{\partial v} \right) dv + \left(\operatorname{grad} w \cdot \frac{\partial \vec{r}}{\partial w} \right) dw. \tag{1.2.21}$$

By comparing like terms in equations (1.2.20) and (1.2.21) we find

$$\vec{E}^2 \cdot \vec{E}_1 = 0, \qquad \vec{E}^2 \cdot \vec{E}_2 = 1, \qquad \vec{E}^2 \cdot \vec{E}_3 = 0$$
$$\vec{E}^3 \cdot \vec{E}_1 = 0, \qquad \vec{E}^3 \cdot \vec{E}_2 = 0, \qquad \vec{E}^3 \cdot \vec{E}_3 = 1. \qquad (1.2.22)$$

The equations (1.2.22) and (1.2.19) show us that the basis vectors defined by equations (1.2.12) and (1.2.13) are reciprocal.

Introducing the notation

$$(x^1, x^2, x^3) = (u, v, w) \qquad (y^1, y^2, y^3) = (x, y, z) \qquad (1.2.23)$$

where the $x's$ denote the generalized coordinates and the $y's$ denote the rectangular Cartesian coordinates, the above equations can be expressed in a more concise form with the index notation. For example, if

$$x^i = x^i(x, y, z) = x^i(y^1, y^2, y^3), \quad \text{and} \quad y^i = y^i(u, v, w) = y^i(x^1, x^2, x^3), \quad i = 1, 2, 3 \qquad (1.2.24)$$

then the reciprocal basis vectors can be represented

$$\vec{E}^i = \operatorname{grad} x^i, \qquad i = 1, 2, 3 \qquad (1.2.25)$$

and

$$\vec{E}_i = \frac{\partial \vec{r}}{\partial x^i}, \qquad i = 1, 2, 3. \qquad (1.2.26)$$

We now show that these basis vectors are reciprocal. Observe that $\vec{r} = \vec{r}(x^1, x^2, x^3)$ with

$$d\vec{r} = \frac{\partial \vec{r}}{\partial x^m} dx^m \qquad (1.2.27)$$

and consequently

$$dx^i = \operatorname{grad} x^i \cdot d\vec{r} = \operatorname{grad} x^i \cdot \frac{\partial \vec{r}}{\partial x^m} dx^m = \left(\vec{E}^i \cdot \vec{E}_m \right) dx^m = \delta^i_m \, dx^m, \qquad i = 1, 2, 3 \qquad (1.2.28)$$

Comparing like terms in this last equation establishes the result that

$$\vec{E}^i \cdot \vec{E}_m = \delta^i_m, \quad i, m = 1, 2, 3 \qquad (1.2.29)$$

which demonstrates that the basis vectors are reciprocal.

Scalars, Vectors and Tensors

Tensors are quantities which obey certain transformation laws. That is, scalars, vectors, matrices and higher order arrays can be thought of as components of a tensor quantity. We shall be interested in finding how these components are represented in various coordinate systems. We desire knowledge of these transformation laws in order that we can represent various physical laws in a form which is independent of the coordinate system chosen. Before defining different types of tensors let us examine what we mean by a coordinate transformation.

Coordinate transformations of the type found in equations (1.2.8) and (1.2.9) can be generalized to higher dimensions. Let x^i, $i = 1, 2, \ldots, N$ denote N variables. These quantities can be thought of as representing a variable point $(x^1, x^2, \ldots, x^N)$ in an N dimensional space V_N. Another set of N quantities, call them barred quantities, $\overline{x}^i$, $i = 1, 2, \ldots, N$, can be used to represent a variable point $(\overline{x}^1, \overline{x}^2, \ldots, \overline{x}^N)$ in an N dimensional space $\overline{V}_N$. When the $x's$ are related to the $\overline{x}'s$ by equations of the form

$$x^i = x^i(\overline{x}^1, \overline{x}^2, \ldots, \overline{x}^N), \quad i = 1, 2, \ldots, N \tag{1.2.30}$$

then a transformation is said to exist between the coordinates x^i and $\overline{x}^i$, $i = 1, 2, \ldots, N$. Whenever the relations (1.2.30) are functionally independent, single valued and possess partial derivatives such that the Jacobian of the transformation

$$J\left(\frac{x}{\overline{x}}\right) = J\left(\frac{x^1, x^2, \ldots, x^N}{\overline{x}^1, \overline{x}^2, \ldots, \overline{x}^N}\right) = \begin{vmatrix} \frac{\partial x^1}{\partial \overline{x}^1} & \frac{\partial x^1}{\partial \overline{x}^2} & \cdots & \frac{\partial x^1}{\partial \overline{x}^N} \\ \vdots & \vdots & \cdots & \vdots \\ \frac{\partial x^N}{\partial \overline{x}^1} & \frac{\partial x^N}{\partial \overline{x}^2} & \cdots & \frac{\partial x^N}{\partial \overline{x}^N} \end{vmatrix} \tag{1.2.31}$$

is different from zero, then there exists an inverse transformation

$$\overline{x}^i = \overline{x}^i(x^1, x^2, \ldots, x^N), \quad i = 1, 2, \ldots, N. \tag{1.2.32}$$

For brevity the transformation equations (1.2.30) and (1.2.32) are sometimes expressed by the notation

$$x^i = x^i(\overline{x}), \, i = 1, \ldots, N \quad \text{and} \quad \overline{x}^i = \overline{x}^i(x), \, i = 1, \ldots, N. \tag{1.2.33}$$

Consider a sequence of transformations from x to $\overline{x}$ and then from $\overline{x}$ to $\overline{\overline{x}}$ coordinates. For simplicity let $\overline{x} = y$ and $\overline{\overline{x}} = z$. If we denote by T_1, T_2 and T_3 the transformations

$$T_1: \quad y^i = y^i(x^1, \ldots, x^N) \quad i = 1, \ldots, N \quad \text{or} \quad T_1 x = y$$

$$T_2: \quad z^i = z^i(y^1, \ldots, y^N) \quad i = 1, \ldots, N \quad \text{or} \quad T_2 y = z$$

Then the transformation T_3 obtained by substituting T_1 into T_2 is called the product of two successive transformations and is written

$$T_3: \quad z^i = z^i(y^1(x^1, \ldots, x^N), \ldots, y^N(x^1, \ldots, x^N)) \quad i = 1, \ldots, N \quad \text{or} \quad T_3 x = T_2 T_1 x = z.$$

This product transformation is denoted symbolically by $T_3 = T_2 T_1$.

The Jacobian of the product transformation is equal to the product of Jacobians associated with the product transformation and $J_3 = J_2 J_1$.

Transformations Form a Group

A group G is a nonempty set of elements together with a law, for combining the elements. The combined elements are denoted by a product. Thus, if a and b are elements in G then no matter how you define the law for combining elements, the product combination is denoted ab. The set G and combining law forms a group if the following properties are satisfied:

(i) For all $a, b \in G$, then $ab \in G$. This is called the closure property.

(ii) There exists an identity element I such that for all $a \in G$ we have $Ia = aI = a$.

(iii) There exists an inverse element. That is, for all $a \in G$ there exists an inverse element a^{-1} such that $a\,a^{-1} = a^{-1}a = I$.

(iv) The associative law holds under the combining law and $a(bc) = (ab)c$ for all $a, b, c \in G$.

For example, the set of elements $G = \{1, -1, i, -i\}$, where $i^2 = -1$ together with the combining law of ordinary multiplication, forms a group. This can be seen from the multiplication table.

$\times$	1	-1	i	-i
1	1	-1	i	-i
-1	-1	1	-i	i
-i	-i	i	1	-1
i	i	-i	-1	1

The set of all coordinate transformations of the form found in equation (1.2.30), with Jacobian different from zero, forms a group because:

(i) The product transformation, which consists of two successive transformations, belongs to the set of transformations. (closure)

(ii) The identity transformation exists in the special case that $\bar{x}$ and x are the same coordinates.

(iii) The inverse transformation exists because the Jacobian of each individual transformation is different from zero.

(iv) The associative law is satisfied in that the transformations satisfy the property $T_3(T_2T_1) = (T_3T_2)T_1$.

When the given transformation equations contain a parameter the combining law is often times represented as a product of symbolic operators. For example, we denote by T_α a transformation of coordinates having a parameter α. The inverse transformation can be denoted by T_α^{-1} and one can write $T_\alpha x = \bar{x}$ or $x = T_\alpha^{-1}\bar{x}$. We let T_β denote the same transformation, but with a parameter β, then the transitive property is expressed symbolically by $T_\alpha T_\beta = T_\gamma$ where the product $T_\alpha T_\beta$ represents the result of performing two successive transformations. The first coordinate transformation uses the given transformation equations and uses the parameter α in these equations. This transformation is then followed by another coordinate transformation using the same set of transformation equations, but this time the parameter value is β. The above symbolic product is used to demonstrate that the result of applying two successive transformations produces a result which is equivalent to performing a single transformation of coordinates having the parameter value of γ. Usually some relationship can then be established between the parameter values α, β and γ.

In this symbolic notation, we let T_θ denote the identity transformation. That is, using the parameter value of θ in the given set of transformation equations produces the identity transformation. The inverse transformation can then be expressed in the form of finding the parameter value β such that $T_\alpha T_\beta = T_\theta$.

42

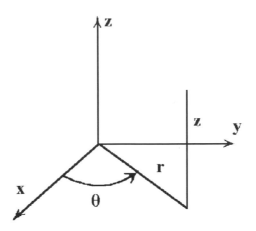

Figure 1.2-2. Cylindrical coordinates.

Cartesian Coordinates

At times it is convenient to introduce an orthogonal Cartesian coordinate system having coordinates y^i, $i = 1, 2, \ldots, N$. This space is denoted E_N and represents an N-dimensional Euclidean space. Whenever the generalized independent coordinates x^i, $i = 1, \ldots, N$ are functions of the $y's$, and these equations are functionally independent, then there exists independent transformation equations

$$y^i = y^i(x^1, x^2, \ldots, x^N), \quad i = 1, 2, \ldots, N, \tag{1.2.34}$$

with Jacobian different from zero. Similarly, if there is some other set of generalized coordinates, say a barred system $\overline{x}^i$, $i = 1, \ldots, N$ where the $\overline{x}'s$ are independent functions of the $y's$, then there will exist another set of independent transformation equations

$$y^i = y^i(\overline{x}^1, \overline{x}^2, \ldots, \overline{x}^N), \quad i = 1, 2, \ldots, N, \tag{1.2.35}$$

with Jacobian different from zero. The transformations found in the equations (1.2.34) and (1.2.35) imply that there exists relations between the $x's$ and $\overline{x}'s$ of the form (1.2.30) with inverse transformations of the form (1.2.32). It should be remembered that the concepts and ideas developed in this section can be applied to a space V_N of any finite dimension. Two dimensional surfaces $(N = 2)$ and three dimensional spaces $(N = 3)$ will occupy most of our applications. In relativity, one must consider spaces where $N = 4$.

EXAMPLE 1.2-1. (cylindrical coordinates (r, θ, z)) Consider the transformation

$$x = x(r, \theta, z) = r \cos \theta \qquad y = y(r, \theta, z) = r \sin \theta \qquad z = z(r, \theta, z) = z$$

from rectangular coordinates (x, y, z) to cylindrical coordinates (r, θ, z), illustrated in the figure 1.2-2. By letting

$$y^1 = x, \quad y^2 = y, \quad y^3 = z \qquad x^1 = r, \quad x^2 = \theta, \quad x^3 = z$$

the above set of equations are examples of the transformation equations (1.2.8) with $u = r$, $v = \theta$, $w = z$ as the generalized coordinates.

■

EXAMPLE 1.2.2. (Spherical Coordinates) (ρ, θ, ϕ)

Consider the transformation

$$x = x(\rho, \theta, \phi) = \rho \sin \theta \cos \phi \qquad y = y(\rho, \theta, \phi) = \rho \sin \theta \sin \phi \qquad z = z(\rho, \theta, \phi) = \rho \cos \theta$$

from rectangular coordinates (x, y, z) to spherical coordinates (ρ, θ, ϕ). By letting

$$y^1 = x, \; y^2 = y, \; y^3 = z \qquad x^1 = \rho, \; x^2 = \theta, \; x^3 = \phi$$

the above set of equations has the form found in equation (1.2.8) with $u = \rho$, $v = \theta$, $w = \phi$ the generalized coordinates. One could place bars over the $x's$ in this example in order to distinguish these coordinates from the $x's$ of the previous example. The spherical coordinates (ρ, θ, ϕ) are illustrated in the figure 1.2-3.

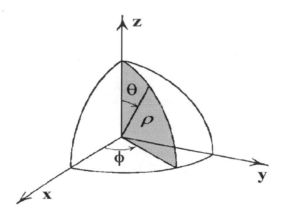

Figure 1.2-3. Spherical coordinates.

■

Scalar Functions and Invariance

We are now at a point where we can begin to define what tensor quantities are. The first definition is for a scalar invariant or tensor of order zero.

Definition: (Absolute scalar field) Assume there exists a coordinate transformation of the type (1.2.30) with Jacobian J different from zero. Let the scalar function

$$f = f(x^1, x^2, \ldots, x^N) \tag{1.2.36}$$

be a function of the coordinates x^i, $i = 1, \ldots, N$ in a space V_N. Whenever there exists a function

$$\overline{f} = \overline{f}(\overline{x}^1, \overline{x}^2, \ldots, \overline{x}^N) \tag{1.2.37}$$

which is a function of the coordinates $\overline{x}^i$, $i = 1, \ldots, N$ such that $\overline{f} = J^W f$, then f is called a tensor of rank or order zero of weight W in the space V_N. Whenever $W = 0$, the scalar f is called the component of an absolute scalar field and is referred to as an absolute tensor of rank or order zero.

44

That is, an absolute scalar field is an invariant object in the space V_N with respect to the group of coordinate transformations. It has a single component in each coordinate system. For any scalar function of the type defined by equation (1.2.36), we can substitute the transformation equations (1.2.30) and obtain

$$f = f(x^1, \ldots, x^N) = f(x^1(\overline{x}), \ldots, x^N(\overline{x})) = \overline{f}(\overline{x}^1, \ldots, \overline{x}^N). \tag{1.2.38}$$

Vector Transformation, Contravariant Components

In V_N consider a curve C defined by the set of parametric equations

$$C: \quad x^i = x^i(t), \quad i = 1, \ldots, N$$

where t is a parameter. The tangent vector to the curve C is the vector

$$\vec{T} = \left(\frac{dx^1}{dt}, \frac{dx^2}{dt}, \ldots, \frac{dx^N}{dt} \right).$$

In index notation, which focuses attention on the components, this tangent vector is denoted

$$T^i = \frac{dx^i}{dt}, \quad i = 1, \ldots, N.$$

For a coordinate transformation of the type defined by equation (1.2.30) with its inverse transformation defined by equation (1.2.32), the curve C is represented in the barred space by

$$\overline{x}^i = \overline{x}^i(x^1(t), x^2(t), \ldots, x^N(t)) = \overline{x}^i(t), \quad i = 1, \ldots, N,$$

with t unchanged. The tangent to the curve in the barred system of coordinates is represented by

$$\frac{d\overline{x}^i}{dt} = \frac{\partial \overline{x}^i}{\partial x^j} \frac{dx^j}{dt}, \quad i = 1, \ldots, N. \tag{1.2.39}$$

Letting $\overline{T}^i$, $i = 1, \ldots, N$ denote the components of this tangent vector in the barred system of coordinates, the equation (1.2.39) can then be expressed in the form

$$\overline{T}^i = \frac{\partial \overline{x}^i}{\partial x^j} T^j, \quad i, j = 1, \ldots, N. \tag{1.2.40}$$

This equation is said to define the transformation law associated with an absolute contravariant tensor of rank or order one. In the case $N = 3$ the matrix form of this transformation is represented

$$\begin{pmatrix} \overline{T}^1 \\ \overline{T}^2 \\ \overline{T}^3 \end{pmatrix} = \begin{pmatrix} \frac{\partial \overline{x}^1}{\partial x^1} & \frac{\partial \overline{x}^1}{\partial x^2} & \frac{\partial \overline{x}^1}{\partial x^3} \\ \frac{\partial \overline{x}^2}{\partial x^1} & \frac{\partial \overline{x}^2}{\partial x^2} & \frac{\partial \overline{x}^2}{\partial x^3} \\ \frac{\partial \overline{x}^3}{\partial x^1} & \frac{\partial \overline{x}^3}{\partial x^2} & \frac{\partial \overline{x}^3}{\partial x^3} \end{pmatrix} \begin{pmatrix} T^1 \\ T^2 \\ T^3 \end{pmatrix} \tag{1.2.41}$$

A more general definition is

Definition: (Contravariant tensor) Whenever N quantities A^i in a coordinate system $(x^1, \ldots, x^N)$ are related to N quantities $\overline{A}^i$ in a coordinate system $(\overline{x}^1, \ldots, \overline{x}^N)$ such that the Jacobian J is different from zero, then if the transformation law

$$\overline{A}^i = J^W \frac{\partial \overline{x}^i}{\partial x^j} A^j$$

is satisfied, these quantities are called the components of a relative tensor of rank or order one with weight W. Whenever $W = 0$ these quantities are called the components of an absolute tensor of rank or order one.

We see that the above transformation law satisfies the group properties.

EXAMPLE 1.2-3. (Transitive Property of Contravariant Transformation)

Show that successive contravariant transformations is also a contravariant transformation.

Solution: Consider the transformation of a vector from an unbarred to a barred system of coordinates. A vector or absolute tensor of rank one $A^i = A^i(x)$, $i = 1, \ldots, N$ will transform like the equation (1.2.40) and

$$\overline{A}^i(\overline{x}) = \frac{\partial \overline{x}^i}{\partial x^j} A^j(x). \tag{1.2.42}$$

Another transformation from $\overline{x} \to \overline{\overline{x}}$ coordinates will produce the components

$$\overline{\overline{A}}^i(\overline{\overline{x}}) = \frac{\partial \overline{\overline{x}}^i}{\partial \overline{x}^j} \overline{A}^j(\overline{x}) \tag{1.2.43}$$

Here we have used the notation $A^j(x)$ to emphasize the dependence of the components A^j upon the x coordinates. Changing indices and substituting equation (1.2.42) into (1.2.43) we find

$$\overline{\overline{A}}^i(\overline{\overline{x}}) = \frac{\partial \overline{\overline{x}}^i}{\partial \overline{x}^j} \frac{\partial \overline{x}^j}{\partial x^m} A^m(x). \tag{1.2.44}$$

From the fact that

$$\frac{\partial \overline{\overline{x}}^i}{\partial \overline{x}^j} \frac{\partial \overline{x}^j}{\partial x^m} = \frac{\partial \overline{\overline{x}}^i}{\partial x^m},$$

the equation (1.2.44) simplifies to

$$\overline{\overline{A}}^i(\overline{\overline{x}}) = \frac{\partial \overline{\overline{x}}^i}{\partial x^m} A^m(x) \tag{1.2.45}$$

and hence this transformation is also contravariant. We express this by saying that the above are transitive with respect to the group of coordinate transformations.

Note that from the chain rule one can write

$$\frac{\partial x^m}{\partial \overline{x}^j} \frac{\partial \overline{x}^j}{\partial x^n} = \frac{\partial x^m}{\partial \overline{x}^1} \frac{\partial \overline{x}^1}{\partial x^n} + \frac{\partial x^m}{\partial \overline{x}^2} \frac{\partial \overline{x}^2}{\partial x^n} + \frac{\partial x^m}{\partial \overline{x}^3} \frac{\partial \overline{x}^3}{\partial x^n} = \frac{\partial x^m}{\partial x^n} = \delta_n^m.$$

Do not make the mistake of writing

$$\frac{\partial x^m}{\partial \overline{x}^2} \frac{\partial \overline{x}^2}{\partial x^n} = \frac{\partial x^m}{\partial x^n} \qquad \text{or} \qquad \frac{\partial x^m}{\partial \overline{x}^3} \frac{\partial \overline{x}^3}{\partial x^n} = \frac{\partial x^m}{\partial x^n}$$

as these expressions are incorrect. Note that there are no summations in these terms, whereas there is a summation index in the representation of the chain rule.

∎

Vector Transformation, Covariant Components

Consider a scalar invariant $A(x) = \overline{A}(\overline{x})$ which is a shorthand notation for the equation

$$A(x^1, x^2, \ldots, x^n) = \overline{A}(\overline{x}^1, \overline{x}^2, \ldots, \overline{x}^n)$$

involving the coordinate transformation of equation (1.2.30). By the chain rule we differentiate this invariant and find that the components of the gradient must satisfy

$$\frac{\partial \overline{A}}{\partial \overline{x}^i} = \frac{\partial A}{\partial x^j} \frac{\partial x^j}{\partial \overline{x}^i}. \tag{1.2.46}$$

Let

$$A_j = \frac{\partial A}{\partial x^j} \qquad \text{and} \qquad \overline{A}_i = \frac{\partial \overline{A}}{\partial \overline{x}^i},$$

then equation (1.2.46) can be expressed as the transformation law

$$\overline{A}_i = A_j \frac{\partial x^j}{\partial \overline{x}^i}. \tag{1.2.47}$$

This is the transformation law for an absolute covariant tensor of rank or order one. A more general definition is

---

Definition: (Covariant tensor) Whenever N quantities A_i in a coordinate system $(x^1, \ldots, x^N)$ are related to N quantities $\overline{A}_i$ in a coordinate system $(\overline{x}^1, \ldots, \overline{x}^N)$, with Jacobian J different from zero, such that the transformation law

$$\overline{A}_i = J^W \frac{\partial x^j}{\partial \overline{x}^i} A_j \tag{1.2.48}$$

is satisfied, then these quantities are called the components of a relative covariant tensor of rank or order one having a weight of W. Whenever $W = 0$, these quantities are called the components of an absolute covariant tensor of rank or order one.

Again we note that the above transformation satisfies the group properties. Absolute tensors of rank or order one are referred to as vectors while absolute tensors of rank or order zero are referred to as scalars.

EXAMPLE 1.2-4. (Transitive Property of Covariant Transformation)

Consider a sequence of transformation laws of the type defined by the equation (1.2.47)

$$x \to \overline{x} \qquad \overline{A}_i(\overline{x}) = A_j(x) \frac{\partial x^j}{\partial \overline{x}^i}$$

$$\overline{x} \to \overline{\overline{x}} \qquad \overline{\overline{A}}_k(\overline{\overline{x}}) = \overline{A}_m(\overline{x}) \frac{\partial \overline{x}^m}{\partial \overline{\overline{x}}^k}$$

We can therefore express the transformation of the components associated with the coordinate transformation $x \to \overline{\overline{x}}$ and

$$\overline{\overline{A}}_k(\overline{\overline{x}}) = \left(A_j(x) \frac{\partial x^j}{\partial \overline{x}^m} \right) \frac{\partial \overline{x}^m}{\partial \overline{\overline{x}}^k} = A_j(x) \frac{\partial x^j}{\partial \overline{\overline{x}}^k},$$

which demonstrates the transitive property of a covariant transformation.

∎

Higher Order Tensors

We have shown that first order tensors are quantities which obey certain transformation laws. Higher order tensors are defined in a similar manner and also satisfy the group properties. We assume that we are given transformations of the type illustrated in equations (1.2.30) and (1.2.32) which are single valued and continuous with Jacobian J different from zero. Further, the quantities x^i and $\overline{x}^i$, $i = 1, \ldots, n$ represent the coordinates in any two coordinate systems. The following transformation laws define second order and third order tensors.

Definition: (Second order contravariant tensor) Whenever N-squared quantities A^{ij} in a coordinate system $(x^1, \ldots, x^N)$ are related to N-squared quantities $\overline{A}^{mn}$ in a coordinate system $(\overline{x}^1, \ldots, \overline{x}^N)$ such that the transformation law

$$\overline{A}^{mn}(\overline{x}) = A^{ij}(x)J^W \frac{\partial \overline{x}^m}{\partial x^i} \frac{\partial \overline{x}^n}{\partial x^j} \tag{1.2.49}$$

is satisfied, then these quantities are called components of a relative contravariant tensor of rank or order two with weight W. Whenever $W = 0$ these quantities are called the components of an absolute contravariant tensor of rank or order two.

Definition: (Second order covariant tensor) Whenever N-squared quantities A_{ij} in a coordinate system $(x^1, \ldots, x^N)$ are related to N-squared quantities $\overline{A}_{mn}$ in a coordinate system $(\overline{x}^1, \ldots, \overline{x}^N)$ such that the transformation law

$$\overline{A}_{mn}(\overline{x}) = A_{ij}(x)J^W \frac{\partial x^i}{\partial \overline{x}^m} \frac{\partial x^j}{\partial \overline{x}^n} \tag{1.2.50}$$

is satisfied, then these quantities are called components of a relative covariant tensor of rank or order two with weight W. Whenever $W = 0$ these quantities are called the components of an absolute covariant tensor of rank or order two.

Definition: (Second order mixed tensor) Whenever N-squared quantities A^i_j in a coordinate system $(x^1, \ldots, x^N)$ are related to N-squared quantities $\overline{A}^m_n$ in a coordinate system $(\overline{x}^1, \ldots, \overline{x}^N)$ such that the transformation law

$$\overline{A}^m_n(\overline{x}) = A^i_j(x)J^W \frac{\partial \overline{x}^m}{\partial x^i} \frac{\partial x^j}{\partial \overline{x}^n} \tag{1.2.51}$$

is satisfied, then these quantities are called components of a relative mixed tensor of rank or order two with weight W. Whenever $W = 0$ these quantities are called the components of an absolute mixed tensor of rank or order two. It is contravariant of order one and covariant of order one.

Higher order tensors are defined in a similar manner. For example, if we can find N-cubed quantities A_{np}^{m} such that

$$\overline{A}_{jk}^{i}(\overline{x}) = A_{\alpha\beta}^{\gamma}(x)J^{W} \frac{\partial \overline{x}^{i}}{\partial x^{\gamma}} \frac{\partial x^{\alpha}}{\partial \overline{x}^{j}} \frac{\partial x^{\beta}}{\partial \overline{x}^{k}} \tag{1.2.52}$$

then this is a relative mixed tensor of order three with weight W. It is contravariant of order one and covariant of order two.

General Definition

In general a mixed tensor of rank or order $(m + n)$

$$T_{j_1 j_2 \ldots j_n}^{i_1 i_2 \ldots i_m} \tag{1.2.53}$$

is contravariant of order m and covariant of order n if it obeys the transformation law

$$\overline{T}_{j_1 j_2 \ldots j_n}^{i_1 i_2 \ldots i_m} = \left[J\left(\frac{x}{\overline{x}}\right) \right]^{W} T_{b_1 b_2 \ldots b_n}^{a_1 a_2 \ldots a_m} \frac{\partial \overline{x}^{i_1}}{\partial x^{a_1}} \frac{\partial \overline{x}^{i_2}}{\partial x^{a_2}} \cdots \frac{\partial \overline{x}^{i_m}}{\partial x^{a_m}} \cdot \frac{\partial x^{b_1}}{\partial \overline{x}^{j_1}} \frac{\partial x^{b_2}}{\partial \overline{x}^{j_2}} \cdots \frac{\partial x^{b_n}}{\partial \overline{x}^{j_n}} \tag{1.2.54}$$

where

$$J\left(\frac{x}{\overline{x}}\right) = \left| \frac{\partial x}{\partial \overline{x}} \right| = \frac{\partial(x^1, x^2, \ldots, x^N)}{\partial(\overline{x}^1, \overline{x}^2, \ldots, \overline{x}^N)}$$

is the Jacobian of the transformation. When $W = 0$ the tensor is called an absolute tensor, otherwise it is called a relative tensor of weight W.

Here superscripts are used to denote contravariant components and subscripts are used to denote covariant components. Thus, if we are given the tensor components in one coordinate system, then the components in any other coordinate system are determined by the transformation law of equation (1.2.54). Throughout the remainder of this text one should treat all tensors as absolute tensors unless specified otherwise.

Dyads and Polyads

Note that vectors can be represented in bold face type with the notation

$$\mathbf{A} = A_i \mathbf{E}^i$$

This notation can also be generalized to tensor quantities. Higher order tensors can also be denoted by bold face type. For example the tensor components T_{ij} and B_{ijk} can be represented in terms of the basis vectors $\mathbf{E}^i, i = 1, \ldots, N$ by using a notation which is similar to that for the representation of vectors. For example,

$$\mathbf{T} = T_{ij} \mathbf{E}^i \mathbf{E}^j$$

$$\mathbf{B} = B_{ijk} \mathbf{E}^i \mathbf{E}^j \mathbf{E}^k.$$

Here $\mathbf{T}$ denotes a tensor with components T_{ij} and $\mathbf{B}$ denotes a tensor with components B_{ijk}. The quantities $\mathbf{E}^i \mathbf{E}^j$ are called unit dyads and $\mathbf{E}^i \mathbf{E}^j \mathbf{E}^k$ are called unit triads. There is no multiplication sign between the basis vectors. This notation is called a polyad notation. A further generalization of this notation is the representation of an arbitrary tensor using the basis and reciprocal basis vectors in bold type. For example, a mixed tensor would have the polyadic representation

$$\mathbf{T} = T_{lm \ldots n}^{ij \ldots k} \mathbf{E}_i \mathbf{E}_j \ldots \mathbf{E}_k \mathbf{E}^l \mathbf{E}^m \ldots \mathbf{E}^n.$$

A dyadic is formed by the outer or direct product of two vectors. For example, the outer product of the vectors

$$\mathbf{a} = a_1\mathbf{E}^1 + a_2\mathbf{E}^2 + a_3\mathbf{E}^3 \quad \text{and} \quad \mathbf{b} = b_1\mathbf{E}^1 + b_2\mathbf{E}^2 + b_3\mathbf{E}^3$$

gives the dyad

$$\mathbf{ab} = a_1b_1\mathbf{E}^1\mathbf{E}^1 + a_1b_2\mathbf{E}^1\mathbf{E}^2 + a_1b_3\mathbf{E}^1\mathbf{E}^3$$
$$a_2b_1\mathbf{E}^2\mathbf{E}^1 + a_2b_2\mathbf{E}^2\mathbf{E}^2 + a_2b_3\mathbf{E}^2\mathbf{E}^3$$
$$a_3b_1\mathbf{E}^3\mathbf{E}^1 + a_3b_2\mathbf{E}^3\mathbf{E}^2 + a_3b_3\mathbf{E}^3\mathbf{E}^3.$$

In general, a dyad can be represented

$$\mathbf{A} = A_{ij}\mathbf{E}^i\mathbf{E}^j \qquad i,j = 1,\ldots,N$$

where the summation convention is in effect for the repeated indices. The coefficients A_{ij} are called the coefficients of the dyad. When the coefficients are written as an $N \times N$ array it is called a matrix. Every second order tensor can be written as a linear combination of dyads. The dyads form a basis for the second order tensors. As the example above illustrates, the nine dyads $\{\mathbf{E}^1\mathbf{E}^1, \mathbf{E}^1\mathbf{E}^2, \ldots, \mathbf{E}^3\mathbf{E}^3\}$, associated with the outer products of three dimensional base vectors, constitute a basis for the second order tensor $\mathbf{A} = \mathbf{ab}$ having the components $A_{ij} = a_ib_j$ with $i,j = 1,2,3$. Similarly, a triad has the form

$$\mathbf{T} = T_{ijk}\mathbf{E}^i\mathbf{E}^j\mathbf{E}^k \quad \text{Sum on repeated indices}$$

where i,j,k have the range $1,2,\ldots,N$. The set of outer or direct products $\{\mathbf{E}^i\mathbf{E}^j\mathbf{E}^k\}$, with $i,j,k = 1,\ldots,N$ constitutes a basis for all third order tensors. Tensor components with mixed suffixes like C^i_{jk} are associated with triad basis of the form

$$\mathbf{C} = C^i_{jk}\mathbf{E}_i\mathbf{E}^j\mathbf{E}^k$$

where i,j,k have the range $1,2,\ldots N$. Dyads are associated with the outer product of two vectors, while triads, tetrads,... are associated with higher-order outer products. These higher-order outer or direct products are referred to as polyads.

The polyad notation is a generalization of the vector notation. The subject of how polyad components transform between coordinate systems is the subject of tensor calculus.

In Cartesian coordinates we have $\mathbf{E}^i = \mathbf{E}_i = \widehat{\mathbf{e}}_i$ and a dyadic with components called dyads is written $\mathbf{A} = A_{ij}\widehat{\mathbf{e}}_i\widehat{\mathbf{e}}_j$ or

$$\mathbf{A} = A_{11}\widehat{\mathbf{e}}_1\widehat{\mathbf{e}}_1 + A_{12}\widehat{\mathbf{e}}_1\widehat{\mathbf{e}}_2 + A_{13}\widehat{\mathbf{e}}_1\widehat{\mathbf{e}}_3$$
$$A_{21}\widehat{\mathbf{e}}_2\widehat{\mathbf{e}}_1 + A_{22}\widehat{\mathbf{e}}_2\widehat{\mathbf{e}}_2 + A_{23}\widehat{\mathbf{e}}_2\widehat{\mathbf{e}}_3$$
$$A_{31}\widehat{\mathbf{e}}_3\widehat{\mathbf{e}}_1 + A_{32}\widehat{\mathbf{e}}_3\widehat{\mathbf{e}}_2 + A_{33}\widehat{\mathbf{e}}_3\widehat{\mathbf{e}}_3$$

where the terms $\widehat{\mathbf{e}}_i\widehat{\mathbf{e}}_j$ are called unit dyads. Note that a dyadic has nine components as compared with a vector which has only three components. The conjugate dyadic $\mathbf{A}_c$ is defined by a transposition of the unit vectors in $\mathbf{A}$, to obtain

$$\mathbf{A}_c = A_{11}\widehat{\mathbf{e}}_1\widehat{\mathbf{e}}_1 + A_{12}\widehat{\mathbf{e}}_2\widehat{\mathbf{e}}_1 + A_{13}\widehat{\mathbf{e}}_3\widehat{\mathbf{e}}_1$$
$$A_{21}\widehat{\mathbf{e}}_1\widehat{\mathbf{e}}_2 + A_{22}\widehat{\mathbf{e}}_2\widehat{\mathbf{e}}_2 + A_{23}\widehat{\mathbf{e}}_3\widehat{\mathbf{e}}_2$$
$$A_{31}\widehat{\mathbf{e}}_1\widehat{\mathbf{e}}_3 + A_{32}\widehat{\mathbf{e}}_2\widehat{\mathbf{e}}_3 + A_{33}\widehat{\mathbf{e}}_3\widehat{\mathbf{e}}_3$$

If a dyadic equals its conjugate $\mathbf{A} = \mathbf{A}_c$, then $A_{ij} = A_{ji}$ and the dyadic is called symmetric. If a dyadic equals the negative of its conjugate $\mathbf{A} = -\mathbf{A}_c$, then $A_{ij} = -A_{ji}$ and the dyadic is called skew-symmetric. A special dyadic called the identical dyadic or idemfactor is defined by

$$J = \widehat{\mathbf{e}}_1\,\widehat{\mathbf{e}}_1 + \widehat{\mathbf{e}}_2\,\widehat{\mathbf{e}}_2 + \widehat{\mathbf{e}}_3\,\widehat{\mathbf{e}}_3.$$

This dyadic has the property that pre or post dot product multiplication of J with a vector $\vec{V}$ produces the same vector $\vec{V}$. For example,

$$\vec{V} \cdot J = (V_1\,\widehat{\mathbf{e}}_1 + V_2\,\widehat{\mathbf{e}}_2 + V_3\,\widehat{\mathbf{e}}_3) \cdot J$$
$$= V_1\,\widehat{\mathbf{e}}_1 \cdot \widehat{\mathbf{e}}_1\,\widehat{\mathbf{e}}_1 + V_2\,\widehat{\mathbf{e}}_2 \cdot \widehat{\mathbf{e}}_2\,\widehat{\mathbf{e}}_2 + V_3\,\widehat{\mathbf{e}}_3 \cdot \widehat{\mathbf{e}}_3\,\widehat{\mathbf{e}}_3 = \vec{V}$$

and $\quad J \cdot \vec{V} = J \cdot (V_1\,\widehat{\mathbf{e}}_1 + V_2\,\widehat{\mathbf{e}}_2 + V_3\,\widehat{\mathbf{e}}_3)$

$$= V_1\,\widehat{\mathbf{e}}_1\,\widehat{\mathbf{e}}_1 \cdot \widehat{\mathbf{e}}_1 + V_2\,\widehat{\mathbf{e}}_2\,\widehat{\mathbf{e}}_2 \cdot \widehat{\mathbf{e}}_2 + V_3\,\widehat{\mathbf{e}}_3\,\widehat{\mathbf{e}}_3 \cdot \widehat{\mathbf{e}}_3 = \vec{V}$$

A dyadic operation often used in physics and chemistry is the double dot product $\mathbf{A} : \mathbf{B}$ where $\mathbf{A}$ and $\mathbf{B}$ are both dyadics. Here both dyadics are expanded using the distributive law of multiplication, and then each unit dyad pair $\widehat{\mathbf{e}}_i\,\widehat{\mathbf{e}}_j : \widehat{\mathbf{e}}_m\,\widehat{\mathbf{e}}_n$ are combined according to the rule

$$\widehat{\mathbf{e}}_i\,\widehat{\mathbf{e}}_j : \widehat{\mathbf{e}}_m\,\widehat{\mathbf{e}}_n = (\widehat{\mathbf{e}}_i \cdot \widehat{\mathbf{e}}_m)(\widehat{\mathbf{e}}_j \cdot \widehat{\mathbf{e}}_n).$$

For example, if $\mathbf{A} = A_{ij}\,\widehat{\mathbf{e}}_i\,\widehat{\mathbf{e}}_j$ and $\mathbf{B} = B_{ij}\,\widehat{\mathbf{e}}_i\,\widehat{\mathbf{e}}_j$, then the double dot product $\mathbf{A} : \mathbf{B}$ is calculated as follows.

$$\mathbf{A} : \mathbf{B} = (A_{ij}\,\widehat{\mathbf{e}}_i\,\widehat{\mathbf{e}}_j) : (B_{mn}\,\widehat{\mathbf{e}}_m\,\widehat{\mathbf{e}}_n) = A_{ij}B_{mn}(\widehat{\mathbf{e}}_i\,\widehat{\mathbf{e}}_j : \widehat{\mathbf{e}}_m\,\widehat{\mathbf{e}}_n) = A_{ij}B_{mn}(\widehat{\mathbf{e}}_i \cdot \widehat{\mathbf{e}}_m)(\widehat{\mathbf{e}}_j \cdot \widehat{\mathbf{e}}_n)$$

$$= A_{ij}B_{mn}\delta_{im}\delta_{jn} = A_{mj}B_{mj}$$

$$= A_{11}B_{11} + A_{12}B_{12} + A_{13}B_{13}$$

$$+ A_{21}B_{21} + A_{22}B_{22} + A_{23}B_{23}$$

$$+ A_{31}B_{31} + A_{32}B_{32} + A_{33}B_{33}$$

When operating with dyads, triads and polyads, there is a definite order to the way vectors and polyad components are represented. For example, for $\vec{A} = A_i\,\widehat{\mathbf{e}}_i$ and $\vec{B} = B_i\,\widehat{\mathbf{e}}_i$ vectors with outer product

$$\vec{A}\vec{B} = A_m B_n\,\widehat{\mathbf{e}}_m\,\widehat{\mathbf{e}}_n = \phi$$

there is produced the dyadic ϕ with components $A_m B_n$. In comparison, the outer product

$$\vec{B}\vec{A} = B_m A_n\,\widehat{\mathbf{e}}_m\,\widehat{\mathbf{e}}_n = \psi$$

produces the dyadic ψ with components $B_m A_n$. That is

$$\phi = \vec{A}\vec{B} = A_1 B_1\,\widehat{\mathbf{e}}_1\,\widehat{\mathbf{e}}_1 + A_1 B_2\,\widehat{\mathbf{e}}_1\,\widehat{\mathbf{e}}_2 + A_1 B_3\,\widehat{\mathbf{e}}_1\,\widehat{\mathbf{e}}_3$$

$$A_2 B_1\,\widehat{\mathbf{e}}_2\,\widehat{\mathbf{e}}_1 + A_2 B_2\,\widehat{\mathbf{e}}_2\,\widehat{\mathbf{e}}_2 + A_2 B_3\,\widehat{\mathbf{e}}_2\,\widehat{\mathbf{e}}_3$$

$$A_3 B_1\,\widehat{\mathbf{e}}_3\,\widehat{\mathbf{e}}_1 + A_3 B_2\,\widehat{\mathbf{e}}_3\,\widehat{\mathbf{e}}_2 + A_3 B_3\,\widehat{\mathbf{e}}_3\,\widehat{\mathbf{e}}_3$$

and $\quad \psi = \vec{B}\vec{A} = B_1 A_1\,\widehat{\mathbf{e}}_1\,\widehat{\mathbf{e}}_1 + B_1 A_2\,\widehat{\mathbf{e}}_1\,\widehat{\mathbf{e}}_2 + B_1 A_3\,\widehat{\mathbf{e}}_1\,\widehat{\mathbf{e}}_3$

$$B_2 A_1\,\widehat{\mathbf{e}}_2\,\widehat{\mathbf{e}}_1 + B_2 A_2\,\widehat{\mathbf{e}}_2\,\widehat{\mathbf{e}}_2 + B_2 A_3\,\widehat{\mathbf{e}}_2\,\widehat{\mathbf{e}}_3$$

$$B_3 A_1\,\widehat{\mathbf{e}}_3\,\widehat{\mathbf{e}}_1 + B_3 A_2\,\widehat{\mathbf{e}}_3\,\widehat{\mathbf{e}}_2 + B_3 A_3\,\widehat{\mathbf{e}}_3\,\widehat{\mathbf{e}}_3$$

are different dyadics.

The scalar dot product of a dyad with a vector $\vec{C}$ is defined for both pre and post multiplication as

$$\phi \cdot \vec{C} = \vec{A}\vec{B} \cdot \vec{C} = \vec{A}(\vec{B} \cdot \vec{C})$$

$$\vec{C} \cdot \phi = \vec{C} \cdot \vec{A}\vec{B} = (\vec{C} \cdot \vec{A})\vec{B}$$

These products are, in general, not equal.

Operations Using Tensors

The following are some important tensor operations which are used to derive special equations and to prove various identities.

Addition and Subtraction

Tensors of the same type and weight can be added or subtracted. For example, two third order mixed tensors, when added, produce another third order mixed tensor. Let A^i_{jk} and B^i_{jk} denote two third order mixed tensors. Their sum is denoted

$$C^i_{jk} = A^i_{jk} + B^i_{jk}.$$

That is, like components are added. The sum is also a mixed tensor as we now verify. By hypothesis A^i_{jk} and B^i_{jk} are third order mixed tensors and hence must obey the transformation laws

$$\overline{A}^i_{jk} = A^m_{np} \frac{\partial \overline{x}^i}{\partial x^m} \frac{\partial x^n}{\partial \overline{x}^j} \frac{\partial x^p}{\partial \overline{x}^k}$$

$$\overline{B}^i_{jk} = B^m_{np} \frac{\partial \overline{x}^i}{\partial x^m} \frac{\partial x^n}{\partial \overline{x}^j} \frac{\partial x^p}{\partial \overline{x}^k}.$$

We let $\overline{C}^i_{jk} = \overline{A}^i_{jk} + \overline{B}^i_{jk}$ denote the sum in the transformed coordinates. Then the addition of the above transformation equations produces

$$\overline{C}^i_{jk} = \left(\overline{A}^i_{jk} + \overline{B}^i_{jk} \right) = \left(A^m_{np} + B^m_{np} \right) \frac{\partial \overline{x}^i}{\partial x^m} \frac{\partial x^n}{\partial \overline{x}^j} \frac{\partial x^p}{\partial \overline{x}^k} = C^m_{np} \frac{\partial \overline{x}^i}{\partial x^m} \frac{\partial x^n}{\partial \overline{x}^j} \frac{\partial x^p}{\partial \overline{x}^k}.$$

Consequently, the sum transforms as a mixed third order tensor.

Multiplication (Outer Product)

The product of two tensors is also a tensor. The rank or order of the resulting tensor is the sum of the ranks of the tensors occurring in the multiplication. As an example, let A^i_{jk} denote a mixed third order tensor and let B^l_m denote a mixed second order tensor. The outer product of these two tensors is the fifth order tensor

$$C^{il}_{jkm} = A^i_{jk} B^l_m, \quad i,j,k,l,m = 1,2,\ldots,N.$$

Here all indices are free indices as i,j,k,l,m take on any of the integer values $1,2,\ldots,N$. Let $\overline{A}^i_{jk}$ and $\overline{B}^l_m$ denote the components of the given tensors in the barred system of coordinates. We define $\overline{C}^{il}_{jkm}$ as the outer product of these components. Observe that C^{il}_{jkm} is a tensor for by hypothesis A^i_{jk} and B^l_m are tensors and hence obey the transformation laws

$$\overline{A}^\alpha_{\beta\gamma} = A^i_{jk} \frac{\partial \overline{x}^\alpha}{\partial x^i} \frac{\partial x^j}{\partial \overline{x}^\beta} \frac{\partial x^k}{\partial \overline{x}^\gamma}$$

$$\overline{B}^\delta_\epsilon = B^l_m \frac{\partial \overline{x}^\delta}{\partial x^l} \frac{\partial x^m}{\partial \overline{x}^\epsilon}.$$

(1.2.55)

The outer product of these components produces

$$\overline{C}^{\alpha\delta}_{\beta\gamma\epsilon} = \overline{A}^\alpha_{\beta\gamma} \overline{B}^\delta_\epsilon = A^i_{jk} B^l_m \frac{\partial \overline{x}^\alpha}{\partial x^i} \frac{\partial x^j}{\partial \overline{x}^\beta} \frac{\partial x^k}{\partial \overline{x}^\gamma} \frac{\partial \overline{x}^\delta}{\partial x^l} \frac{\partial x^m}{\partial \overline{x}^\epsilon}$$

$$= C^{il}_{jkm} \frac{\partial \overline{x}^\alpha}{\partial x^i} \frac{\partial x^j}{\partial \overline{x}^\beta} \frac{\partial x^k}{\partial \overline{x}^\gamma} \frac{\partial \overline{x}^\delta}{\partial x^l} \frac{\partial x^m}{\partial \overline{x}^\epsilon}$$

(1.2.56)

which demonstrates that C^{il}_{jkm} transforms as a mixed fifth order absolute tensor. Other outer products are analyzed in a similar way.

52

Contraction

The operation of contraction on any mixed tensor of rank m is performed when an upper index is set equal to a lower index and the summation convention is invoked. When the summation is performed over the repeated indices the resulting quantity is also a tensor of rank or order $(m-2)$. For example, let A^i_{jk}, $i,j,k = 1,2,\ldots,N$ denote a mixed tensor and perform a contraction by setting j equal to i. We obtain

$$A^i_{ik} = A^1_{1k} + A^2_{2k} + \cdots + A^N_{Nk} = A_k \qquad (1.2.57)$$

where k is a free index. To show that A_k is a tensor, we let $\overline{A}^i_{ik} = \overline{A}_k$ denote the contraction on the transformed components of A^i_{jk}. By hypothesis A^i_{jk} is a mixed tensor and hence the components must satisfy the transformation law

$$\overline{A}^i_{jk} = A^m_{np}\frac{\partial \overline{x}^i}{\partial x^m}\frac{\partial x^n}{\partial \overline{x}^j}\frac{\partial x^p}{\partial \overline{x}^k}.$$

Now execute a contraction by setting j equal to i and perform a summation over the repeated index. We find

$$\begin{aligned}\overline{A}^i_{ik} = \overline{A}_k &= A^m_{np}\frac{\partial \overline{x}^i}{\partial x^m}\frac{\partial x^n}{\partial \overline{x}^i}\frac{\partial x^p}{\partial \overline{x}^k} = A^m_{np}\frac{\partial x^n}{\partial x^m}\frac{\partial x^p}{\partial \overline{x}^k}\\ &= A^m_{np}\delta^n_m\frac{\partial x^p}{\partial \overline{x}^k} = A^n_{np}\frac{\partial x^p}{\partial \overline{x}^k} = A_p\frac{\partial x^p}{\partial \overline{x}^k}.\end{aligned} \qquad (1.2.58)$$

Hence, the contraction produces a tensor of rank two less than the original tensor. Contractions on other mixed tensors can be analyzed in a similar manner.

New tensors can be constructed from old tensors by performing a contraction on an upper and lower index. This process can be repeated as long as there is an upper and lower index upon which to perform the contraction. Each time a contraction is performed the rank of the resulting tensor is two less than the rank of the original tensor.

Multiplication (Inner Product)

The inner product of two tensors is obtained by:
(i) first taking the outer product of the given tensors and
(ii) performing a contraction on two of the indices.

EXAMPLE 1.2-5. (Inner product)

Let A^i and B_j denote the components of two first order tensors (vectors). The outer product of these tensors is

$$C^i_j = A^i B_j,\ i,j = 1,2,\ldots,N.$$

The inner product of these tensors is the scalar

$$C = A^i B_i = A^1 B_1 + A^2 B_2 + \cdots + A^N B_N.$$

Note that in some situations the inner product is performed by employing only subscript indices. For example, the above inner product is sometimes expressed as

$$C = A_i B_i = A_1 B_1 + A_2 B_2 + \cdots A_N B_N.$$

This notation is discussed later when Cartesian tensors are considered.

Quotient Law

Assume B_r^{qs} and C_p^s are arbitrary absolute tensors. Further assume we have a quantity $A(ijk)$ which we think might be a third order mixed tensor A_{jk}^i. By showing that the equation

$$A_{qp}^r B_r^{qs} = C_p^s$$

is satisfied, then it follows that A_{qp}^r must be a tensor. This is an example of the quotient law. Obviously, this result can be generalized to apply to tensors of any order or rank. To prove the above assertion we shall show from the above equation that A_{jk}^i is a tensor. Let x^i and $\overline{x}^i$ denote a barred and unbarred system of coordinates which are related by transformations of the form defined by equation (1.2.30). In the barred system, we assume that

$$\overline{A}_{qp}^r \overline{B}_r^{qs} = \overline{C}_p^s \qquad (1.2.59)$$

where by hypothesis B_k^{ij} and C_m^l are arbitrary absolute tensors and therefore must satisfy the transformation equations

$$\overline{B}_r^{qs} = B_k^{ij} \frac{\partial \overline{x}^q}{\partial x^i} \frac{\partial \overline{x}^s}{\partial x^j} \frac{\partial x^k}{\partial \overline{x}^r}$$

$$\overline{C}_p^s = C_m^l \frac{\partial \overline{x}^s}{\partial x^l} \frac{\partial x^m}{\partial \overline{x}^p}.$$

We substitute for $\overline{B}_r^{qs}$ and $\overline{C}_p^s$ in the equation (1.2.59) and obtain the equation

$$\overline{A}_{qp}^r \left(B_k^{ij} \frac{\partial \overline{x}^q}{\partial x^i} \frac{\partial \overline{x}^s}{\partial x^j} \frac{\partial x^k}{\partial \overline{x}^r} \right) = \left(C_m^l \frac{\partial \overline{x}^s}{\partial x^l} \frac{\partial x^m}{\partial \overline{x}^p} \right)$$

$$= A_{qm}^r B_r^{ql} \frac{\partial \overline{x}^s}{\partial x^l} \frac{\partial x^m}{\partial \overline{x}^p}.$$

Since the summation indices are dummy indices they can be replaced by other symbols. We change l to j, q to i and r to k and write the above equation as

$$\frac{\partial \overline{x}^s}{\partial x^j} \left(\overline{A}_{qp}^r \frac{\partial \overline{x}^q}{\partial x^i} \frac{\partial x^k}{\partial \overline{x}^r} - A_{im}^k \frac{\partial x^m}{\partial \overline{x}^p} \right) B_k^{ij} = 0.$$

Use inner multiplication by $\frac{\partial x^n}{\partial \overline{x}^s}$ and simplify this equation to the form

$$\delta_j^n \left[\overline{A}_{qp}^r \frac{\partial \overline{x}^q}{\partial x^i} \frac{\partial x^k}{\partial \overline{x}^r} - A_{im}^k \frac{\partial x^m}{\partial \overline{x}^p} \right] B_k^{ij} = 0 \qquad \text{or}$$

$$\left[\overline{A}_{qp}^r \frac{\partial \overline{x}^q}{\partial x^i} \frac{\partial x^k}{\partial \overline{x}^r} - A_{im}^k \frac{\partial x^m}{\partial \overline{x}^p} \right] B_k^{in} = 0.$$

Because B_k^{in} is an arbitrary tensor, the quantity inside the brackets is zero and therefore

$$\overline{A}_{qp}^r \frac{\partial \overline{x}^q}{\partial x^i} \frac{\partial x^k}{\partial \overline{x}^r} - A_{im}^k \frac{\partial x^m}{\partial \overline{x}^p} = 0.$$

This equation is simplified by inner multiplication by $\frac{\partial x^i}{\partial \overline{x}^j} \frac{\partial \overline{x}^l}{\partial x^k}$ to obtain

$$\delta_j^q \delta_r^l \overline{A}_{qp}^r - A_{im}^k \frac{\partial x^m}{\partial \overline{x}^p} \frac{\partial x^i}{\partial \overline{x}^j} \frac{\partial \overline{x}^l}{\partial x^k} = 0 \qquad \text{or}$$

$$\overline{A}_{jp}^l = A_{im}^k \frac{\partial x^m}{\partial \overline{x}^p} \frac{\partial x^i}{\partial \overline{x}^j} \frac{\partial \overline{x}^l}{\partial x^k}$$

which is the transformation law for a third order mixed tensor.

EXERCISE 1.2

▶ **1.**   Consider the transformation equations representing a rotation of axes through an angle α.

$$T_\alpha : \begin{cases} x^1 &= \overline{x}^1 \cos\alpha - \overline{x}^2 \sin\alpha \\ x^2 &= \overline{x}^1 \sin\alpha + \overline{x}^2 \cos\alpha \end{cases}$$

Treat α as a parameter and show this set of transformations constitutes a group by finding the value of α which:

  (i) gives the identity transformation.

  (ii) gives the inverse transformation.

  (iii) show the transformation is transitive in that a transformation with $\alpha = \theta_1$ followed by a transformation with $\alpha = \theta_2$ is equivalent to the transformation using $\alpha = \theta_1 + \theta_2$.

▶ **2.**   Show the transformation

$$T_\alpha : \begin{cases} \overline{x}^1 &= \alpha x^1 \\ \overline{x}^2 &= \frac{1}{\alpha} x^2 \end{cases}$$

forms a group with α as a parameter. Find the value of α such that:

  (i) the identity transformation exists.

  (ii) the inverse transformation exists.

  (iii) the transitive property is satisfied.

▶ **3.**   Show the given transformation forms a group with parameter α.

$$T_\alpha : \begin{cases} \overline{x}^1 &= \frac{x^1}{1 - \alpha x^1} \\ \overline{x}^2 &= \frac{x^2}{1 - \alpha x^1} \end{cases}$$

▶ **4.**   Consider the Lorentz transformation from relativity theory having the velocity parameter V, c is the speed of light and $x_4 = t$ is time.

$$T_V : \begin{cases} \overline{x}^1 &= \dfrac{x^1 - V x^4}{\sqrt{1 - \frac{V^2}{c^2}}} \\ \overline{x}^2 &= x^2 \\ \overline{x}^3 &= x^3 \\ \overline{x}^4 &= \dfrac{x^4 - \frac{V x^1}{c^2}}{\sqrt{1 - \frac{V^2}{c^2}}} \end{cases}$$

Show this set of transformations constitutes a group, by establishing:

  (i) $V = 0$ gives the identity transformation T_0.

  (ii) $T_{V_2} \cdot T_{V_1} = T_0$ requires that $V_2 = -V_1$.

  (iii) $T_{V_2} \cdot T_{V_1} = T_{V_3}$ requires that

$$V_3 = \frac{V_1 + V_2}{1 + \frac{V_1 V_2}{c^2}}.$$

▶ **5.**   For $(\vec{E}_1, \vec{E}_2, \vec{E}_3)$ an arbitrary independent basis, (a) Verify that

$$\vec{E}^1 = \frac{1}{V} \vec{E}_2 \times \vec{E}_3, \qquad \vec{E}^2 = \frac{1}{V} \vec{E}_3 \times \vec{E}_1, \qquad \vec{E}^3 = \frac{1}{V} \vec{E}_1 \times \vec{E}_2$$

is a reciprocal basis, where $V = \vec{E}_1 \cdot (\vec{E}_2 \times \vec{E}_3)$ (b) Show that $\vec{E}^j = g^{ij} \vec{E}_i$.

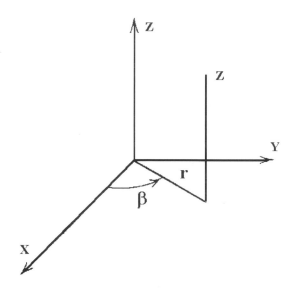

Figure 1.2-4. Cylindrical coordinates (r, β, z).

6. For the cylindrical coordinates (r, β, z) illustrated in the figure 1.2-4.

(a) Write out the transformation equations from rectangular (x, y, z) coordinates to cylindrical (r, β, z) coordinates. Also write out the inverse transformation.

(b) Determine the following basis vectors in cylindrical coordinates and represent your results in terms of cylindrical coordinates.

$\qquad$ (i) The tangential basis $\vec{E}_1, \vec{E}_2, \vec{E}_3$. $\quad$ (ii)The normal basis $\vec{E}^1, \vec{E}^2, \vec{E}^3$. $\quad$ (iii) $\hat{\mathbf{e}}_r, \hat{\mathbf{e}}_\beta, \hat{\mathbf{e}}_z$

where $\hat{\mathbf{e}}_r, \hat{\mathbf{e}}_\beta, \hat{\mathbf{e}}_z$ are normalized vectors in the directions of the tangential basis.

(c) A vector $\vec{A} = A_x\,\hat{\mathbf{e}}_1 + A_y\,\hat{\mathbf{e}}_2 + A_z\,\hat{\mathbf{e}}_3$ can be represented in any of the forms:

$$\vec{A} = A^1\vec{E}_1 + A^2\vec{E}_2 + A^3\vec{E}_3$$
$$\vec{A} = A_1\vec{E}^1 + A_2\vec{E}^2 + A_3\vec{E}^3$$
$$\vec{A} = A_r\hat{\mathbf{e}}_r + A_\beta\hat{\mathbf{e}}_\beta + A_z\hat{\mathbf{e}}_z$$

depending upon the basis vectors selected . In terms of the components A_x, A_y, A_z

(i) Solve for the contravariant components A^1, A^2, A^3.

(ii) Solve for the covariant components A_1, A_2, A_3.

(iii) Solve for the components A_r, A_β, A_z. Express all results in cylindrical coordinates. (Note the components A_r, A_β, A_z are referred to as physical components. Physical components are considered in more detail in a later section.)

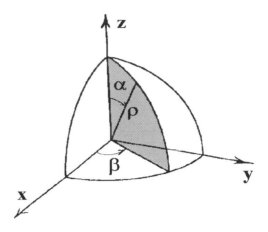

Figure 1.2-5. Spherical coordinates (ρ, α, β).

▶ **7.** For the spherical coordinates (ρ, α, β) illustrated in the figure 1.2-5.

(a) Write out the transformation equations from rectangular (x, y, z) coordinates to spherical (ρ, α, β) co-ordinates. Also write out the equations which describe the inverse transformation.

(b) Determine the following basis vectors in spherical coordinates

(i) The tangential basis $\vec{E}_1, \vec{E}_2, \vec{E}_3$.

(ii) The normal basis $\vec{E}^1, \vec{E}^2, \vec{E}^3$.

(iii) $\hat{\mathbf{e}}_\rho, \hat{\mathbf{e}}_\alpha, \hat{\mathbf{e}}_\beta$ which are normalized vectors in the directions of the tangential basis. Express all results in terms of spherical coordinates.

(c) A vector $\vec{A} = A_x\,\hat{\mathbf{e}}_1 + A_y\,\hat{\mathbf{e}}_2 + A_z\,\hat{\mathbf{e}}_3$ can be represented in any of the forms:

$$\vec{A} = A^1\vec{E}_1 + A^2\vec{E}_2 + A^3\vec{E}_3$$
$$\vec{A} = A_1\vec{E}^1 + A_2\vec{E}^2 + A_3\vec{E}^3$$
$$\vec{A} = A_\rho\hat{\mathbf{e}}_\rho + A_\alpha\hat{\mathbf{e}}_\alpha + A_\beta\hat{\mathbf{e}}_\beta$$

depending upon the basis vectors selected . Calculate, in terms of the coordinates (ρ, α, β) and the components A_x, A_y, A_z

(i) The contravariant components A^1, A^2, A^3.

(ii) The covariant components A_1, A_2, A_3.

(iii) The components $A_\rho, A_\alpha, A_\beta$ which are called physical components.

▶ **8.** Work the problems 6,7 and then let $(x^1, x^2, x^3) = (r, \beta, z)$ denote the coordinates in the cylindrical system and let $(\overline{x}^1, \overline{x}^2, \overline{x}^3) = (\rho, \alpha, \beta)$ denote the coordinates in the spherical system.

(a) Write the transformation equations $x \to \overline{x}$ from cylindrical to spherical coordinates. Also find the inverse transformations. (Hint: See the figures 1.2-4 and 1.2-5.)

(b) Use the results from part (a) and the results from problems 6,7 to verify that

$$\overline{A}_i = A_j\frac{\partial x^j}{\partial \overline{x}^i} \quad \text{for} \quad i = 1, 2, 3.$$

(i.e. Substitute A_j from problem 6 to get $\bar{A}_i$ given in problem 7.)

(c) Use the results from part (a) and the results from problems 6,7 to verify that

$$\overline{A}^i = A^j \frac{\partial \overline{x}^i}{\partial x^j} \quad \text{for} \quad i = 1, 2, 3.$$

(i.e. Substitute A^j from problem 6 to get $\overline{A}^i$ given by problem 7.)

9. Pick two arbitrary noncolinear vectors in the x, y plane, say

$$\vec{V_1} = 5\,\hat{e}_1 + \hat{e}_2 \quad \text{and} \quad \vec{V_2} = \hat{e}_1 + 5\,\hat{e}_2$$

and let $\vec{V_3} = \hat{e}_3$ be a unit vector perpendicular to both $\vec{V_1}$ and $\vec{V_2}$. The vectors $\vec{V_1}$ and $\vec{V_2}$ can be thought of as defining an oblique coordinate system, as illustrated in the figure 1.2-6.

(a) Find the reciprocal basis $(\vec{V}^1, \vec{V}^2, \vec{V}^3)$.

(b) Let

$$\vec{r} = x\,\hat{e}_1 + y\,\hat{e}_2 + z\,\hat{e}_3 = \alpha \vec{V_1} + \beta \vec{V_2} + \gamma \vec{V_3}$$

and show that

$$\alpha = \frac{5x}{24} - \frac{y}{24}$$
$$\beta = -\frac{x}{24} + \frac{5y}{24}$$
$$\gamma = z$$

(c) Show

$$x = 5\alpha + \beta$$
$$y = \alpha + 5\beta$$
$$z = \gamma$$

(d) For $\gamma = \gamma_0$ constant, show the coordinate lines are described by $\alpha = $ constant and $\beta = $ constant, and sketch some of these coordinate lines. (See figure 1.2-6.)

(e) Find the metrics g_{ij} and conjugate metrices g^{ij} associated with the (α, β, γ) space.

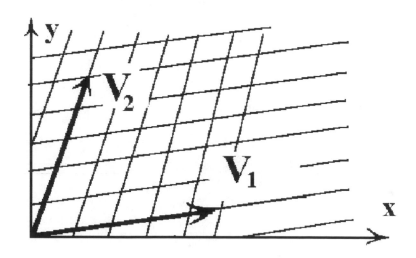

Figure 1.2-6. Oblique coordinates.

▶ **10.** Consider the transformation equations

$$x = x(u, v, w)$$
$$y = y(u, v, w)$$
$$z = z(u, v, w)$$

substituted into the position vector

$$\vec{r} = x\,\hat{\mathbf{e}}_1 + y\,\hat{\mathbf{e}}_2 + z\,\hat{\mathbf{e}}_3.$$

Define the basis vectors

$$(\vec{E}_1, \vec{E}_2, \vec{E}_3) = \left(\frac{\partial \vec{r}}{\partial u}, \frac{\partial \vec{r}}{\partial v}, \frac{\partial \vec{r}}{\partial w} \right)$$

with the reciprocal basis

$$\vec{E}^1 = \frac{1}{V}\vec{E}_2 \times \vec{E}_3, \qquad \vec{E}^2 = \frac{1}{V}\vec{E}_3 \times \vec{E}_1, \qquad \vec{E}^3 = \frac{1}{V}\vec{E}_1 \times \vec{E}_2.$$

where

$$V = \vec{E}_1 \cdot (\vec{E}_2 \times \vec{E}_3).$$

Let $v = \vec{E}^1 \cdot (\vec{E}^2 \times \vec{E}^3)$ and show that $v \cdot V = 1$.

▶ **11.** Given the coordinate transformation

$$x = -u - 2v \qquad y = -u - v \qquad z = z$$

(a) Find and illustrate graphically some of the coordinate curves.

(b) For $\vec{r} = \vec{r}(u, v, z)$ a position vector, define the basis vectors

$$\vec{E}_1 = \frac{\partial \vec{r}}{\partial u}, \qquad \vec{E}_2 = \frac{\partial \vec{r}}{\partial v}, \qquad \vec{E}_3 = \frac{\partial \vec{r}}{\partial z}.$$

Calculate these vectors and then calculate the reciprocal basis $\vec{E}^1, \vec{E}^2, \vec{E}^3$.

(c) With respect to the basis vectors in (b) find the contravariant components A^i associated with the vector

$$\vec{A} = \alpha_1 \,\hat{\mathbf{e}}_1 + \alpha_2 \,\hat{\mathbf{e}}_2 + \alpha_3 \,\hat{\mathbf{e}}_3$$

where $(\alpha_1, \alpha_2, \alpha_3)$ are constants.

(d) Find the covariant components A_i associated with the vector $\vec{A}$ given in part (c).

(e) Calculate the metric tensor g_{ij} and conjugate metric tensor g^{ij}.

(f) From the results (e), verify that $g_{ij}g^{jk} = \delta_i^k$

(g) Use the results from (c)(d) and (e) to verify that $A_i = g_{ik}A^k$

(h) Use the results from (c)(d) and (e) to verify that $A^i = g^{ik}A_k$

(i) Find the projection of the vector $\vec{A}$ on unit vectors in the directions $\vec{E}_1, \vec{E}_2, \vec{E}_3$.

(j) Find the projection of the vector $\vec{A}$ on unit vectors the directions $\vec{E}^1, \vec{E}^2, \vec{E}^3$.

12. For $\vec{r} = y^i\,\widehat{\mathbf{e}}_i$ where $y^i = y^i(x^1, x^2, x^3)$, $i = 1, 2, 3$ we have by definition

$$\vec{E}_j = \frac{\partial\vec{r}}{\partial x^j} = \frac{\partial y^i}{\partial x^j}\,\widehat{\mathbf{e}}_i.\quad \text{From this relation show that}\quad \vec{E}^m = \frac{\partial x^m}{\partial y^j}\,\widehat{\mathbf{e}}_j$$

and consequently

$$g_{ij} = \vec{E}_i \cdot \vec{E}_j = \frac{\partial y^m}{\partial x^i}\frac{\partial y^m}{\partial x^j}, \quad\text{and}\quad g^{ij} = \vec{E}^i \cdot \vec{E}^j = \frac{\partial x^i}{\partial y^m}\frac{\partial x^j}{\partial y^m}, \qquad i, j, m = 1, \ldots, 3$$

13. Consider the set of all coordinate transformations of the form

$$y^i = a^i_j x^j + b^i$$

where a^i_j and b^i are constants and the determinant of a^i_j is different from zero. Show this set of transformations forms a group.

14. For α_i, β_i constants and t a parameter, $x^i = \alpha_i + t\,\beta_i, i = 1, 2, 3$ is the parametric representation of a straight line. Find the parametric equation of the line which passes through the two points $(1, 2, 3)$ and $(14, 7, -3)$. What does the vector $\frac{d\vec{r}}{dt}$ represent?

15. A surface can be represented using two parameters u, v by introducing the parametric equations

$$x^i = x^i(u, v), \quad i = 1, 2, 3, \quad a < u < b \quad\text{and}\quad c < v < d.$$

The parameters u, v are called the curvilinear coordinates of a point on the surface. A point on the surface can be represented by the position vector $\vec{r} = \vec{r}(u, v) = x^1(u, v)\,\widehat{\mathbf{e}}_1 + x^2(u, v)\,\widehat{\mathbf{e}}_2 + x^3(u, v)\,\widehat{\mathbf{e}}_3$. The vectors $\frac{\partial\vec{r}}{\partial u}$ and $\frac{\partial\vec{r}}{\partial v}$ are tangent vectors to the coordinate surface curves $\vec{r}(u, c_2)$ and $\vec{r}(c_1, v)$ respectively. An element of surface area dS on the surface is defined as the area of the elemental parallelogram having the vector sides $\frac{\partial\vec{r}}{\partial u}du$ and $\frac{\partial\vec{r}}{\partial v}dv$. Show that

$$dS = \left|\frac{\partial\vec{r}}{\partial u} \times \frac{\partial\vec{r}}{\partial v}\right| du\,dv = \sqrt{g_{11}g_{22} - (g_{12})^2}\,du\,dv$$

where

$$g_{11} = \frac{\partial\vec{r}}{\partial u} \cdot \frac{\partial\vec{r}}{\partial u} \qquad g_{12} = \frac{\partial\vec{r}}{\partial u} \cdot \frac{\partial\vec{r}}{\partial v} \qquad g_{22} = \frac{\partial\vec{r}}{\partial v} \cdot \frac{\partial\vec{r}}{\partial v}.$$

Hint: $(\vec{A} \times \vec{B}) \cdot (\vec{A} \times \vec{B}) = |\vec{A} \times \vec{B}|^2$ See Exercise 1.1, problem 9(c).

16.

(a) Use the results from problem 15 and find the element of surface area of the circular cone

$$x = u\sin\alpha\cos v \qquad y = u\sin\alpha\sin v \qquad z = u\cos\alpha$$

$$\alpha \text{ a constant} \qquad 0 \le u \le b \qquad 0 \le v \le 2\pi$$

(b) Find the surface area of the above cone.

▶ **17.** The equation of a plane is defined in terms of two parameters u and v and has the form

$$x^i = \alpha_i \, u + \beta_i \, v + \gamma_i \qquad i = 1, 2, 3,$$

where α_i β_i and γ_i are constants. Find the equation of the plane which passes through the points $(1, 2, 3)$, $(14, 7, -3)$ and $(5, 5, 5)$. What does this problem have to do with the position vector $\vec{r}(u, v)$, the vectors $\frac{\partial \vec{r}}{\partial u}, \frac{\partial \vec{r}}{\partial v}$ and $\vec{r}(0, 0)$? Hint: See problem 15.

▶ **18.** Determine the points of intersection of the curve $x^1 = t$, $x^2 = (t)^2$, $x^3 = (t)^3$ with the plane

$$8\,x^1 - 5\,x^2 + x^3 - 4 = 0.$$

▶ **19.** Verify the relations $V \, e_{ijk} \vec{E}^k = \vec{E}_i \times \vec{E}_j$ and $v^{-1} e^{ijk} \vec{E}_k = \vec{E}^i \times \vec{E}^j$ where $v = \vec{E}^1 \cdot (\vec{E}^2 \times \vec{E}^3)$ and $V = \vec{E}_1 \cdot (\vec{E}_2 \times \vec{E}_3)$..

▶ **20.** Let $\bar{x}^i$ and x^i, $i = 1, 2, 3$ be related by the linear transformation $\bar{x}^i = c^i_j x^j$, where c^i_j are constants such that the determinant $c = det(c^i_j)$ is different from zero. Let γ^n_m denote the cofactor of c^m_n divided by the determinant c.

(a) Show that $c^i_j \gamma^j_k = \gamma^i_j c^j_k = \delta^i_k$.

(b) Show the inverse transformation can be expressed $x^i = \gamma^i_j \bar{x}^j$.

(c) Show that if A^i is a contravariant vector, then its transformed components are $\bar{A}^p = c^p_q A^q$.

(d) Show that if A_i is a covariant vector, then its transformed components are $\bar{A}_i = \gamma^p_i A_p$.

▶ **21.** Show that the outer product of two contravariant vectors A^i and B^i, $i = 1, 2, 3$ results in a second order contravariant tensor.

▶ **22.** Show that for the position vector $\vec{r} = y^i(x^1, x^2, x^3)\,\widehat{\mathbf{e}}_i$ the element of arc length squared is $ds^2 = d\vec{r} \cdot d\vec{r} = g_{ij}dx^i dx^j$ where $g_{ij} = \vec{E}_i \cdot \vec{E}_j = \dfrac{\partial y^m}{\partial x^i} \dfrac{\partial y^m}{\partial x^j}$.

▶ **23.** For A^i_{jk}, B^m_n and C^p_{tq} absolute tensors, show that if $A^i_{jk} B^k_n = C^i_{jn}$ then $\bar{A}^i_{jk} \bar{B}^k_n = \bar{C}^i_{jn}$.

▶ **24.** Let A_{ij} denote an absolute covariant tensor of order 2. Show that the determinant $A = det(A_{ij})$ is an invariant of weight 2 and $\sqrt{(A)}$ is an invariant of weight 1.

▶ **25.** Let B^{ij} denote an absolute contravariant tensor of order 2. Show that the determinant $B = det(B^{ij})$ is an invariant of weight -2 and $\sqrt{B}$ is an invariant of weight -1.

▶ **26.**

(a) Write out the contravariant components of the following vectors

$$(i) \quad \vec{E}_1 \qquad (ii) \quad \vec{E}_2 \qquad (iii) \quad \vec{E}_3 \quad \text{where} \quad \vec{E}_i = \frac{\partial \vec{r}}{\partial x^i} \quad \text{for} \quad i = 1, 2, 3.$$

(b) Write out the covariant components of the following vectors

$$(i) \quad \vec{E}^1 \qquad (ii) \quad \vec{E}^2 \qquad (ii) \quad \vec{E}^3 \quad \text{where} \quad \vec{E}^i = \text{grad}\, x^i, \quad \text{for} \quad i = 1, 2, 3.$$

27. Let A_{ij} and A^{ij} denote absolute second order tensors. Show that $\lambda = A_{ij}A^{ij}$ is a scalar invariant.

28. Assume that a_{ij}, $i, j = 1, 2, 3, 4$ is a skew-symmetric second order absolute tensor. (a) Show that

$$b_{ijk} = \frac{\partial a_{jk}}{\partial x^i} + \frac{\partial a_{ki}}{\partial x^j} + \frac{\partial a_{ij}}{\partial x^k}$$

is a third order tensor. (b) Show b_{ijk} is skew-symmetric in all pairs of indices and (c) determine the number of independent components this tensor has.

29. Show the linear forms $A_1x + B_1y + C_1$ and $A_2x + B_2y + C_2$, with respect to the group of rotations and translations $x = \overline{x}\cos\theta - \overline{y}\sin\theta + h$ and $y = \overline{x}\sin\theta + \overline{y}\cos\theta + k$, have the forms $\overline{A}_1\overline{x} + \overline{B}_1\overline{y} + \overline{C}_1$ and $\overline{A}_2\overline{x} + \overline{B}_2\overline{y} + \overline{C}_2$. Also show that the quantities $A_1B_2 - A_2B_1$ and $A_1A_2 + B_1B_2$ are invariants.

30. Show that the curvature of a curve $y = f(x)$ is $\kappa = \pm y''(1+y'^2)^{-3/2}$ and that this curvature remains invariant under the group of rotations given in the problem 1. Hint: Calculate $\frac{dy}{dx} = \frac{dy}{d\overline{x}}\frac{d\overline{x}}{dx}$.

31. Show that when the equation of a curve is given in the parametric form $x = x(t)$, $y = y(t)$, then the curvature is $\kappa = \pm\dfrac{\dot{x}\ddot{y} - \dot{y}\ddot{x}}{(\dot{x}^2 + \dot{y}^2)^{3/2}}$ and remains invariant under the change of parameter $t = t(\overline{t})$, where $\dot{x} = \frac{dx}{dt}$, etc.

32. Let A^{ij}_k denote a third order mixed tensor. (a) Show that the contraction A^{ij}_i is a first order contravariant tensor. (b) Show that contraction of i and j produces A^{ii}_k which is not a tensor. This shows that in general, the process of contraction does not always apply to indices at the same level.

33. Let $\phi = \phi(x^1, x^2, \ldots, x^N)$ denote an absolute scalar invariant. (a) Is the quantity $\frac{\partial\phi}{\partial x^i}$ a tensor? (b) Is the quantity $\frac{\partial^2\phi}{\partial x^i\partial x^j}$ a tensor?

34. Consider the second order absolute tensor a_{ij}, $i, j = 1, 2$ where $a_{11} = 1, a_{12} = 2, a_{21} = 3$ and $a_{22} = 4$. Find the components of $\overline{a}_{ij}$ under the transformation of coordinates $\overline{x}^1 = x^1 + x^2$ and $\overline{x}^2 = x^1 - x^2$.

35. Let A_i, B_i denote the components of two covariant absolute tensors of order one. Show that $C_{ij} = A_iB_j$ is an absolute second order covariant tensor.

36. Let A^i denote the components of an absolute contravariant tensor of order one and let B_i denote the components of an absolute covariant tensor of order one, show that $C^i_j = A^iB_j$ transforms as an absolute mixed tensor of order two.

37. (a) Show the sum and difference of two tensors of the same kind is also a tensor of this kind. (b) Show that the outer product of two tensors is a tensor. Do parts (a) (b) in the special case where one tensor A^i is a relative tensor of weight 4 and the other tensor B^j_k is a relative tensor of weight 3. What is the weight of the outer product tensor $T^{ij}_k = A^iB^j_k$ in this special case?

38. Let A^{ij}_{km} denote the components of a mixed tensor of weight M. Form the contraction $B^j_m = A^{ij}_{im}$ and determine how B^j_m transforms. What is its weight?

39. Let A^i_j denote the components of an absolute mixed tensor of order two. Show that the scalar contraction $S = A^i_i$ is an invariant.

▶ **40.** Let $A^i = A^i(x^1, x^2, \ldots, x^N)$ denote the components of an absolute contravariant tensor. Form the quantity $B^i_j = \frac{\partial A^i}{\partial x^j}$ and determine if B^i_j transforms like a tensor.

▶ **41.** Let A_i denote the components of a covariant vector. (a) Show that $a_{ij} = \dfrac{\partial A_i}{\partial x^j} - \dfrac{\partial A_j}{\partial x^i}$ are the components of a second order tensor. (b) Show that $\dfrac{\partial a_{ij}}{\partial x^k} + \dfrac{\partial a_{jk}}{\partial x^i} + \dfrac{\partial a_{ki}}{\partial x^j} = 0$.

▶ **42.** Show that $x^i = K\, e^{ijk} A_j B_k$, with $K \neq 0$ and arbitrary, is a general solution of the system of equations $A_i x^i = 0$, $B_i x^i = 0$, $i = 1, 2, 3$. Give a geometric interpretation of this result in terms of vectors.

▶ **43.** Given the vector $\vec{A} = y\,\widehat{\mathbf{e}}_1 + z\,\widehat{\mathbf{e}}_2 + x\,\widehat{\mathbf{e}}_3$ where $\widehat{\mathbf{e}}_1, \widehat{\mathbf{e}}_2, \widehat{\mathbf{e}}_3$ denote a set of unit basis vectors which define a set of orthogonal x, y, z axes. Let $\vec{E}_1 = 3\,\widehat{\mathbf{e}}_1 + 4\,\widehat{\mathbf{e}}_2$, $\vec{E}_2 = 4\,\widehat{\mathbf{e}}_1 + 7\,\widehat{\mathbf{e}}_2$ and $\vec{E}_3 = \widehat{\mathbf{e}}_3$ denote a set of basis vectors which define a set of u, v, w axes. (a) Find the coordinate transformation between these two sets of axes. (b) Find a set of reciprocal vectors $\vec{E}^1, \vec{E}^3, \vec{E}^3$. (c) Calculate the covariant components of $\vec{A}$. (d) Calculate the contravariant components of $\vec{A}$.

▶ **44.** Let $\mathbf{A} = A_{ij}\,\widehat{\mathbf{e}}_i\,\widehat{\mathbf{e}}_j$ denote a dyadic. Show that

$$\mathbf{A} : \mathbf{A}_c = A_{11}A_{11} + A_{12}A_{21} + A_{13}A_{31} + A_{21}A_{12} + A_{22}A_{22} + A_{23}A_{32} + A_{31}A_{13} + A_{32}A_{23} + A_{23}A_{33}$$

▶ **45.** Let $\vec{A} = A_i\,\widehat{\mathbf{e}}_i$, $\vec{B} = B_i\,\widehat{\mathbf{e}}_i$, $\vec{C} = C_i\,\widehat{\mathbf{e}}_i$, $\vec{D} = D_i\,\widehat{\mathbf{e}}_i$ denote vectors and let $\phi = \vec{A}\vec{B}$, $\psi = \vec{C}\vec{D}$ denote dyadics which are the outer products involving the above vectors. Show that the double dot product satisfies

$$\phi : \psi = \vec{A}\vec{B} : \vec{C}\vec{D} = (\vec{A} \cdot \vec{C})(\vec{B} \cdot \vec{D})$$

▶ **46.** Show that if a_{ij} is a symmetric tensor in one coordinate system, then it is symmetric in all coordinate systems.

▶ **47.** Write the transformation laws for the given tensors. $(a)\quad A^k_{ij} \qquad (b)\quad A^{ij}_k \qquad (c)\quad A^{ijk}_m$

▶ **48.** Show that if $\overline{A}_i = A_j \dfrac{\partial x^j}{\partial \overline{x}^i}$, then $A_i = \overline{A}_j \dfrac{\partial \overline{x}^j}{\partial x^i}$. Note that this is equivalent to interchanging the bar and unbarred systems.

▶ **49.**

(a) Show that under the linear homogeneous transformation

$$x_1 = a^1_1 \overline{x}_1 + a^2_1 \overline{x}_2$$
$$x_2 = a^1_2 \overline{x}_1 + a^2_2 \overline{x}_2$$

the quadratic form

$$Q(x_1, x_2) = g_{11}(x_1)^2 + 2g_{12}x_1 x_2 + g_{22}(x_2)^2 \quad \text{becomes} \quad \overline{Q}(\overline{x}_1, \overline{x}_2) = \overline{g}_{11}(\overline{x}_1)^2 + 2\overline{g}_{12}\overline{x}_1\overline{x}_2 + \overline{g}_{22}(\overline{x}_2)^2$$

where $\overline{g}_{ij} = g_{11}a^j_1 a^i_1 + g_{12}(a^i_1 a^j_2 + a^j_1 a^i_2) + g_{22}a^i_2 a^j_2$.

(b) Show $F = g_{11}g_{22} - (g_{12})^2$ is a relative invariant of weight 2 of the quadratic form $Q(x_1, x_2)$ with respect to the group of linear homogeneous transformations. i.e. Show that $\overline{F} = \Delta^2 F$ where $\overline{F} = \overline{g}_{11}\overline{g}_{22} - (\overline{g}_{12})^2$ and $\Delta = (a^1_1 a^2_2 - a^2_1 a^1_2)$.

50. Let $\mathbf{a}_i$ and $\mathbf{b}_i$ for $i = 1, \ldots, n$ denote arbitrary vectors and form the dyadic

$$\Phi = \mathbf{a}_1\mathbf{b}_1 + \mathbf{a}_2\mathbf{b}_2 + \cdots + \mathbf{a}_n\mathbf{b}_n.$$

By definition the first scalar invariant of Φ is

$$\phi_1 = \mathbf{a}_1 \cdot \mathbf{b}_1 + \mathbf{a}_2 \cdot \mathbf{b}_2 + \cdots + \mathbf{a}_n \cdot \mathbf{b}_n$$

where a dot product operator has been placed between the vectors. The first vector invariant of Φ is defined

$$\vec{\phi} = \mathbf{a}_1 \times \mathbf{b}_1 + \mathbf{a}_2 \times \mathbf{b}_2 + \cdots + \mathbf{a}_n \times \mathbf{b}_n$$

where a vector cross product operator has been placed between the vectors.

(a) Show that the first scalar and vector invariant of

$$\Phi = \widehat{\mathbf{e}}_1\widehat{\mathbf{e}}_2 + \widehat{\mathbf{e}}_2\widehat{\mathbf{e}}_3 + \widehat{\mathbf{e}}_3\widehat{\mathbf{e}}_3$$

are respectively 1 and $\widehat{\mathbf{e}}_1 + \widehat{\mathbf{e}}_3$.

(b) From the vector $\mathbf{f} = f_1\widehat{\mathbf{e}}_1 + f_2\widehat{\mathbf{e}}_2 + f_3\widehat{\mathbf{e}}_3$ one can form the dyadic $\nabla\mathbf{f}$ having the matrix components

$$\nabla\mathbf{f} = \begin{pmatrix} \frac{\partial f_1}{\partial x} & \frac{\partial f_2}{\partial x} & \frac{\partial f_3}{\partial x} \\ \frac{\partial f_1}{\partial y} & \frac{\partial f_2}{\partial y} & \frac{\partial f_3}{\partial y} \\ \frac{\partial f_1}{\partial z} & \frac{\partial f_2}{\partial z} & \frac{\partial f_3}{\partial z} \end{pmatrix}.$$

Show that this dyadic has the first scalar and vector invariants given by

$$\nabla \cdot \mathbf{f} = \frac{\partial f_1}{\partial x} + \frac{\partial f_2}{\partial y} + \frac{\partial f_3}{\partial z}$$

$$\nabla \times \mathbf{f} = \left(\frac{\partial f_3}{\partial y} - \frac{\partial f_2}{\partial z}\right)\widehat{\mathbf{e}}_1 + \left(\frac{\partial f_1}{\partial z} - \frac{\partial f_3}{\partial x}\right)\widehat{\mathbf{e}}_2 + \left(\frac{\partial f_2}{\partial x} - \frac{\partial f_1}{\partial y}\right)\widehat{\mathbf{e}}_3$$

51. Let Φ denote the dyadic given in problem 50. The dyadic Φ_2 defined by

$$\Phi_2 = \frac{1}{2}\sum_{i,j} \mathbf{a}_i \times \mathbf{a}_j\mathbf{b}_i \times \mathbf{b}_j$$

is called the Gibbs second dyadic of Φ, where the summation is taken over all permutations of i and j. When $i = j$ the dyad vanishes. Note that the permutations i, j and j, i give the same dyad and so occurs twice in the final sum. The factor $1/2$ removes this doubling. Associated with the Gibbs dyad Φ_2 are the scalar invariants

$$\phi_2 = \frac{1}{2}\sum_{i,j}(\mathbf{a}_i \times \mathbf{a}_j) \cdot (\mathbf{b}_i \times \mathbf{b}_j)$$

$$\phi_3 = \frac{1}{6}\sum_{i,j,k}(\mathbf{a}_i \times \mathbf{a}_j \cdot \mathbf{a}_k)(\mathbf{b}_i \times \mathbf{b}_j \cdot \mathbf{b}_k)$$

Show that the dyad

$$\Phi = \mathbf{a}\mathbf{s} + \mathbf{t}\mathbf{q} + \mathbf{c}\mathbf{u}$$

has

the first scalar invariant $\quad \phi_1 = \mathbf{a} \cdot \mathbf{s} + \mathbf{b} \cdot \mathbf{t} + \mathbf{c} \cdot \mathbf{u}$

the first vector invariant $\quad \vec{\phi} = \mathbf{a} \times \mathbf{s} + \mathbf{b} \times \mathbf{t} + \mathbf{c} \times \mathbf{u}$

Gibbs second dyad $\quad \Phi_2 = \mathbf{b} \times \mathbf{c}\mathbf{t} \times \mathbf{u} + \mathbf{c} \times \mathbf{a}\mathbf{u} \times \mathbf{s} + \mathbf{a} \times \mathbf{b}\mathbf{s} \times \mathbf{t}$

second scalar of $\Phi \quad \phi_2 = (\mathbf{b} \times \mathbf{c}) \cdot (\mathbf{t} \times \mathbf{u}) + (\mathbf{c} \times \mathbf{a}) \cdot (\mathbf{u} \times \mathbf{s}) + (\mathbf{a} \times \mathbf{b}) \cdot (\mathbf{s} \times \mathbf{t})$

third scalar of $\Phi \quad \phi_3 = (\mathbf{a} \times \mathbf{b} \cdot \mathbf{c})(\mathbf{s} \times \mathbf{t} \cdot \mathbf{u})$

▶ **52.** (**Spherical Trigonometry**) Construct a spherical triangle ABC on the surface of a unit sphere with sides and angles less than 180 degrees. Denote by $\mathbf{a}, \mathbf{b}\ \mathbf{c}$ the unit vectors from the origin of the sphere to the vertices A,B and C. Make the construction such that $\mathbf{a} \cdot (\mathbf{b} \times \mathbf{c})$ is positive with $\mathbf{a}, \mathbf{b}, \mathbf{c}$ forming a right-handed system. Let α, β, γ denote the angles between these unit vectors such that

$$\mathbf{a} \cdot \mathbf{b} = \cos \gamma \qquad \mathbf{c} \cdot \mathbf{a} = \cos \beta \qquad \mathbf{b} \cdot \mathbf{c} = \cos \alpha. \tag{1}$$

The great circles through the vertices A,B,C then make up the sides of the spherical triangle where side α is opposite vertex A, side β is opposite vertex B and side γ is opposite the vertex C. The angles A,B and C between the various planes formed by the vectors $\mathbf{a}, \mathbf{b}$ and $\mathbf{c}$ are called the interior dihedral angles of the spherical triangle. Note that the cross products

$$\mathbf{a} \times \mathbf{b} = \sin \gamma\, \overline{\mathbf{c}} \qquad \mathbf{b} \times \mathbf{c} = \sin \alpha\, \overline{\mathbf{a}} \qquad \mathbf{c} \times \mathbf{a} = \sin \beta\, \overline{\mathbf{b}} \tag{2}$$

define unit vectors $\overline{\mathbf{a}}, \overline{\mathbf{b}}$ and $\overline{\mathbf{c}}$ perpendicular to the planes determined by the unit vectors $\mathbf{a}, \mathbf{b}$ and $\mathbf{c}$. The dot products

$$\overline{\mathbf{a}} \cdot \overline{\mathbf{b}} = \cos \overline{\gamma} \qquad \overline{\mathbf{b}} \cdot \overline{\mathbf{c}} = \cos \overline{\alpha} \qquad \overline{\mathbf{c}} \cdot \overline{\mathbf{a}} = \cos \overline{\beta} \tag{3}$$

define the angles $\overline{\alpha}, \overline{\beta}$ and $\overline{\gamma}$ which are called the exterior dihedral angles at the vertices A,B and C and are such that

$$\overline{\alpha} = \pi - A \qquad \overline{\beta} = \pi - B \qquad \overline{\gamma} = \pi - C. \tag{4}$$

(a) Using appropriate scaling, show that the vectors $\mathbf{a}, \mathbf{b}, \mathbf{c}$ and $\overline{\mathbf{a}}, \overline{\mathbf{b}}, \overline{\mathbf{c}}$ form a reciprocal set.

(b) Show that $\mathbf{a} \cdot (\mathbf{b} \times \mathbf{c}) = \sin \alpha\, \mathbf{a} \cdot \overline{\mathbf{a}} = \sin \beta\, \mathbf{b} \cdot \overline{\mathbf{b}} = \sin \gamma\, \mathbf{c} \cdot \overline{\mathbf{c}}$

(c) Show that $\overline{\mathbf{a}} \cdot (\overline{\mathbf{b}} \times \overline{\mathbf{c}}) = \sin \overline{\alpha}\, \mathbf{a} \cdot \overline{\mathbf{a}} = \sin \overline{\beta}\, \mathbf{b} \cdot \overline{\mathbf{b}} = \sin \overline{\gamma}\, \mathbf{c} \cdot \overline{\mathbf{c}}$

(d) Using parts (b) and (c) show that

$$\frac{\sin \alpha}{\sin \overline{\alpha}} = \frac{\sin \beta}{\sin \overline{\beta}} = \frac{\sin \gamma}{\sin \overline{\gamma}}$$

(e) Use the results from equation (4) to derive the law of sines for spherical triangles

$$\frac{\sin \alpha}{\sin A} = \frac{\sin \beta}{\sin B} = \frac{\sin \gamma}{\sin C}$$

(f) Using the equations (2) show that

$$\sin \beta \sin \gamma\, \mathbf{b} \cdot \mathbf{c} = (\mathbf{c} \times \mathbf{a}) \cdot (\mathbf{a} \times \mathbf{b}) = (\mathbf{c} \cdot \mathbf{a})(\mathbf{a} \cdot \mathbf{b}) - \mathbf{b} \cdot \mathbf{c}$$

and hence show that

$$\cos \alpha = \cos \beta \cos \gamma - \sin \beta \sin \gamma \cos \overline{\alpha}.$$

In a similar manner show also that

$$\cos \overline{\alpha} = \cos \overline{\beta} \cos \overline{\gamma} - \sin \overline{\beta} \sin \overline{\gamma} \cos \alpha.$$

(g) Using part (f) derive the law of cosines for spherical triangles

$$\cos \alpha = \cos \beta \cos \gamma + \sin \beta \sin \gamma \cos A$$

$$\cos A = - \cos B \cos C + \sin B \sin C \cos \alpha$$

A cyclic permutation of the symbols produces similar results involving the other angles and sides of the spherical triangle.

§1.3 SPECIAL TENSORS

Knowing how tensors are defined and recognizing a tensor when it pops up in front of you are two different things. Some quantities, which are tensors, frequently arise in applied problems and you should learn to recognize these special tensors when they occur. In this section some important tensor quantities are defined. We also consider how these special tensors can in turn be used to define other tensors.

Metric Tensor

Define y^i, $i = 1, \ldots, N$ as independent coordinates in an N dimensional orthogonal Cartesian coordinate system. The distance squared between two points y^i and $y^i + dy^i$, $i = 1, \ldots, N$ is defined by the expression

$$ds^2 = dy^m dy^m = (dy^1)^2 + (dy^2)^2 + \cdots + (dy^N)^2. \tag{1.3.1}$$

Assume that the coordinates y^i are related to a set of independent generalized coordinates x^i, $i = 1, \ldots, N$ by a set of transformation equations

$$y^i = y^i(x^1, x^2, \ldots, x^N), \quad i = 1, \ldots, N. \tag{1.3.2}$$

To emphasize that each y^i depends upon the x coordinates we sometimes use the notation $y^i = y^i(x)$, for $i = 1, \ldots, N$. The differential of each coordinate can be written as

$$dy^m = \frac{\partial y^m}{\partial x^j} dx^j, \quad m = 1, \ldots, N, \tag{1.3.3}$$

and consequently in the x-generalized coordinates the distance squared, found from the equation (1.3.1), becomes a quadratic form. Substituting equation (1.3.3) into equation (1.3.1) we find

$$ds^2 = \frac{\partial y^m}{\partial x^i} \frac{\partial y^m}{\partial x^j} dx^i dx^j = g_{ij} dx^i dx^j \tag{1.3.4}$$

where

$$g_{ij} = \frac{\partial y^m}{\partial x^i} \frac{\partial y^m}{\partial x^j}, \quad i, j = 1, \ldots, N \tag{1.3.5}$$

are called the metrices of the space defined by the coordinates x^i, $i = 1, \ldots, N$. Here the g_{ij} are functions of the x coordinates and is sometimes written as $g_{ij} = g_{ij}(x)$. Further, the metrices g_{ij} are symmetric in the indices i and j so that $g_{ij} = g_{ji}$ for all values of i and j over the range of the indices. If we transform to another coordinate system, say $\overline{x}^i$, $i = 1, \ldots, N$, then the element of arc length squared is expressed in terms of the barred coordinates and $ds^2 = \overline{g}_{ij} d\overline{x}^i d\overline{x}^j$, where $\overline{g}_{ij} = \overline{g}_{ij}(\overline{x})$ is a function of the barred coordinates. The following example demonstrates that these metrices are second order covariant tensors.

EXAMPLE 1.3-1. Show the metric components g_{ij} are covariant tensors of the second order.

Solution: In a coordinate system x^i, $i = 1, \ldots, N$ the element of arc length squared is

$$ds^2 = g_{ij} dx^i dx^j \tag{1.3.6}$$

while in a coordinate system $\overline{x}^i$, $i = 1, \ldots, N$ the element of arc length squared is represented in the form

$$ds^2 = \overline{g}_{mn} d\overline{x}^m d\overline{x}^n. \tag{1.3.7}$$

The element of arc length squared is to be an invariant and so we require that

$$\overline{g}_{mn} d\overline{x}^m d\overline{x}^n = g_{ij} dx^i dx^j \tag{1.3.8}$$

Here it is assumed that there exists a coordinate transformation of the form defined by equation (1.2.30) together with an inverse transformation, as in equation (1.2.32), which relates the barred and unbarred coordinates. In general, if $x^i = x^i(\overline{x})$, then for $i = 1, \ldots, N$ we have

$$dx^i = \frac{\partial x^i}{\partial \overline{x}^m} d\overline{x}^m \quad \text{and} \quad dx^j = \frac{\partial x^j}{\partial \overline{x}^n} d\overline{x}^n \tag{1.3.9}$$

Substituting these differentials in equation (1.3.8) gives us the result

$$\overline{g}_{mn} d\overline{x}^m d\overline{x}^n = g_{ij} \frac{\partial x^i}{\partial \overline{x}^m} \frac{\partial x^j}{\partial \overline{x}^n} d\overline{x}^m d\overline{x}^n \quad \text{or} \quad \left(\overline{g}_{mn} - g_{ij} \frac{\partial x^i}{\partial \overline{x}^m} \frac{\partial x^j}{\partial \overline{x}^n} \right) d\overline{x}^m d\overline{x}^n = 0$$

For arbitrary changes in $d\overline{x}^m$ this equation implies that $\overline{g}_{mn} = g_{ij} \dfrac{\partial x^i}{\partial \overline{x}^m} \dfrac{\partial x^j}{\partial \overline{x}^n}$ and therefore g_{ij} transforms as a second order absolute covariant tensor.

∎

EXAMPLE 1.3-2. (Curvilinear coordinates) Consider a set of general transformation equations from rectangular coordinates (x, y, z) to curvilinear coordinates (u, v, w). These transformation equations and the corresponding inverse transformations are represented

$$\begin{aligned}
x &= x(u, v, w) & u &= u(x, y, z) \\
y &= y(u, v, w) & v &= v(x, y, z) \\
z &= z(u, v, w). & w &= w(x, y, z)
\end{aligned} \tag{1.3.10}$$

Here $y^1 = x$, $y^2 = y$, $y^3 = z$ and $x^1 = u$, $x^2 = v$, $x^3 = w$ are the Cartesian and generalized coordinates and $N = 3$. The intersection of the coordinate surfaces $u = c_1, v = c_2$ and $w = c_3$ define coordinate curves of the curvilinear coordinate system. The substitution of the given transformation equations (1.3.10) into the position vector $\vec{r} = x\,\widehat{\mathbf{e}}_1 + y\,\widehat{\mathbf{e}}_2 + z\,\widehat{\mathbf{e}}_3$ produces the position vector which is a function of the generalized coordinates and

$$\vec{r} = \vec{r}(u, v, w) = x(u, v, w)\,\widehat{\mathbf{e}}_1 + y(u, v, w)\,\widehat{\mathbf{e}}_2 + z(u, v, w)\,\widehat{\mathbf{e}}_3$$

and consequently $d\vec{r} = \dfrac{\partial \vec{r}}{\partial u}\, du + \dfrac{\partial \vec{r}}{\partial v}\, dv + \dfrac{\partial \vec{r}}{\partial w}\, dw$, where

$$\begin{aligned}
\vec{E}_1 &= \frac{\partial \vec{r}}{\partial u} = \frac{\partial x}{\partial u}\,\widehat{\mathbf{e}}_1 + \frac{\partial y}{\partial u}\,\widehat{\mathbf{e}}_2 + \frac{\partial z}{\partial u}\,\widehat{\mathbf{e}}_3 \\
\vec{E}_2 &= \frac{\partial \vec{r}}{\partial v} = \frac{\partial x}{\partial v}\,\widehat{\mathbf{e}}_1 + \frac{\partial y}{\partial v}\,\widehat{\mathbf{e}}_2 + \frac{\partial z}{\partial v}\,\widehat{\mathbf{e}}_3 \\
\vec{E}_3 &= \frac{\partial \vec{r}}{\partial w} = \frac{\partial x}{\partial w}\,\widehat{\mathbf{e}}_1 + \frac{\partial y}{\partial w}\,\widehat{\mathbf{e}}_2 + \frac{\partial z}{\partial w}\,\widehat{\mathbf{e}}_3.
\end{aligned} \tag{1.3.11}$$

are tangent vectors to the coordinate curves. The element of arc length in the curvilinear coordinates is

$$ds^2 = d\vec{r} \cdot d\vec{r} = \frac{\partial \vec{r}}{\partial u} \cdot \frac{\partial \vec{r}}{\partial u} \, dudu + \frac{\partial \vec{r}}{\partial u} \cdot \frac{\partial \vec{r}}{\partial v} \, dudv + \frac{\partial \vec{r}}{\partial u} \cdot \frac{\partial \vec{r}}{\partial w} \, dudw$$

$$+ \frac{\partial \vec{r}}{\partial v} \cdot \frac{\partial \vec{r}}{\partial u} \, dvdu + \frac{\partial \vec{r}}{\partial v} \cdot \frac{\partial \vec{r}}{\partial v} \, dvdv + \frac{\partial \vec{r}}{\partial v} \cdot \frac{\partial \vec{r}}{\partial w} \, dvdw \qquad (1.3.12)$$

$$+ \frac{\partial \vec{r}}{\partial w} \cdot \frac{\partial \vec{r}}{\partial u} \, dwdu + \frac{\partial \vec{r}}{\partial w} \cdot \frac{\partial \vec{r}}{\partial v} \, dwdv + \frac{\partial \vec{r}}{\partial w} \cdot \frac{\partial \vec{r}}{\partial w} \, dwdw.$$

Utilizing the summation convention, the above can be expressed in the index notation. Define the quantities

$$g_{11} = \frac{\partial \vec{r}}{\partial u} \cdot \frac{\partial \vec{r}}{\partial u} \qquad g_{12} = \frac{\partial \vec{r}}{\partial u} \cdot \frac{\partial \vec{r}}{\partial v} \qquad g_{13} = \frac{\partial \vec{r}}{\partial u} \cdot \frac{\partial \vec{r}}{\partial w}$$

$$g_{21} = \frac{\partial \vec{r}}{\partial v} \cdot \frac{\partial \vec{r}}{\partial u} \qquad g_{22} = \frac{\partial \vec{r}}{\partial v} \cdot \frac{\partial \vec{r}}{\partial v} \qquad g_{23} = \frac{\partial \vec{r}}{\partial v} \cdot \frac{\partial \vec{r}}{\partial w}$$

$$g_{31} = \frac{\partial \vec{r}}{\partial w} \cdot \frac{\partial \vec{r}}{\partial u} \qquad g_{32} = \frac{\partial \vec{r}}{\partial w} \cdot \frac{\partial \vec{r}}{\partial v} \qquad g_{33} = \frac{\partial \vec{r}}{\partial w} \cdot \frac{\partial \vec{r}}{\partial w}$$

and let $x^1 = u, \quad x^2 = v, \quad x^3 = w$. Then the above element of arc length can be expressed as

$$ds^2 = \vec{E}_i \cdot \vec{E}_j \, dx^i dx^j = g_{ij} dx^i dx^j, \qquad i,j = 1,2,3$$

where

$$g_{ij} = \vec{E}_i \cdot \vec{E}_j = \frac{\partial \vec{r}}{\partial x^i} \cdot \frac{\partial \vec{r}}{\partial x^j} = \frac{\partial y^m}{\partial x^i} \frac{\partial y^m}{\partial x^j}, \qquad i,j \text{ free indices} \qquad (1.3.13)$$

are called the metric components of the curvilinear coordinate system. The metric components may be thought of as the elements of a symmetric matrix, since $g_{ij} = g_{ji}$. The coordinate system is orthogonal if $g_{ij} = 0$ for $i \neq j$, this means that the three vectors $\frac{\partial \vec{r}}{\partial u}, \frac{\partial \vec{r}}{\partial v}, \frac{\partial \vec{r}}{\partial w}$ are orthogonal vectors.

Note that an orthogonal coordinate system is right-handed if the cross product of the first two coordinate directions points in the third coordinate direction. Otherwise it is called a left-handed coordinate system. We shall use right-handed coordinates for our applications.

A special case of the above is the rectangular coordinate system x, y, z, with element of arc length squared is $ds^2 = dx^2 + dy^2 + dz^2$. In this space the metric components are

$$g_{ij} = \begin{pmatrix} 1 & 0 & 0 \\ 0 & 1 & 0 \\ 0 & 0 & 1 \end{pmatrix}.$$

∎

EXAMPLE 1.3-3. (Cylindrical coordinates (r, θ, z))

The transformation equations from rectangular coordinates to cylindrical coordinates can be expressed as $x = r\cos\theta, \quad y = r\sin\theta, \quad z = z$. Here $y^1 = x, y^2 = y, y^3 = z$ and $x^1 = r, x^2 = \theta, x^3 = z$, and the position vector can be expressed $\vec{r} = \vec{r}(r, \theta, z) = r\cos\theta \,\hat{e}_1 + r\sin\theta \,\hat{e}_2 + z\hat{e}_3$. The derivatives of this position vector are calculated and we find

$$\vec{E}_1 = \frac{\partial \vec{r}}{\partial r} = \cos\theta \,\hat{e}_1 + \sin\theta \,\hat{e}_2, \quad \vec{E}_2 = \frac{\partial \vec{r}}{\partial \theta} = -r\sin\theta \,\hat{e}_1 + r\cos\theta \,\hat{e}_2, \quad \vec{E}_3 = \frac{\partial \vec{r}}{\partial z} = \hat{e}_3.$$

From the results in equation (1.3.13), the metric components of this space are

$$g_{ij} = \begin{pmatrix} 1 & 0 & 0 \\ 0 & r^2 & 0 \\ 0 & 0 & 1 \end{pmatrix}.$$

We note that since $g_{ij} = 0$ when $i \neq j$, the coordinate system is orthogonal.

∎

68

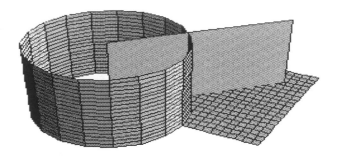

Figure 1.3-1. Cylindrical coordinates.

Orthogonal Coordinate Systems

Given a set of transformations of the form found in equation (1.3.10), one can readily determine the metric components associated with the generalized coordinates. For future reference we list several different coordinate systems together with their metric components. Each of the listed coordinate systems are orthogonal and so $g_{ij} = 0$ for $i \neq j$. The metric components of these orthogonal systems have the form

$$g_{ij} = \begin{pmatrix} h_1^2 & 0 & 0 \\ 0 & h_2^2 & 0 \\ 0 & 0 & h_3^2 \end{pmatrix}$$

and the element of arc length squared is

$$ds^2 = h_1^2 (dx^1)^2 + h_2^2 (dx^2)^2 + h_3^2 (dx^3)^2.$$

1. Cartesian coordinates (x, y, z)

$$x = x \qquad h_1 = 1$$
$$y = y \qquad h_2 = 1$$
$$z = z \qquad h_3 = 1$$

The coordinate curves are formed by the intersection of the coordinate surfaces
$$x = \text{Constant},\ y = \text{Constant and } z = \text{Constant}.$$

2. Cylindrical coordinates (r, θ, z)

$$x = r \cos\theta \quad r \geq 0 \qquad\qquad h_1 = 1$$
$$y = r \sin\theta \quad 0 \leq \theta \leq 2\pi \qquad h_2 = r$$
$$z = z \qquad -\infty < z < \infty \quad h_3 = 1$$

The coordinate curves, illustrated in the figure 1.3-1, are formed by the intersection of the coordinate surfaces
$$x^2 + y^2 = r^2, \qquad \text{Cylinders}$$
$$y/x = \tan\theta \qquad \text{Planes}$$
$$z = Constant \qquad \text{Planes}.$$

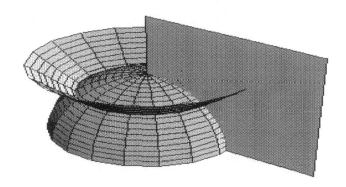

Figure 1.3-2. Spherical coordinates.

3. Spherical coordinatesf (ρ, θ, ϕ)

$$x = \rho \sin \theta \cos \phi \qquad \rho \geq 0 \qquad \qquad h_1 = 1$$

$$y = \rho \sin \theta \sin \phi \qquad 0 \leq \theta \leq \pi \qquad h_2 = \rho$$

$$z = \rho \cos \theta \qquad \qquad 0 \leq \phi \leq 2\pi \qquad h_3 = \rho \sin \theta$$

The coordinate curves, illustrated in the figure 1.3-2, are formed by the intersection of the coordinate surfaces

$$x^2 + y^2 + z^2 = \rho^2 \qquad \text{Spheres}$$

$$x^2 + y^2 = \tan^2 \theta \, z^2 \qquad \text{Cones}$$

$$y = x \tan \phi \quad \text{Planes.}$$

4. Parabolic cylindrical coordinates (ξ, η, z)

$$x = \xi \eta \qquad \qquad -\infty < \xi < \infty \qquad h_1 = \sqrt{\xi^2 + \eta^2}$$

$$y = \frac{1}{2}(\xi^2 - \eta^2) \qquad -\infty < z < \infty \qquad h_2 = \sqrt{\xi^2 + \eta^2}$$

$$z = z \qquad \qquad \eta \geq 0 \qquad \qquad h_3 = 1$$

The coordinate curves, illustrated in the figure 1.3-3, are formed by the intersection of the coordinate surfaces

$$x^2 = -2\xi^2(y - \frac{\xi^2}{2}) \qquad \text{Parabolic cylinders}$$

$$x^2 = 2\eta^2(y + \frac{\eta^2}{2}) \qquad \text{Parabolic cylinders}$$

$$z = Constant \qquad \text{Planes.}$$

70

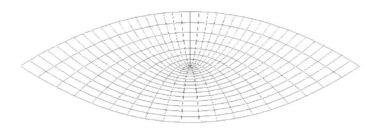

Figure 1.3-3. Parabolic cylindrical coordinates in plane $z = 0$.

5. Parabolic coordinates (ξ, η, ϕ)

$$
\begin{array}{lll}
x = \xi\eta\cos\phi & \xi \geq 0 & h_1 = \sqrt{\xi^2 + \eta^2} \\
y = \xi\eta\sin\phi & \eta \geq 0 & h_2 = \sqrt{\xi^2 + \eta^2} \\
z = \dfrac{1}{2}(\xi^2 - \eta^2) & 0 < \phi < 2\pi & h_3 = \xi\eta
\end{array}
$$

The coordinate curves, illustrated in the figure 1.3-4, are formed by the intersection of the coordinate surfaces

$$
\begin{array}{ll}
x^2 + y^2 = -2\xi^2(z - \dfrac{\xi^2}{2}) & \text{Paraboloids} \\
x^2 + y^2 = 2\eta^2(z + \dfrac{\eta^2}{2}) & \text{Paraboloids} \\
y = x\tan\phi & \text{Planes.}
\end{array}
$$

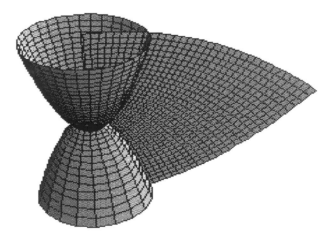

Figure 1.3-4. Parabolic coordinates, $\phi = \pi/4$.

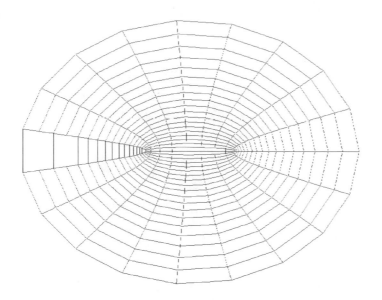

Figure 1.3-5. Elliptic cylindrical coordinates in the plane $z = 0$.

6. Elliptic cylindrical coordinates (ξ, η, z)

$$x = \cosh \xi \cos \eta \quad \xi \geq 0 \qquad\qquad h_1 = \sqrt{\sinh^2 \xi + \sin^2 \eta}$$

$$y = \sinh \xi \sin \eta \quad 0 \leq \eta \leq 2\pi \qquad h_2 = \sqrt{\sinh^2 \xi + \sin^2 \eta}$$

$$z = z \qquad\qquad -\infty < z < \infty \qquad h_3 = 1$$

The coordinate curves, illustrated in the figure 1.3-5, are formed by the intersection of the coordinate surfaces

$$\frac{x^2}{\cosh^2 \xi} + \frac{y^2}{\sinh^2 \xi} = 1 \qquad \text{Elliptic cylinders}$$

$$\frac{x^2}{\cos^2 \eta} - \frac{y^2}{\sin^2 \eta} = 1 \qquad \text{Hyperbolic cylinders}$$

$$z = Constant \qquad \text{Planes.}$$

7. Elliptic coordinates (ξ, η, ϕ)

$$x = \sqrt{(1 - \eta^2)(\xi^2 - 1)} \cos \phi \quad 1 \leq \xi < \infty \qquad h_1 = \sqrt{\frac{\xi^2 - \eta^2}{\xi^2 - 1}}$$

$$y = \sqrt{(1 - \eta^2)(\xi^2 - 1)} \sin \phi \quad -1 \leq \eta \leq 1 \qquad h_2 = \sqrt{\frac{\xi^2 - \eta^2}{1 - \eta^2}}$$

$$z = \xi\eta \qquad\qquad 0 \leq \phi < 2\pi \qquad h_3 = \sqrt{(1 - \eta^2)(\xi^2 - 1)}$$

The coordinate curves, illustrated in the figure 1.3-6, are formed by the intersection of the coordinate surfaces

$$\frac{x^2}{\xi^2 - 1} + \frac{y^2}{\xi^2 - 1} + \frac{z^2}{\xi^2} = 1 \qquad \text{Prolate ellipsoid}$$

$$\frac{z^2}{\eta^2} - \frac{x^2}{1 - \eta^2} - \frac{y^2}{1 - \eta^2} = 1 \qquad \text{Two-sheeted hyperboloid}$$

$$y = x \tan \phi \qquad \text{Planes}$$

72

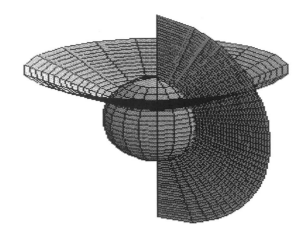

Figure 1.3-6. Elliptic coordinates $\phi = \pi/4$.

8. Bipolar coordinates (u, v, z)

$$x = \frac{a \sinh v}{\cosh v - \cos u}, \qquad 0 \le u < 2\pi \qquad h_1^2 = h_2^2$$

$$y = \frac{a \sin u}{\cosh v - \cos u}, \qquad -\infty < v < \infty \qquad h_2^2 = \frac{a^2}{(\cosh v - \cos u)^2}$$

$$z = z \qquad -\infty < z < \infty \qquad h_3^2 = 1$$

The coordinate curves, illustrated in the figure 1.3-7, are formed by the intersection of the coordinate surfaces

$$(x - a \coth v)^2 + y^2 = \frac{a^2}{\sinh^2 v} \qquad \text{Cylinders}$$

$$x^2 + (y - a \cot u)^2 = \frac{a^2}{\sin^2 u} \qquad \text{Cylinders}$$

$$z = Constant \qquad \text{Planes.}$$

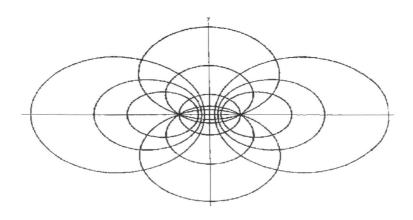

Figure 1.3-7. Bipolar coordinates.

9. Conical coordinates (u, v, w)

$$x = \frac{uvw}{ab}, \qquad b^2 > v^2 > a^2 > w^2, \quad u \geq 0 \qquad h_1^2 = 1$$

$$y = \frac{u}{a}\sqrt{\frac{(v^2 - a^2)(w^2 - a^2)}{a^2 - b^2}} \qquad\qquad h_2^2 = \frac{u^2(v^2 - w^2)}{(v^2 - a^2)(b^2 - v^2)}$$

$$z = \frac{u}{b}\sqrt{\frac{(v^2 - b^2)(w^2 - b^2)}{b^2 - a^2}} \qquad\qquad h_3^2 = \frac{u^2(v^2 - w^2)}{(w^2 - a^2)(w^2 - b^2)}$$

The coordinate curves, illustrated in the figure 1.3-8, are formed by the intersection of the coordinate surfaces

$$x^2 + y^2 + z^2 = u^2 \qquad \text{Spheres}$$

$$\frac{x^2}{v^2} + \frac{y^2}{v^2 - a^2} + \frac{z^2}{v^2 - b^2} = 0, \qquad \text{Cones}$$

$$\frac{x^2}{w^2} + \frac{y^2}{w^2 - a^2} + \frac{z^2}{w^2 - b^2} = 0, \qquad \text{Cones.}$$

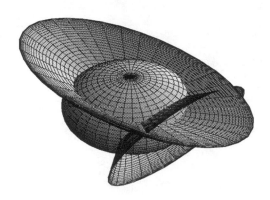

Figure 1.3-8. Conical coordinates.

10. Prolate spheroidal coordinates (u, v, ϕ)

$$x = a \sinh u \sin v \cos \phi, \quad u \geq 0 \qquad h_1^2 = h_2^2$$

$$y = a \sinh u \sin v \sin \phi, \quad 0 \leq v \leq \pi \qquad h_2^2 = a^2(\sinh^2 u + \sin^2 v)$$

$$z = a \cosh u \cos v, \quad 0 \leq \phi < 2\pi \qquad h_3^2 = a^2 \sinh^2 u \sin^2 v$$

The coordinate curves, illustrated in the figure 1.3-9, are formed by the intersection of the coordinate surfaces

$$\frac{x^2}{(a\sinh u)^2} + \frac{y^2}{(a\sinh u)^2} + \frac{z^2}{(a\cosh u)^2} = 1, \qquad \text{Prolate ellipsoids}$$

$$\frac{z^2}{(a\cos v)^2} - \frac{x^2}{(a\sin v)^2} - \frac{y^2}{(a\sin v)^2} = 1, \qquad \text{Two-sheeted hyperboloid}$$

$$y = x \tan \phi, \qquad \text{Planes.}$$

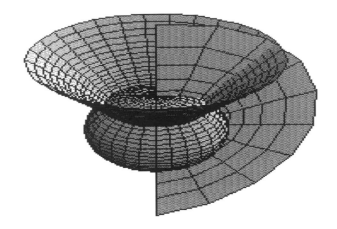

Figure 1.3-9. Prolate spheroidal coordinates

11. Oblate spheroidal coordinates (ξ, η, ϕ)

$$x = a \cosh \xi \cos \eta \cos \phi, \qquad \xi \geq 0 \qquad\qquad h_1^2 = h_2^2$$
$$y = a \cosh \xi \cos \eta \sin \phi, \qquad -\frac{\pi}{2} \leq \eta \leq \frac{\pi}{2} \qquad h_2^2 = a^2(\sinh^2 \xi + \sin^2 \eta)$$
$$z = a \sinh \xi \sin \eta, \quad 0 \leq \phi \leq 2\pi \qquad\qquad h_3^2 = a^2 \cosh^2 \xi \cos^2 \eta$$

The coordinate curves, illustrated in the figure 1.3-10, are formed by the intersection of the coordinate surfaces

$$\frac{x^2}{(a \cosh \xi)^2} + \frac{y^2}{(a \cosh \xi)^2} + \frac{z^2}{(a \sinh \xi)^2} = 1, \qquad \text{Oblate ellipsoids}$$

$$\frac{x^2}{(a \cos \eta)^2} + \frac{y^2}{(a \cos \eta)^2} - \frac{z^2}{(a \sin \eta)^2} = 1, \qquad \text{One-sheet hyperboloids}$$

$$y = x \tan \phi, \qquad \text{Planes.}$$

12. Toroidal coordinates (u, v, ϕ)

$$x = \frac{a \sinh v \cos \phi}{\cosh v - \cos u}, \quad 0 \leq u < 2\pi \qquad h_1^2 = h_2^2$$
$$y = \frac{a \sinh v \sin \phi}{\cosh v - \cos u}, \quad -\infty < v < \infty \qquad h_2^2 = \frac{a^2}{(\cosh v - \cos u)^2}$$
$$z = \frac{a \sin u}{\cosh v - \cos u}, \quad 0 \leq \phi < 2\pi \qquad h_3^2 = \frac{a^2 \sinh^2 v}{(\cosh v - \cos u)^2}$$

The coordinate curves, illustrated in the figure 1.3-11, are formed by the intersection of the coordinate surfaces

$$x^2 + y^2 + \left(z - \frac{a \cos u}{\sin u}\right)^2 = \frac{a^2}{\sin^2 u}, \qquad \text{Spheres}$$

$$\left(\sqrt{x^2 + y^2} - a\frac{\cosh v}{\sinh v}\right)^2 + z^2 = \frac{a^2}{\sinh^2 v}, \qquad \text{Torus}$$

$$y = x \tan \phi, \qquad \text{planes}$$

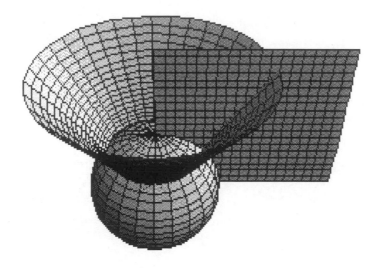

Figure 1.3-10. Oblate spheroidal coordinates

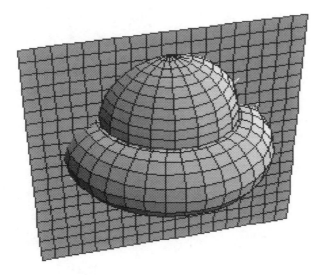

Figure 1.3-11. Toroidal coordinates

13. Dipole coordinate system (μ, χ, ϕ)

The standard dipole coordinate system (μ, χ, ϕ) is an orthogonal coordinate system used whenever the Earth's dipolar magnetic field is involved in modeling physical problems. In terms of spherical coordinates (r, θ, ϕ) the dipole coordinates μ and χ are defined

$$\mu = -\frac{\cos\theta}{r^2}, \qquad \chi = \frac{\sin^2\theta}{r}$$

where r is a normalized length from the Earth's center, ($r = 1$ corresponds to one Earth radii), θ is a colatitude and ϕ is the longitude. Here μ is a potential function of the dipole field. The dipole field lines are represented by the curves χ equal to a constant in a meridan plane ϕ equal to a constant. In this coordinate system the element of arc length squared is given by $ds^2 = h_\mu^2 \, d\mu^2 + h_\chi^2 \, d\chi^2 + h_\phi^2 \, d\phi^2$ where

$$h_\mu = \frac{1}{|\nabla\mu|} = \frac{r^3}{\sqrt{1 + 3\cos^2\theta}}, \qquad h_\chi = \frac{1}{|\nabla\chi|} = \frac{r^2}{\sin\theta\sqrt{1 + 3\cos^2\theta}}, \qquad h_\phi = \frac{1}{|\nabla\phi|} = r\sin\theta$$

The metric h_μ varies dramatically along the field lines and consequently it is sometimes more desirable to use coordinates $(\bar\mu, \chi, \phi)$ where $\bar\mu = -\frac{\sqrt{\cos\theta}}{r}$ for $\theta < \pi/2$. The coordinates $(\bar\mu, \chi, \phi)$ produce an orthogonal coordinate system with

$$h_{\bar\mu} = \frac{1}{|\nabla\bar\mu|} = 2r^2 \frac{\sqrt{\cos\theta}}{\sqrt{1 + 3\cos^2\theta}}.$$

This coordinate system has the disadvantage that $h_{\bar\mu}$ vanishes at the equator with $h_{\bar\mu}\big|_{\bar\mu=0} = 0$ and consequently cannot be applied for all latitudes. The standard dipole coordinate system is not recommended for use with numerical methods.

14. General Curvilinear coordinates (u, v, w)

$$x = x(u, v, w)$$
$$y = y(u, v, w)$$
$$z = z(u, v, w)$$

Let $\vec{r} = \vec{r}(u, v, w) = x\,\widehat{\mathbf{e}}_1 + y\,\widehat{\mathbf{e}}_2 + z\,\widehat{\mathbf{e}}_3$, then

$$\frac{\partial\vec{r}}{\partial u} = h_1\,\widehat{\mathbf{e}}_u, \qquad \frac{\partial\vec{r}}{\partial v} = h_2\,\widehat{\mathbf{e}}_v, \qquad \frac{\partial\vec{r}}{\partial w} = h_3\,\widehat{\mathbf{e}}_w$$

where $\widehat{\mathbf{e}}_u, \widehat{\mathbf{e}}_v, \widehat{\mathbf{e}}_w$ are unit tangent vectors to the coordinate curves $\vec{r}(u, c_2, c_3)$, $\vec{r}(c_1, v, c_3)$ and $\vec{r}(c_1, c_2, w)$. The scale factors h_1, h_2, h_3 are given by

$$h_1 = \left|\frac{\partial\vec{r}}{\partial u}\right|, \qquad h_2 = \left|\frac{\partial\vec{r}}{\partial v}\right|, \qquad h_3 = \left|\frac{\partial\vec{r}}{\partial w}\right|$$

Note that the curvilinear coordinate system is called orthogonal if $\widehat{\mathbf{e}}_u, \widehat{\mathbf{e}}_v, \widehat{\mathbf{e}}_w$ are mutually perpendicular. In this case the element of arc length squared is given by

$$ds^2 = h_1^2 du^2 + h_2^2 dv^2 + h_3^2 dw^2$$

and an element of volume is found to be

$$d\tau = \left| \frac{\partial \vec{r}}{\partial u} \cdot \frac{\partial \vec{r}}{\partial v} \times \frac{\partial \vec{r}}{\partial w} \right| du\, dv\, dw = \left| \frac{\partial(x,y,z)}{\partial(u,v,w)} \right| du\, dv\, dw$$

where

$$\left| \frac{\partial(x,y,z)}{\partial(u,v,w)} \right| = \begin{vmatrix} \frac{\partial x}{\partial u} & \frac{\partial x}{\partial v} & \frac{\partial x}{\partial w} \\ \frac{\partial y}{\partial u} & \frac{\partial y}{\partial v} & \frac{\partial y}{\partial w} \\ \frac{\partial z}{\partial u} & \frac{\partial z}{\partial v} & \frac{\partial z}{\partial w} \end{vmatrix}$$

is the Jacobian of the transformation.

EXAMPLE 1.3-4. Show the Kronecker delta δ_j^i is a mixed second order tensor.

Solution: Assume we have a coordinate transformation $x^i = x^i(\overline{x}), i = 1, \ldots, N$ of the form (1.2.30) and possessing an inverse transformation of the form (1.2.32). Let $\overline{\delta}_j^i$ and δ_j^i denote the Kronecker delta in the barred and unbarred system of coordinates. By definition the Kronecker delta is defined

$$\overline{\delta}_j^i = \delta_j^i = \begin{cases} 0, & \text{if } i \neq j \\ 1, & \text{if } i = j \end{cases}.$$

Employing the chain rule we write

$$\frac{\partial \overline{x}^m}{\partial \overline{x}^n} = \frac{\partial \overline{x}^m}{\partial x^i} \frac{\partial x^i}{\partial \overline{x}^n} = \frac{\partial \overline{x}^m}{\partial x^i} \frac{\partial x^k}{\partial \overline{x}^n} \delta_k^i \tag{1.3.14}$$

By hypothesis, the $\overline{x}^i, i = 1, \ldots, N$ are independent coordinates and therefore we have $\frac{\partial \overline{x}^m}{\partial \overline{x}^n} = \overline{\delta}_n^m$ and (1.3.14) simplifies to

$$\overline{\delta}_n^m = \delta_k^i \frac{\partial \overline{x}^m}{\partial x^i} \frac{\partial x^k}{\partial \overline{x}^n}.$$

Therefore, the Kronecker delta transforms as a mixed second order tensor. ∎

Conjugate Metric Tensor

Let g denote the determinant of the matrix having the metric tensor $g_{ij}, i, j = 1, \ldots, N$ as its elements. In our study of cofactor elements of a matrix we have shown that

$$cof(g_{1j})g_{1k} + cof(g_{2j})g_{2k} + \ldots + cof(g_{Nj})g_{Nk} = g\delta_k^j. \tag{1.3.15}$$

We can use this fact to find the elements in the inverse matrix associated with the matrix having the components g_{ij}. The elements of this inverse matrix are

$$g^{ij} = \frac{1}{g} cof(g_{ij}) \tag{1.3.16}$$

and are called the conjugate metric components. We examine the summation $g^{ij}g_{ik}$ and find:

$$g^{ij}g_{ik} = g^{1j}g_{1k} + g^{2j}g_{2k} + \ldots + g^{Nj}g_{Nk}$$
$$= \frac{1}{g} \left[cof(g_{1j})g_{1k} + cof(g_{2j})g_{2k} + \ldots + cof(g_{Nj})g_{Nk} \right]$$
$$= \frac{1}{g} \left[g\delta_k^j \right] = \delta_k^j$$

The equation

$$g^{ij}g_{ik} = \delta^j_k \tag{1.3.17}$$

is an example where we can use the quotient law to show g^{ij} is a second order contravariant tensor. Because of the symmetry of g^{ij} and g_{ij} the equation (1.3.17) can be represented in other forms.

EXAMPLE 1.3-5. Let A_i and A^i denote respectively the covariant and contravariant components of a vector $\vec{A}$. Show these components are related by the equations

$$A_i = g_{ij}A^j \tag{1.3.18}$$

$$A^k = g^{jk}A_j \tag{1.3.19}$$

where g_{ij} and g^{ij} are the metric and conjugate metric components of the space.

Solution: We multiply the equation (1.3.18) by g^{im} (inner product) and use equation (1.3.17) to simplify the results. This produces the equation $g^{im}A_i = g^{im}g_{ij}A^j = \delta^m_j A^j = A^m$. Changing indices produces the result given in equation (1.3.19). Conversely, if we start with equation (1.3.19) and multiply by g_{km} (inner product) we obtain $g_{km}A^k = g_{km}g^{jk}A_j = \delta^j_m A_j = A_m$ which is another form of the equation (1.3.18) with the indices changed.

Notice the consequences of what the equations (1.3.18) and (1.3.19) imply when we are in an orthogonal Cartesian coordinate system where

$$g_{ij} = \begin{pmatrix} 1 & 0 & 0 \\ 0 & 1 & 0 \\ 0 & 0 & 1 \end{pmatrix} \quad \text{and} \quad g^{ij} = \begin{pmatrix} 1 & 0 & 0 \\ 0 & 1 & 0 \\ 0 & 0 & 1 \end{pmatrix}.$$

In this special case, we have

$$A_1 = g_{11}A^1 + g_{12}A^2 + g_{13}A^3 = A^1$$
$$A_2 = g_{21}A^1 + g_{22}A^2 + g_{23}A^3 = A^2$$
$$A_3 = g_{31}A^1 + g_{32}A^2 + g_{33}A^3 = A^3.$$

These equations tell us that in a Cartesian coordinate system the contravariant and covariant components are identically the same.

∎

EXAMPLE 1.3-6. We have previously shown that if A_i is a covariant tensor of rank 1 its components in a barred system of coordinates are

$$\overline{A}_i = A_j \frac{\partial x^j}{\partial \overline{x}^i}. \tag{1.3.20}$$

Solve for the A_j in terms of the $\overline{A}_j$. (i.e. find the inverse transformation).

Solution: Multiply equation (1.3.20) by $\frac{\partial \overline{x}^i}{\partial x^m}$ (inner product) and obtain

$$\overline{A}_i \frac{\partial \overline{x}^i}{\partial x^m} = A_j \frac{\partial x^j}{\partial \overline{x}^i} \frac{\partial \overline{x}^i}{\partial x^m}. \tag{1.3.21}$$

In the above product we have $\dfrac{\partial x^j}{\partial \overline{x}^i}\dfrac{\partial \overline{x}^i}{\partial x^m} = \dfrac{\partial x^j}{\partial x^m} = \delta_m^j$ since x^j and x^m are assumed to be independent coordinates. This reduces equation (1.3.21) to the form

$$\overline{A}_i \frac{\partial \overline{x}^i}{\partial x^m} = A_j \delta_m^j = A_m \tag{1.3.22}$$

which is the desired inverse transformation.

This result can be obtained in another way. Examine the transformation equation (1.3.20) and ask the question, "When we have two coordinate systems, say a barred and an unbarred system, does it matter which system we call the barred system?" With some thought it should be obvious that it doesn't matter which system you label as the barred system. Therefore, we can interchange the barred and unbarred symbols in equation (1.3.20) and obtain the result $A_i = \overline{A}_j \dfrac{\partial \overline{x}^j}{\partial x^i}$ which is the same form as equation (1.3.22), but with a different set of indices.

■

Associated Tensors

Associated tensors can be constructed by taking the inner product of known tensors with either the metric or conjugate metric tensor.

> **Definition: (Associated tensor)**   Any tensor constructed by multiplying (inner product) a given tensor with the metric or conjugate metric tensor is called an associated tensor.

Associated tensors are different ways of representing a tensor. The multiplication of a tensor by the metric or conjugate metric tensor has the effect of lowering or raising indices. For example the covariant and contravariant components of a vector are different representations of the same vector in different forms. These forms are associated with one another by way of the metric and conjugate metric tensor and

$$g^{ij} A_i = A^j \qquad g_{ij} A^j = A_i.$$

EXAMPLE 1.3-7.   The following are some examples of associated tensors.

$$A^j = g^{ij} A_i \qquad\qquad A_j = g_{ij} A^i$$
$$A^m_{.jk} = g^{mi} A_{ijk} \qquad\qquad A^{i.k}_m = g_{mj} A^{ijk}$$
$$A^{.nm}_{i..} = g^{mk} g^{nj} A_{ijk} \qquad\qquad A_{mjk} = g_{im} A^i_{.jk}$$

Sometimes 'dots' are used as indices in order to represent the location of the index that was raised or lowered. If a tensor is symmetric, the position of the index is immaterial and so a dot is not needed. For example, if A_{nm} is a symmetric tensor, then it is easy to show that $A^n_{.m}$ and $A^{.n}_m$ are equal and therefore can be written as A^n_m without confusion.

Higher order tensors are similarly related. For example, if we find a fourth order covariant tensor T_{ijkm} we can then construct the fourth order contravariant tensor T^{pqrs} from the relation

$$T^{pqrs} = g^{pi} g^{qj} g^{rk} g^{sm} T_{ijkm}.$$

This fourth order tensor can also be expressed as a mixed tensor. Some mixed tensors associated with the given fourth order covariant tensor are:

$$T^{p}_{.jkm} = g^{pi} T_{ijkm}, \qquad T^{pq}_{..km} = g^{qj} T^{p}_{.jkm}.$$

∎

Riemann Space V_N

A Riemannian space V_N is said to exist if the element of arc length squared has the form

$$ds^2 = g_{ij} dx^i dx^j \qquad (1.3.23)$$

where the metrices $g_{ij} = g_{ij}(x^1, x^2, \ldots, x^N)$ are continuous functions of the coordinates and are different from constants. In the special case $g_{ij} = \delta_{ij}$ the Riemannian space V_N reduces to a Euclidean space E_N. The element of arc length squared defined by equation (1.3.23) is called the Riemannian metric and any geometry which results by using this metric is called a Riemannian geometry. A space V_N is called flat if it is possible to find a coordinate transformation where the element of arc length squared is $ds^2 = \epsilon_i (dx^i)^2$ where each ϵ_i is either $+1$ or -1. A space which is not flat is called curved.

Geometry in V_N

Given two vectors $\vec{A} = A^i \vec{E}_i$ and $\vec{B} = B^j \vec{E}_j$, then their dot product can be represented

$$\vec{A} \cdot \vec{B} = A^i B^j \vec{E}_i \cdot \vec{E}_j = g_{ij} A^i B^j = A_j B^j = A^i B_i = g^{ij} A_j B_i = |\vec{A}||\vec{B}| \cos\theta. \qquad (1.3.24)$$

In an N dimensional Riemannian space V_N the dot or inner product of two vectors $\vec{A}$ and $\vec{B}$ is defined:

$$g_{ij} A^i B^j = A_j B^j = A^i B_i = g^{ij} A_j B_i = AB \cos\theta. \qquad (1.3.25)$$

In this definition A is the magnitude of the vector A^i, the quantity B is the magnitude of the vector B_i and θ is the angle between the vectors when their origins are made to coincide. In the special case that $\theta = 90°$ we have $g_{ij} A^i B^j = 0$ as the condition that must be satisfied in order that the given vectors A^i and B^i are orthogonal to one another. Consider also the special case of equation (1.3.25) when $A^i = B^i$ and $\theta = 0$. In this case the equations (1.3.25) inform us that

$$g^{in} A_n A_i = A^i A_i = g_{in} A^i A^n = (A)^2. \qquad (1.3.26)$$

From this equation one can determine the magnitude of the vector A^i. The magnitudes A and B can be written $A = (g_{in} A^i A^n)^{\frac{1}{2}}$ and $B = (g_{pq} B^p B^q)^{\frac{1}{2}}$ and so we can express equation (1.3.24) in the form

$$\cos\theta = \frac{g_{ij} A^i B^j}{(g_{mn} A^m A^n)^{\frac{1}{2}} (g_{pq} B^p B^q)^{\frac{1}{2}}}. \qquad (1.3.27)$$

An import application of the above concepts arises in the dynamics of rigid body motion. Note that if a vector A^i has constant magnitude and the magnitude of $\frac{dA^i}{dt}$ is different from zero, then the vectors A^i and $\frac{dA^i}{dt}$ must be orthogonal to one another due to the fact that $g_{ij}A^i\frac{dA^j}{dt}=0$. As an example, consider the unit vectors $\widehat{\mathbf{e}}_1, \widehat{\mathbf{e}}_2$ and $\widehat{\mathbf{e}}_3$ on a rotating system of Cartesian axes. We have for constants c_i, $i=1,6$ that

$$\frac{d\widehat{\mathbf{e}}_1}{dt}=c_1\,\widehat{\mathbf{e}}_2+c_2\,\widehat{\mathbf{e}}_3, \qquad \frac{d\widehat{\mathbf{e}}_2}{dt}=c_3\,\widehat{\mathbf{e}}_3+c_4\,\widehat{\mathbf{e}}_1, \qquad \frac{d\widehat{\mathbf{e}}_3}{dt}=c_5\,\widehat{\mathbf{e}}_1+c_6\,\widehat{\mathbf{e}}_2$$

because the derivative of any $\widehat{\mathbf{e}}_i$ (i fixed) constant vector must lie in a plane containing the vectors $\widehat{\mathbf{e}}_j$ and $\widehat{\mathbf{e}}_k$, ($j \neq i$, $k \neq i$ and $j \neq k$), since any vector in this plane must be perpendicular to $\widehat{\mathbf{e}}_i$.

The above definition of a dot product in V_N can be used to define unit vectors in V_N.

Definition: (Unit vector) Whenever the magnitude of a vector A^i is unity, the vector is called a unit vector. In this case we have

$$g_{ij}A^iA^j=1. \qquad (1.3.28)$$

EXAMPLE 1.3-8. (Unit vectors)

In V_N the element of arc length squared is expressed $ds^2 = g_{ij}\,dx^i dx^j$ which can be expressed in the form $1 = g_{ij}\frac{dx^i}{ds}\frac{dx^j}{ds}$. This equation states that the vector $\frac{dx^i}{ds}$, $i=1,\dots,N$ is a unit vector. One application of this equation is to consider a particle moving along a curve in V_N which is described by the parametric equations $x^i = x^i(t)$, for $i=1,\dots,N$. The vector $V^i = \frac{dx^i}{dt}$, $i=1,\dots,N$ represents a velocity vector of the particle. By chain rule differentiation we have

$$V^i=\frac{dx^i}{dt}=\frac{dx^i}{ds}\frac{ds}{dt}=V\frac{dx^i}{ds}, \qquad (1.3.29)$$

where $V=\frac{ds}{dt}$ is the scalar speed of the particle and $\frac{dx^i}{ds}$ is a unit tangent vector to the curve. The equation (1.3.29) shows that the velocity is directed along the tangent to the curve and has a magnitude V. That is

$$\left(\frac{ds}{dt}\right)^2=(V)^2=g_{ij}V^iV^j.$$

∎

EXAMPLE 1.3-9. (Curvilinear coordinates)

Find an expression for the cosine of the angles between the coordinate curves associated with the transformation equations

$$x=x(u,v,w), \qquad y=y(u,v,w), \qquad z=z(u,v,w).$$

Solution: Let $y^1=x$, $y^2=y$, $y^3=z$ and $x^1=u$, $x^2=v$, $x^3=w$ denote the Cartesian and curvilinear coordinates respectively. With reference to the figure 1.3-12 we can interpret the intersection of the surfaces

$v = c_2$ and $w = c_3$ as the curve $\vec{r} = \vec{r}(u, c_2, c_3)$ which is a function of the parameter u. By moving only along this curve we have $d\vec{r} = \dfrac{\partial \vec{r}}{\partial u}\, du$ and consequently

$$ds^2 = d\vec{r} \cdot d\vec{r} = \frac{\partial \vec{r}}{\partial u} \cdot \frac{\partial \vec{r}}{\partial u}\, du\, du = g_{11}(dx^1)^2,$$

or

$$1 = \frac{d\vec{r}}{ds} \cdot \frac{d\vec{r}}{ds} = g_{11}\left(\frac{dx^1}{ds}\right)^2.$$

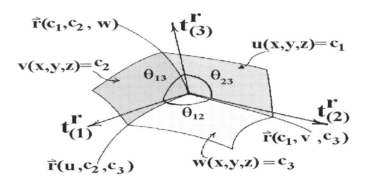

Figure 1.3-12. Angles between curvilinear coordinates.

This equation shows that the vector $\frac{dx^1}{ds} = \frac{1}{\sqrt{g_{11}}}$ is a unit vector along this curve. This tangent vector can be represented by $t^r_{(1)} = \frac{1}{\sqrt{g_{11}}}\delta^r_1$. The curve which is defined by the intersection of the surfaces $u = c_1$ and $w = c_3$ has the unit tangent vector $t^r_{(2)} = \frac{1}{\sqrt{g_{22}}}\delta^r_2$. Similarly, the curve which is defined as the intersection of the surfaces $u = c_1$ and $v = c_2$ has the unit tangent vector $t^r_{(3)} = \frac{1}{\sqrt{g_{33}}}\delta^r_3$. The cosine of the angle θ_{12}, which is the angle between the unit vectors $t^r_{(1)}$ and $t^r_{(2)}$, is obtained from the result of equation (1.3.25). We find

$$\cos\theta_{12} = g_{pq}t^p_{(1)}t^q_{(2)} = g_{pq}\frac{1}{\sqrt{g_{11}}}\delta^p_1 \frac{1}{\sqrt{g_{22}}}\delta^q_2 = \frac{g_{12}}{\sqrt{g_{11}}\sqrt{g_{22}}}.$$

For θ_{13} the angle between the directions $t^i_{(1)}$ and $t^i_{(3)}$ we find

$$\cos\theta_{13} = \frac{g_{13}}{\sqrt{g_{11}}\sqrt{g_{33}}}.$$

Finally, for θ_{23} the angle between the directions $t^i_{(2)}$ and $t^i_{(3)}$ we find

$$\cos\theta_{23} = \frac{g_{23}}{\sqrt{g_{22}}\sqrt{g_{33}}}.$$

When $\theta_{13} = \theta_{12} = \theta_{23} = 90°$, we have $g_{12} = g_{13} = g_{23} = 0$ and the coordinate curves which make up the curvilinear coordinate system are orthogonal to one another.

In an orthogonal coordinate system we adopt the notation

$$g_{11} = (h_1)^2, \qquad g_{22} = (h_2)^2, \qquad g_{33} = (h_3)^2 \qquad \text{and} \qquad g_{ij} = 0,\ i \neq j.$$

Epsilon Permutation Symbol

Associated with the $e-$permutation symbols there are the epsilon permutation symbols defined by the relations

$$\epsilon_{ijk} = \sqrt{g}\, e_{ijk} \qquad \text{and} \qquad \epsilon^{ijk} = \frac{1}{\sqrt{g}} e^{ijk} \tag{1.3.30}$$

where g is the determinant of the metrices g_{ij}.

It can be demonstrated that the e_{ijk} permutation symbol is a relative tensor of weight -1 whereas the ϵ_{ijk} permutation symbol is an absolute tensor. Similarly, the e^{ijk} permutation symbol is a relative tensor of weight $+1$ and the corresponding ϵ^{ijk} permutation symbol is an absolute tensor.

EXAMPLE 1.3-10. (ϵ permutation symbol)

Show that e_{ijk} is a relative tensor of weight -1 and the corresponding ϵ_{ijk} permutation symbol is an absolute tensor.

Solution: Examine the Jacobian

$$J\left(\frac{x}{\overline{x}}\right) = \begin{vmatrix} \frac{\partial x^1}{\partial \overline{x}^1} & \frac{\partial x^1}{\partial \overline{x}^2} & \frac{\partial x^1}{\partial \overline{x}^3} \\ \frac{\partial x^2}{\partial \overline{x}^1} & \frac{\partial x^2}{\partial \overline{x}^2} & \frac{\partial x^2}{\partial \overline{x}^3} \\ \frac{\partial x^3}{\partial \overline{x}^1} & \frac{\partial x^3}{\partial \overline{x}^2} & \frac{\partial x^3}{\partial \overline{x}^3} \end{vmatrix}$$

and make the substitution

$$a^i_j = \frac{\partial x^i}{\partial \overline{x}^j},\ i,j=1,2,3.$$

From the definition of a determinant we may write

$$e_{ijk} a^i_m a^j_n a^k_p = J\left(\frac{x}{\overline{x}}\right) e_{mnp}. \tag{1.3.31}$$

By definition, $\overline{e}_{mnp} = e_{mnp}$ in all coordinate systems and hence equation (1.3.31) can be expressed in the form

$$\left[J\left(\frac{x}{\overline{x}}\right)\right]^{-1} e_{ijk} \frac{\partial x^i}{\partial \overline{x}^m} \frac{\partial x^j}{\partial \overline{x}^n} \frac{\partial x^k}{\partial \overline{x}^p} = \overline{e}_{mnp} \tag{1.3.32}$$

which demonstrates that e_{ijk} transforms as a relative tensor of weight -1.

We have previously shown the metric tensor g_{ij} is a second order covariant tensor and transforms according to the rule $\overline{g}_{ij} = g_{mn} \frac{\partial x^m}{\partial \overline{x}^i} \frac{\partial x^n}{\partial \overline{x}^j}$. Taking the determinant of this result we find

$$\overline{g} = |\overline{g}_{ij}| = |g_{mn}| \left|\frac{\partial x^m}{\partial \overline{x}^i}\right|^2 = g\left[J\left(\frac{x}{\overline{x}}\right)\right]^2 \tag{1.3.33}$$

where g is the determinant of (g_{ij}) and $\overline{g}$ is the determinant of $(\overline{g}_{ij})$. This result demonstrates that g is a scalar invariant of weight $+2$. Taking the square root of this result we find that

$$\sqrt{\overline{g}} = \sqrt{g}\, J\left(\frac{x}{\overline{x}}\right). \tag{1.3.34}$$

This shows that $\sqrt{g}$ a scalar invariant of weight $+1$. Now multiply both sides of equation (1.3.32) by $\sqrt{\overline{g}}$ and use (1.3.34) to verify the relation

$$\sqrt{g}\, e_{ijk} \frac{\partial x^i}{\partial \overline{x}^m} \frac{\partial x^j}{\partial \overline{x}^n} \frac{\partial x^k}{\partial \overline{x}^p} = \sqrt{\overline{g}}\, \overline{e}_{mnp}. \tag{1.3.35}$$

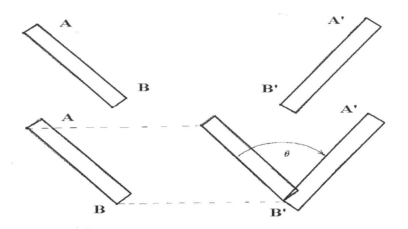

Figure 1.3-13. Motion of rigid rod

This equation demonstrates that the quantity $\epsilon_{ijk} = \sqrt{g}\, e_{ijk}$ transforms like an absolute tensor.

In a similar manner one can show e^{ijk} is a relative tensor of weight $+1$ and $\epsilon^{ijk} = \frac{1}{\sqrt{g}} e^{ijk}$ is an absolute tensor. This is left as an exercise.

$\blacksquare$

Another exercise found at the end of this section is to show that a generalization of the $e - \delta$ identity is the epsilon identity

$$g^{ij}\epsilon_{ipt}\epsilon_{jrs} = g_{pr}g_{ts} - g_{ps}g_{tr}. \tag{1.3.36}$$

Cartesian Tensors

Consider the motion of a rigid rod in two dimensions. No matter how complicated the movement of the rod is we can describe the motion as a translation followed by a rotation. Consider the rigid rod $\overline{AB}$ illustrated in the figure 1.3-13.

In this figure there is a before and after picture of the rod's position. By moving the point B to B' we have a translation. This is then followed by a rotation holding B fixed.

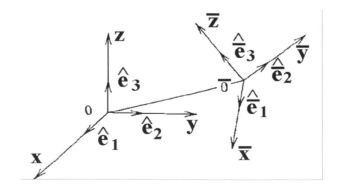

Figure 1.3-14. Translation followed by rotation of axes

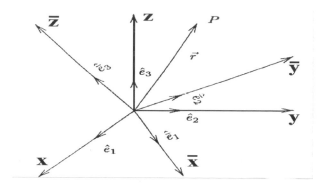

Figure 1.3-15. Rotation of axes

A similar situation exists in three dimensions. Consider two sets of Cartesian axes, say a barred and unbarred system as illustrated in the figure 1.3-14. Let us translate the origin 0 to $\overline{0}$ and then rotate the (x, y, z) axes until they coincide with the $(\overline{x}, \overline{y}, \overline{z})$ axes. We consider first the rotation of axes when the origins 0 and $\overline{0}$ coincide as the translational distance can be represented by a vector b^k, $k = 1, 2, 3$. When the origin 0 is translated to $\overline{0}$ we have the situation illustrated in the figure 1.3-15, where the barred axes can be thought of as a transformation due to rotation.

Let

$$\vec{r} = x\,\widehat{\mathbf{e}}_1 + y\,\widehat{\mathbf{e}}_2 + z\,\widehat{\mathbf{e}}_3 \tag{1.3.37}$$

denote the position vector of a variable point P with coordinates (x, y, z) with respect to the origin 0 and the unit vectors $\widehat{\mathbf{e}}_1$, $\widehat{\mathbf{e}}_2$, $\widehat{\mathbf{e}}_3$. This same point, when referenced with respect to the origin $\overline{0}$ and the unit vectors $\widehat{\overline{\mathbf{e}}}_1$, $\widehat{\overline{\mathbf{e}}}_2$, $\widehat{\overline{\mathbf{e}}}_3$, has the representation

$$\vec{r} = \overline{x}\,\widehat{\overline{\mathbf{e}}}_1 + \overline{y}\,\widehat{\overline{\mathbf{e}}}_2 + \overline{z}\,\widehat{\overline{\mathbf{e}}}_3. \tag{1.3.38}$$

By considering the projections of $\vec{r}$ upon the barred and unbarred axes we can construct the transformation equations relating the barred and unbarred axes. We calculate the projections of $\vec{r}$ onto the x, y and z axes and find:

$$\vec{r} \cdot \widehat{\mathbf{e}}_1 = x = \overline{x}(\widehat{\overline{\mathbf{e}}}_1 \cdot \widehat{\mathbf{e}}_1) + \overline{y}(\widehat{\overline{\mathbf{e}}}_2 \cdot \widehat{\mathbf{e}}_1) + \overline{z}(\widehat{\overline{\mathbf{e}}}_3 \cdot \widehat{\mathbf{e}}_1)$$
$$\vec{r} \cdot \widehat{\mathbf{e}}_2 = y = \overline{x}(\widehat{\overline{\mathbf{e}}}_1 \cdot \widehat{\mathbf{e}}_2) + \overline{y}(\widehat{\overline{\mathbf{e}}}_2 \cdot \widehat{\mathbf{e}}_2) + \overline{z}(\widehat{\overline{\mathbf{e}}}_3 \cdot \widehat{\mathbf{e}}_2) \tag{1.3.39}$$
$$\vec{r} \cdot \widehat{\mathbf{e}}_3 = z = \overline{x}(\widehat{\overline{\mathbf{e}}}_1 \cdot \widehat{\mathbf{e}}_3) + \overline{y}(\widehat{\overline{\mathbf{e}}}_2 \cdot \widehat{\mathbf{e}}_3) + \overline{z}(\widehat{\overline{\mathbf{e}}}_3 \cdot \widehat{\mathbf{e}}_3).$$

We also calculate the projection of $\vec{r}$ onto the $\overline{x}, \overline{y}, \overline{z}$ axes and find:

$$\vec{r} \cdot \widehat{\overline{\mathbf{e}}}_1 = \overline{x} = x(\widehat{\mathbf{e}}_1 \cdot \widehat{\overline{\mathbf{e}}}_1) + y(\widehat{\mathbf{e}}_2 \cdot \widehat{\overline{\mathbf{e}}}_1) + z(\widehat{\mathbf{e}}_3 \cdot \widehat{\overline{\mathbf{e}}}_1)$$
$$\vec{r} \cdot \widehat{\overline{\mathbf{e}}}_2 = \overline{y} = x(\widehat{\mathbf{e}}_1 \cdot \widehat{\overline{\mathbf{e}}}_2) + y(\widehat{\mathbf{e}}_2 \cdot \widehat{\overline{\mathbf{e}}}_2) + z(\widehat{\mathbf{e}}_3 \cdot \widehat{\overline{\mathbf{e}}}_2) \tag{1.3.40}$$
$$\vec{r} \cdot \widehat{\overline{\mathbf{e}}}_3 = \overline{z} = x(\widehat{\mathbf{e}}_1 \cdot \widehat{\overline{\mathbf{e}}}_3) + y(\widehat{\mathbf{e}}_2 \cdot \widehat{\overline{\mathbf{e}}}_3) + z(\widehat{\mathbf{e}}_3 \cdot \widehat{\overline{\mathbf{e}}}_3).$$

By introducing the notation $(y_1, y_2, y_3) = (x, y, z)$ $(\overline{y}_1, \overline{y}_2, \overline{y}_3) = (\overline{x}, \overline{y}, \overline{z})$ and defining θ_{ij} as the angle between the unit vectors $\widehat{\mathbf{e}}_i$ and $\widehat{\overline{\mathbf{e}}}_j$, we can represent the above transformation equations in a more concise form. We observe that the direction cosines can be written as

$$\ell_{11} = \widehat{\mathbf{e}}_1 \cdot \widehat{\overline{\mathbf{e}}}_1 = \cos\theta_{11} \qquad \ell_{12} = \widehat{\mathbf{e}}_1 \cdot \widehat{\overline{\mathbf{e}}}_2 = \cos\theta_{12} \qquad \ell_{13} = \widehat{\mathbf{e}}_1 \cdot \widehat{\overline{\mathbf{e}}}_3 = \cos\theta_{13}$$
$$\ell_{21} = \widehat{\mathbf{e}}_2 \cdot \widehat{\overline{\mathbf{e}}}_1 = \cos\theta_{21} \qquad \ell_{22} = \widehat{\mathbf{e}}_2 \cdot \widehat{\overline{\mathbf{e}}}_2 = \cos\theta_{22} \qquad \ell_{23} = \widehat{\mathbf{e}}_2 \cdot \widehat{\overline{\mathbf{e}}}_3 = \cos\theta_{23} \qquad (1.3.41)$$
$$\ell_{31} = \widehat{\mathbf{e}}_3 \cdot \widehat{\overline{\mathbf{e}}}_1 = \cos\theta_{31} \qquad \ell_{32} = \widehat{\mathbf{e}}_3 \cdot \widehat{\overline{\mathbf{e}}}_2 = \cos\theta_{32} \qquad \ell_{33} = \widehat{\mathbf{e}}_3 \cdot \widehat{\overline{\mathbf{e}}}_3 = \cos\theta_{33}$$

which enables us to write the equations (1.3.39) and (1.3.40) in the form

$$y_i = \ell_{ij}\overline{y}_j \qquad \text{and} \qquad \overline{y}_i = \ell_{ji}y_j. \qquad (1.3.42)$$

Using the index notation we represent the unit vectors as:

$$\widehat{\overline{\mathbf{e}}}_r = \ell_{pr}\widehat{\mathbf{e}}_p \qquad \text{or} \qquad \widehat{\mathbf{e}}_p = \ell_{pr}\widehat{\overline{\mathbf{e}}}_r \qquad (1.3.43)$$

where ℓ_{pr} are the direction cosines. In both the barred and unbarred system the unit vectors are orthogonal and therefore we must have the dot products

$$\widehat{\overline{\mathbf{e}}}_r \cdot \widehat{\overline{\mathbf{e}}}_p = \delta_{rp} \qquad \text{and} \qquad \widehat{\mathbf{e}}_m \cdot \widehat{\mathbf{e}}_n = \delta_{mn} \qquad (1.3.44)$$

where δ_{ij} is the Kronecker delta. Substituting equation (1.3.43) into equation (1.3.44) we find the direction cosines ℓ_{ij} must satisfy the relations:

$$\widehat{\overline{\mathbf{e}}}_r \cdot \widehat{\overline{\mathbf{e}}}_s = \ell_{pr}\widehat{\mathbf{e}}_p \cdot \ell_{ms}\widehat{\mathbf{e}}_m = \ell_{pr}\ell_{ms}\,\widehat{\mathbf{e}}_p \cdot \widehat{\mathbf{e}}_m = \ell_{pr}\ell_{ms}\delta_{pm} = \ell_{mr}\ell_{ms} = \delta_{rs}$$
$$\text{and} \qquad \widehat{\mathbf{e}}_r \cdot \widehat{\mathbf{e}}_s = \ell_{rm}\widehat{\overline{\mathbf{e}}}_m \cdot \ell_{sn}\widehat{\overline{\mathbf{e}}}_n = \ell_{rm}\ell_{sn}\,\widehat{\overline{\mathbf{e}}}_m \cdot \widehat{\overline{\mathbf{e}}}_n = \ell_{rm}\ell_{sn}\delta_{mn} = \ell_{rm}\ell_{sm} = \delta_{rs}.$$

The relations

$$\ell_{mr}\ell_{ms} = \delta_{rs} \qquad \text{and} \qquad \ell_{rm}\ell_{sm} = \delta_{rs}, \qquad (1.3.45)$$

with summation index m, are important relations which are satisfied by the direction cosines associated with a rotation of axes.

Combining the rotation and translation equations we find

$$y_i = \underbrace{\ell_{ij}\overline{y}_j}_{rotation} + \underbrace{b_i}_{translation} \qquad . \qquad (1.3.46)$$

We multiply this equation by ℓ_{ik} and make use of the relations (1.3.45) to find the inverse transformation

$$\overline{y}_k = \ell_{ik}(y_i - b_i). \qquad (1.3.47)$$

These transformations are called linear or affine transformations.

Consider the $\overline{x}_i$ axes as fixed, while the x_i axes are rotating with respect to the $\overline{x}_i$ axes where both sets of axes have a common origin. Let $\vec{A} = A^i\widehat{\mathbf{e}}_i$ denote a vector fixed in and rotating with the x_i axes. We denote by $\left.\dfrac{d\vec{A}}{dt}\right|_f$ and $\left.\dfrac{d\vec{A}}{dt}\right|_r$ the derivatives of $\vec{A}$ with respect to the fixed (f) and rotating (r) axes. We can write, with respect to the fixed axes, that $\left.\dfrac{d\vec{A}}{dt}\right|_f = \dfrac{dA^i}{dt}\,\widehat{\mathbf{e}}_i + A^i\dfrac{d\widehat{\mathbf{e}}_i}{dt}$. Note that $\dfrac{d\widehat{\mathbf{e}}_i}{dt}$ is the derivative of a vector with constant magnitude. Therefore there exists constants ω_i, $i = 1,\ldots,6$ such that

$$\frac{d\widehat{\mathbf{e}}_1}{dt} = \omega_3\,\widehat{\mathbf{e}}_2 - \omega_2\,\widehat{\mathbf{e}}_3, \qquad \frac{d\widehat{\mathbf{e}}_2}{dt} = \omega_1\,\widehat{\mathbf{e}}_3 - \omega_4\,\widehat{\mathbf{e}}_1, \qquad \frac{d\widehat{\mathbf{e}}_3}{dt} = \omega_5\,\widehat{\mathbf{e}}_1 - \omega_6\,\widehat{\mathbf{e}}_2$$

i.e. see page 80. From the dot product $\widehat{\mathbf{e}}_1 \cdot \widehat{\mathbf{e}}_2 = 0$ we obtain by differentiation $\widehat{\mathbf{e}}_1 \cdot \dfrac{d\widehat{\mathbf{e}}_2}{dt} + \dfrac{d\widehat{\mathbf{e}}_1}{dt} \cdot \widehat{\mathbf{e}}_2 = 0$ which implies $\omega_4 = \omega_3$. Similarly, from the dot products $\widehat{\mathbf{e}}_1 \cdot \widehat{\mathbf{e}}_3$ and $\widehat{\mathbf{e}}_2 \cdot \widehat{\mathbf{e}}_3$ we obtain by differentiation the additional relations $\omega_5 = \omega_2$ and $\omega_6 = \omega_1$. The derivative of $\vec{A}$ with respect to the fixed axes can now be represented

$$\left.\frac{d\vec{A}}{dt}\right|_f = \frac{dA^i}{dt}\,\widehat{\mathbf{e}}_i + (\omega_2 A_3 - \omega_3 A_2)\widehat{\mathbf{e}}_1 + (\omega_3 A_1 - \omega_1 A_3)\widehat{\mathbf{e}}_2 + (\omega_1 A_2 - \omega_2 A_1)\widehat{\mathbf{e}}_3 = \left.\frac{d\vec{A}}{dt}\right|_r + \vec{\omega} \times \vec{A}$$

where $\vec{\omega} = \omega_i\,\widehat{\mathbf{e}}_i$ is called an angular velocity vector of the rotating system. The term $\vec{\omega} \times \vec{A}$ represents the velocity of the rotating system relative to the fixed system and $\left.\dfrac{d\vec{A}}{dt}\right|_r = \dfrac{dA^i}{dt}\,\widehat{\mathbf{e}}_i$ represents the derivative with respect to the rotating system.

Employing the special transformation equations (1.3.46) let us examine how tensor quantities transform when subjected to a translation and rotation of axes. These are our special transformation laws for Cartesian tensors. We examine only the transformation laws for first and second order Cartesian tensor as higher order transformation laws are easily discerned. We have previously shown that in general the first and second order tensor quantities satisfy the transformation laws:

$$\overline{A}_i = A_j\frac{\partial y_j}{\partial \overline{y}_i} \tag{1.3.48}$$

$$\overline{A}^i = A^j\frac{\partial \overline{y}_i}{\partial y_j} \tag{1.3.49}$$

$$\overline{A}^{mn} = A^{ij}\frac{\partial \overline{y}_m}{\partial y_i}\frac{\partial \overline{y}_n}{\partial y_j} \tag{1.3.50}$$

$$\overline{A}_{mn} = A_{ij}\frac{\partial y_i}{\partial \overline{y}_m}\frac{\partial y_j}{\partial \overline{y}_n} \tag{1.3.51}$$

$$\overline{A}^m_n = A^i_j\frac{\partial \overline{y}_m}{\partial y_i}\frac{\partial y_j}{\partial \overline{y}_n} \tag{1.3.52}$$

For the special case of Cartesian tensors we assume that y_i and $\overline{y}_i$, $i = 1,2,3$ are linearly independent. We differentiate the equations (1.3.46) and (1.3.47) and find

$$\frac{\partial y_i}{\partial \overline{y}_k} = \ell_{ij}\frac{\partial \overline{y}_j}{\partial \overline{y}_k} = \ell_{ij}\delta_{jk} = \ell_{ik}, \qquad \text{and} \qquad \frac{\partial \overline{y}_k}{\partial y_m} = \ell_{ik}\frac{\partial y_i}{\partial y_m} = \ell_{ik}\delta_{im} = \ell_{mk}.$$

Substituting these derivatives into the transformation equations (1.3.48) through (1.3.52) we produce the transformation equations

$$\overline{A}_i = A_j\ell_{ji}$$

$$\overline{A}^i = A^j\ell_{ji}$$

$$\overline{A}^{mn} = A^{ij}\ell_{im}\ell_{jn}$$

$$\overline{A}_{mn} = A_{ij}\ell_{im}\ell_{jn}$$

$$\overline{A}^m_n = A^i_j\ell_{im}\ell_{jn}.$$

88

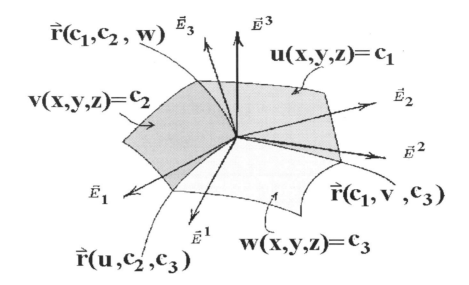

Figure 1.3-16. Transformation to curvilinear coordinates

These are the transformation laws when moving from one orthogonal system to another. In this case the direction cosines ℓ_{im} are constants and satisfy the relations given in equation (1.3.45). The transformation laws for higher ordered tensors are similar in nature to those given above.

In the unbarred system (y_1, y_2, y_3) the metric tensor and conjugate metric tensor are:

$$g_{ij} = \delta_{ij} \qquad \text{and} \qquad g^{ij} = \delta_{ij}$$

where δ_{ij} is the Kronecker delta. In the barred system of coordinates, which is also orthogonal, we have

$$\overline{g}_{ij} = \frac{\partial y_m}{\partial \overline{y}_i} \frac{\partial y_m}{\partial \overline{y}_j}.$$

From the orthogonality relations (1.3.45) we find

$$\overline{g}_{ij} = \ell_{mi}\ell_{mj} = \delta_{ij} \qquad \text{and} \qquad \overline{g}^{ij} = \delta_{ij}.$$

We examine the associated tensors

$$A^i = g^{ij} A_j \qquad\qquad A_i = g_{ij} A^j$$
$$A^{ij} = g^{im} g^{jn} A_{mn} \qquad A_{mn} = g_{mi} g_{nj} A^{ij}$$
$$A^i_n = g^{im} A_{mn} \qquad\qquad A^i_n = g_{nj} A^{ij}$$

and find that the contravariant and covariant components are identical to one another. This holds also in the barred system of coordinates. Also note that these special circumstances allow the representation of contractions using subscript quantities only. This type of a contraction is not allowed for general tensors. It is left as an exercise to try a contraction on a general tensor using only subscripts to see what happens. Note that such a contraction does not produce a tensor. These special situations are considered in the exercises.

Physical Components

We have previously shown an arbitrary vector $\vec{A}$ can be represented in many forms depending upon the coordinate system and basis vectors selected. For example, consider the figure 1.3-16 which illustrates a Cartesian coordinate system and a curvilinear coordinate system.

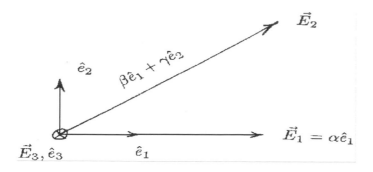

Figure 1.3-17. Physical components

In the Cartesian coordinate system we can represent a vector $\vec{A}$ as

$$\vec{A} = A_x\,\widehat{\mathbf{e}}_1 + A_y\,\widehat{\mathbf{e}}_2 + A_z\,\widehat{\mathbf{e}}_3$$

where $(\widehat{\mathbf{e}}_1, \widehat{\mathbf{e}}_2, \widehat{\mathbf{e}}_3)$ are the basis vectors. Consider a coordinate transformation to a more general coordinate system, say (x^1, x^2, x^3). The vector $\vec{A}$ can be represented with contravariant components as

$$\vec{A} = A^1\vec{E}_1 + A^2\vec{E}_2 + A^3\vec{E}_3 \qquad (1.3.53)$$

with respect to the tangential basis vectors $(\vec{E}_1, \vec{E}_2, \vec{E}_3)$. Alternatively, the same vector $\vec{A}$ can be represented in the form

$$\vec{A} = A_1\vec{E}^1 + A_2\vec{E}^2 + A_3\vec{E}^3 \qquad (1.3.54)$$

having covariant components with respect to the gradient basis vectors $(\vec{E}^1, \vec{E}^2, \vec{E}^3)$. These equations are just different ways of representing the same vector. In the above representations the basis vectors need not be orthogonal and they need not be unit vectors. In general, the physical dimensions of the components A^i and A_j are not the same.

The physical components of the vector $\vec{A}$ in a direction is defined as the projection of $\vec{A}$ upon a unit vector in the desired direction. For example, the physical component of $\vec{A}$ in the direction $\vec{E}_1$ is

$$\vec{A} \cdot \frac{\vec{E}_1}{|\vec{E}_1|} = \frac{A_1}{|\vec{E}_1|} = \text{projection of } \vec{A} \text{ on } \vec{E}_1. \qquad (1.3.58)$$

Similarly, the physical component of $\vec{A}$ in the direction $\vec{E}^1$ is

$$\vec{A} \cdot \frac{\vec{E}^1}{|\vec{E}^1|} = \frac{A^1}{|\vec{E}^1|} = \text{projection of } \vec{A} \text{ on } \vec{E}^1. \qquad (1.3.59)$$

EXAMPLE 1.3-11. (Physical components) Let α, β, γ denote nonzero positive constants such that the product relation $\alpha\gamma = 1$ is satisfied. Consider the nonorthogonal basis vectors

$$\vec{E}_1 = \alpha\,\widehat{\mathbf{e}}_1, \qquad \vec{E}_2 = \beta\,\widehat{\mathbf{e}}_1 + \gamma\,\widehat{\mathbf{e}}_2, \qquad \vec{E}_3 = \widehat{\mathbf{e}}_3$$

illustrated in the figure 1.3-17.

It is readily verified that the reciprocal basis is

$$\vec{E}^1 = \gamma\,\widehat{\mathbf{e}}_1 - \beta\,\widehat{\mathbf{e}}_2, \qquad \vec{E}^2 = \alpha\,\widehat{\mathbf{e}}_2, \qquad \vec{E}^3 = \widehat{\mathbf{e}}_3.$$

Consider the problem of representing the vector $\vec{A} = A_x\,\widehat{\mathbf{e}}_1 + A_y\,\widehat{\mathbf{e}}_2$ in the contravariant vector form

$$\vec{A} = A^1\vec{E}_1 + A^2\vec{E}_2 \quad \text{or tensor form} \quad A^i,\ i = 1, 2.$$

This vector has the contravariant components

$$A^1 = \vec{A}\cdot\vec{E}^1 = \gamma A_x - \beta A_y \qquad \text{and} \qquad A^2 = \vec{A}\cdot\vec{E}^2 = \alpha A_y.$$

Alternatively, this same vector can be represented as the covariant vector

$$\vec{A} = A_1\vec{E}^1 + A_2\vec{E}^2 \quad \text{which has the tensor form} \quad A_i,\ i = 1, 2.$$

The covariant components are found from the relations

$$A_1 = \vec{A}\cdot\vec{E}_1 = \alpha A_x \qquad\qquad A_2 = \vec{A}\cdot\vec{E}_2 = \beta A_x + \gamma A_y.$$

The physical components of $\vec{A}$ in the directions $\vec{E}^1$ and $\vec{E}^2$ are found to be:

$$\vec{A}\cdot\frac{\vec{E}^1}{|\vec{E}^1|} = \frac{A^1}{|\vec{E}^1|} = \frac{\gamma A_x - \beta A_y}{\sqrt{\gamma^2 + \beta^2}} = A(1)$$

$$\vec{A}\cdot\frac{\vec{E}^2}{|\vec{E}^2|} = \frac{A^2}{|\vec{E}^2|} = \frac{\alpha A_y}{\alpha} = A_y = A(2).$$

Note that these same results are obtained from the dot product relations using either form of the vector $\vec{A}$. For example, we can write

$$\vec{A}\cdot\frac{\vec{E}^1}{|\vec{E}^1|} = \frac{A_1(\vec{E}^1\cdot\vec{E}^1) + A_2(\vec{E}^2\cdot\vec{E}^1)}{|\vec{E}^1|} = A(1)$$

$$\text{and} \qquad \vec{A}\cdot\frac{\vec{E}^2}{|\vec{E}^2|} = \frac{A_1(\vec{E}^1\cdot\vec{E}^2) + A_2(\vec{E}^2\cdot\vec{E}^2)}{|\vec{E}^2|} = A(2).$$

In general, the physical components of a vector $\vec{A}$ in a direction of a unit vector λ^i is the generalized dot product in V_N. This dot product is an invariant and can be expressed

$$g_{ij}A^i\lambda^j = A^i\lambda_i = A_i\lambda^i = \text{projection of } \vec{A} \text{ in direction of } \lambda^i$$

$\blacksquare$

Physical Components For Orthogonal Coordinates

In orthogonal coordinates observe the element of arc length squared in V_3 is

$$ds^2 = g_{ij}dx^i dx^j = (h_1)^2(dx^1)^2 + (h_2)^2(dx^2)^2 + (h_3)^2(dx^3)^2$$

where

$$g_{ij} = \begin{pmatrix} (h_1)^2 & 0 & 0 \\ 0 & (h_2)^2 & 0 \\ 0 & 0 & (h_3)^2 \end{pmatrix}. \tag{1.3.60}$$

In this case the curvilinear coordinates are orthogonal and

$$h_{(i)}^2 = g_{(i)(i)} \quad i \text{ not summed and} \quad g_{ij} = 0, \ i \neq j.$$

At an arbitrary point in this coordinate system we take $\lambda^i, i = 1, 2, 3$ as a unit vector in the direction of the coordinate x^1. We then obtain

$$\lambda^1 = \frac{dx^1}{ds}, \quad \lambda^2 = 0, \quad \lambda^3 = 0.$$

This is a unit vector since

$$1 = g_{ij}\lambda^i\lambda^j = g_{11}\lambda^1\lambda^1 = h_1^2(\lambda^1)^2$$

or $\lambda^1 = \frac{1}{h_1}$. Here the curvilinear coordinate system is orthogonal and in this case the physical component of a vector A^i, in the direction x^i, is the projection of A^i on λ^i in V_3. The projection in the x^1 direction is determined from

$$A(1) = g_{ij}A^i\lambda^j = g_{11}A^1\lambda^1 = h_1^2 A^1 \frac{1}{h_1} = h_1 A^1.$$

Similarly, we choose unit vectors μ^i and ν^i, $i = 1, 2, 3$ in the x^2 and x^3 directions. These unit vectors can be represented

$$\mu^1 = 0, \quad \mu^2 = \frac{dx^2}{ds} = \frac{1}{h_2}, \quad \mu^3 = 0$$
$$\nu^1 = 0, \quad \nu^2 = 0, \quad \nu^3 = \frac{dx^3}{ds} = \frac{1}{h_3}$$

and the physical components of the vector A^i in these directions are calculated as

$$A(2) = h_2 A^2 \quad \text{and} \quad A(3) = h_3 A^3.$$

In summary, we can say that in an orthogonal coordinate system the physical components of a contravariant tensor of order one can be determined from the equations

$$A(i) = h_{(i)}A^{(i)} = \sqrt{g_{(i)(i)}}A^{(i)}, \quad i = 1, 2 \text{ or } 3 \quad \text{no summation on i,}$$

which is a short hand notation for the physical components $(h_1 A^1, h_2 A^2, h_3 A^3)$. In an orthogonal coordinate system the nonzero conjugate metric components are

$$g^{(i)(i)} = \frac{1}{g_{(i)(i)}}, \quad i = 1, 2, \text{ or } 3 \quad \text{no summation on i.}$$

These components are needed to calculate the physical components associated with a covariant tensor of order one. For example, in the x^1−direction, we have the covariant components

$$\lambda_1 = g_{11}\lambda^1 = h_1^2 \frac{1}{h_1} = h_1, \quad \lambda_2 = 0, \quad \lambda_3 = 0$$

92

and therefore the projection in V_3 can be represented

$$g_{ij}A^i\lambda^j = g_{ij}A^i g^{jm}\lambda_m = A_j g^{jm}\lambda_m = A_1\lambda_1 g^{11} = A_1 h_1 \frac{1}{h_1^2} = \frac{A_1}{h_1} = A(1).$$

In a similar manner we calculate the relations

$$A(2) = \frac{A_2}{h_2} \quad \text{and} \quad A(3) = \frac{A_3}{h_3}$$

for the other physical components in the directions x^2 and x^3. These physical components can be represented in the short hand notation

$$A(i) = \frac{A_{(i)}}{h_{(i)}} = \frac{A_{(i)}}{\sqrt{g_{(i)(i)}}}, \quad i = 1, 2 \, or \, 3 \qquad \text{no summation on } i.$$

In an orthogonal coordinate system the physical components associated with both the contravariant and covariant components are the same. To show this we note that when $A^i g_{ij} = A_j$ is summed on i we obtain

$$A^1 g_{1j} + A^2 g_{2j} + A^3 g_{3j} = A_j.$$

Since $g_{ij} = 0$ for $i \neq j$ this equation reduces to

$$A^{(i)} g_{(i)(i)} = A_{(i)}, \quad i \text{ not summed.}$$

Another form for this equation is

$$A(i) = A^{(i)}\sqrt{g_{(i)(i)}} = \frac{A_{(i)}}{\sqrt{g_{(i)(i)}}} \quad i \text{ not summed,}$$

which demonstrates that the physical components associated with the contravariant and covariant components are identical.

Notation The physical components are sometimes expressed by symbols with subscripts which represent the coordinate curve along which the projection is taken. For example, let H^i denote the contravariant components of a first order tensor. The following are some examples of the representation of the physical components of H^i in various coordinate systems:

orthogonal coordinates	coordinate system	tensor components	physical components
general	(x^1, x^2, x^3)	H^i	$H(1), H(2), H(3)$
rectangular	(x, y, z)	H^i	H_x, H_y, H_z
cylindrical	(r, θ, z)	H^i	H_r, H_θ, H_z
spherical	(ρ, θ, ϕ)	H^i	H_ρ, H_θ, H_ϕ
general	(u, v, w)	H^i	H_u, H_v, H_w

Higher Order Tensors

The physical components associated with higher ordered tensors are defined by projections in V_N just like the case with first order tensors. For an nth ordered tensor $T_{ij...k}$ we can select n unit vectors $\lambda^i, \mu^i, \ldots, \nu^i$ and form the inner product (projection)

$$T_{ij...k}\lambda^i \mu^j \ldots \nu^k.$$

When projecting the tensor components onto the coordinate curves, there are N choices for each of the unit vectors. This produces N^n physical components.

The above inner product represents the physical component of the tensor $T_{ij...k}$ along the directions of the unit vectors $\lambda^i, \mu^i, \ldots, \nu^i$. The selected unit vectors may or may not be orthogonal. In the cases where the selected unit vectors are all orthogonal to one another, the calculation of the physical components is greatly simplified. By relabeling the unit vectors $\lambda^i_{(m)}, \lambda^i_{(n)}, \ldots, \lambda^i_{(p)}$ where $(m), (n), ..., (p)$ represent one of the N directions, the physical components of a general nth order tensor is represented

$$T(m\,n\ldots p) = T_{ij...k}\lambda^i_{(m)}\lambda^j_{(n)}\ldots\lambda^k_{(p)}$$

EXAMPLE 1.3-12. (Physical components)

In an orthogonal curvilinear coordinate system V_3 with metric g_{ij}, $i, j = 1, 2, 3$, find the physical components of

(i) the second order tensor A_{ij}. (ii) the second order tensor A^{ij}. (iii) the second order tensor A^i_j.
Solution: The physical components of A_{mn}, $m, n = 1, 2, 3$ along the directions of two unit vectors λ^i and μ^i is defined as the inner product in V_3. These physical components can be expressed

$$A(ij) = A_{mn}\lambda^m_{(i)}\mu^n_{(j)} \qquad i, j = 1, 2, 3,$$

where the subscripts (i) and (j) represent one of the coordinate directions. Dropping the subscripts (i) and (j), we make the observation that in an orthogonal curvilinear coordinate system there are three choices for the direction of the unit vector λ^i and also three choices for the direction of the unit vector μ^i. These three choices represent the directions along the x^1, x^2 or x^3 coordinate curves which emanate from a point of the curvilinear coordinate system. This produces a total of nine possible physical components associated with the tensor A_{mn}.

For example, we can obtain the components of the unit vector λ^i, $i = 1, 2, 3$ in the x^1 direction directly from an examination of the element of arc length squared

$$ds^2 = (h_1)^2(dx^1)^2 + (h_2)^2(dx^2)^2 + (h_3)^2(dx^3)^2.$$

By setting $dx^2 = dx^3 = 0$, we find

$$\frac{dx^1}{ds} = \frac{1}{h_1} = \lambda^1, \qquad \lambda^2 = 0, \qquad \lambda^3 = 0.$$

This is the vector $\lambda^i_{(1)}$, $i = 1, 2, 3$. Similarly, if we choose to select the unit vector λ^i, $i = 1, 2, 3$ in the x^2 direction, we set $dx^1 = dx^3 = 0$ in the element of arc length squared and find the components

$$\lambda^1 = 0, \qquad \lambda^2 = \frac{dx^2}{ds} = \frac{1}{h_2}, \qquad \lambda^3 = 0.$$

This is the vector $\lambda^i_{(2)}$, $i = 1, 2, 3$. Finally, if we select λ^i, $i = 1, 2, 3$ in the x^3 direction, we set $dx^1 = dx^2 = 0$ in the element of arc length squared and determine the unit vector

$$\lambda^1 = 0, \qquad \lambda^2 = 0, \qquad \lambda^3 = \frac{dx^3}{ds} = \frac{1}{h_3}.$$

This is the vector $\lambda^i_{(3)}$, $i = 1, 2, 3$. Similarly, the unit vector μ^i can be selected as one of the above three directions. Examining all nine possible combinations for selecting the unit vectors, we calculate the physical components in an orthogonal coordinate system as:

$$A(11) = \frac{A_{11}}{h_1 h_1} \qquad A(12) = \frac{A_{12}}{h_1 h_2} \qquad A(13) = \frac{A_{13}}{h_1 h_3}$$

$$A(21) = \frac{A_{21}}{h_1 h_2} \qquad A(22) = \frac{A_{22}}{h_2 h_2} \qquad A(23) = \frac{A_{23}}{h_2 h_3}$$

$$A(31) = \frac{A_{31}}{h_3 h_1} \qquad A(32) = \frac{A_{32}}{h_3 h_2} \qquad A(33) = \frac{A_{33}}{h_3 h_3}$$

These results can be written in the more compact form

$$A(ij) = \frac{A_{(i)(j)}}{h_{(i)} h_{(j)}} \qquad \text{no summation on } i \text{ or } j . \tag{1.3.61}$$

For mixed tensors we have

$$A^i_j = g^{im} A_{mj} = g^{i1} A_{1j} + g^{i2} A_{2j} + g^{i3} A_{3j}. \tag{1.3.62}$$

From the fact $g^{ij} = 0$ for $i \neq j$, together with the physical components from equation (1.3.61), the equation (1.3.62) reduces to

$$A^{(i)}_{(j)} = g^{(i)(i)} A_{(i)(j)} = \frac{1}{h^2_{(i)}} \cdot h_{(i)} h_{(j)} A(ij) \qquad \text{no summation on } i \text{ and } i, j = 1, 2 \text{ or } 3.$$

This can also be written in the form

$$A(ij) = A^{(i)}_{(j)} \frac{h_{(i)}}{h_{(j)}} \qquad \text{no summation on } i \text{ or } j. \tag{1.3.63}$$

Hence, the physical components associated with the mixed tensor A^i_j in an orthogonal coordinate system can be expressed as

$$A(11) = A^1_1 \qquad A(12) = A^1_2 \frac{h_1}{h_2} \qquad A(13) = A^1_3 \frac{h_1}{h_3}$$

$$A(21) = A^2_1 \frac{h_2}{h_1} \qquad A(22) = A^2_2 \qquad A(23) = A^2_3 \frac{h_2}{h_3}$$

$$A(31) = A^3_1 \frac{h_3}{h_1} \qquad A(32) = A^3_2 \frac{h_3}{h_2} \qquad A(33) = A^3_3.$$

For second order contravariant tensors we may write

$$A^{ij} g_{jm} = A^i_m = A^{i1} g_{1m} + A^{i2} g_{2m} + A^{i3} g_{3m}.$$

We use the fact $g_{ij} = 0$ for $i \neq j$ together with the physical components from equation (1.3.63) to reduce the above equation to the form $A_{(m)}^{(i)} = A^{(i)(m)} g_{(m)(m)}$ no summation on m. In terms of physical components we have

$$\frac{h_{(m)}}{h_{(i)}} A(im) = A^{(i)(m)} h_{(m)}^2 \quad \text{or} \quad A(im) = A^{(i)(m)} h_{(i)} h_{(m)}. \quad \text{no summation} \quad i, m = 1, 2, 3 \qquad (1.3.64)$$

Examining the results from equation (1.3.64) we find that the physical components associated with the contravariant tensor A^{ij}, in an orthogonal coordinate system, can be written as:

$$A(11) = A^{11} h_1 h_1 \qquad A(12) = A^{12} h_1 h_2 \qquad A(13) = A^{13} h_1 h_3$$
$$A(21) = A^{21} h_2 h_1 \qquad A(22) = A^{22} h_2 h_2 \qquad A(23) = A^{23} h_2 h_3$$
$$A(31) = A^{31} h_3 h_1 \qquad A(32) = A^{32} h_3 h_2 \qquad A(33) = A^{33} h_3 h_3.$$

∎

Physical Components in General

In an orthogonal curvilinear coordinate system, the physical components associated with the nth order tensor $T_{ij...kl}$ along the curvilinear coordinate directions can be represented:

$$T(ij \ldots kl) = \frac{T_{(i)(j)...(k)(l)}}{h_{(i)} h_{(j)} \ldots h_{(k)} h_{(l)}} \quad \text{no summations.}$$

These physical components can be related to the various tensors associated with $T_{ij...kl}$. For example, in an orthogonal coordinate system, the physical components associated with the mixed tensor $T_{n...kl}^{ij...m}$ can be expressed as:

$$T(ij \ldots mn \ldots kl) = T_{(n)...(k)(l)}^{(i)(j)...(m)} \frac{h_{(i)} h_{(j)} \ldots h_{(m)}}{h_{(n)} \ldots h_{(k)} h_{(l)}} \quad \text{no summations.} \qquad (1.3.65)$$

EXAMPLE 1.3-13. (Physical components) Let $x^i = x^i(t), i = 1, 2, 3$ denote the position vector of a particle which moves as a function of time t. Assume there exists a coordinate transformation $\overline{x}^i = \overline{x}^i(x)$, for $i = 1, 2, 3$, of the form given by equations (1.2.33). The position of the particle when referenced with respect to the barred system of coordinates can be found by substitution. The generalized velocity of the particle in the unbarred system is a vector with components

$$v^i = \frac{dx^i}{dt}, i = 1, 2, 3.$$

The generalized velocity components of the same particle in the barred system is obtained from the chain rule. We find this velocity is represented by

$$\overline{v}^i = \frac{d\overline{x}^i}{dt} = \frac{\partial \overline{x}^i}{\partial x^j} \frac{dx^j}{dt} = \frac{\partial \overline{x}^i}{\partial x^j} v^j.$$

This equation implies that the contravariant quantities $(v^1, v^2, v^3) = (\frac{dx^1}{dt}, \frac{dx^2}{dt}, \frac{dx^3}{dt})$ are tensor quantities. These quantities are called the components of the generalized velocity. The coordinates x^1, x^2, x^3 are generalized coordinates. This means we can select any set of three independent variables for the representation of the motion. The variables selected might not have the same dimensions. For example, in cylindrical coordinates we let $(x^1 = r, x^2 = \theta, x^3 = z)$. Here x^1 and x^3 have dimensions of distance but x^2 has dimensions of angular displacement. The generalized velocities are

$$v^1 = \frac{dx^1}{dt} = \frac{dr}{dt}, \qquad v^2 = \frac{dx^2}{dt} = \frac{d\theta}{dt}, \qquad v^3 = \frac{dx^3}{dt} = \frac{dz}{dt}.$$

Here v^1 and v^3 have units of length divided by time while v^2 has the units of angular velocity or angular change divided by time. Clearly, these dimensions are not all the same. Let us examine the physical components of the generalized velocities. We find in cylindrical coordinates $h_1 = 1$, $h_2 = r$, $h_3 = 1$ and the physical components of the velocity have the forms:

$$v_r = v(1) = v^1 h_1 = \frac{dr}{dt}, \qquad v_\theta = v(2) = v^2 h_2 = r\frac{d\theta}{dt}, \qquad v_z = v(3) = v^3 h_3 = \frac{dz}{dt}.$$

Now the physical components of the velocity all have the same units of length divided by time.

■

Additional examples of the use of physical components are considered later. For the time being, just remember that when tensor equations are derived, the equations are valid in any generalized coordinate system. In particular, we are interested in the representation of physical laws which are to be invariant and independent of the coordinate system used to represent these laws. Once a tensor equation is derived, we can chose any type of generalized coordinates and expand the tensor equations. Before using any expanded tensor equations we must replace all the tensor components by their corresponding physical components in order that the equations are dimensionally homogeneous. It is these expanded equations, expressed in terms of the physical components, which are used to solve applied problems.

Tensors and Multilinear Forms

Tensors can be thought of as being created by multilinear forms defined on some vector space V. Let us define on a vector space V a linear form, a bilinear form and a general multilinear form. We can then illustrate how tensors are created from these forms.

Definition: (Linear form) Let V denote a vector space which contains vectors $\vec{x}, \vec{x}_1, \vec{x}_2, \ldots$ A linear form in $\vec{x}$ is a scalar function $\varphi(\vec{x})$ having a single vector argument $\vec{x}$ which satisfies the linearity properties:

$$\begin{aligned} (i) & \quad \varphi(\vec{x}_1 + \vec{x}_2) = \varphi(\vec{x}_1) + \varphi(\vec{x}_2) \\ (ii) & \quad \varphi(\mu\vec{x}_1) = \mu\varphi(\vec{x}_1) \end{aligned} \qquad (1.3.66)$$

for all arbitrary vectors $\vec{x}_1, \vec{x}_2$ in V and all real numbers μ.

An example of a linear form is the dot product relation

$$\varphi(\vec{x}) = \vec{A} \cdot \vec{x} \qquad (1.3.67)$$

where $\vec{A}$ is a constant vector and $\vec{x}$ is an arbitrary vector belonging to the vector space V.

Note that a linear form in $\vec{x}$ can be expressed in terms of the components of the vector $\vec{x}$ and the base vectors $(\widehat{\mathbf{e}}_1, \widehat{\mathbf{e}}_2, \widehat{\mathbf{e}}_3)$ used to represent $\vec{x}$. To show this, we write the vector $\vec{x}$ in the component form

$$\vec{x} = x^i \widehat{\mathbf{e}}_i = x^1 \widehat{\mathbf{e}}_1 + x^2 \widehat{\mathbf{e}}_2 + x^3 \widehat{\mathbf{e}}_3,$$

where $x^i, i = 1, 2, 3$ are the components of $\vec{x}$ with respect to the basis vectors $(\widehat{\mathbf{e}}_1, \widehat{\mathbf{e}}_2, \widehat{\mathbf{e}}_3)$. By the linearity property of φ we can write

$$\begin{aligned}
\varphi(\vec{x}) = \varphi(x^i \widehat{\mathbf{e}}_i) &= \varphi(x^1 \widehat{\mathbf{e}}_1 + x^2 \widehat{\mathbf{e}}_2 + x^3 \widehat{\mathbf{e}}_3) \\
&= \varphi(x^1 \widehat{\mathbf{e}}_1) + \varphi(x^2 \widehat{\mathbf{e}}_2) + \varphi(x^3 \widehat{\mathbf{e}}_3) \\
&= x^1 \varphi(\widehat{\mathbf{e}}_1) + x^2 \varphi(\widehat{\mathbf{e}}_2) + x^3 \varphi(\widehat{\mathbf{e}}_3) = x^i \varphi(\widehat{\mathbf{e}}_i)
\end{aligned}$$

Thus we can write $\varphi(\vec{x}) = x^i \varphi(\widehat{\mathbf{e}}_i)$ and by defining the quantity $\varphi(\widehat{\mathbf{e}}_i) = a_i$ as a tensor we obtain $\varphi(\vec{x}) = x^i a_i$. Note that if we change basis from $(\widehat{\mathbf{e}}_1, \widehat{\mathbf{e}}_2, \widehat{\mathbf{e}}_3)$ to $(\vec{E}_1, \vec{E}_2, \vec{E}_3)$ then the components of $\vec{x}$ also must change. Letting $\overline{x}^i$ denote the components of $\vec{x}$ with respect to the new basis, we would have

$$\vec{x} = \overline{x}^i \vec{E}_i \quad \text{and} \quad \varphi(\vec{x}) = \varphi(\overline{x}^i \vec{E}_i) = \overline{x}^i \varphi(\vec{E}_i).$$

The linear form φ defines a new tensor $\overline{a}_i = \varphi(\vec{E}_i)$ so that $\varphi(\vec{x}) = \overline{x}^i \overline{a}_i$. Whenever there is a definite relation between the basis vectors $(\widehat{\mathbf{e}}_1, \widehat{\mathbf{e}}_2, \widehat{\mathbf{e}}_3)$ and $(\vec{E}_1, \vec{E}_2, \vec{E}_3)$, say,

$$\vec{E}_i = \frac{\partial x^j}{\partial \overline{x}^i} \widehat{\mathbf{e}}_j,$$

then there exists a definite relation between the tensors a_i and $\overline{a}_i$. This relation is

$$\overline{a}_i = \varphi(\vec{E}_i) = \varphi(\frac{\partial x^j}{\partial \overline{x}^i} \widehat{\mathbf{e}}_j) = \frac{\partial x^j}{\partial \overline{x}^i} \varphi(\widehat{\mathbf{e}}_j) = \frac{\partial x^j}{\partial \overline{x}^i} a_j.$$

This is the transformation law for an absolute covariant tensor of rank or order one.

The above idea is now extended to higher order tensors.

> **Definition: (Bilinear form)** A bilinear form in $\vec{x}$ and $\vec{y}$ is a scalar function $\varphi(\vec{x}, \vec{y})$ with two vector arguments, which satisfies the linearity properties:
>
> $$\begin{aligned}
(i) \quad & \varphi(\vec{x}_1 + \vec{x}_2, \vec{y}_1) = \varphi(\vec{x}_1, \vec{y}_1) + \varphi(\vec{x}_2, \vec{y}_1) \\
(ii) \quad & \varphi(\vec{x}_1, \vec{y}_1 + \vec{y}_2) = \varphi(\vec{x}_1, \vec{y}_1) + \varphi(\vec{x}_1, \vec{y}_2) \\
(iii) \quad & \varphi(\mu \vec{x}_1, \vec{y}_1) = \mu \varphi(\vec{x}_1, \vec{y}_1) \\
(iv) \quad & \varphi(\vec{x}_1, \mu \vec{y}_1) = \mu \varphi(\vec{x}_1, \vec{y}_1)
\end{aligned}$$
>
> (1.3.68)
>
> for arbitrary vectors $\vec{x}_1, \vec{x}_2, \vec{y}_1, \vec{y}_2$ in the vector space V and for all real numbers μ.

Note in the definition of a bilinear form that the scalar function φ is linear in both the arguments $\vec{x}$ and $\vec{y}$. An example of a bilinear form is the dot product relation

$$\varphi(\vec{x}, \vec{y}) = \vec{x} \cdot \vec{y} \tag{1.3.69}$$

where both $\vec{x}$ and $\vec{y}$ belong to the same vector space V.

The definition of a bilinear form suggests how multilinear forms can be defined.

Definition: (Multilinear forms) A multilinear form of degree M or a M degree linear form in the vector arguments

$$\vec{x}_1, \vec{x}_2, \ldots, \vec{x}_M$$

is a scalar function

$$\varphi(\vec{x}_1, \vec{x}_2, \ldots, \vec{x}_M)$$

of M vector arguments which satisfies the property that it is a linear form in each of its arguments. That is, φ must satisfy for each $j = 1, 2, \ldots, M$ the properties:

$(i) \quad \varphi(\vec{x}_1, \ldots, \vec{x}_{j1} + \vec{x}_{j2}, \ldots \vec{x}_M) = \varphi(\vec{x}_1, \ldots, \vec{x}_{j1}, \ldots, \vec{x}_M) + \varphi(\vec{x}_1, \ldots, \vec{x}_{j2}, \ldots, \vec{x}_M)$

$(ii) \quad \varphi(\vec{x}_1, \ldots, \mu\vec{x}_j, \ldots, \vec{x}_M) = \mu\varphi(\vec{x}_1, \ldots, \vec{x}_j, \ldots, \vec{x}_M)$

$$\tag{1.3.70}$$

for all arbitrary vectors $\vec{x}_1, \ldots, \vec{x}_M$ in the vector space V and all real numbers μ.

An example of a third degree multilinear form or trilinear form is the triple scalar product

$$\varphi(\vec{x}, \vec{y}, \vec{z}) = \vec{x} \cdot (\vec{y} \times \vec{z}). \tag{1.3.71}$$

Note that multilinear forms are independent of the coordinate system selected and depend only upon the vector arguments. In a three dimensional vector space we select the basis vectors $(\widehat{\mathbf{e}}_1, \widehat{\mathbf{e}}_2, \widehat{\mathbf{e}}_3)$ and represent all vectors with respect to this basis set. For example, if $\vec{x}, \vec{y}, \vec{z}$ are three vectors we can represent these vectors in the component forms

$$\vec{x} = x^i\,\widehat{\mathbf{e}}_i, \qquad \vec{y} = y^j\,\widehat{\mathbf{e}}_j, \qquad \vec{z} = z^k\,\widehat{\mathbf{e}}_k \tag{1.3.72}$$

where we have employed the summation convention on the repeated indices i, j and k. Substituting equations (1.3.72) into equation (1.3.71) we obtain

$$\varphi(x^i\,\widehat{\mathbf{e}}_i, y^j\,\widehat{\mathbf{e}}_j, z^k\,\widehat{\mathbf{e}}_k) = x^i y^j z^k \varphi(\widehat{\mathbf{e}}_i, \widehat{\mathbf{e}}_j, \widehat{\mathbf{e}}_k), \tag{1.3.73}$$

since φ is linear in all its arguments. By defining the tensor quantity

$$\varphi(\widehat{\mathbf{e}}_i, \widehat{\mathbf{e}}_j, \widehat{\mathbf{e}}_k) = e_{ijk} \tag{1.3.74}$$

(See exercise 1.1, problem 15) the trilinear form, given by equation (1.3.71), with vectors from equations (1.3.72), can be expressed as

$$\varphi(\vec{x}, \vec{y}, \vec{z}) = e_{ijk} x^i y^j z^k, \quad i, j, k = 1, 2, 3. \tag{1.3.75}$$

The coefficients e_{ijk} of the trilinear form is called a third order tensor. It is the familiar permutation symbol considered earlier.

In a multilinear form of degree M, $\varphi(\vec{x}, \vec{y}, \ldots, \vec{z})$, the M arguments can be represented in a component form with respect to a set of basis vectors $(\widehat{\mathbf{e}}_1, \widehat{\mathbf{e}}_2, \widehat{\mathbf{e}}_3)$. Let these vectors have components $x^i, y^i, z^i, i = 1, 2, 3$ with respect to the selected basis vectors. We then can write

$$\vec{x} = x^i \widehat{\mathbf{e}}_i, \qquad \vec{y} = y^j \widehat{\mathbf{e}}_j, \qquad \vec{z} = z^k \widehat{\mathbf{e}}_k.$$

Substituting these vectors into the M degree multilinear form produces

$$\varphi(x^i \widehat{\mathbf{e}}_i, y^j \widehat{\mathbf{e}}_j, \ldots, z^k \widehat{\mathbf{e}}_k) = x^i y^j \cdots z^k \varphi(\widehat{\mathbf{e}}_i, \widehat{\mathbf{e}}_j, \ldots, \widehat{\mathbf{e}}_k). \tag{1.3.76}$$

The multilinear form defines a set of coefficients

$$a_{ij\ldots k} = \varphi(\widehat{\mathbf{e}}_i, \widehat{\mathbf{e}}_j, \ldots, \widehat{\mathbf{e}}_k) \tag{1.3.77}$$

which are referred to as the components of a tensor of order M. The tensor is thus created by the multilinear form and has M indices if φ is of degree M.

Note that if we change to a different set of basis vectors, say, $(\vec{E}_1, \vec{E}_2, \vec{E}_3)$ the multilinear form defines a new tensor

$$\overline{a}_{ij\ldots k} = \varphi(\vec{E}_i, \vec{E}_j, \ldots, \vec{E}_k). \tag{1.3.78}$$

This new tensor has a bar over it to distinguish it from the previous tensor. A definite relation exists between the new and old basis vectors and so there exists a definite relation between the components of the barred and unbarred tensor components. Recall that if we are given a set of transformation equations

$$y^i = y^i(x^1, x^2, x^3), i = 1, 2, 3, \tag{1.3.79}$$

from rectangular to generalized curvilinear coordinates, we can express the basis vectors in the new system by the equations

$$\vec{E}_i = \frac{\partial y^j}{\partial x^i} \widehat{\mathbf{e}}_j, \quad i = 1, 2, 3. \tag{1.3.80}$$

For example, see equations (1.3.11) with $y^1 = x, y^2 = y, y^3 = z, x^1 = u, x^2 = v, x^3 = w$. Substituting equations (1.3.80) into equations (1.3.78) we obtain

$$\overline{a}_{ij\ldots k} = \varphi\left(\frac{\partial y^\alpha}{\partial x^i} \widehat{\mathbf{e}}_\alpha, \frac{\partial y^\beta}{\partial x^j} \widehat{\mathbf{e}}_\beta, \ldots, \frac{\partial y^\gamma}{\partial x^k} \widehat{\mathbf{e}}_\gamma\right).$$

By the linearity property of φ, this equation is expressible in the form

$$\overline{a}_{ij\ldots k} = \frac{\partial y^\alpha}{\partial x^i} \frac{\partial y^\beta}{\partial x^j} \cdots \frac{\partial y^\gamma}{\partial x^k} \varphi(\widehat{\mathbf{e}}_\alpha, \widehat{\mathbf{e}}_\beta, \ldots, \widehat{\mathbf{e}}_\gamma)$$

$$\overline{a}_{ij\ldots k} = \frac{\partial y^\alpha}{\partial x^i} \frac{\partial y^\beta}{\partial x^j} \cdots \frac{\partial y^\gamma}{\partial x^k} a_{\alpha\beta\ldots\gamma}$$

This is the familiar transformation law for a covariant tensor of degree M. By selecting reciprocal basis vectors the corresponding transformation laws for contravariant vectors can be determined.

The above examples illustrate that tensors can be considered as quantities derivable from multilinear forms defined on some vector space.

Dual Tensors

The e-permutation symbol is often used to generate new tensors from given tensors. For $T_{i_1 i_2 \ldots i_m}$ a skew-symmetric tensor, we define the tensor

$$\hat{T}^{j_1 j_2 \ldots j_{n-m}} = \frac{1}{m!} e^{j_1 j_2 \ldots j_{n-m} i_1 i_2 \ldots i_m} T_{i_1 i_2 \ldots i_m} \qquad m \leq n \tag{1.3.81}$$

as the dual tensor associated with $T_{i_1 i_2 \ldots i_m}$. Note that the e-permutation symbol or alternating tensor has a weight of $+1$ and consequently the dual tensor will have a higher weight than the original tensor.

The e-permutation symbol has the following properties

$$e^{i_1 i_2 \ldots i_N} e_{i_1 i_2 \ldots i_N} = N!$$

$$e^{i_1 i_2 \ldots i_N} e_{j_1 j_2 \ldots j_N} = \delta^{i_1 i_2 \ldots i_N}_{j_1 j_2 \ldots j_N}$$

$$e_{k_1 k_2 \ldots k_m i_1 i_2 \ldots i_{N-m}} e^{j_1 j_2 \ldots j_m i_1 i_2 \ldots i_{N-m}} = (N-m)! \delta^{j_1 j_2 \ldots j_m}_{k_1 k_2 \ldots k_m} \tag{1.3.82}$$

$$\delta^{j_1 j_2 \ldots j_m}_{k_1 k_2 \ldots k_m} T_{j_1 j_2 \ldots j_m} = m! T_{k_1 k_2 \ldots k_m}.$$

Using the above properties we can solve for the skew-symmetric tensor in terms of the dual tensor. We find

$$T_{i_1 i_2 \ldots i_m} = \frac{1}{(n-m)!} e_{i_1 i_2 \ldots i_m j_1 j_2 \ldots j_{n-m}} \hat{T}^{j_1 j_2 \ldots j_{n-m}}. \tag{1.3.83}$$

For example, if $A_{ij}\ i,j = 1, 2, 3$ is a skew-symmetric tensor, we may associate with it the dual tensor

$$V^i = \frac{1}{2!} e^{ijk} A_{jk},$$

which is a first order tensor or vector. Note that A_{ij} has the components

$$\begin{pmatrix} 0 & A_{12} & A_{13} \\ -A_{12} & 0 & A_{23} \\ -A_{13} & -A_{23} & 0 \end{pmatrix} \tag{1.3.84}$$

which implies the components of the vector $\vec{V}$ are

$$(V^1, V^2, V^3) = (A_{23}, A_{31}, A_{12}). \tag{1.3.85}$$

Note that the vector components have a cyclic order to the indices which comes from the cyclic properties of the e-permutation symbol.

As another example, consider the fourth order skew-symmetric tensor A_{ijkl}, $i, j, k, l = 1, \ldots, n$. We can associate with this tensor any of the dual tensor quantities

$$V = \frac{1}{4!} e^{ijkl} A_{ijkl}$$

$$V^i = \frac{1}{4!} e^{ijklm} A_{jklm}$$

$$V^{ij} = \frac{1}{4!} e^{ijklmn} A_{klmn} \tag{1.3.86}$$

$$V^{ijk} = \frac{1}{4!} e^{ijklmnp} A_{lmnp}$$

$$V^{ijkl} = \frac{1}{4!} e^{ijklmnpr} A_{mnpr}$$

Applications of dual tensors can be found in section 2.2.

EXERCISE 1.3

▶ **1.**

(a) From the transformation law for the second order tensor $\overline{g}_{ij} = g_{ab}\dfrac{\partial x^a}{\partial \overline{x}^i}\dfrac{\partial x^b}{\partial \overline{x}^j}$

    solve for the g_{ab} in terms of $\overline{g}_{ij}$.

(b) Show that if g_{ij} is symmetric in one coordinate system it is symmetric in all coordinate systems.

(c) Let $\overline{g} = det(\overline{g}_{ij})$ and $g = det(g_{ij})$ and show that $\overline{g} = gJ^2(\frac{x}{\overline{x}})$ and therefore $\sqrt{\overline{g}} = \sqrt{g}J(\frac{x}{\overline{x}})$. This shows

    that g is a scalar invariant of weight 2 and $\sqrt{g}$ is a scalar invariant of weight 1.

▶ **2.**   For $g_{ij} = \dfrac{\partial y^m}{\partial x^i}\dfrac{\partial y^m}{\partial x^j}$ show that $g^{ij} = \dfrac{\partial x^i}{\partial y^m}\dfrac{\partial x^j}{\partial y^m}$

▶ **3.**   Show that in a curvilinear coordinate system which is orthogonal we have:

$$(a) \qquad g = det(g_{ij}) = g_{11}g_{22}g_{33}$$

$$(b) \qquad g_{mn} = g^{mn} = 0 \qquad \text{for } m \neq n$$

$$(c) \qquad g^{NN} = \frac{1}{g_{NN}} \quad \text{for } N = 1,2,3 \quad \text{(no summation on N)}$$

▶ **4.**   Show that $g = det(g_{ij}) = \left|\dfrac{\partial y^i}{\partial x^j}\right|^2 = J^2$, where J is the Jacobian.

▶ **5.**   Define the quantities $h_1 = h_u = |\dfrac{\partial \vec{r}}{\partial u}|$, $h_2 = h_v = |\dfrac{\partial \vec{r}}{\partial v}|$, $h_3 = h_w = |\dfrac{\partial \vec{r}}{\partial w}|$ and construct the unit

vectors

$$\widehat{e}_u = \frac{1}{h_1}\frac{\partial \vec{r}}{\partial u}, \qquad \widehat{e}_v = \frac{1}{h_2}\frac{\partial \vec{r}}{\partial v}, \qquad \widehat{e}_w = \frac{1}{h_3}\frac{\partial \vec{r}}{\partial w}.$$

(a) Assume the coordinate system is orthogonal and show that

$$g_{11} = h_1^2 = \left(\frac{\partial x}{\partial u}\right)^2 + \left(\frac{\partial y}{\partial u}\right)^2 + \left(\frac{\partial z}{\partial u}\right)^2,$$

$$g_{22} = h_2^2 = \left(\frac{\partial x}{\partial v}\right)^2 + \left(\frac{\partial y}{\partial v}\right)^2 + \left(\frac{\partial z}{\partial v}\right)^2,$$

$$g_{33} = h_3^2 = \left(\frac{\partial x}{\partial w}\right)^2 + \left(\frac{\partial y}{\partial w}\right)^2 + \left(\frac{\partial z}{\partial w}\right)^2.$$

(b) Show that $d\vec{r}$ can be expressed in the form $d\vec{r} = h_1\,\widehat{e}_u\,du + h_2\,\widehat{e}_v\,dv + h_3\,\widehat{e}_w\,dw$.

(c) Show that the volume of the elemental parallelepiped having $d\vec{r}$ as diagonal can be represented

$$d\tau = \sqrt{g}\,dudvdw = J\,dudvdw = \frac{\partial(x,y,z)}{\partial(u,v,w)}\,dudvdw.$$

    Hint:

$$|\vec{A}\cdot(\vec{B}\times\vec{C})| = \begin{vmatrix} A_1 & A_2 & A_3 \\ B_1 & B_2 & B_3 \\ C_1 & C_2 & C_3 \end{vmatrix}$$

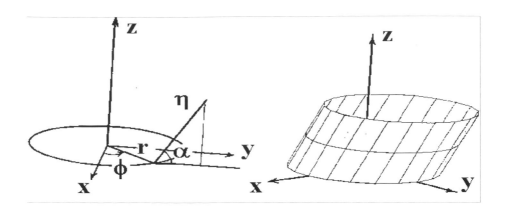

Figure 1.3-18 Oblique cylindrical coordinates.

▶ **6.** Calculate the metric and conjugate metric components for the oblique cylindrical coordinates (r, ϕ, η), illustrated in figure 1.3-18, where $x = r\cos\phi$, $y = r\sin\phi + \eta\cos\alpha$, $z = \eta\sin\alpha$ and α is a parameter $0 < \alpha \le \frac{\pi}{2}$. Note: When $\alpha = \frac{\pi}{2}$ cylindrical coordinates result.

▶ **7.** In Cartesian coordinates you are given the affine transformation. $\bar{x}_i = \ell_{ij}x_j$ where

$$\bar{x}_1 = \frac{1}{15}(5x_1 - 14x_2 + 2x_3), \qquad \bar{x}_2 = -\frac{1}{3}(2x_1 + x_2 + 2x_3), \qquad \bar{x}_3 = \frac{1}{15}(10x_1 + 2x_2 - 11x_3)$$

(a) Show the transformation is orthogonal.

(b) A vector $\vec{A}(x_1, x_2, x_3)$ in the unbarred system has the components

$$A_1 = (x_1)^2, \qquad A_2 = (x_2)^2 \qquad A_3 = (x_3)^2.$$

Find the components of this vector in the barred system of coordinates.

▶ **8.** Calculate the metric and conjugate metric tensors in cylindrical coordinates (r, θ, z).

▶ **9.** Calculate the metric and conjugate metric tensors in spherical coordinates (ρ, θ, ϕ).

▶ **10.** Calculate the metric and conjugate metric tensors in parabolic cylindrical coordinates (ξ, η, z).

▶ **11.** Calculate the metric and conjugate metric components in elliptic cylindrical coordinates (ξ, η, z).

▶ **12.** For the change $d\vec{r}$ given in problem 5, show the elemental parallelepiped with diagonal $d\vec{r}$ has:

(a) the element of area $dS_1 = \sqrt{g_{22}g_{33} - g_{23}^2}\, dvdw$ in the $u =$constant surface.

(b) The element of area $dS_2 = \sqrt{g_{33}g_{11} - g_{13}^2}\, dudw$ in the $v =$constant surface.

(c) the element of area $dS_3 = \sqrt{g_{11}g_{22} - g_{12}^2}\, dudv$ in the $w =$constant surface.

(d) Simplify the above elements of area in the special case where the curvilinear coordinates are orthogonal?
Hint: $|\vec{A} \times \vec{B}| = \sqrt{(\vec{A} \times \vec{B}) \cdot (\vec{A} \times \vec{B})} = \sqrt{(\vec{A} \cdot \vec{A})(\vec{B} \cdot \vec{B}) - (\vec{A} \cdot \vec{B})(\vec{A} \cdot \vec{B})}.$

13. Calculate the metric and conjugate metric tensor associated with the toroidal surface coordinates (ξ, η) illustrated in the figure 1.3-19, where

$$x = (a + b\cos\xi)\cos\eta \qquad a > b > 0$$
$$y = (a + b\cos\xi)\sin\eta \qquad 0 < \xi < 2\pi$$
$$z = b\sin\xi \qquad 0 < \eta < 2\pi$$

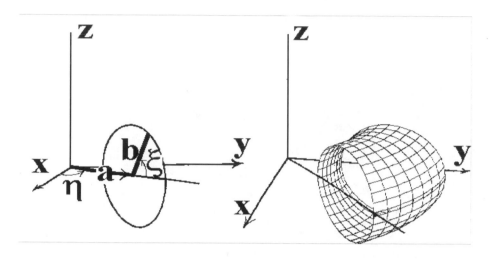

Figure 1.3-19. Toroidal surface coordinates

14. Calculate the metric and conjugate metric tensor associated with the spherical surface coordinates (θ, ϕ), illustrated in the figure 1.3-20, where

$$x = a\sin\theta\cos\phi \qquad a > 0 \quad \text{is constant}$$
$$y = a\sin\theta\sin\phi \qquad 0 < \phi < 2\pi$$
$$z = a\cos\theta \qquad 0 < \theta < \frac{\pi}{2}$$

15. Consider g_{ij}, $i, j = 1, 2$
(a) Show that $g^{11} = \dfrac{g_{22}}{\Delta}$, $g^{12} = g^{21} = \dfrac{-g_{12}}{\Delta}$, $g^{22} = \dfrac{g_{11}}{\Delta}$ where $\Delta = g_{11}g_{22} - g_{12}g_{21}$.
(b) Use the results in part (a) and verify that $g_{ij}g^{ik} = \delta_j^k$, $i, j, k = 1, 2$.

16. Let A_x, A_y, A_z denote the constant components of a vector in Cartesian coordinates. Using the transformation laws (1.2.42) and (1.2.47) to find the contravariant and covariant components of this vector upon changing to (a) cylindrical coordinates (r, θ, z). (b) spherical coordinates (ρ, θ, ϕ) and (c) Parabolic cylindrical coordinates.

17. Find the relationship which exists between the given associated tensors.

(a) $A_{r.}^{pqk}$ and A_{rs}^{pq} (c) $A_{.l.m}^{i.j.}$ and $A_{r.t.}^{.s.p}$

(b) $A_{.mrs}^{p}$ and $A_{..rs}^{pq}$ (d) A_{mnk} and $A_{..k}^{ij}$

104

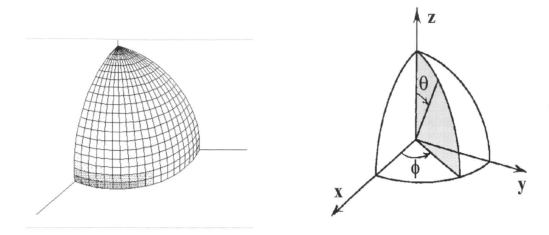

Figure 1.3-20. Spherical surface coordinates

▶ **18.** Given the fourth order tensor $C_{ikmp} = \lambda\delta_{ik}\delta_{mp} + \mu(\delta_{im}\delta_{kp} + \delta_{ip}\delta_{km}) + \nu(\delta_{im}\delta_{kp} - \delta_{ip}\delta_{km})$ where λ, μ and ν are scalars and δ_{ij} is the Kronecker delta. Show that under an orthogonal transformation of rotation of axes with $\overline{x}_i = \ell_{ij}x_j$ where $\ell_{rs}\ell_{is} = \ell_{mr}\ell_{mi} = \delta_{ri}$ the components of the above tensor are unaltered. Any tensor whose components are unaltered under an orthogonal transformation is called an 'isotropic' tensor. Another way of stating this problem is to say "Show C_{ikmp} is an isotropic tensor."

▶ **19.** Assume A_{ijl} is a third order covariant tensor and B^{pqmn} is a fourth order contravariant tensor. Prove that $A_{ikl}B^{klmn}$ is a mixed tensor of order three, with one covariant and two contravariant indices.

▶ **20.** Assume that T_{mnrs} is an absolute tensor. Show that if $T_{ijkl} + T_{ijlk} = 0$ in the coordinate system x^r then $\overline{T}_{ijkl} + \overline{T}_{ijlk} = 0$ in any other coordinate system $\overline{x}^r$.

▶ **21.** Show that $\epsilon_{ijk}\epsilon_{rst} = \begin{vmatrix} g_{ir} & g_{is} & g_{it} \\ g_{jr} & g_{js} & g_{jt} \\ g_{kr} & g_{ks} & g_{kt} \end{vmatrix}$.   Hint: See problem 38, Exercise 1.1

▶ **22.** Determine if the tensor equation $\epsilon_{mnp}\epsilon_{mij} + \epsilon_{mnj}\epsilon_{mpi} = \epsilon_{mni}\epsilon_{mpj}$ is true or false. Justify your answer.

▶ **23.** Prove the epsilon identity $g^{ij}\epsilon_{ipt}\epsilon_{jrs} = g_{pr}g_{ts} - g_{ps}g_{tr}$.   Hint: See problem 38, Exercise 1.1

▶ **24.** Let A^{rs} denote a skew-symmetric contravariant tensor and let $c_r = \frac{1}{2}\epsilon_{rmn}A^{mn}$ where $\epsilon_{rmn} = \sqrt{g}\,e_{rmn}$. Show that c_r are the components of a covariant tensor. Write out all the components.

▶ **25.** Let A_{rs} denote a skew-symmetric covariant tensor and let $c^r = \frac{1}{2}\epsilon^{rmn}A_{mn}$ where $\epsilon^{rmn} = \frac{1}{\sqrt{g}}e^{rmn}$. Show that c^r are the components of a contravariant tensor. Write out all the components.

▶ **26.** Let $A_{pq}B_r^{qs} = C_{pr}^s$ where B_r^{qs} is a relative tensor of weight ω_1 and C_{pr}^s is a relative tensor of weight ω_2. Prove that A_{pq} is a relative tensor of weight $(\omega_2 - \omega_1)$.

▶ **27.** When A_j^i is an absolute tensor prove that $\sqrt{g}A_j^i$ is a relative tensor of weight $+1$.

▶ **28.** When A_j^i is an absolute tensor prove that $\frac{1}{\sqrt{g}}A_j^i$ is a relative tensor of weight -1.

29.

(a) Show e^{ijk} is a relative tensor of weight $+1$.

(b) Show $\epsilon^{ijk} = \frac{1}{\sqrt{g}} e^{ijk}$ is an absolute tensor. Hint: See example 1.1-26.

30. The equation of a surface can be represented by an equation of the form $\Phi(x^1, x^2, x^3) = constant$. Show that a unit normal vector to the surface can be represented by the vector

$$n^i = \frac{g^{ij}\frac{\partial \Phi}{\partial x^j}}{(g^{mn}\frac{\partial \Phi}{\partial x^m}\frac{\partial \Phi}{\partial x^n})^{\frac{1}{2}}}.$$

31. Assume that $\bar{g}_{ij} = \lambda g_{ij}$ with λ a nonzero constant. Find and calculate $\bar{g}^{ij}$ in terms of g^{ij}.

32. Determine if the following tensor equation is true. Justify your answer.

$$\epsilon_{rjk}A_i^r + \epsilon_{irk}A_j^r + \epsilon_{ijr}A_k^r = \epsilon_{ijk}A_r^r.$$

Hint: See problem 21, Exercise 1.1.

33. Show that for C_i and C^i associated tensors, and $C^i = \epsilon^{ijk}A_j B_k$, then $C_i = \epsilon_{ijk}A^j B^k$

34. Prove that ϵ^{ijk} and ϵ_{ijk} are associated tensors. Hint: Consider the determinant of g_{ij}.

35. Show $\epsilon^{ijk}A_i B_j C_k = \epsilon_{ijk}A^i B^j C^k$.

36. Let T_j^i, $i, j = 1, 2, 3$ denote a second order mixed tensor. Show that the given quantities are scalar invariants. (i) $I_1 = T_i^i$ (ii) $I_2 = \frac{1}{2}\left[(T_i^i)^2 - T_m^i T_i^m\right]$ (iii) $I_3 = det|T_j^i|$

37.

(a) Assume A^{ij} and B^{ij}, $i, j = 1, 2, 3$ are absolute contravariant tensors, and determine if the inner product $C^{ik} = A^{ij}B^{jk}$ is an absolute tensor?

(b) Assume that the condition $\frac{\partial \bar{x}^j}{\partial x^n}\frac{\partial \bar{x}^j}{\partial x^m} = \delta_{nm}$ is satisfied, and determine whether the inner product in part (a) is a tensor?

(c) Consider only transformations which are a rotation and translation of axes $\bar{y}_i = \ell_{ij}y_j + b_i$, where ℓ_{ij} are direction cosines for the rotation of axes. Show that $\frac{\partial \bar{y}_j}{\partial y_n}\frac{\partial \bar{y}_j}{\partial y_m} = \delta_{nm}$

38. For A_{ijk} a Cartesian tensor, determine if a contraction on the indices i and j is allowed. That is, determine if the quantity $A_k = A_{iik}$, (summation on i) is a tensor. Hint: See part(c) of the previous problem.

39. Prove the e-δ identity $e^{ijk}e_{imn} = \delta_m^j\delta_n^k - \delta_n^j\delta_m^k$.

40. Consider the vector V_k, $k = 1, 2, 3$ and define the matrix (a_{ij}) having the elements $a_{ij} = e_{ijk}V_k$, where e_{ijk} is the e-permutation symbol.

(a) Solve for V_i in terms of a_{mn} by multiplying both sides of the given equation by e^{ijl} and note the $e - \delta$ identity allows us to simplify the result.

(b) Sum the given expression on k and then assign values to the free indices (i,j=1,2,3) and compare your results with part (a).

(c) Is a_{ij} symmetric, skew-symmetric, or neither?

▶ **41.** It can be shown that the continuity equation of fluid dynamics can be expressed in the tensor form

$$\frac{1}{\sqrt{g}}\frac{\partial}{\partial x^r}\left(\sqrt{g}\varrho V^r\right) + \frac{\partial\varrho}{\partial t} = 0,$$

where ϱ is the density of the fluid, t is time, V^r, with $r = 1, 2, 3$ are the velocity components and $g = |g_{ij}|$ is the determinant of the metric tensor. Employing the summation convention and replacing the tensor components of velocity by their physical components, express the continuity equation in

    (a) Cartesian coordinates (x, y, z) with physical components V_x, V_y, V_z.

    (b) Cylindrical coordinates (r, θ, z) with physical components V_r, V_θ, V_z.

    (c) Spherical coordinates (ρ, θ, ϕ) with physical components V_ρ, V_θ, V_ϕ.

▶ **42.** Let x^1, x^2, x^3 denote a set of skewed coordinates with respect to the Cartesian coordinates y^1, y^2, y^3. Assume that $\vec{E}_1, \vec{E}_2, \vec{E}_3$ are unit vectors in the directions of the x^1, x^2 and x^3 axes respectively. If the unit vectors satisfy the relations

$$\begin{aligned}
\vec{E}_1 \cdot \vec{E}_1 &= 1 & \vec{E}_1 \cdot \vec{E}_2 &= \cos\theta_{12} \\
\vec{E}_2 \cdot \vec{E}_2 &= 1 & \vec{E}_1 \cdot \vec{E}_3 &= \cos\theta_{13} \\
\vec{E}_3 \cdot \vec{E}_3 &= 1 & \vec{E}_2 \cdot \vec{E}_3 &= \cos\theta_{23},
\end{aligned}$$

then calculate the metrices g_{ij} and conjugate metrices g^{ij}.

▶ **43.** Let A_{ij}, $i, j = 1, 2, 3, 4$ denote the skew-symmetric second rank tensor

$$A_{ij} = \begin{pmatrix} 0 & a & b & c \\ -a & 0 & d & e \\ -b & -d & 0 & f \\ -c & -e & -f & 0 \end{pmatrix},$$

where a, b, c, d, e, f are complex constants. Calculate the components of the dual tensor

$$V^{ij} = \frac{1}{2}e^{ijkl}A_{kl}.$$

▶ **44.** In Cartesian coordinates the vorticity tensor at a point in a fluid medium is defined

$$\omega_{ij} = \frac{1}{2}\left(\frac{\partial V_j}{\partial x^i} - \frac{\partial V_i}{\partial x^j}\right)$$

where V_i are the velocity components of the fluid at the point. The vorticity vector at a point in a fluid medium in Cartesian coordinates is defined by $\omega^i = \frac{1}{2}e^{ijk}\omega_{jk}$. Show that these tensors are dual tensors.

▶ **45.** Write out the relation between each of the components of the dual tensors

$$\hat{T}^{ij} = \frac{1}{2}e^{ijkl}T_{kl} \quad i, j, k, l = 1, 2, 3, 4$$

and show that if $ijkl$ is an even permutation of 1234, then $\hat{T}^{ij} = T_{kl}$.

46. Consider the general affine transformation $\bar{x}_i = a_{ij}x_j$ where $(x^1, x^2, x^3) = (x, y, z)$ with inverse transformation $x_i = b_{ij}\bar{x}_j$. Determine (a) the image of the plane $Ax + By + Cz + D = 0$ under this transformation and (b) the image of a second degree conic section

$$Ax^2 + 2Bxy + Cy^2 + Dx + Ey + F = 0.$$

47. Using a multilinear form of degree M, derive the transformation law for a contravariant vector of degree M.

48. Let g denote the determinant of g_{ij} and show that $\dfrac{\partial g}{\partial x^k} = gg^{ij}\dfrac{\partial g_{ij}}{\partial x^k}$.

49. We have shown that for a rotation of xyz axes with respect to a set of fixed $\bar{x}\bar{y}\bar{z}$ axes, the derivative of a vector $\vec{A}$ with respect to an observer on the barred axes is given by $\dfrac{d\vec{A}}{dt}\bigg|_f = \dfrac{d\vec{A}}{dt}\bigg|_r + \vec{\omega} \times \vec{A}$. Introduce the operators $D_f\vec{A} = \dfrac{d\vec{A}}{dt}\bigg|_f =$ derivative in fixed system and $D_r\vec{A} = \dfrac{d\vec{A}}{dt}\bigg|_r =$ derivative in rotating system.

(a) Show that $D_f\vec{A} = (D_r + \vec{\omega}\times)\vec{A}$.

(b) Consider the special case that the vector $\vec{A}$ is the position vector $\vec{r}$. Show that $D_f\vec{r} = (D_r + \vec{\omega}\times)\vec{r}$ produces $\vec{V}\bigg|_f = \vec{V}\bigg|_r + \vec{\omega} \times \vec{r}$ where $\vec{V}\bigg|_f$ represents the velocity of a particle relative to the fixed system and $\vec{V}\bigg|_r$ represents the velocity of a particle with respect to the rotating system of coordinates.

(c) Show that $\vec{a}\bigg|_f = \vec{a}\bigg|_r + \vec{\omega} \times (\vec{\omega} \times \vec{r})$ where $\vec{a}\bigg|_f$ represents the acceleration of a particle relative to the fixed system and $\vec{a}\bigg|_r$ represents the acceleration of a particle with respect to the rotating system.

(d) Show in the special case $\vec{\omega}$ is a constant that $\vec{a}\bigg|_f = 2\vec{\omega} \times \vec{V} + \vec{\omega} \times (\vec{\omega} \times \vec{r})$. where $\vec{V}$ is the velocity of the particle relative to the rotating system. The term $2\vec{\omega} \times \vec{V}$ is referred to as the Coriolis acceleration and the term $\vec{\omega} \times (\vec{\omega} \times \vec{r})$ is referred to as the centripetal acceleration.

50. For the confocal paraboloidal coordinates (u, v, w) with transformation equations

$$x^2 = \frac{(b^2 - u)(b^2 - v)(b^2 - w)}{a^2 - b^2} \qquad -\infty < u < a^2$$

$$y^2 = \frac{(a^2 - u)(a^2 - v)(a^2 - w)}{b^2 - a^2} \qquad a^2 < v < b^2$$

$$2z = u + v + w - a^2 - b^2 \qquad a^2 < w < \infty$$

show that

$$h_1^2 = \frac{(v - u)(w - u)}{4(a^2 - u)(b^2 - u)}$$

$$h_2^2 = \frac{(w - v)(u - v)}{4(a^2 - v)(b^2 - v)}$$

$$h_3^2 = \frac{(u - w)(v - w)}{16(a^2 - w)(b^2 - w)}$$

§1.4 DERIVATIVE OF A TENSOR

In this section we develop some additional operations associated with tensors. Historically, one of the basic problems of the tensor calculus was to try and find a tensor quantity which is a function of the metric tensor g_{ij} and some of its derivatives $\dfrac{\partial g_{ij}}{\partial x^m}$, $\dfrac{\partial^2 g_{ij}}{\partial x^m \partial x^n}$, ... A solution of this problem is the fourth order Riemann Christoffel tensor R_{ijkl} to be developed shortly. In order to understand how this tensor was arrived at, we must first develop some preliminary relationships involving Christoffel symbols.

Christoffel Symbols

Let us consider the metric tensor g_{ij} which we know satisfies the transformation law $\overline{g}_{\alpha\beta} = g_{ab} \dfrac{\partial x^a}{\partial \overline{x}^\alpha} \dfrac{\partial x^b}{\partial \overline{x}^\beta}$. Define the quantity

$$(\alpha, \beta, \gamma) = \frac{\partial \overline{g}_{\alpha\beta}}{\partial \overline{x}^\gamma} = \frac{\partial g_{ab}}{\partial x^c} \frac{\partial x^c}{\partial \overline{x}^\gamma} \frac{\partial x^a}{\partial \overline{x}^\alpha} \frac{\partial x^b}{\partial \overline{x}^\beta} + g_{ab} \frac{\partial^2 x^a}{\partial \overline{x}^\alpha \partial \overline{x}^\gamma} \frac{\partial x^b}{\partial \overline{x}^\beta} + g_{ab} \frac{\partial x^a}{\partial \overline{x}^\alpha} \frac{\partial^2 x^b}{\partial \overline{x}^\beta \partial \overline{x}^\gamma}$$

and form the combination of terms $\dfrac{1}{2}\left[(\alpha, \beta, \gamma) + (\beta, \gamma, \alpha) - (\gamma, \alpha, \beta)\right]$ to obtain the result

$$\frac{1}{2}\left[\frac{\partial \overline{g}_{\alpha\beta}}{\partial \overline{x}^\gamma} + \frac{\partial \overline{g}_{\beta\gamma}}{\partial \overline{x}^\alpha} - \frac{\partial \overline{g}_{\gamma\alpha}}{\partial \overline{x}^\beta}\right] = \frac{1}{2}\left[\frac{\partial g_{ab}}{\partial x^c} + \frac{\partial g_{bc}}{\partial x^a} - \frac{\partial g_{ca}}{\partial x^b}\right]\frac{\partial x^a}{\partial \overline{x}^\alpha} \frac{\partial x^b}{\partial \overline{x}^\beta} \frac{\partial x^c}{\partial \overline{x}^\gamma} + g_{ab}\frac{\partial x^b}{\partial \overline{x}^\beta} \frac{\partial^2 x^a}{\partial \overline{x}^\alpha \partial \overline{x}^\gamma}. \tag{1.4.1}$$

In this equation the combination of derivatives occurring inside the brackets is called a Christoffel symbol of the first kind and is defined by the notation

$$[ac, b] = [ca, b] = \frac{1}{2}\left[\frac{\partial g_{ab}}{\partial x^c} + \frac{\partial g_{bc}}{\partial x^a} - \frac{\partial g_{ac}}{\partial x^b}\right]. \tag{1.4.2}$$

The equation (1.4.1) defines the transformation for a Christoffel symbol of the first kind and can be expressed as

$$\overline{[\alpha\,\gamma, \beta]} = [ac, b]\frac{\partial x^a}{\partial \overline{x}^\alpha} \frac{\partial x^b}{\partial \overline{x}^\beta} \frac{\partial x^c}{\partial \overline{x}^\gamma} + g_{ab}\frac{\partial^2 x^a}{\partial \overline{x}^\alpha \partial \overline{x}^\gamma} \frac{\partial x^b}{\partial \overline{x}^\beta}. \tag{1.4.3}$$

Observe that the Christoffel symbol of the first kind $[ac, b]$ does not transform like a tensor. However, it is symmetric in the indices a and c.

At this time it is convenient to use the equation (1.4.3) to develop an expression for the second derivative term which occurs in that equation as this second derivative term arises in some of our future considerations. To solve for this second derivative we can multiply equation (1.4.3) by $\dfrac{\partial \overline{x}^\beta}{\partial x^d} g^{de}$ and simplify the result to the form

$$\frac{\partial^2 x^e}{\partial \overline{x}^\alpha \partial \overline{x}^\gamma} = -g^{de}[ac, d]\frac{\partial x^a}{\partial \overline{x}^\alpha} \frac{\partial x^c}{\partial \overline{x}^\gamma} + \overline{[\alpha\,\gamma, \beta]}\frac{\partial \overline{x}^\beta}{\partial x^d}g^{de}. \tag{1.4.4}$$

The transformation $g^{de} = \overline{g}^{\lambda\mu}\dfrac{\partial x^d}{\partial \overline{x}^\lambda} \dfrac{\partial x^e}{\partial \overline{x}^\mu}$ allows us to express the equation (1.4.4) in the form

$$\frac{\partial^2 x^e}{\partial \overline{x}^\alpha \partial \overline{x}^\gamma} = -g^{de}[ac, d]\frac{\partial x^a}{\partial \overline{x}^\alpha} \frac{\partial x^c}{\partial \overline{x}^\gamma} + \overline{g}^{\beta\mu}\overline{[\alpha\,\gamma, \beta]}\frac{\partial x^e}{\partial \overline{x}^\mu}. \tag{1.4.5}$$

Define the Christoffel symbol of the second kind as

$$\left\{\begin{matrix} i \\ j\,k \end{matrix}\right\} = \left\{\begin{matrix} i \\ k\,j \end{matrix}\right\} = g^{i\alpha}[jk, \alpha] = \frac{1}{2}g^{i\alpha}\left(\frac{\partial g_{k\alpha}}{\partial x^j} + \frac{\partial g_{j\alpha}}{\partial x^k} - \frac{\partial g_{jk}}{\partial x^\alpha}\right). \tag{1.4.6}$$

This Christoffel symbol of the second kind is symmetric in the indices j and k and from equation (1.4.5) we see that it satisfies the transformation law

$$\overline{\left\{ \begin{array}{c} \mu \\ \alpha\,\gamma \end{array} \right\}} \frac{\partial x^e}{\partial \overline{x}^\mu} = \left\{ \begin{array}{c} e \\ a\,c \end{array} \right\} \frac{\partial x^a}{\partial \overline{x}^\alpha} \frac{\partial x^c}{\partial \overline{x}^\gamma} + \frac{\partial^2 x^e}{\partial \overline{x}^\alpha \partial \overline{x}^\gamma}. \tag{1.4.7}$$

Observe that the Christoffel symbol of the second kind does not transform like a tensor quantity. We can use the relation defined by equation (1.4.7) to express the second derivative of the transformation equations in terms of the Christoffel symbols of the second kind. At times it will be convenient to represent the Christoffel symbols with a subscript to indicate the metric from which they are calculated. Thus, an alternative notation for $\left\{ \begin{array}{c} i \\ j\,k \end{array} \right\}$ is the notation $\left\{ \begin{array}{c} i \\ j\,k \end{array} \right\}_{\overline{g}}$.

EXAMPLE 1.4-1. (Christoffel symbols) Solve for the Christoffel symbol of the first kind in terms of the Christoffel symbol of the second kind.

Solution: By the definition from equation (1.4.6) we have $\left\{ \begin{array}{c} i \\ j\,k \end{array} \right\} = g^{i\alpha}[jk,\alpha]$. We multiply this equation by $g_{\beta i}$ and find $g_{\beta i} \left\{ \begin{array}{c} i \\ j\,k \end{array} \right\} = \delta_\beta^\alpha [jk,\alpha] = [jk,\beta]$ and so $[jk,\alpha] = g_{\alpha i} \left\{ \begin{array}{c} i \\ j\,k \end{array} \right\} = g_{\alpha 1} \left\{ \begin{array}{c} 1 \\ j\,k \end{array} \right\} + \cdots + g_{\alpha N} \left\{ \begin{array}{c} N \\ j\,k \end{array} \right\}$. ∎

EXAMPLE 1.4-2. (Christoffel symbols of first kind)

Derive formulas to find the Christoffel symbols of the first kind in a generalized orthogonal coordinate system with metric coefficients

$$g_{ij} = 0 \quad \text{for} \quad i \neq j \qquad \text{and} \qquad g_{(i)(i)} = h_{(i)}^2, \quad i = 1,2,3$$

where i is not summed.

Solution: In an orthogonal coordinate system where $g_{ij} = 0$ for $i \neq j$ we observe that

$$[ab,c] = \frac{1}{2}\left(\frac{\partial g_{ac}}{\partial x^b} + \frac{\partial g_{bc}}{\partial x^a} - \frac{\partial g_{ab}}{\partial x^c} \right). \tag{1.4.8}$$

Here there are $3^3 = 27$ quantities to calculate. We consider the following cases:

CASE I Let $a = b = c = i$, then the equation (1.4.8) simplifies to

$$[ab,c] = [ii,i] = \frac{1}{2}\frac{\partial g_{ii}}{\partial x^i} \quad \text{(no summation on } i\text{)}. \tag{1.4.9}$$

From this equation we can calculate any of the Christoffel symbols

$$[11,1], \quad [22,2], \quad \text{or} \quad [33,3].$$

CASE II Let $a = b = i \neq c$, then the equation (1.4.8) simplifies to the form

$$[ab,c] = [ii,c] = -\frac{1}{2}\frac{\partial g_{ii}}{\partial x^c} \quad \text{(no summation on } i \text{ and } i \neq c\text{)}. \tag{1.4.10}$$

since, $g_{ic} = 0$ for $i \neq c$. This equation shows how we may calculate any of the six Christoffel symbols

$$[11,2], \quad [11,3], \quad [22,1], \quad [22,3], \quad [33,1], \quad [33,2].$$

110

CASE III Let $a = c = i \neq b$, and noting that $g_{ib} = 0$ for $i \neq b$, it can be verified that the equation (1.4.8) simplifies to the form

$$[ab, c] = [ib, i] = [bi, i] = \frac{1}{2}\frac{\partial g_{ii}}{\partial x^b} \quad \text{(no summation on } i \text{ and } i \neq b).} \tag{1.4.11}$$

From this equation we can calculate any of the twelve Christoffel symbols

$$[12, 1] = [21, 1] \qquad\qquad [31, 3] = [13, 3]$$
$$[32, 3] = [23, 3] \qquad\qquad [21, 2] = [12, 2]$$
$$[13, 1] = [31, 1] \qquad\qquad [23, 2] = [32, 2]$$

CASE IV Let $a \neq b \neq c$ and show that the equation (1.4.8) reduces to

$$[ab, c] = 0, \qquad (a \neq b \neq c.)$$

This represents the six Christoffel symbols

$$[12, 3] = [21, 3] = [23, 1] = [32, 1] = [31, 2] = [13, 2] = 0.$$

From the Cases I,II,III,IV all twenty seven Christoffel symbols of the first kind can be determined. In practice, only the nonzero Christoffel symbols are listed.

■

EXAMPLE 1.4-3. (Christoffel symbols of the first kind) Find the nonzero Christoffel symbols of the first kind in cylindrical coordinates.
Solution: From the results of example 1.4-2 we find that for $x^1 = r$, $x^2 = \theta$, $x^3 = z$ and

$$g_{11} = 1, \qquad g_{22} = (x^1)^2 = r^2, \qquad g_{33} = 1$$

the nonzero Christoffel symbols of the first kind in cylindrical coordinates are:

$$[22, 1] = -\frac{1}{2}\frac{\partial g_{22}}{\partial x^1} = -x^1 = -r$$
$$[21, 2] = [12, 2] = \frac{1}{2}\frac{\partial g_{22}}{\partial x^1} = x^1 = r.$$

■

EXAMPLE 1.4-4. (Christoffel symbols of the second kind)
Find formulas for the calculation of the Christoffel symbols of the second kind in a generalized orthogonal coordinate system with metric coefficients

$$g_{ij} = 0 \quad \text{for} \quad i \neq j \quad \text{and} \quad g_{(i)(i)} = h_{(i)}^2, \quad i = 1, 2, 3$$

where i is not summed.
Solution: By definition we have

$$\left\{ \begin{matrix} i \\ j\,k \end{matrix} \right\} = g^{im}[jk, m] = g^{i1}[jk, 1] + g^{i2}[jk, 2] + g^{i3}[jk, 3] \tag{1.4.12}$$

By hypothesis the coordinate system is orthogonal and so

$$g^{ij} = 0 \quad \text{for} \quad i \neq j \quad \text{and} \quad g^{ii} = \frac{1}{g_{ii}} \quad i \text{ not summed.}$$

The only nonzero term in the equation (1.4.12) occurs when $m = i$ and consequently

$$\begin{Bmatrix} i \\ j\,k \end{Bmatrix} = g^{ii}[jk, i] = \frac{[jk, i]}{g_{ii}} \quad \text{no summation on } i. \tag{1.4.13}$$

We can now consider the four cases considered in the example 1.4-2.

CASE I Let $j = k = i$ and show

$$\begin{Bmatrix} i \\ i\,i \end{Bmatrix} = \frac{[ii, i]}{g_{ii}} = \frac{1}{2g_{ii}} \frac{\partial g_{ii}}{\partial x^i} = \frac{1}{2} \frac{\partial}{\partial x^i} \ln g_{ii} \quad \text{no summation on } i. \tag{1.4.14}$$

CASE II Let $k = j \neq i$ and show

$$\begin{Bmatrix} i \\ j\,j \end{Bmatrix} = \frac{[jj, i]}{g_{ii}} = \frac{-1}{2g_{ii}} \frac{\partial g_{jj}}{\partial x^i} \quad \text{no summation on } i \text{ or } j. \tag{1.4.15}$$

CASE III Let $i = j \neq k$ and verify that

$$\begin{Bmatrix} j \\ j\,k \end{Bmatrix} = \begin{Bmatrix} j \\ k\,j \end{Bmatrix} = \frac{[jk, j]}{g_{jj}} = \frac{1}{2g_{jj}} \frac{\partial g_{jj}}{\partial x^k} = \frac{1}{2} \frac{\partial}{\partial x^k} \ln g_{jj} \quad \text{no summation on } i \text{ or } j. \tag{1.4.16}$$

CASE IV For the case $i \neq j \neq k$ we find

$$\begin{Bmatrix} i \\ j\,k \end{Bmatrix} = \frac{[jk, i]}{g_{ii}} = 0, \quad i \neq j \neq k \quad \text{no summation on } i.$$

The above cases represent all 27 terms.

■

EXAMPLE 1.4-5. (Notation) In the case of cylindrical coordinates we can use the above relations and find the nonzero Christoffel symbols of the second kind:

$$\begin{Bmatrix} 1 \\ 2\,2 \end{Bmatrix} = -\frac{1}{2g_{11}} \frac{\partial g_{22}}{\partial x^1} = -x^1 = -r$$

$$\begin{Bmatrix} 2 \\ 1\,2 \end{Bmatrix} = \begin{Bmatrix} 2 \\ 2\,1 \end{Bmatrix} = \frac{1}{2g_{22}} \frac{\partial g_{22}}{\partial x^1} = \frac{1}{x^1} = \frac{1}{r}$$

Note 1: The notation for the above Christoffel symbols are based upon the assumption that $x^1 = r, x^2 = \theta$ and $x^3 = z$. However, in tensor calculus the choice of the coordinates can be arbitrary. We could just as well have defined $x^1 = z, x^2 = r$ and $x^3 = \theta$. In this latter case, the numbering system of the Christoffel symbols changes. To avoid confusion, an alternate method of writing the Christoffel symbols is to use coordinates in place of the integers 1,2 and 3. For example, in cylindrical coordinates we can write

$$\begin{Bmatrix} \theta \\ r\,\theta \end{Bmatrix} = \begin{Bmatrix} \theta \\ \theta\,r \end{Bmatrix} = \frac{1}{r} \quad \text{and} \quad \begin{Bmatrix} r \\ \theta\,\theta \end{Bmatrix} = -r.$$

112

If we define $x^1 = r, x^2 = \theta, x^3 = z$, then the nonzero Christoffel symbols are written as

$$\left\{ \begin{matrix} 2 \\ 1\,2 \end{matrix} \right\} = \left\{ \begin{matrix} 2 \\ 2\,1 \end{matrix} \right\} = \frac{1}{r} \quad \text{and} \quad \left\{ \begin{matrix} 1 \\ 2\,2 \end{matrix} \right\} = -r.$$

In contrast, if we define $x^1 = z, x^2 = r, x^3 = \theta$, then the nonzero Christoffel symbols are written

$$\left\{ \begin{matrix} 3 \\ 2\,3 \end{matrix} \right\} = \left\{ \begin{matrix} 3 \\ 3\,2 \end{matrix} \right\} = \frac{1}{r} \quad \text{and} \quad \left\{ \begin{matrix} 2 \\ 3\,3 \end{matrix} \right\} = -r.$$

Note 2: Some textbooks use the notation $\Gamma_{a,bc}$ for Christoffel symbols of the first kind and $\Gamma^d_{bc} = g^{da}\Gamma_{a,bc}$ for Christoffel symbols of the second kind. This notation is not used in these notes since the notation suggests that the Christoffel symbols are third order tensors, which is not true. The Christoffel symbols of the first and second kind are not tensors. This fact is clearly illustrated by the transformation equations (1.4.3) and (1.4.7).

■

EXAMPLE 1.4-6.(Surface coordinates) Let $z = f(x, y)$ denote a smooth surface where $\vec{r}(x, y) = x\,\hat{\mathbf{e}}_1 + y\,\hat{\mathbf{e}}_2 + f(x, y)\,\hat{\mathbf{e}}_3$ denotes a position vector to a point on the surface. We let $u^1 = x$ and $u^2 = y$ denote surface coordinates then

$$\frac{\partial \vec{r}}{\partial u^1} = \hat{\mathbf{e}}_1 + \frac{\partial f}{\partial u^1}\,\hat{\mathbf{e}}_3 \qquad \frac{\partial \vec{r}}{\partial u^2} = \hat{\mathbf{e}}_2 + \frac{\partial f}{\partial u^2}\,\hat{\mathbf{e}}_3$$

The metric associated with the 2-dimensional surface coordinates (u^1, u^2) is found from the arc length squared

$$ds^2 = d\vec{r} \cdot d\vec{r} = \left(\frac{\partial \vec{r}}{\partial u^1}du^1 + \frac{\partial \vec{r}}{\partial u^2}du^2 \right) \cdot \left(\frac{\partial \vec{r}}{\partial u^1}du^1 + \frac{\partial \vec{r}}{\partial u^2}du^2 \right) = g_{ij}du^i du^j$$

We find the surface metrics

$$g_{11} = 1 + \left(\frac{\partial f}{\partial u^1} \right)^2 \qquad g_{12} = \frac{\partial f}{\partial u^1}\frac{\partial f}{\partial u^2}$$

$$g_{21} = \frac{\partial f}{\partial u^2}\frac{\partial f}{\partial u^1} \qquad g_{22} = 1 + \left(\frac{\partial f}{\partial u^2} \right)^2$$

Using the notation

$$f_i = \frac{\partial f}{\partial u^i} \qquad f_{ij} = \frac{\partial^2 f}{\partial u^i \partial u^j} \qquad i, j = 1, 2$$

we write the above surface metrics in the form

$$g_{11} = 1 + f_1^2 \qquad g_{12} = f_1 f_2 \qquad g_{21} = f_2 f_1 \qquad g_{22} = 1 + f_2^2$$

then the conjugate metric tensor is found to be

$$g^{11} = \frac{1 + f_2^2}{1 + f_1^2 + f_2^2} \qquad g^{12} = \frac{-f_1 f_2}{1 + f_1^2 + f_2^2} \qquad g^{21} = \frac{-f_2 f_1}{1 + f_1^2 + f_2^2} \qquad g^{22} = \frac{1 + f_1^2}{1 + f_1^2 + f_2^2}.$$

Note that this 2-dimensional curvilinear coordinate system is not orthogonal (unless $g_{12} = g_{21} = 0$). For this nonorthogonal 2-dimensional curvilinear coordinate system we have from equation (1.4.8) that the Christoffel symbols of the first kind are given by

$$[11,1] = \frac{1}{2}\frac{\partial g_{11}}{\partial u^1} = f_1 f_{11} \qquad [11,2] = \frac{\partial g_{12}}{\partial u^1} - \frac{1}{2}\frac{\partial g_{11}}{\partial u^2} = f_2 f_{11}$$

$$[12,1] = \frac{1}{2}\frac{\partial g_{11}}{\partial u^2} = f_1 f_{12} \qquad [12,2] = \frac{1}{2}\frac{\partial g_{22}}{\partial u^1} = f_2 f_{21}$$

$$[22,1] = \frac{\partial g_{12}}{\partial u^2} - \frac{1}{2}\frac{\partial g_{22}}{\partial u^1} = f_1 f_{22} \qquad [22,2] = \frac{1}{2}\frac{\partial g_{22}}{\partial u^2} = f_2 f_{22}$$

We find from equation (1.4.12) the Christoffel symbols of the second kind are given by

$$\left\{ {i \atop j\,k} \right\} = g^{im}[jk,m] = g^{i1}[jk,1] + g^{i2}[jk,2].$$

Consequently,

$$\left\{ {1 \atop 1\,1} \right\} = g^{11}[11,1] + g^{12}[11,2] = \frac{f_1 f_{11}}{1 + f_1^2 + f_2^2} \qquad \left\{ {2 \atop 1\,1} \right\} = g^{21}[11,1] + g^{22}[11,2] = \frac{f_2 f_{11}}{1 + f_1^2 + f_2^2}$$

$$\left\{ {1 \atop 1\,2} \right\} = g^{11}[12,1] + g^{12}[12,2] = \frac{f_1 f_{12}}{1 + f_1^2 + f_2^2} \qquad \left\{ {2 \atop 1\,2} \right\} = g^{21}[12,1] + g^{22}[12,2] = \frac{f_2 f_{21}}{1 + f_1^2 + f_2^2}$$

$$\left\{ {1 \atop 2\,2} \right\} = g^{11}[22,1] + g^{12}[22,2] = \frac{f_1 f_{22}}{1 + f_1^2 + f_2^2} \qquad \left\{ {2 \atop 2\,2} \right\} = g^{21}[22,1] + g^{22}[22,2] = \frac{f_2 f_{22}}{1 + f_1^2 + f_2^2}$$

∎

Covariant Differentiation

Let A_i denote a covariant tensor of rank 1 which obeys the transformation law

$$\overline{A}_\alpha = A_i \frac{\partial x^i}{\partial \overline{x}^\alpha}. \tag{1.4.17}$$

Differentiate this relation with respect to $\overline{x}^\beta$ and show

$$\frac{\partial \overline{A}_\alpha}{\partial \overline{x}^\beta} = A_i \frac{\partial^2 x^i}{\partial \overline{x}^\alpha \partial \overline{x}^\beta} + \frac{\partial A_i}{\partial x^j}\frac{\partial x^j}{\partial \overline{x}^\beta}\frac{\partial x^i}{\partial \overline{x}^\alpha}. \tag{1.4.18}$$

Now use the relation from equation (1.4.7) to eliminate the second derivative term from (1.4.18) and express it in the form

$$\frac{\partial \overline{A}_\alpha}{\partial \overline{x}^\beta} = A_i \left[\left\{ {\sigma \atop \alpha\,\beta} \right\} \frac{\partial x^i}{\partial \overline{x}^\sigma} - \left\{ {i \atop j\,k} \right\} \frac{\partial x^j}{\partial \overline{x}^\alpha}\frac{\partial x^k}{\partial \overline{x}^\beta} \right] + \frac{\partial A_i}{\partial x^j}\frac{\partial x^j}{\partial \overline{x}^\beta}\frac{\partial x^i}{\partial \overline{x}^\alpha}. \tag{1.4.19}$$

Employing the equation (1.4.17), with α replaced by σ, the equation (1.4.19) is expressible in the form

$$\frac{\partial \overline{A}_\alpha}{\partial \overline{x}^\beta} - \overline{A}_\sigma \left\{ {\sigma \atop \alpha\,\beta} \right\} = \frac{\partial A_j}{\partial x^k}\frac{\partial x^j}{\partial \overline{x}^\alpha}\frac{\partial x^k}{\partial \overline{x}^\beta} - A_i \left\{ {i \atop j\,k} \right\} \frac{\partial x^j}{\partial \overline{x}^\alpha}\frac{\partial x^k}{\partial \overline{x}^\beta} \tag{1.4.20}$$

or alternatively

$$\left[\frac{\partial \overline{A}_\alpha}{\partial \overline{x}^\beta} - \overline{A}_\sigma \left\{ {\sigma \atop \alpha\,\beta} \right\} \right] = \left[\frac{\partial A_j}{\partial x^k} - A_i \left\{ {i \atop j\,k} \right\} \right] \frac{\partial x^j}{\partial \overline{x}^\alpha}\frac{\partial x^k}{\partial \overline{x}^\beta}. \tag{1.4.21}$$

Define the quantity

$$A_{j,k} = \frac{\partial A_j}{\partial x^k} - A_i \begin{Bmatrix} i \\ j\,k \end{Bmatrix} \tag{1.4.22}$$

as the covariant derivative of A_j with respect to x^k. The equation (1.4.21) demonstrates that the covariant derivative of a covariant tensor produces a second order tensor which satisfies the transformation law

$$\overline{A}_{\alpha,\beta} = A_{j,k} \frac{\partial x^j}{\partial \overline{x}^\alpha} \frac{\partial x^k}{\partial \overline{x}^\beta}. \tag{1.4.23}$$

Other notations frequently used to denote the covariant derivative are:

$$A_{j,k} = A_{j;k} = A_{j/k} = \nabla_k A_j = A_j|_k. \tag{1.4.24}$$

In the special case where g_{ij} are constants the Christoffel symbols of the second kind are zero, and so the covariant derivative reduces to $A_{j,k} = \dfrac{\partial A_j}{\partial x^k}$. That is, under the special circumstances where the Christoffel symbols of the second kind are zero, the covariant derivative reduces to an ordinary derivative.

Covariant Derivative of Contravariant Tensor

A contravariant tensor A^i obeys the transformation law $\overline{A}^i = A^\alpha \dfrac{\partial \overline{x}^i}{\partial x^\alpha}$ which can be expressed in the form

$$A^i = \overline{A}^\alpha \frac{\partial x^i}{\partial \overline{x}^\alpha} \tag{1.4.24}$$

by interchanging the barred and unbarred quantities. We write the transformation law in the form of equation (1.4.24) in order to make use of the second derivative relation from the previously derived equation (1.4.7). Differentiate equation (1.4.24) with respect to x^j to obtain the relation

$$\frac{\partial A^i}{\partial x^j} = \overline{A}^\alpha \frac{\partial^2 x^i}{\partial \overline{x}^\alpha \partial \overline{x}^\beta} \frac{\partial \overline{x}^\beta}{\partial x^j} + \frac{\partial \overline{A}^\alpha}{\partial \overline{x}^\beta} \frac{\partial \overline{x}^\beta}{\partial x^j} \frac{\partial x^i}{\partial \overline{x}^\alpha}. \tag{1.4.25}$$

Changing the indices in equation (1.4.25) and substituting for the second derivative term, using the relation from equation (1.4.7), produces the equation

$$\frac{\partial A^i}{\partial x^j} = \overline{A}^\alpha \left[\begin{Bmatrix} \sigma \\ \alpha\,\beta \end{Bmatrix} \frac{\partial x^i}{\partial \overline{x}^\sigma} - \begin{Bmatrix} i \\ m\,k \end{Bmatrix} \frac{\partial x^m}{\partial \overline{x}^\alpha} \frac{\partial x^k}{\partial \overline{x}^\beta} \right] \frac{\partial \overline{x}^\beta}{\partial x^j} + \frac{\partial \overline{A}^\alpha}{\partial \overline{x}^\beta} \frac{\partial \overline{x}^\beta}{\partial x^j} \frac{\partial x^i}{\partial \overline{x}^\alpha}. \tag{1.4.26}$$

Applying the relation found in equation (1.4.24), with i replaced by m, together with the relation

$$\frac{\partial \overline{x}^\beta}{\partial x^j} \frac{\partial x^k}{\partial \overline{x}^\beta} = \delta_j^k,$$

we simplify equation (1.4.26) to the form

$$\left[\frac{\partial A^i}{\partial x^j} + \begin{Bmatrix} i \\ m\,j \end{Bmatrix} A^m \right] = \left[\frac{\partial \overline{A}^\sigma}{\partial \overline{x}^\beta} + \overline{\begin{Bmatrix} \sigma \\ \alpha\,\beta \end{Bmatrix}} \overline{A}^\alpha \right] \frac{\partial \overline{x}^\beta}{\partial x^j} \frac{\partial x^i}{\partial \overline{x}^\sigma}. \tag{1.4.27}$$

Define the quantity

$$A^i{}_{,j} = \frac{\partial A^i}{\partial x^j} + \begin{Bmatrix} i \\ m\,j \end{Bmatrix} A^m \tag{1.4.28}$$

as the covariant derivative of the contravariant tensor A^i. The equation (1.4.27) demonstrates that a covariant derivative of a contravariant tensor will transform like a mixed second order tensor and

$$A^i_{\,,j} = \overline{A}^\sigma_{\,,\beta} \frac{\partial \overline{x}^\beta}{\partial x^j} \frac{\partial x^i}{\partial \overline{x}^\sigma}. \tag{1.4.29}$$

Again it should be observed that for the condition where g_{ij} are constants we have $A^i_{\,,j} = \dfrac{\partial A^i}{\partial x^j}$ and the covariant derivative of a contravariant tensor reduces to an ordinary derivative in this special case.

In a similar manner the covariant derivative of second rank tensors can be derived. We find these derivatives have the forms:

$$A_{ij,k} = \frac{\partial A_{ij}}{\partial x^k} - A_{\sigma j}\begin{Bmatrix} \sigma \\ i\,k \end{Bmatrix} - A_{i\sigma}\begin{Bmatrix} \sigma \\ j\,k \end{Bmatrix}$$

$$A^i_{j\,,k} = \frac{\partial A^i_j}{\partial x^k} + A^\sigma_j\begin{Bmatrix} i \\ \sigma\,k \end{Bmatrix} - A^i_\sigma\begin{Bmatrix} \sigma \\ j\,k \end{Bmatrix} \tag{1.4.30}$$

$$A^{ij}_{\,\,,k} = \frac{\partial A^{ij}}{\partial x^k} + A^{\sigma j}\begin{Bmatrix} i \\ \sigma\,k \end{Bmatrix} + A^{i\sigma}\begin{Bmatrix} j \\ \sigma\,k \end{Bmatrix}.$$

In general, the covariant derivative of a mixed tensor $A^{ij\ldots k}_{lm\ldots p}$ of rank n has the form

$$A^{ij\ldots k}_{lm\ldots p,q} = \frac{\partial A^{ij\ldots k}_{lm\ldots p}}{\partial x^q} + A^{\sigma j\ldots k}_{lm\ldots p}\begin{Bmatrix} i \\ \sigma\,q \end{Bmatrix} + A^{i\sigma\ldots k}_{lm\ldots p}\begin{Bmatrix} j \\ \sigma\,q \end{Bmatrix} + \cdots + A^{ij\ldots \sigma}_{lm\ldots p}\begin{Bmatrix} k \\ \sigma\,q \end{Bmatrix}$$
$$- A^{ij\ldots k}_{\sigma m\ldots p}\begin{Bmatrix} \sigma \\ l\,q \end{Bmatrix} - A^{ij\ldots k}_{l\sigma\ldots p}\begin{Bmatrix} \sigma \\ m\,q \end{Bmatrix} - \cdots - A^{ij\ldots k}_{lm\ldots \sigma}\begin{Bmatrix} \sigma \\ p\,q \end{Bmatrix} \tag{1.4.31}$$

and this derivative is a tensor of rank $n + 1$. Note the pattern of the $+$ signs for the contravariant indices and the $-$ signs for the covariant indices.

Observe that the covariant derivative of an nth order tensor produces an $n+1$st order tensor, the indices of these higher order tensors can also be raised and lowered by multiplication by the metric or conjugate metric tensor. For example we can write

$$g^{im}A_{jk}|_m = A_{jk}|^i \quad \text{and} \quad g^{im}A^{jk}|_m = A^{jk}|^i$$

Rules for Covariant Differentiation

The rules for covariant differentiation are the same as for ordinary differentiation. That is:

(i) The covariant derivative of a sum is the sum of the covariant derivatives.

(ii) The covariant derivative of a product of tensors is the first times the covariant derivative of the second plus the second times the covariant derivative of the first.

(iii) Higher derivatives are defined as derivatives of derivatives. Be careful in calculating higher order derivatives as in general $A_{i,jk} \neq A_{i,kj}$.

EXAMPLE 1.4-7. (Covariant differentiation) Calculate the second covariant derivative $A_{i,jk}$.
Solution: The covariant derivative of A_i is

$$A_{i,j} = \frac{\partial A_i}{\partial x^j} - A_\sigma\begin{Bmatrix} \sigma \\ i\,j \end{Bmatrix}.$$

By definition, the second covariant derivative is the covariant derivative of a covariant derivative and hence

$$A_{i,jk} = (A_{i,j})_{,k} = \frac{\partial}{\partial x^k}\left[\frac{\partial A_i}{\partial x^j} - A_\sigma \begin{Bmatrix} \sigma \\ i\,j \end{Bmatrix}\right] - A_{m,j}\begin{Bmatrix} m \\ i\,k \end{Bmatrix} - A_{i,m}\begin{Bmatrix} m \\ j\,k \end{Bmatrix}.$$

Simplifying this expression one obtains

$$A_{i,jk} = \frac{\partial^2 A_i}{\partial x^j \partial x^k} - \frac{\partial A_\sigma}{\partial x^k}\begin{Bmatrix} \sigma \\ i\,j \end{Bmatrix} - A_\sigma \frac{\partial}{\partial x^k}\begin{Bmatrix} \sigma \\ i\,j \end{Bmatrix}$$
$$- \left[\frac{\partial A_m}{\partial x^j} - A_\sigma \begin{Bmatrix} \sigma \\ m\,j \end{Bmatrix}\right]\begin{Bmatrix} m \\ i\,k \end{Bmatrix} - \left[\frac{\partial A_i}{\partial x^m} - A_\sigma \begin{Bmatrix} \sigma \\ i\,m \end{Bmatrix}\right]\begin{Bmatrix} m \\ j\,k \end{Bmatrix}.$$

Rearranging terms, the second covariant derivative can be expressed in the form

$$A_{i,jk} = \frac{\partial^2 A_i}{\partial x^j \partial x^k} - \frac{\partial A_\sigma}{\partial x^k}\begin{Bmatrix} \sigma \\ i\,j \end{Bmatrix} - \frac{\partial A_m}{\partial x^j}\begin{Bmatrix} m \\ i\,k \end{Bmatrix} - \frac{\partial A_i}{\partial x^m}\begin{Bmatrix} m \\ j\,k \end{Bmatrix}$$
$$- A_\sigma \left[\frac{\partial}{\partial x^k}\begin{Bmatrix} \sigma \\ i\,j \end{Bmatrix} - \begin{Bmatrix} \sigma \\ i\,m \end{Bmatrix}\begin{Bmatrix} m \\ j\,k \end{Bmatrix} - \begin{Bmatrix} m \\ i\,k \end{Bmatrix}\begin{Bmatrix} \sigma \\ m\,j \end{Bmatrix}\right].$$

(1.4.32)

∎

Riemann Christoffel Tensor

Utilizing the equation (1.4.32), it is left as an exercise to show that

$$A_{i,jk} - A_{i,kj} = A_\sigma R^\sigma_{ijk}$$

where

$$R^\sigma_{ijk} = \frac{\partial}{\partial x^j}\begin{Bmatrix} \sigma \\ i\,k \end{Bmatrix} - \frac{\partial}{\partial x^k}\begin{Bmatrix} \sigma \\ i\,j \end{Bmatrix} + \begin{Bmatrix} m \\ i\,k \end{Bmatrix}\begin{Bmatrix} \sigma \\ m\,j \end{Bmatrix} - \begin{Bmatrix} m \\ i\,j \end{Bmatrix}\begin{Bmatrix} \sigma \\ m\,k \end{Bmatrix}$$

(1.4.33)

is called the Riemann Christoffel tensor. The covariant form of this tensor is

$$R_{hjkl} = g_{ih}R^i_{jkl}.$$

(1.4.34)

It is an easy exercise to show that this covariant form can be expressed in either of the forms

$$R_{injk} = \frac{\partial}{\partial x^j}[nk,i] - \frac{\partial}{\partial x^k}[nj,i] + [ik,s]\begin{Bmatrix} s \\ n\,j \end{Bmatrix} - [ij,s]\begin{Bmatrix} s \\ n\,k \end{Bmatrix}$$

or

$$R_{ijkl} = \frac{1}{2}\left(\frac{\partial^2 g_{il}}{\partial x^j \partial x^k} - \frac{\partial^2 g_{jl}}{\partial x^i \partial x^k} - \frac{\partial^2 g_{ik}}{\partial x^j \partial x^l} + \frac{\partial^2 g_{jk}}{\partial x^i \partial x^l}\right) + g^{\alpha\beta}\left([jk,\beta][il,\alpha] - [jl,\beta][ik,\alpha]\right).$$

From these forms we find that the Riemann Christoffel tensor is skew symmetric in the first two indices and the last two indices as well as being symmetric in the interchange of the first pair and last pairs of indices and consequently

$$R_{jikl} = -R_{ijkl} \qquad R_{ijlk} = -R_{ijkl} \qquad R_{klij} = R_{ijkl}.$$

In a two dimensional space there are only four components of the Riemann Christoffel tensor to consider. These four components are either $+R_{1212}$ or $-R_{1212}$ since they are all related by

$$R_{1212} = -R_{2112} = R_{2121} = -R_{1221}.$$

In a Cartesian coordinate system $R_{hijk} = 0$. The Riemann Christoffel tensor is important because it occurs in differential geometry and relativity which are two areas of interest to be considered later. Additional properties of this tensor are found in the exercises of section 1.5.

Physical Interpretation of Covariant Differentiation

In a system of generalized coordinates (x^1, x^2, x^3) we can construct the basis vectors $(\vec{E}_1, \vec{E}_2, \vec{E}_3)$. These basis vectors change with position. That is, each basis vector is a function of the coordinates at which they are evaluated. We can emphasize this dependence by writing

$$\vec{E}_i = \vec{E}_i(x^1, x^2, x^3) = \frac{\partial \vec{r}}{\partial x^i} \qquad i = 1, 2, 3.$$

Associated with these basis vectors we have the reciprocal basis vectors

$$\vec{E}^i = \vec{E}^i(x^1, x^2, x^3), \qquad i = 1, 2, 3$$

which are also functions of position. A vector $\vec{A}$ can be represented in terms of contravariant components as

$$\vec{A} = A^1 \vec{E}_1 + A^2 \vec{E}_2 + A^3 \vec{E}_3 = A^j \vec{E}_j \tag{1.4.35}$$

or it can be represented in terms of covariant components as

$$\vec{A} = A_1 \vec{E}^1 + A_2 \vec{E}^2 + A_3 \vec{E}^3 = A_j \vec{E}^j. \tag{1.4.36}$$

A change in the vector $\vec{A}$ is represented as $d\vec{A} = \frac{\partial \vec{A}}{\partial x^k} dx^k$ where from equation (1.4.35) we find

$$\frac{\partial \vec{A}}{\partial x^k} = A^j \frac{\partial \vec{E}_j}{\partial x^k} + \frac{\partial A^j}{\partial x^k} \vec{E}_j \tag{1.4.37}$$

or alternatively from equation (1.4.36) we may write

$$\frac{\partial \vec{A}}{\partial x^k} = A_j \frac{\partial \vec{E}^j}{\partial x^k} + \frac{\partial A_j}{\partial x^k} \vec{E}^j. \tag{1.4.38}$$

We define the covariant derivative of the covariant components as

$$A_{i,k} = \frac{\partial \vec{A}}{\partial x^k} \cdot \vec{E}_i = \frac{\partial A_i}{\partial x^k} + A_j \frac{\partial \vec{E}^j}{\partial x^k} \cdot \vec{E}_i. \tag{1.4.39}$$

The covariant derivative of the contravariant components are defined by the relation

$$A^i_{\ ,k} = \frac{\partial \vec{A}}{\partial x^k} \cdot \vec{E}^i = \frac{\partial A^i}{\partial x^k} + A^j \frac{\partial \vec{E}_j}{\partial x^k} \cdot \vec{E}^i. \tag{1.4.40}$$

Introduce the notation

$$\frac{\partial \vec{E}_j}{\partial x^k} = \left\{ \begin{matrix} m \\ j\,k \end{matrix} \right\} \vec{E}_m \qquad \text{and} \qquad \frac{\partial \vec{E}^j}{\partial x^k} = -\left\{ \begin{matrix} j \\ m\,k \end{matrix} \right\} \vec{E}^m. \tag{1.4.41}$$

We then have

$$\vec{E}^i \cdot \frac{\partial \vec{E}_j}{\partial x^k} = \left\{ \begin{matrix} m \\ j\,k \end{matrix} \right\} \vec{E}_m \cdot \vec{E}^i = \left\{ \begin{matrix} m \\ j\,k \end{matrix} \right\} \delta^i_m = \left\{ \begin{matrix} i \\ j\,k \end{matrix} \right\} \tag{1.4.42}$$

and

$$\vec{E}_i \cdot \frac{\partial \vec{E}^j}{\partial x^k} = - \left\{ \begin{matrix} j \\ m\,k \end{matrix} \right\} \vec{E}^m \cdot \vec{E}_i = - \left\{ \begin{matrix} j \\ m\,k \end{matrix} \right\} \delta_i^m = - \left\{ \begin{matrix} j \\ i\,k \end{matrix} \right\}. \tag{1.4.43}$$

Then equations (1.4.39) and (1.4.40) become

$$A_{i,k} = \frac{\partial A_i}{\partial x^k} - \left\{ \begin{matrix} j \\ i\,k \end{matrix} \right\} A_j \qquad A^i{}_{,k} = \frac{\partial A^i}{\partial x^k} + \left\{ \begin{matrix} i \\ j\,k \end{matrix} \right\} A^j$$

which is consistent with our earlier definitions from equations (1.4.22) and (1.4.28). Here the first term of the covariant derivative represents the rate of change of the tensor field as we move along a coordinate curve. The second term in the covariant derivative represents the change in the local basis vectors as we move along the coordinate curves. This is the physical interpretation associated with the Christoffel symbols of the second kind.

We make the observation that the derivatives of the basis vectors in equations (1.4.39) and (1.4.40) are related since

$$\vec{E}_i \cdot \vec{E}^j = \delta_i^j$$

and therefore

$$\frac{\partial}{\partial x^k}(\vec{E}_i \cdot \vec{E}^j) = \vec{E}_i \cdot \frac{\partial \vec{E}^j}{\partial x^k} + \frac{\partial \vec{E}_i}{\partial x^k} \cdot \vec{E}^j = 0$$

$$\text{or} \qquad \vec{E}_i \cdot \frac{\partial \vec{E}^j}{\partial x^k} = -\vec{E}^j \cdot \frac{\partial \vec{E}_i}{\partial x^k}$$

Hence we can express equation (1.4.39) in the form

$$A_{i,k} = \frac{\partial A_i}{\partial x^k} - A_j \vec{E}^j \cdot \frac{\partial \vec{E}_i}{\partial x^k}. \tag{1.4.44}$$

We write the first equation in (1.4.41) in the form

$$\frac{\partial \vec{E}_j}{\partial x^k} = \left\{ \begin{matrix} m \\ j\,k \end{matrix} \right\} g_{im} \vec{E}^i = [jk, i] \vec{E}^i \tag{1.4.45}$$

and therefore

$$\frac{\partial \vec{E}_j}{\partial x^k} \cdot \vec{E}^m = \left\{ \begin{matrix} i \\ j\,k \end{matrix} \right\} \vec{E}_i \cdot \vec{E}^m = \left\{ \begin{matrix} i \\ j\,k \end{matrix} \right\} \delta_i^m = \left\{ \begin{matrix} m \\ j\,k \end{matrix} \right\}$$

$$\text{and} \qquad \frac{\partial \vec{E}_j}{\partial x^k} \cdot \vec{E}_m = [jk, i] \vec{E}^i \cdot \vec{E}_m = [jk, i] \delta_m^i = [jk, m]. \tag{1.4.46}$$

These results also reduce the equations (1.4.40) and (1.4.44) to our previous forms for the covariant derivatives.

The equations (1.4.41) are representations of the vectors $\frac{\partial \vec{E}_i}{\partial x^k}$ and $\frac{\partial \vec{E}^j}{\partial x^k}$ in terms of the basis vectors and reciprocal basis vectors of the space. The covariant derivative relations then take into account how these vectors change with position and affect changes in the tensor field.

The Christoffel symbols in equations (1.4.46) are symmetric in the indices j and k since

$$\frac{\partial \vec{E}_j}{\partial x^k} = \frac{\partial}{\partial x^k}\left(\frac{\partial \vec{r}}{\partial x^j} \right) = \frac{\partial}{\partial x^j}\left(\frac{\partial \vec{r}}{\partial x^k} \right) = \frac{\partial \vec{E}_k}{\partial x^j}. \tag{1.4.47}$$

The equations (1.4.46) and (1.4.47) enable us to write

$$
\begin{aligned}
[jk,m] =& \vec{E}_m \cdot \frac{\partial \vec{E}_j}{\partial x^k} = \frac{1}{2}\left[\vec{E}_m \cdot \frac{\partial \vec{E}_j}{\partial x^k} + \vec{E}_m \cdot \frac{\partial \vec{E}_k}{\partial x^j}\right] \\
=& \frac{1}{2}\left[\frac{\partial}{\partial x^k}\left(\vec{E}_m \cdot \vec{E}_j\right) + \frac{\partial}{\partial x^j}\left(\vec{E}_m \cdot \vec{E}_k\right) - \vec{E}_j \cdot \frac{\partial \vec{E}_m}{\partial x^k} - \vec{E}_k \cdot \frac{\partial \vec{E}_m}{\partial x^j}\right] \\
=& \frac{1}{2}\left[\frac{\partial}{\partial x^k}\left(\vec{E}_m \cdot \vec{E}_j\right) + \frac{\partial}{\partial x^j}\left(\vec{E}_m \cdot \vec{E}_k\right) - \vec{E}_j \cdot \frac{\partial \vec{E}_k}{\partial x^m} - \vec{E}_k \cdot \frac{\partial \vec{E}_j}{\partial x^m}\right] \\
=& \frac{1}{2}\left[\frac{\partial}{\partial x^k}\left(\vec{E}_m \cdot \vec{E}_j\right) + \frac{\partial}{\partial x^j}\left(\vec{E}_m \cdot \vec{E}_k\right) - \frac{\partial}{\partial x^m}\left(\vec{E}_j \cdot \vec{E}_k\right)\right] \\
=& \frac{1}{2}\left[\frac{\partial g_{mj}}{\partial x^k} + \frac{\partial g_{mk}}{\partial x^j} - \frac{\partial g_{jk}}{\partial x^m}\right] = [kj,m]
\end{aligned}
$$

which again agrees with our previous result.

For future reference we make the observation that if the vector $\vec{A}$ is represented in the form $\vec{A} = A^j \vec{E}_j$, involving contravariant components, then we may write

$$
\begin{aligned}
d\vec{A} = \frac{\partial \vec{A}}{\partial x^k}\,dx^k =& \left(\frac{\partial A^j}{\partial x^k}\vec{E}_j + A^j \frac{\partial \vec{E}_j}{\partial x^k}\right)dx^k \\
=& \left(\frac{\partial A^j}{\partial x^k}\vec{E}_j + A^j \begin{Bmatrix} i \\ j\,k \end{Bmatrix}\vec{E}_i\right)dx^k \\
=& \left(\frac{\partial A^j}{\partial x^k} + \begin{Bmatrix} j \\ m\,k \end{Bmatrix}A^m\right)\vec{E}_j\,dx^k = A^j{}_{,k}\,dx^k\,\vec{E}_j.
\end{aligned}
\tag{1.4.48}
$$

Similarly, if the vector $\vec{A}$ is represented in the form $\vec{A} = A_j \vec{E}^j$ involving covariant components it is left as an exercise to show that

$$
d\vec{A} = A_{j,k}\,dx^k\,\vec{E}^j
\tag{1.4.49}
$$

Ricci's Theorem

Ricci's theorem states that the covariant derivative of the metric tensor vanishes and $g_{ik,l} = 0$.
Proof: We have

$$
\begin{aligned}
g_{ik,l} &= \frac{\partial g_{ik}}{\partial x^l} - \begin{Bmatrix} m \\ k\,l \end{Bmatrix}g_{im} - \begin{Bmatrix} m \\ i\,l \end{Bmatrix}g_{mk} \\
g_{ik,l} &= \frac{\partial g_{ik}}{\partial x^l} - [kl,i] - [il,k] \\
g_{ik,l} &= \frac{\partial g_{ik}}{\partial x^l} - \frac{1}{2}\left[\frac{\partial g_{ik}}{\partial x^l} + \frac{\partial g_{il}}{\partial x^k} - \frac{\partial g_{kl}}{\partial x^i}\right] - \frac{1}{2}\left[\frac{\partial g_{ik}}{\partial x^l} + \frac{\partial g_{kl}}{\partial x^i} - \frac{\partial g_{il}}{\partial x^k}\right] = 0.
\end{aligned}
$$

Because of Ricci's theorem the components of the metric tensor can be regarded as constants during covariant differentiation.

EXAMPLE 1.4-8. (Covariant differentiation) Show that $\delta^i_{j,k} = 0$.
Solution

$$\delta^i_{j,k} = \frac{\partial \delta^i_j}{\partial x^k} + \delta^\sigma_j \begin{Bmatrix} i \\ \sigma \, k \end{Bmatrix} - \delta^i_\sigma \begin{Bmatrix} \sigma \\ j \, k \end{Bmatrix} = \begin{Bmatrix} i \\ j \, k \end{Bmatrix} - \begin{Bmatrix} i \\ j \, k \end{Bmatrix} = 0.$$

∎

EXAMPLE 1.4-9. (Covariant differentiation) Show that $g^{ij}{}_{,k} = 0$.
Solution: Since $g_{ij}g^{jk} = \delta^k_i$ we take the covariant derivative of this expression and find

$$(g_{ij}g^{jk})_{,l} = \delta^k_{i,l} = 0$$

$$g_{ij}g^{jk}{}_{,l} + g_{ij,l}g^{jk} = 0.$$

But $g_{ij,l} = 0$ by Ricci's theorem and hence $g_{ij}g^{jk}{}_{,l} = 0$. We multiply this expression by g^{im} and obtain

$$g^{im}g_{ij}g^{jk}{}_{,l} = \delta^m_j g^{jk}{}_{,l} = g^{mk}{}_{,l} = 0$$

which demonstrates that the covariant derivative of the conjugate metric tensor is also zero.

∎

EXAMPLE 1.4-10. (Covariant differentiation) Some additional examples of covariant differentiation are:

$$(i) \quad (g_{il}A^l)_{,k} = g_{il}A^l{}_{,k} = A_{i,k} \qquad\qquad (ii) \quad (g_{im}g_{jn}A^{ij})_{,k} = g_{im}g_{jn}A^{ij}{}_{,k} = A_{mn,k}$$

∎

Intrinsic or Absolute Differentiation

The intrinsic or absolute derivative of a covariant vector A_i taken along a curve $x^i = x^i(t), i = 1, \ldots, N$ is defined as the inner product of the covariant derivative with the tangent vector to the curve. The intrinsic derivative is represented

$$\frac{\delta A_i}{\delta t} = A_{i,j}\frac{dx^j}{dt} = \left[\frac{\partial A_i}{\partial x^j} - A_\alpha \begin{Bmatrix} \alpha \\ i \, j \end{Bmatrix} \right] \frac{dx^j}{dt}$$

$$\frac{\delta A_i}{\delta t} = \frac{dA_i}{dt} - A_\alpha \begin{Bmatrix} \alpha \\ i \, j \end{Bmatrix} \frac{dx^j}{dt}.$$

(1.4.50)

Similarly, the absolute or intrinsic derivative of a contravariant tensor A^i is represented

$$\frac{\delta A^i}{\delta t} = A^i{}_{,j}\frac{dx^j}{dt} = \frac{dA^i}{dt} + \begin{Bmatrix} i \\ j \, k \end{Bmatrix} A^k \frac{dx^j}{dt}.$$

The intrinsic or absolute derivative is used to differentiate sums and products in the same manner as used in ordinary differentiation. Also if the coordinate system is Cartesian the intrinsic derivative becomes an ordinary derivative.

The intrinsic derivative of higher order tensors is similarly defined as an inner product of the covariant derivative with the tangent vector to the given curve. For example,

$$\frac{\delta A^{ij}_{klm}}{\delta t} = A^{ij}_{klm,p}\frac{dx^p}{dt}$$

is the intrinsic derivative of the fifth order mixed tensor A^{ij}_{klm}.

EXAMPLE 1.4-11. (Generalized velocity and acceleration) Let t denote time and let $x^i = x^i(t)$ for $i = 1, \ldots, N$, denote the position vector of a particle in the generalized coordinates $(x^1, \ldots, x^N)$. From the transformation equations (1.2.30), the position vector of the same particle in the barred system of coordinates, $(\overline{x}^1, \overline{x}^2, \ldots, \overline{x}^N)$, is

$$\overline{x}^i = \overline{x}^i(x^1(t), x^2(t), \ldots, x^N(t)) = \overline{x}^i(t), \quad i = 1, \ldots, N.$$

The generalized velocity is $v^i = \frac{dx^i}{dt}$, $i = 1, \ldots, N$. The quantity v^i transforms as a tensor since by definition

$$\overline{v}^i = \frac{d\overline{x}^i}{dt} = \frac{\partial \overline{x}^i}{\partial x^j}\frac{dx^j}{dt} = \frac{\partial \overline{x}^i}{\partial x^j}v^j. \tag{1.4.51}$$

Let us now find an expression for the generalized acceleration. Write equation (1.4.51) in the form

$$v^j = \overline{v}^i\frac{\partial x^j}{\partial \overline{x}^i} \tag{1.4.52}$$

and differentiate with respect to time to obtain

$$\frac{dv^j}{dt} = \overline{v}^i\frac{\partial^2 x^j}{\partial \overline{x}^i \partial \overline{x}^k}\frac{d\overline{x}^k}{dt} + \frac{d\overline{v}^i}{dt}\frac{\partial x^j}{\partial \overline{x}^i} \tag{1.4.53}$$

The equation (1.4.53) demonstrates that $\frac{dv^i}{dt}$ does not transform like a tensor. From the equation (1.4.7) previously derived, we change indices and write equation (1.4.53) in the form

$$\frac{dv^j}{dt} = \overline{v}^i\frac{d\overline{x}^k}{dt}\left[\begin{Bmatrix}\sigma\\i\,k\end{Bmatrix}\frac{\partial x^j}{\partial \overline{x}^\sigma} - \begin{Bmatrix}j\\a\,c\end{Bmatrix}\frac{\partial x^a}{\partial \overline{x}^i}\frac{\partial x^c}{\partial \overline{x}^k}\right] + \frac{\partial x^j}{\partial \overline{x}^i}\frac{d\overline{v}^i}{dt}.$$

Rearranging terms we find

$$\frac{\partial v^j}{\partial x^k}\frac{dx^k}{dt} + \begin{Bmatrix}j\\a\,c\end{Bmatrix}\left(\frac{\partial x^a}{\partial \overline{x}^i}\overline{v}^i\right)\left(\frac{\partial x^c}{\partial \overline{x}^k}\frac{d\overline{x}^k}{dt}\right) = \frac{\partial x^j}{\partial \overline{x}^i}\frac{\partial \overline{v}^i}{\partial \overline{x}^k}\frac{d\overline{x}^k}{dt} + \overline{\begin{Bmatrix}\sigma\\i\,k\end{Bmatrix}}\overline{v}^i\frac{\partial x^j}{\partial \overline{x}^\sigma}\frac{d\overline{x}^k}{dt} \qquad \text{or}$$

$$\left[\frac{\partial v^j}{\partial x^k} + \begin{Bmatrix}j\\a\,k\end{Bmatrix}v^a\right]\frac{dx^k}{dt} = \left[\frac{\partial \overline{v}^\sigma}{\partial \overline{x}^k} + \overline{\begin{Bmatrix}\sigma\\i\,k\end{Bmatrix}}\overline{v}^i\right]\frac{d\overline{x}^k}{dt}\frac{\partial x^j}{\partial \overline{x}^\sigma}$$

$$\frac{\delta v^j}{\delta t} = \frac{\delta \overline{v}^\sigma}{\delta t}\frac{\partial x^j}{\partial \overline{x}^\sigma}.$$

The above equation illustrates that the intrinsic derivative of the velocity is a tensor quantity. This derivative is called the generalized acceleration and is denoted

$$f^i = \frac{\delta v^i}{\delta t} = v^i_{,j}\frac{dx^j}{dt} = \frac{dv^i}{dt} + \begin{Bmatrix}i\\m\,n\end{Bmatrix}v^m v^n = \frac{d^2 x^i}{dt^2} + \begin{Bmatrix}i\\m\,n\end{Bmatrix}\frac{dx^m}{dt}\frac{dx^n}{dt}, \quad i = 1, \ldots, N \tag{1.4.54}$$

To summarize, we have shown that if

$$x^i = x^i(t), \quad i = 1, \ldots, N \quad \text{is the generalized position vector, then}$$

$$v^i = \frac{dx^i}{dt}, \quad i = 1, \ldots, N \quad \text{is the generalized velocity, and}$$

$$f^i = \frac{\delta v^i}{\delta t} = v^i_{,j}\frac{dx^j}{dt}, \quad i = 1, \ldots, N \quad \text{is the generalized acceleration.}$$

Parallel Vector Fields

Let $y^i = y^i(t)$, $i = 1, 2, 3$ denote a space curve C in a Cartesian coordinate system and let Y^i define a constant vector in this system. Construct at each point of the curve C the vector Y^i. This produces a field of parallel vectors along the curve C. What happens to the curve and the field of parallel vectors when we transform to an arbitrary coordinate system using the transformation equations

$$y^i = y^i(x^1, x^2, x^3), \quad i = 1, 2, 3$$

with inverse transformation

$$x^i = x^i(y^1, y^2, y^3), \quad i = 1, 2, 3?$$

The space curve C in the new coordinates is obtained directly from the transformation equations and can be written

$$x^i = x^i(y^1(t), y^2(t), y^3(t)) = x^i(t), \quad i = 1, 2, 3.$$

The field of parallel vectors Y^i become X^i in the new coordinates where

$$Y^i = X^j \frac{\partial y^i}{\partial x^j}. \tag{1.4.55}$$

Since the components of Y^i are constants, their derivatives will be zero and so we obtain by differentiating the equation (1.4.55), with respect to the parameter t, that the field of parallel vectors X^i must satisfy the differential equation

$$\frac{dX^j}{dt} \frac{\partial y^i}{\partial x^j} + X^j \frac{\partial^2 y^i}{\partial x^j \partial x^m} \frac{dx^m}{dt} = \frac{dY^i}{dt} = 0. \tag{1.4.56}$$

Changing symbols in the equation (1.4.7) and setting the Christoffel symbol to zero in the Cartesian system of coordinates, we represent equation (1.4.7) in the form

$$\frac{\partial^2 y^i}{\partial x^j \partial x^m} = \left\{ \begin{matrix} \alpha \\ j\, m \end{matrix} \right\} \frac{\partial y^i}{\partial x^\alpha}$$

and therefore, the equation (1.4.56) can be reduced to the form

$$\frac{\delta X^j}{\delta t} = \frac{dX^j}{dt} + \left\{ \begin{matrix} j \\ k\, m \end{matrix} \right\} X^k \frac{dx^m}{dt} = 0. \tag{1.4.57}$$

The equation (1.4.57) is the differential equation which must be satisfied by a parallel field of vectors X^i along an arbitrary curve $x^i(t)$.

EXERCISE 1.4

▶ **1.** Find the nonzero Christoffel symbols of the first and second kind in cylindrical coordinates $(x^1, x^2, x^3) = (r, \theta, z)$, where $x = r\cos\theta, \qquad y = r\sin\theta, \qquad z = z.$

▶ **2.** Find the nonzero Christoffel symbols of the first and second kind in spherical coordinates $(x^1, x^2, x^3) = (\rho, \theta, \phi)$, where $x = \rho\sin\theta\cos\phi, \quad y = \rho\sin\theta\sin\phi, \quad z = \rho\cos\theta.$

▶ **3.** Find the nonzero Christoffel symbols of the first and second kind in parabolic cylindrical coordinates $(x^1, x^2, x^3) = (\xi, \eta, z)$, where $x = \xi\eta, \quad y = \dfrac{1}{2}(\xi^2 - \eta^2), \quad z = z.$

▶ **4.** Find the nonzero Christoffel symbols of the first and second kind in parabolic coordinates $(x^1, x^2, x^3) = (\xi, \eta, \phi)$, where $x = \xi\eta\cos\phi, \quad y = \xi\eta\sin\phi, \quad z = \dfrac{1}{2}(\xi^2 - \eta^2).$

▶ **5.** Find the nonzero Christoffel symbols of the first and second kind in elliptic cylindrical coordinates $(x^1, x^2, x^3) = (\xi, \eta, z)$, where $x = \cosh\xi\cos\eta, \quad y = \sinh\xi\sin\eta, \quad z = z.$

▶ **6.** Find the nonzero Christoffel symbols of the first and second kind for the oblique cylindrical coordinates $(x^1, x^2, x^3) = (r, \phi, \eta)$, where $x = r\cos\phi, \quad y = r\sin\phi + \eta\cos\alpha, \quad z = \eta\sin\alpha$ with $0 < \alpha < \frac{\pi}{2}$ and α constant.
Hint: See figure 1.3-18 and exercise 1.3, problem 6.

▶ **7.** Show $[ij, k] + [kj, i] = \dfrac{\partial g_{ik}}{\partial x^j}$.

▶ **8.**

(a) Let $\begin{Bmatrix} r \\ st \end{Bmatrix} = g^{ri}[st, i]$ and solve for the Christoffel symbol of the first kind in terms of the Christoffel symbol of the second kind.

(b) Assume $[st, i] = g_{ni}\begin{Bmatrix} n \\ st \end{Bmatrix}$ and solve for the Christoffel symbol of the second kind in terms of the Christoffel symbol of the first kind.

▶ **9.**

(a) Write down the transformation law satisfied by the fourth order tensor $\epsilon_{ijk,m}$.

(b) Show that $\epsilon_{ijk,m} = 0$ in all coordinate systems.

(c) Show that $(\sqrt{g})_{,k} = 0$.

▶ **10.** Show $\epsilon^{ijk}_{,m} = 0$.

▶ **11.** Calculate the second covariant derivative $A_{i,kj}$.

▶ **12.** The gradient of a scalar field $\phi(x^1, x^2, x^3)$ is the vector $\operatorname{grad}\phi = \vec{E}^i \dfrac{\partial\phi}{\partial x^i}$.

(a) Find the physical components associated with the covariant components $\phi_{,i}$

(b) Show the directional derivative of ϕ in a direction A^i is $\dfrac{d\phi}{dA} = \dfrac{A^i\phi_{,i}}{(g_{mn}A^mA^n)^{1/2}}$.

► **13.**

  (a) Show $\sqrt{g}$ is a relative scalar of weight $+1$.

  (b) Use the results from problem 9(c) and problem 44, Exercise 1.4, to show that
$$(\sqrt{g})_{,k} = \frac{\partial \sqrt{g}}{\partial x^k} - \left\{ \begin{matrix} m \\ k\,m \end{matrix} \right\} \sqrt{g} = 0.$$

  (c) Show that $\left\{ \begin{matrix} m \\ k\,m \end{matrix} \right\} = \dfrac{\partial}{\partial x^k} \ln(\sqrt{g}) = \dfrac{1}{2g}\dfrac{\partial g}{\partial x^k}.$

► **14.**   Use the result from problem 9(b) to show $\left\{ \begin{matrix} m \\ k\,m \end{matrix} \right\} = \dfrac{\partial}{\partial x^k} \ln(\sqrt{g}) = \dfrac{1}{2g}\dfrac{\partial g}{\partial x^k}.$

     Hint: Expand the covariant derivative $\epsilon_{rst,p}$ and then substitute $\epsilon_{rst} = \sqrt{g}e_{rst}$. Simplify by inner multiplication with $\dfrac{e^{rst}}{\sqrt{g}}$ and note the Exercise 1.1, problem 26.

► **15.**   Calculate the covariant derivative $A^i{}_{,m}$ and then contract on m and i to show that

$$A^i{}_{,i} = \frac{1}{\sqrt{g}}\frac{\partial}{\partial x^i}\left(\sqrt{g}A^i \right).$$

► **16.**   Show $\dfrac{1}{\sqrt{g}}\dfrac{\partial}{\partial x^j}\left(\sqrt{g}g^{ij} \right) + \left\{ \begin{matrix} i \\ p\,q \end{matrix} \right\}g^{pq} = 0.$ Hint: See problem 14.

► **17.**   Prove that the covariant derivative of a sum equals the sum of the covariant derivatives.

     Hint: Assume $C_i = A_i + B_i$ and write out the covariant derivative for $C_{i,j}$.

► **18.**   Let $C^i_j = A^i B_j$ and prove that the covariant derivative of a product equals the first term times the covariant derivative of the second term plus the second term times the covariant derivative of the first term.

► **19.**   Start with the transformation law $\bar{A}_{ij} = A_{\alpha\beta}\dfrac{\partial x^\alpha}{\partial \bar{x}^i}\dfrac{\partial x^\beta}{\partial \bar{x}^j}$ and take an ordinary derivative of both sides with respect to $\bar{x}^k$ and hence derive the relation for $A_{ij,k}$ given in (1.4.30).

► **20.**   Start with the transformation law $A^{ij} = \bar{A}^{\alpha\,\beta}\dfrac{\partial x^i}{\partial \bar{x}^\alpha}\dfrac{\partial x^j}{\partial \bar{x}^\beta}$ and take an ordinary derivative of both sides with respect to x^k and hence derive the relation for $A^{ij}{}_{,k}$ given in (1.4.30).

► **21.**   Find the covariant derivatives of

$$(a) \quad A^{ijk} \qquad (b) \quad A^{ij}{}_k \qquad (c) \quad A^i{}_{jk} \qquad (d) \quad A_{ijk}$$

► **22.**   Find the intrinsic derivative along the curve $x^i = x^i(t), \quad i = 1,\ldots,N$ for

$$(a) \quad A^{ijk} \qquad (b) \quad A^{ij}{}_k \qquad (c) \quad A^i{}_{jk} \qquad (d) \quad A_{ijk}$$

► **23.**

  (a) Assume $\vec{A} = A^i \vec{E}_i$ and show that $d\vec{A} = A^i{}_{,k}\,dx^k\,\vec{E}_i$.

  (b) Assume $\vec{A} = A_i \vec{E}^i$ and show that $d\vec{A} = A_{i,k}\,dx^k\,\vec{E}^i$.

▶ **24.** (parallel vector field) Imagine a vector field $A^i = A^i(x^1, x^2, x^3)$ which is a function of position. Assume that at all points along a curve $x^i = x^i(t), i = 1, 2, 3$ the vector field points in the same direction, we would then have a parallel vector field or homogeneous vector field. Assume $\vec{A}$ is a constant, then $d\vec{A} = \frac{\partial \vec{A}}{\partial x^k} dx^k = 0$. Show that for a parallel vector field the condition $A_{i,k} = 0$ must be satisfied.

▶ **25.** Show that $\dfrac{\partial [ik, n]}{\partial x^j} = g_{n\sigma} \dfrac{\partial}{\partial x^j} \left\{ \begin{matrix} \sigma \\ i\, k \end{matrix} \right\} + ([nj, \sigma] + [\sigma j, n]) \left\{ \begin{matrix} \sigma \\ i\, k \end{matrix} \right\}$.

▶ **26.** Show $A_{r,s} - A_{s,r} = \dfrac{\partial A_r}{\partial x^s} - \dfrac{\partial A_s}{\partial x^r}$.

▶ **27.** In cylindrical coordinates you are given the contravariant vector components

$$A^1 = r \qquad A^2 = \cos\theta \qquad A^3 = z\sin\theta$$

(a) Find the physical components A_r, A_θ, and A_z.

(b) Denote the physical components of $A^i_{\ ,j}$, $i, j = 1, 2, 3$, by
$$\begin{matrix} A_{rr} & A_{r\theta} & A_{rz} \\ A_{\theta r} & A_{\theta\theta} & A_{\theta z} \\ A_{zr} & A_{z\theta} & A_{zz}. \end{matrix}$$
Find these physical components.

▶ **28.** Find the covariant form of the contravariant tensor $C^i = \epsilon^{ijk} A_{k,j}$. Express your answer in terms of $A^k_{\ ,j}$.

▶ **29.** In Cartesian coordinates let x denote the magnitude of the position vector x_i. Show that (a) $x_{,j} = \dfrac{1}{x} x_j$
(b) $x_{,ij} = \dfrac{1}{x}\delta_{ij} - \dfrac{1}{x^3} x_i x_j$ (c) $x_{,ii} = \dfrac{2}{x}$. (d) Let $U = \dfrac{1}{x}$, $x \neq 0$, and show that $U_{,ij} = \dfrac{-\delta_{ij}}{x^3} + \dfrac{3x_i x_j}{x^5}$ and $U_{,ii} = 0$.

▶ **30.** Consider a two dimensional space with element of arc length squared

$$ds^2 = g_{11}(du^1)^2 + g_{22}(du^2)^2 \quad \text{and metric} \quad g_{ij} = \begin{pmatrix} g_{11} & 0 \\ 0 & g_{22} \end{pmatrix}$$

where u^1, u^2 are surface coordinates.
(a) Find formulas to calculate the Christoffel symbols of the first kind.
(b) Find formulas to calculate the Christoffel symbols of the second kind.

▶ **31.** Find the metric tensor and Christoffel symbols of the first and second kind associated with the two dimensional space describing points on a cylinder of radius a. Let $u^1 = \theta$ and $u^2 = z$ denote surface coordinates where

$$x = a\cos\theta = a\cos u^1$$
$$y = a\sin\theta = a\sin u^1$$
$$z = z = u^2$$

▶ **32.** Find the metric tensor and Christoffel symbols of the first and second kind associated with the two dimensional space describing points on a sphere of radius a. Let $u^1 = \theta$ and $u^2 = \phi$ denote surface coordinates where

$$x = a\sin\theta\cos\phi = a\sin u^1\cos u^2$$
$$y = a\sin\theta\sin\phi = a\sin u^1\sin u^2$$
$$z = a\cos\theta = a\cos u^1$$

▶ **33.** Find the metric tensor and Christoffel symbols of the first and second kind associated with the two dimensional space describing points on a torus having the parameters a and b and surface coordinates $u^1 = \xi$, $u^2 = \eta$. illustrated in the figure 1.3-19. The points on the surface of the torus are given in terms of the surface coordinates by the equations

$$x = (a + b\cos\xi)\cos\eta$$
$$y = (a + b\cos\xi)\sin\eta$$
$$z = b\sin\xi$$

▶ **34.** Prove that $e_{ijk}a^m b^j c^k u^i_{,m} + e_{ijk}a^i b^m c^k u^j_{,m} + e_{ijk}a^i b^j c^m u^k_{,m} = u^r_{,r} e_{ijk}a^i b^j c^k$. Hint: See Exercise 1.3, problem 32 and Exercise 1.1, problem 21.

▶ **35.** Calculate the second covariant derivative $A^i_{,jk}$.

▶ **36.** Show that $\sigma^{ij}_{,j} = \dfrac{1}{\sqrt{g}}\dfrac{\partial}{\partial x^j}\left(\sqrt{g}\sigma^{ij}\right) + \sigma^{mn}\begin{Bmatrix} i \\ m\,n \end{Bmatrix}$

▶ **37.** Find the contravariant, covariant and physical components of velocity and acceleration in (a) Cartesian coordinates and (b) cylindrical coordinates.

▶ **38.** Find the contravariant, covariant and physical components of velocity and acceleration in spherical coordinates.

▶ **39.** In spherical coordinates (ρ, θ, ϕ) show that the acceleration components can be represented in terms of the velocity components as

$$f_\rho = \dot{v}_\rho - \frac{v_\theta^2 + v_\phi^2}{\rho}, \qquad f_\theta = \dot{v}_\theta + \frac{v_\rho v_\theta}{\rho} - \frac{v_\phi^2}{\rho\tan\theta}, \qquad f_\phi = \dot{v}_\phi + \frac{v_\rho v_\phi}{\rho} + \frac{v_\theta v_\phi}{\rho\tan\theta}$$

Hint: Calculate $\dot{v}_\rho, \dot{v}_\theta, \dot{v}_\phi$.

▶ **40.** The divergence of a vector A^i is $A^i_{,i}$. That is, perform a contraction on the covariant derivative $A^i_{,j}$ to obtain $A^i_{,i}$. Calculate the divergence in (a) Cartesian coordinates (b) cylindrical coordinates and (c) spherical coordinates.

▶ **41.** If S is a scalar invariant of weight one and A^i_{jk} is a third order relative tensor of weight W, show that $S^{-W}A^i_{jk}$ is an absolute tensor.

42. Let $\bar{Y}^i, i = 1, 2, 3$ denote the components of a field of parallel vectors along the curve $\overline{C}$ defined by the equations $\bar{y}^i = \bar{y}^i(t)$, $i = 1, 2, 3$ in a space with metric tensor $\bar{g}_{ij}$, $i, j = 1, 2, 3$. Assume that $\bar{Y}^i$ and $\frac{d\bar{y}^i}{dt}$ are unit vectors such that at each point of the curve $\overline{C}$ we have

$$\bar{g}_{ij}\bar{Y}^i\frac{d\bar{y}^j}{dt} = \cos\theta = \text{Constant}.$$

(i.e. The field of parallel vectors makes a constant angle θ with the tangent to each point of the curve $\overline{C}$.) Show that if $\bar{Y}^i$ and $\bar{y}^i(t)$ undergo a transformation $x^i = x^i(\bar{y}^1, \bar{y}^2, \bar{y}^3)$, $i = 1, 2, 3$ then the transformed vector $X^m = \bar{Y}^i\frac{\partial x^m}{\partial \bar{y}^i}$ makes a constant angle with the tangent vector to the transformed curve C given by $x^i = x^i(\bar{y}^1(t), \bar{y}^2(t), \bar{y}^3(t))$.

▶ **43.** Let J denote the Jacobian determinant $|\frac{\partial x^i}{\partial \bar{x}^j}|$. Differentiate J with respect to x^m and show that

$$\frac{\partial J}{\partial x^m} = J\overline{\left\{\begin{matrix} \alpha \\ \alpha\,p \end{matrix}\right\}}\frac{\partial \bar{x}^p}{\partial x^m} - J\left\{\begin{matrix} r \\ r\,m \end{matrix}\right\}.$$

Hint: See Exercise 1.1, problem 27 and (1.4.7).

▶ **44.** Assume that ϕ is a relative scalar of weight W so that $\bar{\phi} = J^W\phi$. Differentiate this relation with respect to $\bar{x}^k$. Use the result from problem 43 to obtain the transformation law:

$$\left[\frac{\partial \bar{\phi}}{\partial \bar{x}^k} - W\overline{\left\{\begin{matrix} \alpha \\ \alpha\,k \end{matrix}\right\}}\bar{\phi}\right] - J^W\left[\frac{\partial \phi}{\partial x^m} - W\left\{\begin{matrix} r \\ m\,r \end{matrix}\right\}\phi\right]\frac{\partial x^m}{\partial \bar{x}^k}.$$

The quantity inside the brackets is called the covariant derivative of a relative scalar of weight W. The covariant derivative of a relative scalar of weight W is defined as

$$\phi_{,k} = \frac{\partial \phi}{\partial x^k} - W\left\{\begin{matrix} r \\ k\,r \end{matrix}\right\}\phi$$

and this definition has an extra term involving the weight.

It can be shown that similar results hold for relative tensors of weight W. For example, the covariant derivative of first and second order relative tensors of weight W have the forms

$$T^i_{,k} = \frac{\partial T^i}{\partial x^k} + \left\{\begin{matrix} i \\ k\,m \end{matrix}\right\}T^m - W\left\{\begin{matrix} r \\ k\,r \end{matrix}\right\}T^i$$

$$T^i_{j,k} = \frac{\partial T^i_j}{\partial x^k} + \left\{\begin{matrix} i \\ k\,\sigma \end{matrix}\right\}T^\sigma_j - \left\{\begin{matrix} \sigma \\ j\,k \end{matrix}\right\}T^i_\sigma - W\left\{\begin{matrix} r \\ k\,r \end{matrix}\right\}T^i_j$$

When the weight term is zero these covariant derivatives reduce to the results given in our previous definitions.

▶ **45.** Let $\frac{dx^i}{dt} = v^i$ denote a generalized velocity and define the scalar function of kinetic energy T of a particle with mass m as

$$T = \frac{1}{2}m\,g_{ij}\,v^i\,v^j = \frac{1}{2}m\,g_{ij}\,\dot{x}^i\,\dot{x}^j.$$

Show that the intrinsic derivative of T is the same as an ordinary derivative of T. (i.e. Show that $\frac{\delta T}{\delta T} = \frac{dT}{dt}$.)

▶ **46.** Verify the relations

$$\frac{\partial g_{ij}}{\partial x^k} = -g_{mj}\, g_{ni}\, \frac{\partial g^{nm}}{\partial x^k}$$

$$\frac{\partial g^{in}}{\partial x^k} = -g^{mn}\, g^{ij}\, \frac{\partial g_{jm}}{\partial x^k}$$

▶ **47.** Assume that B^{ijk} is an absolute tensor. Is the quantity $T^{jk} = \dfrac{1}{\sqrt{g}}\dfrac{\partial}{\partial x^i}\left(\sqrt{g}B^{ijk}\right)$ a tensor? Justify your answer. If your answer is "no", explain your answer and determine if there any conditions you can impose upon B^{ijk} such that the above quantity will be a tensor?

▶ **48.** The e-permutation symbol can be used to define various vector products. Let A_i, B_i, C_i, D_i $i = 1,\ldots, N$ denote vectors, then expand and verify the following products:

(a) In two dimensions

$$R = e_{ij}A_iB_j \quad \text{a scalar determinant.}$$

$$R_i = e_{ij}A_j \quad \text{a vector (rotation).}$$

(b) In three dimensions

$$S = e_{ijk}A_iB_jC_k \quad \text{a scalar determinant.}$$

$$S_i = e_{ijk}B_jC_k \quad \text{a vector cross product.}$$

$$S_{ij} = e_{ijk}C_k \quad \text{a skew-symmetric matrix}$$

(c) In four dimensions

$$T = e_{ijkm}A_iB_jC_kD_m \quad \text{a scalar determinant.}$$

$$T_i = e_{ijkm}B_jC_kD_m \quad \text{4-dimensional cross product.}$$

$$T_{ij} = e_{ijkm}C_kD_m \quad \text{skew-symmetric matrix.}$$

$$T_{ijk} = e_{ijkm}D_m \quad \text{skew-symmetric tensor.}$$

with similar products in higher dimensions.

▶ **49.** Expand the curl operator for:

(a) Two dimensions $B = e_{ij}A_{j,i}$

(b) Three dimensions $B_i = e_{ijk}A_{k,j}$

(c) Four dimensions $B_{ij} = e_{ijkm}A_{m,k}$

▶ **50.** Assume A_i has continuous second order derivatives. Show in Cartesian coordinates that $e_{ijk}A_{i,kj} = 0$.

§1.5 DIFFERENTIAL GEOMETRY AND RELATIVITY

In this section we will examine some fundamental properties of curves and surfaces. In particular, at each point of a space curve we can construct a moving coordinate system consisting of a tangent vector, a normal vector and a binormal vector which is perpendicular to both the tangent and normal vectors. How these vectors change as we move along the space curve brings up the subjects of curvature and torsion associated with a space curve. The curvature is a measure of how the tangent vector to the curve is changing and the torsion is a measure of the twisting of the curve out of a plane. We will find that straight lines have zero curvature and plane curves have zero torsion.

In a similar fashion, associated with every smooth surface there are two coordinate surface curves and a normal surface vector through each point on the surface. The coordinate surface curves have tangent vectors which together with the normal surface vectors create a set of basis vectors. These vectors can be used to define such things as a two dimensional surface metric and a second order curvature tensor. The coordinate curves have tangent vectors which together with the surface normal form a coordinate system at each point of the surface. How these surface vectors change brings into consideration two different curvatures. A normal curvature and a tangential curvature (geodesic curvature). How these curvatures are related to the curvature tensor and to the Riemann Christoffel tensor, introduced in the last section, as well as other interesting relationships between the various surface vectors and curvatures, is the subject area of differential geometry.

Also presented in this section is a brief introduction to relativity where again the Riemann Christoffel tensor will occur. Properties of this important tensor are developed in the exercises of this section.

Space Curves and Curvature

For $x^i = x^i(s)$, $i = 1, 2, 3$, a 3-dimensional space curve in a Riemannian space V_n with metric tensor g_{ij}, and arc length parameter s, the vector $T^i = \frac{dx^i}{ds}$ represents a tangent vector to the curve at a point P on the curve. The vector T^i is a unit vector because

$$g_{ij}T^iT^j = g_{ij}\frac{dx^i}{ds}\frac{dx^j}{ds} = 1. \tag{1.5.1}$$

Differentiate intrinsically, with respect to arc length, the relation (1.5.1) and verify that

$$g_{ij}T^i\frac{\delta T^j}{\delta s} + g_{ij}\frac{\delta T^i}{\delta s}T^j = 0, \tag{1.5.2}$$

which implies that

$$g_{ij}T^j\frac{\delta T^i}{\delta s} = 0. \tag{1.5.3}$$

Hence, the vector $\frac{\delta T^i}{\delta s}$ is perpendicular to the tangent vector T^i. Define the unit normal vector N^i to the space curve to be in the same direction as the vector $\frac{\delta T^i}{\delta s}$ and write

$$N^i = \frac{1}{\kappa}\frac{\delta T^i}{\delta s} \tag{1.5.4}$$

where κ is a scale factor, called the curvature, and is selected such that

$$g_{ij}N^iN^j = 1 \quad \text{which implies} \quad g_{ij}\frac{\delta T^i}{\delta s}\frac{\delta T^j}{\delta s} = \kappa^2. \tag{1.5.5}$$

The reciprocal of curvature is called the radius of curvature. The curvature measures the rate of change of the tangent vector to the curve as the arc length varies. By differentiating intrinsically, with respect to arc length s, the relation $g_{ij}T^iN^j = 0$ we find that

$$g_{ij}T^i\frac{\delta N^j}{\delta s} + g_{ij}\frac{\delta T^i}{\delta s}N^j = 0. \tag{1.5.6}$$

Consequently, the curvature κ can be determined from the relation

$$g_{ij}T^i\frac{\delta N^j}{\delta s} = -g_{ij}\frac{\delta T^i}{\delta s}N^j = -g_{ij}\kappa N^iN^j = -\kappa \tag{1.5.7}$$

which defines the sign of the curvature. In a similar fashion we differentiate the relation (1.5.5) and find that

$$g_{ij}N^i\frac{\delta N^j}{\delta s} = 0. \tag{1.5.8}$$

This later equation indicates that the vector $\frac{\delta N^j}{\delta s}$ is perpendicular to the unit normal N^i. The equation (1.5.3) indicates that T^i is also perpendicular to N^i and hence any linear combination of these vectors will also be perpendicular to N^i. The unit binormal vector is defined by selecting the linear combination

$$\frac{\delta N^j}{\delta s} + \kappa T^j \tag{1.5.9}$$

and then scaling it into a unit vector by defining

$$B^j = \frac{1}{\tau}\left(\frac{\delta N^j}{\delta s} + \kappa T^j\right) \tag{1.5.10}$$

where τ is a scalar called the torsion. The sign of τ is selected such that the vectors T^i, N^i and B^i form a right handed system with $\epsilon_{ijk}T^iN^jB^k = 1$ and the magnitude of τ is selected such that B^i is a unit vector satisfying

$$g_{ij}B^iB^j = 1. \tag{1.5.11}$$

The triad of vectors T^i, N^i, B^i at a point on the curve form three planes. The plane containing T^i and B^i is called the rectifying plane. The plane containing N^i and B^i is called the normal plane. The plane containing T^i and N^i is called the osculating plane. The reciprocal of the torsion is called the radius of torsion. The torsion measures the rate of change of the osculating plane. The vectors T^i, N^i and B^i form a right-handed orthogonal system at a point on the space curve and satisfy the relation

$$B^i = \epsilon^{ijk}T_jN_k. \tag{1.5.12}$$

By using the equation (1.5.10) it can be shown that B^i is perpendicular to both the vectors T^i and N^i since

$$g_{ij}B^iT^j = 0 \quad \text{and} \quad g_{ij}B^iN^j = 0.$$

It is left as an exercise to show that the binormal vector B^i satisfies the relation $\frac{\delta B^i}{\delta s} = -\tau N^i$. The three relations

$$\frac{\delta T^i}{\delta s} = \kappa N^i \qquad \frac{\delta N^i}{\delta s} = \tau B^i - \kappa T^i \qquad \frac{\delta B^i}{\delta s} = -\tau N^i \tag{1.5.13}$$

are known as the Frenet-Serret formulas of differential geometry.

Surfaces and Curvature

Let us examine surfaces in a Cartesian frame of reference and then later we can generalize our results to other coordinate systems. A surface in Euclidean 3-dimensional space can be defined in several different ways. Explicitly, $z = f(x, y)$, implicitly, $F(x, y, z) = 0$ or parametrically by defining a set of parametric equations of the form

$$x = x(u, v), \qquad y = y(u, v), \qquad z = z(u, v)$$

which contain two independent parameters u, v called surface coordinates. For example, the equations

$$x = a \sin \theta \cos \phi, \qquad y = a \sin \theta \sin \phi, \qquad z = a \cos \theta$$

are the parametric equations which define a spherical surface of radius a with parameters $u = \theta$ and $v = \phi$. See for example figure 1.3-20 in section 1.3. By eliminating the parameters u, v one can derive the implicit form of the surface and by solving for z one obtains the explicit form of the surface. Using the parametric form of a surface we can define the position vector to a point on the surface which is then represented in terms of the parameters u, v as

$$\vec{r} = \vec{r}(u, v) = x(u, v)\,\widehat{\mathbf{e}}_1 + y(u, v)\,\widehat{\mathbf{e}}_2 + z(u, v)\,\widehat{\mathbf{e}}_3. \tag{1.5.14}$$

The coordinates (u, v) are called the curvilinear coordinates of a point on the surface. The functions $x(u, v), y(u, v), z(u, v)$ are assumed to be real and differentiable such that $\frac{\partial \vec{r}}{\partial u} \times \frac{\partial \vec{r}}{\partial v} \neq 0$. The curves

$$\vec{r}(u, c_2) \qquad \text{and} \qquad \vec{r}(c_1, v) \tag{1.5.15}$$

with c_1, c_2 constants, then define two surface curves called coordinate curves, which intersect at the surface coordinates (c_1, c_2). The family of curves defined by equations (1.5.15) with equally spaced constant values $c_i, c_i + \Delta c_i, c_i + 2\Delta c_i, \ldots$ define a surface coordinate grid system. The vectors $\frac{\partial \vec{r}}{\partial u}$ and $\frac{\partial \vec{r}}{\partial v}$ evaluated at the surface coordinates (c_1, c_2) on the surface, are tangent vectors to the coordinate curves through the point and are basis vectors for any vector lying in the surface. Letting $(x, y, z) = (y^1, y^2, y^3)$ and $(u, v) = (u^1, u^2)$ and utilizing the summation convention, we can write the position vector in the form

$$\vec{r} = \vec{r}(u^1, u^2) = y^i(u^1, u^2)\,\widehat{\mathbf{e}}_i. \tag{1.5.16}$$

The tangent vectors to the coordinate curves at a point P can then be represented as the basis vectors

$$\vec{E}_\alpha = \frac{\partial \vec{r}}{\partial u^\alpha} = \frac{\partial y^i}{\partial u^\alpha}\,\widehat{\mathbf{e}}_i, \quad \alpha = 1, 2 \tag{1.5.17}$$

where the partial derivatives are to be evaluated at the point P where the coordinate curves on the surface intersect. From these basis vectors we construct a unit normal vector to the surface at the point P by calculating the cross product of the tangent vector $\vec{r}_u = \frac{\partial \vec{r}}{\partial u}$ and $\vec{r}_v = \frac{\partial \vec{r}}{\partial v}$. A unit normal is then

$$\widehat{n} = \widehat{n}(u, v) = \frac{\vec{E}_1 \times \vec{E}_2}{|\vec{E}_1 \times \vec{E}_2|} = \frac{\vec{r}_u \times \vec{r}_v}{|\vec{r}_u \times \vec{r}_v|} \tag{1.5.18}$$

and is such that the vectors $\vec{E}_1, \vec{E}_2$ and $\widehat{n}$ form a right-handed system of coordinates.

If we transform from one set of curvilinear coordinates (u, v) to another set $(\bar{u}, \bar{v})$, which are determined by a set of transformation laws

$$u = u(\bar{u}, \bar{v}), \qquad v = v(\bar{u}, \bar{v}),$$

the equation of the surface becomes

$$\vec{r} = \vec{r}(\bar{u}, \bar{v}) = x(u(\bar{u}, \bar{v}), v(\bar{u}, \bar{v}))\,\widehat{\mathbf{e}}_1 + y(u(\bar{u}, \bar{v}), v(\bar{u}, \bar{v}))\,\widehat{\mathbf{e}}_2 + z(u(\bar{u}, \bar{v}), v(\bar{u}, \bar{v}))\,\widehat{\mathbf{e}}_3$$

and the tangent vectors to the new coordinate curves are

$$\frac{\partial \vec{r}}{\partial \bar{u}} = \frac{\partial \vec{r}}{\partial u}\frac{\partial u}{\partial \bar{u}} + \frac{\partial \vec{r}}{\partial v}\frac{\partial v}{\partial \bar{u}} \quad \text{and} \quad \frac{\partial \vec{r}}{\partial \bar{v}} = \frac{\partial \vec{r}}{\partial u}\frac{\partial u}{\partial \bar{v}} + \frac{\partial \vec{r}}{\partial v}\frac{\partial v}{\partial \bar{v}}.$$

Using the indicial notation this result can be represented as

$$\frac{\partial y^i}{\partial \bar{u}^\alpha} = \frac{\partial y^i}{\partial u^\beta}\frac{\partial u^\beta}{\partial \bar{u}^\alpha}.$$

This is the transformation law connecting the two systems of basis vectors on the surface.

A curve on the surface is defined by a relation $f(u, v) = 0$ between the curvilinear coordinates. Another way to represent a curve on the surface is to represent it in a parametric form where $u = u(t)$ and $v = v(t)$, where t is a parameter. The vector

$$\frac{d\vec{r}}{dt} = \frac{\partial \vec{r}}{\partial u}\frac{du}{dt} + \frac{\partial \vec{r}}{\partial v}\frac{dv}{dt}$$

is tangent to the curve on the surface.

An element of arc length with respect to the surface coordinates is represented by

$$ds^2 = d\vec{r} \cdot d\vec{r} = \frac{\partial \vec{r}}{\partial u^\alpha} \cdot \frac{\partial \vec{r}}{\partial u^\beta}\, du^\alpha\, du^\beta = a_{\alpha\beta} du^\alpha du^\beta \tag{1.5.19}$$

where $a_{\alpha\beta} = \frac{\partial \vec{r}}{\partial u^\alpha} \cdot \frac{\partial \vec{r}}{\partial u^\beta}$ with $\alpha, \beta = 1, 2$ defines a surface metric. This element of arc length on the surface is often written as the quadratic form

$$A = ds^2 = E(du)^2 + 2F\,du\,dv + G(dv)^2 = \frac{1}{E}(E\,du + F\,dv)^2 + \frac{EG - F^2}{E}(dv)^2 \tag{1.5.20}$$

and called the first fundamental form of the surface. Observe that for ds^2 to be positive definite the quantities E and $EG - F^2$ must be positive.

The surface metric associated with the two dimensional surface is defined by

$$a_{\alpha\beta} = \vec{E}_\alpha \cdot \vec{E}_\beta = \frac{\partial \vec{r}}{\partial u^\alpha} \cdot \frac{\partial \vec{r}}{\partial u^\beta} = \frac{\partial y^i}{\partial u^\alpha}\frac{\partial y^i}{\partial u^\beta}, \quad \alpha, \beta = 1, 2 \tag{1.5.21}$$

with conjugate metric tensor $a^{\alpha\beta}$ defined such that $a^{\alpha\beta}a_{\beta\gamma} = \delta^\alpha_\gamma$. Here the surface is embedded in a three dimensional space with metric g_{ij} and $a_{\alpha\beta}$ is the two dimensional surface metric. In the equation (1.5.20) the quantities E, F, G are functions of the surface coordinates u, v and are determined from the relations

$$E = a_{11} = \frac{\partial \vec{r}}{\partial u} \cdot \frac{\partial \vec{r}}{\partial u} = \frac{\partial y^i}{\partial u^1}\frac{\partial y^i}{\partial u^1}$$

$$F = a_{12} = \frac{\partial \vec{r}}{\partial u} \cdot \frac{\partial \vec{r}}{\partial v} = \frac{\partial y^i}{\partial u^1}\frac{\partial y^i}{\partial u^2} \tag{1.5.22}$$

$$G = a_{22} = \frac{\partial \vec{r}}{\partial v} \cdot \frac{\partial \vec{r}}{\partial v} = \frac{\partial y^i}{\partial u^2}\frac{\partial y^i}{\partial u^2}$$

Here and throughout the remainder of this section, we adopt the convention that Greek letters have the range 1,2, while Latin letters have the range 1,2,3.

Construct at a general point P on the surface the unit normal vector $\widehat{n}$ at this point. Also construct a plane which contains this unit surface normal vector $\widehat{n}$. Observe that there are an infinite number of planes which contain this unit surface normal. For now, select one of these planes, then later on we will consider all such planes. Let $\vec{r} = \vec{r}(s)$ denote the position vector defining a curve C which is the intersection of the selected plane with the surface, where s is the arc length along the curve, which is measured from some fixed point on the curve. Let us find the curvature of this curve of intersection. The vector $\widehat{T} = \frac{d\vec{r}}{ds}$, evaluated at the point P, is a unit tangent vector to the curve C and lies in the tangent plane to the surface at the point P. Here we are using ordinary differentiation rather than intrinsic differentiation because we are in a Cartesian system of coordinates. Differentiating the relation $\widehat{T} \cdot \widehat{T} = 1$, with respect to arc length s we find that $\widehat{T} \cdot \frac{d\widehat{T}}{ds} = 0$ which implies that the vector $\frac{d\widehat{T}}{ds}$ is perpendicular to the tangent vector $\widehat{T}$. Since the coordinate system is Cartesian we can treat the curve of intersection C as a space curve, then the vector $\vec{K} = \frac{d\widehat{T}}{ds}$, evaluated at point P, is defined as the curvature vector with curvature $|\vec{K}| = \kappa$ and radius of curvature $R = 1/\kappa$. A unit normal $\widehat{N}$ to the space curve is taken in the same direction as $\frac{d\widehat{T}}{ds}$ so that the curvature will always be positive. We can then write $\vec{K} = \kappa\widehat{N} = \dfrac{d\widehat{T}}{ds}$. Consider the geometry of figure 1.5-1 and define on the surface a unit vector $\widehat{u} = \widehat{n} \times \widehat{T}$ which is perpendicular to both the surface tangent vector $\widehat{T}$ and the surface normal vector $\widehat{n}$, such that the vectors T^i, u^i and n^i forms a right-handed system.

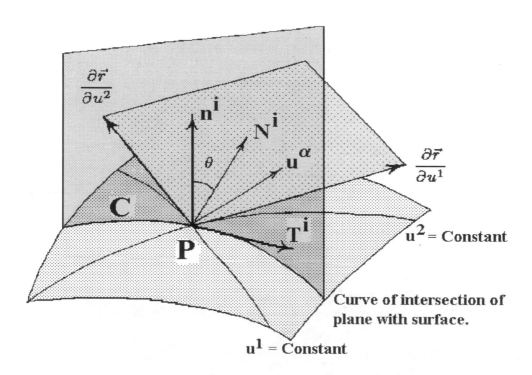

Figure 1.5-1 Surface curve with tangent plane and a normal plane.

134

The direction of $\widehat{u}$ in relation to $\widehat{T}$ is in the same sense as the surface tangents $\vec{E}_1$ and $\vec{E}_2$. Note that the vector $\frac{d\widehat{T}}{ds}$ is perpendicular to the tangent vector $\widehat{T}$ and lies in the plane which contains the vectors $\widehat{n}$ and $\widehat{u}$. We can therefore write the curvature vector $\vec{K}$ in the component form

$$\vec{K} = \frac{d\widehat{T}}{ds} = \kappa_{(n)}\,\widehat{n} + \kappa_{(g)}\,\widehat{u} = \vec{K}_n + \vec{K}_g \tag{1.5.23}$$

where $\kappa_{(n)}$ is called the normal curvature and $\kappa_{(g)}$ is called the geodesic curvature. The subscripts are not indices. These curvatures can be calculated as follows. From the orthogonality condition $\widehat{n}\cdot\widehat{T}=0$ we obtain by differentiation with respect to arc length s the result $\widehat{n}\cdot\frac{d\widehat{T}}{ds} + \widehat{T}\cdot\frac{d\widehat{n}}{ds} = 0$. The normal curvature is determined from the dot product relation

$$\widehat{n}\cdot\vec{K} = \kappa_{(n)} = -\widehat{T}\cdot\frac{d\widehat{n}}{ds} = -\frac{d\vec{r}}{ds}\cdot\frac{d\widehat{n}}{ds}. \tag{1.5.24}$$

By taking the dot product of $\widehat{u}$ with equation (1.5.23) we find that the geodesic curvature is determined from the triple scalar product relation

$$\kappa_{(g)} = \widehat{u}\cdot\frac{d\widehat{T}}{ds} = (\widehat{n}\times\widehat{T})\cdot\frac{d\widehat{T}}{ds}. \tag{1.5.25}$$

Normal Curvature

The equation (1.5.24) can be expressed in terms of a quadratic form by writing

$$\kappa_{(n)}\,ds^2 = -d\vec{r}\cdot d\widehat{n}. \tag{1.5.26}$$

The unit normal to the surface $\widehat{n}$ and position vector $\vec{r}$ are functions of the surface coordinates u,v with

$$d\vec{r} = \frac{\partial\vec{r}}{\partial u}du + \frac{\partial\vec{r}}{\partial v}dv \quad\text{and}\quad d\widehat{n} = \frac{\partial\widehat{n}}{\partial u}du + \frac{\partial\widehat{n}}{\partial v}dv. \tag{1.5.27}$$

We define the quadratic form

$$B = -d\vec{r}\cdot d\widehat{n} = -\left(\frac{\partial\vec{r}}{\partial u}du + \frac{\partial\vec{r}}{\partial v}dv\right)\cdot\left(\frac{\partial\widehat{n}}{\partial u}du + \frac{\partial\widehat{n}}{\partial v}dv\right)$$
$$B = e(du)^2 + 2f\,du\,dv + g(dv)^2 = b_{\alpha\beta}\,du^\alpha du^\beta \tag{1.5.28}$$

where

$$e = -\frac{\partial\vec{r}}{\partial u}\cdot\frac{\partial\widehat{n}}{\partial u}, \quad 2f = -\left(\frac{\partial\vec{r}}{\partial u}\cdot\frac{\partial\widehat{n}}{\partial v} + \frac{\partial\widehat{n}}{\partial u}\cdot\frac{\partial\vec{r}}{\partial v}\right), \quad g = -\frac{\partial\vec{r}}{\partial v}\cdot\frac{\partial\widehat{n}}{\partial v} \tag{1.5.29}$$

and $b_{\alpha\beta}$ $\alpha,\beta = 1,2$ is called the curvature tensor and $a^{\alpha\gamma}b_{\alpha\beta} = b_\beta^\gamma$ is an associated curvature tensor. The quadratic form of equation (1.5.28) is called the second fundamental form of the surface. Alternative methods for calculating the coefficients of this quadratic form result from the following considerations. The unit surface normal is perpendicular to the tangent vectors to the coordinate curves at the point P and therefore we have the orthogonality relationships

$$\frac{\partial\vec{r}}{\partial u}\cdot\widehat{n} = 0 \quad\text{and}\quad \frac{\partial\vec{r}}{\partial v}\cdot\widehat{n} = 0. \tag{1.5.30}$$

Observe that by differentiating the relations in equation (1.5.30), with respect to both u and v, one can derive the results

$$e = \frac{\partial^2 \vec{r}}{\partial u^2} \cdot \widehat{n} = -\frac{\partial \vec{r}}{\partial u} \cdot \frac{\partial \widehat{n}}{\partial u} = b_{11}$$

$$f = \frac{\partial^2 \vec{r}}{\partial u \partial v} \cdot \widehat{n} = -\frac{\partial \vec{r}}{\partial u} \cdot \frac{\partial \widehat{n}}{\partial v} = -\frac{\partial \widehat{n}}{\partial u} \cdot \frac{\partial \vec{r}}{\partial v} = b_{21} = b_{12} \qquad (1.5.31)$$

$$g = \frac{\partial^2 \vec{r}}{\partial v^2} \cdot \widehat{n} = -\frac{\partial \vec{r}}{\partial v} \cdot \frac{\partial \widehat{n}}{\partial v} = b_{22}$$

and consequently the curvature tensor can be expressed as

$$b_{\alpha\beta} = -\frac{\partial \vec{r}}{\partial u^\alpha} \cdot \frac{\partial \widehat{n}}{\partial u^\beta}. \qquad (1.5.32)$$

The quadratic forms from equations (1.5.20) and (1.5.28) enable us to represent the normal curvature in the form of a ratio of quadratic forms. We find from equation (1.5.26) that the normal curvature in the direction $\frac{du}{dv}$ is

$$\kappa_{(n)} = \frac{B}{A} = \frac{e(du)^2 + 2f\,du\,dv + g(dv)^2}{E(du)^2 + 2F\,du\,dv + G(dv)^2}. \qquad (1.5.33)$$

If we write the unit tangent vector to the curve in the form $\widehat{T} = \frac{d\vec{r}}{ds} = \frac{\partial \vec{r}}{\partial u^\alpha} \frac{du^\alpha}{ds}$ and express the derivative of the unit surface normal with respect to arc length as $\frac{d\widehat{n}}{ds} = \frac{\partial \widehat{n}}{\partial u^\beta} \frac{du^\beta}{ds}$, then the normal curvature can be expressed in the form

$$\begin{aligned}\kappa_{(n)} &= -\widehat{T} \cdot \frac{d\widehat{n}}{ds} = -\left(\frac{\partial \vec{r}}{\partial u^\alpha} \cdot \frac{\partial \widehat{n}}{\partial u^\beta}\right) \frac{du^\alpha}{ds} \frac{du^\beta}{ds} \\ &= \frac{b_{\alpha\beta} du^\alpha du^\beta}{ds^2} = \frac{b_{\alpha\beta} du^\alpha du^\beta}{a_{\alpha\beta} du^\alpha du^\beta}.\end{aligned} \qquad (1.5.34)$$

Observe that the curvature tensor is a second order symmetric tensor.

In the previous discussions, the plane containing the unit normal vector was arbitrary. Let us now consider all such planes that pass through this unit surface normal. As we vary the plane containing the unit surface normal $\widehat{n}$ at P we get different curves of intersection with the surface. Each curve has a curvature associated with it. By examining all such planes we can find the maximum and minimum normal curvatures associated with the surface when $\frac{d\kappa}{d\lambda} = 0$. We write equation (1.5.33) in the form

$$\kappa_{(n)} = \frac{e + 2f\lambda + g\lambda^2}{E + 2F\lambda + G\lambda^2} \qquad (1.5.35)$$

where $\lambda = \frac{dv}{du}$. The condition $\frac{d\kappa}{d\lambda} = 0$ implies, using the theory of proportions, that we can write equation (1.5.35) in the form

$$\kappa_{(n)} = \frac{(e + f\lambda) + \lambda(f + g\lambda)}{(E + F\lambda) + \lambda(F + G\lambda)} = \frac{f + g\lambda}{F + G\lambda} = \frac{e + f\lambda}{E + F\lambda}. \qquad (1.5.36)$$

Consequently, the curvature κ will satisfy the differential equations

$$(e - \kappa E)du + (f - \kappa F)dv = 0 \quad \text{and} \quad (f - \kappa F)du + (g - \kappa G)dv = 0. \qquad (1.5.37)$$

The maximum and minimum curvatures occur in those directions λ where $\frac{d\kappa_{(n)}}{d\lambda} = 0$. Calculating the derivative of $\kappa_{(n)}$ with respect to λ and setting the derivative to zero we obtain a quadratic equation in λ

$$(Fg - Gf)\lambda^2 + (Eg - Ge)\lambda + (Ef - Fe) = 0, \qquad (Fg - Gf) \neq 0.$$

This equation has two roots λ_1 and λ_2 which satisfy

$$\lambda_1 + \lambda_2 = -\frac{Eg - Ge}{Fg - Gf} \quad \text{and} \quad \lambda_1 \lambda_2 = \frac{Ef - Fe}{Fg - Gf}, \tag{1.5.38}$$

where $Fg - Gf \neq 0$. The curvatures $\kappa_{(1)}, \kappa_{(2)}$ corresponding to the roots λ_1 and λ_2 are called the principal curvatures at the point P. Several quantities of interest that are related to $\kappa_{(1)}$ and $\kappa_{(2)}$ are: (1) the principal radii of curvature $R_i = 1/\kappa_i, i = 1, 2$; (2) $H = \frac{1}{2}(\kappa_{(1)} + \kappa_{(2)})$ called the mean curvature and $K = \kappa_{(1)}\kappa_{(2)}$ called the total curvature or Gaussian curvature of the surface. Observe that the roots λ_1 and λ_2 determine two directions on the surface

$$\frac{d\vec{r}_1}{du} = \frac{\partial \vec{r}}{\partial u} + \frac{\partial \vec{r}}{\partial v}\lambda_1 \quad \text{and} \quad \frac{d\vec{r}_2}{du} = \frac{\partial \vec{r}}{\partial u} + \frac{\partial \vec{r}}{\partial v}\lambda_2.$$

If these directions are orthogonal we will have

$$\frac{d\vec{r}_1}{du} \cdot \frac{d\vec{r}_2}{du} = (\frac{\partial \vec{r}}{\partial u} + \frac{\partial \vec{r}}{\partial v}\lambda_1)(\frac{\partial \vec{r}}{\partial u} + \frac{\partial \vec{r}}{\partial v}\lambda_2) = 0.$$

This requires that

$$G\lambda_1\lambda_2 + F(\lambda_1 + \lambda_2) + E = 0. \tag{1.5.39}$$

It is left as an exercise to verify that this is indeed the case and so the directions determined by the principal curvatures must be orthogonal. In the case where $Fg - Gf = 0$ we have that $F = 0$ and $f = 0$ because the coordinate curves are orthogonal and G must be positive. In this special case there are still two directions determined by the differential equations (1.5.37) with $dv = 0$, du arbitrary, and $du = 0$, dv arbitrary. From the differential equations (1.5.37) we find these directions correspond to

$$\kappa_{(1)} = \frac{e}{E} \quad \text{and} \quad \kappa_{(2)} = \frac{g}{G}.$$

We let $\lambda^\alpha = \frac{du^\alpha}{ds}$ denote a unit vector on the surface satisfying $a_{\alpha\beta}\lambda^\alpha\lambda^\beta = 1$. Then the equation (1.5.34) can be written as $\kappa_{(n)} = b_{\alpha\beta}\lambda^\alpha\lambda^\beta$ or we can write $(b_{\alpha\beta} - \kappa_{(n)} a_{\alpha\beta})\lambda^\alpha\lambda^\beta = 0$. The maximum and minimum normal curvature occurs in those directions λ^α where

$$(b_{\alpha\beta} - \kappa_{(n)} a_{\alpha\beta})\lambda^\alpha = 0$$

and so $\kappa_{(n)}$ must be a root of the determinant equation $|b_{\alpha\beta} - \kappa_{(n)} a_{\alpha\beta}| = 0$ or

$$|a^{\alpha\gamma}b_{\alpha\beta} - \kappa_{(n)}\delta_\beta^\gamma| = \begin{vmatrix} b_1^1 - \kappa_{(n)} & b_2^1 \\ b_1^2 & b_2^2 - \kappa_{(n)} \end{vmatrix} = \kappa_{(n)}^2 - b_{\alpha\beta}a^{\alpha\beta}\kappa_{(n)} + \frac{b}{a} = 0. \tag{1.5.40}$$

This is a quadratic equation in $\kappa_{(n)}$ of the form $\kappa_{(n)}^2 - (\kappa_{(1)} + \kappa_{(2)})\kappa_{(n)} + \kappa_{(1)}\kappa_{(2)} = 0$. In other words the principal curvatures $\kappa_{(1)}$ and $\kappa_{(2)}$ are the eigenvalues of the matrix with elements $b_\beta^\gamma = a^{\alpha\gamma}b_{\alpha\beta}$. Observe that from the determinant equation in $\kappa_{(n)}$ we can directly find the total curvature or Gaussian curvature which is an invariant given by $K = \kappa_{(1)}\kappa_{(2)} = |b_\beta^\alpha| = |a^{\alpha\gamma}b_{\gamma\beta}| = b/a$. The mean curvature is also an invariant obtained from $H = \frac{1}{2}(\kappa_{(1)} + \kappa_{(2)}) = \frac{1}{2}a^{\alpha\beta}b_{\alpha\beta}$, where $a = a_{11}a_{22} - a_{12}a_{21}$ and $b = b_{11}b_{22} - b_{12}b_{21}$ are the determinants formed from the surface metric tensor and curvature tensor components.

The equations of Gauss, Weingarten and Codazzi

At each point on a space curve we can construct a unit tangent $\vec{T}$, a unit normal $\vec{N}$ and unit binormal $\vec{B}$. The derivatives of these vectors, with respect to arc length, can also be represented as linear combinations of the base vectors $\vec{T}, \vec{N}, \vec{B}$. See for example the Frenet-Serret formulas from equations (1.5.13). In a similar fashion the surface vectors $\vec{r}_u, \vec{r}_v, \hat{n}$ form a basis and the derivatives of these basis vectors with respect to the surface coordinates u, v can also be expressed as linear combinations of the basis vectors $\vec{r}_u, \vec{r}_v, \hat{n}$. For example, the derivatives $\vec{r}_{uu}, \vec{r}_{uv}, \vec{r}_{vv}$ can be expressed as linear combinations of $\vec{r}_u, \vec{r}_v, \hat{n}$. We can write

$$\vec{r}_{uu} = c_1 \vec{r}_u + c_2 \vec{r}_v + c_3 \hat{n}$$
$$\vec{r}_{uv} = c_4 \vec{r}_u + c_5 \vec{r}_v + c_6 \hat{n} \qquad (1.5.41)$$
$$\vec{r}_{vv} = c_7 \vec{r}_u + c_8 \vec{r}_v + c_9 \hat{n}$$

where $c_1, \ldots, c_9$ are constants to be determined. It is an easy exercise (see exercise 1.5, problem 8) to show that these equations can be written in the indicial notation as

$$\frac{\partial^2 \vec{r}}{\partial u^\alpha \partial u^\beta} = \left\{ \begin{matrix} \gamma \\ \alpha\,\beta \end{matrix} \right\} \frac{\partial \vec{r}}{\partial u^\gamma} + b_{\alpha\beta} \hat{n}. \qquad (1.5.42)$$

These equations are known as the Gauss equations.

In a similar fashion the derivatives of the normal vector can be represented as linear combinations of the surface basis vectors. If we write

$$\frac{\partial \hat{n}}{\partial u} = c_1 \vec{r}_u + c_2 \vec{r}_v \qquad \frac{\partial \vec{r}}{\partial u} = c_1^* \frac{\partial \hat{n}}{\partial u} + c_2^* \frac{\partial \hat{n}}{\partial v}$$
$$\text{or} \qquad (1.5.43)$$
$$\frac{\partial \hat{n}}{\partial v} = c_3 \vec{r}_u + c_4 \vec{r}_v \qquad \frac{\partial \vec{r}}{\partial v} = c_3^* \frac{\partial \hat{n}}{\partial u} + c_4^* \frac{\partial \hat{n}}{\partial v}$$

where $c_1, \ldots, c_4$ and $c_1^*, \ldots, c_4^*$ are constants. These equations are known as the Weingarten equations. It is easily demonstrated (see exercise 1.5, problem 9) that the Weingarten equations can be written in the indicial form

$$\frac{\partial \hat{n}}{\partial u^\alpha} = -b_\alpha^\beta \frac{\partial \vec{r}}{\partial u^\beta} \qquad (1.5.44)$$

where $b_\alpha^\beta = a^{\beta\gamma} b_{\gamma\alpha}$ is the mixed second order form of the curvature tensor.

The equations of Gauss produce a system of partial differential equations defining the surface coordinates x^i as a function of the curvilinear coordinates u and v. The equations are not independent as certain compatibility conditions must be satisfied. In particular, it is required that the mixed partial derivatives must satisfy

$$\frac{\partial^3 \vec{r}}{\partial u^\alpha \partial u^\beta \partial u^\delta} = \frac{\partial^3 \vec{r}}{\partial u^\alpha \partial u^\delta \partial u^\beta}.$$

We calculate

$$\frac{\partial^3 \vec{r}}{\partial u^\alpha \partial u^\beta \partial u^\delta} = \left\{ \begin{matrix} \gamma \\ \alpha\,\beta \end{matrix} \right\} \frac{\partial^2 \vec{r}}{\partial u^\gamma \partial u^\delta} + \frac{\partial \left\{ \begin{matrix} \gamma \\ \alpha\,\beta \end{matrix} \right\}}{\partial u^\delta} \frac{\partial \vec{r}}{\partial u^\gamma} + b_{\alpha\beta} \frac{\partial \hat{n}}{\partial u^\delta} + \frac{\partial b_{\alpha\beta}}{\partial u^\delta} \hat{n}$$

and use the equations of Gauss and Weingarten to express this derivative in the form

$$\frac{\partial^3 \vec{r}}{\partial u^\alpha \partial u^\beta \partial u^\delta} = \left[\frac{\partial \left\{ \begin{matrix} \omega \\ \alpha\,\beta \end{matrix} \right\}}{\partial u^\delta} + \left\{ \begin{matrix} \gamma \\ \alpha\,\beta \end{matrix} \right\} \left\{ \begin{matrix} \omega \\ \gamma\,\delta \end{matrix} \right\} - b_{\alpha\beta} b_\delta^\omega \right] \frac{\partial \vec{r}}{\partial u^\omega} + \left[\left\{ \begin{matrix} \gamma \\ \alpha\,\beta \end{matrix} \right\} b_{\gamma\delta} + \frac{\partial b_{\alpha\beta}}{\partial u^\delta} \right] \hat{n}.$$

Forming the difference

$$\frac{\partial^3 \vec{r}}{\partial u^\alpha \partial u^\beta \partial u^\delta} - \frac{\partial^3 \vec{r}}{\partial u^\alpha \partial u^\delta \partial u^\beta} = 0$$

we find that the coefficients of the independent vectors $\hat{n}$ and $\frac{\partial \vec{r}}{\partial u^\omega}$ must be zero. Setting the coefficient of $\hat{n}$ equal to zero produces the Codazzi equations

$$\begin{Bmatrix} \gamma \\ \alpha\,\beta \end{Bmatrix} b_{\gamma\delta} - \begin{Bmatrix} \gamma \\ \alpha\,\delta \end{Bmatrix} b_{\gamma\beta} + \frac{\partial b_{\alpha\beta}}{\partial u^\delta} - \frac{\partial b_{\alpha\delta}}{\partial u^\beta} = 0. \tag{1.5.45}$$

These equations are sometimes referred to as the Mainardi-Codazzi equations. Equating to zero the coefficient of $\frac{\partial \vec{r}}{\partial u^\omega}$ we find that $R^\delta_{\ \alpha\gamma\beta} = b_{\alpha\beta} b^\delta_\gamma - b_{\alpha\gamma} b^\delta_\beta$ or changing indices we have the covariant form

$$a_{\omega\delta} R^\delta_{\ \alpha\beta\gamma} = R_{\omega\alpha\beta\gamma} = b_{\omega\beta} b_{\alpha\gamma} - b_{\omega\gamma} b_{\alpha\beta}, \tag{1.5.46}$$

where

$$R^\delta_{\ \alpha\gamma\beta} = \frac{\partial}{\partial u^\gamma} \begin{Bmatrix} \delta \\ \alpha\,\beta \end{Bmatrix} - \frac{\partial}{\partial u^\beta} \begin{Bmatrix} \delta \\ \alpha\,\gamma \end{Bmatrix} + \begin{Bmatrix} \omega \\ \alpha\,\beta \end{Bmatrix} \begin{Bmatrix} \delta \\ \omega\,\gamma \end{Bmatrix} - \begin{Bmatrix} \omega \\ \alpha\,\gamma \end{Bmatrix} \begin{Bmatrix} \delta \\ \omega\,\beta \end{Bmatrix} \tag{1.5.47}$$

is the mixed Riemann curvature tensor.

EXAMPLE 1.5-1

Show that the Gaussian or total curvature $K = \kappa_{(1)}\kappa_{(2)}$ depends only upon the metric $a_{\alpha\beta}$ and is $K = \dfrac{R_{1212}}{a}$ where $a = det(a_{\alpha\beta})$.

Solution:

Utilizing the two-dimensional alternating tensor $e^{\alpha\beta}$ and the property of determinants we can write $e^{\gamma\delta} K = e^{\alpha\beta} b^\gamma_\alpha b^\delta_\beta$ where from page 137, $K = |b^\gamma_\beta| = |a^{\alpha\gamma} b_{\alpha\beta}|$. Now multiply by $e_{\gamma\zeta}$ and then contract on ζ and δ to obtain

$$e_{\gamma\delta} e^{\gamma\delta} K = e_{\gamma\delta} e^{\alpha\beta} b^\gamma_\alpha b^\delta_\beta = 2K$$

$$2K = e_{\gamma\delta} e^{\alpha\beta} \left(a^{\gamma\mu} b_{\alpha u}\right) \left(a^{\delta\nu} b_{\beta\nu}\right)$$

But $e_{\gamma\delta} a^{\gamma\mu} a^{\delta\nu} = a e^{\mu\nu}$ so that $2K = e^{\alpha\beta} a\, e^{\mu\nu} b_{\alpha\mu} b_{\beta\nu}$. Using $\sqrt{a}\, e^{\mu\nu} = \epsilon^{\mu\nu}$ we have $2K = \epsilon^{\mu\nu} \epsilon^{\alpha\beta} b_{\alpha\mu} b_{\beta\nu}$. Interchanging indices we can write

$$2K = \epsilon^{\beta\gamma} \epsilon^{\omega\alpha} b_{\omega\beta} b_{\alpha\gamma} \quad \text{and} \quad 2K = \epsilon^{\gamma\beta} \epsilon^{\omega\alpha} b_{\omega\gamma} b_{\alpha\beta}.$$

Adding these last two results we find that $4K = \epsilon^{\beta\gamma} \epsilon^{\omega\gamma} (b_{\omega\beta} b_{\alpha\gamma} - b_{\omega\gamma} b_{\alpha\beta}) = \epsilon^{\beta\gamma} \epsilon^{\omega\gamma} R_{\omega\alpha\beta\gamma}$. Now multiply both sides by $\epsilon_{\sigma\tau} \epsilon_{\lambda\nu}$ to obtain $4K\epsilon_{\sigma\tau} \epsilon_{\lambda\nu} = \delta^{\beta\gamma}_{\sigma\tau} \delta^{\omega\alpha}_{\lambda\nu} R_{\omega\alpha\beta\gamma}$. From exercise 1.5, problem 16, the Riemann curvature tensor R_{ijkl} is skew symmetric in the $(i,j), (k,l)$ as well as being symmetric in the $(ij), (kl)$ pair of indices. Consequently, $\delta^{\beta\gamma}_{\sigma\tau} \delta^{\omega\alpha}_{\lambda\nu} R_{\omega\alpha\beta\gamma} = 4R_{\lambda\nu\sigma\tau}$ and hence $R_{\lambda\nu\sigma\tau} = K\epsilon_{\sigma\tau} \epsilon_{\lambda\nu}$ and we have the special case where $K\sqrt{a}\, e_{12} \sqrt{a}\, e_{12} = R_{1212}$ or $K = \dfrac{R_{1212}}{a}$. A much simpler way to obtain this result is to observe $K = \dfrac{b}{a}$ (bottom of page 136) and note from equation (1.5.46) that $R_{1212} = b_{11}b_{22} - b_{12}b_{21} = b$. ∎

Note that on a surface $ds^2 = a_{\alpha\beta} du^\alpha du^\beta$ where $a_{\alpha\beta}$ are the metrices for the surface. This metric is a tensor and satisfies $\bar{a}_{\gamma\delta} = a_{\alpha\beta} \dfrac{\partial u^\alpha}{\partial \bar{u}^\gamma} \dfrac{\partial u^\beta}{\partial \bar{u}^\delta}$ and by taking determinants we find

$$\bar{a} = \left|\bar{a}_{\gamma\delta}\right| \left|\frac{\partial u^\alpha}{\partial \bar{u}^\gamma}\right| \left|\frac{\partial u^\beta}{\partial \bar{u}^\delta}\right| = aJ^2$$

where J is the Jacobian of the surface coordinate transformation. Here the curvature tensor for the surface $R_{\alpha\beta\gamma\delta}$ has only one independent component since $R_{1212} = R_{2121} = -R_{1221} = -R_{2112}$ (See exercises 20,21). From the transformation law

$$\bar{R}_{\epsilon\eta\lambda\mu} = R_{\alpha\beta\gamma\delta}\frac{\partial u^\alpha}{\partial \bar{u}^\epsilon}\frac{\partial u^\beta}{\partial \bar{u}^\eta}\frac{\partial u^\gamma}{\partial \bar{u}^\lambda}\frac{\partial u^\delta}{\partial \bar{u}^\mu}$$

one can sum over the repeated indices and show that $\bar{R}_{1212} = R_{1212}J^2$ and therefore

$$\frac{\bar{R}_{1212}}{\bar{a}} = \frac{R_{1212}}{a} = K$$

which shows that the Gaussian curvature is a scalar invariant in V_2.

Geodesic Curvature

For C an arbitrary curve on a given surface the curvature vector $\vec{K}$, associated with this curve, is the vector sum of the normal curvature $\kappa_{(n)}\,\hat{n}$ and geodesic curvature $\kappa_{(g)}\,\hat{u}$ and lies in a plane which is perpendicular to the tangent vector to the given curve on the surface. The geodesic curvature $\kappa_{(g)}$ is obtained from the equation (1.5.25) and can be represented

$$\kappa_{(g)} = \hat{u}\cdot\vec{K} = \hat{u}\cdot\frac{d\vec{T}}{ds} = (\hat{n}\times\vec{T})\cdot\frac{d\vec{T}}{ds} = \left(\vec{T}\times\frac{d\vec{T}}{ds}\right)\cdot\hat{n}.$$

Substituting into this expression the vectors

$$\vec{T} = \frac{d\vec{r}}{ds} = \vec{r}_u\frac{du}{ds} + \vec{r}_v\frac{dv}{ds}$$

$$\frac{d\vec{T}}{ds} = \vec{K} = \vec{r}_{uu}(u')^2 + 2\vec{r}_{uv}u'v' + \vec{r}_{vv}(v')^2 + \vec{r}_u u'' + \vec{r}_v v'',$$

where $' = \frac{d}{ds}$, and by utilizing the results from problem 10 of the exercises following this section, we find that the geodesic curvature can be represented as

$$\kappa_{(g)} = \left[\left\{{2 \atop 1\,1}\right\}(u')^3 + \left(2\left\{{2 \atop 1\,2}\right\} - \left\{{1 \atop 1\,1}\right\}\right)(u')^2 v' + \right.$$
$$\left. \left(\left\{{2 \atop 2\,2}\right\} - 2\left\{{1 \atop 1\,2}\right\}\right)u'(v')^2 - \left\{{1 \atop 2\,2}\right\}(v')^3 + (u'v'' - u''v')\right]\sqrt{EG - F^2}. \tag{1.5.48}$$

This equation indicates that the geodesic curvature is only a function of the surface metrices E, F, G and the derivatives u', v', u'', v''. When the geodesic curvature is zero the curve is called a geodesic curve. Such curves are often times, but not always, the lines of shortest distance between two points on a surface. For example, the great circle on a sphere which passes through two given points on the sphere is a geodesic curve. If you erase that part of the circle which represents the shortest distance between two points on the circle you are left with a geodesic curve connecting the two points, however, the path is not the shortest distance between the two points.

For plane curves we let $u = x$ and $v = y$ so that the geodesic curvature reduces to

$$k_g = u'v'' - u''v' = \frac{d\phi}{ds}$$

140

where ϕ is the angle between the tangent $\vec{T}$ to the curve and the unit vector $\widehat{e}_1$.

Geodesics are curves on the surface where the geodesic curvature is zero. Since $k_g = 0$ along a geodesic surface curve, then at every point on this surface curve the normal $\vec{N}$ to the curve will be in the same direction as the normal $\widehat{n}$ to the surface. In this case, we have $\vec{r}_u \cdot \widehat{n} = 0$ and $\vec{r}_v \cdot \widehat{n} = 0$ which reduces to

$$\frac{d\vec{T}}{ds} \cdot \vec{r}_u = 0 \quad \text{and} \quad \frac{d\vec{T}}{ds} \cdot \vec{r}_v = 0, \tag{1.5.49}$$

since the vectors $\widehat{n}$ and $\frac{d\vec{T}}{ds}$ have the same direction. In particular, we may write

$$\vec{T} = \frac{d\vec{r}}{ds} = \frac{\partial \vec{r}}{\partial u}\frac{du}{ds} + \frac{\partial \vec{r}}{\partial v}\frac{dv}{ds} = \vec{r}_u\, u' + \vec{r}_v\, v'$$
$$\frac{d\vec{T}}{ds} = \vec{r}_{uu}\,(u')^2 + 2\vec{r}_{uv}\,u'v' + \vec{r}_{vv}\,(v')^2 + \vec{r}_u\,u'' + \vec{r}_v\,v''$$

Consequently, the equations (1.5.49) become

$$\frac{d\vec{T}}{ds} \cdot \vec{r}_u = (\vec{r}_{uu} \cdot \vec{r}_u)\,(u')^2 + 2(\vec{r}_{uv} \cdot \vec{r}_u)\,u'v' + (\vec{r}_{vv} \cdot \vec{r}_u)\,(v')^2 + Eu'' + Fv'' = 0$$
$$\frac{d\vec{T}}{ds} \cdot \vec{r}_v = (\vec{r}_{uu} \cdot \vec{r}_v)\,(u')^2 + 2(\vec{r}_{uv} \cdot \vec{r}_v)\,u'v' + (\vec{r}_{vv} \cdot \vec{r}_v)\,(v')^2 + Fu'' + Gv'' = 0. \tag{1.5.50}$$

Utilizing the results from exercise 1.5,(See problems 4,5 and 6), we can eliminate v'' from the equations (1.5.50) to obtain

$$\frac{d^2u}{ds^2} + \left\{\begin{matrix}1\\1\,1\end{matrix}\right\}\left(\frac{du}{ds}\right)^2 + 2\left\{\begin{matrix}1\\1\,2\end{matrix}\right\}\frac{du}{ds}\frac{dv}{ds} + \left\{\begin{matrix}1\\2\,2\end{matrix}\right\}\left(\frac{dv}{ds}\right)^2 = 0$$

and eliminating u'' from the equations (1.5.50) produces the equation

$$\frac{d^2v}{ds^2} + \left\{\begin{matrix}2\\1\,1\end{matrix}\right\}\left(\frac{du}{ds}\right)^2 + 2\left\{\begin{matrix}2\\1\,2\end{matrix}\right\}\frac{du}{ds}\frac{dv}{ds} + \left\{\begin{matrix}2\\2\,2\end{matrix}\right\}\left(\frac{dv}{ds}\right)^2 = 0.$$

In tensor form, these last two equations are written

$$\frac{d^2u^\alpha}{ds^2} + \left\{\begin{matrix}\alpha\\\beta\,\gamma\end{matrix}\right\}_a \frac{du^\beta}{ds}\frac{du^\gamma}{ds} = 0, \quad \alpha,\beta,\gamma = 1,2 \tag{1.5.51}$$

where $u = u^1$ and $v = u^2$. The equations (1.5.51) are the differential equations defining a geodesic curve on a surface. We will find that these same type of equations arise in considering the shortest distance between two points in a generalized coordinate system. See for example problem 18 in exercise 2.2.

Tensor Derivatives

Let $u^\alpha = u^\alpha(t)$ denote the parametric equations of a curve on the surface defined by the parametric equations $x^i = x^i(u^1, u^2)$. We can then represent the surface curve in the spatial geometry since the surface curve can be represented in the spatial coordinates through the representation $x^i = x^i(u^1(t), u^2(t)) = x^i(t)$. Recall that for $x^i = x^i(t)$ a given curve C , the intrinsic derivative of a vector field A^i along C is defined as the inner product of the covariant derivative of the vector field with the tangent vector to the curve. This intrinsic derivative is written

$$\frac{\delta A^i}{\delta t} = A^i_{,j}\frac{dx^j}{dt} = \left[\frac{\partial A^i}{\partial x^j} + \left\{\begin{matrix} i \\ j\,k \end{matrix}\right\}_g A^k\right]\frac{dx^j}{dt}$$

or

$$\frac{\delta A^i}{\delta t} = \frac{dA^i}{dt} + \left\{\begin{matrix} i \\ j\,k \end{matrix}\right\}_g A^k\frac{dx^j}{dt}$$

where the subscript g indicates that the Christoffel symbol is formed from the spatial metric g_{ij}. If A^α is a surface vector defined along the curve C, the intrinsic derivative is represented

$$\frac{\delta A^\alpha}{\delta t} = A^\alpha_{,\beta}\frac{du^\beta}{dt} = \left[\frac{\partial A^\alpha}{\partial u^\beta} + \left\{\begin{matrix} \alpha \\ \beta\,\gamma \end{matrix}\right\}_a A^\gamma\right]\frac{du^\beta}{dt}$$

or

$$\frac{\delta A^\alpha}{\delta t} = \frac{dA^\alpha}{dt} + \left\{\begin{matrix} \alpha \\ \beta\,\gamma \end{matrix}\right\}_a A^\gamma\frac{du^\beta}{dt}$$

where the subscript a denotes that the Christoffel is formed from the surface metric $a_{\alpha\beta}$.

Similarly, the formulas for the intrinsic derivative of a covariant spatial vector A_i or covariant surface vector A_α are given by

$$\frac{\delta A_i}{\delta t} = \frac{dA_i}{dt} - \left\{\begin{matrix} k \\ i\,j \end{matrix}\right\}_g A_k\frac{dx^j}{dt}$$

and

$$\frac{\delta A_\alpha}{\delta t} = \frac{dA_\alpha}{dt} - \left\{\begin{matrix} \gamma \\ \alpha\,\beta \end{matrix}\right\}_a A_\alpha\frac{du^\beta}{dt}.$$

Consider a mixed tensor T^i_α which is contravariant with respect to a transformation of space coordinates x^i and covariant with respect to a transformation of surface coordinates u^α. For T^i_α defined over the surface curve C, which can also be viewed as a space curve C, define the scalar invariant $\Psi = \Psi(t) = T^i_\alpha A_i B^\alpha$ where A_i is a parallel vector field along the curve C when it is viewed as a space curve and B^α is also a parallel vector field along the curve C when it is viewed as a surface curve. Recall that these parallel vector fields must satisfy the differential equations

$$\frac{\delta A_i}{\delta t} = \frac{dA_i}{dt} - \left\{\begin{matrix} k \\ i\,j \end{matrix}\right\}_g A_k\frac{dx^j}{dt} = 0 \quad\text{and}\quad \frac{\delta B^\alpha}{\delta t} = \frac{dB^\alpha}{dt} + \left\{\begin{matrix} \alpha \\ \beta\,\gamma \end{matrix}\right\}_a B^\gamma\frac{du^\beta}{dt} = 0. \qquad (1.5.52)$$

The scalar invariant Ψ is a function of the parameter t of the space curve since both the tensor and the parallel vector fields are to be evaluated along the curve C. By differentiating the function Ψ with respect to the parameter t there results

$$\frac{d\Psi}{dt} = \frac{dT^i_\alpha}{dt}A_i B^\alpha + T^i_\alpha\frac{dA_i}{dt}B^\alpha + T^i_\alpha A_i\frac{dB^\alpha}{dt}. \qquad (1.5.53)$$

But the vectors A_i and B^α are parallel vector fields and must satisfy the relations given by equations (1.5.52). This implies that equation (1.5.53) can be written in the form

$$\frac{d\Psi}{dt} = \left[\frac{dT_\alpha^i}{dt} + \left\{{i \atop k\,j}\right\}_g T_\alpha^k \frac{dx^j}{dt} - \left\{{\gamma \atop \beta\,\alpha}\right\}_a T_\gamma^i \frac{du^\beta}{dt}\right] A_i B^\alpha. \qquad (1.5.54)$$

The quantity inside the brackets of equation (1.5.54) is defined as the intrinsic tensor derivative with respect to the parameter t along the curve C. This intrinsic tensor derivative is written

$$\frac{\delta T_\alpha^i}{dt} = \frac{dT_\alpha^i}{dt} + \left\{{i \atop k\,j}\right\}_g T_\alpha^k \frac{dx^j}{dt} - \left\{{\gamma \atop \beta\,\alpha}\right\}_a T_\gamma^i \frac{du^\beta}{dt}. \qquad (1.5.55)$$

The spatial representation of the curve C is related to the surface representation of the curve C through the defining equations. Therefore, we can express the equation (1.5.55) in the form

$$\frac{\delta T_\alpha^i}{dt} = \left[\frac{\partial T_\alpha^i}{\partial u^\beta} + \left\{{i \atop k\,j}\right\}_g T_\alpha^k \frac{\partial x^j}{\partial u^\beta} - \left\{{\gamma \atop \beta\,\alpha}\right\}_a T_\gamma^i\right] \frac{du^\beta}{dt} \qquad (1.5.56)$$

The quantity inside the brackets is a mixed tensor which is defined as the tensor derivative of T_α^i with respect to the surface coordinates u^β. The tensor derivative of the mixed tensor T_α^i with respect to the surface coordinates u^β is written

$$T_{\alpha,\beta}^i = \frac{\partial T_\alpha^i}{\partial u^\beta} + \left\{{i \atop k\,j}\right\}_g T_\alpha^k \frac{\partial x^j}{\partial u^\beta} - \left\{{\gamma \atop \beta\,\alpha}\right\}_a T_\gamma^i.$$

In general, given a mixed tensor $T_{\alpha\ldots\beta}^{i\ldots j}$ which is contravariant with respect to transformations of the space coordinates and covariant with respect to transformations of the surface coordinates, then we can define the scalar field along the surface curve C as

$$\Psi(t) = T_{\alpha\ldots\beta}^{i\ldots j} A_i \cdots A_j B^\alpha \cdots B^\beta \qquad (1.5.57)$$

where $A_i, \ldots, A_j$ and $B^\alpha, \ldots, B^\beta$ are parallel vector fields along the curve C. The intrinsic tensor derivative is then derived by differentiating the equation (1.5.57) with respect to the parameter t.

Tensor derivatives of the metric tensors g_{ij}, $a_{\alpha\beta}$ and the alternating tensors ϵ_{ijk}, $\epsilon_{\alpha\beta}$ and their associated tensors are all zero. Hence, they can be treated as constants during the tensor differentiation process.

Generalizations

In a Riemannian space V_n with metric g_{ij} and curvilinear coordinates x^i, $i = 1, 2, 3$, the equations of a surface can be written in the parametric form $x^i = x^i(u^1, u^2)$ where u^α, $\alpha = 1, 2$ are called the curvilinear coordinates of the surface. Since

$$dx^i = \frac{\partial x^i}{\partial u^\alpha} du^\alpha \qquad (1.5.58)$$

then a small change du^α on the surface results in change dx^i in the space coordinates. Hence an element of arc length on the surface can be represented in terms of the curvilinear coordinates of the surface. This same element of arc length can also be represented in terms of the curvilinear coordinates of the space. Thus, an element of arc length squared in terms of the surface coordinates is represented

$$ds^2 = a_{\alpha\beta} du^\alpha du^\beta \qquad (1.5.59)$$

where $a_{\alpha\beta}$ is the metric of the surface. This same element when viewed as a spatial element is represented

$$ds^2 = g_{ij}dx^i dx^j. \tag{1.5.60}$$

By equating the equations (1.5.59) and (1.5.60) we find that

$$g_{ij}dx^i dx^j = g_{ij}\frac{\partial x^i}{\partial u^\alpha}\frac{\partial x^j}{\partial u^\beta}du^\alpha du^\beta = a_{\alpha\beta}du^\alpha du^\beta. \tag{1.5.61}$$

The equation (1.5.61) shows that the surface metric is related to the spatial metric and can be calculated from the relation $a_{\alpha\beta} = g_{ij}\frac{\partial x^i}{\partial u^\alpha}\frac{\partial x^j}{\partial u^\beta}$. This equation reduces to the equation (1.5.21) in the special case of Cartesian coordinates. In the surface coordinates we define the quadratic form $A = a_{\alpha\beta}du^\alpha du^\beta$ as the first fundamental form of the surface. The tangent vector to the coordinate curves defining the surface are given by $\frac{\partial x^i}{\partial u^\alpha}$ and can be viewed as either a covariant surface vector or a contravariant spatial vector. We define this vector as

$$x_\alpha^i = \frac{\partial x^i}{\partial u^\alpha}, \qquad i = 1,2,3, \quad \alpha = 1,2. \tag{1.5.62}$$

Any vector which is a linear combination of the tangent vectors to the coordinate curves is called a surface vector. A surface vector A^α can also be viewed as a spatial vector A^i. The relation between the spatial representation and surface representation is $A^i = A^\alpha x_\alpha^i$. The surface representation $A^\alpha, \alpha = 1,2$ and the spatial representation $A^i, i = 1,2,3$ define the same direction and magnitude since

$$g_{ij}A^i A^j = g_{ij}A^\alpha x_\alpha^i A^\beta x_\beta^j = g_{ij}x_\alpha^i x_\beta^j A^\alpha A^\beta = a_{\alpha\beta}A^\alpha A^\beta.$$

Consider any two surface vectors A^α and B^α and their spatial representations A^i and B^i where

$$A^i = A^\alpha x_\alpha^i \qquad \text{and} \qquad B^i = B^\alpha x_\alpha^i. \tag{1.5.63}$$

These vectors are tangent to the surface and so a unit normal vector to the surface can be defined from the cross product relation

$$n_i AB\sin\theta = \epsilon_{ijk}A^j B^k \tag{1.5.64}$$

where A, B are the magnitudes of A^i, B^i and θ is the angle between the vectors when their origins are made to coincide. Substituting equations (1.5.63) into the equation (1.5.64) we find

$$n_i AB\sin\theta = \epsilon_{ijk}A^\alpha x_\alpha^j B^\beta x_\beta^k. \tag{1.5.65}$$

In terms of the surface metric we have $AB\sin\theta = \epsilon_{\alpha\beta}A^\alpha B^\beta$ so that equation (1.5.65) can be written in the form

$$(n_i\epsilon_{\alpha\beta} - \epsilon_{ijk}x_\alpha^j x_\beta^k)A^\alpha B^\beta = 0 \tag{1.5.66}$$

which for arbitrary surface vectors implies

$$n_i\epsilon_{\alpha\beta} = \epsilon_{ijk}x_\alpha^j x_\beta^k \quad \text{or} \quad n_i = \frac{1}{2}\epsilon^{\alpha\beta}\epsilon_{ijk}x_\alpha^j x_\beta^k. \tag{1.5.67}$$

The equation (1.5.67) defines a unit normal vector to the surface in terms of the tangent vectors to the coordinate curves. This unit normal vector is related to the covariant derivative of the surface tangents as

144

is now demonstrated. By using the results from equation (1.5.50), the tensor derivative of equation (1.5.59), with respect to the surface coordinates, produces

$$x^i_{\alpha,\beta} = \frac{\partial^2 x^i}{\partial u^\alpha \partial u^\beta} + \left\{\begin{matrix} i \\ p\,q \end{matrix}\right\}_g x^p_\alpha x^q_\beta - \left\{\begin{matrix} \sigma \\ \alpha\,\beta \end{matrix}\right\}_a x^i_\sigma \tag{1.5.68}$$

where the subscripts on the Christoffel symbols refer to the metric from which they are calculated. Also the tensor derivative of the equation (1.5.57) produces the result

$$g_{ij}x^i_{\alpha,\gamma}x^j_\beta + g_{ij}x^i_\alpha x^j_{\beta,\gamma} = a_{\alpha\beta,\gamma} = 0. \tag{1.5.69}$$

Interchanging the indices α, β, γ cyclically in the equation (1.5.69) one can verify that

$$g_{ij}x^i_{\alpha,\beta}x^j_\gamma = 0. \tag{1.5.70}$$

The equation (1.5.70) indicates that in terms of the space coordinates, the vector $x^i_{\alpha,\beta}$ is perpendicular to the surface tangent vector x^i_γ and so must have the same direction as the unit surface normal n^i. Therefore, there must exist a second order tensor $b_{\alpha\beta}$ such that

$$b_{\alpha\beta}n^i = x^i_{\alpha,\beta}. \tag{1.5.71}$$

By using the relation $g_{ij}n^i n^j = 1$ we can transform equation (1.5.71) to the form

$$b_{\alpha\beta} = g_{ij}n^j x^i_{\alpha,\beta} = \frac{1}{2}\epsilon^{\gamma\delta}\epsilon_{ijk}x^i_{\alpha,\beta}x^j_\gamma x^k_\delta. \tag{1.5.72}$$

The second order symmetric tensor $b_{\alpha\beta}$ is called the curvature tensor and the quadratic form

$$B = b_{\alpha\beta}du^\alpha du^\beta \tag{1.5.73}$$

is called the second fundamental form of the surface.

Consider also the tensor derivative with respect to the surface coordinates of the unit normal vector to the surface. This derivative is

$$n^i_{,\alpha} = \frac{\partial n^i}{\partial u^\alpha} + \left\{\begin{matrix} i \\ j\,k \end{matrix}\right\}_g n^j x^k_\alpha. \tag{1.5.74}$$

Taking the tensor derivative of $g_{ij}n^i n^j = 1$ with respect to the surface coordinates produces the result $g_{ij}n^i n^j_{,\alpha} = 0$ which shows that the vector $n^j_{,\alpha}$ is perpendicular to n^i and must lie in the tangent plane to the surface. It can therefore be expressed as a linear combination of the surface tangent vectors x^i_α and written in the form

$$n^i_{,\alpha} = \eta^\beta_\alpha x^i_\beta \tag{1.5.75}$$

where the coefficients η^β_α can be written in terms of the surface metric components $a_{\alpha\beta}$ and the curvature components $b_{\alpha\beta}$ as follows. The unit vector n^i is normal to the surface so that

$$g_{ij}n^i x^j_\alpha = 0. \tag{1.5.76}$$

The tensor derivative of this equation with respect to the surface coordinates gives

$$g_{ij}n^i_\beta x^j_\alpha + g_{ij}n^i x^j_{\alpha,\beta} = 0. \tag{1.5.77}$$

Substitute into equation (1.5.77) the relations from equations (1.5.57), (1.5.71) and (1.5.75) and show that

$$b_{\alpha\beta} = -a_{\alpha\gamma}\eta^\gamma_\beta. \tag{1.5.78}$$

Solving the equation (1.5.78) for the coefficients η^γ_β we find

$$\eta^\gamma_\beta = -a^{\alpha\gamma}b_{\alpha\beta}. \tag{1.5.79}$$

Now substituting equation (1.5.79) into the equation (1.5.75) produces the Weingarten formula

$$n^i_{,\alpha} = -a^{\gamma\beta}b_{\gamma\alpha}x^i_\beta. \tag{1.5.80}$$

This is a relation for the derivative of the unit normal in terms of the surface metric, curvature tensor and surface tangents.

A third fundamental form of the surface is given by the quadratic form

$$C = c_{\alpha\beta}du^\alpha du^\beta \tag{1.5.81}$$

where $c_{\alpha\beta}$ is defined as the symmetric surface tensor

$$c_{\alpha\beta} = g_{ij}n^i_{,\alpha}n^j_{,\beta}. \tag{1.5.82}$$

By using the Weingarten formula in the equation (1.5.81) one can verify that

$$c_{\alpha\beta} = a^{\gamma\delta}b_{\alpha\gamma}b_{\beta\delta}. \tag{1.5.83}$$

Geodesic Coordinates

In a Cartesian coordinate system the metric tensor g_{ij} is a constant and so the Christoffel symbols are zero at all points of the space. This is because the Christoffel symbols are dependent upon the derivatives of the metric tensor which is constant. If the space V_N is not Cartesian then the Christoffel symbols do not vanish at all points of the space. However, it is possible to find a coordinate system where the Christoffel symbols will all vanish at a given point P of the space. Such coordinates are called geodesic coordinates of the point P.

Consider a two dimensional surface with surface coordinates u^α and surface metric $a_{\alpha\beta}$. If we transform to some other two dimensional coordinate system, say $\bar{u}^\alpha$ with metric $\bar{a}_{\alpha\beta}$, where the two coordinates are related by transformation equations of the form

$$u^\alpha = u^\alpha(\bar{u}^1, \bar{u}^2), \quad \alpha = 1, 2, \tag{1.5.84}$$

then from the transformation equation (1.4.7) we can write, after changing symbols,

$$\begin{Bmatrix} \delta \\ \beta\gamma \end{Bmatrix}_{\bar{a}} \frac{\partial u^\alpha}{\partial \bar{u}^\delta} = \begin{Bmatrix} \alpha \\ \delta\epsilon \end{Bmatrix}_a \frac{\partial u^\delta}{\partial \bar{u}^\beta} \frac{\partial u^\epsilon}{\partial \bar{u}^\gamma} + \frac{\partial^2 u^\alpha}{\partial \bar{u}^\beta \partial \bar{u}^\gamma}. \tag{1.5.85}$$

This is a relationship between the Christoffel symbols in the two coordinate systems. If $\begin{Bmatrix} \delta \\ \beta\gamma \end{Bmatrix}_{\bar{a}}$ vanishes at a point P, then for that particular point the equation (1.5.85) reduces to

$$\frac{\partial^2 u^\alpha}{\partial \bar{u}^\beta \partial \bar{u}^\gamma} = -\begin{Bmatrix} \alpha \\ \delta\epsilon \end{Bmatrix}_a \frac{\partial u^\delta}{\partial \bar{u}^\beta} \frac{\partial u^\epsilon}{\partial \bar{u}^\gamma} \tag{1.5.86}$$

where all terms are evaluated at the point P. Conversely, if the equation (1.5.86) is satisfied at the point P, then the Christoffel symbol $\begin{Bmatrix} \delta \\ \beta\gamma \end{Bmatrix}_{\bar{a}}$ must be zero at this point. Consider the special coordinate transformation

$$u^\alpha = u_0^\alpha + \bar{u}^\alpha - \frac{1}{2} \begin{Bmatrix} \alpha \\ \beta\gamma \end{Bmatrix}_a \bar{u}^\beta \bar{u}^\alpha \tag{1.5.87}$$

where u_0^α are the surface coordinates of the point P. The point P in the new coordinates is given by $\bar{u}^\alpha = 0$. We now differentiate the relation (1.5.87) to see if it satisfies the equation (1.5.86). We calculate the derivatives

$$\frac{\partial u^\alpha}{\partial \bar{u}^\tau} = \delta_\tau^\alpha - \frac{1}{2} \begin{Bmatrix} \alpha \\ \beta\tau \end{Bmatrix}_a \bar{u}^\beta - \frac{1}{2} \begin{Bmatrix} \alpha \\ \tau\gamma \end{Bmatrix}_a \bar{u}^\gamma \Big|_{u^\alpha=0} \tag{1.5.88}$$

and

$$\frac{\partial^2 u^\alpha}{\partial \bar{u}^\tau \partial \bar{u}^\sigma} = -\begin{Bmatrix} \alpha \\ \tau\sigma \end{Bmatrix}_a \Big|_{u^\alpha=0} \tag{1.5.89}$$

where these derivative are evaluated at $\bar{u}^\alpha = 0$. We find the derivative equations (1.5.88) and (1.5.89) do satisfy the equation (1.5.86) locally at the point P. Hence, the Christoffel symbols will all be zero at this particular point. The new coordinates can then be called geodesic coordinates.

Riemann Christoffel Tensor

Consider the Riemann Christoffel tensor defined by the equation (1.4.33). Various properties of this tensor are derived in the exercises at the end of this section. We will be particularly interested in the Riemann Christoffel tensor in a two dimensional space with metric $a_{\alpha\beta}$ and coordinates u^α. We find the Riemann Christoffel tensor has the form

$$R^\delta_{\cdot\alpha\beta\gamma} = \frac{\partial}{\partial u^\beta} \begin{Bmatrix} \delta \\ \alpha\gamma \end{Bmatrix} - \frac{\partial}{\partial u^\gamma} \begin{Bmatrix} \delta \\ \alpha\beta \end{Bmatrix} + \begin{Bmatrix} \tau \\ \alpha\gamma \end{Bmatrix} \begin{Bmatrix} \delta \\ \beta\tau \end{Bmatrix} - \begin{Bmatrix} \tau \\ \alpha\beta \end{Bmatrix} \begin{Bmatrix} \delta \\ \gamma\tau \end{Bmatrix} \tag{1.5.90}$$

where the Christoffel symbols are evaluated with respect to the surface metric. The above tensor has the associated tensor

$$R_{\sigma\alpha\beta\gamma} = a_{\sigma\delta} R^\delta_{\cdot\alpha\beta\gamma} \tag{1.5.91}$$

which is skew-symmetric in the indices (σ, α) and (β, γ) such that

$$R_{\sigma\alpha\beta\gamma} = -R_{\alpha\sigma\beta\gamma} \qquad \text{and} \qquad R_{\sigma\alpha\beta\gamma} = -R_{\sigma\alpha\gamma\beta}. \tag{1.5.92}$$

The two dimensional alternating tensor is used to define the constant

$$K = \frac{1}{4} \epsilon^{\alpha\beta} \epsilon^{\gamma\delta} R_{\alpha\beta\gamma\delta} \tag{1.5.93}$$

(see example 1.5-1) which is an invariant of the surface and called the Gaussian curvature or total curvature. In the exercises following this section it is shown that the Riemann Christoffel tensor of the surface can be expressed in terms of the total curvature and the alternating tensors as

$$R_{\alpha\beta\gamma\delta} = K\epsilon_{\alpha\beta}\epsilon_{\gamma\delta}. \tag{1.5.94}$$

Consider the second tensor derivative of x_α^r which is given by

$$x_{\alpha,\beta\gamma}^r = \frac{\partial x_{\alpha,\beta}^r}{\partial u^\gamma} + \left\{\begin{matrix} r \\ i\,j \end{matrix}\right\}_g x_{\alpha,\beta}^i x_\gamma^j - \left\{\begin{matrix} \delta \\ \alpha\,\gamma \end{matrix}\right\}_a x_{\delta,\beta}^r - \left\{\begin{matrix} \delta \\ \beta\,\gamma \end{matrix}\right\}_a x_{\alpha,\delta}^r \tag{1.5.95}$$

which can be shown (See A.J. McConnell reference) to satisfy the relation

$$x_{\alpha,\beta\gamma}^r - x_{\alpha,\gamma\beta}^r = R_{.\alpha\beta\gamma}^\delta x_\delta^r. \tag{1.5.96}$$

Using the relation (1.5.96) we can now derive some interesting properties relating to the tensors $a_{\alpha\beta}, b_{\alpha\beta}$, $c_{\alpha\beta}$, $R_{\alpha\beta\gamma\delta}$, the mean curvature H and the total curvature K.

Consider the tensor derivative of the equation (1.5.71) which can be written

$$x_{\alpha,\beta\gamma}^i = b_{\alpha\beta,\gamma}n^i + b_{\alpha\beta}n_{,\gamma}^i \tag{1.5.97}$$

where

$$b_{\alpha\beta,\gamma} = \frac{\partial b_{\alpha\beta}}{\partial u^\alpha} - \left\{\begin{matrix} \sigma \\ \alpha\,\gamma \end{matrix}\right\}_a b_{\sigma\beta} - \left\{\begin{matrix} \sigma \\ \beta\,\gamma \end{matrix}\right\}_a b_{\alpha\sigma}. \tag{1.5.98}$$

By using the Weingarten formula, given in equation (1.5.80), the equation (1.5.97) can be expressed in the form

$$x_{\alpha,\beta\gamma}^i = b_{\alpha\beta,\gamma}n^i - b_{\alpha\beta}a^{\tau\sigma}b_{\tau\gamma}x_\sigma^i \tag{1.5.99}$$

and by using the equations (1.5.98) and (1.5.99) it can be established that

$$x_{\alpha,\beta\gamma}^r - x_{\alpha,\gamma\beta}^r = (b_{\alpha\beta,\gamma} - b_{\alpha\gamma,\beta})n^r - a^{\tau\delta}(b_{\alpha\beta}b_{\tau\gamma} - b_{\alpha\gamma}b_{\tau\beta})x_\delta^r. \tag{1.5.100}$$

Now by equating the results from the equations (1.5.96) and (1.5.100) we arrive at the relation

$$R_{.\alpha\beta\gamma}^\delta x_\delta^r = (b_{\alpha\beta,\gamma} - b_{\alpha\gamma,\beta})n^r - a^{\tau\delta}(b_{\alpha\beta}b_{\tau\gamma} - b_{\alpha\gamma}b_{\tau\beta})x_\delta^r. \tag{1.5.101}$$

Multiplying the equation (1.5.101) by n_r and using the results from the equation (1.5.76) there results the Codazzi equations

$$b_{\alpha\beta,\gamma} - b_{\alpha\gamma,\beta} = 0. \tag{1.5.102}$$

Multiplying the equation (1.5.101) by $g_{rm}x_\sigma^m$ and simplifying one can derive the Gauss equations of the surface

$$R_{\sigma\alpha\beta\gamma} = b_{\alpha\gamma}b_{\sigma\beta} - b_{\alpha\beta}b_{\sigma\gamma}. \tag{1.5.103}$$

By using the Gauss equations (1.5.103) the equation (1.5.94) can be written as

$$K\epsilon_{\sigma\alpha}\epsilon_{\beta\gamma} = b_{\alpha\gamma}b_{\sigma\beta} - b_{\alpha\beta}b_{\sigma\gamma}. \tag{1.5.104}$$

148

Another form of equation (1.5.104) is obtained by using the equation (1.5.83) together with the relation $a_{\alpha\beta} = -a^{\sigma\gamma}\epsilon_{\sigma\alpha}\epsilon_{\beta\gamma}$. It is left as an exercise to verify the resulting form

$$-Ka_{\alpha\beta} = c_{\alpha\beta} - a^{\sigma\gamma}b_{\sigma\gamma}b_{\alpha\beta}. \tag{1.5.106}$$

Define the quantity

$$H = \frac{1}{2}a^{\sigma\gamma}b_{\sigma\gamma} \tag{1.5.107}$$

as the mean curvature of the surface, then the equation (1.5.106) can be written in the form

$$c_{\alpha\beta} - 2H\,b_{\alpha\beta} + K\,a_{\alpha\beta} = 0. \tag{1.5.108}$$

By multiplying the equation (1.5.108) by $du^\alpha du^\beta$ and summing, we find

$$C - 2H\,B + K\,A = 0 \tag{1.5.109}$$

is a relation connecting the first, second and third fundamental forms.

EXAMPLE 1.5-2

In a two dimensional space the Riemann Christoffel tensor has only one nonzero independent component R_{1212}. (See Exercise 1.5, problem number 21.) The equation (1.5.104) can be written in the form $K\sqrt{a}e_{12}\sqrt{a}e_{12} = b_{22}b_{11} - b_{21}b_{12}$ and solving for the Gaussian curvature K we find

$$K = \frac{b_{22}b_{11} - b_{12}b_{21}}{a_{11}a_{22} - a_{12}a_{21}} = \frac{b}{a} = \frac{R_{1212}}{a}. \tag{1.5.110}$$

∎

Surface Curvature

For a surface curve $u^\alpha = u^\alpha(s), \alpha = 1,2$ lying upon a surface $x^i = x^i(u^1,u^2), i = 1,2,3$, we have a two dimensional space embedded in a three dimensional space. Thus, if $t^\alpha = \frac{du^\alpha}{ds}$ is a unit tangent vector to the surface curve then $a_{\alpha\beta}\frac{du^\alpha}{ds}\frac{du^\beta}{ds} = a_{\alpha\beta}t^\alpha t^\beta = 1$. This same vector can be represented as the unit tangent vector to the space curve $x^i = x^i(u^1(s),u^2(s))$ with $T^i = \frac{dx^i}{ds}$. That is we will have $g_{ij}\frac{dx^i}{ds}\frac{dx^j}{ds} = g_{ij}T^iT^j = 1$. The surface vector t^α and the space vector T^i are related by

$$T^i = \frac{\partial x^i}{\partial u^\alpha}\frac{du^\alpha}{ds} = x^i_\alpha t^\alpha. \tag{1.5.111}$$

The surface vector t^α is a unit vector so that $a_{\alpha\beta}t^\alpha t^\beta = 1$. If we differentiate this equation intrinsically with respect to the parameter s, we find that $a_{\alpha\beta}t^\alpha\frac{\delta t^\beta}{\delta s} = 0$. This shows that the surface vector $\frac{\delta t^\alpha}{\delta s}$ is perpendicular to the surface vector t^α. Let u^α denote a unit normal vector in the surface plane which is orthogonal to the tangent vector t^α. The direction of u^α is selected such that $\epsilon_{\alpha\beta}t^\alpha u^\beta = 1$. Therefore, there exists a scalar $\kappa_{(g)}$ such that

$$\frac{\delta t^\alpha}{\delta s} = \kappa_{(g)}u^\alpha \tag{1.5.112}$$

where $\kappa_{(g)}$ is called the geodesic curvature of the curve. In a similar manner it can be shown that $\frac{\delta u^\alpha}{\delta s}$ is a surface vector either colinear to t^α or orthogonal to an intrinsic derivative of t^α. Let $\frac{\delta u^\alpha}{\delta s} = \beta t^\alpha$ where β is a scalar constant to be determined. By differentiating the relation $a_{\alpha\beta}t^\alpha u^\beta = 0$ intrinsically and simplifying we find that $\beta = -\kappa_{(g)}$ and therefore

$$\frac{\delta u^\alpha}{\delta s} = -\kappa_{(g)}t^\alpha. \tag{1.5.113}$$

The equations (1.5.112) and (1.5.113) are sometimes referred to as the Frenet-Serret formula for a curve relative to a surface.

Taking the intrinsic derivative of equation (1.5.111), with respect to the parameter s, we find that

$$\frac{\delta T^i}{\delta s} = x^i_\alpha \frac{\delta t^\alpha}{\delta s} + x^i_{\alpha,\beta}\frac{du^\beta}{ds}t^\alpha. \tag{1.5.114}$$

Treating the curve as a space curve we use the Frenet formulas (1.5.13). If we treat the curve as a surface curve, then we use the Frenet formulas (1.5.112) and (1.5.113). In this way the equation (1.5.114) can be written in the form

$$\kappa N^i = x^i_\alpha \kappa_{(g)} u^\alpha + x^i_{\alpha,\beta} t^\beta t^\alpha. \tag{1.5.115}$$

By using the results from equation (1.5.71) in equation (1.5.115) we obtain

$$\kappa N^i = \kappa_{(g)} u^i + b_{\alpha\beta} n^i t^\alpha t^\beta \tag{1.5.116}$$

where u^i is the space vector counterpart of the surface vector u^α. Let θ denote the angle between the surface normal n^i and the principal normal N^i, then we have that $\cos\theta = n_i N^i$. Hence, by multiplying the equation (1.5.116) by n_i we obtain

$$\kappa \cos\theta = b_{\alpha\beta} t^\alpha t^\beta. \tag{1.5.117}$$

Consequently, for all curves on the surface with the same tangent vector t^α, the quantity $\kappa\cos\theta$ will remain constant. This result is known as Meusnier's theorem. Note also that $\kappa\cos\theta = \kappa_{(n)}$ is the normal component of the curvature and $\kappa\sin\theta = \kappa_{(g)}$ is the geodesic component of the curvature. Therefore, we write the equation (1.5.117) as

$$\kappa_{(n)} = b_{\alpha\beta} t^\alpha t^\beta \tag{1.5.118}$$

which represents the normal curvature of the surface in the direction t^α. The equation (1.5.118) can also be written in the form

$$\kappa_{(n)} = b_{\alpha\beta}\frac{du^\alpha}{ds}\frac{du^\beta}{ds} = \frac{B}{A} \tag{1.5.119}$$

which is a ratio of quadratic forms.

The surface directions for which $\kappa_{(n)}$ has a maximum or minimum value is determined from the equation (1.5.119) which is written as

$$(b_{\alpha\beta} - \kappa_{(n)}a_{\alpha\beta})\lambda^\alpha\lambda^\beta = 0. \tag{1.5.120}$$

The direction giving a maximum or minimum value to $\kappa_{(n)}$ must then satisfy

$$(b_{\alpha\beta} - \kappa_{(n)}a_{\alpha\beta})\lambda^\beta = 0 \tag{1.5.121}$$

150

so that $\kappa_{(n)}$ must be a root of the determinant equation

$$det(b_{\alpha\beta} - \kappa_{(n)}a_{\alpha\beta}) = 0. \tag{1.5.122}$$

The expanded form of equation (1.5.122) can be written as

$$\kappa_{(n)}^2 - a^{\alpha\beta}b_{\alpha\beta}\kappa_{(n)} + \frac{b}{a} = 0 \tag{1.5.123}$$

where $a = a_{11}a_{22} - a_{12}a_{21}$ and $b = b_{11}b_{22} - b_{12}b_{21}$. Using the definition given in equation (1.5.107) and using the result from equation (1.5.110), the equation (1.5.123) can be expressed in the form

$$\kappa_{(n)}^2 - 2H\,\kappa_{(n)} + K = 0. \tag{1.5.124}$$

The roots $\kappa_{(1)}$ and $\kappa_{(2)}$ of the equation (1.5.124) then satisfy the relations

$$H = \frac{1}{2}(\kappa_{(1)} + \kappa_{(2)}) \tag{1.5.125}$$

and

$$K = \kappa_{(1)}\kappa_{(2)}. \tag{1.5.126}$$

Here H is the mean value of the principal curvatures and K is the Gaussian or total curvature which is the product of the principal curvatures. It is readily verified that

$$H = \frac{Eg - 2fF + eG}{2(EG - F^2)} \quad \text{and} \quad K = \frac{eg - f^2}{EG - F^2}$$

are invariants obtained from the surface metric and curvature tensor.

Relativity

Sir Isaac Newton and Albert Einstein viewed the world differently when it came to describing gravity and the motion of the planets. In this brief introduction to relativity we will compare the Newtonian equations with the relativistic equations in describing planetary motion. We begin with an examination of Newtonian systems.

Newton's viewpoint of planetary motion is a multiple bodied problem, but for simplicity we consider only a two body problem, say the sun and some planet where the motion takes place in a plane. Newton's law of gravitation states that two masses m and M are attracted toward each other with a force of magnitude $\frac{GmM}{\rho^2}$, where G is a constant, ρ is the distance between the masses, m is the mass of the planet and M is the mass of the sun. One can construct an x, y plane containing the two masses with the origin located at the center of mass of the sun. Let $\widehat{\mathbf{e}}_\rho = \cos\phi\,\widehat{\mathbf{e}}_1 + \sin\phi\,\widehat{\mathbf{e}}_2$ denote a unit vector at the origin of this coordinate system and pointing in the direction of the mass m. The vector force of attraction of mass M on mass m is given by the relation

$$\vec{F} = \frac{-GmM}{\rho^2}\widehat{\mathbf{e}}_\rho. \tag{1.5.127}$$

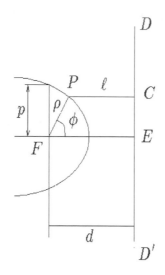

Figure 1.5-2. conic sections

The equation of motion of mass m with respect to mass M is obtained from Newton's second law. Let $\vec{\rho} = \rho\,\widehat{\mathbf{e}}_\rho$ denote the position vector of mass m with respect to the origin. Newton's second law can then be written in any of the forms

$$\vec{F} = \frac{-GmM}{\rho^2}\,\widehat{\mathbf{e}}_\rho = m\frac{d^2\vec{\rho}}{dt^2} = m\frac{d\vec{V}}{dt} = \frac{-GmM}{\rho^3}\,\vec{\rho} \qquad (1.5.128)$$

and from this equation we can show that the motion of the mass m can be described as a conic section.

A conic is the locus of a point P whose distance from a fixed point F, called a focus, and distance from a fixed line DD' not containing the fixed point, called a directrix, is in a constant ratio ϵ called the eccentricity. Let $FP = \rho$ denote the distance of P from the fixed point F and let $PC = \ell$ denote the distance of P from the fixed line DD'. A conic section is then described by the relation $\frac{\rho}{\ell} = \epsilon = $ a constant. From the geometry of figure 1.5-2 we find $\ell = d - \rho\cos\phi$, where $d = FE$ is the distance from the focus F to the fixed line DD'. Consequently, one can write

$$\rho = \epsilon\ell = \epsilon(d - \rho\cos\phi)$$

or solving for ρ one finds that the equation of a conic section can be written in the form

$$\rho = \frac{\epsilon d}{1 + \epsilon\cos\phi}$$

Note that when $\phi = \frac{\pi}{2}$ we find $\rho = p = \epsilon d$. For $\epsilon = 1$ a parabola results; for $0 < \epsilon \le 1$ an ellipse results; for $\epsilon > 1$ a hyperbola results; and if $\epsilon = 0$ the conic section is a circle.

One can write the equation of a conic section as

$$\rho = \frac{p}{1 + \epsilon\cos\phi} \qquad (1.5.129)$$

152

where $p = \epsilon d$ and the angle ϕ is known as the true anomaly associated with the orbit. The quantity p is called the semi-parameter of the conic section. A more general form of the above equation is

$$\rho = \frac{p}{1 + \epsilon \cos(\phi - \phi_0)} \quad \text{or} \quad u = \frac{1}{\rho} = A[1 + \epsilon \cos(\phi - \phi_0)], \tag{1.5.130}$$

where ϕ_0 is an arbitrary starting anomaly and $A = 1/p$ is some new constant. An additional symbol a, known as the semi-major axes of an elliptical orbit can be introduced where q, p, ϵ, a are related by

$$\frac{p}{1 + \epsilon} = q = a(1 - \epsilon) \quad \text{or} \quad p = a(1 - \epsilon^2). \tag{1.5.131}$$

To show that the equation (1.5.128) produces a conic section for the motion of mass m with respect to mass M we will show that one form of the solution of equation (1.5.128) is given by the equation (1.5.129). To verify this we use the following vector identities:

$$\begin{aligned}
\vec{\rho} \times \widehat{\mathbf{e}}_\rho &= 0 \\
\frac{d}{dt}\left(\vec{\rho} \times \frac{d\vec{\rho}}{dt}\right) &= \vec{\rho} \times \frac{d^2\vec{\rho}}{dt^2} \\
\widehat{\mathbf{e}}_\rho \cdot \frac{d\widehat{\mathbf{e}}_\rho}{dt} &= 0 \\
\widehat{\mathbf{e}}_\rho \times \left(\widehat{\mathbf{e}}_\rho \times \frac{d\widehat{\mathbf{e}}_\rho}{dt}\right) &= -\frac{d\widehat{\mathbf{e}}_\rho}{dt}.
\end{aligned} \tag{1.5.132}$$

From the equation (1.5.128) we find that

$$\frac{d}{dt}\left(\vec{\rho} \times \frac{d\vec{\rho}}{dt}\right) = \vec{\rho} \times \frac{d^2\vec{\rho}}{dt^2} = -\frac{GM}{\rho^2}\vec{\rho} \times \widehat{\mathbf{e}}_\rho = \vec{0} \tag{1.5.133}$$

so that an integration of equation (1.5.133) produces

$$\vec{\rho} \times \frac{d\vec{\rho}}{dt} = \vec{h} = \text{constant.} \tag{1.5.134}$$

The quantity $\vec{H} = \vec{\rho} \times m\vec{V} = \vec{\rho} \times m\frac{d\vec{\rho}}{dt}$ is the angular momentum of the mass m so that the quantity $\vec{h}$ represents the angular momentum per unit mass. The equation (1.5.134) tells us that $\vec{h}$ is a constant for our two body system. Note that because $\vec{h}$ is constant we have

$$\begin{aligned}
\frac{d}{dt}\left(\vec{V} \times \vec{h}\right) = \frac{d\vec{V}}{dt} \times \vec{h} &= -\frac{GM}{\rho^2}\widehat{\mathbf{e}}_\rho \times \left(\vec{\rho} \times \frac{d\vec{\rho}}{dt}\right) \\
&= -\frac{GM}{\rho^2}\widehat{\mathbf{e}}_\rho \times [\vec{\rho}\widehat{\mathbf{e}}_\rho \times (\rho\frac{d\widehat{\mathbf{e}}_\rho}{dt} + \frac{d\rho}{dt}\widehat{\mathbf{e}}_\rho)] \\
&= -\frac{GM}{\rho^2}\widehat{\mathbf{e}}_\rho \times (\widehat{\mathbf{e}}_\rho \times \frac{d\widehat{\mathbf{e}}_\rho}{dt})\rho^2 = GM\frac{d\widehat{\mathbf{e}}_\rho}{dt}
\end{aligned}$$

and so an integration produces

$$\vec{V} \times \vec{h} = GM\widehat{\mathbf{e}}_\rho + \vec{C}$$

where $\vec{C}$ is a vector constant of integration. The triple scalar product formula gives us

$$\vec{\rho} \cdot (\vec{V} \times \vec{h}) = \vec{h} \cdot (\vec{\rho} \times \frac{d\vec{\rho}}{dt}) = h^2 = GM\vec{\rho} \cdot \widehat{\mathbf{e}}_\rho + \vec{\rho} \cdot \vec{C}$$

or

$$h^2 = GM\rho + C\rho\cos\phi \tag{1.5.135}$$

where ϕ is the angle between the vectors $\vec{C}$ and $\vec{\rho}$. From the equation (1.5.135) we find that

$$\rho = \frac{p}{1 + \epsilon\cos\phi} \tag{1.5.136}$$

where $p = h^2/GM$ and $\epsilon = C/GM$. This result is known as Kepler's first law and implies that when $\epsilon < 1$ the mass m describes an elliptical orbit with the sun at one focus.

We present now an alternate derivation of equation (1.5.130) for later use. From the equation (1.5.128) we have

$$2\frac{d\vec{\rho}}{dt} \cdot \frac{d^2\vec{\rho}}{dt^2} = \frac{d}{dt}\left(\frac{d\vec{\rho}}{dt} \cdot \frac{d\vec{\rho}}{dt}\right) = -2\frac{GM}{\rho^3}\vec{\rho} \cdot \frac{d\vec{\rho}}{dt} = -\frac{GM}{\rho^3}\frac{d}{dt}(\vec{\rho} \cdot \vec{\rho}). \tag{1.5.137}$$

Consider the equation (1.5.137) in spherical coordinates ρ, θ, ϕ. The tensor velocity components are $V^1 = \frac{d\rho}{dt}$, $V^2 = \frac{d\theta}{dt}$, $V^3 = \frac{d\phi}{dt}$ and the physical components of velocity are given by $V_\rho = \frac{d\rho}{dt}$, $V_\theta = \rho\frac{d\theta}{dt}$, $V_\phi = \rho\sin\theta\frac{d\phi}{dt}$ so that the velocity can be written

$$\vec{V} = \frac{d\vec{\rho}}{dt} = \frac{d\rho}{dt}\,\widehat{\mathbf{e}}_\rho + \rho\frac{d\theta}{dt}\,\widehat{\mathbf{e}}_\theta + \rho\sin\theta\frac{d\phi}{dt}\,\widehat{\mathbf{e}}_\phi. \tag{1.5.138}$$

Substituting equation (1.5.138) into equation (1.5.137) gives the result

$$\frac{d}{dt}\left[\left(\frac{d\rho}{dt}\right)^2 + \rho^2\left(\frac{d\theta}{dt}\right)^2 + \rho^2\sin^2\theta\left(\frac{d\phi}{dt}\right)^2\right] = -\frac{GM}{\rho^3}\frac{d}{dt}(\rho^2) = -\frac{2GM}{\rho^2}\frac{d\rho}{dt} = 2GM\frac{d}{dt}\left(\frac{1}{\rho}\right)$$

which can be integrated directly to give

$$\left(\frac{d\rho}{dt}\right)^2 + \rho^2\left(\frac{d\theta}{dt}\right)^2 + \rho^2\sin^2\theta\left(\frac{d\phi}{dt}\right)^2 = \frac{2GM}{\rho} - E \tag{1.5.139}$$

where $-E$ is a constant of integration. In the special case of a planar orbit we set $\theta = \frac{\pi}{2}$ constant so that the equation (1.5.139) reduces to

$$\begin{aligned}
\left(\frac{d\rho}{dt}\right)^2 + \rho^2\left(\frac{d\phi}{dt}\right)^2 &= \frac{2GM}{\rho} - E \\
\left(\frac{d\rho}{d\phi}\frac{d\phi}{dt}\right)^2 + \rho^2\left(\frac{d\phi}{dt}\right)^2 &= \frac{2GM}{\rho} - E.
\end{aligned} \tag{1.5.140}$$

Also for this special case of planar motion we have

$$\left|\vec{\rho} \times \frac{d\vec{\rho}}{dt}\right| = \rho^2\frac{d\phi}{dt} = h. \tag{1.5.141}$$

By eliminating $\frac{d\phi}{dt}$ from the equation (1.5.140) we obtain the result

$$\left(\frac{d\rho}{d\phi}\right)^2 + \rho^2 = \frac{2GM}{h^2}\rho^3 - \frac{E}{h^2}\rho^4. \tag{1.5.142}$$

The substitution $\rho = \frac{1}{u}$ can be used to represent the equation (1.5.142) in the form

$$\left(\frac{du}{d\phi}\right)^2 + u^2 - \frac{2GM}{h^2}u + \frac{E}{h^2} = 0 \tag{1.5.143}$$

which is a form we will return to later in this section. Note that we can separate the variables in equations (1.5.142) or (1.5.143). The results can then be integrate to produce the equation (1.5.130).

Newton also considered the relative motion of two inertial systems, say S and $\overline{S}$. Consider two such systems as depicted in the figure 1.5-3 where the $\overline{S}$ system is moving in the $x-$direction with speed v relative to the system S.

154

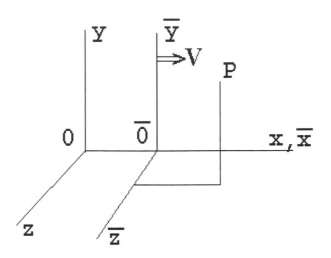

Figure 1.5-3. Relative motion of two inertial systems.

For a Newtonian system, if at time $t = 0$ we have clocks in both systems which coincide, than at time t a point $P(\overline{x}, \overline{y}, \overline{z})$ in the $\overline{S}$ system can be described by the transformation equations

$$
\begin{array}{lll}
x = \overline{x} + v\overline{t} & \qquad & \overline{x} = x - vt \\[4pt]
y = \overline{y} & & \overline{y} = y \\[4pt]
& \text{or} & \\[4pt]
z = \overline{z} & & \overline{z} = z \\[4pt]
t = \overline{t} & & \overline{t} = t.
\end{array}
\tag{1.5.144}
$$

These are the transformation equation of Newton's relativity sometimes referred to as a Galilean transformation.

Before Einstein the principle of relativity required that velocities be additive and obey Galileo's velocity addition rule

$$
V_{P/R} = V_{P/Q} + V_{Q/R}.
\tag{1.5.145}
$$

That is, the velocity of P with respect to R equals the velocity of P with respect to Q plus the velocity of Q with respect to R. For example, a person (P) running north at 3 km/hr on a train (Q) moving north at 60 km/hr with respect to the ground (R) has a velocity of 63 km/hr with respect to the ground. What happens when (P) is a light wave moving on a train (Q) which is moving with velocity V relative to the ground? Are the velocities still additive? This type of question led to the famous Michelson-Morley experiment which has been labeled as the starting point for relativity. Einstein's answer to the above question was "NO" and required that $V_{P/R} = V_{P/Q} = c =$speed of light be a universal constant.

In contrast to the Newtonian equations, Einstein considered the motion of light from the origins 0 and $\overline{0}$ of the systems S and $\overline{S}$. If the $\overline{S}$ system moves with velocity v relative to the S system and at time $t = 0$

a light signal is sent from the S system to the $\overline{S}$ system, then this light signal will move out in a spherical wave front and lie on the sphere

$$x^2 + y^2 + z^2 = c^2 t^2 \tag{1.5.146}$$

where c is the speed of light. Conversely, if a light signal is sent out from the $\overline{S}$ system at time $\bar{t} = 0$, it will lie on the spherical wave front

$$\overline{x}^2 + \overline{y}^2 + \overline{z}^2 = c^2 \overline{t}^2. \tag{1.5.147}$$

Observe that the Newtonian equations (1.5.144) do not satisfy the equations (1.5.146) and (1.5.147) identically. If $y = \overline{y}$ and $z = \overline{z}$ then the space variables $(x, \overline{x})$ and time variables $(t, \overline{t})$ must somehow be related. Einstein suggested the following transformation equations between these variables

$$\overline{x} = \gamma(x - vt) \quad \text{and} \quad x = \gamma(\overline{x} + v\overline{t}) \tag{1.5.148}$$

where γ is a constant to be determined. The differentials of equations (1.5.148) produce

$$d\overline{x} = \gamma(dx - vdt) \quad \text{and} \quad dx = \gamma(d\overline{x} + vd\overline{t}) \tag{1.5.149}$$

from which we obtain the ratios

$$\frac{d\overline{x}}{\gamma(d\overline{x} + v\,d\overline{t})} = \frac{\gamma(dx - v\,dt)}{dx} \quad \text{or} \quad \frac{1}{\gamma(1 + \frac{v}{\frac{d\overline{x}}{d\overline{t}}})} = \gamma(1 - \frac{v}{\frac{dx}{dt}}). \tag{1.5.150}$$

When $\dfrac{d\overline{x}}{d\overline{t}} = \dfrac{dx}{dt} = c$, the speed of light, the equation (1.5.150) requires that

$$\gamma^2 = (1 - \frac{v^2}{c^2})^{-1} \quad \text{or} \quad \gamma = (1 - \frac{v^2}{c^2})^{-1/2}. \tag{1.5.151}$$

From the equations (1.5.148) we eliminate $\overline{x}$ and find

$$\overline{t} = \gamma(t - \frac{v}{c^2}x). \tag{1.5.152}$$

We can now replace the Newtonian equations (1.5.144) by the relativistic transformation equations

$$
\begin{aligned}
x &= \gamma(\overline{x} + v\overline{t}) & \overline{x} &= \gamma(x - vt) \\
y &= \overline{y} & \overline{y} &= y \\
z &= \overline{z} & \overline{z} &= z \\
t &= \gamma(\overline{t} + \frac{v}{c^2}\overline{x}) & \overline{t} &= \gamma(t - \frac{v}{c^2}x)
\end{aligned}
\quad \text{or} \tag{1.5.153}
$$

where γ is given by equation (1.5.151). These equations are also known as the Lorentz transformation. Note that for $v \ll c$, then $\dfrac{v}{c^2} \approx 0$, $\gamma \approx 1$, then the equations (1.5.153) closely approximate the equations (1.5.144). The equations (1.5.153) also satisfy the equations (1.5.146) and (1.5.147) identically as can be readily verified by substitution. Further, by using chain rule differentiation we obtain from the relations (1.5.148) that

$$\frac{dx}{dt} = \frac{\frac{d\overline{x}}{d\overline{t}} + v}{1 + \frac{d\overline{x}}{d\overline{t}}\frac{v}{c}\frac{}{c}}. \tag{1.5.154}$$

The equation (1.5.154) is the Einstein relative velocity addition rule which replaces the previous Newtonian rule given by equation (1.5.145). We can rewrite equation (1.5.154) in the notation of equation (1.5.145) as

$$V_{P/R} = \frac{V_{P/Q} + V_{Q/R}}{1 + \frac{V_{P/Q}}{c}\frac{V_{Q/R}}{c}}. \tag{1.5.155}$$

Observe that when $V_{P/Q} << c$ and $V_{Q/R} << c$ then equation (1.5.155) approximates closely the equation (1.5.145). Also as $V_{P/Q}$ and $V_{Q/R}$ approach the speed of light we have

$$\lim_{\substack{V_{P/Q} \to c \\ V_{Q/R} \to c}} \frac{V_{P/Q} + V_{Q/R}}{1 + \frac{V_{P/Q}}{c}\frac{V_{Q/R}}{c}} = c \tag{1.5.156}$$

which agrees with Einstein's hypothesis that the speed of light is an invariant.

Let us return now to the viewpoint of what gravitation is. Einstein thought of space and time as being related and viewed the motion of the planets as being that of geodesic paths in a space-time continuum. Recall the equations of geodesics are given by

$$\frac{d^2 x^i}{ds^2} + \left\{ \begin{matrix} i \\ j\,k \end{matrix} \right\} \frac{dx^j}{ds}\frac{dx^k}{ds} = 0, \tag{1.5.157}$$

where s is arc length. These equations are to be associated with a 4-dimensional space-time metric g_{ij} where the indices i, j take on the values $1, 2, 3, 4$ and the x^i are generalized coordinates. Einstein asked the question, "Can one introduce a space-time metric g_{ij} such that the equations (1.5.157) can somehow reproduce the law of gravitational attraction $\frac{d^2 \vec{\rho}}{dt^2} + \frac{GM}{\rho^3}\vec{\rho} = 0$?" Then the motion of the planets can be viewed as optimized motion in a space-time continuum where the metrices of the space simulate the law of gravitational attraction. Einstein thought that this motion should be related to the curvature of the space which can be obtained from the Riemann-Christoffel tensor $R^i{}_{jkl}$. The metric we desire g_{ij}, $i, j = 1, 2, 3, 4$ has 16 components. The conjugate metric tensor g^{ij} is defined such that $g^{ij}g_{jk} = \delta^i_k$ and an element of arc length squared is given by $ds^2 = g_{ij}dx^i dx^j$. Einstein thought that the metrices should come from the Riemann-Christoffel curvature tensor which, for $n = 4$ has 256 components, but only 20 of these are linearly independent. This seems like a large number of equations from which to obtain the law of gravitational attraction and so Einstein considered the contracted tensor

$$G_{ij} = R^t{}_{ijt} = \frac{\partial}{\partial x^j}\left\{ \begin{matrix} n \\ i\,n \end{matrix} \right\} - \frac{\partial}{\partial x^n}\left\{ \begin{matrix} n \\ i\,j \end{matrix} \right\} + \left\{ \begin{matrix} m \\ i\,n \end{matrix} \right\}\left\{ \begin{matrix} n \\ m\,j \end{matrix} \right\} - \left\{ \begin{matrix} m \\ i\,j \end{matrix} \right\}\left\{ \begin{matrix} n \\ m\,n \end{matrix} \right\}. \tag{1.5.158}$$

Spherical coordinates (ρ, θ, ϕ) suggests a metric similar to

$$ds^2 = -(d\rho)^2 - \rho^2(d\theta)^2 - \rho^2 \sin^2\theta (d\phi)^2 + c^2(dt)^2$$

where $g_{11} = -1$, $g_{22} = -\rho^2$, $g_{33} = -\rho^2 \sin^2\theta$, $g_{44} = c^2$ and $g_{ij} = 0$ for $i \neq j$. The negative signs are introduced so that $\left(\frac{ds}{dt}\right)^2 = c^2 - v^2$ is positive when $v < c$ and the velocity is not greater than c. However, this metric will not work since the curvature tensor vanishes. The spherical symmetry of the problem suggest that g_{11} and g_{44} change while g_{22} and g_{33} remain fixed. Let $(x^1, x^2, x^3, x^4) = (\rho, \theta, \phi, t)$ and assume

$$g_{11} = -e^u, \quad g_{22} = -\rho^2, \quad g_{33} = -\rho^2 \sin^2\theta, \quad g_{44} = e^v \tag{1.5.159}$$

where u and v are unknown functions of ρ to be determined. This gives the conjugate metric tensor

$$g^{11} = -e^{-u}, \quad g^{22} = \frac{-1}{\rho^2}, \quad g^{33} = \frac{-1}{\rho^2 \sin^2 \theta}, \quad g^{44} = e^{-v} \tag{1.5.160}$$

and $g^{ij} = 0$ for $i \neq j$. This choice of a metric produces

$$ds^2 = -e^u (d\rho)^2 - \rho^2 (d\theta)^2 - \rho^2 \sin^2 \theta (d\phi)^2 + e^v (dt)^2 \tag{1.5.161}$$

together with the nonzero Christoffel symbols

$$\begin{Bmatrix} 1 \\ 1\,1 \end{Bmatrix} = \frac{1}{2}\frac{du}{d\rho} \qquad \begin{Bmatrix} 2 \\ 1\,2 \end{Bmatrix} = \frac{1}{\rho} \qquad \begin{Bmatrix} 3 \\ 1\,3 \end{Bmatrix} = \frac{1}{\rho}$$

$$\begin{Bmatrix} 1 \\ 2\,2 \end{Bmatrix} = -\rho e^{-u} \qquad \begin{Bmatrix} 2 \\ 2\,1 \end{Bmatrix} = \frac{1}{\rho} \qquad \begin{Bmatrix} 3 \\ 2\,3 \end{Bmatrix} = \frac{\cos\theta}{\sin\theta} \qquad \begin{Bmatrix} 4 \\ 1\,4 \end{Bmatrix} = \frac{1}{2}\frac{dv}{d\rho}$$

$$\begin{Bmatrix} 1 \\ 3\,3 \end{Bmatrix} = -\rho e^{-u}\sin^2\theta \qquad \begin{Bmatrix} 2 \\ 3\,3 \end{Bmatrix} = -\sin\theta\cos\theta \qquad \begin{Bmatrix} 3 \\ 3\,1 \end{Bmatrix} = \frac{1}{\rho} \qquad \begin{Bmatrix} 4 \\ 4\,1 \end{Bmatrix} = \frac{1}{2}\frac{dv}{d\rho}.$$

$$\begin{Bmatrix} 1 \\ 4\,4 \end{Bmatrix} = \frac{1}{2}e^{v-u}\frac{dv}{dr} \qquad \begin{Bmatrix} 3 \\ 3\,2 \end{Bmatrix} = \frac{\cos\theta}{\sin\theta} \tag{1.5.162}$$

The equation (1.5.158) is used to calculate the nonzero G_{ij} and we find that

$$G_{11} = \frac{1}{2}\frac{d^2v}{d\rho^2} + \frac{1}{4}\left(\frac{dv}{d\rho}\right)^2 - \frac{1}{4}\frac{du}{d\rho}\frac{dv}{d\rho} - \frac{1}{\rho}\frac{du}{d\rho}$$

$$G_{22} = e^{-u}\left(1 + \frac{1}{2}\rho\frac{dv}{d\rho} - \frac{1}{2}\rho\frac{du}{d\rho} - e^u\right)$$

$$G_{33} = e^{-u}\left(1 + \frac{1}{2}\rho\frac{dv}{d\rho} - \frac{1}{2}\rho\frac{du}{d\rho} - e^u\right)\sin^2\theta \tag{1.5.163}$$

$$G_{44} = -e^{v-u}\left(\frac{1}{2}\frac{d^2v}{d\rho^2} - \frac{1}{4}\frac{du}{d\rho}\frac{dv}{d\rho} + \frac{1}{4}\left(\frac{dv}{d\rho}\right)^2 + \frac{1}{\rho}\frac{dv}{d\rho}\right)$$

and $G_{ij} = 0$ for $i \neq j$. The assumption that $G_{ij} = 0$ for all i, j leads to the differential equations

$$\frac{d^2v}{d\rho^2} + \frac{1}{2}\left(\frac{dv}{d\rho}\right)^2 - \frac{1}{2}\frac{du}{d\rho}\frac{dv}{d\rho} - \frac{2}{\rho}\frac{du}{d\rho} = 0$$

$$1 + \frac{1}{2}\rho\frac{dv}{d\rho} - \frac{1}{2}\rho\frac{du}{d\rho} - e^u = 0 \tag{1.5.164}$$

$$\frac{d^2v}{d\rho^2} + \frac{1}{2}\left(\frac{dv}{d\rho}\right)^2 - \frac{1}{2}\frac{du}{d\rho}\frac{dv}{d\rho} + \frac{2}{\rho}\frac{dv}{d\rho} = 0.$$

Subtracting the first equation from the third equation gives

$$\frac{du}{d\rho} + \frac{dv}{d\rho} = 0 \quad \text{or} \quad u + v = c_1 = \text{constant}. \tag{1.5.165}$$

The second equation in (1.5.164) then becomes

$$\rho\frac{du}{d\rho} = 1 - e^u \tag{1.5.166}$$

158

Separate the variables in equation (1.5.166) and integrate to obtain the result

$$e^u = \frac{1}{1 - \frac{c_2}{\rho}} \tag{1.5.167}$$

where c_2 is a constant of integration and consequently

$$e^v = e^{c_1 - u} = e^{c_1}\left(1 - \frac{c_2}{\rho}\right). \tag{1.5.168}$$

The constant c_1 is selected such that g_{44} approaches c^2 as ρ increases without bound. This produces the metrices

$$g_{11} = \frac{-1}{1 - \frac{c_2}{\rho}}, \quad g_{22} = -\rho^2, \quad g_{33} = -\rho^2 \sin^2\theta, \quad g_{44} = c^2(1 - \frac{c_2}{\rho}) \tag{1.5.169}$$

where c_2 is a constant still to be determined. The metrices given by equation (1.5.169) are now used to expand the equations (1.5.157) representing the geodesics in this four dimensional space. The differential equations representing the geodesics are found to be

$$\frac{d^2\rho}{ds^2} + \frac{1}{2}\frac{du}{d\rho}\left(\frac{d\rho}{ds}\right)^2 - \rho e^{-u}\left(\frac{d\theta}{ds}\right)^2 - \rho e^{-u}\sin^2\theta\left(\frac{d\phi}{ds}\right)^2 + \frac{1}{2}e^{v-u}\frac{dv}{d\rho}\left(\frac{dt}{ds}\right)^2 = 0 \tag{1.5.170}$$

$$\frac{d^2\theta}{ds^2} + \frac{2}{\rho}\frac{d\theta}{ds}\frac{d\rho}{ds} - \sin\theta\cos\theta\left(\frac{d\phi}{ds}\right)^2 = 0 \tag{1.5.171}$$

$$\frac{d^2\phi}{ds^2} + \frac{2}{\rho}\frac{d\phi}{ds}\frac{d\rho}{ds} + 2\frac{\cos\theta}{\sin\theta}\frac{d\phi}{ds}\frac{d\theta}{ds} = 0 \tag{1.5.172}$$

$$\frac{d^2t}{ds^2} + \frac{dv}{d\rho}\frac{dt}{ds}\frac{d\rho}{ds} = 0. \tag{1.5.173}$$

The equation (1.5.171) is identically satisfied if we examine planar orbits where $\theta = \frac{\pi}{2}$ is a constant. This value of θ also simplifies the equations (1.5.170) and (1.5.172). The equation (1.5.172) becomes an exact differential equation

$$\frac{d}{ds}\left(\rho^2\frac{d\phi}{ds}\right) = 0 \quad \text{or} \quad \rho^2\frac{d\phi}{ds} = c_4, \tag{1.5.174}$$

and the equation (1.5.173) also becomes an exact differential

$$\frac{d}{ds}\left(\frac{dt}{ds}e^v\right) = 0 \quad \text{or} \quad \frac{dt}{ds}e^v = c_5, \tag{1.5.175}$$

where c_4 and c_5 are constants of integration. This leaves the equation (1.5.170) which determines ρ. Substituting the results from equations (1.5.174) and (1.5.175), together with the relation (1.5.161), the equation (1.5.170) reduces to

$$\frac{d^2\rho}{ds^2} + \frac{c_2}{2\rho^2} + \frac{c_2 c_4^2}{2\rho^4} - (1 - \frac{c_2}{\rho})\frac{c_4^2}{\rho^3} = 0. \tag{1.5.176}$$

By the chain rule we have

$$\frac{d^2\rho}{ds^2} = \frac{d^2\rho}{d\phi^2}\left(\frac{d\phi}{ds}\right)^2 + \frac{d\rho}{d\phi}\frac{d^2\phi}{ds^2} = \frac{d^2\rho}{d\phi^2}\frac{c_4^2}{\rho^4} + \left(\frac{d\rho}{d\phi}\right)^2\left(\frac{-2c_4^2}{\rho^5}\right)$$

and so equation (1.5.176) can be written in the form

$$\frac{d^2\rho}{d\phi^2} - \frac{2}{\rho}\left(\frac{d\rho}{d\phi}\right)^2 + \frac{c_2}{2}\frac{\rho^2}{c_4^2} + \frac{c_2}{2} - \left(1 - \frac{c_2}{\rho}\right)\rho = 0. \tag{1.5.177}$$

The substitution $\rho = \frac{1}{u}$ reduces the equation (1.5.177) to the form

$$\frac{d^2u}{d\phi^2} + u - \frac{c_2}{2c_4^2} = \frac{3}{2}c_2u^2. \tag{1.5.178}$$

Multiply the equation (1.5.178) by $2\frac{du}{d\phi}$ and integrate with respect to ϕ to obtain

$$\left(\frac{du}{d\phi}\right)^2 + u^2 - \frac{c_2}{c_4^2}u = c_2u^3 + c_6. \tag{1.5.179}$$

where c_6 is a constant of integration. To determine the constant c_6 we write the equation (1.5.161) in the special case $\theta = \frac{\pi}{2}$ and use the substitutions from the equations (1.5.174) and (1.5.175) to obtain

$$e^u\left(\frac{d\rho}{ds}\right)^2 = e^u\left(\frac{d\rho}{d\phi}\frac{d\phi}{ds}\right)^2 = 1 - \rho^2\left(\frac{d\phi}{ds}\right)^2 + e^v\left(\frac{dt}{ds}\right)^2$$

or

$$\left(\frac{d\rho}{d\phi}\right)^2 + \left(1 - \frac{c_2}{\rho}\right)\rho^2 + \left(1 - \frac{c_2}{\rho} - \frac{c_5^2}{c^2}\right)\frac{\rho^4}{c_4^2} = 0. \tag{1.5.180}$$

The substitution $\rho = \frac{1}{u}$ reduces the equation (1.5.180) to the form

$$\left(\frac{du}{d\phi}\right)^2 + u^2 - c_2u^3 + \frac{1}{c_4^2} - \frac{c_2}{c_4^2}u - \frac{c_5^2}{c^2c_4^2} = 0. \tag{1.5.181}$$

Now comparing the equations (1.5.181) and (1.5.179) we select

$$c_6 = \left(\frac{c_5^2}{c^2} - 1\right)\frac{1}{c_4^2}$$

so that the equation (1.5.179) takes on the form

$$\left(\frac{du}{d\phi}\right)^2 + u^2 - \frac{c_2}{c_4^2}u + \left(1 - \frac{c_5^2}{c^2}\right)\frac{1}{c_4^2} = c_2u^3 \tag{1.5.182}$$

Now we can compare our relativistic equation (1.5.182) with our Newtonian equation (1.5.143). In order that the two equations almost agree we select the constants c_2, c_4, c_5 so that

$$\frac{c_2}{c_4^2} = \frac{2GM}{h^2} \quad \text{and} \quad \frac{1 - \frac{c_5^2}{c^2}}{c_4^2} = \frac{E}{h^2}. \tag{1.5.183}$$

The equations (1.5.183) are only two equations in three unknowns and so we use the additional equation

$$\lim_{\rho \to \infty} \rho^2\frac{d\phi}{dt} = \lim_{\rho \to \infty} \rho^2\frac{d\phi}{ds}\frac{ds}{dt} = h \tag{1.5.184}$$

which is obtained from equation (1.5.141). Substituting equations (1.5.174) and (1.5.175) into equation (1.5.184), rearranging terms and taking the limit we find that

$$\frac{c_4c^2}{c_5} = h. \tag{1.5.185}$$

From equations (1.5.183) and (1.5.185) we obtain the results that

$$c_5^2 = \frac{c^2}{1 + \frac{E}{c^2}}, \quad c_2 = \frac{2GM}{c^2}\left(\frac{1}{1 + E/c^2}\right), \quad c_4 = \frac{h}{c\sqrt{1 + E/c^2}} \tag{1.5.186}$$

These values substituted into equation (1.5.181) produce the differential equation

$$\left(\frac{du}{d\phi}\right)^2 + u^2 - \frac{2GM}{h^2}u + \frac{E}{h^2} = \frac{2GM}{h^2}\left(\frac{1}{1 + E/c^2}\right)u^3. \tag{1.5.187}$$

Let $\alpha = \frac{c_2}{c_4^2} = \frac{2GM}{h^2}$ and $\beta = c_2 = \frac{2GM}{c^2}(\frac{1}{1+E/c^2})$ then the differential equation (1.5.178) can be written as

$$\frac{d^2u}{d\phi^2} + u - \frac{\alpha}{2} = \frac{3}{2}\beta u^2. \tag{1.5.188}$$

We know the solution to equation (1.5.143) is given by

$$u = \frac{1}{\rho} = A(1 + \epsilon\cos(\phi - \phi_0)) \tag{1.5.189}$$

and so we assume a solution to equation (1.5.188) of this same general form. We know that A is small and so we make the assumption that the solution of equation (1.5.188) given by equation (1.5.189) is such that ϕ_0 is approximately constant and varies slowly as a function of $A\phi$. Observe that if $\phi_0 = \phi_0(A\phi)$, then $\frac{d\phi_0}{d\phi} = \phi_0'A$ and $\frac{d^2\phi_0}{d\phi^2} = \phi_0''A^2$, where primes denote differentiation with respect to the argument of the function. (i.e. $A\phi$ for this problem.) The derivatives of equation (1.5.189) produce

$$\frac{du}{d\phi} = -\epsilon A\sin(\phi - \phi_0)(1 - \phi_0'A)$$

$$\frac{d^2u}{d\phi^2} = \epsilon A^3\sin(\phi - \phi_0)\phi_0'' - \epsilon A\cos(\phi - \phi_0)(1 - 2A\phi_0' + A^2(\phi_0')^2)$$

$$= -\epsilon A\cos(\phi - \phi_0) + 2\epsilon A^2\phi_0'\cos(\phi - \phi_0) + O(A^3).$$

Substituting these derivatives into the differential equation (1.5.188) produces the equations

$$2\epsilon A^2\phi_0'\cos(\phi - \phi_0) + A - \frac{\alpha}{2} = \frac{3\beta}{2}\left(A^2 + 2\epsilon A^2\cos(\phi - \phi_0) + \epsilon^2 A^2\cos^2(\phi - \phi_0)\right) + O(A^3).$$

Now A is small so that terms $O(A^3)$ can be neglected. Equating the constant terms and the coefficient of the $\cos(\phi - \phi_0)$ terms we obtain the equations

$$A - \frac{\alpha}{2} = \frac{3\beta}{2}A^2 \qquad 2\epsilon A^2\phi_0' = 3\beta\epsilon A^2 + \frac{3\beta}{2}\epsilon^2 A^2\cos(\phi - \phi_0).$$

Treating ϕ_0 as essentially constant, the above system has the approximate solutions

$$A \approx \frac{\alpha}{2} \qquad \phi_0 \approx \frac{3\beta}{2}A\phi + \frac{3\beta}{4}A\epsilon\sin(\phi - \phi_0) \tag{1.5.190}$$

The solutions given by equations (1.5.190) tells us that ϕ_0 varies slowly with time. For ϵ less than 1, the elliptical motion is affected by this change in ϕ_0. It causes the semi-major axis of the ellipse to slowly rotate at a rate given by $\frac{d\phi_0}{dt}$. Using the following values for the planet Mercury

$$G = 6.67(10^{-8})\, \text{dyne cm}^2/\text{g}^2$$

$$c = 3(10^{10})\, \text{cm/sec}$$

$$M = 1.99(10^{33})\, \text{g}$$

$$\beta \approx \frac{2GM}{c^2} = 2.95(10^5)\, \text{cm}$$

$$a = 5.78(10^{12})\, \text{cm}$$

$$h \approx \sqrt{GMa(1-\epsilon^2)} = 2.71(10^{19})\, \text{cm}^2/\text{sec}$$

$$\epsilon = 0.206$$

$$\frac{d\phi}{dt} \approx \left(\frac{GM}{a^3}\right)^{1/2}\, \text{sec}^{-1}\, \text{Kepler's third law}$$

$$(1.5.191)$$

we calculate the slow rate of rotation of the semi-major axis to be approximately

$$\frac{d\phi_0}{dt} = \frac{d\phi_0}{d\phi}\frac{d\phi}{dt} \approx \frac{3}{2}\beta A \frac{d\phi}{dt} \approx 3\left(\frac{GM}{ch}\right)^2\left(\frac{GM}{a^3}\right)^{1/2} = 6.628(10^{-14})\, \text{rad/sec}$$

$$= 43.01 \quad \text{seconds of arc per century.}$$

$$(1.5.192)$$

It has long been known that the planet Mercury's orbit revolves about the Sun with an extremely slow rate of revolution. This slow variation in Mercury's semi-major axis has been observed and measured and is in agreement with the above value. Newtonian mechanics could not account for the changes in Mercury's semi-major axis, but Einstein's theory of relativity does give this prediction. The resulting solution of equation (1.5.188) can be viewed as being caused by the curvature of the space-time continuum.

The contracted curvature tensor G_{ij} set equal to zero is just one of many conditions that can be assumed in order to arrive at a metric for the space-time continuum. Any assumption on the value of G_{ij} relates to imposing some kind of curvature on the space. Within the large expanse of our universe only our imaginations limit us as to how space, time and matter interact. You can also imagine the existence of other tensor metrics in higher dimensional spaces where the geodesics within the space-time continuum give rise to the motion of other physical quantities.

Other predications using Einstein's general theory of relativity are (i) the bending of light beams by gravitational fields. This prediction was verified during varous eclipses that occured in the 1910-1930 time frame. (ii) Gravitational redshift where periodic processes within atoms on the Sun move at a slower rate than on Earth. This was verified by measuring wavelengths of the radiation of a given element from the Sun compared with wavelength measured in a labortory. The wavelengths of elements from the Sun being longer. (iii) The existence of "black holes". These objects are extremely dense gravitational fields which have an event horizon where things can enter but cannot escape. Not even light can escape. This prediction was verified around 1970 with the discovery of a black hole in the vacinity of the constellation Cygnus.

This short introduction to relativity is concluded with a quote from the NASA News@hg.nasa.gov news release, spring 1998, Release:98-51. "An international team of NASA and university researchers has found the first direct evidence of a phenomenon predicted 80 years ago using Einstein's theory of general relativity–that the Earth is dragging space and time around itself as it rotates." The news release explains that the effect is known as frame dragging and goes on to say "Frame dragging is like what happens if a bowling ball spins in a thick fluid such as molasses. As the ball spins, it pulls the molasses around itself. Anything stuck in the molasses will also move around the ball. Similarly, as the Earth rotates it pulls space-time in its vicinity around itself. This will shift the orbits of satellites near the Earth." This research is reported in the journal Science.

EXERCISE 1.5

▶ **1.** Let $\kappa = \frac{\delta \vec{T}}{\delta s} \cdot \vec{N}$ and $\tau = \frac{\delta \vec{N}}{\delta s} \cdot \vec{B}$. Assume in turn that each of the intrinsic derivatives of $\vec{T}, \vec{N}, \vec{B}$ are some linear combination of $\vec{T}, \vec{N}, \vec{B}$ and hence derive the Frenet-Serret formulas of differential geometry.

▶ **2.** Determine the given surfaces. Describe and sketch the curvilinear coordinates upon each surface.

(a) $\vec{r}(u, v) = u \, \widehat{\mathbf{e}}_1 + v \, \widehat{\mathbf{e}}_2$
(b) $\vec{r}(u, v) = u \cos v \, \widehat{\mathbf{e}}_1 + u \sin v \, \widehat{\mathbf{e}}_2$
(c) $\vec{r}(u, v) = \frac{2uv^2}{u^2 + v^2} \, \widehat{\mathbf{e}}_1 + \frac{2u^2 v}{u^2 + v^2} \, \widehat{\mathbf{e}}_2.$

▶ **3.** Determine the given surfaces and describe the curvilinear coordinates upon the surface. Use some graphics package to plot the surface and illustrate the coordinate curves on the surface. Find element of area dS in terms of u and v.

(a) $\vec{r}(u, v) = a \sin u \cos v \, \widehat{\mathbf{e}}_1 + b \sin u \sin v \, \widehat{\mathbf{e}}_2 + c \cos u \, \widehat{\mathbf{e}}_3$ $\quad a, b, c$ constants $\quad 0 \leq u, v \leq 2\pi$

(b) $\vec{r}(u, v) = \left(4 + v \sin \frac{u}{2}\right) \cos u \, \widehat{\mathbf{e}}_1 + \left(4 + v \sin \frac{u}{2}\right) \sin u \, \widehat{\mathbf{e}}_2 + v \cos \frac{u}{2} \, \widehat{\mathbf{e}}_3$ $\quad -1 \leq v \leq 1, \quad 0 \leq u \leq 2\pi$

(c) $\vec{r}(u, v) = au \cos v \, \widehat{\mathbf{e}}_1 + bu \sin v \, \widehat{\mathbf{e}}_2 + cu \, \widehat{\mathbf{e}}_3$

(d) $\vec{r}(u, v) = u \cos v \, \widehat{\mathbf{e}}_1 + u \sin v \, \widehat{\mathbf{e}}_2 + \alpha v \, \widehat{\mathbf{e}}_3$ $\quad \alpha$ constant

(e) $\vec{r}(u, v) = a \cos v \, \widehat{\mathbf{e}}_1 + b \sin v \, \widehat{\mathbf{e}}_2 + u \, \widehat{\mathbf{e}}_3$ $\quad a, b$ constant

(f) $\vec{r}(u, v) = u \cos v \, \widehat{\mathbf{e}}_1 + u \sin v \, \widehat{\mathbf{e}}_2 + u^2 \, \widehat{\mathbf{e}}_3$

▶ **4.** Consider a two dimensional space with metric tensor $(a_{\alpha\beta}) = \begin{pmatrix} E & F \\ F & G \end{pmatrix}$. Assume that the surface is described by equations of the form $y^i = y^i(u, v)$ and that any point on the surface is given by the position vector $\vec{r} = \vec{r}(u, v) = y^i \, \widehat{\mathbf{e}}_i$. Show that the metrices E, F, G are functions of the parameters u, v and are given by

$$E = \vec{r}_u \cdot \vec{r}_u, \qquad F = \vec{r}_u \cdot \vec{r}_v, \quad G = \vec{r}_v \cdot \vec{r}_v \quad \text{where} \quad \vec{r}_u = \frac{\partial \vec{r}}{\partial u} \quad \text{and} \quad \vec{r}_v = \frac{\partial \vec{r}}{\partial v}.$$

▶ **5.** For the metric given in problem 4 show that the Christoffel symbols of the first kind are given by

$$[1\,1, 1] = \vec{r}_u \cdot \vec{r}_{uu} \qquad [1\,2, 1] = [2\,1, 1] = \vec{r}_u \cdot \vec{r}_{uv} \qquad [2\,2, 1] = \vec{r}_u \cdot \vec{r}_{vv}$$

$$[1\,1, 2] = \vec{r}_v \cdot \vec{r}_{uu} \qquad [1\,2, 2] = [2\,1, 2] = \vec{r}_v \cdot \vec{r}_{uv} \qquad [2\,2, 2] = \vec{r}_v \cdot \vec{r}_{vv}$$

which can be represented $[\alpha\,\beta, \gamma] = \frac{\partial^2 \vec{r}}{\partial u^\alpha \partial u^\beta} \cdot \frac{\partial \vec{r}}{\partial u^\gamma}, \quad \alpha, \beta, \gamma = 1, 2.$

▶ **6.** Show that the results in problem 5 can also be written in the form

$$[1\,1, 1] = \frac{1}{2} E_u \qquad [1\,2, 1] = [2\,1, 1] = \frac{1}{2} E_v \qquad [2\,2, 1] = F_v - \frac{1}{2} G_u$$

$$[1\,1, 2] = F_u - \frac{1}{2} E_v \qquad [1\,2, 2] = [2\,1, 2] = \frac{1}{2} G_u \qquad [2\,2, 2] = \frac{1}{2} G_v$$

where the subscripts indicate partial differentiation.

▶ **7.** For the metric given in problem 4, show that the Christoffel symbols of the second kind can be expressed in the form $\left\{ \begin{matrix} \gamma \\ \alpha\,\beta \end{matrix} \right\} = a^{\gamma\delta} [\alpha\,\beta, \delta], \qquad \alpha, \beta, \gamma = 1, 2$ and produce the results

$$\left\{ \begin{matrix} 1 \\ 1\,1 \end{matrix} \right\} = \frac{GE_u - 2FF_u + FE_v}{2(EG - F^2)} \qquad \left\{ \begin{matrix} 1 \\ 1\,2 \end{matrix} \right\} = \left\{ \begin{matrix} 1 \\ 2\,1 \end{matrix} \right\} = \frac{GE_v - FG_u}{2(EG - F^2)} \qquad \left\{ \begin{matrix} 2 \\ 1\,1 \end{matrix} \right\} = \frac{2EF_u - EE_v - FE_u}{2(EG - F^2)}$$

$$\left\{ \begin{matrix} 1 \\ 2\,2 \end{matrix} \right\} = \frac{2GF_v - GG_u - FG_v}{2(EG - F^2)} \qquad \left\{ \begin{matrix} 2 \\ 1\,2 \end{matrix} \right\} = \left\{ \begin{matrix} 2 \\ 2\,1 \end{matrix} \right\} = \frac{EG_u - FE_v}{2(EG - F^2)} \qquad \left\{ \begin{matrix} 2 \\ 2\,2 \end{matrix} \right\} = \frac{EG_v - 2FF_v + FG_u}{2(EG - F^2)}$$

where the subscripts indicate partial differentiation.

8. Derive the Gauss equations by assuming that

$$\vec{r}_{uu} = c_1\vec{r}_u + c_2\vec{r}_v + c_3\widehat{n}, \qquad \vec{r}_{uv} = c_4\vec{r}_u + c_5\vec{r}_v + c_6\widehat{n}, \qquad \vec{r}_{vv} = c_7\vec{r}_u + c_8\vec{r}_v + c_9\widehat{n}$$

where $c_1, \ldots, c_9$ are constants determined by taking dot products of the above vectors with the vectors $\vec{r}_u, \vec{r}_v$, and $\widehat{n}$. Show that $c_1 = \left\{\begin{array}{c} 1 \\ 11 \end{array}\right\}$, $c_2 = \left\{\begin{array}{c} 2 \\ 11 \end{array}\right\}$, $c_3 = e$, $c_4 = \left\{\begin{array}{c} 1 \\ 12 \end{array}\right\}$, $c_5 = \left\{\begin{array}{c} 2 \\ 12 \end{array}\right\}$, $c_6 = f$,

$c_7 = \left\{\begin{array}{c} 1 \\ 22 \end{array}\right\}$, $c_8 = \left\{\begin{array}{c} 2 \\ 22 \end{array}\right\}$, $c_9 = g$ Show the Gauss equations can be written $\dfrac{\partial^2 \vec{r}}{\partial u^\alpha \partial u^\beta} = \left\{\begin{array}{c} \gamma \\ \alpha\,\beta \end{array}\right\} \dfrac{\partial \vec{r}}{\partial u^\gamma} + b_{\alpha\beta}\widehat{n}$.

9. Derive the Weingarten equations

$$\begin{aligned} \widehat{n}_u &= c_1\vec{r}_u + c_2\vec{r}_v \\ \widehat{n}_v &= c_3\vec{r}_u + c_4\vec{r}_v \end{aligned} \qquad \text{and} \qquad \begin{aligned} \vec{r}_u &= c_1^*\widehat{n}_u + c_2^*\widehat{n}_v \\ \vec{r}_v &= c_3^*\widehat{n}_u + c_4^*\widehat{n}_v \end{aligned}$$

and show

$$c_1 = \frac{fF - eG}{EG - F^2} \qquad c_3 = \frac{gF - fG}{EG - F^2} \qquad c_1^* = \frac{fF - gE}{eg - f^2} \qquad c_3^* = \frac{fG - gF}{eg - f^2}$$

$$c_2 = \frac{eF - fE}{EG - F^2} \qquad c_4 = \frac{fF - gE}{EG - F^2} \qquad c_2^* = \frac{fE - eF}{eg - f^2} \qquad c_4^* = \frac{fF - eG}{eg - f^2}$$

The constants in the above equations are determined in a manner similar to that suggested in problem 8. Show that the Weingarten equations can be written in the form

$$\frac{\partial \widehat{n}}{\partial u^\alpha} = -b_\alpha^\beta \frac{\partial \vec{r}}{\partial u^\beta}.$$

10. Using $\widehat{n} = \dfrac{\vec{r}_u \times \vec{r}_v}{\sqrt{EG - F^2}}$, the results from exercise 1.1, problem 9(a), and the results from problem 5, verify that

$$(\vec{r}_u \times \vec{r}_{uu}) \cdot \widehat{n} = \left\{\begin{array}{c} 2 \\ 11 \end{array}\right\} \sqrt{EG - F^2}$$

$$(\vec{r}_u \times \vec{r}_{uv}) \cdot \widehat{n} = \left\{\begin{array}{c} 2 \\ 12 \end{array}\right\} \sqrt{EG - F^2} \qquad (\vec{r}_v \times \vec{r}_{uv}) \cdot \widehat{n} = -\left\{\begin{array}{c} 1 \\ 21 \end{array}\right\} \sqrt{EG - F^2}$$

$$(\vec{r}_v \times \vec{r}_{uu}) \cdot \widehat{n} = -\left\{\begin{array}{c} 1 \\ 11 \end{array}\right\} \sqrt{EG - F^2} \qquad (\vec{r}_v \times \vec{r}_{vv}) \cdot \widehat{n} = -\left\{\begin{array}{c} 1 \\ 22 \end{array}\right\} \sqrt{EG - F^2}$$

$$(\vec{r}_u \times \vec{r}_{vv}) \cdot \widehat{n} = \left\{\begin{array}{c} 2 \\ 22 \end{array}\right\} \sqrt{EG - F^2} \qquad (\vec{r}_u \times \vec{r}_v) \cdot \widehat{n} = \sqrt{EG - F^2}$$

and then derive the formula for the geodesic curvature given by equation (1.5.48).

Hint:$(\widehat{n} \times \vec{T}) \cdot \dfrac{d\vec{T}}{ds} = (\vec{T} \times \dfrac{d\vec{T}}{ds}) \cdot \widehat{n}$ and $a^{\alpha\delta}]\beta\,\gamma, \delta] = \left\{\begin{array}{c} \alpha \\ \beta\,\gamma \end{array}\right\}$.

164

▶ **11.** Verify the equation (1.5.39) which shows that the normal curvature directions are orthogonal. i.e. verify that $G\lambda_1\lambda_2 + F(\lambda_1 + \lambda_2) + E = 0$.

▶ **12.** Verify that $\delta^{\beta\gamma}_{\sigma\tau}\delta^{\omega\alpha}_{\lambda\nu}R_{\omega\alpha\beta\gamma} = 4R_{\lambda\nu\sigma\tau}$.

▶ **13.** Find the first fundamental form and unit normal to the surface defined by $z = f(x, y)$.

▶ **14.** Verify

$$A_{i,jk} - A_{i,kj} = A_\sigma R^\sigma_{.ijk}$$

where

$$R^\sigma_{.ijk} = \frac{\partial}{\partial x^j}\begin{Bmatrix} \sigma \\ i\,k \end{Bmatrix} - \frac{\partial}{\partial x^k}\begin{Bmatrix} \sigma \\ i\,j \end{Bmatrix} + \begin{Bmatrix} n \\ i\,k \end{Bmatrix}\begin{Bmatrix} \sigma \\ n\,j \end{Bmatrix} - \begin{Bmatrix} n \\ i\,j \end{Bmatrix}\begin{Bmatrix} \sigma \\ n\,k \end{Bmatrix}.$$

which is sometimes written

$$R^\sigma_{.ijk} = \begin{vmatrix} \frac{\partial}{\partial x^j} & \frac{\partial}{\partial x^k} \\ \begin{Bmatrix} \sigma \\ i\,j \end{Bmatrix} & \begin{Bmatrix} \sigma \\ i\,k \end{Bmatrix} \end{vmatrix} + \begin{vmatrix} \begin{Bmatrix} n \\ i\,k \end{Bmatrix} & \begin{Bmatrix} n \\ i\,j \end{Bmatrix} \\ \begin{Bmatrix} \sigma \\ n\,k \end{Bmatrix} & \begin{Bmatrix} \sigma \\ n\,j \end{Bmatrix} \end{vmatrix}$$

▶ **15.** For $R_{ijkl} = g_{i\sigma}R^\sigma_{.jkl}$ show

$$R_{injk} = \frac{\partial}{\partial x^j}[nk, i] - \frac{\partial}{\partial x^k}[nj, i] + [ik, s]\begin{Bmatrix} s \\ n\,j \end{Bmatrix} - [ij, s]\begin{Bmatrix} s \\ n\,k \end{Bmatrix}$$

which is sometimes written

$$R_{injk} = \begin{vmatrix} \frac{\partial}{\partial x^j} & \frac{\partial}{\partial x^k} \\ [nj, i] & [nk, i] \end{vmatrix} + \begin{vmatrix} \begin{Bmatrix} s \\ n\,j \end{Bmatrix} & \begin{Bmatrix} s \\ n\,k \end{Bmatrix} \\ [ij, s] & [ik, s] \end{vmatrix}$$

▶ **16.** Show

$$R_{ijkl} = \frac{1}{2}\left(\frac{\partial^2 g_{il}}{\partial x^j \partial x^k} - \frac{\partial^2 g_{jl}}{\partial x^i \partial x^k} - \frac{\partial^2 g_{ik}}{\partial x^j \partial x^l} + \frac{\partial^2 g_{jk}}{\partial x^i \partial x^l}\right) + g^{\alpha\beta}\left([jk, \beta][il, \alpha] - [jl, \beta][ik, \alpha]\right).$$

▶ **17.** Show that

$$(i)\quad R_{jikl} = -R_{ijkl}, \qquad (ii)\quad R_{ijlk} = -R_{ijkl}, \qquad (iii)\quad R_{klij} = R_{ijkl}$$

Hence, the tensor R_{ijkl} is skew-symmetric in the indices i, j and k, l. Also the tensor R_{ijkl} is symmetric with respect to the (ij) and (kl) pair of indices.

▶ **18.** Verify the following cyclic properties of the Riemann Christoffel symbol:

$$(i)\quad R_{nijk} + R_{njki} + R_{nkij} = 0 \qquad \text{first index fixed}$$
$$(ii)\quad R_{injk} + R_{jnki} + R_{knij} = 0 \qquad \text{second index fixed}$$
$$(iii)\quad R_{ijnk} + R_{jkni} + R_{kinj} = 0 \qquad \text{third index fixed}$$
$$(iv)\quad R_{ikjn} + R_{kjin} + R_{jikn} = 0 \qquad \text{fourth index fixed}$$

▶ **19.** By employing the results from the previous problems, show all components of the form:
$R_{iijk},\quad R_{injj},\quad R_{iijj},\quad R_{iiii},\quad$ (no summation on i or j) must be zero.

20. Find the number of independent components associated with the Riemann Christoffel tensor R_{ijkm}, $i, j, k, m = 1, 2, \ldots, N$. There are N^4 components to examine in an $N-$dimensional space. Many of these components are zero and many of the nonzero components are related to one another by symmetries or the cyclic properties. Verify the following cases:

CASE I We examine components of the form R_{inin}, $i \neq n$ with no summation of i or n. The first index can be chosen in N ways and therefore with $i \neq n$ the second index can be chosen in $N - 1$ ways. Observe that $R_{inin} = R_{nini}$, (no summation on i or n) and so one half of the total combinations are repeated. This leaves $M_1 = \frac{1}{2}N(N - 1)$ components of the form R_{inin}. The quantity M_1 can also be thought of as the number of distinct pairs of indices (i, n).

CASE II We next examine components of the form R_{inji}, $i \neq n \neq j$ where there is no summation on the index i. We have previously shown that the first pair of indices can be chosen in M_1 ways. Therefore, the third index can be selected in $N - 2$ ways and therefore there are $M_2 = \frac{1}{2}N(N - 1)(N - 2)$ distinct components of the form R_{inji} with $i \neq n \neq j$.

CASE III Next examine components of the form R_{injk} where $i \neq n \neq j \neq k$. From CASE I the first pairs of indices (i, n) can be chosen in M_1 ways. Taking into account symmetries, it can be shown that the second pair of indices can be chosen in $\frac{1}{2}(N - 2)(N - 3)$ ways. This implies that there are $\frac{1}{4}N(N-1)(N-2)(N-3)$ ways of choosing the indices i, n, j and k with $i \neq n \neq j \neq k$. By symmetry the pairs (i, n) and (j, k) can be interchanged and therefore only one half of these combinations are distinct. This leaves

$$\frac{1}{8}N(N - 1)(N - 2)(N - 3)$$

distinct pairs of indices. Also from the cyclic relations we find that only two thirds of the above components are distinct. This produces

$$M_3 = \frac{N(N - 1)(N - 2)(N - 3)}{12}$$

distinct components of the form R_{injk} with $i \neq n \neq j \neq k$.

Adding the above components from each case we find there are

$$M_4 = M_1 + M_2 + M_3 = \frac{N^2(N^2 - 1)}{12}$$

distinct and independent components.

Verify the entries in the following table:

Dimension of space N	1	2	3	4	5
Number of components N^4	1	16	81	256	625
M_4 = Independent components of R_{ijkm}	0	1	6	20	50

Note 1: A one dimensional space can not be curved and all one dimensional spaces are Euclidean. (i.e. if we have an element of arc length squared given by $ds^2 = f(x)(dx)^2$, we can make the coordinate transformation $\sqrt{f(x)}dx = du$ and reduce the arc length squared to the form $ds^2 = du^2$.)

Note 2: In a two dimensional space, the indices can only take on the values 1 and 2. In this special case there are 16 possible components. It can be shown that the only nonvanishing components are:

$$R_{1212} = -R_{1221} = -R_{2112} = R_{2121}.$$

For these nonvanishing components only one independent component exists. By convention, the component R_{1212} is selected as the single independent component and all other nonzero components are expressed in terms of this component.

Find the nonvanishing independent components R_{ijkl} for $i, j, k, l = 1, 2, 3, 4$ and show that

$$
\begin{array}{cccc}
R_{1212} & R_{3434} & R_{2142} & R_{4124} \\
R_{1313} & R_{1231} & R_{2342} & R_{4314} \\
R_{2323} & R_{1421} & R_{3213} & R_{4234} \\
R_{1414} & R_{1341} & R_{3243} & R_{1324} \\
R_{2424} & R_{2132} & R_{3143} & R_{1432}
\end{array}
$$

can be selected as the twenty independent components.

▶ **21.**

(a) For $N = 2$ show R_{1212} is the only nonzero independent component and
$$R_{1212} = R_{2121} = -R_{1221} = -R_{2112}.$$

(b) Show that on the surface of a sphere of radius r_0 we have $R_{1212} = r_0^2 \sin^2 \theta$.

▶ **22.** Show for $N = 2$ that
$$\overline{R}_{1212} = R_{1212}J^2 = R_{1212}\left|\frac{\partial x}{\partial \overline{x}}\right|^2$$

▶ **23.** Define $R_{ij} = R^s_{.ijs}$ as the Ricci tensor and $G^i_j = R^i_j - \frac{1}{2}\delta^i_j R$ as the Einstein tensor, where $R^i_j = g^{ik}R_{kj}$ and $R = R^i_i$. Show that

$$(a) \quad R_{jk} = g^{ab}R_{jabk}$$

$$(b) \quad R_{ij} = \frac{\partial^2 \log \sqrt{g}}{\partial x^i \partial x^j} - \left\{\begin{matrix} b \\ i\,j \end{matrix}\right\}\frac{\partial \log \sqrt{g}}{\partial x^b} - \frac{\partial}{\partial x^a}\left\{\begin{matrix} a \\ i\,j \end{matrix}\right\} + \left\{\begin{matrix} b \\ i\,a \end{matrix}\right\}\left\{\begin{matrix} a \\ j\,b \end{matrix}\right\}$$

$$(c) \quad R^i_{ijk} = 0$$

▶ **24.** By employing the results from the previous problem show that in the case $N = 2$ we have

$$\frac{R_{11}}{g_{11}} = \frac{R_{22}}{g_{22}} = \frac{R_{12}}{g_{12}} = -\frac{R_{1212}}{g}$$

where g is the determinant of g_{ij}.

▶ **25.** Consider the case $N = 2$ where we have $g_{12} = g_{21} = 0$ and show that

$$(a) \quad R_{12} = R_{21} = 0 \qquad\qquad (c) \quad R = \frac{2R_{1221}}{g_{11}g_{22}}$$

$$(b) \quad R_{11}g_{22} = R_{22}g_{11} = R_{1221} \qquad (d) \quad R_{ij} = \frac{1}{2}Rg_{ij}, \quad \text{where} \quad R = g^{ij}R_{ij}$$

The scalar invariant R is known as the Einstein curvature of the surface and the tensor $G^i_j = R^i_j - \frac{1}{2}\delta^i_j R$ is known as the Einstein tensor.

▶ **26.** For $N = 3$ show that $R_{1212}, R_{1313}, R_{2323}, R_{1213}, R_{2123}, R_{3132}$ are independent components of the Riemann Christoffel tensor.

27. For $N = 2$ and $a_{\alpha\beta} = \begin{pmatrix} a_{11} & 0 \\ 0 & a_{22} \end{pmatrix}$ show that

$$K = \frac{R_{1212}}{a} = -\frac{1}{2\sqrt{a}}\left[\frac{\partial}{\partial u^1}\left(\frac{1}{\sqrt{a}}\frac{\partial a_{22}}{\partial u^1}\right) + \frac{\partial}{\partial u^2}\left(\frac{1}{\sqrt{a}}\frac{\partial a_{11}}{\partial u^2}\right)\right].$$

28. For $N = 2$ and $a_{\alpha\beta} = \begin{pmatrix} a_{11} & a_{12} \\ a_{21} & a_{22} \end{pmatrix}$ show that

$$K = \frac{1}{2\sqrt{a}}\left\{\frac{\partial}{\partial u^1}\left[\frac{a_{12}}{a_{11}\sqrt{a}}\frac{\partial a_{11}}{\partial u^2} - \frac{1}{\sqrt{a}}\frac{\partial a_{22}}{\partial u^1}\right] + \frac{\partial}{\partial u^2}\left[\frac{2}{\sqrt{a}}\frac{\partial a_{12}}{\partial u^1} - \frac{1}{\sqrt{a}}\frac{\partial a_{11}}{\partial u^2} - \frac{a_{12}}{a_{11}\sqrt{a}}\frac{\partial a_{11}}{\partial u^1}\right]\right\}.$$

Check your results by setting $a_{12} = a_{21} = 0$ and comparing this answer with that given in the problem 27.

29. Write out the Frenet-Serret formulas (1.5.112)(1.5.113) for surface curves in terms of Christoffel symbols of the second kind.

30.

(a) Use the fact that for $n = 2$ we have $R_{1212} = R_{2121} = -R_{2112} = -R_{1221}$ together with $e_{\alpha\beta}$, $e^{\alpha\beta}$ the two dimensional alternating tensors to show that the equation (1.5.110) can be written as

$$R_{\alpha\beta\gamma\delta} = K\epsilon_{\alpha\beta}\epsilon_{\gamma\delta} \quad \text{where} \quad \epsilon_{\alpha\beta} = \sqrt{a}\,e_{\alpha\beta} \quad \text{and} \quad \epsilon^{\alpha\beta} = \frac{1}{\sqrt{a}}e^{\alpha\beta}$$

are the corresponding epsilon tensors.

(b) Show that from the result in part (a) we obtain $\frac{1}{4}R_{\alpha\beta\gamma\delta}\epsilon^{\alpha\beta}\epsilon^{\gamma\delta} = K$.

Hint: See equations (1.3.82),(1.5.93) and (1.5.94).

31. Verify the result given by the equation (1.5.100).

32. Show that $a^{\alpha\beta}c_{\alpha\beta} = 4H^2 - 2K$.

33. Find equations for the principal curvatures associated with the surface

$$x = u, \quad y = v, \quad z = f(u, v).$$

34. <u>Geodesics on a sphere</u> Let (θ, ϕ) denote the surface coordinates of the sphere of radius ρ defined by the parametric equations

$$x = \rho\sin\theta\cos\phi, \ y = \rho\sin\theta\sin\phi, \ z = \rho\cos\theta. \tag{1}$$

Consider also a plane which passes through the origin with normal having the direction numbers (n_1, n_2, n_3). This plane is represented by $n_1x + n_2y + n_3z = 0$ and intersects the sphere in a great circle which is described by the relation

$$n_1\sin\theta\cos\phi + n_2\sin\theta\sin\phi + n_3\cos\theta = 0. \tag{2}$$

This is an implicit relation between the surface coordinates θ, ϕ which describes the great circle lying on the sphere. We can write this later equation in the form

$$n_1\cos\phi + n_2\sin\phi = \frac{-n_3}{\tan\theta} \tag{3}$$

and in the special case where $n_1 = \cos\beta$, $n_2 = \sin\beta$, $n_3 = -\tan\alpha$ is expressible in the form

$$\cos(\phi - \beta) = \frac{\tan\alpha}{\tan\theta} \quad \text{or} \quad \phi - \beta = \cos^{-1}\left(\frac{\tan\alpha}{\tan\theta}\right). \tag{4}$$

The above equation defines an explicit relationship between the surface coordinates which defines a great circle on the sphere. The arc length squared relation satisfied by the surface coordinates together with the equation obtained by differentiating equation (4) with respect to arc length s gives the relations

$$\sin^2\theta \frac{d\phi}{ds} = \frac{\tan\alpha}{\sqrt{1 - \frac{\tan^2\alpha}{\tan^2\theta}}} \frac{d\theta}{ds} \tag{5}$$

$$ds^2 = \rho^2\, d\theta^2 + \rho^2 \sin^2\theta\, d\phi^2 \tag{6}$$

The above equations (1)-(6) are needed to consider the following problem.

(a) Show that the differential equations defining the geodesics on the surface of a sphere (equations (1.5.51)) are

$$\frac{d^2\theta}{ds^2} - \sin\theta\cos\theta\left(\frac{d\phi}{ds}\right)^2 = 0 \tag{7}$$

$$\frac{d^2\phi}{ds^2} + 2\cot\theta\frac{d\theta}{ds}\frac{d\phi}{ds} = 0 \tag{8}$$

(b) Multiply equation (8) by $\sin^2\theta$ and integrate to obtain

$$\sin^2\theta \frac{d\phi}{ds} = c_1 \tag{9}$$

where c_1 is a constant of integration.

(c) Multiply equation (7) by $\frac{d\theta}{ds}$ and use the result of equation (9) to show that an integration produces

$$\left(\frac{d\theta}{ds}\right)^2 = \frac{-c_1^2}{\sin^2\theta} + c_2^2 \tag{10}$$

where c_2^2 is a constant of integration.

(d) Use the equations (5)(6) to show that $c_2 = 1/\rho$ and $c_1 = \frac{\sin\alpha}{\rho}$.

(e) Show that equations (9) and (10) imply that

$$\frac{d\phi}{d\theta} = \frac{\tan\alpha}{\tan^2\theta}\frac{\sec^2\theta}{\sqrt{1 - \frac{\tan^2\alpha}{\tan^2\theta}}}$$

and making the substitution $u = \frac{\tan\alpha}{\tan\theta}$ this equation can be integrated to obtain the equation (4). We can now expand the equation (4) and express the results in terms of x, y, z to obtain the equation (3). This produces a plane which intersects the sphere in a great circle. This shows that the geodesics on a sphere are great circles.

35. Find the differential equations defining the geodesics on the surface of a cylinder.

36. Find the differential equations defining the geodesics on the surface of a torus. (See problem 13, Exercise 1.3)

37. Find the differential equations defining the geodesics on the surface of revolution

$$x = r \cos \phi, \qquad y = r \sin \phi, \qquad z = f(r).$$

Note the curve $z = f(x)$ gives a profile of the surface. The curves $r = $ Constant are the parallels, while the curves $\phi = $ Constant are the meridians of the surface and

$$ds^2 = (1 + f'^2) \, dr^2 + r^2 d\phi^2.$$

38. Find the unit normal and tangent plane to an arbitrary point on the right circular cone

$$x = u \sin \alpha \cos \phi, \qquad y = u \sin \alpha \sin \phi, \qquad z = u \cos \alpha.$$

This is a surface of revolution with $r = u \sin \alpha$ and $f(r) = r \cot \alpha$ with α constant.

39. Let s denote arc length and assume the position vector $\vec{r}(s)$ is analytic about a point s_0. Show that the Taylor series $\vec{r}(s) = \vec{r}(s_0) + h\vec{r}'(s_0) + \dfrac{h^2}{2!}\vec{r}''(s_0) + \dfrac{h^3}{3!}\vec{r}'''(s_0) + \cdots$ about the point s_0, with $h = s - s_0$ is given by $\vec{r}(s) = \vec{r}(s_0) + h\vec{T} + \frac{1}{2}\kappa h^2 \vec{N} + \frac{1}{6}h^3(-\kappa^2\vec{T} + \kappa'\vec{N} + \kappa\tau\vec{B}) + \cdots$ which is obtained by differentiating the Frenet formulas.

40.

(a) Show that the circular helix defined by $x = a\cos t, \quad y = a\sin t, \quad z = bt$ with a, b constants, has the property that any tangent to the curve makes a constant angle with the line defining the z-axis. (i.e. $\vec{T} \cdot \hat{e}_3 = \cos \alpha = $ constant.)

(b) Show also that $\vec{N} \cdot \hat{e}_3 = 0$ and consequently $\hat{e}_3$ is parallel to the rectifying plane, which implies that $\hat{e}_3 = \vec{T}\cos\alpha + \vec{B}\sin\alpha$.

(c) Differentiate the result in part (b) and show that $\kappa/\tau = \tan\alpha$ is a constant.

41. Consider a space curve $x_i = x_i(s)$ in Cartesian coordinates.

(a) Show that $\kappa = \left| \dfrac{d\vec{T}}{ds} \right| = \sqrt{x_i' x_i'}$

(b) Show that $\tau = \dfrac{1}{\kappa^2} e_{ijk} x_i' x_j'' x_k'''$. Hint: Consider $\vec{r}' \cdot \vec{r}'' \times \vec{r}'''$

42.

(a) Find the direction cosines of a normal to a surface $z = f(x, y)$.

(b) Find the direction cosines of a normal to a surface $F(x, y, z) = 0$.

(c) Find the direction cosines of a normal to a surface $x = x(u, v), y = y(u, v), z = z(u, v)$.

43. Show that for a smooth surface $z = f(x, y)$ the Gaussian curvature at a point on the surface is given by

$$K = \frac{f_{xx}f_{yy} - f_{xy}^2}{(f_x^2 + f_y^2 + 1)^2}.$$

170

▶ **44.** Show that for a smooth surface $z = f(x, y)$ the mean curvature at a point on the surface is given by

$$H = \frac{(1 + f_y^2)f_{xx} - 2f_x f_y f_{xy} + (1 + f_x^2)f_{yy}}{2(f_x^2 + f_y^2 + 1)^{3/2}}.$$

▶ **45.** Express the Frenet-Serret formulas (1.5.13) in terms of Christoffel symbols of the second kind.

▶ **46.** Verify the relation (1.5.106).

▶ **47.** In V_n assume that $R_{ij} = \rho g_{ij}$ and show that $\rho = \frac{R}{n}$ where $R = g^{ij}R_{ij}$. This result is known as Einstein's gravitational equation at points where matter is present. It is analogous to the Poisson equation $\nabla^2 V = \rho$ from the Newtonian theory of gravitation.

▶ **48.** In V_n assume that $R_{ijkl} = K(g_{ik}g_{jl} - g_{il}g_{jk})$ and show that $R = Kn(1 - n)$. (Hint: See problem 23.)

▶ **49.** Assume $g_{ij} = 0$ for $i \neq j$ and verify the following.

(a) $R_{hijk} = 0$ for $h \neq i \neq j \neq k$

(b) $R_{hiik} = \sqrt{g_{ii}} \left(\frac{\partial^2 \sqrt{g_{ii}}}{\partial x^h \partial x^k} - \frac{\partial \sqrt{g_{ii}}}{\partial x^h} \frac{\partial \log \sqrt{g_{hh}}}{\partial x^k} - \frac{\partial \sqrt{g_{ii}}}{\partial x^k} \frac{\partial \log \sqrt{g_{kk}}}{\partial x^h} \right)$ for h, i, k unequal.

(c) $R_{hiih} = \sqrt{g_{ii}}\sqrt{g_{hh}} \left[\frac{\partial}{\partial x^h} \left(\frac{1}{\sqrt{g_{hh}}} \frac{\partial \sqrt{g_{ii}}}{\partial x^h} \right) + \frac{\partial}{\partial x^i} \left(\frac{1}{\sqrt{g_{ii}}} \frac{\partial \sqrt{g_{hh}}}{\partial x^i} \right) + \sum_{\substack{m=1 \\ m \neq h \ m \neq i}}^{n} \frac{1}{g_{mm}} \frac{\partial \sqrt{g_{ii}}}{\partial x^m} \frac{\partial \sqrt{g_{hh}}}{\partial x^m} \right]$ for $h \neq i$.

▶ **50.** Consider a surface of revolution where $x = r\cos\theta$, $y = r\sin\theta$ and $z = f(r)$ is a given function of r.

(a) Show in this V_2 we have $ds^2 = (1 + (f')^2)dr^2 + r^2 d\theta^2$ where $' = \frac{d}{ds}$.

(b) Show the geodesic equations in this V_2 are

$$\frac{d^2 r}{ds^2} + \frac{f' f''}{1 + (f')^2} \left(\frac{dr}{ds} \right)^2 - \frac{r}{1 + (f')^2} \left(\frac{d\theta}{ds} \right)^2 = 0$$

$$\frac{d^2 \theta}{ds^2} + \frac{2}{r} \frac{d\theta}{ds} \frac{dr}{ds} = 0$$

(c) Solve the second equation in part (b) to obtain $\frac{d\theta}{ds} = \frac{a}{r^2}$. Substitute this result for ds in part (a) to show
$d\theta = \pm \frac{a\sqrt{1 + (f')^2}}{r\sqrt{r^2 - a^2}} dr$ which theoretically can be integrated.

▶ **51.**

A curve $y = f(x)$ rotated about the x-axis generates a surface of revolution.

(a) Show that the position vector $\vec{r}$ to a point on the surface is given by

$\vec{r} = x\,\hat{e}_1 + f(x)\cos\theta\,\hat{e}_2 + f(x)\sin\theta\,\hat{e}_3$

(b) Find functions $f(x)$ which produce

(i) a cone   (ii) a cylinder   (iii) a sphere   (iv) a para

(c) What are the surface coordinates?

Describe the surface curves θ=constant.

Describe the surface curves x =constant.

Are these surface curves geodesics?

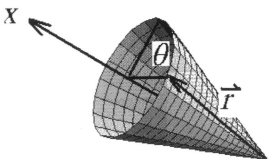

PART 2: INTRODUCTION TO CONTINUUM MECHANICS

In the following sections we develop some applications of tensor calculus in the areas of dynamics, elasticity, fluids, electricity and magnetism. We begin by first developing generalized expressions for the vector operations of gradient, divergence, and curl. Also generalized expressions for other vector operators are considered in order that tensor equations can be converted to vector equations. We construct a table to aid in the translating of generalized tensor equations to vector form and vice versa.

The basic equations of continuum mechanics are developed in the later sections. These equations are developed in both Cartesian and generalized tensor form and then converted to vector form.

§2.1 TENSOR NOTATION FOR SCALAR AND VECTOR QUANTITIES

We consider the tensor representation of some vector expressions. Our goal is to develop the ability to convert vector equations to tensor form as well as being able to represent tensor equations in vector form. In this section the basic equations of continuum mechanics are represented using both a vector notation and the indicial notation which focuses attention on the tensor components. In order to move back and forth between these notations, the representation of vector quantities in tensor form is now considered.

Gradient

For $\Phi = \Phi(x^1, x^2, \ldots, x^N)$ a scalar function of the coordinates $x^i, i = 1, \ldots, N$, the gradient of Φ is defined as the covariant vector

$$\Phi_{,i} = \frac{\partial \Phi}{\partial x^i}, \quad i = 1, \ldots, N. \tag{2.1.1}$$

The contravariant form of the gradient is

$$g^{im} \Phi_{,m}. \tag{2.1.2}$$

Note, if $C^i = g^{im} \Phi_{,m}$, $i = 1, 2, 3$ are the tensor components of the gradient then in an orthogonal coordinate system we will have

$$C^1 = g^{11} \Phi_{,1}, \qquad C^2 = g^{22} \Phi_{,2}, \qquad C^3 = g^{33} \Phi_{,3}.$$

We note that in an orthogonal coordinate system that $g^{ii} = 1/h_i^2$, (no sum on i), $i = 1, 2, 3$ and hence replacing the tensor components by their equivalent physical components there results the equations

$$\frac{C(1)}{h_1} = \frac{1}{h_1^2} \frac{\partial \Phi}{\partial x^1}, \qquad \frac{C(2)}{h_2} = \frac{1}{h_2^2} \frac{\partial \Phi}{\partial x^2}, \qquad \frac{C(3)}{h_3} = \frac{1}{h_3^2} \frac{\partial \Phi}{\partial x^3}.$$

Simplifying, we find the physical components of the gradient are

$$C(1) = \frac{1}{h_1} \frac{\partial \Phi}{\partial x^1}, \qquad C(2) = \frac{1}{h_2} \frac{\partial \Phi}{\partial x^2}, \qquad C(3) = \frac{1}{h_3} \frac{\partial \Phi}{\partial x^3}.$$

These results are only valid when the coordinate system is orthogonal and $g_{ij} = 0$ for $i \neq j$ and $g_{ii} = h_i^2$, with $i = 1, 2, 3$, and where i is not summed.

Divergence

The divergence of a contravariant tensor A^r is obtained by taking the covariant derivative with respect to x^k and then performing a contraction. This produces

$$\text{div } A^r = A^r{}_{,r}. \tag{2.1.3}$$

Still another form for the divergence is obtained by simplifying the expression (2.1.3). The covariant derivative can be represented

$$A^r{}_{,k} = \frac{\partial A^r}{\partial x^k} + \left\{ \begin{matrix} r \\ m\,k \end{matrix} \right\} A^m.$$

Upon contracting the indices r and k and using the result from Exercise 1.4, problem 13, we obtain

$$A^r{}_{,r} = \frac{\partial A^r}{\partial x^r} + \frac{1}{\sqrt{g}} \frac{\partial(\sqrt{g})}{\partial x^m} A^m$$

$$A^r{}_{,r} = \frac{1}{\sqrt{g}} \left(\sqrt{g} \frac{\partial A^r}{\partial x^r} + A^r \frac{\partial \sqrt{g}}{\partial x^r} \right) \tag{2.1.4}$$

$$A^r{}_{,r} = \frac{1}{\sqrt{g}} \frac{\partial}{\partial x^r} \left(\sqrt{g} A^r \right).$$

EXAMPLE 2.1-1. (Divergence)   Find the representation of the divergence of a vector A^r in spherical coordinates (ρ, θ, ϕ). **Solution:**   In spherical coordinates we have

$$x^1 = \rho, \quad x^2 = \theta, \quad x^3 = \phi \quad \text{with} \quad g_{ij} = 0 \quad \text{for} \quad i \neq j \quad \text{and}$$

$$g_{11} = h_1^2 = 1, \qquad g_{22} = h_2^2 = \rho^2, \qquad g_{33} = h_3^2 = \rho^2 \sin^2 \theta.$$

The determinant of g_{ij} is $g = |g_{ij}| = \rho^4 \sin^2 \theta$ and $\sqrt{g} = \rho^2 \sin \theta$. Employing the relation (2.1.4) we find

$$\text{div } A^r = \frac{1}{\sqrt{g}} \left[\frac{\partial}{\partial x^1}(\sqrt{g} A^1) + \frac{\partial}{\partial x^2}(\sqrt{g} A^2) + \frac{\partial}{\partial x^3}(\sqrt{g} A^3) \right].$$

In terms of the physical components this equation becomes

$$\text{div } A^r = \frac{1}{\sqrt{g}} \left[\frac{\partial}{\partial \rho}(\sqrt{g} \frac{A(1)}{h_1}) + \frac{\partial}{\partial \theta}(\sqrt{g} \frac{A(2)}{h_2}) + \frac{\partial}{\partial \phi}(\sqrt{g} \frac{A(3)}{h_3}) \right].$$

By using the notation

$$A(1) = A_\rho, \qquad A(2) = A_\theta, \qquad A(3) = A_\phi$$

for the physical components, the divergence can be expressed in either of the forms:

$$\text{div } A^r = \frac{1}{\rho^2 \sin \theta} \left[\frac{\partial}{\partial \rho}(\rho^2 \sin \theta A_\rho) + \frac{\partial}{\partial \theta}(\rho^2 \sin \theta \frac{A_\theta}{\rho}) + \frac{\partial}{\partial \phi}(\rho^2 \sin \theta \frac{A_\phi}{\rho \sin \theta}) \right] \quad \text{or}$$

$$\text{div } A^r = \frac{1}{\rho^2} \frac{\partial}{\partial \rho}(\rho^2 A_\rho) + \frac{1}{\rho \sin \theta} \frac{\partial}{\partial \theta}(\sin \theta A_\theta) + \frac{1}{\rho \sin \theta} \frac{\partial A_\phi}{\partial \phi}.$$

Curl

The contravariant components of the vector $\vec{C} = \text{curl } \vec{A}$ are represented

$$C^i = \epsilon^{ijk} A_{k,j}. \tag{2.1.5}$$

In expanded form this representation becomes:

$$C^1 = \frac{1}{\sqrt{g}}\left(\frac{\partial A_3}{\partial x^2} - \frac{\partial A_2}{\partial x^3}\right)$$

$$C^2 = \frac{1}{\sqrt{g}}\left(\frac{\partial A_1}{\partial x^3} - \frac{\partial A_3}{\partial x^1}\right) \tag{2.1.6}$$

$$C^3 = \frac{1}{\sqrt{g}}\left(\frac{\partial A_2}{\partial x^1} - \frac{\partial A_1}{\partial x^2}\right).$$

EXAMPLE 2.1-2. (Curl) Find the representation for the components of curl $\vec{A}$ in spherical coordinates (ρ, θ, ϕ).

Solution: In spherical coordinates we have : $x^1 = \rho$, $x^2 = \theta$, $x^3 = \phi$ with $g_{ij} = 0$ for $i \neq j$ and

$$g_{11} = h_1^2 = 1, \qquad g_{22} = h_2^2 = \rho^2, \qquad g_{33} = h_3^2 = \rho^2 \sin^2\theta.$$

The determinant of g_{ij} is $g = |g_{ij}| = \rho^4 \sin^2\theta$ with $\sqrt{g} = \rho^2 \sin\theta$. The relations (2.1.6) are tensor equations representing the components of the vector curl $\vec{A}$. To find the components of curl $\vec{A}$ in spherical components we write the equations (2.1.6) in terms of their physical components. These equations take on the form:

$$\frac{C(1)}{h_1} = \frac{1}{\sqrt{g}}\left[\frac{\partial}{\partial\theta}(h_3 A(3)) - \frac{\partial}{\partial\phi}(h_2 A(2))\right]$$

$$\frac{C(2)}{h_2} = \frac{1}{\sqrt{g}}\left[\frac{\partial}{\partial\phi}(h_1 A(1)) - \frac{\partial}{\partial\rho}(h_3 A(3))\right] \tag{2.1.7}$$

$$\frac{C(3)}{h_3} = \frac{1}{\sqrt{g}}\left[\frac{\partial}{\partial\rho}(h_2 A(2)) - \frac{\partial}{\partial\theta}(h_1 A(1))\right].$$

We employ the notations

$$C(1) = C_\rho, \quad C(2) = C_\theta, \quad C(3) = C_\phi, \quad A(1) = A_\rho, \quad A(2) = A_\theta, \quad A(3) = A_\phi$$

to denote the physical components, and find the components of the vector curl $\vec{A}$, in spherical coordinates, are expressible in the form:

$$C_\rho = \frac{1}{\rho^2 \sin\theta}\left[\frac{\partial}{\partial\theta}(\rho\sin\theta A_\phi) - \frac{\partial}{\partial\phi}(\rho A_\theta)\right]$$

$$C_\theta = \frac{1}{\rho\sin\theta}\left[\frac{\partial}{\partial\phi}(A_\rho) - \frac{\partial}{\partial\rho}(\rho\sin\theta A_\phi)\right] \tag{2.1.8}$$

$$C_\phi = \frac{1}{\rho}\left[\frac{\partial}{\partial\rho}(\rho A_\theta) - \frac{\partial}{\partial\theta}(A_\rho)\right].$$

Laplacian

The Laplacian $\nabla^2 U$ has the contravariant form

$$\nabla^2 U = g^{ij}U_{,ij} = (g^{ij}U_{,i})_{,j} = \left(g^{ij}\frac{\partial U}{\partial x^i}\right)_{,j}. \tag{2.1.9}$$

Expanding this expression produces the equations:

$$\nabla^2 U = \frac{\partial}{\partial x^j}\left(g^{ij}\frac{\partial U}{\partial x^i}\right) + g^{im}\frac{\partial U}{\partial x^i}\begin{Bmatrix} j \\ m\,j \end{Bmatrix}$$

$$\nabla^2 U = \frac{\partial}{\partial x^j}\left(g^{ij}\frac{\partial U}{\partial x^i}\right) + \frac{1}{\sqrt{g}}\frac{\partial\sqrt{g}}{\partial x^j}g^{ij}\frac{\partial U}{\partial x^i}$$

$$\nabla^2 U = \frac{1}{\sqrt{g}}\left[\sqrt{g}\frac{\partial}{\partial x^j}\left(g^{ij}\frac{\partial U}{\partial x^i}\right) + g^{ij}\frac{\partial U}{\partial x^i}\frac{\partial\sqrt{g}}{\partial x^j}\right] \tag{2.1.10}$$

$$\nabla^2 U = \frac{1}{\sqrt{g}}\frac{\partial}{\partial x^j}\left(\sqrt{g}g^{ij}\frac{\partial U}{\partial x^i}\right).$$

In orthogonal coordinates we have $g^{ij} = 0$ for $i \neq j$ and

$$g_{11} = h_1^2, \qquad g_{22} = h_2^2, \qquad g_{33} = h_3^2$$

and so (2.1.10) when expanded reduces to the form

$$\nabla^2 U = \frac{1}{h_1 h_2 h_3}\left[\frac{\partial}{\partial x^1}\left(\frac{h_2 h_3}{h_1}\frac{\partial U}{\partial x^1}\right) + \frac{\partial}{\partial x^2}\left(\frac{h_1 h_3}{h_2}\frac{\partial U}{\partial x^2}\right) + \frac{\partial}{\partial x^3}\left(\frac{h_1 h_2}{h_3}\frac{\partial U}{\partial x^3}\right)\right]. \tag{2.1.11}$$

This representation is only valid in an orthogonal system of coordinates.

EXAMPLE 2.1-3. (Laplacian)   Find the Laplacian in spherical coordinates.

Solution: Utilizing the results given in the previous example we find the Laplacian in spherical coordinates has the form

$$\nabla^2 U = \frac{1}{\rho^2 \sin\theta}\left[\frac{\partial}{\partial\rho}\left(\rho^2 \sin\theta\frac{\partial U}{\partial\rho}\right) + \frac{\partial}{\partial\theta}\left(\sin\theta\frac{\partial U}{\partial\theta}\right) + \frac{\partial}{\partial\phi}\left(\frac{1}{\sin\theta}\frac{\partial U}{\partial\phi}\right)\right]. \tag{2.1.12}$$

This simplifies to

$$\nabla^2 U = \frac{\partial^2 U}{\partial\rho^2} + \frac{2}{\rho}\frac{\partial U}{\partial\rho} + \frac{1}{\rho^2}\frac{\partial^2 U}{\partial\theta^2} + \frac{\cot\theta}{\rho^2}\frac{\partial U}{\partial\theta} + \frac{1}{\rho^2\sin^2\theta}\frac{\partial^2 U}{\partial\phi^2}. \tag{2.1.13}$$

■

The table 1 gives the vector and tensor representation for various quantities of interest.

VECTOR	GENERAL TENSOR	CARTESIAN TENSOR
$\vec{A}$	A^i or A_i	A_i
$\vec{A} \cdot \vec{B}$	$A^i B_i = g_{ij} A^i B^j = A_i B^i$ $A^i B_i = g^{ij} A_i B_j$	$A_i B_i$
$\vec{C} = \vec{A} \times \vec{B}$	$C^i = \dfrac{1}{\sqrt{g}} e^{ijk} A_j B_k$	$C_i = e_{ijk} A_j B_k$
$\nabla \Phi = \operatorname{grad} \Phi$	$g^{im} \Phi_{,m}$	$\Phi_{,i} = \dfrac{\partial \Phi}{\partial x^i}$
$\nabla \cdot \vec{A} = \operatorname{div} \vec{A}$	$g^{mn} A_{m,n} = A^r{}_{,r} = \dfrac{1}{\sqrt{g}} \dfrac{\partial}{\partial x^r} \left(\sqrt{g} A^r \right)$	$A_{i,i} = \dfrac{\partial A_i}{\partial x^i}$
$\nabla \times \vec{A} = \vec{C} = \operatorname{curl} \vec{A}$	$C^i = \epsilon^{ijk} A_{k,j}$	$C_i = e_{ijk} \dfrac{\partial A_k}{\partial x^j}$
$\nabla^2 U$	$g^{mn} U_{,mn} = \dfrac{1}{\sqrt{g}} \dfrac{\partial}{\partial x^j} \left(\sqrt{g} g^{ij} \dfrac{\partial U}{\partial x^i} \right)$	$\dfrac{\partial}{\partial x^i} \left(\dfrac{\partial U}{\partial x^i} \right)$
$\vec{C} = (\vec{A} \cdot \nabla)\vec{B}$	$C^i = A^m B^i{}_{,m}$	$C_i = A_m \dfrac{\partial B_i}{\partial x^m}$
$\vec{C} = \vec{A}(\nabla \cdot \vec{B})$	$C^i = A^i B^j{}_{,j}$	$C_i = A_i \dfrac{\partial B_m}{\partial x^m}$
$\vec{C} = \nabla^2 \vec{A}$	$C^i = g^{jm} A^i{}_{,mj}$ or $C_i = g^{jm} A_{i,mj}$	$C_i = \dfrac{\partial}{\partial x^m} \left(\dfrac{\partial A_i}{\partial x^m} \right)$
$\left(\vec{A} \cdot \nabla \right) \phi$	$g^{im} A^i \phi_{,m}$	$A_i \phi_{,i}$
$\nabla \left(\nabla \cdot \vec{A} \right)$	$g^{im} \left(A^r{}_{,r} \right)_{,m}$	$\dfrac{\partial^2 A_r}{\partial x_i \partial x_r}$
$\nabla \times \left(\nabla \times \vec{A} \right)$	$\epsilon_{ijk} g^{jm} \left(\epsilon^{kst} A_{t,s} \right)_{,m}$	$\dfrac{\partial^2 A_j}{\partial x_j \partial x_i} - \dfrac{\partial^2 A_i}{\partial x_j \partial x_j}$

Table 1 Vector and tensor representations.

EXAMPLE 2.1-4. (Maxwell's equations) In the study of electrodynamics there arises the following vectors and scalars:

$$\vec{E} = \text{Electric force vector}, [\vec{E}] = \text{Newton/coulomb}$$

$$\vec{B} = \text{Magnetic force vector}, [\vec{B}] = \text{Weber/m}^2$$

$$\vec{D} = \text{Displacement vector}, [\vec{D}] = \text{coulomb/m}^2$$

$$\vec{H} = \text{Auxilary magnetic force vector}, [\vec{H}] = \text{ampere/m}$$

$$\vec{J} = \text{Free current density}, [\vec{J}] = \text{ampere/m}^2$$

$$\varrho = \text{free charge density}, [\varrho] = \text{coulomb/m}^3$$

The above quantities arise in the representation of the following laws:

Faraday's Law This law states the line integral of the electromagnetic force around a loop is proportional to the rate of flux of magnetic induction through the loop. This gives rise to the first electromagnetic field equation:

$$\nabla \times \vec{E} = -\frac{\partial \vec{B}}{\partial t} \qquad \text{or} \qquad \epsilon^{ijk} E_{k,j} = -\frac{\partial B^i}{\partial t}. \tag{2.1.15}$$

Ampere's Law This law states the line integral of the magnetic force vector around a closed loop is proportional to the sum of the current through the loop and the rate of flux of the displacement vector through the loop. This produces the second electromagnetic field equation:

$$\nabla \times \vec{H} = \vec{J} + \frac{\partial \vec{D}}{\partial t} \qquad \text{or} \qquad \epsilon^{ijk} H_{k,j} = J^i + \frac{\partial D^i}{\partial t}. \tag{2.1.16}$$

Gauss's Law for Electricity This law states that the flux of the electric force vector through a closed surface is proportional to the total charge enclosed by the surface. This results in the third electromagnetic field equation:

$$\nabla \cdot \vec{D} = \varrho \qquad \text{or} \qquad \frac{1}{\sqrt{g}} \frac{\partial}{\partial x^i} \left(\sqrt{g} D^i \right) = \varrho. \tag{2.1.17}$$

Gauss's Law for Magnetism This law states the magnetic flux through any closed volume is zero. This produces the fourth electromagnetic field equation:

$$\nabla \cdot \vec{B} = 0 \qquad \text{or} \qquad \frac{1}{\sqrt{g}} \frac{\partial}{\partial x^i} \left(\sqrt{g} B^i \right) = 0. \tag{2.1.18}$$

The four electromagnetic field equations are referred to as Maxwell's equations. These equations arise in the study of electrodynamics and can be represented in other forms. These other forms will depend upon such things as the material assumptions and units of measurements used. Note that the tensor equations (2.1.15) through (2.1.18) are representations of Maxwell's equations in a form which is independent of the coordinate system chosen.

In applications, the tensor quantities must be expressed in terms of their physical components. In a general orthogonal curvilinear coordinate system we will have

$$g_{11} = h_1^2, \quad g_{22} = h_2^2, \quad g_{33} = h_3^2, \quad \text{and} \quad g_{ij} = 0 \quad \text{for} \quad i \neq j.$$

This produces the result $\sqrt{g} = h_1 h_2 h_3$. Further, if we represent the physical components of

$$D_i, B_i, E_i, H_i \quad \text{by} \quad D(i), B(i), E(i), \text{ and } H(i)$$

the Maxwell equations can be represented by the equations in table 2. The tables 3, 4 and 5 are the representation of Maxwell's equations in rectangular, cylindrical, and spherical coordinates. These latter tables are special cases associated with the more general table 2.

$$\frac{1}{h_1h_2h_3}\left[\frac{\partial}{\partial x^2}(h_3E(3)) - \frac{\partial}{\partial x^3}(h_2E(2))\right] = -\frac{1}{h_1}\frac{\partial B(1)}{\partial t}$$

$$\frac{1}{h_1h_2h_3}\left[\frac{\partial}{\partial x^3}(h_1E(1)) - \frac{\partial}{\partial x^1}(h_3E(3))\right] = -\frac{1}{h_2}\frac{\partial B(2)}{\partial t}$$

$$\frac{1}{h_1h_2h_3}\left[\frac{\partial}{\partial x^1}(h_2E(2)) - \frac{\partial}{\partial x^2}(h_1E(1))\right] = -\frac{1}{h_3}\frac{\partial B(3)}{\partial t}$$

$$\frac{1}{h_1h_2h_3}\left[\frac{\partial}{\partial x^2}(h_3H(3)) - \frac{\partial}{\partial x^3}(h_2H(2))\right] = \frac{J(1)}{h_1} + \frac{1}{h_1}\frac{\partial D(1)}{\partial t}$$

$$\frac{1}{h_1h_2h_3}\left[\frac{\partial}{\partial x^3}(h_1H(1)) - \frac{\partial}{\partial x^1}(h_3H(3))\right] = \frac{J(2)}{h_2} + \frac{1}{h_2}\frac{\partial D(2)}{\partial t}$$

$$\frac{1}{h_1h_2h_3}\left[\frac{\partial}{\partial x^1}(h_2H(2)) - \frac{\partial}{\partial x^2}(h_1H(1))\right] = \frac{J(3)}{h_3} + \frac{1}{h_3}\frac{\partial D(3)}{\partial t}$$

$$\frac{1}{h_1h_2h_3}\left[\frac{\partial}{\partial x^1}\left(h_1h_2h_3\frac{D(1)}{h_1}\right) + \frac{\partial}{\partial x^2}\left(h_1h_2h_3\frac{D(2)}{h_2}\right) + \frac{\partial}{\partial x^3}\left(h_1h_2h_3\frac{D(3)}{h_3}\right)\right] = \varrho$$

$$\frac{1}{h_1h_2h_3}\left[\frac{\partial}{\partial x^1}\left(h_1h_2h_3\frac{B(1)}{h_1}\right) + \frac{\partial}{\partial x^2}\left(h_1h_2h_3\frac{B(2)}{h_2}\right) + \frac{\partial}{\partial x^3}\left(h_1h_2h_3\frac{B(3)}{h_3}\right)\right] = 0$$

Table 2 Maxwell's equations in generalized orthogonal coordinates.
Note that all the tensor components have been replaced by their physical components.

$$\frac{\partial E_z}{\partial y} - \frac{\partial E_y}{\partial z} = -\frac{\partial B_x}{\partial t} \qquad \frac{\partial H_z}{\partial y} - \frac{\partial H_y}{\partial z} = J_x + \frac{\partial D_x}{\partial t} \qquad \frac{\partial D_x}{\partial x} + \frac{\partial D_y}{\partial y} + \frac{\partial D_z}{\partial z} = \varrho$$

$$\frac{\partial E_x}{\partial z} - \frac{\partial E_z}{\partial x} = -\frac{\partial B_y}{\partial t} \qquad \frac{\partial H_x}{\partial z} - \frac{\partial H_z}{\partial x} = J_y + \frac{\partial D_y}{\partial t}$$

$$\frac{\partial E_y}{\partial x} - \frac{\partial E_x}{\partial y} = -\frac{\partial B_z}{\partial t} \qquad \frac{\partial H_y}{\partial x} - \frac{\partial H_x}{\partial y} = J_z + \frac{\partial D_z}{\partial t} \qquad \frac{\partial B_x}{\partial x} + \frac{\partial B_y}{\partial y} + \frac{\partial B_z}{\partial z} = 0$$

Here we have introduced the notations:

$$D_x = D(1) \quad B_x = B(1) \quad H_x = H(1) \quad J_x = J(1) \quad E_x = E(1)$$

$$D_y = D(2) \quad B_y = B(2) \quad H_y = H(2) \quad J_y = J(2) \quad E_y = E(2)$$

$$D_z = D(3) \quad B_z = B(3) \quad H_z = H(3) \quad J_z = J(3) \quad E_z = E(3)$$

with $x^1 = x,\quad x^2 = y,\quad x^3 = z,\quad h_1 = h_2 = h_3 = 1$

Table 3 Maxwell's equations Cartesian coordinates

$$\frac{1}{r}\frac{\partial E_z}{\partial \theta} - \frac{\partial E_\theta}{\partial z} = -\frac{\partial B_r}{\partial t} \qquad \frac{1}{r}\frac{\partial H_z}{\partial \theta} - \frac{\partial H_\theta}{\partial z} = J_r + \frac{\partial D_r}{\partial t}$$

$$\frac{\partial E_r}{\partial z} - \frac{\partial E_z}{\partial r} = -\frac{\partial B_\theta}{\partial t} \qquad \frac{\partial H_r}{\partial z} - \frac{\partial H_z}{\partial r} = J_\theta + \frac{\partial D_\theta}{\partial t}$$

$$\frac{1}{r}\frac{\partial}{\partial r}(rE_\theta) - \frac{1}{r}\frac{\partial E_r}{\partial \theta} = -\frac{\partial B_z}{\partial t} \qquad \frac{1}{r}\frac{\partial}{\partial r}(rH_\theta) - \frac{1}{r}\frac{\partial H_r}{\partial \theta} = J_z + \frac{\partial D_z}{\partial t}$$

$$\frac{1}{r}\frac{\partial}{\partial r}(rD_r) + \frac{1}{r}\frac{\partial D_\theta}{\partial \theta} + \frac{\partial D_z}{\partial z} = \varrho \qquad \frac{1}{r}\frac{\partial}{\partial r}(rB_r) + \frac{1}{r}\frac{\partial B_\theta}{\partial \theta} + \frac{\partial B_z}{\partial z} = 0$$

Here we have introduced the notations:

$$D_r = D(1) \quad B_r = B(1) \quad H_r = H(1) \quad J_r = J(1) \quad E_r = E(1)$$

$$D_\theta = D(2) \quad B_\theta = B(2) \quad H_\theta = H(2) \quad J_\theta = J(2) \quad E_\theta = E(2)$$

$$D_z = D(3) \quad B_z = B(3) \quad H_z = H(3) \quad J_z = J(3) \quad E_z = E(3)$$

with $x^1 = r,\quad x^2 = \theta,\quad x^3 = z,\quad h_1 = 1,\quad h_2 = r,\quad h_3 = 1.$

Table 4 Maxwell's equations in cylindrical coordinates.

$$\frac{1}{\rho\sin\theta}\left[\frac{\partial}{\partial\theta}\left(\sin\theta E_\phi\right) - \frac{\partial E_\theta}{\partial\phi}\right] = -\frac{\partial B_\rho}{\partial t} \qquad \frac{1}{\rho\sin\theta}\left[\frac{\partial}{\partial\theta}\left(\sin\theta H_\phi\right) - \frac{\partial H_\theta}{\partial\phi}\right] = J_\rho + \frac{\partial D_\rho}{\partial t}$$

$$\frac{1}{\rho\sin\theta}\frac{\partial E_\rho}{\partial\phi} - \frac{1}{\rho}\frac{\partial}{\partial\rho}(\rho E_\phi) = -\frac{\partial B_\theta}{\partial t} \qquad \frac{1}{\rho\sin\theta}\frac{\partial H_\rho}{\partial\phi} - \frac{1}{\rho}\frac{\partial}{\partial\rho}(\rho H_\phi) = J_\theta + \frac{\partial D_\theta}{\partial t}$$

$$\frac{1}{\rho}\frac{\partial}{\partial\rho}(\rho E_\theta) - \frac{1}{\rho}\frac{\partial E_\rho}{\partial\theta} = -\frac{\partial B_\phi}{\partial t} \qquad \frac{1}{\rho}\frac{\partial}{\partial\rho}(\rho H_\theta) - \frac{1}{\rho}\frac{\partial H_\rho}{\partial\theta} = J_\phi + \frac{\partial D_\phi}{\partial t}$$

$$\frac{1}{\rho^2}\frac{\partial}{\partial\rho}(\rho^2 D_\rho) + \frac{1}{\rho\sin\theta}\frac{\partial}{\partial\theta}(\sin\theta D_\theta) + \frac{1}{\rho\sin\theta}\frac{\partial D_\phi}{\partial\phi} = \varrho$$

$$\frac{1}{\rho^2}\frac{\partial}{\partial\rho}(\rho^2 B_\rho) + \frac{1}{\rho\sin\theta}\frac{\partial}{\partial\theta}(\sin\theta B_\theta) + \frac{1}{\rho\sin\theta}\frac{\partial B_\phi}{\partial\phi} = 0$$

Here we have introduced the notations:

$$D_\rho = D(1) \qquad B_\rho = B(1) \qquad H_\rho = H(1) \qquad J_\rho = J(1) \qquad E_\rho = E(1)$$

$$D_\theta = D(2) \qquad B_\theta = B(2) \qquad H_\theta = H(2) \qquad J_\theta = J(2) \qquad E_\theta = E(2)$$

$$D_\phi = D(3) \qquad B_\phi = B(3) \qquad H_\phi = H(3) \qquad J_\phi = J(3) \qquad E_\phi = E(3)$$

with $x^1 = \rho, \quad x^2 = \theta, \quad x^3 = \phi, \quad h_1 = 1, \quad h_2 = \rho, \quad h_3 = \rho\sin\theta$

Table 5 Maxwell's equations spherical coordinates.

Eigenvalues and Eigenvectors of Symmetric Tensors

Consider the equation

$$T_{ij}A_j = \lambda A_i, \quad i, j = 1, 2, 3, \tag{2.1.19}$$

where $T_{ij} = T_{ji}$ is symmetric, A_i are the components of a vector and λ is a scalar. Any nonzero solution A_i of equation (2.1.19) is called an eigenvector of the tensor T_{ij} and the associated scalar λ is called an eigenvalue. When expanded these equations have the form

$$(T_{11} - \lambda)A_1 + T_{12}A_2 + T_{13}A_3 = 0$$

$$T_{21}A_1 + (T_{22} - \lambda)A_2 + T_{23}A_3 = 0$$

$$T_{31}A_1 + T_{32}A_2 + (T_{33} - \lambda)A_3 = 0.$$

The condition for equation (2.1.19) to have a nonzero solution A_i is that the characteristic equation should be zero. This equation is found from the determinant equation

$$f(\lambda) = \begin{vmatrix} T_{11} - \lambda & T_{12} & T_{13} \\ T_{21} & T_{22} - \lambda & T_{23} \\ T_{31} & T_{32} & T_{33} - \lambda \end{vmatrix} = 0, \tag{2.1.20}$$

which when expanded is a cubic equation of the form

$$f(\lambda) = -\lambda^3 + I_1\lambda^2 - I_2\lambda + I_3 = 0, \tag{2.1.21}$$

where I_1, I_2 and I_3 are invariants defined by the relations

$$\begin{aligned} I_1 &= T_{ii} \\ I_2 &= \frac{1}{2}T_{ii}T_{jj} - \frac{1}{2}T_{ij}T_{ij} \\ I_3 &= e_{ijk}T_{i1}T_{j2}T_{k3}. \end{aligned} \tag{2.1.22}$$

When T_{ij} is subjected to an orthogonal transformation, where $\bar{T}_{mn} = T_{ij}\ell_{im}\ell_{jn}$, then

$$\ell_{im}\ell_{jn}(T_{mn} - \lambda\,\delta_{mn}) = \bar{T}_{ij} - \lambda\,\delta_{ij} \quad \text{and} \quad \det(T_{mn} - \lambda\,\delta_{mn}) = \det\left(\bar{T}_{ij} - \lambda\,\delta_{ij}\right).$$

Hence, the eigenvalues of a second order tensor remain invariant under an orthogonal transformation.

If T_{ij} is real and symmetric then

- the eigenvalues of T_{ij} will be real, and
- the eigenvectors corresponding to distinct eigenvalues will be orthogonal.

Proof: To show a quantity is real we show that the conjugate of the quantity equals the given quantity. If (2.1.19) is satisfied, we multiply by the conjugate $\bar{A}_i$ and obtain

$$\bar{A}_i T_{ij} A_j = \lambda A_i \bar{A}_i. \tag{2.1.25}$$

The right hand side of this equation has the inner product $A_i\bar{A}_i$ which is real. It remains to show the left hand side of equation (2.1.25) is also real. Consider the conjugate of this left hand side and write

$$\overline{\bar{A}_i T_{ij} A_j} = A_i \bar{T}_{ij} \bar{A}_j = A_i T_{ji} \bar{A}_j = \bar{A}_i T_{ij} A_j.$$

Consequently, the left hand side of equation (2.1.25) is real and the eigenvalue λ can be represented as the ratio of two real quantities.

Assume that $\lambda_{(1)}$ and $\lambda_{(2)}$ are two distinct eigenvalues which produce the unit eigenvectors $\hat{L}_1$ and $\hat{L}_2$ with components ℓ_{i1} and $\ell_{i2}, i = 1, 2, 3$ respectively. We then have

$$T_{ij}\ell_{j1} = \lambda_{(1)}\ell_{i1} \quad \text{and} \quad T_{ij}\ell_{j2} = \lambda_{(2)}\ell_{i2}. \tag{2.1.26}$$

Consider the products

$$\begin{aligned} \lambda_{(1)}\ell_{i1}\ell_{i2} &= T_{ij}\ell_{j1}\ell_{i2}, \\ \lambda_{(2)}\ell_{i1}\ell_{i2} &= \ell_{i1}T_{ij}\ell_{j2} = \ell_{j1}T_{ji}\ell_{i2}. \end{aligned} \tag{2.1.27}$$

and subtract these equations. We find that

$$[\lambda_{(1)} - \lambda_{(2)}]\ell_{i1}\ell_{i2} = 0. \tag{2.1.28}$$

By hypothesis, $\lambda_{(1)}$ is different from $\lambda_{(2)}$ and therefore the inner product $\ell_{i1}\ell_{i2}$ must be zero. This shows that the eigenvectors corresponding to distinct eigenvalues are orthogonal.

Associated with distinct eigenvalues $\lambda_{(i)}, i = 1, 2, 3$ there are unit eigenvectors

$$\hat{L}_{(i)} = \ell_{i1}\,\hat{\mathbf{e}}_1 + \ell_{i2}\,\hat{\mathbf{e}}_2 + \ell_{i3}\,\hat{\mathbf{e}}_3$$

with components $\ell_{im}, m = 1, 2, 3$ which are direction cosines and satisfy

$$\ell_{in}\ell_{im} = \delta_{mn} \qquad \text{and} \qquad \ell_{ij}\ell_{jm} = \delta_{im}. \tag{2.1.23}$$

The unit eigenvectors satisfy the relations

$$T_{ij}\ell_{j1} = \lambda_{(1)}\ell_{i1} \qquad T_{ij}\ell_{j2} = \lambda_{(2)}\ell_{i2} \qquad T_{ij}\ell_{j3} = \lambda_{(3)}\ell_{i3}$$

and can be written as the single equation

$$T_{ij}\ell_{jm} = \lambda_{(m)}\ell_{im}, \quad m = 1, 2, \textit{or }3 \quad \text{m not summed.}$$

Consider the transformation

$$\overline{x}_i = \ell_{ij}x_j \qquad \text{or} \qquad x_m = \ell_{mj}\overline{x}_j$$

which represents a rotation of axes, where ℓ_{ij} are the direction cosines from the eigenvectors of T_{ij}. This is a linear transformation where the ℓ_{ij} satisfy equation (2.1.23). Such a transformation is called an orthogonal transformation. In the new $\overline{x}$ coordinate system, called principal axes, we have

$$\overline{T}_{mn} = T_{ij}\frac{\partial x^i}{\partial \overline{x}^m}\frac{\partial x^j}{\partial \overline{x}^n} = T_{ij}\ell_{im}\ell_{jn} = \lambda_{(n)}\ell_{in}\ell_{im} = \lambda_{(n)}\delta_{mn} \quad \text{(no sum on } n). \tag{2.1.24}$$

This equation shows that in the barred coordinate system there are the components

$$(\overline{T}_{mn}) = \begin{bmatrix} \lambda_{(1)} & 0 & 0 \\ 0 & \lambda_{(2)} & 0 \\ 0 & 0 & \lambda_{(3)} \end{bmatrix}.$$

That is, along the principal axes the tensor components T_{ij} are transformed to the components $\overline{T}_{ij}$ where $\overline{T}_{ij} = 0$ for $i \neq j$. The elements $\overline{T}_{(i)(i)}$, i not summed, represent the eigenvalues of the transformation (2.1.19).

EXERCISE 2.1

▶ **1.** In cylindrical coordinates (r, θ, z) with $f = f(r, \theta, z)$ find the gradient of f.

▶ **2.** In cylindrical coordinates (r, θ, z) with $\vec{A} = \vec{A}(r, \theta, z)$ find div $\vec{A}$.

▶ **3.** In cylindrical coordinates (r, θ, z) for $\vec{A} = \vec{A}(r, \theta, z)$ find curl $\vec{A}$.

▶ **4.** In cylindrical coordinates (r, θ, z) for $f = f(r, \theta, z)$ find $\nabla^2 f$.

▶ **5.** In spherical coordinates (ρ, θ, ϕ) with $f = f(\rho, \theta, \phi)$ find the gradient of f.

▶ **6.** In spherical coordinates (ρ, θ, ϕ) with $\vec{A} = \vec{A}(\rho, \theta, \phi)$ find div $\vec{A}$.

▶ **7.** In spherical coordinates (ρ, θ, ϕ) for $\vec{A} = \vec{A}(\rho, \theta, \phi)$ find curl $\vec{A}$.

▶ **8.** In spherical coordinates (ρ, θ, ϕ) for $f = f(\rho, \theta, \phi)$ find $\nabla^2 f$.

▶ **9.** Let $\vec{r} = x\,\hat{\mathbf{e}}_1 + y\,\hat{\mathbf{e}}_2 + z\,\hat{\mathbf{e}}_3$ denote the position vector of a variable point (x, y, z) in Cartesian coordinates. Let $r = |\vec{r}|$ denote the distance of this point from the origin. Find in terms of $\vec{r}$ and r:

(a) grad (r) $\quad$ (b) grad (r^m) $\quad$ (c) grad $(\dfrac{1}{r})$ $\quad$ (d) grad $(\ln r)$ $\quad$ (e) grad (ϕ)

where $\phi = \phi(r)$ is an arbitrary function of r.

▶ **10.** Let $\vec{r} = x\,\hat{\mathbf{e}}_1 + y\,\hat{\mathbf{e}}_2 + z\,\hat{\mathbf{e}}_3$ denote the position vector of a variable point (x, y, z) in Cartesian coordinates. Let $r = |\vec{r}|$ denote the distance of this point from the origin. Find:

(a) div $(\vec{r})$ $\quad$ (b) div $(r^m \vec{r})$ $\quad$ (c) div $(r^{-3}\vec{r})$ $\quad$ (d) div $(\phi \vec{r})$

where $\phi = \phi(r)$ is an arbitrary function or r.

▶ **11.** Let $\vec{r} = x\,\hat{\mathbf{e}}_1 + y\,\hat{\mathbf{e}}_2 + z\,\hat{\mathbf{e}}_3$ denote the position vector of a variable point (x, y, z) in Cartesian coordinates. Let $r = |\vec{r}|$ denote the distance of this point from the origin. Find: (a) curl $\vec{r}$ $\quad$ (b) curl $(\phi \vec{r})$ where $\phi = \phi(r)$ is an arbitrary function of r.

▶ **12.** Expand and simplify the representation for curl (curl $\vec{A}$).

▶ **13.** Show that the curl of the gradient is zero in generalized coordinates.

▶ **14.** Write out the physical components associated with the gradient of $\phi = \phi(x^1, x^2, x^3)$.

▶ **15.** Show that

$$g^{im} A_{i,m} = \frac{1}{\sqrt{g}} \frac{\partial}{\partial x^i} \left[\sqrt{g} g^{im} A_m \right] = A^i_{,i} = \frac{1}{\sqrt{g}} \frac{\partial}{\partial x^i} \left[\sqrt{g} A^i \right].$$

16. Let $r = (\vec{r} \cdot \vec{r})^{1/2} = \sqrt{x^2 + y^2 + z^2})$ and calculate (a) $\nabla^2(r)$ (b) $\nabla^2(1/r)$ (c) $\nabla^2(r^2)$ (d) $\nabla^2(1/r^2)$

17. Given the tensor equations $D_{ij} = \frac{1}{2}(v_{i,j} + v_{j,i})$, $i, j = 1, 2, 3$. Let $v(1), v(2), v(3)$ denote the physical components of v_1, v_2, v_3 and let $D(ij)$ denote the physical components associated with D_{ij}. Assume the coordinate system (x^1, x^2, x^3) is orthogonal with metric coefficients $g_{(i)(i)} = h_i^2$, $i = 1, 2, 3$ and $g_{ij} = 0$ for $i \neq j$.

(a) Find expressions for the physical components $D(11), D(22)$ and $D(33)$ in terms of the physical components $v(i), i = 1, 2, 3$. Answer: $D(ii) = \frac{1}{h_i} \frac{\partial V(i)}{\partial x^i} + \sum_{j \neq i} \frac{V(j)}{h_i h_j} \frac{\partial h_i}{\partial x^j}$ no sum on i.

(b) Find expressions for the physical components $D(12), D(13)$ and $D(23)$ in terms of the physical components $v(i), i = 1, 2, 3$. Answer: $D(ij) = \frac{1}{2} \left[\frac{h_i}{h_j} \frac{\partial}{\partial x^j} \left(\frac{V(i)}{h_i} \right) + \frac{h_j}{h_i} \frac{\partial}{\partial x^i} \left(\frac{V(j)}{h_j} \right) \right]$

▶ **18.** Write out the tensor equations in problem 17 in Cartesian coordinates.

▶ **19.** Write out the tensor equations in problem 17 in cylindrical coordinates.

20. Write out the tensor equations in problem 17 in spherical coordinates.

21. Express the vector equation $(\lambda + 2\mu)\nabla\Phi - 2\mu\nabla \times \vec{\omega} + \vec{F} = \vec{0}$ in tensor form.

▶ **22.** Write out the equations in problem 21 for a generalized orthogonal coordinate system in terms of physical components.

▶ **23.** Write out the equations in problem 22 for cylindrical coordinates.

▶ **24.** Write out the equations in problem 22 for spherical coordinates.

▶ **25.** Use equation (2.1.4) to represent the divergence in parabolic cylindrical coordinates (ξ, η, z).

▶ **26.** Use equation (2.1.4) to represent the divergence in parabolic coordinates (ξ, η, ϕ).

▶ **27.** Use equation (2.1.4) to represent the divergence in elliptic cylindrical coordinates (ξ, η, z).

Change the given equations from a vector notation to a tensor notation.

▶ **28.** $\vec{B} = \vec{v}\nabla \cdot \vec{A} + (\nabla \cdot \vec{v})\vec{A}$

▶ **29.** $\frac{d}{dt}[\vec{A} \cdot (\vec{B} \times \vec{C})] = \frac{d\vec{A}}{dt} \cdot (\vec{B} \times \vec{C}) + \vec{A} \cdot (\frac{d\vec{B}}{dt} \times \vec{C}) + \vec{A} \cdot (\vec{B} \times \frac{d\vec{C}}{dt})$

▶ **30.** $\frac{d\vec{v}}{dt} = \frac{\partial \vec{v}}{\partial t} + (\vec{v} \cdot \nabla)\vec{v}$

▶ **31.** $\frac{1}{c} \frac{\partial \vec{H}}{\partial t} = -\text{curl } \vec{E}$

▶ **32.** $\frac{d\vec{B}}{dt} - (\vec{B} \cdot \nabla)\vec{v} + \vec{B}(\nabla \cdot \vec{v}) = \vec{0}$

Change the given equations from a tensor notation to a vector notation.

▶ 33. $\epsilon^{ijk} B_{k,j} + F^i = 0$

▶ 34. $g_{ij} \epsilon^{jkl} B_{l,k} + F_i = 0$

▶ 35. $\dfrac{\partial \varrho}{\partial t} + (\varrho v_i), i = 0$

▶ 36. $\varrho \left(\dfrac{\partial v_i}{\partial t} + v_m \dfrac{\partial v_i}{\partial x^m} \right) = -\dfrac{\partial P}{\partial x^i} + \mu \dfrac{\partial^2 v_i}{\partial x^m \partial x^m} + F_i$

▶ 37. The moment of inertia of an area or second moment of area is defined by $I_{ij} = \displaystyle\int\!\!\int_A (y_m y_m \delta_{ij} - y_i y_j)\, dA$ where dA is an element of area. Calculate the moment of inertia I_{ij}, $i,j = 1,2$ for the triangle illustrated in the figure 2.1-1 and show that $I_{ij} = \begin{pmatrix} \frac{1}{12} bh^3 & -\frac{1}{24} b^2 h^2 \\ -\frac{1}{24} b^2 h^2 & \frac{1}{12} b^3 h \end{pmatrix}$.

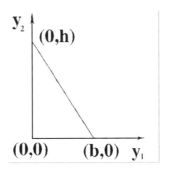

Figure 2.1-1 Moments of inertia for a triangle

▶ 38. Use the results from problem 37 and rotate the axes in figure 2.1-1 through an angle θ to a barred system of coordinates.

(a) Show that in the barred system of coordinates

$$\overline{I}_{11} = \left(\frac{I_{11} + I_{22}}{2} \right) + \left(\frac{I_{11} - I_{22}}{2} \right) \cos 2\theta + I_{12} \sin 2\theta$$

$$\overline{I}_{12} = \overline{I}_{21} = -\left(\frac{I_{11} - I_{22}}{2} \right) \sin 2\theta + I_{12} \cos 2\theta$$

$$\overline{I}_{22} = \left(\frac{I_{11} + I_{22}}{2} \right) - \left(\frac{I_{11} - I_{22}}{2} \right) \cos 2\theta - I_{12} \sin 2\theta$$

(b) For what value of θ will $\overline{I}_{11}$ have a maximum value?

(c) Show that when $\overline{I}_{11}$ is a maximum, we will have $\overline{I}_{22}$ a minimum and $\overline{I}_{12} = \overline{I}_{21} = 0$.

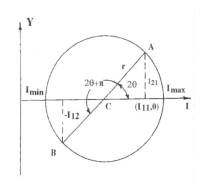

Figure 2.1-2 Mohr's circle

39. Otto Mohr[1] gave the following physical interpretation to the results obtained in problem 38:

• Plot the points $A(I_{11}, I_{12})$ and $B(I_{22}, -I_{12})$ as illustrated in the figure 2.1-2

• Draw the line $\overline{AB}$ and calculate the point C where this line intersects the I axes. Show the point C has the coordinates

$$\left(\frac{I_{11} + I_{22}}{2}, 0\right)$$

• Calculate the radius of the circle with center at the point C and with diagonal $\overline{AB}$ and show this radius is

$$r = \sqrt{\left(\frac{I_{11} - I_{22}}{2}\right)^2 + I_{12}^2}$$

• Show the maximum and minimum values of I occur where the constructed circle intersects the I axes. Show that $I_{max} = \bar{I}_{11} = \dfrac{I_{11} + I_{22}}{2} + r \qquad I_{min} = \bar{I}_{22} = \dfrac{I_{11} + I_{22}}{2} - r.$

40. Show directly that the eigenvalues of the symmetric matrix $I_{ij} = \begin{pmatrix} I_{11} & I_{12} \\ I_{21} & I_{22} \end{pmatrix}$ are $\lambda_1 = I_{max}$ and $\lambda_2 = I_{min}$ where I_{max} and I_{min} are given in problem 39.

41. Find the principal axes and moments of inertia for the triangle given in problem 37 and summarize your results from problems 37, 38, 39, and 40.

42. Verify for orthogonal coordinates the relations

$$\left[\nabla \times \vec{A}\right] \cdot \hat{e}_{(i)} = \sum_{k=1}^{3} \frac{e_{(i)jk}}{h_1 h_2 h_3} h_{(i)} \frac{\partial(h_{(k)}A(k))}{\partial x_j}$$

or

$$\nabla \times \vec{A} = \frac{1}{h_1 h_2 h_3} \begin{vmatrix} h_1\,\hat{e}_1 & h_2\,\hat{e}_2 & h_3\,\hat{e}_3 \\ \frac{\partial}{\partial x_1} & \frac{\partial}{\partial x_2} & \frac{\partial}{\partial x_3} \\ h_1 A(1) & h_2 A(2) & h_3 A(3) \end{vmatrix}.$$

43. Verify for orthogonal coordinates the relation

$$\left[\nabla \times (\nabla \times \vec{A})\right] \cdot \hat{e}_{(i)} = \sum_{m=1}^{3} e_{(i)jr} e_{rsm} \frac{h_{(i)}}{h_1 h_2 h_3} \frac{\partial}{\partial x_j} \left[\frac{h_{(r)}^2}{h_1 h_2 h_3} \frac{\partial(h_{(m)}A(m))}{\partial x_s}\right]$$

---

[1]Christian Otto Mohr (1835-1918) German civil engineer.

▶ **44.** Verify for orthogonal coordinates the relation

$$\left[\nabla\left(\nabla\cdot\vec{A}\right)\right]\cdot\hat{e}_{(i)} = \frac{1}{h_{(i)}}\frac{\partial}{\partial x_{(i)}}\left\{\frac{1}{h_1 h_2 h_3}\left[\frac{\partial(h_2 h_3 A(1))}{\partial x_1} + \frac{\partial(h_1 h_3 A(2))}{\partial x_2} + \frac{\partial(h_1 h_2 A(3))}{\partial x_3}\right]\right\}$$

▶ **45.** Verify the relation

$$\left[(\vec{A}\cdot\nabla)\vec{B}\right]\cdot\hat{e}_{(i)} = \sum_{k=1}^{3}\frac{A(k)}{h_{(k)}}\frac{\partial B(i)}{\partial x_k} + \sum_{k\neq i}\frac{B(k)}{h_k h_{(i)}}\left(A(i)\frac{\partial h_{(i)}}{\partial x_k} - A(k)\frac{\partial h_k}{\partial x_{(i)}}\right)$$

▶ **46.** The Gauss divergence theorem is written

$$\iiint_V\left(\frac{\partial F^1}{\partial x} + \frac{\partial F^2}{\partial y} + \frac{\partial F^3}{\partial z}\right)d\tau = \iint_S\left(n_1 F^1 + n_2 F^2 + n_3 F^3\right)d\sigma$$

where V is the volume within a simple closed surface S. Here it is assumed that $F^i = F^i(x, y, z)$ are continuous functions with continuous first order derivatives throughout V and n_i are the direction cosines of the outward normal to S, $d\tau$ is an element of volume and $d\sigma$ is an element of surface area.

(a) Show that in a Cartesian coordinate system

$$F^i_{,i} = \frac{\partial F^1}{\partial x} + \frac{\partial F^2}{\partial y} + \frac{\partial F^3}{\partial z}$$

and that the tensor form of this theorem is $\iiint_V F^i_{,i}\,d\tau = \iint_S F^i n_i\,d\sigma$.

(b) Write the vector form of this theorem.

(c) Show that if we define

$$u_r = \frac{\partial u}{\partial x^r}, \quad v_r = \frac{\partial v}{\partial x^r} \quad \text{and} \quad F_r = g_{rm}F^m = uv_r$$

then $F^i_{,i} = g^{im}F_{i,m} = g^{im}\left(uv_{i,m} + u_m v_i\right)$

(d) Show that another form of the Gauss divergence theorem is

$$\iiint_V g^{im}u_m v_i\,d\tau = \iint_S uv_m n^m\,d\sigma - \iiint_V ug^{im}v_{i,m}\,d\tau$$

Write out the above equation in Cartesian coordinates.

▶ **47.** Find the eigenvalues and eigenvectors associated with the matrix $A = \begin{pmatrix} 1 & 1 & 2 \\ 1 & 2 & 1 \\ 2 & 1 & 1 \end{pmatrix}$.

Show that the eigenvectors are orthogonal.

▶ **48.** Find the eigenvalues and eigenvectors associated with the matrix $A = \begin{pmatrix} 1 & 2 & 1 \\ 2 & 1 & 0 \\ 1 & 0 & 1 \end{pmatrix}$.

Show that the eigenvectors are orthogonal.

▶ **49.** Find the eigenvalues and eigenvectors associated with the matrix $A = \begin{pmatrix} 1 & 1 & 0 \\ 1 & 1 & 1 \\ 0 & 1 & 1 \end{pmatrix}$.

Show that the eigenvectors are orthogonal.

▶ **50.** The harmonic and biharmonic functions or potential functions occur in the mathematical modeling of many physical problems. Any solution of Laplace's equation $\nabla^2\Phi = 0$ is called a harmonic function and any solution of the biharmonic equation $\nabla^4\Phi = 0$ is called a biharmonic function.

(a) Expand the Laplace equation in Cartesian, cylindrical and spherical coordinates.

(b) Expand the biharmonic equation in two dimensional Cartesian and polar coordinates.

   Hint: Consider $\nabla^4\Phi = \nabla^2(\nabla^2\Phi)$. In Cartesian coordinates $\nabla^2\Phi = \Phi_{,ii}$ and $\nabla^4\Phi = \Phi_{,iijj}$.

§2.2 DYNAMICS

Dynamics is concerned with studying the motion of particles and rigid bodies. By studying the motion of a single hypothetical particle, one can discern the motion of a system of particles. This in turn leads to the study of the motion of individual points in a continuous deformable medium.

Particle Movement

The trajectory of a particle in a generalized coordinate system is described by the parametric equations

$$x^i = x^i(t), \quad i = 1, \ldots, N \tag{2.2.1}$$

where t is a time parameter. If the coordinates are changed to a barred system by introducing a coordinate transformation $\bar{x}^i = \bar{x}^i(x^1, x^2, \ldots, x^N), \quad i = 1, \ldots, N$ then the trajectory of the particle in the barred system of coordinates is

$$\bar{x}^i = \bar{x}^i(x^1(t), x^2(t), \ldots, x^N(t)), \quad i = 1, \ldots, N. \tag{2.2.2}$$

The generalized velocity of the particle in the unbarred system is defined by

$$v^i = \frac{dx^i}{dt}, \quad i = 1, \ldots, N. \tag{2.2.3}$$

By the chain rule differentiation of the transformation equations (2.2.2) one can verify that the velocity in the barred system is

$$\bar{v}^r = \frac{d\bar{x}^r}{dt} = \frac{\partial \bar{x}^r}{\partial x^j}\frac{dx^j}{dt} = \frac{\partial \bar{x}^r}{\partial x^j}v^j, \quad r = 1, \ldots, N. \tag{2.2.4}$$

This show that the generalized velocity v^i is a first order contravariant tensor. The speed of the particle is obtained from the magnitude of the velocity and is $v^2 = g_{ij}v^iv^j$. The generalized acceleration f^i of the particle is defined as the intrinsic derivative of the generalized velocity. The generalized acceleration has the form

$$f^i = \frac{\delta v^i}{\delta t} = v^i{}_{,n}\frac{dx^n}{dt} = \frac{dv^i}{dt} + \begin{Bmatrix} i \\ m\,n \end{Bmatrix}v^mv^n = \frac{d^2x^i}{dt^2} + \begin{Bmatrix} i \\ m\,n \end{Bmatrix}\frac{dx^m}{dt}\frac{dx^n}{dt} \tag{2.2.5}$$

and the magnitude of the acceleration is $f^2 = g_{ij}f^if^j$.

Frenet-Serret Formulas

The parametric equations (2.2.1) describe a curve in our generalized space. With reference to the figure 2.2-1 we wish to define at each point P of the curve the following orthogonal unit vectors:

$$T^i = \text{unit tangent vector at each point } P.$$
$$N^i = \text{unit normal vector at each point } P.$$
$$B^i = \text{unit binormal vector at each point } P.$$

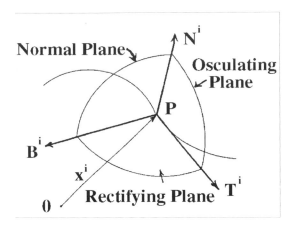

Figure 2.2-1 Tangent, normal and binormal to point P on curve.

These vectors define the osculating, normal and rectifying planes illustrated in the figure 2.2-1.

In the generalized coordinates the arc length squared is $ds^2 = g_{ij}dx^i dx^j$. Define $T^i = \frac{dx^i}{ds}$ as the tangent vector to the parametric curve defined by equation (2.2.1). This vector is a unit tangent vector because if we write the element of arc length squared in the form

$$1 = g_{ij}\frac{dx^i}{ds}\frac{dx^j}{ds} = g_{ij}T^i T^j, \tag{2.2.6}$$

we obtain the generalized dot product for T^i. This generalized dot product implies that the tangent vector is a unit vector. Differentiating the equation (2.2.6) intrinsically with respect to arc length s along the curve produces

$$g_{mn}\frac{\delta T^m}{\delta s}T^n + g_{mn}T^m\frac{\delta T^n}{\delta s} = 0,$$

which simplifies to

$$g_{mn}T^n\frac{\delta T^m}{\delta s} = 0. \tag{2.2.7}$$

The equation (2.2.7) is a statement that the vector $\frac{\delta T^m}{\delta s}$ is orthogonal to the vector T^m. The unit normal vector is defined as

$$N^i = \frac{1}{\kappa}\frac{\delta T^i}{\delta s} \qquad \text{or} \qquad N_i = \frac{1}{\kappa}\frac{\delta T_i}{\delta s}, \tag{2.2.8}$$

where κ is a scalar called the curvature and is chosen such that the magnitude of N^i is unity. The reciprocal of the curvature is $R = \frac{1}{\kappa}$, which is called the radius of curvature. The curvature of a straight line is zero while the curvature of a circle is a constant. The curvature measures the rate of change of the tangent vector as the arc length varies.

The equation (2.2.7) can be expressed in the form

$$g_{ij}T^i N^j = 0. \tag{2.2.9}$$

Taking the intrinsic derivative of equation (2.2.9) with respect to the arc length s produces

$$g_{ij}T^i\frac{\delta N^j}{\delta s} + g_{ij}\frac{\delta T^i}{\delta s}N^j = 0$$

or

$$g_{ij}T^i\frac{\delta N^j}{\delta s} = -g_{ij}\frac{\delta T^i}{\delta s}N^j = -\kappa g_{ij}N^iN^j = -\kappa. \tag{2.2.10}$$

The generalized dot product can be written $g_{ij}T^iT^j = 1$, and so we can express equation (2.2.10) in the form

$$g_{ij}T^i\frac{\delta N^j}{\delta s} = -\kappa g_{ij}T^iT^j \quad\text{or}\quad g_{ij}T^i\left(\frac{\delta N^j}{\delta s} + \kappa T^j\right) = 0. \tag{2.2.11}$$

therefore, the vector

$$\frac{\delta N^j}{\delta s} + \kappa T^j \tag{2.2.12}$$

is orthogonal to T^i. In a similar manner, we can use the relation $g_{ij}N^iN^j = 1$ and differentiate intrinsically with respect to the arc length s to show that $g_{ij}N^i\frac{\delta N^j}{\delta s} = 0$. This in turn can be expressed in the form $g_{ij}N^i\left(\frac{\delta N^j}{\delta s} + \kappa T^j\right) = 0$. This form of the equation implies that the vector represented in equation (2.2.12) is also orthogonal to the unit normal N^i. We define the unit binormal vector as

$$B^i = \frac{1}{\tau}\left(\frac{\delta N^i}{\delta s} + \kappa T^i\right) \quad\text{or}\quad B_i = \frac{1}{\tau}\left(\frac{\delta N_i}{\delta s} + \kappa T_i\right) \tag{2.2.13}$$

where τ is a scalar called the torsion. The torsion is chosen such that the binormal vector is a unit vector. The torsion τ measures the rate of change of the osculating plane and is a measure of the twisting of the curve out of a plane. The value $\tau = 0$ corresponds to a plane curve. The vectors $T^i, N^i, B^i, i = 1,2,3$ satisfy the cross product relation $B^i = \epsilon^{ijk}T_jN_k$. If we differentiate this relation intrinsically with respect to arc length s we find

$$\begin{aligned}\frac{\delta B^i}{\delta s} &= \epsilon^{ijk}\left(T_j\frac{\delta N_k}{\delta s} + \frac{\delta T_j}{\delta s}N_k\right)\\ &= \epsilon^{ijk}[T_j(\tau B_k - \kappa T_k) + \kappa N_jN_k]\\ &= \tau\epsilon^{ijk}T_jB_k = -\tau\epsilon^{ikj}B_kT_j = -\tau N^i.\end{aligned} \tag{2.2.14}$$

The relations (2.2.8),(2.2.13) and (2.2.14) are now summarized and written

$$\frac{\delta T^i}{\delta s} = \kappa N^i \qquad \frac{\delta N^i}{\delta s} = \tau B^i - \kappa T^i \qquad \frac{\delta B^i}{\delta s} = -\tau N^i. \tag{2.2.15}$$

These equations are known as the Frenet-Serret formulas of differential geometry.

Velocity and Acceleration

Chain rule differentiation of the generalized velocity is expressible in the form

$$v^i = \frac{dx^i}{dt} = \frac{dx^i}{ds}\frac{ds}{dt} = T^iv, \tag{2.2.16}$$

where $v = \frac{ds}{dt}$ is the speed of the particle and is the magnitude of v^i. The vector T^i is the unit tangent vector to the trajectory curve at the time t. The equation (2.2.16) is a statement of the fact that the velocity of a particle is always in the direction of the tangent vector to the curve and has the speed v.

190

By chain rule differentiation, the generalized acceleration is expressible in the form

$$
\begin{aligned}
f^r = \frac{\delta v^r}{\delta t} &= \frac{dv}{dt}T^r + v\frac{\delta T^r}{\delta t} \\
&= \frac{dv}{dt}T^r + v\frac{\delta T^r}{\delta s}\frac{ds}{dt} \\
&= \frac{dv}{dt}T^r + \kappa v^2 N^r.
\end{aligned}
\tag{2.2.17}
$$

The equation (2.2.17) states that the acceleration lies in the osculating plane. Further, the equation (2.2.17) indicates that the tangential component of the acceleration is $\frac{dv}{dt}$, while the normal component of the acceleration is κv^2.

Work and Potential Energy

Define M as the constant mass of the particle as it moves along the curve defined by equation (2.2.1). Also let Q^r denote the components of a force vector (in appropriate units of measurements) which acts upon the particle. Newton's second law of motion can then be expressed in the form

$$
Q^r = Mf^r \qquad \text{or} \qquad Q_r = Mf_r.
\tag{2.2.18}
$$

The work done W in moving a particle from a point P_0 to a point P_1 along a curve $x^r = x^r(t), r = 1,2,3$, with parameter t, is represented by a summation of the tangential components of the forces acting along the path and is defined as the line integral

$$
W = \int_{P_0}^{P_1} Q_r \frac{dx^r}{ds}ds = \int_{P_0}^{P_1} Q_r\, dx^r = \int_{t_0}^{t_1} Q_r \frac{dx^r}{dt}dt = \int_{t_0}^{t_1} Q_r v^r\, dt
\tag{2.2.19}
$$

where $Q_r = g_{rs}Q^s$ is the covariant form of the force vector, t is the time parameter and s is arc length along the curve.

Conservative Systems

If the force vector is conservative it means that the force is derivable from a scalar potential function

$$
V = V(x^1, x^2, \ldots, x^N) \qquad \text{such that} \qquad Q_r = -V_{,r} = -\frac{\partial V}{\partial x^r}, \qquad r = 1, \ldots, N.
\tag{2.2.20}
$$

In this case the equation (2.2.19) can be integrated and we find that to within an additive constant we will have $V = -W$. The potential function V is called the potential energy of the particle and the work done becomes the change in potential energy between the starting and end points and is independent of the path connecting the points.

Lagrange's Equations of Motion

The kinetic energy T of the particle is defined as one half the mass times the velocity squared and can be expressed in any of the forms

$$T = \frac{1}{2}M\left(\frac{ds}{dt}\right)^2 = \frac{1}{2}Mv^2 = \frac{1}{2}Mg_{mn}v^m v^n = \frac{1}{2}Mg_{mn}\dot{x}^m \dot{x}^n, \qquad (2.2.21)$$

where the dot notation denotes differentiation with respect to time. It is an easy exercise to calculate the derivatives

$$\frac{\partial T}{\partial \dot{x}^r} = Mg_{rm}\dot{x}^m$$
$$\frac{d}{dt}\left(\frac{\partial T}{\partial \dot{x}^r}\right) = M\left[g_{rm}\ddot{x}^m + \frac{\partial g_{rm}}{\partial x^n}\dot{x}^n \dot{x}^m\right] \qquad (2.2.22)$$
$$\frac{\partial T}{\partial x^r} = \frac{1}{2}M\frac{\partial g_{mn}}{\partial x^r}\dot{x}^m \dot{x}^n,$$

and thereby verify the relation

$$\frac{d}{dt}\left(\frac{\partial T}{\partial \dot{x}^r}\right) - \frac{\partial T}{\partial x^r} = Mf_r = Q_r, \quad r = 1, \dots, N. \qquad (2.2.23)$$

This equation is called the Lagrange's form of the equations of motion.

EXAMPLE 2.2-1. (Equations of motion in spherical coordinates) Find the Lagrange's form of the equations of motion in spherical coordinates.

Solution: Let $x^1 = \rho$, $x^2 = \theta$, $x^3 = \phi$ then the element of arc length squared in spherical coordinates has the form

$$ds^2 = (d\rho)^2 + \rho^2(d\theta)^2 + \rho^2\sin^2\theta(d\phi)^2.$$

The element of arc length squared can be used to construct the kinetic energy. For example,

$$T = \frac{1}{2}M\left(\frac{ds}{dt}\right)^2 = \frac{1}{2}M\left[(\dot\rho)^2 + \rho^2(\dot\theta)^2 + \rho^2\sin^2\theta(\dot\phi)^2\right].$$

The Lagrange form of the equations of motion of a particle are found from the relations (2.2.23) and are calculated to be:

$$Mf_1 = Q_1 = \frac{d}{dt}\left(\frac{\partial T}{\partial \dot\rho}\right) - \frac{\partial T}{\partial \rho} = M\left[\ddot\rho - \rho(\dot\theta)^2 - \rho\sin^2\theta(\dot\phi)^2\right]$$
$$Mf_2 = Q_2 = \frac{d}{dt}\left(\frac{\partial T}{\partial \dot\theta}\right) - \frac{\partial T}{\partial \theta} = M\left[\frac{d}{dt}\left(\rho^2\dot\theta\right) - \rho^2\sin\theta\cos\theta(\dot\phi)^2\right]$$
$$Mf_3 = Q_3 = \frac{d}{dt}\left(\frac{\partial T}{\partial \dot\phi}\right) - \frac{\partial T}{\partial \phi} = M\left[\frac{d}{dt}\left(\rho^2\sin^2\theta\dot\phi\right)\right].$$

In terms of physical components we have

$$Q_\rho = M\left[\ddot\rho - \rho(\dot\theta)^2 - \rho\sin^2\theta(\dot\phi)^2\right]$$
$$Q_\theta = \frac{M}{\rho}\left[\frac{d}{dt}\left(\rho^2\dot\theta\right) - \rho^2\sin\theta\cos\theta(\dot\phi)^2\right]$$
$$Q_\phi = \frac{M}{\rho\sin\theta}\left[\frac{d}{dt}\left(\rho^2\sin^2\theta\dot\phi\right)\right].$$

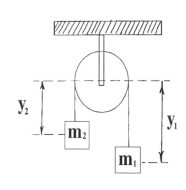

Figure 2.2-2 Simply pulley system

Euler-Lagrange Equations of Motion

Starting with the Lagrange's form of the equations of motion from equation (2.2.23), we assume that the external force Q_r is derivable from a potential function V as specified by the equation (2.2.20). That is, we assume the system is conservative and express the equations of motion in the form

$$\frac{d}{dt}\left(\frac{\partial T}{\partial \dot{x}^r}\right) - \frac{\partial T}{\partial x^r} = -\frac{\partial V}{\partial x^r} = Q_r, \quad r = 1, \ldots, N \qquad (2.2.24)$$

The Lagrangian is defined by the equation

$$L = T - V = T(x^1, \ldots, x^N, \dot{x}^1, \ldots, \dot{x}^N) - V(x^1, \ldots, x^N) = L(x^i, \dot{x}^i). \qquad (2.2.25)$$

Employing the defining equation (2.2.25), it is readily verified that the equations of motion are expressible in the form

$$\frac{d}{dt}\left(\frac{\partial L}{\partial \dot{x}^r}\right) - \frac{\partial L}{\partial x^r} = 0, \qquad r = 1, \ldots, N, \qquad (2.2.26)$$

which are called the Euler-Lagrange form for the equations of motion.

EXAMPLE 2.2-2. (Simple pulley system) Find the equation of motion for the simply pulley system illustrated in the figure 2.2-2.

Solution: The given system has only one degree of freedom, say y_1. It is assumed that $y_1 + y_2 = \ell =$ a constant. The kinetic energy of the system is $T = \frac{1}{2}(m_1 + m_2)\dot{y}_1^2$. Let y_1 increase by an amount dy_1 and show the work done by gravity can be expressed as

$$dW = m_1 g\, dy_1 + m_2 g\, dy_2 = m_1 g\, dy_1 - m_2 g\, dy_1 = (m_1 - m_2)g\, dy_1 = Q_1\, dy_1.$$

Here $Q_1 = (m_1 - m_2)g$ is the external force acting on the system where g is the acceleration of gravity. The Lagrange equation of motion is

$$\frac{d}{dt}\left(\frac{\partial T}{\partial \dot{y}_1}\right) - \frac{\partial T}{\partial y_1} = Q_1 \qquad \text{or} \qquad (m_1 + m_2)\ddot{y}_1 = (m_1 - m_2)g.$$

Initial conditions must be applied to y_1 and $\dot{y}_1$ before this equation can be solved.

193

EXAMPLE 2.2-3. (Simple pendulum) Find the equation of motion for the pendulum system illustrated in the figure 2.2-3.

Solution: Choose the angle θ illustrated in the figure 2.2-3 as the generalized coordinate. If the pendulum is moved from a vertical position through an angle θ, we observe that the mass m moves up a distance $h = \ell - \ell\cos\theta$. The work done in moving this mass a vertical distance h is

$$W = -mgh = -mg\ell(1 - \cos\theta),$$

since the force is $-mg$ in this coordinate system. In moving the pendulum through an angle θ, the arc length s swept out by the mass m is $s = \ell\theta$. This implies that the kinetic energy can be expressed

$$T = \frac{1}{2}m\left(\frac{ds}{dt}\right)^2 = \frac{1}{2}m\left(\ell\dot\theta\right)^2 = \frac{1}{2}m\ell^2(\dot\theta)^2.$$

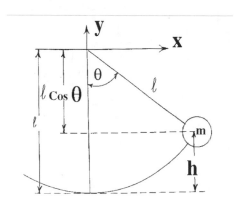

Figure 2.2-3 Simple pendulum system

The Lagrangian of the system is $L = T - V = \frac{1}{2}m\ell^2(\dot\theta)^2 - mg\ell(1 - \cos\theta)$ and from this we find the equation of motion

$$\frac{d}{dt}\left(\frac{\partial L}{\partial \dot\theta}\right) - \frac{\partial L}{\partial \theta} = 0 \quad \text{or} \quad \frac{d}{dt}\left(m\ell^2\dot\theta\right) - mg\ell(-\sin\theta) = 0.$$

This in turn simplifies to the equation

$$\ddot\theta + \frac{g}{\ell}\sin\theta = 0.$$

This equation together with a set of initial conditions for θ and $\dot\theta$ represents the nonlinear differential equation which describes the motion of a pendulum without damping.

■

EXAMPLE 2.2-4. (Compound pendulum) Find the equations of motion for the compound pendulum illustrated in the figure 2.2-4.

Solution: Choose for the generalized coordinates the angles $x^1 = \theta_1$ and $x^2 = \theta_2$ illustrated in the figure 2.2-4. To find the potential function V for this system we consider the work done as the masses m_1 and m_2 are moved. Consider independent motions of the angles θ_1 and θ_2. Imagine the compound pendulum initially in the vertical position as illustrated in the figure 2.2-4(a). Now let m_1 be displaced due to a change in θ_1 and obtain the figure 2.2-4(b). The work done to achieve this position is

$$W_1 = -(m_1 + m_2)gh_1 = -(m_1 + m_2)gL_1(1 - \cos\theta_1).$$

Starting from the position in figure 2.2-4(b) we now let θ_2 undergo a displacement and achieve the configuration in the figure 2.2-4(c).

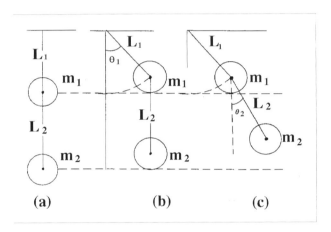

Figure 2.2-4 Compound pendulum

The work done due to the displacement θ_2 can be represented

$$W_2 = -m_2gh_2 = -m_2gL_2(1 - \cos\theta_2).$$

Since the potential energy V satisfies $V = -W$ to within an additive constant, we can write

$$V = -W = -W_1 - W_2 = -(m_1 + m_2)gL_1\cos\theta_1 - m_2gL_2\cos\theta_2 + constant,$$

where the constant term in the potential energy has been neglected since it does not contribute anything to the equations of motion. (i.e. the derivative of a constant is zero.)

The kinetic energy term for this system can be represented

$$T = \frac{1}{2}m_1\left(\frac{ds_1}{dt}\right)^2 + \frac{1}{2}m_2\left(\frac{ds_2}{dt}\right)^2$$
$$T = \frac{1}{2}m_1(\dot{x}_1^2 + \dot{y}_1^2) + \frac{1}{2}m_2(\dot{x}_2^2 + \dot{y}_2^2),$$

(2.2.27)

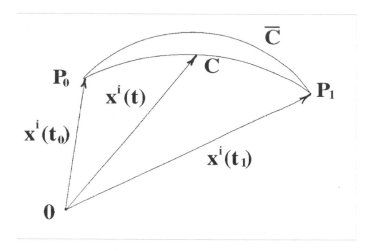

Figure 2.2-5. Motion along curves c and $\bar{c}$

where

$$(x_1, y_1) = (L_1 \sin\theta_1 , \, -L_1 \cos\theta_1)$$
$$(x_2, y_2) = (L_1 \sin\theta_1 + L_2 \sin\theta_2, -L_1 \cos\theta_1 - L_2 \cos\theta_2) \tag{2.2.28}$$

are the coordinates of the masses m_1 and m_2 respectively. Substituting the equations (2.2.28) into equation (2.2.27) and simplifying produces the kinetic energy expression

$$T = \frac{1}{2}(m_1 + m_2)L_1^2\dot{\theta}_1^2 + m_2 L_1 L_2 \dot{\theta}_1 \dot{\theta}_2 \cos(\theta_1 - \theta_2) + \frac{1}{2}m_2 L_2^2 \dot{\theta}_2^2. \tag{2.2.29}$$

Writing the Lagrangian as $L = T - V$, the equations describing the motion of the compound pendulum are obtained from the Lagrangian equations

$$\frac{d}{dt}\left(\frac{\partial L}{\partial \dot{\theta}_1}\right) - \frac{\partial L}{\partial \theta_1} = 0 \quad \text{and} \quad \frac{d}{dt}\left(\frac{\partial L}{\partial \dot{\theta}_2}\right) - \frac{\partial L}{\partial \theta_2} = 0.$$

Calculating the necessary derivatives, substituting them into the Lagrangian equations of motion and then simplifying we derive the equations of motion

$$L_1\ddot{\theta}_1 + \frac{m_2}{m_1 + m_2}L_2\ddot{\theta}_2 \cos(\theta_1 - \theta_2) + \frac{m_2}{m_1 + m_2}L_2(\dot{\theta}_2)^2 \sin(\theta_1 - \theta_2) + g\sin\theta_1 = 0$$
$$L_1\ddot{\theta}_1 \cos(\theta_1 - \theta_2) + L_2\ddot{\theta}_2 - L_1(\dot{\theta}_1)^2 \sin(\theta_1 - \theta_2) + g\sin\theta_2 = 0.$$

These equations are a set of coupled, second order nonlinear ordinary differential equations. These equations are subject to initial conditions being imposed upon the angular displacements (θ_1, θ_2) and the angular velocities $(\dot{\theta}_1, \dot{\theta}_2)$.

■

Alternative Derivation of Lagrange's Equations of Motion

Let c denote a given curve represented in the parametric form $x^i = x^i(t)$, $\quad i = 1, \ldots, N$, $\quad t_0 \le t \le t_1$ and let P_0, P_1 denote two points on this curve corresponding to the parameter values t_0 and t_1 respectively. Let $\bar{c}$ denote another curve which also passes through the two points P_0 and P_1 as illustrated in the figure 2.2-5.

The curve $\bar{c}$ is represented in the parametric form $\bar{x}^i = \bar{x}^i(t) = x^i(t) + \epsilon \eta^i(t)$, $\quad i = 1, \ldots, N$, $\quad t_0 \le t \le t_1$ in terms of a parameter ϵ. In this representation the function $\eta^i(t)$ must satisfy the end conditions

$$\eta^i(t_0) = 0 \quad \text{and} \quad \eta^i(t_1) = 0 \quad i = 1, \ldots, N$$

since the curve $\bar{c}$ is assumed to pass through the end points P_0 and P_1.

Consider the line integral

$$I(\epsilon) = \int_{t_0}^{t_1} L(t, x^i + \epsilon \eta^i, \dot{x}^i + \epsilon \dot{\eta}^i) \, dt, \tag{2.2.30}$$

where

$$L = T - V = L(t, \bar{x}^i, \dot{\bar{x}}^i)$$

is the Lagrangian evaluated along the curve $\bar{c}$. We ask the question, "What conditions must be satisfied by the curve c in order that the integral $I(\epsilon)$ have an extremum value when ϵ is zero?" If the integral $I(\epsilon)$ has a minimum value when ϵ is zero it follows that its derivative with respect to ϵ will be zero at this value and we will have

$$\left. \frac{dI(\epsilon)}{d\epsilon} \right|_{\epsilon=0} = 0.$$

Employing the definition

$$\left. \frac{dI}{d\epsilon} \right|_{\epsilon=0} = \lim_{\epsilon \to 0} \frac{I(\epsilon) - I(0)}{\epsilon} = I'(0) = 0$$

we expand the Lagrangian in equation (2.2.30) in a series about the point $\epsilon = 0$. Substituting the expansion

$$L(t, x^i + \epsilon \eta^i, \dot{x}^i + \epsilon \dot{\eta}^i) = L(t, x^i, \dot{x}^i) + \epsilon \left[\frac{\partial L}{\partial x^i} \eta^i + \frac{\partial L}{\partial \dot{x}^i} \dot{\eta}^i \right] + \epsilon^2 [\quad] + \cdots$$

into equation (2.2.30) we calculate the derivative

$$I'(0) = \lim_{\epsilon \to 0} \frac{I(\epsilon) - I(0)}{\epsilon} = \lim_{\epsilon \to 0} \int_{t_0}^{t_1} \left[\frac{\partial L}{\partial x^i} \eta^i(t) + \frac{\partial L}{\partial \dot{x}^i} \dot{\eta}^i(t) \right] dt + \epsilon [\quad] + \cdots = 0,$$

where we have neglected higher order powers of ϵ since ϵ is approaching zero. Analysis of this equation informs us that the integral I has a minimum value at $\epsilon = 0$ provided that the integral

$$\delta I = \int_{t_0}^{t_1} \left[\frac{\partial L}{\partial x^i} \eta^i(t) + \frac{\partial L}{\partial \dot{x}^i} \dot{\eta}^i(t) \right] dt = 0 \tag{2.2.31}$$

is satisfied. Integrating the second term of this integral by parts we find

$$\delta I = \int_{t_0}^{t_1} \frac{\partial L}{\partial x^i} \eta^i \, dt + \left[\frac{\partial L}{\partial \dot{x}^i} \eta^i(t) \right]_{t_0}^{t_1} - \int_{t_0}^{t_1} \frac{d}{dt} \left(\frac{\partial L}{\partial \dot{x}^i} \right) \eta^i(t) \, dt = 0. \tag{2.2.32}$$

The end condition on $\eta^i(t)$ makes the middle term in equation (2.2.32) vanish and we are left with the integral

$$\delta I = \int_{t_0}^{t_1} \eta^i(t) \left[\frac{\partial L}{\partial x^i} - \frac{d}{dt} \left(\frac{\partial L}{\partial \dot{x}^i} \right) \right] dt = 0, \tag{2.2.33}$$

which must equal zero for all $\eta^i(t)$. Since $\eta^i(t)$ is arbitrary, the only way the integral in equation (2.2.33) can be zero for all $\eta^i(t)$ is for the term inside the brackets to vanish. This produces the result that the integral of the Lagrangian is an extremum when the Euler-Lagrange equations

$$\frac{d}{dt}\left(\frac{\partial L}{\partial \dot{x}^i}\right) - \frac{\partial L}{\partial x^i} = 0, \quad i = 1, \ldots, N \tag{2.2.34}$$

are satisfied. This is a necessary condition for the integral $I(\epsilon)$ to have a minimum value.

In general, any line integral of the form

$$I = \int_{t_0}^{t_1} \phi(t, x^i, \dot{x}^i)\, dt \tag{2.2.35}$$

has an extremum value if the curve c defined by $x^i = x^i(t)$, $i = 1, \ldots, N$ satisfies the Euler-Lagrange equations

$$\frac{d}{dt}\left(\frac{\partial \phi}{\partial \dot{x}^i}\right) - \frac{\partial \phi}{\partial x^i} = 0, \quad i = 1, \ldots, N. \tag{2.2.36}$$

The above derivation is a special case of (2.2.36) when $\phi = L$. Note that the equations of motion equations (2.2.34) are just another form of the equations (2.2.24). Note also that

$$\frac{\delta T}{\delta t} = \frac{\delta}{\delta t}\left(\frac{1}{2} m g_{ij} v^i v^j\right) = m g_{ij} v^i f^j = m f_i v^i = m f_i \dot{x}^i$$

and if we assume that the force Q_i is derivable from a potential function V, then $m f_i = Q_i = -\dfrac{\partial V}{\partial x^i}$, so that $\dfrac{\delta T}{\delta t} = m f_i \dot{x}^i = Q_i \dot{x}^i = -\dfrac{\partial V}{\partial x^i}\dot{x}^i = -\dfrac{\delta V}{\delta t}$ or $\dfrac{\delta}{\delta t}(T + V) = 0$ or $T + V = h = $ constant called the energy constant of the system.

Action Integral

The equations of motion (2.2.34) or (2.2.24) are interpreted as describing geodesics in a space whose line-element is

$$ds^2 = 2m(h - V)g_{jk}dx^j dx^k$$

where V is the potential function for the force system and $T + V = h$ is the energy constant of the motion. The integral of ds along a curve C between two points P_1 and P_2 is called an action integral and is

$$A = \sqrt{2m} \int_{P_1}^{P_2} \left\{(h - V)g_{jk}\frac{dx^j}{d\tau}\frac{dx^k}{d\tau}\right\}^{1/2} d\tau$$

where τ is a parameter used to describe the curve C. The principle of stationary action states that of all curves through the points P_1 and P_2 the one which makes the action an extremum is the curve specified by Newton's second law. The extremum is usually a minimum. To show this let

$$\phi = \sqrt{2m} \left\{(h - V)g_{jk}\frac{dx^j}{d\tau}\frac{dx^k}{d\tau}\right\}^{1/2}$$

in equation (2.2.36). Using the notation $\dot{x}^k = \frac{dx^k}{d\tau}$ we find that

$$\frac{\partial \phi}{\partial \dot{x}^i} = \frac{2m}{\phi}(h - V)g_{ik}\dot{x}^k$$

$$\frac{\partial \phi}{\partial x^i} = \frac{2m}{2\phi}(h - V)\frac{\partial g_{jk}}{\partial x^i}\dot{x}^j\dot{x}^k - \frac{2m}{2\phi}\frac{\partial V}{\partial x^i}g_{jk}\dot{x}^j\dot{x}^k.$$

The equation (2.2.36) which describe the extremum trajectories are found to be

$$\frac{d}{dt}\left[\frac{2m}{\phi}(h - V)g_{ik}\dot{x}^k\right] - \frac{2m}{2\phi}(h - V)\frac{\partial g_{jk}}{\partial x^i}\dot{x}^j\dot{x}^k + \frac{2m}{\phi}\frac{\partial V}{\partial x^i}g_{jk}\dot{x}^j\dot{x}^k = 0.$$

By changing variables from τ to t where $\frac{dt}{d\tau} = \frac{\sqrt{m}\phi}{\sqrt{2}(h-V)}$ we find that the trajectory for an extremum must satisfy the equation

$$m\frac{d}{dt}\left(g_{ik}\frac{dx^k}{dt}\right) - \frac{m}{2}\frac{\partial g_{jk}}{\partial x^i}\frac{dx^j}{dt}\frac{dx^k}{dt} + \frac{\partial V}{\partial x^i} = 0$$

which are the same equations as (2.2.24). (i.e. See also the equations (2.2.22).)

Dynamics of Rigid Body Motion

Let us derive the equations of motion of a rigid body which is rotating due to external forces acting upon it. We neglect any translational motion of the body since this type of motion can be discerned using our knowledge of particle dynamics. The derivation of the equations of motion is restricted to Cartesian tensors and rotational motion.

Consider a system of N particles rotating with angular velocity ω_i, $i = 1, 2, 3$, about a line L through the center of mass of the system. Let $\vec{V}^{(\alpha)}$ denote the velocity of the αth particle which has mass $m_{(\alpha)}$ and position $x_i^{(\alpha)}$, $i = 1, 2, 3$ with respect to an origin on the line L. Without loss of generality we can assume that the origin of the coordinate system is also at the center of mass of the system of particles, as this choice of an origin simplifies the derivation. The velocity components for each particle is obtained by taking cross products and we can write

$$\vec{V}^{(\alpha)} = \vec{\omega} \times \vec{r}^{(\alpha)} \qquad \text{or} \qquad V_i^{(\alpha)} = e_{ijk}\omega_j x_k^{(\alpha)}. \tag{2.2.37}$$

The kinetic energy of the system of particles is written as the sum of the kinetic energies of each individual particle and is

$$T = \frac{1}{2}\sum_{\alpha=1}^{N} m_{(\alpha)} V_i^{(\alpha)} V_i^{(\alpha)} = \frac{1}{2}\sum_{\alpha=1}^{N} m_{(\alpha)} e_{ijk}\omega_j x_k^{(\alpha)} e_{imn}\omega_m x_n^{(\alpha)}. \tag{2.2.38}$$

Employing the $e - \delta$ identity the equation (2.2.38) can be simplified to the form

$$T = \frac{1}{2}\sum_{\alpha=1}^{N} m_{(\alpha)} \left(\omega_m\omega_m x_k^{(\alpha)} x_k^{(\alpha)} - \omega_n\omega_k x_k^{(\alpha)} x_n^{(\alpha)}\right).$$

Define the second moments and products of inertia by the equation

$$I_{ij} = \sum_{\alpha=1}^{N} m_{(\alpha)} \left(x_k^{(\alpha)} x_k^{(\alpha)}\delta_{ij} - x_i^{(\alpha)} x_j^{(\alpha)}\right) \tag{2.2.39}$$

and write the kinetic energy in the form

$$T = \frac{1}{2}I_{ij}\omega_i\omega_j. \tag{2.2.40}$$

Similarly, the angular momentum of the system of particles can also be represented in terms of the second moments and products of inertia. The angular momentum of a system of particles is defined as a summation of the moments of the linear momentum of each individual particle and is

$$H_i = \sum_{\alpha=1}^{N} m_{(\alpha)}e_{ijk}x_j^{(\alpha)}v_k^{(\alpha)} = \sum_{\alpha=1}^{N} m_{(\alpha)}e_{ijk}x_j^{(\alpha)}e_{kmn}\omega_m x_n^{(\alpha)}. \tag{2.2.41}$$

The $e-\delta$ identity simplifies the equation (2.2.41) to the form

$$H_i = \omega_j \sum_{\alpha=1}^{N} m_{(\alpha)}\left(x_n^{(\alpha)}x_n^{(\alpha)}\delta_{ij} - x_j^{(\alpha)}x_i^{(\alpha)}\right) = \omega_j I_{ji}. \tag{2.2.42}$$

The equations of motion of a rigid body is obtained by applying Newton's second law of motion to the system of N particles. The equation of motion of the αth particle is written

$$m_{(\alpha)}\ddot{x}_i^{(\alpha)} = F_i^{(\alpha)}. \tag{2.2.43}$$

Summing equation (2.2.43) over all particles gives the result

$$\sum_{\alpha=1}^{N} m_{(\alpha)}\ddot{x}_i^{(\alpha)} = \sum_{\alpha=1}^{N} F_i^{(\alpha)}. \tag{2.2.44}$$

This represents the translational equations of motion of the rigid body. The equation (2.2.44) represents the rate of change of linear momentum being equal to the total external force acting upon the system. Taking the cross product of equation (2.2.43) with the position vector $x_j^{(\alpha)}$ produces

$$m_{(\alpha)}\ddot{x}_t^{(\alpha)}e_{rst}x_s^{(\alpha)} = e_{rst}x_s^{(\alpha)}F_t^{(\alpha)}$$

and summing over all particles we find the equation

$$\sum_{\alpha=1}^{N} m_{(\alpha)}e_{rst}x_s^{(\alpha)}\ddot{x}_t^{(\alpha)} = \sum_{\alpha=1}^{N} e_{rst}x_s^{(\alpha)}F_t^{(\alpha)}. \tag{2.2.45}$$

The equations (2.2.44) and (2.2.45) represent the conservation of linear and angular momentum and can be written in the forms

$$\frac{d}{dt}\left(\sum_{\alpha=1}^{N} m_{(\alpha)}\dot{x}_r^{(\alpha)}\right) = \sum_{\alpha=1}^{N} F_r^{(\alpha)} \tag{2.2.46}$$

and

$$\frac{d}{dt}\left(\sum_{\alpha=1}^{N} m_{(\alpha)}e_{rst}x_s^{(\alpha)}\dot{x}_t^{(\alpha)}\right) = \sum_{\alpha=1}^{N} e_{rst}x_s^{(\alpha)}F_t^{(\alpha)}. \tag{2.2.47}$$

By definition we have $G_r = \sum m_{(\alpha)}\dot{x}_r^{(\alpha)}$ representing the linear momentum, $F_r = \sum F_r^{(\alpha)}$ the total force acting on the system of particles, $H_r = \sum m_{(\alpha)}e_{rst}x_s^{(\alpha)}\dot{x}_t^{(\alpha)}$ is the angular momentum of the system relative

200

to the origin, and $M_r = \sum e_{rst} x_s^{(\alpha)} F_t^{(\alpha)}$ is the total moment of the system relative to the origin. We can therefore express the equations (2.2.46) and (2.2.47) in the form

$$\frac{dG_r}{dt} = F_r \qquad (2.2.48)$$

and

$$\frac{dH_r}{dt} = M_r. \qquad (2.2.49)$$

The equation (2.2.49) expresses the fact that the rate of change of angular momentum is equal to the moment of the external forces about the origin. These equations show that the motion of a system of particles can be studied by considering the motion of the center of mass of the system (translational motion) and simultaneously considering the motion of points about the center of mass (rotational motion).

We now develop some relations in order to express the equations (2.2.49) in an alternate form. Toward this purpose we consider first the concepts of relative motion and angular velocity.

Relative Motion and Angular Velocity

Consider two different reference frames denoted by $\overline{S}$ and S. Both reference frames are Cartesian coordinates with axes $\overline{x}_i$ and x_i, $i = 1, 2, 3$, respectively. The reference frame S is fixed in space and is called an inertial reference frame or space-fixed reference system of axes. The reference frame $\overline{S}$ is fixed to and rotates with the rigid body and is called a body-fixed system of axes. Again, for convenience, it is assumed that the origins of both reference systems are fixed at the center of mass of the rigid body. Further, we let the system $\overline{S}$ have the basis vectors $\widehat{\overline{e}}_i$, $i = 1, 2, 3$, while the reference system S has the basis vectors $\hat{e}_i$, $i = 1, 2, 3$. The transformation equations between the two sets of reference axes are the affine transformations

$$\overline{x}_i = \ell_{ji} x_j \qquad \text{and} \qquad x_i = \ell_{ij} \overline{x}_j \qquad (2.2.50)$$

where $\ell_{ij} = \ell_{ij}(t)$ are direction cosines which are functions of time t (i.e. the ℓ_{ij} are the cosines of the angles between the barred and unbarred axes where the barred axes are rotating relative to the space-fixed unbarred axes.) The direction cosines satisfy the relations

$$\ell_{ij} \ell_{ik} = \delta_{jk} \qquad \text{and} \qquad \ell_{ij} \ell_{kj} = \delta_{ik}. \qquad (2.2.51)$$

EXAMPLE 2.2-5. (Euler angles ϕ, θ, ψ) Consider the following sequence of transformations which are used in celestial mechanics. First a rotation about the x_3 axis taking the x_i axes to the y_i axes

$$\begin{pmatrix} y_1 \\ y_2 \\ y_3 \end{pmatrix} = \begin{pmatrix} \cos\phi & \sin\phi & 0 \\ -\sin\phi & \cos\phi & 0 \\ 0 & 0 & 1 \end{pmatrix} \begin{pmatrix} x_1 \\ x_2 \\ x_3 \end{pmatrix}$$

where the rotation angle ϕ is called the longitude of the ascending node. Second, a rotation about the y_1 axis taking the y_i axes to the y_i' axes

$$\begin{pmatrix} y_1' \\ y_2' \\ y_3' \end{pmatrix} = \begin{pmatrix} 1 & 0 & 0 \\ 0 & \cos\theta & \sin\theta \\ 0 & -\sin\theta & \cos\theta \end{pmatrix} \begin{pmatrix} y_1 \\ y_2 \\ y_3 \end{pmatrix}$$

where the rotation angle θ is called the angle of inclination of the orbital plane. Finally, a rotation about the y'_3 axis taking the y'_i axes to the $\bar{x}_i$ axes

$$\begin{pmatrix} \bar{x}_1 \\ \bar{x}_2 \\ \bar{x}_3 \end{pmatrix} = \begin{pmatrix} \cos\psi & \sin\psi & 0 \\ -\sin\psi & \cos\psi & 0 \\ 0 & 0 & 1 \end{pmatrix} \begin{pmatrix} y'_1 \\ y'_2 \\ y'_3 \end{pmatrix}$$

where the rotation angle ψ is called the argument of perigee. The Euler angle θ is the angle $\bar{x}_3 0 x_3$, the angle ϕ is the angle $x_1 0 y_1$ and ψ is the angle $y_1 0 \bar{x}_1$. These angles are illustrated in the figure 2.2-6. Note also that the rotation vectors associated with these transformations are vectors of magnitude $\dot{\phi}, \dot{\theta}, \dot{\psi}$ in the directions indicated in the figure 2.2-6.

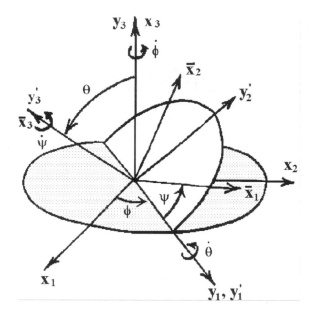

Figure 2.2-6. Euler angles.

By combining the above transformations there results the transformation equations (2.2.50)

$$\begin{pmatrix} \bar{x}_1 \\ \bar{x}_2 \\ \bar{x}_3 \end{pmatrix} = \begin{pmatrix} \cos\psi\cos\phi - \cos\theta\sin\phi\sin\psi & \cos\psi\sin\phi + \cos\theta\cos\phi\sin\psi & \sin\psi\sin\theta \\ -\sin\psi\cos\phi - \cos\theta\sin\phi\cos\psi & -\sin\psi\sin\phi + \cos\theta\cos\phi\cos\psi & \cos\psi\sin\theta \\ \sin\theta\sin\phi & -\sin\theta\cos\phi & \cos\theta \end{pmatrix} \begin{pmatrix} x_1 \\ x_2 \\ x_3 \end{pmatrix}.$$

It is left as an exercise to verify that the transformation matrix is orthogonal and the components ℓ_{ji} satisfy the relations (2.2.51).

■

Consider the velocity of a point which is rotating with the rigid body. Denote by $v_i = v_i(S)$, for $i = 1, 2, 3$, the velocity components relative to the S reference frame and by $\overline{v}_i = \overline{v}_i(\overline{S})$, $i = 1, 2, 3$ the velocity components of the same point relative to the body-fixed axes. In terms of the basis vectors we can write

$$\vec{V} = v_1(S)\,\hat{\mathbf{e}}_1 + v_2(S)\,\hat{\mathbf{e}}_2 + v_3(S)\,\hat{\mathbf{e}}_3 = \frac{dx_i}{dt}\,\hat{\mathbf{e}}_i \tag{2.2.52}$$

as the velocity in the S reference frame. Similarly, we write

$$\vec{\overline{V}} = \overline{v}_1(\overline{S})\widehat{\overline{\mathbf{e}}}_1 + \overline{v}_2(\overline{S})\widehat{\overline{\mathbf{e}}}_2 + \overline{v}_3(\overline{S})\widehat{\overline{\mathbf{e}}}_3 = \frac{d\overline{x}_i}{dt}\widehat{\overline{\mathbf{e}}}_i \tag{2.2.53}$$

as the velocity components relative to the body-fixed reference frame. There are occasions when it is desirable to represent $\vec{\overline{V}}$ in the S frame of reference and $\vec{V}$ in the $\overline{S}$ frame of reference. In these instances we can write

$$\vec{V} = v_1(\overline{S})\widehat{\overline{\mathbf{e}}}_1 + v_2(\overline{S})\widehat{\overline{\mathbf{e}}}_2 + v_3(\overline{S})\widehat{\overline{\mathbf{e}}}_3 \tag{2.2.54}$$

and

$$\vec{\overline{V}} = \overline{v}_1(S)\,\hat{\mathbf{e}}_1 + \overline{v}_2(S)\,\hat{\mathbf{e}}_2 + \overline{v}_3(S)\,\hat{\mathbf{e}}_3. \tag{2.2.55}$$

Here we have adopted the notation that $v_i(S)$ are the velocity components relative to the S reference frame and $v_i(\overline{S})$ are the same velocity components relative to the $\overline{S}$ reference frame. Similarly, $\overline{v}_i(\overline{S})$ denotes the velocity components relative to the $\overline{S}$ reference frame, while $\overline{v}_i(S)$ denotes the same velocity components relative to the S reference frame.

Here both $\vec{V}$ and $\vec{\overline{V}}$ are vectors and so their components are first order tensors and satisfy the transformation laws

$$\overline{v}_i(S) = \ell_{ji}v_j(S) = \ell_{ji}\dot{x}_j \qquad \text{and} \qquad v_i(\overline{S}) = \ell_{ij}\overline{v}_j(\overline{S}) = \ell_{ij}\dot{\overline{x}}_j. \tag{2.2.56}$$

The equations (2.2.56) define the relative velocity components as functions of time t. By differentiating the equations (2.2.50) we obtain

$$\frac{d\overline{x}_i}{dt} = \overline{v}_i(\overline{S}) = \ell_{ji}\dot{x}_j + \dot{\ell}_{ji}x_j \tag{2.2.57}$$

and

$$\frac{dx_i}{dt} = v_i(S) = \ell_{ij}\dot{\overline{x}}_j + \dot{\ell}_{ij}\overline{x}_j. \tag{2.2.58}$$

Multiply the equation (2.2.57) by ℓ_{mi} and multiply the equation (2.2.58) by ℓ_{im} and derive the relations

$$v_m(\overline{S}) = v_m(S) + \ell_{mi}\dot{\ell}_{ji}x_j \tag{2.2.59}$$

and

$$\overline{v}_m(S) = \overline{v}_m(\overline{S}) + \ell_{im}\dot{\ell}_{ij}\overline{x}_j. \tag{2.2.60}$$

The equations (2.2.59) and (2.2.60) describe the transformation laws of the velocity components upon changing from the S to the $\overline{S}$ reference frame. These equations can be expressed in terms of the angular velocity by making certain substitutions which are now defined.

The first order angular velocity vector ω_i is related to the second order skew-symmetric angular velocity tensor ω_{ij} by the defining equation

$$\omega_{mn} = e_{imn}\omega_i. \tag{2.2.61}$$

The equation (2.2.61) implies that ω_i and ω_{ij} are dual tensors and $\omega_i = \frac{1}{2}e_{ijk}\omega_{jk}$. Also the velocity of a point which is rotating about the origin relative to the S frame of reference is given by $v_i(S) = e_{ijk}\omega_j x_k$ which can also be written in the form $v_m(S) = -\omega_{mk}x_k$. Since the barred axes rotate with the rigid body, then a particle in the barred reference frame will have $v_m(\overline{S}) = 0$, since the coordinates of a point in the rigid body will be constants with respect to this reference frame. We therefore can write equation (2.2.59) in the form $0 = v_m(S) + \ell_{mi}\dot{\ell}_{ji}x_j$ which implies that

$$v_m(S) = -\ell_{mi}\dot{\ell}_{ji}x_j = -\omega_{mk}x_k \quad \text{or} \quad \omega_{mj} = \omega_{mj}(\overline{S}, S) = \ell_{mi}\dot{\ell}_{ji}.$$

This equation is interpreted as describing the angular velocity tensor of $\overline{S}$ relative to S. Since ω_{ij} is a tensor, it can be represented in the barred system by

$$\overline{\omega}_{mn}(\overline{S}, S) = \ell_{im}\ell_{jn}\omega_{ij}(\overline{S}, S) = \ell_{im}\ell_{jn}\ell_{is}\dot{\ell}_{js} = \delta_{ms}\ell_{jn}\dot{\ell}_{js} = \ell_{jn}\dot{\ell}_{jm} \tag{2.2.62}$$

By differentiating the equations (2.2.51) it is an easy exercise to show that ω_{ij} is skew-symmetric. The second order angular velocity tensor can be used to write the equations (2.2.59) and (2.2.60) in the forms

$$\begin{aligned} v_m(\overline{S}) &= v_m(S) + \omega_{mj}(\overline{S}, S)x_j \\ \overline{v}_m(S) &= \overline{v}_m(\overline{S}) + \overline{\omega}_{jm}(\overline{S}, S)\overline{x}_j \end{aligned} \tag{2.2.63}$$

The above relations are now employed to derive the celebrated Euler's equations of motion of a rigid body.

Euler's Equations of Motion

We desire to find the equations of motion of a rigid body which is subjected to external forces. These equations are the formulas (2.2.49), and we now proceed to write these equations in a slightly different form. Similar to the introduction of the angular velocity tensor, given in equation (2.2.61), we now introduce the following tensors

1. The fourth order moment of inertia tensor I_{mnst} which is related to the second order moment of inertia tensor I_{ij} by the equations

$$I_{mnst} = \frac{1}{2}e_{jmn}e_{ist}I_{ij} \quad \text{or} \quad I_{ij} = \frac{1}{2}I_{pqrs}e_{ipq}e_{jrs} \tag{2.2.64}$$

2. The second order angular momentum tensor H_{jk} which is related to the angular momentum vector H_i by the equation

$$H_i = \frac{1}{2}e_{ijk}H_{jk} \quad \text{or} \quad H_{jk} = e_{ijk}H_i \tag{2.2.65}$$

3. The second order moment tensor M_{jk} which is related to the moment M_i by the relation

$$M_i = \frac{1}{2}e_{ijk}M_{jk} \quad \text{or} \quad M_{jk} = e_{ijk}M_i. \tag{2.2.66}$$

Now if we multiply equation (2.2.49) by e_{rjk}, then it can be written in the form

$$\frac{dH_{ij}}{dt} = M_{ij}. \tag{2.2.67}$$

204

Similarly, if we multiply the equation (2.2.42) by e_{imn}, then it can be expressed in the alternate form

$$H_{mn} = e_{imn}\omega_j I_{ji} = I_{mnst}\omega_{st}$$

and because of this relation the equation (2.2.67) can be expressed as

$$\frac{d}{dt}\left(I_{ijst}\omega_{st}\right) = M_{ij}. \tag{2.2.68}$$

We write this equation in the barred system of coordinates where $\overline{I}_{pqrs}$ will be a constant and consequently its derivative will be zero. We employ the transformation equations

$$I_{ijst} = \ell_{ip}\ell_{jq}\ell_{sr}\ell_{tk}\overline{I}_{pqrk}$$
$$\overline{\omega}_{ij} = \ell_{si}\ell_{tj}\omega_{st}$$
$$\overline{M}_{pq} = \ell_{ip}\ell_{jq}M_{ij}$$

and then multiply the equation (2.2.68) by $\ell_{ip}\ell_{jq}$ and simplify to obtain

$$\ell_{ip}\ell_{jq}\frac{d}{dt}\left(\ell_{i\alpha}\ell_{j\beta}\overline{I}_{\alpha\beta rk}\overline{\omega}_{rk}\right) = \overline{M}_{pq}.$$

Expand all terms in this equation and take note that the derivative of the $\overline{I}_{\alpha\beta rk}$ is zero. The expanded equation then simplifies to

$$\overline{I}_{pqrk}\frac{d\overline{\omega}_{rk}}{dt} + (\delta_{\alpha u}\delta_{pv}\delta_{\beta q} + \delta_{p\alpha}\delta_{\beta u}\delta_{qv})\overline{I}_{\alpha\beta rk}\overline{\omega}_{rk}\overline{\omega}_{uv} = \overline{M}_{pq}. \tag{2.2.69}$$

Substitute into equation (2.2.69) the relations from equations (2.2.61),(2.2.64) and (2.2.66), and then multiply by e_{mpq} and simplify to obtain the Euler's equations of motion

$$\overline{I}_{im}\frac{d\overline{\omega}_i}{dt} - e_{tmj}\overline{I}_{ij}\overline{\omega}_i\overline{\omega}_t = \overline{M}_m. \tag{2.2.70}$$

Dropping the bar notation and performing the indicated summations over the range 1,2,3 we find the Euler equations have the form

$$I_{11}\frac{d\omega_1}{dt} + I_{21}\frac{d\omega_2}{dt} + I_{31}\frac{d\omega_3}{dt} + (I_{13}\omega_1 + I_{23}\omega_2 + I_{33}\omega_3)\,\omega_2 - (I_{12}\omega_1 + I_{22}\omega_2 + I_{32}\omega_3)\,\omega_3 = M_1$$
$$I_{12}\frac{d\omega_1}{dt} + I_{22}\frac{d\omega_2}{dt} + I_{32}\frac{d\omega_3}{dt} + (I_{11}\omega_1 + I_{21}\omega_2 + I_{31}\omega_3)\,\omega_3 - (I_{13}\omega_1 + I_{23}\omega_2 + I_{33}\omega_3)\,\omega_1 = M_2 \tag{2.2.71}$$
$$I_{13}\frac{d\omega_1}{dt} + I_{23}\frac{d\omega_2}{dt} + I_{33}\frac{d\omega_3}{dt} + (I_{12}\omega_1 + I_{22}\omega_2 + I_{32}\omega_3)\,\omega_1 - (I_{11}\omega_1 + I_{21}\omega_2 + I_{31}\omega_3)\,\omega_2 = M_3.$$

In the special case where the barred axes are principal axes, then $I_{ij} = 0$ for $i \neq j$ and the Euler's equations reduces to the system of nonlinear differential equations

$$I_{11}\frac{d\omega_1}{dt} + (I_{33} - I_{22})\omega_2\omega_3 = M_1$$
$$I_{22}\frac{d\omega_2}{dt} + (I_{11} - I_{33})\omega_3\omega_1 = M_2 \tag{2.2.72}$$
$$I_{33}\frac{d\omega_3}{dt} + (I_{22} - I_{11})\omega_1\omega_2 = M_3.$$

In the case of constant coefficients and constant moments the solutions of the above differential equations can be expressed in terms of Jacobi elliptic functions.

EXERCISE 2.2

▸ **1.** Find a set of parametric equations for the straight line which passes through the points $P_1(1,1,1)$ and $P_2(2,3,4)$. Find the unit tangent vector to any point on this line.

▸ **2.** Consider the space curve $x = \frac{1}{2}\sin^2 t$, $y = \frac{1}{2}t - \frac{1}{4}\sin 2t$, $z = \sin t$ where t is a parameter. Find the unit vectors $T^i, B^i, N^i, i = 1, 2, 3$ at the point where $t = \pi$.

3. A claim has been made that the space curve $x = t$, $y = t^2$, $z = t^3$ intersects the plane 11x-6y+z=6 in three distinct points. Determine if this claim is true or false. Justify your answer and find the three points of intersection if they exist.

▸ **4.** Find a set of parametric equations $x_i = x_i(s_1, s_2), i = 1, 2, 3$ for the plane which passes through the points $P_1(3,0,0)$, $P_2(0,4,0)$ and $P_3(0,0,5)$. Find a unit normal to this plane.

▸ **5.** For the helix $x = \sin t \quad y = \cos t \quad z = \frac{2}{\pi}t$ find the equation of the tangent plane to the curve at the point where $t = \pi/4$. Find the equation of the tangent line to the curve at the point where $t = \pi/4$. Find the scalar curvature κ.

6. Verify the derivative $\dfrac{\partial T}{\partial \dot{x}^r} = M g_{rm} \dot{x}^m$.

▸ **7.** Verify the derivative $\dfrac{d}{dt}\left(\dfrac{\partial T}{\partial \dot{x}^r} \right) = M\left[g_{rm}\ddot{x}^m + \dfrac{\partial g_{rm}}{\partial x^n}\dot{x}^n\dot{x}^m \right]$.

▸ **8.** Verify the derivative $\dfrac{\partial T}{\partial x^r} = \dfrac{1}{2}M\dfrac{\partial g_{mn}}{\partial x^r}\dot{x}^m\dot{x}^n$.

9. Use the results from problems 6,7 and 8 to derive the Lagrange's form for the equations of motion defined by equation (2.2.23).

10. Express the generalized velocity and acceleration in cylindrical coordinates $(x^1, x^2, x^3) = (r, \theta, z)$ and show

$$
\begin{aligned}
V^1 &= \frac{dx^1}{dt} = \frac{dr}{dt} & f^1 &= \frac{\delta V^1}{\delta t} = \frac{d^2r}{dt^2} - r\left(\frac{d\theta}{dt}\right)^2 \\
V^2 &= \frac{dx^2}{dt} = \frac{d\theta}{dt} & f^2 &= \frac{\delta V^2}{\delta t} = \frac{d^2\theta}{dt^2} + \frac{2}{r}\frac{dr}{dt}\frac{d\theta}{dt} \\
V^3 &= \frac{dx^3}{dt} = \frac{dz}{dt} & f^3 &= \frac{\delta V^3}{\delta t} = \frac{d^2z}{dt^2}
\end{aligned}
$$

Find the physical components of velocity and acceleration in cylindrical coordinates and show

$$
\begin{aligned}
V_r &= \frac{dr}{dt} & f_r &= \frac{d^2r}{dt^2} - r\left(\frac{d\theta}{dt}\right)^2 \\
V_\theta &= r\frac{d\theta}{dt} & f_\theta &= r\frac{d^2\theta}{dt^2} + 2\frac{dr}{dt}\frac{d\theta}{dt} \\
V_z &= \frac{dz}{dt} & f_z &= \frac{d^2z}{dt^2}
\end{aligned}
$$

► **11.** Express the generalized velocity and acceleration in spherical coordinates $(x^2, x^2, x^3) = (\rho, \theta, \phi)$ and show

$$V^1 = \frac{dx^1}{dt} = \frac{d\rho}{dt} \qquad f^1 = \frac{\delta V^1}{\delta t} = \frac{d^2\rho}{dt^2} - \rho\left(\frac{d\theta}{dt}\right)^2 - \rho\sin^2\theta\left(\frac{d\phi}{dt}\right)^2$$

$$V^2 = \frac{dx^2}{dt} = \frac{d\theta}{dt} \qquad f^2 = \frac{\delta V^2}{\delta t} = \frac{d^2\theta}{dt^2} - \sin\theta\cos\theta\left(\frac{d\phi}{dt}\right)^2 + \frac{2}{\rho}\frac{d\rho}{dt}\frac{d\theta}{dt}$$

$$V^3 = \frac{dx^3}{dt} = \frac{d\phi}{dt} \qquad f^3 = \frac{\delta V^3}{\delta t} = \frac{d^2\phi}{dt^2} + \frac{2}{\rho}\frac{d\rho}{dt}\frac{d\phi}{dt} + 2\cot\theta\frac{d\theta}{dt}\frac{d\phi}{dt}$$

Find the physical components of velocity and acceleration in spherical coordinates and show

$$V_\rho = \frac{d\rho}{dt} \qquad f_\rho = \frac{d^2\rho}{dt^2} - \rho\left(\frac{d\theta}{dt}\right)^2 - \rho\sin^2\theta\left(\frac{d\phi}{dt}\right)^2$$

$$V_\theta = \rho\frac{d\theta}{dt} \qquad f_\theta = \rho\frac{d^2\theta}{dt^2} - \rho\sin\theta\cos\theta\left(\frac{d\phi}{dt}\right)^2 + 2\frac{d\rho}{dt}\frac{d\theta}{dt}$$

$$V_\phi = \rho\sin\theta\frac{d\phi}{dt} \qquad f_\phi = \rho\sin\theta\frac{d^2\phi}{dt^2} + 2\sin\theta\frac{d\rho}{dt}\frac{d\phi}{dt} + 2\rho\cos\theta\frac{d\theta}{dt}\frac{d\phi}{dt}$$

► **12.** Expand equation (2.2.39) and write out all the components of the moment of inertia tensor I_{ij}.

► **13.** For ρ the density of a continuous material and $d\tau$ an element of volume inside a region R where the material is situated, we write $\rho d\tau$ as an element of mass inside R. Find an equation which describes the center of mass of the region R.

► **14.** Use the equation (2.2.68) to derive the equation (2.2.69).

► **15.** Drop the bar notation and expand the equation (2.2.70) and derive the equations (2.2.71).

► **16.** Verify the Euler transformation, given in example 2.2-5, is orthogonal.

► **17.** For the pulley and mass system illustrated in the figure 2.2-7 let

$$a = \text{the radius of each pulley.}$$
$$\ell_1 = \text{the length of the upper chord.}$$
$$\ell_2 = \text{the length of the lower chord.}$$

Neglect the weight of the pulley and find the equations of motion for the pulley mass system.

► **18.** For $T = \frac{1}{2}mg_{ij}v^iv^j$ the kinetic energy of a particle and V the potential energy of the particle show that $T + V = constant$.

Hint: $mf_i = Q_i = -\frac{\partial V}{\partial x^i}, \quad i = 1, 2, 3 \quad \text{and} \quad \frac{dx^i}{dt} = \dot{x}^i = v^i, i = 1, 2, 3.$

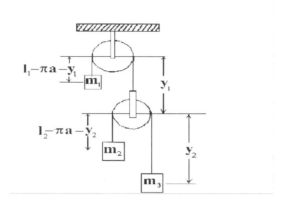

Figure 2.2-7. Pulley and mass system

19. Let $\phi = \frac{ds}{dt}$, where s is the arc length between two points on a curve in generalized coordinates.

(a) Write the arc length in general coordinates as $ds = \sqrt{g_{mn}\dot{x}^m\dot{x}^n}dt$ and show the integral I, defined by equation (2.2.35), represents the distance between two points on a curve.

(b) Using the Euler-Lagrange equations (2.2.36) show that the shortest distance between two points in a generalized space is the curve defined by the equations: $\ddot{x}^i + \begin{Bmatrix} i \\ j\,k \end{Bmatrix}\dot{x}^j\dot{x}^k = \dot{x}^i\dfrac{\frac{d^2s}{dt^2}}{\frac{ds}{dt}}$

(c) Show in the special case $t = s$ the equations in part (b) reduce to $\dfrac{d^2x^i}{ds^2} + \begin{Bmatrix} i \\ j\,k \end{Bmatrix}\dfrac{dx^j}{ds}\dfrac{dx^k}{ds} = 0$, for $i = 1, \ldots, N$. An examination of equation (1.5.51) shows that the above curves are geodesic curves.

(d) Show that the shortest distance between two points in a plane is a straight line.

(e) Consider two points on the surface of a cylinder of radius a. Let $u^1 = \theta$ and $u^2 = z$ denote surface coordinates in the two dimensional space defined by the surface of the cylinder. Show that the shortest distance between the points where $\theta = 0$, $z = 0$ and $\theta = \pi$, $z = H$ is $L = \sqrt{a^2\pi^2 + H^2}$.

▶ **20.** Define $H = T + V$ as the sum of the kinetic energy and potential energy of a particle. The quantity $H = H(x^r, p_r)$ is called the Hamiltonian of the particle and it is expressed in terms of:

- the particle position x^i and
- the particle momentum $p_i = mv_i = mg_{ij}\dot{x}^j$. Here x^r and p_r are treated as independent variables.

(a) Show that the particle momentum is a covariant tensor of rank 1.

(b) Express the kinetic energy T in terms of the particle momentum.

(c) Show that $p_i = \dfrac{\partial T}{\partial \dot{x}^i}$.

(d) Show that $\dfrac{dx^i}{dt} = \dfrac{\partial H}{\partial p_i}$ and $\dfrac{dp_i}{dt} = -\dfrac{\partial H}{\partial x^i}$. These are a set of differential equations describing the position change and momentum change of the particle and are known as Hamilton's equations of motion for a particle.

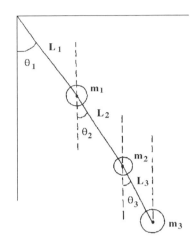

Figure 2.2-8. Compound pendulum

▶ **21.** Let $\frac{\delta T^i}{\delta s} = \kappa N^i$ and $\frac{\delta N^i}{\delta s} = \tau B^i - \kappa T^i$ and calculate the intrinsic derivative of the cross product $B^i = \epsilon^{ijk} T_j N_k$ and find $\frac{\delta B^i}{\delta s}$ in terms of the unit normal vector.

▶ **22.** For T the kinetic energy of a particle and V the potential energy of a particle, define the Lagrangian $L = L(x^i, \dot{x}^i) = T - V = \frac{1}{2} M g_{ij} \dot{x}^i \dot{x}^j - V$ as a function of the independent variables $x^i, \dot{x}^i$. Define the Hamiltonian $H = H(x^i, p_i) = T + V = \frac{1}{2M} g^{ij} p_i p_j + V$, as a function of the independent variables x^i, p_i, where p_i is the momentum vector of the particle and M is the mass of the particle.

(a) Show that $p_i = \frac{\partial T}{\partial \dot{x}^i}$.

(b) Show that $\frac{\partial H}{\partial x^i} = -\frac{\partial L}{\partial x^i}$

▶ **23.** When the Euler angles, figure 2.2-6, are applied to the motion of rotating objects, θ is the angle of nutation, ϕ is the angle of precession and ψ is the angle of spin. Take projections and show that the time derivative of the Euler angles are related to the angular velocity vector components $\omega_x, \omega_y, \omega_z$ by the relations

$$\omega_x = \dot{\theta} \cos\psi + \dot{\phi} \sin\theta \sin\psi, \qquad \omega_y = -\dot{\theta} \sin\psi + \dot{\phi} \sin\theta \cos\psi, \qquad \omega_z = \dot{\psi} + \dot{\phi} \cos\theta$$

where $\omega_x, \omega_y, \omega_z$ are the angular velocity components along the $\overline{x}_1, \overline{x}_2, \overline{x}_3$ axes.

▶ **24.** Find the equations of motion for the compound pendulum illustrated in the figure 2.2-8.

▶ **25.** Let $\vec{F} = -\frac{GMm}{r^3} \vec{r}$ denote the inverse square law force of attraction between the earth and sun, with G a universal constant, M the mass of the sun, m the mass of the earth and $\frac{\vec{r}}{r}$ a unit vector from origin at the center of the sun pointing toward the earth. (a) Write down Newton's second law, in both vector and tensor form, which describes the motion of the earth about the sun. (b) Show that $\frac{d}{dt} (\vec{r} \times \vec{v}) = \vec{0}$ and therefore $\vec{r} \times \vec{v} = \vec{r} \times \frac{d\vec{r}}{dt} = \vec{h} =$ a constant.

26. Construct a set of axes fixed and attached to an airplane. Let the x axis be a longitudinal axis running from the rear to the front of the plane along its center line. Let the y axis run between the wing tips and let the z axis form a right-handed system of coordinates. The y axis is called a lateral axis and the z axis is called a normal axis. Define *pitch* as any angular motion about the lateral axis. Define *roll* as any angular motion about the longitudinal axis. Define *yaw* as any angular motion about the normal axis. Consider two sets of axes. One set is the x, y, z axes attached to and moving with the aircraft. The other set of axes is denoted X, Y, Z and is fixed in space (an inertial set of axes). Describe the pitch, roll and yaw of an aircraft with respect to the inertial set of axes. Show the transformation is orthogonal. Hint: Consider pitch with respect to the fixed axes, then consider roll with respect to the pitch axes and finally consider yaw with respect to the roll axes. This produces three separate transformation matrices which can then be combined to describe the motions of pitch, roll and yaw of an aircraft.

27. In Cartesian coordinates let $F_i = F_i(x^1, x^2, x^3)$ denote a force field and let $x^i = x^i(t)$ denote a space curve C. (a) Show Newton's second law implies that along the curve C we have
$$\frac{d}{dt}\left(\frac{1}{2}m\left(\frac{dx^i}{dt}\right)^2\right) = F_i(x^1, x^2, x^3)\frac{dx^i}{dt} \text{ (no summation on i) and hence}$$

$$\frac{d}{dt}\left[\frac{1}{2}m\left(\left(\frac{dx^1}{dt}\right)^2 + \left(\frac{dx^2}{dt}\right)^2 + \left(\frac{dx^3}{dt}\right)^2\right)\right] = \frac{d}{dt}\left[\frac{1}{2}mv^2\right] = F_1\frac{dx^1}{dt} + F_2\frac{dx^2}{dt} + F_3\frac{dx^3}{dt}$$

(b) Consider two points on the curve C, say point A, $x^i(t_A)$ and point B, $x^i(t_B)$ and show that the work done in moving from A to B in the force field F_i is

$$\frac{1}{2}mv^2\bigg]_{t_A}^{t_B} = \int_A^B F_1dx^1 + F_2dx^2 + F_3dx^3$$

where the right hand side is a line integral along the path C from A to B. (c) Show that if the force field is derivable from a potential function $U(x^1, x^2, x^3)$ by taking the gradient, then the work done is independent of the path C and depends only upon the end points A and B.

28. Find the Lagrangian equations of motion of a spherical pendulum which consists of a bob of mass m suspended at the end of a wire of length ℓ, which is free to swing in any direction subject to the constraint that the wire length is constant. Neglect the weight of the wire and show that for the wire attached to the origin of a right handed x, y, z coordinate system, with the z axis downward, ϕ the angle between the wire and the z axis and θ the angle of rotation of the bob from the y axis, that there results the equations of motion $\quad \frac{d}{dt}\left(\sin^2\phi\frac{d\theta}{dt}\right) = 0 \quad$ and $\quad \frac{d^2\phi}{dt^2} - \left(\frac{d\theta}{dt}\right)^2\sin\phi\cos\phi + \frac{g}{\ell}\sin\phi = 0$

210

▶ **29.** In Cartesian coordinates show the Frenet formulas can be written

$$\frac{d\vec{T}}{ds} = \vec{\delta} \times \vec{T}, \qquad \frac{d\vec{N}}{ds} = \vec{\delta} \times \vec{N}, \qquad \frac{d\vec{B}}{ds} = \vec{\delta} \times \vec{B}$$

where $\vec{\delta} = \tau \vec{T} + \kappa \vec{B}$ is known as the Darboux vector.

▶ **30.** Consider the following two cases for rigid body rotation.

Case 1: Rigid body rotation about a fixed line which is called the fixed axis of rotation. Select a point 0 on this fixed axis and denote by $\hat{\mathbf{e}}$ a unit vector from 0 in the direction of the fixed line and denote by $\hat{\mathbf{e}}_R$ a unit vector which is perpendicular to the fixed axis of rotation. The position vector of a general point in the rigid body can then be represented by a position vector from the point 0 given by $\vec{r} = h\,\hat{\mathbf{e}} + r_0\,\hat{\mathbf{e}}_R$ where h, r_0 and $\hat{\mathbf{e}}$ are all constants and the vector $\hat{\mathbf{e}}_R$ is fixed in and rotating with the rigid body. Denote by $\omega = \dfrac{d\theta}{dt}$ the scalar angular change with respect to time of the vector $\hat{\mathbf{e}}_R$ as it rotates about the fixed line and define the vector angular velocity as $\vec{\omega} = \dfrac{d}{dt}(\theta\,\hat{\mathbf{e}}) = \dfrac{d\theta}{dt}\,\hat{\mathbf{e}}$ where $\theta\,\hat{\mathbf{e}}$ is defined as the vector angle of rotation.

(a) Show that $\dfrac{d\,\hat{\mathbf{e}}_R}{d\theta} = \hat{\mathbf{e}} \times \hat{\mathbf{e}}_R$.

(b) Show that $\vec{V} = \dfrac{d\vec{r}}{dt} = r_0 \dfrac{d\,\hat{\mathbf{e}}_R}{dt} = r_0 \dfrac{d\,\hat{\mathbf{e}}_R}{d\theta}\dfrac{d\theta}{dt} = \vec{\omega} \times (r_0\,\hat{\mathbf{e}}_R) = \vec{\omega} \times (h\,\hat{\mathbf{e}} + r_0\,\hat{\mathbf{e}}_R) = \vec{\omega} \times \vec{r}$.

Case 2: Rigid body rotation about a fixed point 0. Construct at point 0 the unit vector $\hat{\mathbf{e}}_1$ which is fixed in and rotating with the rigid body. From pages 80,87 we know that $\dfrac{d\,\hat{\mathbf{e}}_1}{dt}$ must be perpendicular to $\hat{\mathbf{e}}_1$ and so we can define the vector $\hat{\mathbf{e}}_2$ as a unit vector which is in the direction of $\dfrac{d\,\hat{\mathbf{e}}_1}{dt}$ such that $\dfrac{d\,\hat{\mathbf{e}}_1}{dt} = \alpha\,\hat{\mathbf{e}}_2$ for some constant α. We can then define the unit vector $\hat{\mathbf{e}}_3$ from $\hat{\mathbf{e}}_3 = \hat{\mathbf{e}}_1 \times \hat{\mathbf{e}}_2$.

(a) Show that $\dfrac{d\,\hat{\mathbf{e}}_3}{dt}$, which must be perpendicular to $\hat{\mathbf{e}}_3$, is also perpendicular to $\hat{\mathbf{e}}_1$.

(b) Show that $\dfrac{d\,\hat{\mathbf{e}}_3}{dt}$ can be written as $\dfrac{d\,\hat{\mathbf{e}}_3}{dt} = \beta\,\hat{\mathbf{e}}_2$ for some constant β.

(c) From $\hat{\mathbf{e}}_2 = \hat{\mathbf{e}}_3 \times \hat{\mathbf{e}}_1$ show that $\dfrac{d\,\hat{\mathbf{e}}_2}{dt} = (\alpha\,\hat{\mathbf{e}}_3 - \beta\,\hat{\mathbf{e}}_1) \times \hat{\mathbf{e}}_2$

(d) Define $\vec{\omega} = \alpha\,\hat{\mathbf{e}}_3 - \beta\,\hat{\mathbf{e}}_1$ and show that $\dfrac{d\,\hat{\mathbf{e}}_1}{dt} = \vec{\omega} \times \hat{\mathbf{e}}_1$, $\qquad \dfrac{d\,\hat{\mathbf{e}}_2}{dt} = \vec{\omega} \times \hat{\mathbf{e}}_2$, $\qquad \dfrac{d\,\hat{\mathbf{e}}_3}{dt} = \vec{\omega} \times \hat{\mathbf{e}}_3$

(e) Let $\vec{r} = x\,\hat{\mathbf{e}}_1 + y\,\hat{\mathbf{e}}_2 + z\,\hat{\mathbf{e}}_3$ denote an arbitrary point within the rigid body with respect to the point 0. Show that $\dfrac{d\vec{r}}{dt} = \vec{\omega} \times \vec{r}$.

Note that in Case 2 the direction of $\vec{\omega}$ is not fixed as the unit vectors $\hat{\mathbf{e}}_3$ and $\hat{\mathbf{e}}_1$ are constantly changing. In this case the direction $\vec{\omega}$ is called an instantaneous axis of rotation and $\vec{\omega}$, which also can change in magnitude and direction, is called the instantaneous angular velocity.

▶ **31.** A particle of mass m moves on the surface of the cone $x_1 = z \sin\alpha\cos\theta$, $x_2 = z\sin\alpha\sin\theta$, $x_3 = z\cos\alpha$ where $ds^2 = dz^2 + z^2\sin^2\alpha\,d\theta^2$, with α constant. Assume a potential function due to gravity $V = mgz\cos\alpha$. Show the equations of motion are given by $\frac{d}{dt}\left(mz^2\sin^2\alpha\frac{d\theta}{dt}\right) = 0$ and $m\frac{d^2z}{dt^2} - mz\sin^2\alpha\left(\frac{d\theta}{dt}\right)^2 + mg\cos\alpha = 0$.

§2.3 BASIC EQUATIONS OF CONTINUUM MECHANICS

Continuum mechanics is the study of how materials behave when subjected to external influences. External influences which affect the properties of a substance are such things as forces, temperature, chemical reactions, and electric phenomena. Examples of forces are gravitational forces, electromagnetic forces, and mechanical forces. Solids deform under external forces and so deformations are studied. Fluids move under external forces and so the velocity of the fluid is studied.

A material is considered to be a continuous media which is a collection of material points interconnected by internal forces (forces between the atoms making up the material). We concentrate upon the macroscopic properties rather than the microscopic properties of the material. We treat the material as a body which is homogeneous and continuous in its makeup.

In this introduction we will only consider solid media and liquid media. In general, most of the ideas and concepts developed in this section can be applied to any type of material which is assumed to be a collection of material points held together by some kind of internal forces.

An elastic material is one which deforms under applied forces in such a way that it will return to its original unloaded state when the applied forces are removed. When a linear relation exists between the applied forces and material displacements, then the material is called a linear elastic material. In contrast, a plastic material is one which deforms under applied forces in such a way that it does not return to its original state after removal of the applied forces. Plastic materials will always exhibit some permanent deformation after removal of the applied forces. An elastic material is called homogeneous if it has the same properties throughout. An isotropic material has the same properties, at a point, in all directions about the point.

In this introduction we develop the basic mathematical equations which describe how a continuum behaves when subjected to external forces. We shall discover that there exists a set of basic equations associated with all continuous material media. These basic equations are developed for linear elastic materials and applied to solids and fluids in later sections.

Introduction to Elasticity

Take a rubber band, which has a rectangular cross section, and mark on it a parallelepiped having a length ℓ, a width w and a height h, as illustrated in the figure 2.3-1.

Now apply a force F to both ends of the parallelepiped cross section on the rubber band and examine what happens to the parallelepiped. You will see that:

1. ℓ increases by an amount $\Delta\ell$.
2. w decreases by an amount Δw.
3. h decreases by an amount Δh.

There are many materials which behave in a manner very similar to the rubber band. Most materials, when subjected to tension forces will break if the change $\Delta\ell$ is only one or two percent of the original length. The above example introduces us to several concepts which arise in the study of materials when they are subjected to external forces. The first concept is that of strain which is defined as

$$strain = \frac{\text{change in length}}{\text{original length}}, \qquad \text{(dimensionless)}.$$

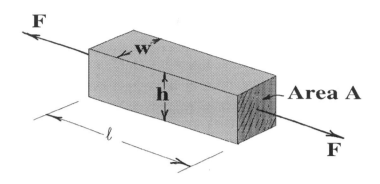

Figure 2.3-1. Section of a rubber band

When the force F is applied to our rubber band example there arises the strains

$$\frac{\Delta\ell}{\ell}, \qquad \frac{\Delta w}{w}, \qquad \frac{\Delta h}{h}.$$

The second concept introduced by our simple example is stress. Stress is defined as a force per unit area. In particular,

$$stress = \frac{\text{Force}}{\text{Area over which force acts}}, \qquad \text{with dimension of} \quad \frac{\text{force}}{\text{unit area}}.$$

We will be interested in studying stress and strain in homogeneous, isotropic materials which are in equilibrium with respect to the force system acting on the material.

Hooke's Law

For linear elastic materials, where the forces are all one dimensional, the stress and strains are related by Hooke's law which has two parts. The Hooke's law, part one, states that stress is proportional to strain in the stretch direction, where the Young's modulus E is the proportionality constant. This is written

$$\text{Hooke's law part 1} \qquad \frac{F}{A} = E\left(\frac{\Delta\ell}{\ell}\right). \tag{2.3.1}$$

A graph of stress vs strain is a straight line with slope E in the linear elastic range of the material.

The Hooke's law, part two, involves the fact that there is a strain contraction perpendicular to the stretch direction. The strain contraction is the same for both the width and height and is proportional to the strain in the stretch direction. The proportionality constant being the Poisson's ratio ν.

$$\text{Hooke's law part 2} \qquad \frac{\Delta w}{w} = \frac{\Delta h}{h} = -\nu\frac{\Delta\ell}{\ell}, \qquad 0 < \nu < \frac{1}{2}. \tag{2.3.2}$$

The proportionality constants E and ν depend upon the material being considered. The constant ν is called the Poisson's ratio and it is always a positive number which is less than one half. Some representative values for E and ν are as follows.

Various types of steel	$28\,(10)^6\,\text{psi} \le E \le 30\,(10)^6\,\text{psi}$	$0.26 \le \nu \le 0.31$
Various types of aluminium	$9.0\,(10)^6\,\text{psi} \le E \le 11.0\,(10)^6\,\text{psi}$	$0.3 \le \nu \le 0.35$

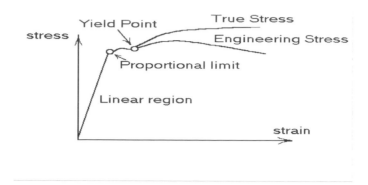

Figure 2.3-2. Typical Stress-strain curve.

Consider a typical stress-strain curve, such as the one illustrated in the figure 2.3-2, which is obtained by placing a material in the shape of a rod or wire in a machine capable of performing tensile straining at a low rate. The engineering stress is the tensile force F divided by the original cross sectional area A_0. Note that during a tensile straining the cross sectional area A of the sample is continually changing and getting smaller so that the actual stress will be larger than the engineering stress. Observe in the figure 2.3-2 that the stress-strain relation remains linear up to a point labeled the proportional limit. For stress-strain points in this linear region the Hooke's law holds and the material will return to its original shape when the loading is removed. For points beyond the proportional limit, but less than the yield point, the material no longer obeys Hooke's law. In this nonlinear region the material still returns to its original shape when the loading is removed. The region beyond the yield point is called the plastic region. At the yield point and beyond, there is a great deal of material deformation while the loading undergoes only small changes. For points in this plastic region, the material undergoes a permanent deformation and does not return to its original shape when the loading is removed. In the plastic region there usually occurs deformation due to slipping of atomic planes within the material. In this introductory section we will restrict our discussions of material stress-strain properties to the linear region.

EXAMPLE 2.3-1. (One dimensional elasticity) Consider a circular rod with cross sectional area A which is subjected to an external force F applied to both ends. The figure 2.3-3 illustrates what happens to the rod after the tension force F is applied. Consider two neighboring points P and Q on the rod, where P is at the point x and Q is at the point $x + \Delta x$. When the force F is applied to the rod it is stretched and P moves to P' and Q moves to Q'. We assume that when F is applied to the rod there is a displacement function $u = u(x,t)$ which describes how each point in the rod moves as a function of time t. If we know the displacement function $u = u(x,t)$ we would then be able to calculate the following distances in terms of the displacement function

$$\overline{PP'} = u(x,t), \qquad \overline{0P'} = x + u(x,t), \qquad \overline{QQ'} = u(x + \Delta x, t) \qquad \overline{0Q'} = x + \Delta x + u(x + \Delta x, t).$$

214

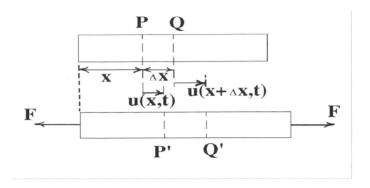

Figure 2.3-3. One dimensional rod subjected to tension force

The strain associated with the distance $\ell = \Delta x = \overline{PQ}$ is

$$e = \frac{\Delta\ell}{\ell} = \frac{\overline{P'Q'} - \overline{PQ}}{\overline{PQ}} = \frac{(\overline{0Q'} - \overline{0P'}) - (\overline{0Q} - \overline{0P})}{\overline{PQ}}$$

$$e = \frac{[x + \Delta x + u(x + \Delta x, t) - (x + u(x,t))] - [(x + \Delta x) - x]}{\Delta x}$$

$$e = \frac{u(x + \Delta x, t) - u(x,t)}{\Delta x}.$$

Use the Hooke's law part(i) and write

$$\frac{F}{A} = E \frac{u(x + \Delta x, t) - u(x,t)}{\Delta x}.$$

Taking the limit as Δx approaches zero we find that

$$\frac{F}{A} = E \frac{\partial u(x,t)}{\partial x}.$$

Hence, the stress is proportional to the spatial derivative of the displacement function.

∎

Normal and Shearing Stresses

Let us consider a more general situation in which we have some material which can be described as having a surface area S which encloses a volume V. Assume that the density of the material is ϱ and the material is homogeneous and isotropic. Further assume that the material is subjected to the forces $\vec{b}$ and $\vec{t}^{(n)}$ where $\vec{b}$ is a body force per unit mass $[force/mass]$, and $\vec{t}^{(n)}$ is a surface traction per unit area $[force/area]$. The superscript (n) denotes the normal to the surface upon which the traction is acting. We will neglect body couples, surface couples, and concentrated forces or couples that act at a single point. If the forces described above are everywhere continuous we can calculate the resultant force $\vec{F}$ and resultant moment $\vec{M}$ acting on the material by constructing various surface and volume integrals which sum the forces acting upon the material. In particular, the resultant force $\vec{F}$ acting on our material can be described by the surface and volume integrals:

$$\vec{F} = \iint_S \vec{t}^{(n)} \, dS + \iiint_V \varrho \vec{b} \, d\tau \tag{2.3.3}$$

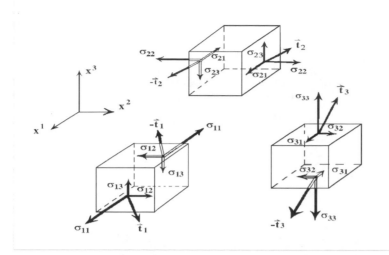

Figure 2.3-4. Stress vectors acting upon an element of volume

which is a summation of all the body forces and surface tractions acting upon our material. Here ϱ is the density of the material, dS is an element of surface area, and $d\tau$ is an element of volume.

The resultant moment $\vec{M}$ about the origin is similarly expressed as

$$\vec{M} = \iint_S \vec{r} \times \vec{t}^{(n)}\, dS + \iiint_V \varrho(\vec{r} \times \vec{b})\, d\tau. \tag{2.3.4}$$

The global motion of the material is governed by the Euler equations of motion.

- The time rate of change of linear momentum equals the resultant force or

$$\frac{d}{dt}\left[\iiint_V \varrho\vec{v}\, d\tau\right] = \vec{F} = \iint_S \vec{t}^{(n)}\, dS + \iiint_V \varrho\vec{b}\, d\tau. \tag{2.3.5}$$

This is a statement concerning the conservation of linear momentum.

- The time rate of change of angular momentum equals the resultant moment or

$$\frac{d}{dt}\left[\iiint_V \varrho\vec{r} \times \vec{v}\, d\tau\right] = \vec{M} = \iint_S \vec{r} \times \vec{t}^{(n)}\, dS + \iiint_V \varrho(\vec{r} \times \vec{b})\, d\tau. \tag{2.3.6}$$

This is a statement concerning conservation of angular momentum.

The Stress Tensor

Define the stress vectors

$$\vec{t}^1 = \sigma^{11}\,\hat{\mathbf{e}}_1 + \sigma^{12}\,\hat{\mathbf{e}}_2 + \sigma^{13}\,\hat{\mathbf{e}}_3$$
$$\vec{t}^2 = \sigma^{21}\,\hat{\mathbf{e}}_1 + \sigma^{22}\,\hat{\mathbf{e}}_2 + \sigma^{23}\,\hat{\mathbf{e}}_3 \tag{2.3.7}$$
$$\vec{t}^3 = \sigma^{31}\,\hat{\mathbf{e}}_1 + \sigma^{32}\,\hat{\mathbf{e}}_2 + \sigma^{33}\,\hat{\mathbf{e}}_3,$$

where σ^{ij}, $i,j = 1,2,3$ is the stress tensor acting at each point of the material. The index i indicates the coordinate surface $x^i = $ a constant, upon which $\vec{t}^i$ acts. The second index j denotes the direction associated with the components of $\vec{t}^i$.

216

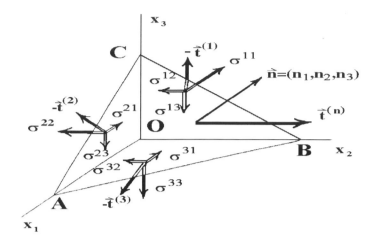

Figure 2.3-5. Stress distribution at a point

For $i = 1, 2, 3$ we adopt the convention of sketching the components of $\vec{t}^i$ in the positive directions if the exterior normal to the surface $x^i =$ constant also points in the positive direction. This gives rise to the figure 2.3-4 which illustrates the stress vectors acting upon an element of volume in rectangular Cartesian coordinates. The components $\sigma^{11}, \sigma^{22}, \sigma^{33}$ are called normal stresses while the components σ^{ij}, $i \neq j$ are called shearing stresses. The equations (2.3.7) can be written in the more compact form using the indicial notation as

$$\vec{t}^i = \sigma^{ij}\,\hat{\mathbf{e}}_j, \quad i, j = 1, 2, 3. \tag{2.3.8}$$

If we know the stress distribution at three orthogonal interfaces at a point P in a solid body, we can then determine the stress at the point P with respect to any plane passing through the point P. With reference to the figure 2.3-5, consider an arbitrary plane passing through the point P which lies within the material body being considered. Construct the elemental tetrahedron with orthogonal axes parallel to the $x^1 = x, x^2 = y$ and $x^3 = z$ axes. In this figure we have the following surface tractions:[1]

$$-\vec{t}^1 \quad \text{on the surface 0BC}$$
$$-\vec{t}^2 \quad \text{on the surface 0AC}$$
$$-\vec{t}^3 \quad \text{on the surface 0AB}$$
$$\vec{t}^{(n)} \quad \text{on the surface ABC}$$

The superscript parenthesis n is to remind you that this surface traction depends upon the orientation of the plane ABC which is determined by a unit normal vector having the direction cosines n_1, n_2 and n_3.

---

[1] From equilibrium considerations note that $\vec{t}^{(-i)} = -\vec{t}^{(i)}$.

Let
$$\Delta S_1 = \text{the surface area 0BC}$$
$$\Delta S_2 = \text{the surface area 0AC}$$
$$\Delta S_3 = \text{the surface area 0AB}$$
$$\Delta S = \text{the surface area ABC .}$$

These surface areas are related by the relations

$$\Delta S_1 = n_1 \Delta S, \qquad \Delta S_2 = n_2 \Delta S, \qquad \Delta S_3 = n_3 \Delta S \qquad (2.3.9)$$

which can be thought of as projections of ΔS upon the planes $x_i =$ constant for $i = 1, 2, 3$.

Cauchy Stress Law

Let $t^{j\,(n)}$ denote the components of the surface traction on the surface ABC. That is, we let

$$\vec{t}^{\,(n)} = t^{1\,(n)}\,\hat{\mathbf{e}}_1 + t^{2\,(n)}\,\hat{\mathbf{e}}_2 + t^{3\,(n)}\,\hat{\mathbf{e}}_3 = t^{j\,(n)}\,\hat{\mathbf{e}}_j. \qquad (2.3.10)$$

It will be demonstrated that the components $t^{j\,(n)}$ of the surface traction forces $\vec{t}^{\,(n)}$ associated with a plane through P and having the unit normal with direction cosines n_1, n_2 and n_3, must satisfy the relations

$$t^{j\,(n)} = n_i\,\sigma^{ij}, \quad i, j = 1, 2, 3. \qquad (2.3.11)$$

This relation is known as the Cauchy stress law.

Proof: Sum the forces acting on the elemental tetrahedron in the figure 2.3-5. If the body is in equilibrium, then the sum of these forces must equal zero or

$$(-\vec{t}^{\,1}\,\Delta S_1) + (-\vec{t}^{\,2}\,\Delta S_2) + (-\vec{t}^{\,3}\,\Delta S_3) + \vec{t}^{\,(n)}\,\Delta S = 0. \qquad (2.3.12)$$

The relations in the equations (2.3.9) are used to simplify the sum of forces in the equation (2.3.12). It is readily verified that the sum of forces simplifies to

$$\vec{t}^{\,(n)} = n_1\vec{t}^{\,1} + n_2\vec{t}^{\,2} + n_3\vec{t}^{\,3} = n_i\vec{t}^{\,i}. \qquad (2.3.13)$$

Substituting in the relations from equation (2.3.8) we find

$$\vec{t}^{\,(n)} = t^{j\,(n)}\,\hat{\mathbf{e}}_j = n_i\sigma^{ij}\,\hat{\mathbf{e}}_j, \quad i, j = 1, 2, 3 \qquad (2.3.14)$$

or in component form
$$t^{j\,(n)} = n_i\sigma^{ij} \qquad (2.3.15)$$

which is the Cauchy stress law.

Conservation of Linear Momentum

Let R denote a region in space where there exists a material volume with density ϱ having surface tractions and body forces acting upon it. Let v^i denote the velocity of the material volume and use Newton's second law to set the time rate of change of linear momentum equal to the forces acting upon the volume as in (2.3.5). We find

$$\frac{\delta}{\delta t}\left[\iiint_R \varrho v^j \, d\tau\right] = \iint_S \sigma^{ij} n_i \, dS + \iiint_R \varrho b^j \, d\tau.$$

Here $d\tau$ is an element of volume, dS is an element of surface area, b^j are body forces per unit mass, and σ^{ij} are the stresses. Employing the Gauss divergence theorem, the surface integral term is replaced by a volume integral and Newton's second law is expressed in the form

$$\iiint_R \left[\varrho f^j - \varrho b^j - \sigma^{ij}{}_{,i}\right] d\tau = 0, \tag{2.3.16}$$

where f^j is the acceleration from equation (1.4.54). Since R is an arbitrary region, the equation (2.3.16) implies that

$$\sigma^{ij}{}_{,i} + \varrho b^j = \varrho f^j. \tag{2.3.17}$$

This equation arises from a balance of linear momentum and represents the equations of motion for material in a continuum. If there is no velocity term, then equation (2.3.17) reduces to an equilibrium equation which can be written

$$\sigma^{ij}{}_{,i} + \varrho b^j = 0. \tag{2.3.18}$$

This equation can also be written in the covariant form

$$g^{si}\sigma_{ms,i} + \varrho b_m = 0,$$

which reduces to $\sigma_{ij,j} + \varrho b_i = 0$ in Cartesian coordinates. The equation (2.3.18) is an equilibrium equation and is one of our fundamental equations describing a continuum.

Conservation of Angular Momentum

The conservation of angular momentum equation (2.3.6) has the Cartesian tensors representation

$$\frac{d}{dt}\left[\iiint_R \varrho e_{ijk} x_j v_k \, d\tau\right] = \iint_S e_{ijk} x_j \sigma_{pk} n_p \, dS + \iiint_R \varrho e_{ijk} x_j b_k \, d\tau. \tag{2.3.19}$$

Employing the Gauss divergence theorem, the surface integral term is replaced by a volume integral to obtain

$$\iiint_R \left[e_{ijk}\varrho\frac{d}{dt}(x_j v_k) - e_{ijk}\left\{\varrho x_j b_k + \frac{\partial}{\partial x^p}(x_j \sigma_{pk})\right\}\right] d\tau = 0. \tag{2.3.20}$$

Since equation (2.3.20) must hold for all arbitrary volumes R we conclude that

$$e_{ijk}\varrho\frac{d}{dt}(x_j v_k) = e_{ijk}\left\{\varrho x_j b_k + x_j \frac{\partial \sigma_{pk}}{\partial x^p} + \sigma_{jk}\right\}$$

219

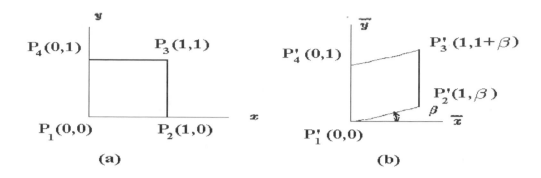

Figure 2.3-6. Shearing parallel to the y axis

which can be rewritten in the form

$$e_{ijk}\left[\sigma_{jk} + x_j\left(\frac{\partial \sigma_{pk}}{\partial x^p} + \varrho b_k - \varrho\frac{dv_k}{dt}\right) - \varrho v_j v_k\right] = 0. \tag{2.3.21}$$

In the equation (2.3.21) the middle term is zero because of the equation (2.3.17). Also the last term in (2.3.21) is zero because $e_{ijk}v_j v_k$ represents the cross product of a vector with itself. The equation (2.3.21) therefore reduces to

$$e_{ijk}\sigma_{jk} = 0, \tag{2.3.22}$$

which implies (see exercise 1.1, problem 22) that $\sigma_{ij} = \sigma_{ji}$ for all i and j. Thus, the conservation of angular momentum requires that the stress tensor be symmetric and so there are only 6 independent stress components to be determined. This is another fundamental law for a continuum.

Strain in Two Dimensions

Consider the matrix equation

$$\begin{pmatrix} \overline{x} \\ \overline{y} \end{pmatrix} = \begin{pmatrix} 1 & 0 \\ \beta & 1 \end{pmatrix}\begin{pmatrix} x \\ y \end{pmatrix} \tag{2.3.23}$$

which can be used to transform points (x,y) to points $(\overline{x},\overline{y})$. When this transformation is applied to the unit square illustrated in the figure 2.3-6(a) we obtain the geometry illustrated in the figure 2.3-6(b) which represents a shearing parallel to the y axis. If β is very small, we can use the approximation $\tan\beta \approx \beta$ and then this transformation can be thought of as a rotation of the element $\overline{P_1 P_2}$ through an angle β to the position $\overline{P_1' P_2'}$ when the barred axes are placed atop the unbarred axes.

Similarly, the matrix equation

$$\begin{pmatrix} \overline{x} \\ \overline{y} \end{pmatrix} = \begin{pmatrix} 1 & \alpha \\ 0 & 1 \end{pmatrix}\begin{pmatrix} x \\ y \end{pmatrix} \tag{2.3.24}$$

can be used to represent a shearing of the unit square parallel to the x axis as illustrated in the figure 2.3-7(b).

220

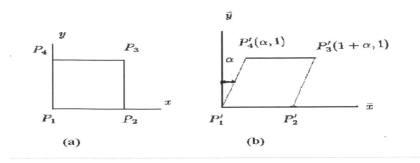

Figure 2.3-7. Shearing parallel to the x axis

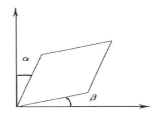

Figure 2.3-8. Shearing parallel to x and y axes

Again, if α is very small, we may use the approximation $\tan\alpha \approx \alpha$ and interpret α as an angular rotation of the element $\overline{P_1 P_4}$ to the position $\overline{P_1' P_4'}$. Now let us multiply the matrices given in equations (2.3.23) and (2.3.24). Note that the order of multiplication is important as can be seen by an examination of the products

$$
\begin{aligned}
\binom{\overline{x}}{\overline{y}} &= \begin{pmatrix} 1 & 0 \\ \beta & 1 \end{pmatrix}\begin{pmatrix} 1 & \alpha \\ 0 & 1 \end{pmatrix}\binom{x}{y} = \begin{pmatrix} 1 & \alpha \\ \beta & 1+\alpha\beta \end{pmatrix}\binom{x}{y} \\
\binom{\overline{x}}{\overline{y}} &= \begin{pmatrix} 1 & \alpha \\ 0 & 1 \end{pmatrix}\begin{pmatrix} 1 & 0 \\ \beta & 1 \end{pmatrix}\binom{x}{y} = \begin{pmatrix} 1+\alpha\beta & \alpha \\ \beta & 1 \end{pmatrix}\binom{x}{y}.
\end{aligned}
\tag{2.3.25}
$$

In equation (2.3.25) we will assume that the product $\alpha\beta$ is very, very small and can be neglected. Then the order of matrix multiplication will be immaterial and the transformation equation (2.3.25) will reduce to

$$
\binom{\overline{x}}{\overline{y}} = \begin{pmatrix} 1 & \alpha \\ \beta & 1 \end{pmatrix}\binom{x}{y}.
\tag{2.3.26}
$$

Applying this transformation to our unit square we obtain the simultaneous shearing parallel to both the x and y axes as illustrated in the figure 2.3-8.

This transformation can then be interpreted as the superposition of the two shearing elements depicted in the figure 2.3-9.

For comparison, we consider also the transformation equation

$$
\binom{\overline{x}}{\overline{y}} = \begin{pmatrix} 1 & 0 \\ -\alpha & 1 \end{pmatrix}\binom{x}{y}
\tag{2.3.27}
$$

221

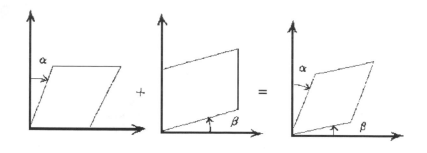

Figure 2.3-9. Superposition of shearing elements

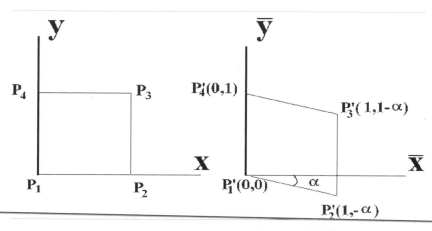

Figure 2.3-10. Rotation of element $\overline{P_1P_2}$

where α is very small. Applying this transformation to the unit square previously considered we obtain the results illustrated in the figure 2.3-10.

Note the difference in the direction of shearing associated with the transformation equations (2.3.27) and (2.3.23) illustrated in the figures 2.3-6 and 2.3-10. If the matrices appearing in the equations (2.3.24) and (2.3.27) are multiplied and we neglect product terms because α is assumed to be very small, we obtain the matrix equation

$$\begin{pmatrix} \overline{x} \\ \overline{y} \end{pmatrix} = \begin{pmatrix} 1 & \alpha \\ -\alpha & 1 \end{pmatrix}\begin{pmatrix} x \\ y \end{pmatrix} = \underbrace{\begin{pmatrix} 1 & 0 \\ 0 & 1 \end{pmatrix}\begin{pmatrix} x \\ y \end{pmatrix}}_{\text{identity}} + \underbrace{\begin{pmatrix} 0 & \alpha \\ -\alpha & 0 \end{pmatrix}\begin{pmatrix} x \\ y \end{pmatrix}}_{\text{rotation}}. \tag{2.3.28}$$

This can be interpreted as a superposition of the transformation equations (2.3.24) and (2.3.27) which represents a rotation of the unit square as illustrated in the figure 2.3-11.

The matrix on the right-hand side of equation (2.3.28) is referred to as a rotation matrix. The ideas illustrated by the above simple transformations will appear again when we consider the transformation of an arbitrary small element in a continuum when it under goes a strain. In particular, we will be interested in extracting the rigid body rotation from a deformed element and treating this rotation separately from the strain displacement.

222

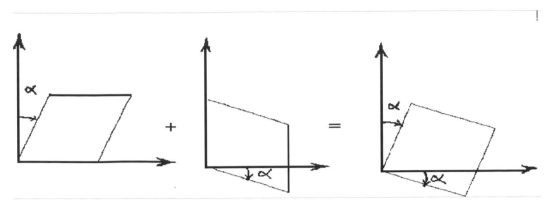

Figure 2.3-11. Rotation of unit square

Transformation of an Arbitrary Element

In two dimensions, we consider a rectangular element ABCD as illustrated in the figure 2.3-12.

Let the points $ABCD$ have the coordinates

$$A(x, y), \qquad B(x + \Delta x, y), \qquad C(x, y + \Delta y), \qquad D(x + \Delta x, y + \Delta y) \tag{2.3.29}$$

and denote by

$$u = u(x, y), \qquad v = v(x, y)$$

the displacement field associated with each of the points in the material continuum when it undergoes a deformation. Assume that the deformation of the element $ABCD$ in figure 2.3-12 can be represented by the matrix equation

$$\begin{pmatrix} \overline{x} \\ \overline{y} \end{pmatrix} = \begin{pmatrix} b_{11} & b_{12} \\ b_{21} & b_{22} \end{pmatrix} \begin{pmatrix} x \\ y \end{pmatrix} \tag{2.3.30}$$

where the coefficients $b_{ij}, i, j = 1, 2, 3$ are to be determined. Let us define $u = u(x, y)$ as the horizontal displacement of the point (x, y) and $v = v(x, y)$ as the vertical displacement of the same point. We can now express the displacement of each of the points A, B, C and D in terms of the displacement field $u = u(x, y)$ and $v = v(x, y)$. Consider first the displacement of the point A to A'. Here the coordinates (x, y) deform to the new coordinates

$$\overline{x} = x + u, \qquad \overline{y} = y + v.$$

That is, the coefficients b_{ij} must be chosen such that the equation

$$\begin{pmatrix} x + u \\ y + v \end{pmatrix} = \begin{pmatrix} b_{11} & b_{12} \\ b_{21} & b_{22} \end{pmatrix} \begin{pmatrix} x \\ y \end{pmatrix} \tag{2.3.31}$$

is satisfied. We next examine the displacement of the point B to B'. This displacement is described by the coordinates $(x + \Delta x, y)$ transforming to $(\overline{x}, \overline{y})$, where

$$\overline{x} = x + \Delta x + u(x + \Delta x, y), \qquad \overline{y} = y + v(x + \Delta x, y). \tag{2.3.32}$$

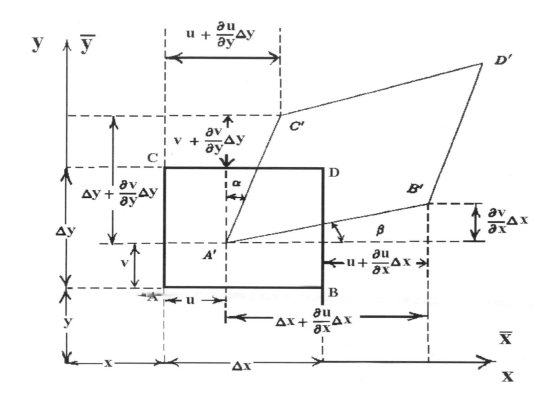

Figure 2.3-12. Displacement of element $ABCD$ to $A'B'C'D'$

Expanding u and v in (2.3.32) in Taylor series about the point (x, y) we find

$$\overline{x} = x + \Delta x + u + \frac{\partial u}{\partial x}\Delta x + h.o.t.$$

$$\overline{y} = y + v + \frac{\partial v}{\partial x}\Delta x + h.o.t.,$$

(2.3.33)

where $h.o.t.$ denotes higher order terms which have been neglected. The equations (2.3.33) require that the coefficients b_{ij} satisfy the matrix equation

$$\begin{pmatrix} x + u + \Delta x + \frac{\partial u}{\partial x}\Delta x \\ y + v + \frac{\partial v}{\partial x}\Delta x \end{pmatrix} = \begin{pmatrix} b_{11} & b_{12} \\ b_{21} & b_{22} \end{pmatrix} \begin{pmatrix} x + \Delta x \\ y \end{pmatrix}.$$

(2.3.34)

The displacement of the point C to C' is described by the coordinates $(x, y + \Delta y)$ transforming to $(\bar{x}, \bar{y})$ where

$$\bar{x} = x + u(x, y + \Delta y), \qquad \bar{y} = y + \Delta y + v(x, y + \Delta y). \tag{2.3.35}$$

Again we expand the displacement field components u and v in a Taylor series about the point (x, y) and find

$$\bar{x} = x + u + \frac{\partial u}{\partial y} \Delta y + h.o.t.$$
$$\bar{y} = y + \Delta y + v + \frac{\partial v}{\partial y} \Delta y + h.o.t. \tag{2.3.36}$$

This equation implies that the coefficients b_{ij} must be chosen such that

$$\begin{pmatrix} x + u + \frac{\partial u}{\partial y} \Delta y \\ y + v + \Delta y + \frac{\partial v}{\partial y} \Delta y \end{pmatrix} = \begin{pmatrix} b_{11} & b_{12} \\ b_{21} & b_{22} \end{pmatrix} \begin{pmatrix} x \\ y + \Delta y \end{pmatrix}. \tag{2.3.37}$$

Finally, it can be verified that the point D with coordinates $(x + \Delta x, y + \Delta y)$ moves to the point D' with coordinates

$$\bar{x} = x + \Delta x + u(x + \Delta x, y + \Delta y), \qquad \bar{y} = y + \Delta y + v(x + \Delta x, y + \Delta y). \tag{2.3.38}$$

Expanding u and v in a Taylor series about the point (x, y) we find the coefficients b_{ij} must be chosen to satisfy the matrix equation

$$\begin{pmatrix} x + \Delta x + u + \frac{\partial u}{\partial x} \Delta x + \frac{\partial u}{\partial y} \Delta y \\ y + \Delta y + v + \frac{\partial v}{\partial x} \Delta x + \frac{\partial v}{\partial y} \Delta y \end{pmatrix} = \begin{pmatrix} b_{11} & b_{12} \\ b_{21} & b_{22} \end{pmatrix} \begin{pmatrix} x + \Delta x \\ y + \Delta y \end{pmatrix}. \tag{2.3.39}$$

The equations (2.3.31),(2.3.34),(2.3.37) and (2.3.39) give rise to the simultaneous equations

$$b_{11}x + b_{12}y = x + u$$
$$b_{21}x + b_{22}y = y + v$$
$$b_{11}(x + \Delta x) + b_{12}y = x + u + \Delta x + \frac{\partial u}{\partial x} \Delta x$$
$$b_{21}(x + \Delta x) + b_{22}y = y + v + \frac{\partial v}{\partial x} \Delta x$$
$$b_{11}x + b_{12}(y + \Delta y) = x + u + \frac{\partial u}{\partial y} \Delta y \tag{2.3.40}$$
$$b_{21}x + b_{22}(y + \Delta y) = y + v + \Delta y + \frac{\partial v}{\partial y} \Delta y$$
$$b_{11}(x + \Delta x) + b_{12}(y + \Delta y) = x + \Delta x + u + \frac{\partial u}{\partial x} \Delta x + \frac{\partial u}{\partial y} \Delta y$$
$$b_{21}(x + \Delta x) + b_{22}(y + \Delta y) = y + \Delta y + v + \frac{\partial v}{\partial x} \Delta x + \frac{\partial v}{\partial y} \Delta y.$$

It is readily verified that the system of equations (2.3.40) has the solution

$$b_{11} = 1 + \frac{\partial u}{\partial x} \qquad b_{12} = \frac{\partial u}{\partial y}$$
$$b_{21} = \frac{\partial v}{\partial x} \qquad b_{22} = 1 + \frac{\partial v}{\partial y}. \tag{2.3.41}$$

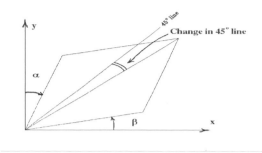

Figure 2.3-13. Change in 45° line

Hence the transformation equation (2.3.30) can be written as

$$\begin{pmatrix} \overline{x} \\ \overline{y} \end{pmatrix} = \begin{pmatrix} 1 + \frac{\partial u}{\partial x} & \frac{\partial u}{\partial y} \\ \frac{\partial v}{\partial x} & 1 + \frac{\partial v}{\partial y} \end{pmatrix} \begin{pmatrix} x \\ y \end{pmatrix}. \tag{2.3.42}$$

A physical interpretation associated with this transformation is obtained by writing it in the form:

$$\begin{pmatrix} \overline{x} \\ \overline{y} \end{pmatrix} = \underbrace{\begin{pmatrix} 1 & 0 \\ 0 & 1 \end{pmatrix} \begin{pmatrix} x \\ y \end{pmatrix}}_{\text{identity}} + \underbrace{\begin{pmatrix} e_{11} & e_{12} \\ e_{21} & e_{22} \end{pmatrix} \begin{pmatrix} x \\ y \end{pmatrix}}_{\text{strain matrix}} + \underbrace{\begin{pmatrix} \omega_{11} & \omega_{12} \\ \omega_{21} & \omega_{22} \end{pmatrix} \begin{pmatrix} x \\ y \end{pmatrix}}_{\text{rotation matrix}}, \tag{2.3.43}$$

where

$$e_{11} = \frac{\partial u}{\partial x} \qquad\qquad e_{21} = \frac{1}{2}\left(\frac{\partial u}{\partial y} + \frac{\partial v}{\partial x}\right)$$
$$e_{12} = \frac{1}{2}\left(\frac{\partial v}{\partial x} + \frac{\partial u}{\partial y}\right) \qquad e_{22} = \frac{\partial v}{\partial y} \tag{2.3.44}$$

are the elements of a symmetric matrix called the strain matrix and

$$\omega_{11} = 0$$
$$\omega_{12} = \frac{1}{2}\left(\frac{\partial u}{\partial y} - \frac{\partial v}{\partial x}\right)$$
$$\omega_{21} = \frac{1}{2}\left(\frac{\partial v}{\partial x} - \frac{\partial u}{\partial y}\right)$$
$$\omega_{22} = 0 \tag{2.3.45}$$

are the elements of a skew symmetric matrix called the rotation matrix.

The strain per unit length in the x-direction associated with the point A in the figure 2.3-12 is

$$e_{11} = \frac{\Delta x + \frac{\partial u}{\partial x}\Delta x - \Delta x}{\Delta x} = \frac{\partial u}{\partial x} \tag{2.3.46}$$

and the strain per unit length of the point A in the y direction is

$$e_{22} = \frac{\Delta y + \frac{\partial v}{\partial y}\Delta y - \Delta y}{\Delta y} = \frac{\partial v}{\partial y}. \tag{2.3.47}$$

These are the terms along the main diagonal in the strain matrix. The geometry of the figure 2.3-12 implies that

$$\tan\beta = \frac{\frac{\partial v}{\partial x}\Delta x}{\Delta x + \frac{\partial u}{\partial x}\Delta x}, \qquad \text{and} \qquad \tan\alpha = \frac{\frac{\partial u}{\partial y}\Delta y}{\Delta y + \frac{\partial v}{\partial y}\Delta y}. \tag{2.3.48}$$

For small derivatives associated with the displacements u and v it is assumed that the angles α and β are small and the equations (2.3.48) therefore reduce to the approximate equations

$$\tan\beta \approx \beta = \frac{\partial v}{\partial x} \qquad \tan\alpha \approx \alpha = \frac{\partial u}{\partial y}. \tag{2.3.49}$$

For a physical interpretation of these terms we consider the deformation of a small rectangular element which undergoes a shearing as illustrated in the figure 2.3-13.

226

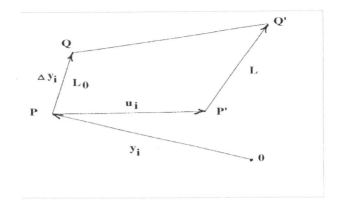

Figure 2.3-14. Displacement field due to state of strain

The quantity

$$\alpha + \beta = \left(\frac{\partial u}{\partial y} + \frac{\partial v}{\partial x}\right) = 2e_{12} = 2e_{21} \qquad (2.3.50)$$

is the change from a ninety degree angle due to the deformation and hence we can write $\frac{1}{2}(\alpha+\beta) = e_{12} = e_{21}$ as representing a change from a 45° angle due to the deformation. The quantities e_{21}, e_{12} are called the shear strains and the quantity

$$\gamma_{12} = 2e_{12} \qquad (2.3.51)$$

is called the shear angle. We will find that in general $\gamma_{ij} = 2e_{ij}, \ i \neq j$.

In the equation (2.3.45), the quantities $\omega_{21} = -\omega_{12}$ are the elements of the rigid body rotation matrix and are interpreted as angles associated with a rotation. The situation is analogous to the transformations and figures for the deformation of the unit square which was considered earlier.

Strain in Three Dimensions

The development of strain in three dimensions is approached from two different viewpoints. The first approach considers the derivation using Cartesian tensors and the second approach considers the derivation of strain using generalized tensors.

Cartesian Tensor Derivation of Strain.

Consider a material which is subjected to external forces such that all the points in the material undergo a deformation. Let (y_1, y_2, y_3) denote a set of orthogonal Cartesian coordinates, fixed in space, which is used to describe the deformations within the material. Further, let $u_i = u_i(y_1, y_2, y_3), i = 1, 2, 3$ denote a displacement field which describes the displacement of each point within the material. With reference to the figure 2.3-14 let P and Q denote two neighboring points within the material while it is in an unstrained state. These points move to the points P' and Q' when the material is in a state of strain. We let $y_i, i = 1, 2, 3$ represent the position vector to the general point P in the material, which is in an unstrained state, and denote by $y_i + u_i, i = 1, 2, 3$ the position vector of the point P' when the material is in a state of strain.

For Q a neighboring point of P which moves to Q' when the material is in a state of strain, we have from the figure 2.3-14 the following vectors:

$$\text{position of } P: \quad y_i, \quad i = 1, 2, 3$$
$$\text{position of } P': \quad y_i + u_i(y_1, y_2, y_3), \quad i = 1, 2, 3$$
$$\text{position of } Q: \quad y_i + \Delta y_i, \quad i = 1, 2, 3 \tag{2.3.52}$$
$$\text{position of } Q': \quad y_i + \Delta y_i + u_i(y_1 + \Delta y_1, y_2 + \Delta y_2, y_3 + \Delta y_3), \quad i = 1, 2, 3$$

Employing our earlier one dimensional definition of strain, we define the strain associated with the point P in the direction $\overline{PQ}$ as $e = \dfrac{L - L_0}{L_0}$, where $L_0 = \overline{PQ}$ and $L = \overline{P'Q'}$. To calculate the strain we need to first calculate the distances L_0 and L. The quantities L_0^2 and L^2 are easily calculated by considering dot products of vectors. For example, we have $L_0^2 = \Delta y_i \Delta y_i$, and the distance $L = \overline{P'Q'}$ is the magnitude of the vector

$$y_i + \Delta y_i + u_i(y_1 + \Delta y_1, y_2 + \Delta y_2, y_3 + \Delta y_3) - (y_i + u_i(y_1, y_2, y_3)), \quad i = 1, 2, 3.$$

Expanding the quantity $u_i(y_1 + \Delta y_1, y_2 + \Delta y_2, y_3 + \Delta y_3)$ in a Taylor series about the point P and neglecting higher order terms of the expansion we find that

$$L^2 = (\Delta y_i + \frac{\partial u_i}{\partial y_m} \Delta y_m)(\Delta y_i + \frac{\partial u_i}{\partial y_n} \Delta y_n).$$

Expanding the terms in this expression produces the equation

$$L^2 = \Delta y_i \Delta y_i + \frac{\partial u_i}{\partial y_n} \Delta y_i \Delta y_n + \frac{\partial u_i}{\partial y_m} \Delta y_m \Delta y_i + \frac{\partial u_i}{\partial y_m} \frac{\partial u_i}{\partial y_n} \Delta y_m \Delta y_n.$$

Note that L and L_0 are very small and so we express the difference $L^2 - L_0^2$ in terms of the strain e. We can write

$$L^2 - L_0^2 = (L + L_0)(L - L_0) = (L - L_0 + 2L_0)(L - L_0) = (e + 2)eL_0^2.$$

Now for e very small, and e^2 negligible, the above equation produces the approximation

$$eL_0^2 \approx \frac{L^2 - L_0^2}{2} = \frac{1}{2} \left[\frac{\partial u_m}{\partial y_n} + \frac{\partial u_n}{\partial y_m} + \frac{\partial u_r}{\partial y_m} \frac{\partial u_r}{\partial y_n} \right] \Delta y_m \Delta y_n.$$

The quantities

$$e_{mn} = \frac{1}{2} \left[\frac{\partial u_m}{\partial y_n} + \frac{\partial u_n}{\partial y_m} + \frac{\partial u_r}{\partial y_m} \frac{\partial u_r}{\partial y_n} \right] \tag{2.3.53}$$

is called the Green strain tensor or Lagrangian strain tensor. To show that e_{ij} is indeed a tensor, we consider the transformation $y_i = \ell_{ij}\overline{y}_j + b_i$, where $\ell_{ji}\ell_{ki} = \delta_{jk} = \ell_{ij}\ell_{ik}$. Note that from the derivative relation $\frac{\partial y_i}{\partial \overline{y}_j} = \ell_{ij}$ and the transformation equations $\overline{u}_i = \ell_{ij}u_j, i = 1, 2, 3$ we can express the strain in the barred system of coordinates. Performing the necessary calculations produces

$$\overline{e}_{ij} = \frac{1}{2} \left[\frac{\partial \overline{u}_i}{\partial \overline{y}_j} + \frac{\partial \overline{u}_j}{\partial \overline{y}_i} + \frac{\partial \overline{u}_r}{\partial \overline{y}_i} \frac{\partial \overline{u}_r}{\partial \overline{y}_j} \right]$$

$$= \frac{1}{2} \left[\frac{\partial}{\partial y_n}(\ell_{ik}u_k)\frac{\partial y_n}{\partial \overline{y}_j} + \frac{\partial}{\partial y_m}(\ell_{jk}u_k)\frac{\partial y_m}{\partial \overline{y}_i} + \frac{\partial}{\partial y_k}(\ell_{rs}u_s)\frac{\partial y_k}{\partial \overline{y}_i} \frac{\partial}{\partial y_t}(\ell_{rm}u_m)\frac{\partial y_t}{\partial \overline{y}_j} \right]$$

$$= \frac{1}{2} \left[\ell_{im}\ell_{nj}\frac{\partial u_m}{\partial y_n} + \ell_{jk}\ell_{mi}\frac{\partial u_k}{\partial y_m} + \ell_{rs}\ell_{rp}\ell_{ki}\ell_{tj}\frac{\partial u_s}{\partial y_k} \frac{\partial u_p}{\partial y_t} \right]$$

$$= \frac{1}{2} \left[\frac{\partial u_m}{\partial y_n} + \frac{\partial u_n}{\partial y_m} + \frac{\partial u_s}{\partial y_m} \frac{\partial u_s}{\partial y_n} \right] \ell_{im}\ell_{nj}$$

or $\quad \overline{e}_{ij} = e_{mn}\ell_{im}\ell_{nj}.$ This shows the strain e_{ij} transforms like a second order Cartesian tensor.

228

Lagrangian and Eulerian Systems

Let $\overline{x}^i$ denote the initial position of a material particle in a continuum. Assume that at a later time the particle has moved to another point whose coordinates are x^i. Both sets of coordinates are referred to the same coordinate system. When the final position can be expressed as a function of the initial position and time we can write $x^i = x^i(\overline{x}^1, \overline{x}^2, \overline{x}^3, t)$. Whenever the changes of any physical quantity is represented in terms of its initial position and time, the representation is referred to as a Lagrangian or material representation of the quantity. This can be thought of as a transformation of the coordinates. When the Jacobian $J(\frac{x}{\overline{x}})$ of this transformation is different from zero, the above set of equations have a unique inverse $\overline{x}^i = \overline{x}^i(x^1, x^2, x^3, t)$, where the position of the particle is now expressed in terms of its instantaneous position and time. Such a representation is referred to as an Eulerian or spatial description of the motion.

Let $(\overline{x}_1, \overline{x}_2, \overline{x}_3)$ denote the initial position of a particle whose motion is described by $x_i = x_i(\overline{x}_1, \overline{x}_2, \overline{x}_3, t)$, then $u_i = x_i - \overline{x}_i$ denotes the displacement vector which can by represented in a Lagrangian or Eulerian form. For example, if

$$x_1 = 2(\overline{x}_1 - \overline{x}_2)(e^t - 1) + (\overline{x}_2 - \overline{x}_1)(e^{-t} - 1) + \overline{x}_1$$
$$x_2 = (\overline{x}_1 - \overline{x}_2)(e^t - 1) + (\overline{x}_2 - \overline{x}_1)(e^{-t} - 1) + \overline{x}_2$$
$$x_3 = \overline{x}_3$$

then the displacement vector can be represented in the Lagrangian form

$$u_1 = 2(\overline{x}_1 - \overline{x}_2)(e^t - 1) + (\overline{x}_2 - \overline{x}_1)(e^{-t} - 1)$$
$$u_2 = (\overline{x}_1 - \overline{x}_2)(e^t - 1) + (\overline{x}_2 - \overline{x}_1)(e^{-t} - 1)$$
$$u_3 = 0$$

or the Eulerian form

$$u_1 = x_1 - (2x_2 - x_1)(1 - e^{-t}) - (x_1 - x_2)(e^{-2t} - e^{-t}) - x_1 e^{-t}$$
$$u_2 = x_2 - (2x_2 - x_1)(1 - e^{-t}) - (x_2 - x_1)(e^{-2t} - e^{-t}) - x_2 e^{-t}$$
$$u_3 = 0.$$

Note that in the Lagrangian system the displacements are expressed in terms of the initial position and time, while in the Eulerian system the independent variables are the position coordinates and time. Euler equations describe, as a function of time, how such things as density, pressure, and fluid velocity change at a fixed point in the medium. In contrast, the Lagrangian viewpoint follows the time history of a moving individual fluid particle as it moves through the medium.

General Tensor Derivation of Strain.

With reference to the figure 2.3-15 consider the deformation of a point P within a continuum. Let (y^1, y^2, y^3) denote a Cartesian coordinate system which is fixed in space. We can introduce a coordinate transformation $y^i = y^i(x^1, x^2, x^3)$, $i = 1, 2, 3$ and represent all points within the continuum with respect to a set of generalized coordinates (x^1, x^2, x^3). Let P denote a general point in the continuum while it is in an unstrained state and assume that this point gets transformed to a point P' when the continuum experiences external forces. If P moves to P', then all points Q which are near P will move to points Q' near P'. We can imagine that in the unstrained state all the points of the continuum are referenced with respect to the set of generalized coordinates (x^1, x^2, x^3). After the strain occurs, we can imagine that it will be convenient to represent all points of the continuum with respect to a new barred system of coordinates $(\overline{x}^1, \overline{x}^2, \overline{x}^3)$. We call the original set of coordinates the Lagrangian system of coordinates and the new set of barred coordinates the Eulerian coordinates. The Eulerian coordinates are assumed to be described by a set of coordinate transformation equations $\overline{x}^i = \overline{x}^i(x^1, x^2, x^3)$, $i = 1, 2, 3$ with inverse transformations $x^i = x^i(\overline{x}^1, \overline{x}^2, \overline{x}^3)$, $i = 1, 2, 3$, which are assumed to exist. The barred and unbarred coordinates can be related to a fixed set of Cartesian coordinates $y^i, i = 1, 2, 3$, and we may assume that there exists transformation equations

$$y^i = f^i(x^1, x^2, x^3), \quad i = 1, 2, 3 \quad \text{and} \quad y^i = g^i(\overline{x}^1, \overline{x}^2, \overline{x}^3), \quad i = 1, 2, 3$$

which relate the barred and unbarred coordinates to the Cartesian axes. In the discussion that follows be sure to note whether there is a bar over a symbol, as we will be jumping back and forth between the Lagrangian and Eulerian reference frames.

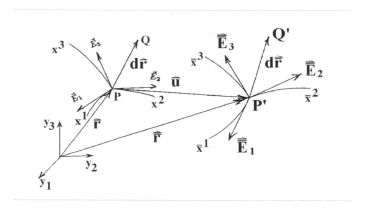

Figure 2.3-15. Strain in generalized coordinates

In the Lagrangian system of unbarred coordinates we have the basis vectors $\vec{E}_i = \dfrac{\partial \vec{r}}{\partial x^i}$ which produce the metrices $g_{ij} = \vec{E}_i \cdot \vec{E}_j$. Similarly, in the Eulerian system of barred coordinates we have the basis vectors $\overline{\vec{E}}_i = \dfrac{\partial \overline{\vec{r}}}{\partial \overline{x}^i}$ which produces the metrices $\overline{G}_{ij} = \overline{\vec{E}}_i \cdot \overline{\vec{E}}_j$. These basis vectors are illustrated in the figure 2.3-15.

We assume that an element of arc length squared ds^2 in the unstrained state is deformed to the element of arc length squared $d\bar{s}^2$ in the strained state. An element of arc length squared can be expressed in terms of the barred or unbarred coordinates. For example, in the Lagrangian system, let $d\vec{r} = \overline{PQ}$ so that

$$L_0^2 = d\vec{r} \cdot d\vec{r} = ds^2 = g_{ij}dx^i dx^j, \qquad (2.3.54)$$

where g_{ij} are the metrices in the Lagrangian coordinate system. This same element of arc length squared can be expressed in the barred system by

$$L_0^2 = ds^2 = \bar{g}_{ij}d\bar{x}^i d\bar{x}^j, \quad \text{where} \quad \bar{g}_{ij} = g_{mn}\frac{\partial x^m}{\partial \bar{x}^i}\frac{\partial x^n}{\partial \bar{x}^j}. \qquad (2.3.55)$$

Similarly, in the Eulerian system of coordinates the deformed arc length squared is

$$L^2 = d\vec{\bar{r}} \cdot d\vec{\bar{r}} = d\bar{s}^2 = \overline{G}_{ij}d\bar{x}^i d\bar{x}^j, \qquad (2.3.56)$$

where $\overline{G}_{ij}$ are the metrices in the Eulerian system of coordinates. This same element of arc length squared can be expressed in the Lagrangian system by the relation

$$L^2 = d\bar{s}^2 = G_{ij}dx^i dx^j, \quad \text{where} \quad G_{ij} = \overline{G}_{mn}\frac{\partial \bar{x}^m}{\partial x^i}\frac{\partial \bar{x}^n}{\partial x^j}. \qquad (2.3.57)$$

In the Lagrangian system we have

$$d\bar{s}^2 - ds^2 = (G_{ij} - g_{ij})dx^i dx^j = 2e_{ij}dx^i dx^j$$

where

$$e_{ij} = \frac{1}{2}\left(G_{ij} - g_{ij}\right) \qquad (2.3.58)$$

is called the Green strain tensor or Lagrangian strain tensor. Alternatively, in the Eulerian system of coordinates we may write

$$d\bar{s}^2 - ds^2 = \left(\overline{G}_{ij} - \bar{g}_{ij}\right)d\bar{x}^i d\bar{x}^j = 2\bar{e}_{ij}d\bar{x}^i d\bar{x}^j$$

where

$$\bar{e}_{ij} = \frac{1}{2}\left(\overline{G}_{ij} - \bar{g}_{ij}\right) \qquad (2.3.59)$$

is called the Almansi strain tensor or Eulerian strain tensor.

Note also in the figure 2.3-15 there is the displacement vector $\vec{u}$. This vector can be represented in any of the following forms:

$$\vec{u} = u^i \vec{E}_i \quad \text{contravariant, Lagrangian basis}$$
$$\vec{u} = u_i \vec{E}^i \quad \text{covariant, Lagrangian reciprocal basis}$$
$$\vec{u} = \overline{u}^i \vec{\overline{E}}_i \quad \text{contravariant, Eulerian basis}$$
$$\vec{u} = \overline{u}_i \vec{\overline{E}}^i \quad \text{covariant, Eulerian reciprocal basis.}$$

By vector addition we have $\vec{r} + \vec{u} = \vec{\overline{r}}$ and therefore $d\vec{r} + d\vec{u} = d\vec{\overline{r}}$. In the Lagrangian frame of reference at the point P we represent $\vec{u}$ in the contravariant form $\vec{u} = u^i \vec{E}_i$ and write $d\vec{r}$ in the form $d\vec{r} = dx^i \vec{E}_i$. By use of the equation (1.4.48) we can express $d\vec{u}$ in the form $d\vec{u} = u^i{}_{,k} dx^k \vec{E}_i$. These substitutions produce the representation $d\vec{\overline{r}} = (dx^i + u^i{}_{,k} dx^k)\vec{E}_i$ in the Lagrangian coordinate system. We can then express $d\overline{s}^2$ in the Lagrangian system. We find

$$d\vec{\overline{r}} \cdot d\vec{\overline{r}} = d\overline{s}^2 = (dx^i + u^i{}_{,k} dx^k)\vec{E}_i \cdot (dx^j + u^j{}_{,m} dx^m)\vec{E}_j$$
$$= (dx^i dx^j + u^j{}_{,m} dx^m dx^i + u^i{}_{,k} dx^k dx^j + u^i{}_{,k} u^j{}_{,m} dx^k dx^m)g_{ij}$$

and so from the relation (2.3.58) one can derive the representation

$$e_{ij} = \frac{1}{2}\left(u_{i,j} + u_{j,i} + u_{m,i} u^m{}_{,j}\right). \tag{2.3.60}$$

This is the representation of the Lagrangian strain tensor in any system of coordinates. The strain tensor e_{ij} is symmetric. We will restrict our study to small deformations and neglect the product terms in equation (2.3.60). Under these conditions the equation (2.3.60) reduces to $e_{ij} = \frac{1}{2}(u_{i,j} + u_{j,i})$.

If instead, we chose to represent the displacement $\vec{u}$ with respect to the Eulerian basis, then we can write

$$\vec{u} = \overline{u}^i \vec{\overline{E}}_i \quad \text{with} \quad d\vec{u} = \overline{u}^i{}_{,k} d\overline{x}^k \vec{\overline{E}}_i.$$

These relations imply that

$$d\vec{r} = d\vec{\overline{r}} - d\vec{u} = (d\overline{x}^i - \overline{u}^i{}_{,k} d\overline{x}^k)\vec{\overline{E}}_i.$$

This representation of $d\vec{r}$ in the Eulerian frame of reference can be used to calculate the strain $\overline{e}_{ij}$ from the relation $d\overline{s}^2 - ds^2$. It is left as an exercise to show that there results

$$\overline{e}_{ij} = \frac{1}{2}\left(\overline{u}_{i,j} + \overline{u}_{j,i} - \overline{u}_{m,i} \overline{u}^m{}_{,j}\right). \tag{2.3.61}$$

The equation (2.3.61) is the representation of the Eulerian strain tensor in any system of coordinates. Under conditions of small deformations both the equations (2.3.60) and (2.3.61) reduce to the linearized Lagrangian and Eulerian strain tensor $\overline{e}_{ij} = \frac{1}{2}(\overline{u}_{i,j} + \overline{u}_{j,i})$. In the case of large deformations the equations (2.3.60) and (2.3.61) describe the strains. In the case of linear elasticity, where the deformations are very small, the product terms in equations (2.3.60) and (2.3.61) are neglected and the Lagrangian and Eulerian strains reduce to their linearized forms

$$e_{ij} = \frac{1}{2}\left[u_{i,j} + u_{j,i}\right] \qquad \overline{e}_{ij} = \frac{1}{2}\left[\overline{u}_{i,j} + \overline{u}_{j,i}\right]. \tag{2.3.62}$$

232

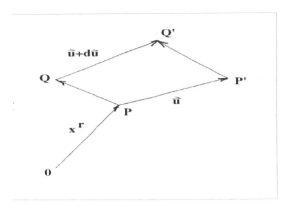

Figure 2.3-16. Displacement due to strain

Compressible and Incompressible Material   With reference to figure 2.3-16, let x^i, $i = 1, 2, 3$ denote the position vector of an arbitrary point P in a continuum before there is a state of strain. Let Q be a neighboring point of P with position vector $x^i + dx^i$, $i = 1, 2, 3$. Also in the figure 2.3-16 there is the displacement vector $\vec{u}$. Here it is assumed that $\vec{u} = \vec{u}(x^1, x^2, x^3)$ denotes the displacement field when the continuum is in a state of strain. The figure 2.3-16 illustrates that in a state of strain P moves to P' and Q moves to Q'. Let us find a relationship between the distance $\overline{PQ}$ before the strain and the distance $\overline{P'Q'}$ when the continuum is in a state of strain. For $\vec{E}_1, \vec{E}_2, \vec{E}_3$ basis functions constructed at P we have previously shown that if

$$\vec{u}(x^1, x^2, x^3) = u^i \vec{E}_i \qquad \text{then} \qquad d\vec{u} = u^i{}_{,j} dx^j \, \vec{E}_i.$$

Now for $\vec{u} + d\vec{u}$ the displacement of the point Q we may use vector addition and write

$$\overline{PQ} + \vec{u} + d\vec{u} = \vec{u} + \overline{P'Q'}. \tag{2.3.63}$$

Let $\overline{PQ} = dx^i \vec{E}_i = a^i \vec{E}_i$ denote an arbitrary small change in the continuum. This arbitrary displacement gets deformed to $\overline{P'Q'} = A^i \vec{E}_i$ due to the state of strain in the continuum. Employing the equation (2.3.63) we write

$$dx^i + u^i{}_{,j} dx^j = a^i + u^i{}_{,j} a^j = A^i$$

which can be written in the form

$$\delta a^i = A^i - a^i = u^i{}_{,j} a^j \quad \text{where} \quad dx^i = a^i, i = 1, 2, 3 \tag{2.3.64}$$

denotes an arbitrary small change. The tensor $u^i{}_{,j}$ and the associated tensor $u_{i,j} = g_{it} u^t{}_{,j}$ are in general not symmetric tensors. However, we know we can express $u_{i,j}$ as the sum of a symmetric (e_{ij}) and skew-symmetric (ω_{ij}) tensor. We therefore write

$$u_{i,j} = e_{ij} + \omega_{ij} \quad \text{or} \quad u^i{}_{,j} = e^i{}_j + \omega^i{}_j,$$

where

$$e_{ij} = \frac{1}{2}(u_{i,j} + u_{j,i}) = \frac{1}{2}(g_{im} u^m{}_{,j} + g_{jm} u^m{}_{,i}) \qquad \text{and} \qquad \omega_{ij} = \frac{1}{2}(u_{i,j} - u_{j,i}) = \frac{1}{2}(g_{im} u^m{}_{,j} - g_{jm} u^m{}_{,i}).$$

The deformation of a small quantity a^i can therefore be represented by a pure strain $A^i - a^i = e^i{}_s a^s$ followed by a rotation $A^i - a^i = \omega^i{}_s a^s$.

Consider now a small element of volume inside a material medium. With reference to the figure 2.3-17(a) we let $\vec{a}, \vec{b}, \vec{c}$ denote three small arbitrary independent vectors constructed at a general point P within the material before any external forces are applied. We imagine $\vec{a}, \vec{b}, \vec{c}$ as representing the sides of a small parallelepiped before any deformation has occurred. When the material is placed in a state of strain the point P will move to P' and the vectors $\vec{a}, \vec{b}, \vec{c}$ will become deformed to the vectors $\vec{A}, \vec{B}, \vec{C}$ as illustrated in the figure 2.3-17(b). The vectors $\vec{A}, \vec{B}, \vec{C}$ represent the sides of the parallelepiped after the deformation.

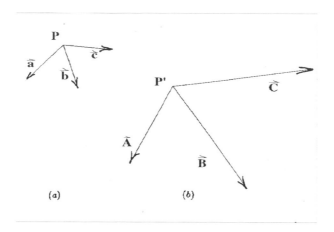

Figure 2.3-17. Deformation of a parallelepiped

Let ΔV denote the volume of the parallelepiped with sides $\vec{a}, \vec{b}, \vec{c}$ at P before the strain and let $\Delta V'$ denote the volume of the deformed parallelepiped after the strain, when it then has sides $\vec{A}, \vec{B}, \vec{C}$ at the point P'. We define the ratio of the change in volume due to the strain divided by the original volume as the dilatation at the point P. The dilatation is thus expressed as

$$\Theta = \frac{\Delta V' - \Delta V}{\Delta V} = dilatation. \qquad (2.3.65)$$

Since $u^i, i = 1, 2, 3$ represents the displacement field due to the strain, we use the result from equation (2.3.64) and represent the displaced vectors $\vec{A}, \vec{B}, \vec{C}$ in the form

$$A^i = a^i + u^i_{,j}a^j$$
$$B^i = b^i + u^i_{,j}b^j \qquad (2.3.66)$$
$$C^i = c^i + u^i_{,j}c^j$$

where $\vec{a}, \vec{b}, \vec{c}$ are arbitrary small vectors emanating from the point P in the unstrained state. The element of volume ΔV, before the strain, is calculated from the triple scalar product relation

$$\Delta V = \vec{a} \cdot (\vec{b} \times \vec{c}) = e_{ijk}a^i b^j c^k.$$

The element of volume $\Delta V'$, which occurs due to the strain, is calculated from the triple scalar product

$$\Delta V' = \vec{A} \cdot (\vec{B} \times \vec{C}) = e_{ijk}A^i B^j C^k.$$

Substituting the relations from the equations (2.3.66) into the triple scalar product gives

$$\Delta V' = e_{ijk}(a^i + u^i_{,m}a^m)(b^j + u^j_{,n}b^n)(c^k + u^k_{,p}c^p).$$

Expanding the triple scalar product and employing the result from Exercise 1.4, problem 34, we find the simplified result gives us the dilatation

$$\Theta = \frac{\Delta V' - \Delta V}{\Delta V} = u^r_{,r} = \text{div}(\vec{u}). \tag{2.3.67}$$

That is, the dilatation is the divergence of the displacement field. If the divergence of the displacement field is zero, there is no volume change and the material is said to be incompressible. If the divergence of the displacement field is different from zero, the material is said to be compressible.

Note that the strain e_{ij} is expressible in terms of the displacement field by the relation

$$e_{ij} = \frac{1}{2}(u_{i,j} + u_{j,i}), \quad \text{and consequently} \quad g^{mn}e_{mn} = u^r_{,r}. \tag{2.3.68}$$

Hence, for an orthogonal system of coordinates the dilatation can be expressed in terms of the strain elements along the main diagonal.

Conservation of Mass

Consider the material in an arbitrary region R of a continuum. Let $\varrho = \varrho(x, y, z, t)$ denote the density of the material within the region. Assume that the dimension of the density ϱ is gm/cm^3 in the cgs system of units. We shall assume that the region R is bounded by a closed surface S with exterior unit normal $\vec{n}$ defined everywhere on the surface. Further, we let $\vec{v} = \vec{v}(x, y, z, t)$ denote a velocity field associated with all points within the continuum. The velocity field has units of cm/sec in the cgs system of units. Neglecting sources and sinks, the law of conservation of mass examines all the material entering and leaving a region R. Enclosed within R is the material mass m where $m = \iiint_R \varrho\, d\tau$ with dimensions of gm in the cgs system of units. Here $d\tau$ denotes an element of volume inside the region R. The change of mass with time is obtained by differentiating the above relation. Differentiating the mass produces the equation

$$\frac{\partial m}{\partial t} = \iiint_R \frac{\partial \varrho}{\partial t}\, d\tau \tag{2.3.69}$$

and has the dimensions of gm/sec.

Consider also the surface integral

$$I = \iint_S \varrho\vec{v} \cdot \hat{n}\, d\sigma \tag{2.3.70}$$

where $d\sigma$ is an element of surface area on the surface S which encloses R and $\hat{n}$ is the exterior unit normal vector to the surface S. The dimensions of the integral I is determined by examining the dimensions of each term in the integrand of I. We find that

$$[I] = \frac{gm}{cm^3} \cdot \frac{cm}{sec} \cdot cm^2 = \frac{gm}{sec}$$

and so the dimension of I is the same as the dimensions for the change of mass within the region R. The surface integral I is the flux rate of material crossing the surface of R and represents the change of mass entering the region if $\vec{v} \cdot \hat{n}$ is negative and the change of mass leaving the region if $\vec{v} \cdot \hat{n}$ is positive, as $\hat{n}$ is always an exterior unit normal vector. Equating the relations from equations (2.3.69) and (2.3.70) we obtain a mathematical statement for mass conservation

$$\frac{\partial m}{\partial t} = \iiint_R \frac{\partial \varrho}{\partial t}\, d\tau = -\iint_S \varrho \vec{v} \cdot \vec{n}\, d\sigma. \qquad (2.3.71)$$

The equation (2.3.71) implies that the rate at which the mass contained in R increases must equal the rate at which the mass flows into R through the surface S. The negative sign changes the direction of the exterior normal so that we consider flow of material into the region. Employing the Gauss divergence theorem, the surface integral in equation (2.3.71) can be replaced by a volume integral and the law of conservation of mass is then expressible in the form

$$\iiint_R \left[\frac{\partial \varrho}{\partial t} + \operatorname{div}(\varrho \vec{v})\right] d\tau = 0. \qquad (2.3.72)$$

Since the region R is an arbitrary volume we conclude that the term inside the brackets must equal zero. This gives us the continuity equation

$$\frac{\partial \varrho}{\partial t} + \operatorname{div}(\varrho \vec{v}) = 0 \qquad (2.3.73)$$

which represents the mass conservation law in terms of velocity components. This is the Eulerian representation of continuity of mass flow.

Equivalent forms of the continuity equation are:

$$\frac{\partial \varrho}{\partial t} + \vec{v} \cdot \operatorname{grad} \varrho + \varrho \operatorname{div} \vec{v} = 0$$

$$\frac{\partial \varrho}{\partial t} + v_i \frac{\partial \varrho}{\partial x^i} + \varrho \frac{\partial v_i}{\partial x^i} = 0$$

$$\frac{D\varrho}{Dt} + \varrho \frac{\partial v_i}{\partial x^i} = 0$$

where $\dfrac{D\varrho}{Dt} = \dfrac{\partial \varrho}{\partial t} + \dfrac{\partial \varrho}{\partial x^i}\dfrac{dx^i}{dt} = \dfrac{\partial \varrho}{\partial t} + \dfrac{\partial \varrho}{\partial x^i} v_i$ is called the material derivative of the density ϱ. Note that the material derivative contains the expression $\frac{\partial \varrho}{\partial x^i} v_i$ which is known as the convective or advection term. If the density $\varrho = \varrho(x, y, z, t)$ is a constant we have

$$\frac{D\varrho}{Dt} = \frac{\partial \varrho}{\partial t} + \frac{\partial \varrho}{\partial x}\frac{dx}{dt} + \frac{\partial \varrho}{\partial y}\frac{dy}{dt} + \frac{\partial \varrho}{\partial z}\frac{dz}{dt} = \frac{\partial \varrho}{\partial t} + \frac{\partial \varrho}{\partial x^i}\frac{dx^i}{dt} = 0 \qquad (2.3.74)$$

and hence the continuity equation reduces to $\operatorname{div}(\vec{v}) = 0$. Thus, if $\operatorname{div}(\vec{v})$ is zero, then the material is incompressible.

EXAMPLE 2.3-2. (Continuity Equation) Find the Lagrangian representation of mass conservation.

Solution: Let (X, Y, Z) denote the initial position of a fluid particle and denote the density of the fluid by $\varrho(X, Y, Z, t)$ so that $\varrho(X, Y, Z, 0)$ denotes the density at the time $t = 0$. Consider a simple closed region in our continuum and denote this region by $R(0)$ at time $t = 0$ and by $R(t)$ at some later time t. That is, all the points in $R(0)$ move in a one-to-one fashion to points in $R(t)$. Initially the mass of material in $R(0)$ is $m(0) = \iiint_{R(0)} \varrho(X, Y, Z, 0)\, d\tau(0)$ where $d\tau(0) = dX\, dY\, dZ$ is an element of volume in $R(0)$. We have after a time t has elapsed the mass of material in the region $R(t)$ given by $m(t) = \iiint_{R(t)} \varrho(X, Y, Z, t)\, d\tau(t)$ where $d\tau(t) = dx\,dy\,dz$ is a deformed element of volume related to the $d\tau(0)$ by $d\tau(t) = J\left(\frac{x,y,z}{X,Y,Z}\right) d\tau(0)$ where J is the Jacobian of the Eulerian (x, y, z) variables with respect to the Lagrangian (X, Y, Z) representation. For mass conservation we require that $m(t) = m(0)$ for all t. This implies that

$$\varrho(X, Y, Z, t)J = \varrho(X, Y, Z, 0) \tag{2.3.75}$$

for all time, since the initial region $R(0)$ is arbitrary. The right hand side of equation (2.3.75) is independent of time and so

$$\frac{d}{dt}\left(\varrho(X, Y, Z, t)J\right) = 0. \tag{2.3.76}$$

This is the Lagrangian form of the continuity equation which expresses mass conservation. Using the result that $\dfrac{dJ}{dt} = J \operatorname{div} \vec{V}$, (see problem 28, Exercise 2.3), the equation (2.3.76) can be expanded and written in the form

$$\frac{D\varrho}{Dt} + \varrho \operatorname{div} \vec{V} = 0 \tag{2.3.77}$$

where $\frac{D\varrho}{Dt}$ is from equation (2.3.74). The form of the continuity equation (2.3.77) is one of the Eulerian forms previously developed.

■

In the Eulerian coordinates the continuity equation is written $\frac{\partial \varrho}{\partial t} + \operatorname{div}(\varrho\vec{v}) = 0$, while in the Lagrangian system the continuity equation is written $\frac{d(\varrho J)}{dt} = 0$. Note that the velocity carries the Lagrangian axes and the density change $\operatorname{grad}\varrho$. This is reflective of the advection term $\vec{v} \cdot \operatorname{grad}\varrho$. Thus, in order for mass to be conserved it need not remain stationary. The mass can flow and the density can change. The material derivative is a transport rule depicting the relation between the Eulerian and Lagrangian viewpoints.

In general, from a Lagrangian viewpoint, any quantity $Q(x, y, z, t)$ which is a function of both position and time is seen as being transported by the fluid velocity (v_1, v_2, v_3) to $Q(x + v_1 dt, y + v_2 dt, z + v_3 dt, t + dt)$. Then the time derivative of Q contains both $\frac{\partial Q}{\partial t}$ and the advection term $\vec{v} \cdot \nabla Q$. In terms of mass flow, the Eulerian viewpoint sees flow into and out of a fixed volume in space, as depicted by the equation (2.3.71), In contrast, the Lagrangian viewpoint sees the same volume moving with the fluid and consequently

$$\frac{D}{Dt} \int\int\int_{R(t)} \rho\, d\tau = 0,$$

where $R(t)$ represents the volume moving with the fluid. Both viewpoints produce the same continuity equation reflecting the conservation of mass.

Summary of Basic Equations

Let us summarize the basic equations which are valid for all types of a continuum. We have derived:

- Conservation of mass (continuity equation)

$$\frac{\partial \varrho}{\partial t} + (\varrho v^i)_{,i} = 0$$

- Conservation of linear momentum sometimes called the Cauchy equation of motion.

$$\sigma^{ij}_{,i} + \varrho b^j = \varrho f^j, \quad j = 1, 2, 3.$$

- Conservation of angular momentum

$$\sigma_{ij} = \sigma_{ji}$$

- Strain tensor for linear elasticity

$$e_{ij} = \frac{1}{2}(u_{i,j} + u_{j,i}).$$

If we assume that the continuum is in equilibrium, and there is no motion, then the velocity and acceleration terms above will be zero. The continuity equation then implies that the density is a constant. The conservation of angular momentum equation requires that the stress tensor be symmetric and we need find only six stresses. The remaining equations reduce to a set of nine equations in the fifteen unknowns:

$$3 \text{ displacements } u_1, u_2, u_3$$

$$6 \text{ strains } \quad e_{11}, e_{12}, e_{13}, e_{22}, e_{23}, e_{33}$$

$$6 \text{ stresses } \quad \sigma_{11}, \sigma_{12}, \sigma_{13}, \sigma_{22}, \sigma_{23}, \sigma_{33}$$

Therefore, we still need additional information if we desire to determine these unknowns.

Note that the above equations do not involve any equations describing the material properties of the continuum. We would expect solid materials to act differently from liquid material when subjected to external forces. An equation or equations which describe the material properties are called constitutive equations. In the following sections we will investigate constitutive equations for solids and liquids. We will restrict our study to linear elastic materials over a range where there is a linear relationship between the stress and strain. We will not consider plastic or viscoelastic materials. Viscoelastic materials have the property that the stress is not only a function of strain but also a function of the rates of change of the stresses and strains and so properties of these types of materials are time dependent.

EXERCISE 2.3

▶ **1.** Assume an orthogonal coordinate system with metric tensor $g_{ij} = 0$ for $i \neq j$ and $g_{(i)(i)} = h_i^2$ (no summation on i). Use the definition of strain

$$e_{rs} = \frac{1}{2}\left(u_{r,s} + u_{s,r}\right) = \frac{1}{2}\left(g_{rt}u^t_{,s} + g_{st}u^t_{,r}\right)$$

and show that in terms of the physical components

$$e(ij) = \frac{e_{ij}}{h_i h_j} \quad \text{no summation on } i \text{ or } j$$

$$u(i) = h_i u^i \quad \text{no summation on } i$$

there results the equations:

$$e_{ii} = g_{it}\left[\frac{\partial u^t}{\partial x^i} + \left\{\begin{matrix} t \\ m\,i \end{matrix}\right\} u^m\right] \quad \text{no summation on } i$$

$$2e_{ij} = g_{it}\frac{\partial u^t}{\partial x^j} + g_{jt}\frac{\partial u^t}{\partial x^i}, \quad i \neq j$$

$$e(ii) = \frac{\partial}{\partial x^i}\left(\frac{u(i)}{h_i}\right) + \frac{1}{2h_i^2}\sum_{m=1}^{3}\frac{u(m)}{h_m}\frac{\partial}{\partial x^m}\left(h_i^2\right) \quad \text{no summation on } i$$

$$2e(ij) = \frac{h_i}{h_j}\frac{\partial}{\partial x^j}\left(\frac{u(i)}{h_i}\right) + \frac{h_j}{h_i}\frac{\partial}{\partial x^i}\left(\frac{u(j)}{h_j}\right), \quad \text{no summation on } i \text{ or } j, \ i \neq j.$$

▶ **2.** Use the results from problem 1 to write out all components of the strain tensor in Cartesian coordinates. Use the notation $u(1) = u$, $u(2) = v$, $u(3) = w$ and

$$e(11) = e_{xx}, \quad e(22) = e_{yy}, \quad e(33) = e_{zz}, \quad e(12) = e_{xy}, \quad e(13) = e_{xz}, \quad e(23) = e_{yz}$$

to verify the relations:

$$e_{xx} = \frac{\partial u}{\partial x} \qquad e_{xy} = \frac{1}{2}\left(\frac{\partial v}{\partial x} + \frac{\partial u}{\partial y}\right)$$

$$e_{yy} = \frac{\partial v}{\partial y} \qquad e_{xz} = \frac{1}{2}\left(\frac{\partial u}{\partial z} + \frac{\partial w}{\partial x}\right)$$

$$e_{zz} = \frac{\partial w}{\partial z} \qquad e_{zy} = \frac{1}{2}\left(\frac{\partial w}{\partial y} + \frac{\partial v}{\partial z}\right)$$

▶ **3.** Use the results from problem 1 to write out all components of the strain tensor in cylindrical coordinates. Use the notation $u(1) = u_r$, $u(2) = u_\theta$, $u(3) = u_z$ and

$$e(11) = e_{rr}, \quad e(22) = e_{\theta\theta}, \quad e(33) = e_{zz}, \quad e(12) = e_{r\theta}, \quad e(13) = e_{rz}, \quad e(23) = e_{\theta z}$$

to verify the relations:

$$e_{rr} = \frac{\partial u_r}{\partial r} \qquad e_{r\theta} = \frac{1}{2}\left(\frac{1}{r}\frac{\partial u_r}{\partial \theta} + \frac{\partial u_\theta}{\partial r} - \frac{u_\theta}{r}\right)$$

$$e_{\theta\theta} = \frac{1}{r}\frac{\partial u_\theta}{\partial \theta} + \frac{u_r}{r} \qquad e_{rz} = \frac{1}{2}\left(\frac{\partial u_z}{\partial r} + \frac{\partial u_r}{\partial z}\right)$$

$$e_{zz} = \frac{\partial u_z}{\partial z} \qquad e_{\theta z} = \frac{1}{2}\left(\frac{\partial u_\theta}{\partial z} + \frac{1}{r}\frac{\partial u_z}{\partial \theta}\right)$$

4. Use the results from problem 1 to write out all components of the strain tensor in spherical coordinates. Use the notation $u(1) = u_\rho, u(2) = u_\theta, u(3) = u_\phi$ and

$$e(11) = e_{\rho\rho}, \quad e(22) = e_{\theta\theta}, \quad e(33) = e_{\phi\phi}, \quad e(12) = e_{\rho\theta}, \quad e(13) = e_{\rho\phi}, \quad e(23) = e_{\theta\phi}$$

to verify the relations

$$e_{\rho\rho} = \frac{\partial u_\rho}{\partial \rho} \qquad\qquad e_{\rho\theta} = \frac{1}{2}\left(\frac{1}{\rho}\frac{\partial u_\rho}{\partial\theta} - \frac{u_\theta}{\rho} + \frac{\partial u_\theta}{\partial\rho}\right)$$

$$e_{00} = \frac{1}{\rho}\frac{\partial u_\theta}{\partial\theta} + \frac{u_\rho}{\rho} \qquad e_{\rho\phi} = \frac{1}{2}\left(\frac{1}{\rho\sin\theta}\frac{\partial u_\rho}{\partial\phi} - \frac{u_\phi}{\rho} + \frac{\partial u_\phi}{\partial\rho}\right)$$

$$e_{\phi\phi} = \frac{1}{\rho\sin\theta}\frac{\partial u_\phi}{\partial\phi} + \frac{u_\rho}{\rho} + \frac{u_\theta}{\rho}\cot\theta \qquad e_{\theta\phi} = \frac{1}{2}\left(\frac{1}{\rho}\frac{\partial u_\phi}{\partial\theta} - \frac{u_\phi}{\rho}\cot\theta + \frac{1}{\rho\sin\theta}\frac{\partial u_\theta}{\partial\phi}\right)$$

5. Expand equation (2.3.67) and find the dilatation in terms of the physical components of an orthogonal system and verify that

$$\Theta = \frac{1}{h_1 h_2 h_3}\left[\frac{\partial(h_2 h_3 u(1))}{\partial x^1} + \frac{\partial(h_1 h_3 u(2))}{\partial x^2} + \frac{\partial(h_1 h_2 u(3))}{\partial x^3}\right]$$

6. Verify that the dilatation in Cartesian coordinates is

$$\Theta = e_{xx} + e_{yy} + e_{zz} = \frac{\partial u}{\partial x} + \frac{\partial v}{\partial y} + \frac{\partial w}{\partial z}.$$

7. Verify that the dilatation in cylindrical coordinates is

$$\Theta = e_{rr} + e_{\theta\theta} + e_{zz} = \frac{\partial u_r}{\partial r} + \frac{1}{r}\frac{\partial u_\theta}{\partial\theta} + \frac{1}{r}u_r + \frac{\partial u_z}{\partial z}.$$

8. Verify that the dilatation in spherical coordinates is

$$\Theta = e_{\rho\rho} + e_{\theta\theta} + e_{\phi\phi} = \frac{\partial u_\rho}{\partial\rho} + \frac{1}{\rho}\frac{\partial u_\theta}{\partial\theta} + \frac{2}{\rho}u_\rho + \frac{1}{\rho\sin\theta}\frac{\partial u_\phi}{\partial\phi} + \frac{u_\theta\cot\theta}{\rho}.$$

9. Show that in an orthogonal set of coordinates the rotation tensor ω_{ij} can be written in terms of physical components in the form

$$\omega(ij) = \frac{1}{2h_i h_j}\left[\frac{\partial(h_i u(i))}{\partial x^j} - \frac{\partial(h_j u(j))}{\partial x^i}\right], \quad \text{no summations}$$

Hint: See problem 1.

10. Use the result from problem 9 to verify that in Cartesian coordinates

$$\omega_{yx} = \frac{1}{2}\left(\frac{\partial v}{\partial x} - \frac{\partial u}{\partial y}\right)$$

$$\omega_{xz} = \frac{1}{2}\left(\frac{\partial u}{\partial z} - \frac{\partial w}{\partial x}\right)$$

$$\omega_{zy} = \frac{1}{2}\left(\frac{\partial w}{\partial y} - \frac{\partial v}{\partial z}\right)$$

240

▶ **11.** Use the results from problem 9 to verify that in cylindrical coordinates

$$\omega_{\theta r} = \frac{1}{2r}\left[\frac{\partial(ru_\theta)}{\partial r} - \frac{\partial u_r}{\partial \theta}\right]$$

$$\omega_{rz} = \frac{1}{2}\left[\frac{\partial u_r}{\partial z} - \frac{\partial u_z}{\partial r}\right]$$

$$\omega_{z\theta} = \frac{1}{2}\left[\frac{1}{r}\frac{\partial u_z}{\partial \theta} - \frac{\partial u_\theta}{\partial z}\right]$$

▶ **12.** Use the results from problem 9 to verify that in spherical coordinates

$$\omega_{\theta\rho} = \frac{1}{2\rho}\left[\frac{\partial(\rho u_\theta)}{\partial \rho} - \frac{\partial u_\rho}{\partial \theta}\right]$$

$$\omega_{\rho\phi} = \frac{1}{2\rho}\left[\frac{1}{\sin\theta}\frac{\partial u_\rho}{\partial \phi} - \frac{\partial(\rho u_\phi)}{\partial \rho}\right]$$

$$\omega_{\phi\theta} = \frac{1}{2\rho\sin\theta}\left[\frac{\partial(u_\phi\sin\theta)}{\partial \theta} - \frac{\partial u_\theta}{\partial \phi}\right]$$

▶ **13.** The conditions for static equilibrium in a linear elastic material are determined from the conservation law

$$\sigma^j_{i\ ,j} + \varrho b_i = 0, \quad i,j = 1,2,3,$$

where σ^i_j are the stress tensor components, b_i are the external body forces per unit mass and ϱ is the density of the material. Assume an orthogonal coordinate system and verify the following results.

(a) Show that

$$\sigma^j_{i\ ,j} = \frac{1}{\sqrt{g}}\frac{\partial}{\partial x^j}(\sqrt{g}\sigma^j_i) - [ij,m]\sigma^{mj}$$

(b) Use the substitutions

$$\sigma(ij) = \sigma^j_i\frac{h_j}{h_i} \quad \text{no summation on } i \text{ or } j$$

$$b(i) = \frac{b_i}{h_i} \quad \text{no summation on } i$$

$$\sigma(ij) = \sigma^{ij}h_i h_j \quad \text{no summation on } i \text{ or } j$$

and express the equilibrium equations in terms of physical components and verify the relations

$$\sum_{j=1}^{3}\frac{1}{\sqrt{g}}\frac{\partial}{\partial x^j}\left(\frac{\sqrt{g}h_i\sigma(ij)}{h_j}\right) - \frac{1}{2}\sum_{j=1}^{3}\frac{\sigma(jj)}{h_j^2}\frac{\partial(h_j^2)}{\partial x^i} + h_i\varrho b(i) = 0,$$

where there is no summation on i.

▶ **14.** Use the results from problem 13 and verify that the equilibrium equations in Cartesian coordinates can be expressed

$$\frac{\partial\sigma_{xx}}{\partial x} + \frac{\partial\sigma_{xy}}{\partial y} + \frac{\partial\sigma_{xz}}{\partial z} + \varrho b_x = 0$$

$$\frac{\partial\sigma_{yx}}{\partial x} + \frac{\partial\sigma_{yy}}{\partial y} + \frac{\partial\sigma_{yz}}{\partial z} + \varrho b_y = 0$$

$$\frac{\partial\sigma_{zx}}{\partial x} + \frac{\partial\sigma_{zy}}{\partial y} + \frac{\partial\sigma_{zz}}{\partial z} + \varrho b_z = 0$$

15. Use the results from problem 13 and verify that the equilibrium equations in cylindrical coordinates can be expressed

$$\frac{\partial \sigma_{rr}}{\partial r} + \frac{1}{r}\frac{\partial \sigma_{r\theta}}{\partial \theta} + \frac{\partial \sigma_{rz}}{\partial z} + \frac{1}{r}(\sigma_{rr} - \sigma_{\theta\theta}) + \varrho b_r = 0$$

$$\frac{\partial \sigma_{\theta r}}{\partial r} + \frac{1}{r}\frac{\partial \sigma_{\theta\theta}}{\partial \theta} + \frac{\partial \sigma_{\theta z}}{\partial z} + \frac{2}{r}\sigma_{\theta r} + \varrho b_\theta = 0$$

$$\frac{\partial \sigma_{zr}}{\partial r} + \frac{1}{r}\frac{\partial \sigma_{z\theta}}{\partial \theta} + \frac{\partial \sigma_{zz}}{\partial z} + \frac{1}{r}\sigma_{zr} + \varrho b_z = 0$$

16. Use the results from problem 13 and verify that the equilibrium equations in spherical coordinates can be expressed

$$\frac{\partial \sigma_{\rho\rho}}{\partial \rho} + \frac{1}{\rho}\frac{\partial \sigma_{\rho\theta}}{\partial \theta} + \frac{1}{\rho\sin\theta}\frac{\partial \sigma_{\rho\phi}}{\partial \phi} + \frac{1}{\rho}(2\sigma_{\rho\rho} - \sigma_{\theta\theta} - \sigma_{\phi\phi} + \sigma_{\rho\theta}\cot\theta) + \varrho b_\rho = 0$$

$$\frac{\partial \sigma_{\theta\rho}}{\partial \rho} + \frac{1}{\rho}\frac{\partial \sigma_{\theta\theta}}{\partial \theta} + \frac{1}{\rho\sin\theta}\frac{\partial \sigma_{\theta\phi}}{\partial \phi} + \frac{1}{\rho}(3\sigma_{\rho\theta} + [\sigma_{\theta\theta} - \sigma_{\phi\phi}]\cot\theta) + \varrho b_\theta = 0$$

$$\frac{\partial \sigma_{\phi\rho}}{\partial \rho} + \frac{1}{\rho}\frac{\partial \sigma_{\phi\theta}}{\partial \theta} + \frac{1}{\rho\sin\theta}\frac{\partial \sigma_{\phi\phi}}{\partial \phi} + \frac{1}{\rho}(3\sigma_{\rho\phi} + 2\sigma_{\theta\phi}\cot\theta) + \varrho b_\phi = 0$$

17. Derive the result for the Lagrangian strain defined by the equation (2.3.60).

18. Derive the result for the Eulerian strain defined by equation (2.3.61).

19. The equation $\delta a^i = u^i_{,j} a^j$, describes the deformation in an elastic solid subjected to forces. The quantity δa^i denotes the difference vector $A^i - a^i$ between the undeformed and deformed states.

(a) Let $|a|$ denote the magnitude of the vector a^i and show that the strain e in the direction a^i can be represented

$$e = \frac{\delta|a|}{|a|} = e_{ij}\left(\frac{a^i}{|a|}\right)\left(\frac{a^j}{|a|}\right) = e_{ij}\lambda^i\lambda^j,$$

where λ^i is a unit vector in the direction a^i.

(b) Show that for $\lambda^1 = 1, \lambda^2 = 0, \lambda^3 = 0$ there results $e = e_{11}$, with similar results applying to vectors λ^i in the y and z directions.

Hint: Consider the magnitude squared $|a|^2 = g_{ij}a^i a^j$.

20. At the point $(1, 2, 3)$ of an elastic solid construct the small vector $\vec{a} = \epsilon(\frac{2}{3}\hat{e}_1 + \frac{2}{3}\hat{e}_2 + \frac{1}{3}\hat{e}_3)$, where $\epsilon > 0$ is a small positive quantity. The solid is subjected to forces such that the following displacement field results.

$$\vec{u} = (xy\,\hat{e}_1 + yz\,\hat{e}_2 + xz\,\hat{e}_3) \times 10^{-2}$$

Calculate the deformed vector $\vec{A}$ after the displacement field has been imposed.

21. For the displacement field

$$\vec{u} = (x^2 + yz)\,\hat{e}_1 + (xy + z^2)\,\hat{e}_2 + xyz\,\hat{e}_3$$

(a) Calculate the strain matrix at the point $(1, 2, 3)$.

(b) Calculate the rotation matrix at the point $(1, 2, 3)$.

▶ **22.** Show that for an orthogonal coordinate system the ith component of the convective operator can be written

$$[(\vec{V} \cdot \nabla)\, \vec{A}]_i = \sum_{m=1}^{3} \frac{V(m)}{h_m} \frac{\partial A(i)}{\partial x^m} + \sum_{\substack{m=1 \\ m \neq i}}^{3} \frac{A(m)}{h_m h_i} \left(V(i) \frac{\partial h_i}{\partial x^m} - V(m) \frac{\partial h_m}{\partial x^i} \right)$$

▶ **23.** Consider a parallelepiped with dimensions ℓ, w, h which has a uniform pressure P applied to each face. Show that the volume strain can be expressed as

$$\frac{\Delta V}{V} = \frac{\Delta \ell}{\ell} + \frac{\Delta w}{w} + \frac{\Delta h}{h} = \frac{-3P(1-2\nu)}{E}.$$

The quantity $k = E/3(1 - 2\nu)$ is called the bulk modulus of elasticity.

▶ **24.** Show in Cartesian coordinates the continuity equation is

$$\frac{\partial \varrho}{\partial t} + \frac{\partial(\varrho u)}{\partial x} + \frac{\partial(\varrho v)}{\partial y} + \frac{\partial(\varrho w)}{\partial z} = 0,$$

where (u, v, w) are the velocity components.

▶ **25.** Show in cylindrical coordinates the continuity equation is

$$\frac{\partial \varrho}{\partial t} + \frac{1}{r} \frac{\partial(r \varrho V_r)}{\partial r} + \frac{1}{r} \frac{\partial(\varrho V_\theta)}{\partial \theta} + \frac{\partial(\varrho V_z)}{\partial z} = 0$$

where V_r, V_θ, V_z are the velocity components.

▶ **26.** Show in spherical coordinates the continuity equation is

$$\frac{\partial \varrho}{\partial t} + \frac{1}{\rho^2} \frac{\partial(\rho^2 \varrho V_\rho)}{\partial \rho} + \frac{1}{\rho \sin\theta} \frac{\partial(\varrho V_\theta \sin\theta)}{\partial \theta} + \frac{1}{\rho \sin\theta} \frac{\partial(\varrho V_\phi)}{\partial \phi} = 0$$

where V_ρ, V_θ, V_ϕ are the velocity components.

▶ **27.** (a) Apply a stress σ_{yy} to both ends of a square element in a x, y continuum. Illustrate and label all changes that occur due to this stress. (b) Apply a stress σ_{xx} to both ends of a square element in a x, y continuum. Illustrate and label all changes that occur due to this stress. (c) Use superposition of your results in parts (a) and (b) and explain each term in the relations

$$e_{xx} = \frac{\sigma_{xx}}{E} - \nu \frac{\sigma_{yy}}{E} \qquad \text{and} \qquad e_{yy} = \frac{\sigma_{yy}}{E} - \nu \frac{\sigma_{xx}}{E}.$$

▶ **28.** Show that the time derivative of the Jacobian $J = J\left(\dfrac{x, y, z}{X, Y, Z}\right)$ satisfies $\dfrac{dJ}{dt} = J \operatorname{div} \vec{V}$ where

$$\operatorname{div} \vec{V} = \frac{\partial V_1}{\partial x} + \frac{\partial V_2}{\partial y} + \frac{\partial V_3}{\partial z} \quad \text{and} \quad V_1 = \frac{dx}{dt}, \quad V_2 = \frac{dy}{dt}, \quad V_3 = \frac{dz}{dt}.$$

Hint: Let $(x, y, z) = (x_1, x_2, x_3)$ and $(X, Y, Z) = (X_1, X_2, X_3)$, then note that

$$e_{ijk} \frac{\partial V_1}{\partial X_i} \frac{\partial x_2}{\partial X_j} \frac{\partial x_3}{\partial X_k} = e_{ijk} \frac{\partial V_1}{\partial x_m} \frac{\partial x_m}{\partial X_i} \frac{\partial x_2}{\partial X_j} \frac{\partial x_3}{\partial X_k} = e_{ijk} \frac{\partial x_1}{\partial X_i} \frac{\partial x_2}{\partial X_j} \frac{\partial x_3}{\partial X_k} \frac{\partial V_1}{\partial x_1}, \quad \text{etc.}$$

§2.4 CONTINUUM MECHANICS (SOLIDS)

In this introduction to continuum mechanics we consider the basic equations describing the physical effects created by external forces acting upon solids and fluids. In addition to the basic equations that are applicable to all continua, there are equations which are constructed to take into account material characteristics. These equations are called constitutive equations. For example, in the study of solids the constitutive equations for a linear elastic material is a set of relations between stress and strain. In the study of fluids, the constitutive equations consists of a set of relations between stress and rate of strain. Constitutive equations are usually constructed from some basic axioms. The resulting equations have unknown material parameters which can be determined from experimental investigations.

One of the basic axioms, used in the study of elastic solids, is that of material invariance. This axiom requires that certain symmetry conditions of solids are to remain invariant under a set of orthogonal transformations and translations. This axiom is employed in the next section to simplify the constitutive equations for elasticity. We begin our study of continuum mechanics by investigating the development of constitutive equations for linear elastic solids.

Generalized Hooke's Law

If the continuum material is a linear elastic material, we introduce the generalized Hooke's law in Cartesian coordinates

$$\sigma_{ij} = c_{ijkl}e_{kl}, \quad i,j,k,l = 1,2,3. \tag{2.4.1}$$

The Hooke's law is a statement that the stress is proportional to the gradient of the deformation occurring in the material. These equations assume a linear relationship exists between the components of the stress tensor and strain tensor and we say stress is a linear function of strain. Such relations are referred to as a set of constitutive equations. Constitutive equations serve to describe the material properties of the medium when it is subjected to external forces.

Constitutive Equations

The equations (2.4.1) are constitutive equations which are applicable for materials exhibiting small deformations when subjected to external forces. The 81 constants c_{ijkl} are called the elastic stiffness of the material. The above relations can also be expressed in the form

$$e_{ij} = s_{ijkl}\sigma_{kl}, \quad i,j,k,l = 1,2,3 \tag{2.4.2}$$

where s_{ijkl} are constants called the elastic compliance of the material. Since the stress σ_{ij} and strain e_{ij} have been shown to be tensors we can conclude that both the elastic stiffness c_{ijkl} and elastic compliance s_{ijkl} are fourth order tensors. Due to the symmetry of the stress and strain tensors we find that the elastic stiffness and elastic compliance tensor must satisfy the relations

$$c_{ijkl} = c_{jikl} = c_{ijlk} = c_{jilk}$$
$$s_{ijkl} = s_{jikl} = s_{ijlk} = s_{jilk} \tag{2.4.3}$$

and so only 36 of the 81 constants are actually independent. If all 36 of the material (crystal) constants are independent the material is called triclinic and there are no material symmetries.

244

Restrictions on Elastic Constants due to Symmetry

The equations (2.4.1) and (2.4.2) can be replaced by an equivalent set of equations which are easier to analyze. This is accomplished by defining the quantities

$$e_1, \quad e_2, \quad e_3, \quad e_4, \quad e_5, \quad e_6$$

$$\sigma_1, \quad \sigma_2, \quad \sigma_3, \quad \sigma_4, \quad \sigma_5, \quad \sigma_6$$

where

$$\begin{pmatrix} e_1 & e_4 & e_5 \\ e_4 & e_2 & e_6 \\ e_5 & e_6 & e_3 \end{pmatrix} = \begin{pmatrix} e_{11} & e_{12} & e_{13} \\ e_{21} & e_{22} & e_{23} \\ e_{31} & e_{32} & e_{33} \end{pmatrix}$$

and

$$\begin{pmatrix} \sigma_1 & \sigma_4 & \sigma_5 \\ \sigma_4 & \sigma_2 & \sigma_6 \\ \sigma_5 & \sigma_6 & \sigma_3 \end{pmatrix} = \begin{pmatrix} \sigma_{11} & \sigma_{12} & \sigma_{13} \\ \sigma_{21} & \sigma_{22} & \sigma_{23} \\ \sigma_{31} & \sigma_{32} & \sigma_{33} \end{pmatrix}.$$

Then the generalized Hooke's law from the equations (2.4.1) and (2.4.2) can be represented in either of the forms

$$\sigma_i = c_{ij}e_j \quad \text{or} \quad e_i = s_{ij}\sigma_j \quad \text{where} \quad i,j = 1,\ldots,6 \tag{2.4.4}$$

where c_{ij} are constants related to the elastic stiffness and s_{ij} are constants related to the elastic compliance. These constants satisfy the relation

$$s_{mi}c_{ij} = \delta_{mj} \quad \text{where} \quad i,m,j = 1,\ldots,6 \tag{2.4.5}$$

Here

$$e_{ij} = \begin{cases} e_i, & i = j = 1,2,3 \\ e_{1+i+j}, & i \neq j, \text{ and } i = 1, \text{ or, } 2 \end{cases}$$

and similarly

$$\sigma_{ij} = \begin{cases} \sigma_i, & i = j = 1,2,3 \\ \sigma_{1+i+j}, & i \neq j, \text{ and } i = 1, \text{ or, } 2. \end{cases}$$

These relations show that the constants c_{ij} are related to the elastic stiffness coefficients c_{pqrs} by the relations

$$c_{m1} = c_{ij11} \qquad c_{m4} = 2c_{ij12}$$

$$c_{m2} = c_{ij22} \qquad c_{m5} = 2c_{ij13}$$

$$c_{m3} = c_{ij33} \qquad c_{m6} = 2c_{ij23}$$

where

$$m = \begin{cases} i, & \text{if } i = j = 1,2, \text{ or } 3 \\ 1+i+j, & \text{if } i \neq j \text{ and } i = 1 \text{ or } 2. \end{cases}$$

A similar type relation holds for the constants s_{ij} and s_{pqrs}. The above relations can be verified by expanding the equations (2.4.1) and (2.4.2) and comparing like terms with the expanded form of the equation (2.4.4).

The generalized Hooke's law can now be expressed in a form where the 36 independent constants can be examined in more detail under special material symmetries. We will examine the form

$$
\begin{pmatrix} e_1 \\ e_2 \\ e_3 \\ e_4 \\ e_5 \\ e_6 \end{pmatrix} - \begin{pmatrix} s_{11} & s_{12} & s_{13} & s_{14} & s_{15} & s_{16} \\ s_{21} & s_{22} & s_{23} & s_{24} & s_{25} & s_{26} \\ s_{31} & s_{32} & s_{33} & s_{34} & s_{35} & s_{36} \\ s_{41} & s_{42} & s_{43} & s_{44} & s_{45} & s_{46} \\ s_{51} & s_{52} & s_{53} & s_{54} & s_{55} & s_{56} \\ s_{61} & s_{62} & s_{63} & s_{64} & s_{65} & s_{66} \end{pmatrix} \begin{pmatrix} \sigma_1 \\ \sigma_2 \\ \sigma_3 \\ \sigma_4 \\ \sigma_5 \\ \sigma_6 \end{pmatrix}.
\tag{2.4.6}
$$

Alternatively, in the arguments that follow, one can examine the equivalent form

$$
\begin{pmatrix} \sigma_1 \\ \sigma_2 \\ \sigma_3 \\ \sigma_4 \\ \sigma_5 \\ \sigma_6 \end{pmatrix} = \begin{pmatrix} c_{11} & c_{12} & c_{13} & c_{14} & c_{15} & c_{16} \\ c_{21} & c_{22} & c_{23} & c_{24} & c_{25} & c_{26} \\ c_{31} & c_{32} & c_{33} & c_{34} & c_{35} & c_{36} \\ c_{41} & c_{42} & c_{43} & c_{44} & c_{45} & c_{46} \\ c_{51} & c_{52} & c_{53} & c_{54} & c_{55} & c_{56} \\ c_{61} & c_{62} & c_{63} & c_{64} & c_{65} & c_{66} \end{pmatrix} \begin{pmatrix} e_1 \\ e_2 \\ e_3 \\ e_4 \\ e_5 \\ e_6 \end{pmatrix}.
$$

Material Symmetries

A material (crystal) with one plane of symmetry is called an aelotropic material. If we let the x_1-x_2 plane be a plane of symmetry then the equations (2.4.6) must remain invariant under the coordinate transformation

$$
\begin{pmatrix} \overline{x}_1 \\ \overline{x}_2 \\ \overline{x}_3 \end{pmatrix} = \begin{pmatrix} 1 & 0 & 0 \\ 0 & 1 & 0 \\ 0 & 0 & -1 \end{pmatrix} \begin{pmatrix} x_1 \\ x_2 \\ x_3 \end{pmatrix}
\tag{2.4.7}
$$

which represents an inversion of the x_3 axis. That is, if the x_1-x_2 plane is a plane of symmetry we should be able to replace x_3 by $-x_3$ and the equations (2.4.6) should remain unchanged. This is equivalent to saying that a transformation of the type from equation (2.4.7) changes the Hooke's law to the form $\overline{e}_i = s_{ij}\overline{\sigma}_j$ where the s_{ij} remain unaltered because it is the same material. Employing the transformation equations

$$
\overline{x}_1 = x_1, \qquad \overline{x}_2 = x_2, \qquad \overline{x}_3 = -x_3
\tag{2.4.8}
$$

we examine the stress and strain transformation equations

$$
\overline{\sigma}_{ij} = \sigma_{pq} \frac{\partial x_p}{\partial \overline{x}_i} \frac{\partial x_q}{\partial \overline{x}_j} \qquad \text{and} \qquad \overline{e}_{ij} = e_{pq} \frac{\partial x_p}{\partial \overline{x}_i} \frac{\partial x_q}{\partial \overline{x}_j}.
\tag{2.4.9}
$$

If we expand both of the equations (2.4.9) and substitute in the nonzero derivatives

$$
\frac{\partial x_1}{\partial \overline{x}_1} = 1, \qquad \frac{\partial x_2}{\partial \overline{x}_2} = 1, \qquad \frac{\partial x_3}{\partial \overline{x}_3} = -1,
\tag{2.4.10}
$$

we obtain the relations

$$
\begin{aligned}
\overline{\sigma}_{11} &= \sigma_{11} & \overline{e}_{11} &= e_{11} \\
\overline{\sigma}_{22} &= \sigma_{22} & \overline{e}_{22} &= e_{22} \\
\overline{\sigma}_{33} &= \sigma_{33} & \overline{e}_{33} &= e_{33} \\
\overline{\sigma}_{21} &= \sigma_{21} & \overline{e}_{21} &= e_{21} \\
\overline{\sigma}_{31} &= -\sigma_{31} & \overline{e}_{31} &= -e_{31} \\
\overline{\sigma}_{23} &= -\sigma_{23} & \overline{e}_{23} &= -e_{23}.
\end{aligned}
\tag{2.4.11}
$$

We conclude that if the material undergoes a strain, with the x_1-x_2 plane as a plane of symmetry then e_5 and e_6 change sign upon reversal of the x_3 axis and e_1, e_2, e_3, e_4 remain unchanged. Similarly, we find σ_5 and σ_6 change sign while $\sigma_1, \sigma_2, \sigma_3, \sigma_4$ remain unchanged. The equation (2.4.6) then becomes

$$
\begin{pmatrix} e_1 \\ e_2 \\ e_3 \\ e_4 \\ -e_5 \\ -e_6 \end{pmatrix} = \begin{pmatrix} s_{11} & s_{12} & s_{13} & s_{14} & s_{15} & s_{16} \\ s_{21} & s_{22} & s_{23} & s_{24} & s_{25} & s_{26} \\ s_{31} & s_{32} & s_{33} & s_{34} & s_{35} & s_{36} \\ s_{41} & s_{42} & s_{43} & s_{44} & s_{45} & s_{46} \\ s_{51} & s_{52} & s_{53} & s_{54} & s_{55} & s_{56} \\ s_{61} & s_{62} & s_{63} & s_{64} & s_{65} & s_{66} \end{pmatrix} \begin{pmatrix} \sigma_1 \\ \sigma_2 \\ \sigma_3 \\ \sigma_4 \\ -\sigma_5 \\ -\sigma_6 \end{pmatrix}. \tag{2.4.12}
$$

If the stress strain relation for the new orientation of the x_3 axis is to have the same form as the old orientation, then the equations (2.4.6) and (2.4.12) must give the same results. Comparison of these equations we find that

$$s_{15} = s_{16} = 0$$

$$s_{25} = s_{26} = 0$$

$$s_{35} = s_{36} = 0$$

$$s_{45} = s_{46} = 0 \tag{2.4.13}$$

$$s_{51} = s_{52} = s_{53} = s_{54} = 0$$

$$s_{61} = s_{62} = s_{63} = s_{64} = 0.$$

In summary, from an examination of the equations (2.4.6) and (2.4.12) we find that for an aelotropic material (crystal), with one plane of symmetry, the 36 constants s_{ij} reduce to 20 constants and the generalized Hooke's law (constitutive equation) has the form

$$
\begin{pmatrix} e_1 \\ e_2 \\ e_3 \\ e_4 \\ e_5 \\ e_6 \end{pmatrix} = \begin{pmatrix} s_{11} & s_{12} & s_{13} & s_{14} & 0 & 0 \\ s_{21} & s_{22} & s_{23} & s_{24} & 0 & 0 \\ s_{31} & s_{32} & s_{33} & s_{34} & 0 & 0 \\ s_{41} & s_{42} & s_{43} & s_{44} & 0 & 0 \\ 0 & 0 & 0 & 0 & s_{55} & s_{56} \\ 0 & 0 & 0 & 0 & s_{65} & s_{66} \end{pmatrix} \begin{pmatrix} \sigma_1 \\ \sigma_2 \\ \sigma_3 \\ \sigma_4 \\ \sigma_5 \\ \sigma_6 \end{pmatrix}. \tag{2.4.14}
$$

Alternatively, the Hooke's law can be represented in the form

$$
\begin{pmatrix} \sigma_1 \\ \sigma_2 \\ \sigma_3 \\ \sigma_4 \\ \sigma_5 \\ \sigma_6 \end{pmatrix} = \begin{pmatrix} c_{11} & c_{12} & c_{13} & c_{14} & 0 & 0 \\ c_{21} & c_{22} & c_{23} & c_{24} & 0 & 0 \\ c_{31} & c_{32} & c_{33} & c_{34} & 0 & 0 \\ c_{41} & c_{42} & c_{43} & c_{44} & 0 & 0 \\ 0 & 0 & 0 & 0 & c_{55} & c_{56} \\ 0 & 0 & 0 & 0 & c_{65} & c_{66} \end{pmatrix} \begin{pmatrix} e_1 \\ e_2 \\ e_3 \\ e_4 \\ e_5 \\ e_6 \end{pmatrix}.
$$

Additional Symmetries

If the material (crystal) is such that there is an additional plane of symmetry, say the x_2-x_3 plane, then reversal of the x_1 axis should leave the equations (2.4.14) unaltered. If there are two planes of symmetry then there will automatically be a third plane of symmetry. Such a material (crystal) is called orthotropic. Introducing the additional transformation

$$\overline{x}_1 = -x_1, \qquad \overline{x}_2 = x_2, \qquad \overline{x}_3 = x_3$$

which represents the reversal of the x_1 axes, the expanded form of equations (2.4.9) are used to calculate the effect of such a transformation upon the stress and strain tensor. We find $\sigma_1, \sigma_2, \sigma_3, \sigma_6, e_1, e_2, e_3, e_6$ remain unchanged while $\sigma_4, \sigma_5, e_4, e_5$ change sign. The equation (2.4.14) then becomes

$$\begin{pmatrix} e_1 \\ e_2 \\ e_3 \\ -e_4 \\ -e_5 \\ e_6 \end{pmatrix} = \begin{pmatrix} s_{11} & s_{12} & s_{13} & s_{14} & 0 & 0 \\ s_{21} & s_{22} & s_{23} & s_{24} & 0 & 0 \\ s_{31} & s_{32} & s_{33} & s_{34} & 0 & 0 \\ s_{41} & s_{42} & s_{43} & s_{44} & 0 & 0 \\ 0 & 0 & 0 & 0 & s_{55} & s_{56} \\ 0 & 0 & 0 & 0 & s_{65} & s_{66} \end{pmatrix} \begin{pmatrix} \sigma_1 \\ \sigma_2 \\ \sigma_3 \\ -\sigma_4 \\ -\sigma_5 \\ \sigma_6 \end{pmatrix}. \qquad (2.4.15)$$

Note that if the constitutive equations (2.4.14) and (2.4.15) are to produce the same results upon reversal of the x_1 axes, then we require that the following coefficients be equated to zero:

$$s_{14} = s_{24} = s_{34} = 0$$
$$s_{41} = s_{42} = s_{43} = 0$$
$$s_{56} = s_{65} = 0.$$

This then produces the constitutive equation

$$\begin{pmatrix} e_1 \\ e_2 \\ e_3 \\ e_4 \\ e_5 \\ e_6 \end{pmatrix} = \begin{pmatrix} s_{11} & s_{12} & s_{13} & 0 & 0 & 0 \\ s_{21} & s_{22} & s_{23} & 0 & 0 & 0 \\ s_{31} & s_{32} & s_{33} & 0 & 0 & 0 \\ 0 & 0 & 0 & s_{44} & 0 & 0 \\ 0 & 0 & 0 & 0 & s_{55} & 0 \\ 0 & 0 & 0 & 0 & 0 & s_{66} \end{pmatrix} \begin{pmatrix} \sigma_1 \\ \sigma_2 \\ \sigma_3 \\ \sigma_4 \\ \sigma_5 \\ \sigma_6 \end{pmatrix} \qquad (2.4.16)$$

or its equivalent form

$$\begin{pmatrix} \sigma_1 \\ \sigma_2 \\ \sigma_3 \\ \sigma_4 \\ \sigma_5 \\ \sigma_6 \end{pmatrix} = \begin{pmatrix} c_{11} & c_{12} & c_{13} & 0 & 0 & 0 \\ c_{21} & c_{22} & c_{23} & 0 & 0 & 0 \\ c_{31} & c_{32} & c_{33} & 0 & 0 & 0 \\ 0 & 0 & 0 & c_{44} & 0 & 0 \\ 0 & 0 & 0 & 0 & c_{55} & 0 \\ 0 & 0 & 0 & 0 & 0 & c_{66} \end{pmatrix} \begin{pmatrix} e_1 \\ e_2 \\ e_3 \\ e_4 \\ e_5 \\ e_6 \end{pmatrix}$$

and the original 36 constants have been reduced to 12 constants. This is the constitutive equation for orthotropic material (crystals).

Axis of Symmetry

If in addition to three planes of symmetry there is an axis of symmetry then the material (crystal) is termed hexagonal. Assume that the x^1 axis is an axis of symmetry and consider the effect of the transformation

$$\bar{x}^1 = x^1, \qquad \bar{x}^2 = x^3 \qquad \bar{x}^3 = -x^2$$

upon the constitutive equations. It is left as an exercise to verify that the constitutive equations reduce to the form where there are 7 independent constants having either of the forms

$$
\begin{pmatrix} e_1 \\ e_2 \\ e_3 \\ e_4 \\ e_5 \\ e_6 \end{pmatrix}
=
\begin{pmatrix}
s_{11} & s_{12} & s_{12} & 0 & 0 & 0 \\
s_{21} & s_{22} & s_{23} & 0 & 0 & 0 \\
s_{21} & s_{23} & s_{22} & 0 & 0 & 0 \\
0 & 0 & 0 & s_{44} & 0 & 0 \\
0 & 0 & 0 & 0 & s_{44} & 0 \\
0 & 0 & 0 & 0 & 0 & s_{66}
\end{pmatrix}
\begin{pmatrix} \sigma_1 \\ \sigma_2 \\ \sigma_3 \\ \sigma_4 \\ \sigma_5 \\ \sigma_6 \end{pmatrix}
$$

or

$$
\begin{pmatrix} \sigma_1 \\ \sigma_2 \\ \sigma_3 \\ \sigma_4 \\ \sigma_5 \\ \sigma_6 \end{pmatrix}
=
\begin{pmatrix}
c_{11} & c_{12} & c_{12} & 0 & 0 & 0 \\
c_{21} & c_{22} & c_{23} & 0 & 0 & 0 \\
c_{21} & c_{23} & c_{22} & 0 & 0 & 0 \\
0 & 0 & 0 & c_{44} & 0 & 0 \\
0 & 0 & 0 & 0 & c_{44} & 0 \\
0 & 0 & 0 & 0 & 0 & c_{66}
\end{pmatrix}
\begin{pmatrix} e_1 \\ e_2 \\ e_3 \\ e_4 \\ e_5 \\ e_6 \end{pmatrix} .
$$

Finally, if the material is completely symmetric, the x^2 axis is also an axis of symmetry and we can consider the effect of the transformation

$$\bar{x}^1 = -x^3, \qquad \bar{x}^2 = x^2, \qquad \bar{x}^3 = x^1$$

upon the constitutive equations. It can be verified that these transformations reduce the Hooke's law constitutive equation to the form

$$
\begin{pmatrix} e_1 \\ e_2 \\ e_3 \\ e_4 \\ e_5 \\ e_6 \end{pmatrix}
=
\begin{pmatrix}
s_{11} & s_{12} & s_{12} & 0 & 0 & 0 \\
s_{12} & s_{11} & s_{12} & 0 & 0 & 0 \\
s_{12} & s_{12} & s_{11} & 0 & 0 & 0 \\
0 & 0 & 0 & s_{44} & 0 & 0 \\
0 & 0 & 0 & 0 & s_{44} & 0 \\
0 & 0 & 0 & 0 & 0 & s_{44}
\end{pmatrix}
\begin{pmatrix} \sigma_1 \\ \sigma_2 \\ \sigma_3 \\ \sigma_4 \\ \sigma_5 \\ \sigma_6 \end{pmatrix} .
\qquad (2.4.17)
$$

Materials (crystals) with atomic arrangements that exhibit the above symmetries are called isotropic materials. An equivalent form of (2.4.17) is the relation

$$
\begin{pmatrix} \sigma_1 \\ \sigma_2 \\ \sigma_3 \\ \sigma_4 \\ \sigma_5 \\ \sigma_6 \end{pmatrix}
=
\begin{pmatrix}
c_{11} & c_{12} & c_{12} & 0 & 0 & 0 \\
c_{12} & c_{11} & c_{12} & 0 & 0 & 0 \\
c_{12} & c_{12} & c_{11} & 0 & 0 & 0 \\
0 & 0 & 0 & c_{44} & 0 & 0 \\
0 & 0 & 0 & 0 & c_{44} & 0 \\
0 & 0 & 0 & 0 & 0 & c_{44}
\end{pmatrix}
\begin{pmatrix} e_1 \\ e_2 \\ e_3 \\ e_4 \\ e_5 \\ e_6 \end{pmatrix} .
$$

The figure 2.4-1 lists values for the elastic stiffness associated with some metals which are isotropic[1]

---

[1]Additional constants are given in "International Tables of Selected Constants", Metals: Thermal and Mechanical Data, Vol. 16, Edited by S. Allard, Pergamon Press, 1969.

Metal	c_{11}	c_{12}	c_{44}
Na	0.074	0.062	0.042
Pb	0.495	0.423	0.149
Cu	1.684	1.214	0.754
Ni	2.508	1.500	1.235
Cr	3.500	0.678	1.008
Mo	4.630	1.610	1.090
W	5.233	2.045	1.607

Figure 2.4-1. Elastic stiffness coefficients for some metals which are cubic.
Constants are given in units of $10^{12}\,dynes/cm^2$

Under these conditions the stress strain constitutive relations can be written as

$$\sigma_1 = \sigma_{11} = (c_{11} - c_{12})e_{11} + c_{12}(e_{11} + e_{22} + e_{33})$$
$$\sigma_2 = \sigma_{22} = (c_{11} - c_{12})e_{22} + c_{12}(e_{11} + e_{22} + e_{33})$$
$$\sigma_3 = \sigma_{33} = (c_{11} - c_{12})e_{33} + c_{12}(e_{11} + e_{22} + e_{33})$$
$$\sigma_4 = \sigma_{12} = c_{44}e_{12}$$
$$\sigma_5 = \sigma_{13} = c_{44}e_{13}$$
$$\sigma_6 = \sigma_{23} = c_{44}e_{23}.$$

(2.4.18)

Isotropic Material

Materials (crystals) which are elastically the same in all directions are called isotropic. We have shown that for a cubic material which exhibits symmetry with respect to all axes and planes, the constitutive stress-strain relation reduces to the form found in equation (2.4.17). Define the quantities

$$s_{11} = \frac{1}{E}, \qquad s_{12} = -\frac{\nu}{E}, \qquad s_{44} = \frac{1}{2\mu}$$

where E is the Young's Modulus of elasticity, ν is the Poisson's ratio, and μ is the shear or rigidity modulus. For isotropic materials the three constants E, ν, μ are not independent as the following example demonstrates.

EXAMPLE 2.4-1. (Elastic constants) For an isotropic material, consider a cross section of material in the x^1-x^2 plane which is subjected to pure shearing so that $\sigma_4 = \sigma_{12}$ is the only nonzero stress as illustrated in the figure 2.4-2.

For the above conditions, the equation (2.4.17) reduces to the single equation

$$e_4 = e_{12} = s_{44}\sigma_4 = s_{44}\sigma_{12} \qquad or \qquad \mu = \frac{\sigma_{12}}{\gamma_{12}}$$

and so the shear modulus is the ratio of the shear stress to the shear angle. Now rotate the axes through a 45 degree angle to a barred system of coordinates where

$$x^1 = \bar{x}^1 \cos\alpha - \bar{x}^2 \sin\alpha \quad x^2 = \bar{x}^1 \sin\alpha + \bar{x}^2 \cos\alpha$$

250

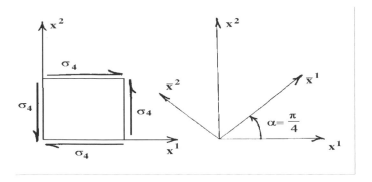

Figure 2.4-2. Element subjected to pure shearing

where $\alpha = \frac{\pi}{4}$. Expanding the transformation equations (2.4.9) we find that

$$\overline{\sigma}_1 = \overline{\sigma}_{11} = \cos\alpha \sin\alpha\, \sigma_{12} + \sin\alpha \cos\alpha\, \sigma_{21} = \sigma_{12} = \sigma_4$$

$$\overline{\sigma}_2 = \overline{\sigma}_{22} = -\sin\alpha \cos\alpha\, \sigma_{12} - \sin\alpha \cos\alpha\, \sigma_{21} = -\sigma_{12} = -\sigma_4,$$

and similarly

$$\overline{e}_1 = \overline{e}_{11} = e_4, \qquad \overline{e}_2 = \overline{e}_{22} = -e_4.$$

In the barred system, the Hooke's law becomes

$$\overline{e}_1 = s_{11}\overline{\sigma}_1 + s_{12}\overline{\sigma}_2 \qquad \text{or}$$

$$e_4 = s_{11}\sigma_4 - s_{12}\sigma_4 = s_{44}\sigma_4.$$

Hence, the constants s_{11}, s_{12}, s_{44} are related by the relation

$$s_{11} - s_{12} = s_{44} \qquad \text{or} \qquad \frac{1}{E} + \frac{\nu}{E} = \frac{1}{2\mu}. \tag{2.4.19}$$

This is an important relation connecting the elastic constants associated with isotropic materials. The above transformation can also be applied to triclinic, aelotropic, orthotropic, and hexagonal materials to find relationships between the elastic constants.

Observe also that some texts postulate the existence of a strain energy function U^* which has the property that $\sigma_{ij} = \frac{\partial U^*}{\partial e_{ij}}$. In this case the strain energy function, in the single index notation, is written $U^* = c_{ij}e_i e_j$ where c_{ij} and consequently s_{ij} are symmetric. In this case the previous discussed symmetries give the following results for the nonzero elastic compliances s_{ij} : 13 nonzero constants instead of 20 for aelotropic material, 9 nonzero constants instead of 12 for orthotropic material, and 6 nonzero constants instead of 7 for hexagonal material. This is because of the additional property that $s_{ij} = s_{ji}$ be symmetric.

■

The previous discussion has shown that for an isotropic material the generalized Hooke's law (constitutive equations) have the form

$$e_{11} = \frac{1}{E}\left[\sigma_{11} - \nu(\sigma_{22} + \sigma_{33})\right]$$

$$e_{22} = \frac{1}{E}\left[\sigma_{22} - \nu(\sigma_{33} + \sigma_{11})\right]$$

$$e_{33} = \frac{1}{E}\left[\sigma_{33} - \nu(\sigma_{11} + \sigma_{22})\right]$$

$$e_{21} = e_{12} = \frac{1+\nu}{E}\sigma_{12}$$

$$e_{32} = e_{23} = \frac{1+\nu}{E}\sigma_{23} \qquad (2.4.20)$$

$$e_{31} = e_{13} = \frac{1+\nu}{E}\sigma_{13}$$

where equation (2.4.19) holds. These equations can be expressed in the indicial notation and have the form

$$e_{ij} = \frac{1+\nu}{E}\sigma_{ij} - \frac{\nu}{E}\sigma_{kk}\delta_{ij}, \qquad (2.4.21)$$

where $\sigma_{kk} = \sigma_{11} + \sigma_{22} + \sigma_{33}$ is a stress invariant and δ_{ij} is the Kronecker delta. We can solve for the stress in terms of the strain by performing a contraction on i and j in equation (2.4.21). This gives the dilatation

$$e_{ii} = \frac{1+\nu}{E}\sigma_{ii} - \frac{3\nu}{E}\sigma_{kk} = \frac{1-2\nu}{E}\sigma_{kk}.$$

Note that from the result in equation (2.4.21) we are now able to solve for the stress in terms of the strain. We find

$$e_{ij} = \frac{1+\nu}{E}\sigma_{ij} - \frac{\nu}{1-2\nu}e_{kk}\delta_{ij}$$

$$\frac{E}{1+\nu}e_{ij} = \sigma_{ij} - \frac{\nu E}{(1+\nu)(1-2\nu)}e_{kk}\delta_{ij} \qquad (2.4.22)$$

$$\text{or} \quad \sigma_{ij} = \frac{E}{1+\nu}e_{ij} + \frac{\nu E}{(1+\nu)(1-2\nu)}e_{kk}\delta_{ij}.$$

The tensor equation (2.4.22) represents the six scalar equations

$$\sigma_{11} = \frac{E}{(1+\nu)(1-2\nu)}\left[(1-\nu)e_{11} + \nu(e_{22}+e_{33})\right] \qquad \sigma_{12} = \frac{E}{1+\nu}e_{12}$$

$$\sigma_{22} = \frac{E}{(1+\nu)(1-2\nu)}\left[(1-\nu)e_{22} + \nu(e_{33}+e_{11})\right] \qquad \sigma_{13} = \frac{E}{1+\nu}e_{13}$$

$$\sigma_{33} = \frac{E}{(1+\nu)(1-2\nu)}\left[(1-\nu)e_{33} + \nu(e_{22}+e_{11})\right] \qquad \sigma_{23} = \frac{E}{1+\nu}e_{23}.$$

Alternative Approach to Constitutive Equations

The constitutive equation defined by Hooke's generalized law for isotropic materials can be approached from another point of view. Consider the generalized Hooke's law

$$\sigma_{ij} = c_{ijkl}e_{kl}, \qquad i,j,k,l = 1,2,3.$$

If we transform to a barred system of coordinates, we will have the new Hooke's law

$$\overline{\sigma}_{ij} = \overline{c}_{ijkl}\overline{e}_{kl}, \qquad i,j,k,l = 1,2,3.$$

For an isotropic material we require that

$$\overline{c}_{ijkl} = c_{ijkl}.$$

Tensors whose components are the same in all coordinate systems are called isotropic tensors. We have previously shown in Exercise 1.3, problem 18, that

$$c_{pqrs} = \lambda\delta_{pq}\delta_{rs} + \mu(\delta_{pr}\delta_{qs} + \delta_{ps}\delta_{qr}) + \kappa(\delta_{pr}\delta_{qs} - \delta_{ps}\delta_{qr})$$

is an isotropic tensor when we consider affine type transformations. If we further require the symmetry conditions found in equations (2.4.3) be satisfied, we find that $\kappa = 0$ so that the generalized Hooke's law takes on the form

$$\sigma_{pq} = c_{pqrs}e_{rs} = \left[\lambda\delta_{pq}\delta_{rs} + \mu(\delta_{pr}\delta_{qs} + \delta_{ps}\delta_{qr})\right]e_{rs}$$

$$\sigma_{pq} = \lambda\delta_{pq}e_{rr} + \mu(e_{pq} + e_{qp}) \qquad (2.4.23)$$

$$\text{or} \qquad \sigma_{pq} = 2\mu e_{pq} + \lambda e_{rr}\delta_{pq},$$

where $e_{rr} = e_{11} + e_{22} + e_{33} = \Theta$ is the dilatation. The constants λ and μ are called Lame's constants. Comparing the equation (2.4.22) with equation (2.4.23) we find that the constants λ and μ satisfy the relations

$$\mu = \frac{E}{2(1+\nu)} \qquad \lambda = \frac{\nu E}{(1+\nu)(1-2\nu)}. \qquad (2.4.24)$$

In addition to the constants E, ν, μ, λ, it is sometimes convenient to introduce the constant k, called the bulk modulus of elasticity, (Exercise 2.3, problem 23), defined by

$$k = \frac{E}{3(1-2\nu)}. \qquad (2.4.25)$$

The stress-strain constitutive equation (2.4.23) was derived using Cartesian tensors. To generalize the equation (2.4.23) we consider a transformation from a Cartesian coordinate system y^i, $i = 1,2,3$ to a general coordinate system $\overline{x}^i$, $i = 1,2,3$. We employ the relations

$$\overline{g}_{ij} = \frac{\partial y^m}{\partial \overline{x}^i}\frac{\partial y^m}{\partial \overline{x}^j}, \qquad \overline{g}^{ij} = \frac{\partial \overline{x}^i}{\partial y^m}\frac{\partial \overline{x}^j}{\partial y^m}$$

and

$$\overline{\sigma}_{mn} = \sigma_{ij}\frac{\partial y^i}{\partial \overline{x}^m}\frac{\partial y^j}{\partial \overline{x}^n}, \qquad \overline{e}_{mn} = e_{ij}\frac{\partial y^i}{\partial \overline{x}^m}\frac{\partial y^j}{\partial \overline{x}^n}, \qquad \text{or} \qquad e_{rq} = \overline{e}_{ij}\frac{\partial \overline{x}^i}{\partial y^r}\frac{\partial \overline{x}^j}{\partial y^q}$$

and convert equation (2.4.23) to a more generalized form. Multiply equation (2.4.23) by $\frac{\partial y^p}{\partial \overline{x}^m}\frac{\partial y^q}{\partial \overline{x}^n}$ and verify the result

$$\overline{\sigma}_{mn} = \lambda \frac{\partial y^q}{\partial \overline{x}^m}\frac{\partial y^q}{\partial \overline{x}^n} e_{rr} + \mu\left(\overline{e}_{mn} + \overline{e}_{nm}\right),$$

which can be simplified to the form

$$\overline{\sigma}_{mn} = \lambda \overline{g}_{mn}\overline{e}_{ij}\overline{g}^{ij} + \mu\left(\overline{e}_{mn} + \overline{e}_{nm}\right).$$

Dropping the bar notation, we have

$$\sigma_{mn} = \lambda g_{mn}g^{ij}e_{ij} + \mu\left(e_{mn} + e_{nm}\right).$$

The contravariant form of this equation is

$$\sigma^{sr} = \lambda g^{sr}g^{ij}e_{ij} + \mu\left(g^{ms}g^{nr} + g^{ns}g^{mr}\right)e_{mn}.$$

Employing the equations (2.4.24) the above result can also be expressed in the form

$$\sigma^{rs} = \frac{E}{2(1+\nu)}\left(g^{ms}g^{nr} + g^{ns}g^{mr} + \frac{2\nu}{1-2\nu}g^{sr}g^{mn}\right)e_{mn}. \tag{2.4.26}$$

This is a more general form for the stress-strain constitutive equations which is valid in all coordinate systems. Multiplying by g_{sk} and employing the use of associative tensors, one can verify

$$\sigma^i_j = \frac{E}{1+\nu}\left(e^i_j + \frac{\nu}{1-2\nu}e^m_m\delta^i_j\right)$$

or $\qquad \sigma^i_j = 2\mu e^i_j + \lambda e^m_m\delta^i_j,$

are alternate forms for the equation (2.4.26). As an exercise, solve for the strains in terms of the stresses and show that

$$Ee^i_j = (1+\nu)\sigma^i_j - \nu\sigma^m_m\delta^i_j.$$

EXAMPLE 2.4-2. (Hooke's law) Let us construct a simple example to test the results we have developed so far. Consider the tension in a cylindrical bar illustrated in the figure 2.4-3.

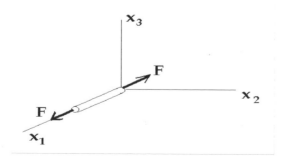

Figure 2.4-3. Stress in a cylindrical bar

Assume that

$$\sigma_{ij} = \begin{pmatrix} \frac{F}{A} & 0 & 0 \\ 0 & 0 & 0 \\ 0 & 0 & 0 \end{pmatrix}$$

where F is the constant applied force and A is the cross sectional area of the cylinder. The generalized Hooke's law (2.4.21) produces the nonzero strains

$$e_{11} = \frac{1+\nu}{E}\sigma_{11} - \frac{\nu}{E}(\sigma_{11} + \sigma_{22} + \sigma_{33}) = \frac{\sigma_{11}}{E}$$

$$e_{22} = \frac{-\nu}{E}\sigma_{11}$$

$$e_{33} = \frac{-\nu}{E}\sigma_{11}$$

From these equations we obtain:

The first part of Hooke's law

$$\sigma_{11} = E e_{11} \text{ or } \frac{F}{A} = E e_{11}.$$

The second part of Hooke's law

$$\frac{\text{lateral contraction}}{\text{longitudinal extension}} = \frac{-e_{22}}{e_{11}} = \frac{-e_{33}}{e_{11}} = \nu = \text{Poisson's ratio}.$$

This example demonstrates that the generalized Hooke's law for homogeneous and isotropic materials reduces to our previous one dimensional result given in (2.3.1) and (2.3.2).

$\blacksquare$

Basic Equations of Elasticity

Assuming the density ϱ is constant, the basic equations of elasticity reduce to the equations representing conservation of linear momentum and angular momentum together with the strain-displacement relations and constitutive equations. In these equations the body forces are assumed known. These basic equations produce 15 equations in 15 unknowns and are a formidable set of equations to solve. Methods for solving these simultaneous equations are: 1) Express the linear momentum equations in terms of the displacements u_i and obtain a system of partial differential equations. Solve the system of partial differential equations for the displacements u_i and then calculate the corresponding strains. The strains can be used to calculate the stresses from the constitutive equations. 2) Solve for the stresses and from the stresses calculate the strains and from the strains calculate the displacements. This converse problem requires some additional considerations which will be addressed shortly.

Basic Equations of Linear Elasticity

- Conservation of linear momentum.

$$\sigma^{ij}_{,i} + \varrho b^j = \varrho f^j \qquad j = 1, 2, 3. \qquad (2.4.27(a))$$

where σ^{ij} is the stress tensor, b^j is the body force per unit mass and f^j is the acceleration. If there is no motion, then $f^j = 0$ and these equations reduce to the equilibrium equations

$$\sigma^{ij}_{,i} + \varrho b^j = 0 \qquad j = 1, 2, 3. \qquad (2.4.27(b))$$

- Conservation of angular momentum. $\qquad \sigma_{ij} = \sigma_{ji}$
- Strain tensor.

$$e_{ij} = \frac{1}{2}(u_{i,j} + u_{j,i}) \qquad (2.4.28)$$

where u_i denotes the displacement field.

- Constitutive equation. For a linear elastic isotropic material we have

$$\sigma^i_j = \frac{E}{1+\nu}e^i_j + \frac{E\nu}{(1+\nu)(1-2\nu)}e^k_k\delta^i_j \qquad i, j = 1, 2, 3 \qquad (2.4.29(a))$$

or its equivalent form

$$\sigma^i_j = 2\mu e^i_j + \lambda e^r_r\delta^i_j \qquad i, j = 1, 2, 3, \qquad (2.4.29(b))$$

where e^r_r is the dilatation. This produces 15 equations for the 15 unknowns

$$u_1, u_2, u_3, \sigma_{11}, \sigma_{12}, \sigma_{13}, \sigma_{22}, \sigma_{23}, \sigma_{33}, e_{11}, e_{12}, e_{13}, e_{22}, e_{23}, e_{33},$$

which represents 3 displacements, 6 strains and 6 stresses. In the above equations it is assumed that the body forces are known.

Navier's Equations

The equations (2.4.27) through (2.4.29) can be combined and written as one set of equations. The resulting equations are known as Navier's equations for the displacements u_i over the range $i = 1, 2, 3$. To derive the Navier's equations in Cartesian coordinates, we write the equations (2.4.27),(2.4.28) and (2.4.29) in Cartesian coordinates. We then calculate $\sigma_{ij,j}$ in terms of the displacements u_i and substitute the results into the momentum equation (2.4.27(a)). Differentiation of the constitutive equations (2.4.29(b)) produces

$$\sigma_{ij,j} = 2\mu e_{ij,j} + \lambda e_{kk,j}\delta_{ij}. \qquad (2.4.30)$$

A contraction of the strain produces the dilatation

$$e_{rr} = \frac{1}{2}\left(u_{r,r} + u_{r,r}\right) = u_{r,r} \tag{2.4.31}$$

From the dilatation we calculate the covariant derivative

$$e_{kk,j} = u_{k,kj}. \tag{2.4.32}$$

Employing the strain relation from equation (2.4.28), we calculate the covariant derivative

$$e_{ij,j} = \frac{1}{2}(u_{i,jj} + u_{j,ij}). \tag{2.4.33}$$

These results allow us to express the covariant derivative of the stress in terms of the displacement field. We find

$$\sigma_{ij,j} = \mu\left[u_{i,jj} + u_{j,ij}\right] + \lambda\delta_{ij}u_{k,kj}$$
$$\text{or} \quad \sigma_{ij,j} = (\lambda + \mu)u_{k,ki} + \mu u_{i,jj}. \tag{2.4.34}$$

Substituting equation (2.4.34) into the linear momentum equation produces the Navier equations:

$$(\lambda + \mu)u_{k,ki} + \mu u_{i,jj} + \varrho b_i = \varrho f_i, \qquad i = 1, 2, 3.$$

In vector form these equations can be expressed

$$(\lambda + \mu)\nabla\left(\nabla \cdot \vec{u}\right) + \mu\nabla^2\vec{u} + \varrho\vec{b} = \varrho\vec{f},$$

where $\vec{u}$ is the displacement vector, $\vec{b}$ is the body force per unit mass and $\vec{f}$ is the acceleration. In Cartesian coordinates these equations have the form:

$$(\lambda + \mu)\left(\frac{\partial^2 u_1}{\partial x_1 \partial x_i} + \frac{\partial^2 u_2}{\partial x_2 \partial x_i} + \frac{\partial^2 u_3}{\partial x_3 \partial x_i}\right) + \mu\nabla^2 u_i + \varrho b_i = \varrho\frac{\partial^2 u_i}{\partial t^2},$$

for $i = 1, 2, 3$, where

$$\nabla^2 u_i = \frac{\partial^2 u_i}{\partial x_1^2} + \frac{\partial^2 u_i}{\partial x_2^2} + \frac{\partial^2 u_i}{\partial x_3^2}.$$

The Navier equations must be satisfied by a set of functions $u_i = u_i(x_1, x_2, x_3)$ which represent the displacement at each point inside some prescribed region R. Knowing the displacement field we can calculate the strain field directly using the equation (2.4.28). Knowledge of the strain field enables us to construct the corresponding stress field from the constitutive equations.

In the absence of body forces, such as gravity, the Navier equation of elastodynamics can be written in the vector form

$$\varrho\frac{\partial^2\vec{u}}{\partial t^2} = (\lambda + \mu)\nabla(\nabla \cdot \vec{u}) + \mu\nabla^2\vec{u}.$$

By using the vector identity $\nabla^2\vec{u} = \nabla(\nabla \cdot \vec{u}) - \nabla \times (\nabla \times \vec{u})$, the Navier equation can also be written in the alternate form

$$\varrho\frac{\partial^2\vec{u}}{\partial t^2} = (\lambda + 2\mu)\nabla(\nabla \cdot \vec{u}) - \mu\nabla \times (\nabla \times \vec{u}) \tag{2.4.35}$$

The displacement field $\vec{u}$ can be represented in the form $\vec{u} = \vec{u}^{(1)} + \vec{u}^{(2)}$, where $\vec{u}^{(1)}$ satisfies $\text{div } \vec{u}^{(1)} = 0$ and the vector $\vec{u}^{(2)}$ satisfies $\text{curl } \vec{u}^{(2)} = \nabla \times \vec{u}^{(2)} = 0$. The vector field $\vec{u}^{(1)}$ is called a solenoidal field, while the vector field $\vec{u}^{(2)}$ is called an irrotational field. Let $\vec{u}^{(1)} = \nabla \times \vec{A}$ and $\vec{u}^{(2)} = \nabla \phi$ so that

$$\vec{u} = \vec{u}^{(1)} + \vec{u}^{(2)} = \nabla \times \vec{A} + \nabla \phi. \tag{2.4.36}$$

This is the Helmholtz theorem where $\vec{A}$ is a vector potential and ϕ is a scalar potential. These quantities are sometimes referred to as Helmholtz potentials. Substituting equation (2.4.36) into equation (2.4.35) gives

$$\varrho \frac{\partial^2}{\partial t^2} \left\{ \nabla \times \vec{A} \right\} = (\lambda + 2\mu)\nabla \nabla \cdot (\nabla \times \vec{A} + \nabla \phi) - \mu \nabla \times \nabla \times \left\{ \nabla \times \vec{A} + \nabla \phi \right\}$$

which simplifies to

$$\nabla \left[(\lambda + 2\mu)\nabla^2 \phi - \varrho \frac{\partial^2 \phi}{\partial t^2} \right] + \nabla \times \left[\mu \nabla^2 \vec{A} - \varrho \frac{\partial^2 \vec{A}}{\partial t^2} \right] = 0. \tag{2.4.37}$$

A sufficient condition for the above equation to hold is for

$$(\lambda + 2\mu)\nabla^2 \phi - \varrho \frac{\partial^2 \phi}{\partial t^2} = 0 \qquad \text{and} \qquad \mu \nabla^2 \vec{A} - \varrho \frac{\partial^2 \vec{A}}{\partial t^2} = 0$$

which implies that $\vec{u}^{(1)}$ and $\vec{u}^{(2)}$ satisfy the wave equations

$$\varrho \frac{\partial^2 \vec{u}^{(2)}}{\partial t^2} = (\lambda + 2\mu)\nabla^2 \vec{u}^{(2)} \qquad \text{and} \qquad \varrho \frac{\partial^2 \vec{u}^{(1)}}{\partial t^2} = \mu \nabla^2 \vec{u}^{(1)}.$$

Here $\vec{u}^{(2)}$ is a compressive wave moving with the speed $\sqrt{(\lambda + 2\mu)/\varrho}$ and the $\vec{u}^{(1)}$ is a shearing wave which moves with speed $\sqrt{\mu/\varrho}$.

The exercises 30 through 38 enable us to write the Navier's equations in Cartesian, cylindrical or spherical coordinates. In particular, we have for Cartesian coordinates

$$(\lambda + \mu)\left(\frac{\partial^2 u}{\partial x^2} + \frac{\partial^2 v}{\partial x \partial y} + \frac{\partial^2 w}{\partial x \partial z}\right) + \mu\left(\frac{\partial^2 u}{\partial x^2} + \frac{\partial^2 u}{\partial y^2} + \frac{\partial^2 u}{\partial z^2}\right) + \varrho b_x = \varrho \frac{\partial^2 u}{\partial t^2}$$

$$(\lambda + \mu)\left(\frac{\partial^2 u}{\partial x \partial y} + \frac{\partial^2 v}{\partial y^2} + \frac{\partial^2 w}{\partial y \partial z}\right) + \mu\left(\frac{\partial^2 v}{\partial x^2} + \frac{\partial^2 v}{\partial y^2} + \frac{\partial^2 v}{\partial z^2}\right) + \varrho b_y = \varrho \frac{\partial^2 v}{\partial t^2}$$

$$(\lambda + \mu)\left(\frac{\partial^2 u}{\partial x \partial z} + \frac{\partial^2 v}{\partial y \partial z} + \frac{\partial^2 w}{\partial z^2}\right) + \mu\left(\frac{\partial^2 w}{\partial x^2} + \frac{\partial^2 w}{\partial y^2} + \frac{\partial^2 w}{\partial z^2}\right) + \varrho b_z = \varrho \frac{\partial^2 w}{\partial t^2}$$

and in cylindrical coordinates

$$(\lambda + \mu)\frac{\partial}{\partial r}\left(\frac{1}{r}\frac{\partial}{\partial r}(r u_r) + \frac{1}{r}\frac{\partial u_\theta}{\partial \theta} + \frac{\partial u_z}{\partial z}\right) +$$

$$\mu\left(\frac{\partial^2 u_r}{\partial r^2} + \frac{1}{r}\frac{\partial u_r}{\partial r} + \frac{1}{r^2}\frac{\partial^2 u_r}{\partial \theta^2} + \frac{\partial^2 u_r}{\partial z^2} - \frac{u_r}{r^2} - \frac{2}{r^2}\frac{\partial u_\theta}{\partial \theta}\right) + \varrho b_r = \varrho \frac{\partial^2 u_r}{\partial t^2}$$

$$(\lambda + \mu)\frac{1}{r}\frac{\partial}{\partial \theta}\left(\frac{1}{r}\frac{\partial}{\partial r}(r u_r) + \frac{1}{r}\frac{\partial u_\theta}{\partial \theta} + \frac{\partial u_z}{\partial z}\right) +$$

$$\mu\left(\frac{\partial^2 u_\theta}{\partial r^2} + \frac{1}{r}\frac{\partial u_\theta}{\partial r} + \frac{1}{r^2}\frac{\partial^2 u_\theta}{\partial \theta^2} + \frac{\partial^2 u_\theta}{\partial z^2} + \frac{2}{r^2}\frac{\partial u_r}{\partial \theta} - \frac{u_\theta}{r^2}\right) + \varrho b_\theta = \varrho \frac{\partial^2 u_\theta}{\partial t^2}$$

$$(\lambda + \mu)\frac{\partial}{\partial z}\left(\frac{1}{r}\frac{\partial}{\partial r}(r u_r) + \frac{1}{r}\frac{\partial u_\theta}{\partial \theta} + \frac{\partial u_z}{\partial z}\right) +$$

$$\mu\left(\frac{\partial^2 u_z}{\partial r^2} + \frac{1}{r}\frac{\partial u_z}{\partial r} + \frac{1}{r^2}\frac{\partial^2 u_z}{\partial \theta^2} + \frac{\partial^2 u_z}{\partial z^2}\right) + \varrho b_z = \varrho \frac{\partial^2 u_z}{\partial t^2}$$

and in spherical coordinates

$$(\lambda + \mu)\frac{\partial}{\partial \rho}\left(\frac{1}{\rho^2}\frac{\partial}{\partial \rho}(\rho^2 u_\rho) + \frac{1}{\rho\sin\theta}\frac{\partial}{\partial\theta}(u_\theta\sin\theta) + \frac{1}{\rho\sin\theta}\frac{\partial u_\phi}{\partial\phi}\right) +$$

$$\mu(\nabla^2 u_\rho - \frac{2}{\rho^2}u_\rho - \frac{2}{\rho^2}\frac{\partial u_\theta}{\partial\theta} - \frac{2u_\theta\cot\theta}{\rho^2} - \frac{2}{\rho^2\sin\theta}\frac{\partial u_\phi}{\partial\phi}) + \varrho b_\rho = \varrho\frac{\partial^2 u_\rho}{\partial t^2}$$

$$(\lambda + \mu)\frac{1}{\rho}\frac{\partial}{\partial\theta}\left(\frac{1}{\rho^2}\frac{\partial}{\partial\rho}(\rho^2 u_\rho) + \frac{1}{\rho\sin\theta}\frac{\partial}{\partial\theta}(u_\theta\sin\theta) + \frac{1}{\rho\sin\theta}\frac{\partial u_\phi}{\partial\phi}\right) +$$

$$\mu(\nabla^2 u_\theta + \frac{2}{\rho^2}\frac{\partial u_\rho}{\partial\theta} - \frac{u_\theta}{\rho^2\sin^2\theta} - \frac{2}{\rho^2}\frac{\cos\theta}{\sin^2\theta}\frac{\partial u_\phi}{\partial\phi}) + \varrho b_\theta = \varrho\frac{\partial^2 u_\theta}{\partial t^2}$$

$$(\lambda + \mu)\frac{1}{\rho\sin\theta}\frac{\partial}{\partial\phi}\left(\frac{1}{\rho^2}\frac{\partial}{\partial\rho}(\rho^2 u_\rho) + \frac{1}{\rho\sin\theta}\frac{\partial}{\partial\theta}(u_\theta\sin\theta) + \frac{1}{\rho\sin\theta}\frac{\partial u_\phi}{\partial\phi}\right) +$$

$$\mu(\nabla^2 u_\phi - \frac{1}{\rho^2\sin^2\theta}u_\phi + \frac{2}{\rho^2\sin\theta}\frac{\partial u_\rho}{\partial\phi} + \frac{2\cos\theta}{\rho^2\sin^2\theta}\frac{\partial u_\theta}{\partial\phi}) + \varrho b_\phi = \varrho\frac{\partial^2 u_\phi}{\partial t^2}$$

where ∇^2 is determined from either equation (2.1.12) or (2.1.13).

Boundary Conditions

In elasticity the body forces per unit mass $(b_i, i = 1, 2, 3)$ are assumed known. In addition one of the following type of boundary conditions is usually prescribed:

- The displacements u_i, $i = 1, 2, 3$ are prescribed on the boundary of the region R over which a solution is desired.
- The stresses (surface tractions) are prescribed on the boundary of the region R over which a solution is desired.
- The displacements $u_i, i = 1, 2, 3$ are given over one portion of the boundary and stresses (surface tractions) are specified over the remaining portion of the boundary. This type of boundary condition is known as a mixed boundary condition.

General Solution of Navier's Equations

There has been derived a general solution to the Navier's equations. It is known as the Papkovich-Neuber solution. In the case of a solid in equilibrium one must solve the equilibrium equations

$$(\lambda + \mu)\nabla(\nabla\cdot\vec{u}) + \mu\nabla^2\vec{u} + \varrho\vec{b} = 0 \quad \text{or}$$
$$\nabla^2\vec{u} + \frac{1}{1-2\nu}\nabla(\nabla\cdot\vec{u}) + \frac{\varrho}{\mu}\vec{b} = 0 \quad (\nu \neq \frac{1}{2}) \tag{2.4.38}$$

THEOREM   A general elastostatic solution of the equation (2.4.38) in terms of harmonic potentials $\phi, \vec{\psi}$ is

$$\vec{u} = \text{grad}\,(\phi + \vec{r}\cdot\vec{\psi}) - 4(1-\nu)\vec{\psi} \tag{2.4.39}$$

where ϕ and $\vec{\psi}$ are continuous solutions of the equations

$$\nabla^2\phi = \frac{-\varrho\vec{r}\cdot\vec{b}}{4\mu(1-\nu)} \qquad \text{and} \qquad \nabla^2\vec{\psi} = \frac{\varrho\vec{b}}{4\mu(1-\nu)} \tag{2.4.40}$$

with $\vec{r} = x\,\hat{e}_1 + y\,\hat{e}_2 + z\,\hat{e}_3$ a position vector to a general point (x,y,z) within the continuum.

Proof: First we write equation (2.4.38) in the tensor form

$$u_{i,kk} + \frac{1}{1-2\nu}(u_{j,j})_{,i} + \frac{\varrho}{\mu}b_i = 0 \tag{2.4.41}$$

Now our problem is to show that equation (2.4.39), in tensor form,

$$u_i = \phi_{,i} + (x_j\psi_j)_{,i} - 4(1-\nu)\psi_i \tag{2.4.42}$$

is a solution of equation (2.4.41). Toward this purpose, we differentiate equation (2.4.42)

$$u_{i,k} = \phi_{,ik} + (x_j\psi_j)_{,ik} - 4(1-\nu)\psi_{i,k} \tag{2.4.43}$$

and then contract on i and k giving

$$u_{i,i} = \phi_{,ii} + (x_j\psi_j)_{,ii} - 4(1-\nu)\psi_{i,i}. \tag{2.4.44}$$

Employing the identity $(x_j\psi_j)_{,ii} = 2\psi_{i,i} + x_i\psi_{i,kk}$ the equation (2.4.44) becomes

$$u_{i,i} = \phi_{,ii} + 2\psi_{i,i} + x_i\psi_{i,kk} - 4(1-\nu)\psi_{i,i}. \tag{2.4.45}$$

By differentiating equation (2.4.43) we establish that

$$\begin{aligned} u_{i,kk} &= \phi_{,ikk} + (x_j\psi_j)_{,ikk} - 4(1-\nu)\psi_{i,kk} \\ &= (\phi_{,kk})_{,i} + ((x_j\psi_j)_{,kk})_{,i} - 4(1-\nu)\psi_{i,kk} \\ &= [\phi_{,kk} + 2\psi_{j,j} + x_j\psi_{j,kk}]_{,i} - 4(1-\nu)\psi_{i,kk}. \end{aligned} \tag{2.4.46}$$

We use the hypothesis

$$\phi_{,kk} = \frac{-\varrho x_j b_j}{4\mu(1-\nu)} \qquad \text{and} \qquad \psi_{j,kk} = \frac{\varrho b_j}{4\mu(1-\nu)},$$

and simplify the equation (2.4.46) to the form

$$u_{i,kk} = 2\psi_{j,ji} - 4(1-\nu)\psi_{i,kk}. \tag{2.4.47}$$

Also by differentiating (2.4.45) one can establish that

$$\begin{aligned} u_{j,ji} &= (\phi_{,jj})_{,i} + 2\psi_{j,ji} + (x_j\psi_{j,kk})_{,i} - 4(1-\nu)\psi_{j,ji} \\ &= \left(\frac{-\varrho x_j b_j}{4\mu(1-\nu)}\right)_{,i} + 2\psi_{j,ji} + \left(\frac{\varrho x_j b_j}{4\mu(1-\nu)}\right)_{,i} - 4(1-\nu)\psi_{j,ji} \\ &= -2(1-2\nu)\psi_{j,ji}. \end{aligned} \tag{2.4.48}$$

Finally, from the equations (2.4.47) and (2.4.48) we obtain the desired result that

$$u_{i,kk} + \frac{1}{1-2\nu} u_{j,ji} + \frac{\varrho b_i}{\mu} = 0.$$

This shows that the equation (2.4.39) is a solution of equation (2.4.38).

As a special case of the above theorem, note that when the body forces are zero, the equations (2.4.40) become

$$\nabla^2 \phi = 0 \quad \text{and} \quad \nabla^2 \vec{\psi} = \vec{0}.$$

In this case, we find that equation (2.4.39) is a solution of equation (2.4.38) provided ϕ and each component of $\vec{\psi}$ are harmonic functions. The Papkovich-Neuber potentials are used together with complex variable theory to solve various two-dimensional elastostatic problems of elasticity. Note also that the Papkovich-Neuber potentials are not unique as different combinations of ϕ and $\vec{\psi}$ can produce the same value for $\vec{u}$.

Compatibility Equations

If we know or can derive the displacement field $u_i, i = 1, 2, 3$ we can then calculate the components of the strain tensor

$$e_{ij} = \frac{1}{2}(u_{i,j} + u_{j,i}). \tag{2.4.49}$$

Knowing the strain components, the stress is found using the constitutive relations.

Consider the converse problem where the strain tensor is given or implied due to the assigned stress field and we are asked to determine the displacement field $u_i, i = 1, 2, 3$. Is this a realistic request? Is it even possible to solve for three displacements given six strain components? It turns out that certain mathematical restrictions must be placed upon the strain components in order that the inverse problem have a solution. These mathematical restrictions are known as compatibility equations. That is, we cannot arbitrarily assign six strain components e_{ij} and expect to find a displacement field $u_i, i = 1, 2, 3$ with three components which satisfies the strain relation as given in equation (2.4.49).

EXAMPLE 2.4-3. Suppose we are given the two partial differential equations,

$$\frac{\partial u}{\partial x} = x + y \quad \text{and} \quad \frac{\partial u}{\partial y} = x^3.$$

Can we solve for $u = u(x, y)$? The answer to this question is "no", because the given equations are inconsistent. The inconsistency is illustrated if we calculate the mixed second derivatives from each equation. We find from the first equation that $\frac{\partial^2 u}{\partial x \partial y} = 1$ and from the second equation we calculate $\frac{\partial^2 u}{\partial y \partial x} = 3x^2$. These mixed second partial derivatives are unequal for all x different from $\sqrt{3}/3$. In general, if we have two first order partial differential equations $\frac{\partial u}{\partial x} = f(x, y)$ and $\frac{\partial u}{\partial y} = g(x, y)$, then for consistency (integrability of the equations) we require that the mixed partial derivatives

$$\frac{\partial^2 u}{\partial x \partial y} = \frac{\partial f}{\partial y} = \frac{\partial^2 u}{\partial y \partial x} = \frac{\partial g}{\partial x}$$

be equal to one another for all x and y values over the domain for which the solution is desired. This is an example of a compatibility equation.

A similar situation occurs in two dimensions for a material in a state of strain where $e_{zz} = e_{zx} = e_{zy} = 0$, called plane strain. In this case, are we allowed to arbitrarily assign values to the strains e_{xx}, e_{yy} and e_{xy} and from these strains determine the displacement field $u = u(x, y)$ and $v = v(x, y)$ in the $x-$ and $y-$directions? Let us try to answer this question. Assume a state of plane strain where $e_{zz} = e_{zx} = e_{zy} = 0$. Further, let us assign 3 arbitrary functional values f, g, h such that

$$e_{xx} = \frac{\partial u}{\partial x} = f(x, y), \quad e_{xy} = \frac{1}{2}\left(\frac{\partial u}{\partial y} + \frac{\partial v}{\partial x}\right) = g(x, y), \quad e_{yy} = \frac{\partial v}{\partial y} = h(x, y).$$

We must now decide whether these equations are consistent. That is, will we be able to solve for the displacement field $u = u(x, y)$ and $v = v(x, y)$? To answer this question, let us derive a compatibility equation (integrability condition). From the given equations we can calculate the following partial derivatives

$$\frac{\partial^2 e_{xx}}{\partial y^2} = \frac{\partial^3 u}{\partial x \partial y^2} = \frac{\partial^2 f}{\partial y^2}$$

$$\frac{\partial^2 e_{yy}}{\partial x^2} = \frac{\partial^3 v}{\partial y \partial x^2} = \frac{\partial^2 h}{\partial x^2}$$

$$2\frac{\partial^2 e_{xy}}{\partial x \partial y} = \frac{\partial^3 u}{\partial x \partial y^2} + \frac{\partial^3 v}{\partial y \partial x^2} = 2\frac{\partial^2 g}{\partial x \partial y}.$$

This last equation gives us the compatibility equation

$$2\frac{\partial^2 e_{xy}}{\partial x \partial y} = \frac{\partial^2 e_{xx}}{\partial y^2} + \frac{\partial^2 e_{yy}}{\partial x^2}$$

or the functions g, f, h must satisfy the relation

$$2\frac{\partial^2 g}{\partial x \partial y} = \frac{\partial^2 f}{\partial y^2} + \frac{\partial^2 h}{\partial x^2}.$$

■

Cartesian Derivation of Compatibility Equations

If the displacement field $u_i, i = 1, 2, 3$ is known we can derive the strain and rotation tensors

$$e_{ij} = \frac{1}{2}(u_{i,j} + u_{j,i}) \qquad \text{and} \qquad \omega_{ij} = \frac{1}{2}(u_{i,j} - u_{j,i}). \tag{2.4.50}$$

Now work backwards. Assume the strain and rotation tensors are given and ask the question, "Is it possible to solve for the displacement field $u_i, i = 1, 2, 3$?" If we view the equation (2.4.50) as a system of equations with unknowns e_{ij}, ω_{ij} and u_i and if by some means we can eliminate the unknowns ω_{ij} and u_i then we will be left with equations which must be satisfied by the strains e_{ij}. These equations are known as the compatibility equations and they represent conditions which the strain components must satisfy in order that a displacement function exist and the equations (2.4.37) are satisfied. Let us see if we can operate upon the equations (2.4.50) to eliminate the quantities u_i and ω_{ij} and hence derive the compatibility equations.

Addition of the equations (2.4.50) produces

$$u_{i,j} = \frac{\partial u_i}{\partial x_j} = e_{ij} + \omega_{ij}. \tag{2.4.51}$$

Differentiate this expression with respect to x_k and verify the result

$$\frac{\partial^2 u_i}{\partial x_j \partial x_k} = \frac{\partial e_{ij}}{\partial x_k} + \frac{\partial \omega_{ij}}{\partial x_k}. \tag{2.4.52}$$

We further assume that the displacement field is continuous so that the mixed partial derivatives are equal and

$$\frac{\partial^2 u_i}{\partial x_j \partial x_k} = \frac{\partial^2 u_i}{\partial x_k \partial x_j}. \tag{2.4.53}$$

Interchanging j and k in equation (2.4.52) gives us

$$\frac{\partial^2 u_i}{\partial x_k \partial x_j} = \frac{\partial e_{ik}}{\partial x_j} + \frac{\partial \omega_{ik}}{\partial x_j}. \tag{2.4.54}$$

Equating the second derivatives from equations (2.4.54) and (2.4.52) and rearranging terms produces the result

$$\frac{\partial e_{ij}}{\partial x_k} - \frac{\partial e_{ik}}{\partial x_j} = \frac{\partial \omega_{ik}}{\partial x_j} - \frac{\partial \omega_{ij}}{\partial x_k} \tag{2.4.55}$$

Making the observation that ω_{ij} satisfies $\dfrac{\partial \omega_{ik}}{\partial x_j} - \dfrac{\partial \omega_{ij}}{\partial x_k} = \dfrac{\partial \omega_{jk}}{\partial x_i}$, the equation (2.4.55) simplifies to the form

$$\frac{\partial e_{ij}}{\partial x_k} - \frac{\partial e_{ik}}{\partial x_j} = \frac{\partial \omega_{jk}}{\partial x_i}. \tag{2.4.56}$$

The term involving ω_{jk} can be eliminated by using the mixed partial derivative relation

$$\frac{\partial^2 \omega_{jk}}{\partial x_i \partial x_m} = \frac{\partial^2 \omega_{jk}}{\partial x_m \partial x_i}. \tag{2.4.57}$$

To derive the compatibility equations we differentiate equation (2.4.56) with respect to x_m and then interchanging the indices i and m and substitute the results into equation (2.4.57). This will produce the compatibility equations

$$\frac{\partial^2 e_{ij}}{\partial x_m \partial x_k} + \frac{\partial^2 e_{mk}}{\partial x_i \partial x_j} - \frac{\partial^2 e_{ik}}{\partial x_m \partial x_j} - \frac{\partial^2 e_{mj}}{\partial x_i \partial x_k} = 0. \tag{2.4.58}$$

This is a set of 81 partial differential equations which must be satisfied by the strain components. Fortunately, due to symmetry considerations only 6 of these 81 equations are distinct. These 6 distinct equations are known as the St. Venant's compatibility equations and can be written as

$$\begin{aligned}
\frac{\partial^2 e_{11}}{\partial x_2 \partial x_3} &= \frac{\partial^2 e_{12}}{\partial x_1 \partial x_3} - \frac{\partial^2 e_{23}}{\partial x_1{}^2} + \frac{\partial^2 e_{31}}{\partial x_1 \partial x_2} \\
\frac{\partial^2 e_{22}}{\partial x_1 \partial x_3} &= \frac{\partial^2 e_{23}}{\partial x_2 \partial x_1} - \frac{\partial^2 e_{31}}{\partial x_2{}^2} + \frac{\partial^2 e_{12}}{\partial x_2 \partial x_3} \\
\frac{\partial^2 e_{33}}{\partial x_1 \partial x_2} &= \frac{\partial^2 e_{31}}{\partial x_3 \partial x_2} - \frac{\partial^2 e_{12}}{\partial x_3{}^2} + \frac{\partial^2 e_{23}}{\partial x_3 \partial x_1} \\
2\frac{\partial^2 e_{12}}{\partial x_1 \partial x_2} &= \frac{\partial^2 e_{11}}{\partial x_2{}^2} + \frac{\partial^2 e_{22}}{\partial x_1{}^2} \\
2\frac{\partial^2 e_{23}}{\partial x_2 \partial x_3} &= \frac{\partial^2 e_{22}}{\partial x_3{}^2} + \frac{\partial^2 e_{33}}{\partial x_2{}^2} \\
2\frac{\partial^2 e_{31}}{\partial x_3 \partial x_1} &= \frac{\partial^2 e_{33}}{\partial x_1{}^2} + \frac{\partial^2 e_{11}}{\partial x_3{}^2}.
\end{aligned} \tag{2.4.59}$$

Observe that the fourth compatibility equation is the same as that derived in the example 2.4-3.

These compatibility equations can also be expressed in the indicial form

$$e_{ij,km} + e_{mk,ji} - e_{ik,jm} - e_{mj,ki} = 0. \tag{2.4.60}$$

Compatibility Equations in Terms of Stress

In the generalized Hooke's law, equation (2.4.29), we can solve for the strain in terms of stress. This in turn will give rise to a representation of the compatibility equations in terms of stress. The resulting equations are known as the Beltrami-Michell equations. Utilizing the strain-stress relation

$$e_{ij} = \frac{1+\nu}{E}\sigma_{ij} - \frac{\nu}{E}\sigma_{kk}\delta_{ij}$$

we substitute for the strain in the equations (2.4.60) and rearrange terms to produce the result

$$\sigma_{ij,km} + \sigma_{mk,ji} - \sigma_{ik,jm} - \sigma_{mj,ki} =$$
$$\frac{\nu}{1+\nu}\left[\delta_{ij}\sigma_{nn,km} + \delta_{mk}\sigma_{nn,ji} - \delta_{ik}\sigma_{nn,jm} - \delta_{mj}\sigma_{nn,ki}\right]. \tag{2.4.61}$$

Now only 6 of these 81 equations are linearly independent. It can be shown that the 6 linearly independent equations are equivalent to the equations obtained by setting $k = m$ and summing over the repeated indices. We then obtain the equations

$$\sigma_{ij,mm} + \sigma_{mm,ij} - (\sigma_{im,m})_{,j} - (\sigma_{mj,m})_{,i} = \frac{\nu}{1+\nu}\left[\delta_{ij}\sigma_{nn,mm} + \sigma_{nn,ij}\right].$$

Employing the equilibrium equation $\sigma_{ij,i} + \varrho b_j = 0$ the above result can be written in the form

$$\sigma_{ij,mm} + \frac{1}{1+\nu}\sigma_{kk,ij} - \frac{\nu}{1+\nu}\delta_{ij}\sigma_{nn,mm} = -(\varrho b_i)_{,j} - (\varrho b_j)_{,i}$$

or

$$\nabla^2\sigma_{ij} + \frac{1}{1+\nu}\sigma_{kk,ij} - \frac{\nu}{1+\nu}\delta_{ij}\sigma_{nn,mm} = -(\varrho b_i)_{,j} - (\varrho b_j)_{,i}.$$

This result can be further simplified by observing that a contraction on the indices k and i in equation (2.4.61) followed by a contraction on the indices m and j produces the result

$$\sigma_{ij,ij} = \frac{1-\nu}{1+\nu}\sigma_{nn,jj}.$$

and so the Beltrami-Michell equations can be written in the form

$$\nabla^2\sigma_{ij} + \frac{1}{1+\nu}\sigma_{pp,ij} = -\frac{\nu}{1-\nu}\delta_{ij}(\varrho b_k)_{,k} - (\varrho b_i)_{,j} - (\varrho b_j)_{,i}. \tag{2.4.62}$$

Their derivation is left as an exercise. The Beltrami-Michell equations together with the linear momentum (equilibrium) equations $\sigma_{ij,i} + \varrho b_j = 0$ represent 9 equations in six unknown stresses. This combinations of equations is difficult to handle. An easier combination of equations in terms of stress functions will be developed shortly.

The Navier equations with boundary conditions are difficult to solve in general. Let us take the momentum equations (2.4.27(a)), the strain relations (2.4.28) and constitutive equations (Hooke's law) (2.4.29) and make simplifying assumptions so that a more tractable systems results.

Plane Strain

The plane strain assumption usually is applied in situations where there is a cylindrical shaped body whose axis is parallel to the z axis and loads are applied along the $z-$direction. In any x-y plane we assume that the surface tractions and body forces are independent of z. We set all strains with a subscript z equal to zero. Further, all solutions for the stresses, strains and displacements are assumed to be only functions of x and y and independent of z. Note that in plane strain the stress σ_{zz} is different from zero.

In Cartesian coordinates the strain tensor is expressible in terms of its physical components which can be represented in the matrix form

$$\begin{pmatrix} e_{11} & e_{12} & e_{13} \\ e_{21} & e_{22} & e_{23} \\ e_{31} & e_{32} & e_{33} \end{pmatrix} = \begin{pmatrix} e_{xx} & e_{xy} & e_{xz} \\ e_{yx} & e_{yy} & e_{yz} \\ e_{zx} & e_{zy} & e_{zz} \end{pmatrix}.$$

If we assume that all strains which contain a subscript z are zero and the remaining strain components are functions of only x and y, we obtain a state of plane strain. For a state of plane strain, the stress components are obtained from the constitutive equations. The condition of plane strain reduces the constitutive equations to the form:

$$e_{xx} = \frac{1}{E}[\sigma_{xx} - \nu(\sigma_{yy} + \sigma_{zz})] \qquad \sigma_{xx} = \frac{E}{(1+\nu)(1-2\nu)}[(1-\nu)e_{xx} + \nu e_{yy}]$$

$$e_{yy} = \frac{1}{E}[\sigma_{yy} - \nu(\sigma_{zz} + \sigma_{xx})] \qquad \sigma_{yy} = \frac{E}{(1+\nu)(1-2\nu)}[(1-\nu)e_{yy} + \nu e_{xx}]$$

$$0 = \frac{1}{E}[\sigma_{zz} - \nu(\sigma_{xx} + \sigma_{yy})] \qquad \sigma_{zz} = \frac{E}{(1+\nu)(1-2\nu)}[\nu(e_{yy} + e_{xx})] \qquad (2.4.63)$$

$$e_{xy} = e_{yx} = \frac{1+\nu}{E}\sigma_{xy}$$

$$\sigma_{xy} = \frac{E}{1+\nu}e_{xy}$$

$$e_{zy} = e_{yz} = \frac{1+\nu}{E}\sigma_{yz} = 0 \qquad \sigma_{xz} = 0$$

$$e_{zx} = e_{xz} = \frac{1+\nu}{E}\sigma_{xz} = 0 \qquad \sigma_{yz} = 0$$

where $\sigma_{xx}, \quad \sigma_{yy}, \quad \sigma_{zz}, \quad \sigma_{xy}, \quad \sigma_{xz}, \quad \sigma_{yz}$ are the physical components of the stress. The above constitutive equations imply that for a state of plane strain we will have

$$\sigma_{zz} = \nu(\sigma_{xx} + \sigma_{yy})$$

$$e_{xx} = \frac{1+\nu}{E}[(1-\nu)\sigma_{xx} - \nu\sigma_{yy}]$$

$$e_{yy} = \frac{1+\nu}{E}[(1-\nu)\sigma_{yy} - \nu\sigma_{xx}]$$

$$e_{xy} = \frac{1+\nu}{E}\sigma_{xy}.$$

Also under these conditions the compatibility equations reduce to

$$\frac{\partial^2 e_{xx}}{\partial y^2} + \frac{\partial^2 e_{yy}}{\partial x^2} = 2\frac{\partial^2 e_{xy}}{\partial x \partial y}.$$

Plane Stress

An assumption of plane stress is usually applied to thin flat plates. The plate thinness is assumed to be in the $z-$direction and loads are applied perpendicular to z. Under these conditions all stress components with a subscript z are assumed to be zero. The remaining stress components are then treated as functions of x and y.

In Cartesian coordinates the stress tensor is expressible in terms of its physical components and can be represented by the matrix

$$\begin{pmatrix} \sigma_{11} & \sigma_{12} & \sigma_{13} \\ \sigma_{21} & \sigma_{22} & \sigma_{23} \\ \sigma_{31} & \sigma_{32} & \sigma_{33} \end{pmatrix} = \begin{pmatrix} \sigma_{xx} & \sigma_{xy} & \sigma_{xz} \\ \sigma_{yx} & \sigma_{yy} & \sigma_{yz} \\ \sigma_{zx} & \sigma_{zy} & \sigma_{zz} \end{pmatrix}.$$

If we assume that all the stresses with a subscript z are zero and the remaining stresses are only functions of x and y we obtain a state of plane stress. The constitutive equations simplify if we assume a state of plane stress. These simplified equations are

$$
\begin{aligned}
e_{xx} &= \frac{1}{E}\sigma_{xx} - \frac{\nu}{E}\sigma_{yy} & \sigma_{xx} &= \frac{E}{1-\nu^2}\left[e_{xx}+\nu e_{yy}\right] \\
e_{yy} &= \frac{1}{E}\sigma_{yy} - \frac{\nu}{E}\sigma_{xx} & \sigma_{yy} &= \frac{E}{1-\nu^2}\left[e_{yy}+\nu e_{xx}\right] \\
e_{zz} &= -\frac{\nu}{E}(\sigma_{xx}+\sigma_{yy}) & \sigma_{zz} &= 0 = (1-\nu)e_{zz}+\nu(e_{xx}+e_{yy}) \\
e_{xy} &= \frac{1+\nu}{E}\sigma_{xy} & \sigma_{xy} &= \frac{E}{1+\nu}e_{xy} \\
e_{xz} &= 0 & \sigma_{yz} &= 0 \\
e_{yz} &= 0. & \sigma_{xz} &= 0
\end{aligned}
\tag{2.4.64}
$$

For a state of plane stress the compatibility equations reduce to

$$\frac{\partial^2 e_{xx}}{\partial y^2} + \frac{\partial^2 e_{yy}}{\partial x^2} = 2\frac{\partial^2 e_{xy}}{\partial x \partial y} \tag{2.4.65}$$

and the three additional equations

$$\frac{\partial^2 e_{zz}}{\partial x^2} = 0, \qquad \frac{\partial^2 e_{zz}}{\partial y^2} = 0, \qquad \frac{\partial^2 e_{zz}}{\partial x \partial y} = 0.$$

These three additional equations complicate the plane stress problem.

Airy Stress Function

In Cartesian coordinates we examine the equilibrium equations (2.4.25(b)) under the conditions of plane strain. In terms of physical components we find that these equations reduce to

$$\frac{\partial \sigma_{xx}}{\partial x} + \frac{\partial \sigma_{xy}}{\partial y} + \varrho b_x = 0, \qquad \frac{\partial \sigma_{yx}}{\partial x} + \frac{\partial \sigma_{yy}}{\partial y} + \varrho b_y = 0, \qquad \frac{\partial \sigma_{zz}}{\partial z} = 0.$$

The last equation is satisfied since σ_{zz} is a function of x and y. If we further assume that the body forces are conservative and derivable from a potential function V by the operation $\varrho\vec{b} = -\text{grad}\, V$ or $\varrho b_i = -V_{,i}$ we can express the above equilibrium equations in the form:

$$
\begin{aligned}
\frac{\partial \sigma_{xx}}{\partial x} + \frac{\partial \sigma_{xy}}{\partial y} - \frac{\partial V}{\partial x} &= 0 \\
\frac{\partial \sigma_{yx}}{\partial x} + \frac{\partial \sigma_{yy}}{\partial y} - \frac{\partial V}{\partial y} &= 0
\end{aligned}
\tag{2.4.66}
$$

We will consider these equations together with the compatibility equations (2.4.65). The equations (2.4.66) will be automatically satisfied if we introduce a scalar function $\phi = \phi(x,y)$ and assume that the stresses are derivable from this function and the potential function V according to the rules:

$$\sigma_{xx} = \frac{\partial^2 \phi}{\partial y^2} + V \qquad \sigma_{xy} = -\frac{\partial^2 \phi}{\partial x \partial y} \qquad \sigma_{yy} = \frac{\partial^2 \phi}{\partial x^2} + V. \qquad (2.4.67)$$

The function $\phi = \phi(x,y)$ is called the Airy stress function after the English astronomer and mathematician Sir George Airy (1801–1892). Since the equations (2.4.67) satisfy the equilibrium equations we need only consider the compatibility equation(s).

For a state of plane strain we substitute the relations (2.4.63) into the compatibility equation (2.4.65) and write the compatibility equation in terms of stresses. We then substitute the relations (2.4.67) and express the compatibility equation in terms of the Airy stress function ϕ. These substitutions are left as exercises. After all these substitutions the compatibility equation, for a state of plane strain, reduces to the form

$$\frac{\partial^4 \phi}{\partial x^4} + 2\frac{\partial^4 \phi}{\partial x^2 \partial y^2} + \frac{\partial^4 \phi}{\partial y^4} + \frac{1-2\nu}{1-\nu}\left(\frac{\partial^2 V}{\partial x^2} + \frac{\partial^2 V}{\partial y^2}\right) = 0. \qquad (2.4.68)$$

In the special case where there are no body forces we have $V = 0$ and equation (2.4.68) is further simplified to the biharmonic equation.

$$\nabla^4 \phi = \frac{\partial^4 \phi}{\partial x^4} + 2\frac{\partial^4 \phi}{\partial x^2 \partial y^2} + \frac{\partial^4 \phi}{\partial y^4} = 0. \qquad (2.4.69)$$

In polar coordinates the biharmonic equation is written

$$\nabla^4 \phi = \nabla^2(\nabla^2 \phi) = \left(\frac{\partial^2}{\partial r^2} + \frac{1}{r}\frac{\partial}{\partial r} + \frac{1}{r^2}\frac{\partial^2}{\partial \theta^2}\right)\left(\frac{\partial^2 \phi}{\partial r^2} + \frac{1}{r}\frac{\partial \phi}{\partial r} + \frac{1}{r^2}\frac{\partial^2 \phi}{\partial \theta^2}\right) = 0.$$

For conditions of plane stress, we can again introduce an Airy stress function using the equations (2.4.67). However, an exact solution of the plane stress problem which satisfies all the compatibility equations is difficult to obtain. By removing the assumptions that $\sigma_{xx}, \sigma_{yy}, \sigma_{xy}$ are independent of z, and neglecting body forces, it can be shown that for symmetrically distributed external loads the stress function ϕ can be represented in the form

$$\phi = \psi - \frac{\nu z^2}{2(1+\nu)}\nabla^2 \psi \qquad (2.4.70)$$

where ψ is a solution of the biharmonic equation $\nabla^4 \psi = 0$. Observe that if z is very small, (the condition of a thin plate), then equation (2.4.70) gives the approximation $\phi \approx \psi$. Under these conditions, we obtain the approximate solution by using only the compatibility equation (2.4.65) together with the stress function defined by equations (2.4.67) with $V = 0$. Note that the solution we obtain from equation (2.4.69) does not satisfy all the compatibility equations, however, it does give an excellent first approximation to the solution in the case where the plate is very thin.

In general, for plane strain or plane stress problems, the equation (2.4.68) or (2.4.69) must be solved for the Airy stress function ϕ which is defined over some region R. In addition to specifying a region of the x,y plane, there are certain boundary conditions which must be satisfied. The boundary conditions specified for the stress will translate through the equations (2.4.67) to boundary conditions being specified for ϕ. In the special case where there are no body forces, both the problems for plane stress and plane strain are governed by the biharmonic differential equation with appropriate boundary conditions.

EXAMPLE 2.4-4 Assume there exist a state of plane strain with zero body forces. For F_{11}, F_{12}, F_{22} constants, consider the function defined by

$$\phi = \phi(x,y) = \frac{1}{2}\left(F_{22}\,x^2 - 2F_{12}\,xy + F_{11}\,y^2\right).$$

This function is an Airy stress function because it satisfies the biharmonic equation $\nabla^4\phi = 0$. The resulting stress field is

$$\sigma_{xx} = \frac{\partial^2\phi}{\partial y^2} = F_{11} \qquad \sigma_{yy} = \frac{\partial^2\phi}{\partial x^2} = F_{22} \qquad \sigma_{xy} = -\frac{\partial^2\phi}{\partial x\partial y} = F_{12}.$$

This example, corresponds to stresses on an infinite flat plate and illustrates a situation where all the stress components are constants for all values of x and y. In this case, we have $\sigma_{zz} = \nu(F_{11}+F_{22})$. The corresponding strain field is obtained from the constitutive equations. We find these strains are

$$e_{xx} = \frac{1+\nu}{E}\left[(1-\nu)F_{11} - \nu F_{22}\right] \qquad e_{yy} = \frac{1+\nu}{E}\left[(1-\nu)F_{22} - \nu F_{11}\right] \qquad e_{xy} = \frac{1+\nu}{E}F_{12}.$$

The displacement field is found to be

$$u = u(x,y) = \frac{1+\nu}{E}\left[(1-\nu)F_{11} - \nu F_{22}\right]x + \left(\frac{1+\nu}{E}\right)F_{12}y + c_1 y + c_2$$

$$v = v(x,y) = \frac{1+\nu}{E}\left[(1-\nu)F_{22} - \nu F_{11}\right]y + \left(\frac{1+\nu}{E}\right)F_{12}x - c_1 x + c_3,$$

with c_1, c_2, c_3 constants, and is obtained by integrating the strain displacement equations given in Exercise 2.3, problem 2.

EXAMPLE 2.4-5. A special case from the previous example is obtained by setting $F_{22} = F_{12} = 0$. This is the situation of an infinite plate with only tension in the x−direction. In this special case we have $\phi = \frac{1}{2}F_{11}y^2$. Changing to polar coordinates we write

$$\phi = \phi(r,\theta) = \frac{F_{11}}{2}r^2\sin^2\theta = \frac{F_{11}}{4}r^2(1-\cos 2\theta).$$

The Exercise 2.4, problem 20, suggests we utilize the Airy equations in polar coordinates and calculate the stresses

$$\sigma_{rr} = \frac{1}{r}\frac{\partial\phi}{\partial r} + \frac{1}{r^2}\frac{\partial^2\phi}{\partial\theta^2} = F_{11}\cos^2\theta = \frac{F_{11}}{2}(1+\cos 2\theta)$$

$$\sigma_{\theta\theta} = \frac{\partial^2\phi}{\partial r^2} = F_{11}\sin^2\theta = \frac{F_{11}}{2}(1-\cos 2\theta)$$

$$\sigma_{r\theta} = \frac{1}{r^2}\frac{\partial\phi}{\partial\theta} - \frac{1}{r}\frac{\partial^2\phi}{\partial r\partial\theta} = -\frac{F_{11}}{2}\sin 2\theta.$$

EXAMPLE 2.4-6. We now consider an infinite plate with a circular hole $x^2 + y^2 = a^2$ which is traction free. Assume the plate has boundary conditions at infinity defined by $\sigma_{xx} = F_{11}$, $\sigma_{yy} = 0$, $\sigma_{xy} = 0$. Find the stress field.

Solution:

The traction boundary condition at $r = a$ is $t_i = \sigma_{mi} n_m$ or

$$t_1 = \sigma_{11} n_1 + \sigma_{12} n_2 \qquad \text{and} \qquad t_2 = \sigma_{12} n_1 + \sigma_{22} n_2.$$

For polar coordinates we have $n_1 = n_r = 1$, $n_2 = n_\theta = 0$ and so the traction free boundary conditions at the surface of the hole are written $\sigma_{rr}|_{r=a} = 0$ and $\sigma_{r\theta}|_{r=a} = 0$. The results from the previous example are used as the boundary conditions at infinity.

Our problem is now to solve for the Airy stress function $\phi = \phi(r, \theta)$ which is a solution of the biharmonic equation. The previous example 2.4-5 and the form of the boundary conditions at infinity suggests that we assume a solution to the biharmonic equation which has the form $\phi = \phi(r, \theta) = f_1(r) + f_2(r) \cos 2\theta$, where f_1, f_2 are unknown functions to be determined. Substituting the assumed solution into the biharmonic equation produces the equation

$$\left(\frac{d^2}{dr^2} + \frac{1}{r} \frac{d}{dr} \right) \left(f_1'' + \frac{1}{r} f_1' \right) + \left(\frac{d^2}{dr^2} + \frac{1}{r} \frac{d}{dr} - \frac{4}{r^2} \right) \left(f_2'' + \frac{1}{r} f_2' - 4 \frac{f_2}{r^2} \right) \cos 2\theta = 0.$$

We therefore require that f_1, f_2 be chosen to satisfy the equations

$$\left(\frac{d^2}{dr^2} + \frac{1}{r} \frac{d}{dr} \right) \left(f_1'' + \frac{1}{r} f_1' \right) = 0 \qquad \left(\frac{d^2}{dr^2} + \frac{1}{r} \frac{d}{dr} - \frac{4}{r^2} \right) \left(f_2'' + \frac{1}{r} f_2' - 4 \frac{f_2}{r^2} \right) = 0$$

or $\qquad r^4 f_1^{(iv)} + 2r^3 f_1''' - r^2 f_1'' + r f_1' = 0 \qquad\qquad r^4 f_2^{(iv)} + 2r^3 f_2''' - 9r^2 f_2'' + 9r f_2' = 0$

These equations are Cauchy type equations. Their solutions are obtained by assuming a solution of the form $f_1 = r^\lambda$ and $f_2 = r^m$ and then solving for the constants λ and m. We find the general solutions of the above equations are

$$f_1 = c_1 r^2 \ln r + c_2 r^2 + c_3 \ln r + c_4 \quad \text{and} \quad f_2 = c_5 r^2 + c_6 r^4 + \frac{c_7}{r^2} + c_8.$$

The constants $c_i, i = 1, \ldots, 8$ are now determined from the boundary conditions. The constant c_4 can be arbitrary since the derivative of a constant is zero. The remaining constants are determined from the stress conditions. Using the results from Exercise 2.4, problem 20, we calculate the stresses

$$\sigma_{rr} = c_1(1 + 2\ln r) + 2c_2 + \frac{c_3}{r^2} - \left(2c_5 + 6\frac{c_7}{r^4} + 4\frac{c_8}{r^2} \right) \cos 2\theta$$

$$\sigma_{\theta\theta} = c_1(3 + 2\ln r) + 2c_2 - \frac{c_3}{r^2} + \left(2c_5 + 12c_6 r^2 + 6\frac{c_7}{r^4} \right) \cos 2\theta$$

$$\sigma_{r\theta} = \left(2c_5 + 6c_6 r^2 - 6\frac{c_7}{r^4} - 2\frac{c_8}{r^2} \right) \sin 2\theta.$$

The stresses are to remain bounded for all values of r which requires c_1 and c_6 to be zero to avoid infinite stresses for large values of r. The stress $\sigma_{rr}|_{r=a} = 0$ requires that

$$2c_2 + \frac{c_3}{a^2} = 0 \quad \text{and} \quad 2c_5 + 6\frac{c_7}{a^4} + 4\frac{c_8}{a^2} = 0.$$

The stress $\sigma_{r\theta}|_{r=a} = 0$ requires that

$$2c_5 - 6\frac{c_7}{a^4} - 2\frac{c_8}{a^2} = 0.$$

In the limit as $r \to \infty$ we require that the stresses must satisfy the boundary conditions from the previous example 2.4-5. This leads to the equations $2c_2 = \frac{F_{11}}{2}$ and $2c_5 = -\frac{F_{11}}{2}$. Solving the above system of equations produces the Airy stress function

$$\phi = \phi(r,\theta) = \frac{F_{11}}{4} + \frac{F_{11}}{4}r^2 - \frac{a^2}{2}F_{11}\ln r + c_4 + \left(\frac{F_{11}a^2}{2} - \frac{F_{11}}{4}r^2 - \frac{F_{11}a^4}{4r^2}\right)\cos 2\theta$$

and the corresponding stress field is

$$\sigma_{rr} = \frac{F_{11}}{2}\left(1 - \frac{a^2}{r^2}\right) + \frac{F_{11}}{2}\left(1 + 3\frac{a^4}{r^4} - 4\frac{a^2}{r^2}\right)\cos 2\theta$$

$$\sigma_{r\theta} = -\frac{F_{11}}{2}\left(1 - 3\frac{a^4}{r^4} + 2\frac{a^2}{r^2}\right)\sin 2\theta$$

$$\sigma_{\theta\theta} = \frac{F_{11}}{2}\left(1 + \frac{a^2}{r^2}\right) - \frac{F_{11}}{2}\left(1 + 3\frac{a^4}{r^4}\right)\cos 2\theta.$$

There is a maximum stress $\sigma_{\theta\theta} = 3F_{11}$ at $\theta = \pi/2, 3\pi/2$ and a minimum stress $\sigma_{\theta\theta} = -F_{11}$ at $\theta = 0, \pi$. The effect of the circular hole has been to magnify the applied stress. The factor of 3 is known as a stress concentration factor. In general, sharp corners and unusually shaped boundaries produce much higher stress concentration factors than rounded boundaries.

EXAMPLE 2.4-7. Consider an infinite cylindrical tube, with inner radius R_1 and the outer radius R_0, which is subjected to an internal pressure P_1 and an external pressure P_0 as illustrated in the figure 2.4-7. Find the stress and displacement fields.

Solution: Let u_r, u_θ, u_z denote the displacement field. We assume that $u_\theta = 0$ and $u_z = 0$ since the cylindrical surface r equal to a constant does not move in the θ or z directions. The displacement $u_r = u_r(r)$ is assumed to depend only upon the radial distance r. Under these conditions the Navier equations become

$$(\lambda + 2\mu)\frac{d}{dr}\left(\frac{1}{r}\frac{d}{dr}(ru_r)\right) = 0.$$

This equation has the solution $u_r = c_1\frac{r}{2} + \frac{c_2}{r}$ and the strain components are found from the relations

$$e_{rr} = \frac{du_r}{dr}, \quad e_{\theta\theta} = \frac{u_r}{r}, \quad e_{zz} = e_{r\theta} = e_{rz} = e_{z\theta} = 0.$$

The stresses are determined from Hooke's law (the constitutive equations) and we write

$$\sigma_{ij} = \lambda\delta_{ij}\Theta + 2\mu e_{ij},$$

where

$$\Theta = \frac{\partial u_r}{\partial r} + \frac{u_r}{r} = \frac{1}{r}\frac{\partial}{\partial r}\left(r u_r\right)$$

is the dilatation. These stresses are found to be

$$\sigma_{rr} = (\lambda + \mu)c_1 - \frac{2\mu}{r^2}c_2 \quad \sigma_{\theta\theta} = (\lambda + \mu)c_1 + \frac{2\mu}{r^2}c_2 \quad \sigma_{zz} = \lambda c_1 \quad \sigma_{r\theta} = \sigma_{rz} = \sigma_{z\theta} = 0.$$

We now apply the boundary conditions

$$\sigma_{rr}|_{r=R_1} n_r = -\left[(\lambda + \mu)c_1 - \frac{2\mu}{R_1^2}c_2\right] = +P_1 \quad \text{and} \quad \sigma_{rr}|_{r=R_0} n_r = \left[(\lambda + \mu)c_1 - \frac{2\mu}{R_0^2}c_2\right] = -P_0.$$

Solving for the constants c_1 and c_2 we find

$$c_1 = \frac{R_1^2 P_1 - R_0^2 P_0}{(\lambda + \mu)(R_0^2 - R_1^2)}, \qquad c_2 = \frac{R_1^2 R_0^2 (P_1 - P_0)}{2\mu(R_0^2 - R_1^2)}.$$

This produces the displacement field

$$u_r = \frac{R_1^2 P_1}{2(R_0^2 - R_1^2)}\left(\frac{r}{\lambda + \mu} + \frac{R_0^2}{\mu r}\right) - \frac{R_0^2 P_0}{2(R_0^2 - R_1^2)}\left(\frac{r}{\lambda + \mu} + \frac{R_1^2}{\mu r}\right), \qquad u_\theta = 0, \qquad u_z = 0,$$

and stress fields

$$\sigma_{rr} = \frac{R_1^2 P_1}{R_0^2 - R_1^2}\left(1 - \frac{R_0^2}{r^2}\right) - \frac{R_0^2 P_0}{R_0^2 - R_1^2}\left(1 - \frac{R_1^2}{r^2}\right)$$

$$\sigma_{\theta\theta} = \frac{R_1^2 P_1}{R_0^2 - R_1^2}\left(1 + \frac{R_0^2}{r^2}\right) - \frac{R_0^2 P_0}{R_0^2 - R_1^2}\left(1 + \frac{R_1^2}{r^2}\right)$$

$$\sigma_{zz} = \left(\frac{\lambda}{\lambda + \mu}\right)\frac{R_1^2 P_1 - R_0^2 P_0}{R_0^2 - R_1^2}$$

$$\sigma_{rz} = \sigma_{z\theta} = \sigma_{r\theta} = 0$$

EXAMPLE 2.4-8. By making simplifying assumptions the Navier equations can be reduced to a more tractable form. For example, we can reduce the Navier equations to a one dimensional problem by making the following assumptions

1. Cartesian coordinates $x_1 = x, \quad x_2 = y, \quad x_3 = z$

2. $u_1 = u_1(x,t), \quad u_2 = u_3 = 0.$

3. There are no body forces.

4. Initial conditions of $\quad u_1(x,0) = 0 \quad$ and $\quad \dfrac{\partial u_1(x,0)}{\partial t} = 0$

5. Boundary conditions of the displacement type $\quad u_1(0,t) = f(t),$

where $f(t)$ is a specified function. These assumptions reduce the Navier equations to the single one dimensional wave equation

$$\frac{\partial^2 u_1}{\partial t^2} = \alpha^2 \frac{\partial^2 u_1}{\partial x^2}, \qquad \alpha^2 = \frac{\lambda + 2\mu}{\rho}.$$

The solution of this equation is

$$u_1(x,t) = \begin{cases} f(t - x/\alpha), & x \le \alpha t \\ 0, & x > \alpha t \end{cases}.$$

The solution represents a longitudinal elastic wave propagating in the $x-$direction with speed α. The stress wave associated with this displacement is determined from the constitutive equations. We find

$$\sigma_{xx} = (\lambda + \mu)e_{xx} = (\lambda + \mu)\frac{\partial u_1}{\partial x}.$$

This produces the stress wave

$$\sigma_{xx} = \begin{cases} -\frac{(\lambda+\mu)}{\alpha}f'(t - x/\alpha), & x \leq \alpha t \\ 0, & x > \alpha t \end{cases}.$$

Here there is a discontinuity in the stress wave front at $x = \alpha t$.

Summary of Basic Equations of Elasticity

The equilibrium equations for a continuum have been shown to have the form $\sigma^{ij}_{,j} + \varrho b^i = 0$, where b^i are the body forces per unit mass and σ^{ij} is the stress tensor. In addition to the above equations we have the constitutive equations $\sigma_{ij} = \lambda e_{kk}\delta_{ij} + 2\mu e_{ij}$ which is a generalized Hooke's law relating stress to strain for a linear elastic isotropic material. The strain tensor is related to the displacement field u_i by the strain equations $e_{ij} = \frac{1}{2}\left(u_{i,j} + u_{j,i}\right)$. These equations can be combined to obtain the Navier equations $\mu u_{i,jj} + (\lambda + \mu)u_{j,ji} + \varrho b_i = 0$.

The above equations must be satisfied at all interior points of the material body. A boundary value problem results when conditions on the displacement of the boundary are specified. That is, the Navier equations must be solved subject to the prescribed displacement boundary conditions. If conditions on the stress at the boundary are specified, then these prescribed stresses are called surface tractions and must satisfy the relations $t^{i\,(n)} = \sigma^{ij}n_j$, where n_i is a unit outward normal vector to the boundary. For surface tractions, we need to use the compatibility equations combined with the constitutive equations and equilibrium equations. This gives rise to the Beltrami-Michell equations of compatibility

$$\sigma_{ij,kk} + \frac{1}{1+\nu}\sigma_{kk,ij} + \varrho(b_{i,j} + b_{j,i}) + \frac{\nu}{1-\nu}\varrho b_{k,k} = 0.$$

Here we must solve for the stress components throughout the continuum where the above equations hold subject to the surface traction boundary conditions. Note that if an elasticity problem is formed in terms of the displacement functions, then the compatibility equations can be ignored.

For mixed boundary value problems we must solve a system of equations consisting of the equilibrium equations, constitutive equations, and strain displacement equations. We must solve these equations subject to conditions where the displacements u_i are prescribed on some portion(s) of the boundary and stresses are prescribed on the remaining portion(s) of the boundary. Mixed boundary value problems are more difficult to solve.

For elastodynamic problems, the equilibrium equations are replaced by equations of motion. In this case we need a set of initial conditions as well as boundary conditions before attempting to solve our basic system of equations.

EXERCISE 2.4

▶ **1.** Verify the generalized Hooke's law constitutive equations for hexagonal materials.

In the following problems the Young's modulus E, Poisson's ratio ν, the shear modulus or modulus of rigidity μ (sometimes denoted by G in Engineering texts), Lame's constant λ and the bulk modulus of elasticity k are assumed to satisfy the equations (2.4.19), (2.4.24) and (2.4.25). Show that these relations imply the additional relations given in the problems 2 through 6.

▶ **2.**

$$E = \frac{\mu(3\lambda + 2\mu)}{\mu + \lambda} \qquad E = \frac{9k(k-\lambda)}{3k-\lambda} \qquad E = \frac{9k\mu}{\mu + 3k}$$

$$E = \frac{\lambda(1+\nu)(1-2\nu)}{\nu} \qquad E = 2\mu(1+\nu) \qquad E = 3(1-2\nu)k$$

▶ **3.**

$$\nu = \frac{3k-E}{6k} \qquad \nu = \frac{\sqrt{(E+\lambda)^2 + 8\lambda^2} - (E+\lambda)}{4\lambda} \qquad \nu = \frac{E-2\mu}{2\mu}$$

$$\nu = \frac{\lambda}{2(\mu+\lambda)} \qquad \nu = \frac{3k-2\mu}{2(\mu+3k)} \qquad \nu = \frac{\lambda}{3k-\lambda}$$

▶ **4.**

$$k = \frac{\sqrt{(E+\lambda)^2 + 8\lambda^2} + (E+3\lambda)}{6} \qquad k = \frac{E}{3(1-2\nu)} \qquad k = \frac{2\mu(1+\nu)}{3(1-2\nu)}$$

$$k = \frac{2\mu + 3\lambda}{3} \qquad k = \frac{\mu E}{3(3\mu - E)} \qquad k = \frac{\lambda(1+\nu)}{3\nu}$$

▶ **5.**

$$\mu = \frac{3(k-\lambda)}{2} \qquad \mu = \frac{3k(1-2\nu)}{2(1+\nu)} \qquad \mu = \frac{\sqrt{(E+\lambda)^2 + 8\lambda^2} + (E-3\lambda)}{4}$$

$$\mu = \frac{\lambda(1-2\nu)}{2\nu} \qquad \mu = \frac{3Ek}{9k-E} \qquad \mu = \frac{E}{2(1+\nu)}$$

▶ **6.**

$$\lambda = \frac{3k\nu}{1+\nu} \qquad \lambda = \frac{3k-2\mu}{3} \qquad \lambda = \frac{\nu E}{(1+\nu)(1-2\nu)}$$

$$\lambda = \frac{\mu(2\mu - E)}{E - 3\mu} \qquad \lambda = \frac{3k(3k-E)}{9k-E} \qquad \lambda = \frac{2\mu\nu}{1-2\nu}$$

▶ **7.** The previous exercises 2 through 6 imply that the generalized Hooke's law

$$\sigma_{ij} = 2\mu e_{ij} + \lambda \delta_{ij} e_{kk}$$

is expressible in a variety of forms. From the set of constants $(\mu, \lambda, \nu, E, k)$ we can select any two constants and then express Hooke's law in terms of these constants.

(a) Express the above Hooke's law in terms of the constants E and ν.

(b) Express the above Hooke's law in terms of the constants k and E.

(c) Express the above Hooke's law in terms of physical components. Hint: The quantity e_{kk} is an invariant hence all you need to know is how second order tensors are represented in terms of physical components. See also problems 10,11,12.

8. Verify the equations defining the stress for plane strain in Cartesian coordinates are

$$\sigma_{xx} = \frac{E}{(1+\nu)(1-2\nu)}[(1-\nu)e_{xx} + \nu e_{yy}]$$

$$\sigma_{yy} = \frac{E}{(1+\nu)(1-2\nu)}[(1-\nu)e_{yy} + \nu e_{xx}]$$

$$\sigma_{zz} = \frac{E\nu}{(1+\nu)(1-2\nu)}[e_{xx} + e_{yy}]$$

$$\sigma_{xy} = \frac{E}{1+\nu}e_{xy}$$

$$\sigma_{yz} = \sigma_{xz} = 0$$

9. Verify the equations defining the stress for plane strain in polar coordinates are

$$\sigma_{rr} = \frac{E}{(1+\nu)(1-2\nu)}[(1-\nu)e_{rr} + \nu e_{\theta\theta}]$$

$$\sigma_{\theta\theta} = \frac{E}{(1+\nu)(1-2\nu)}[(1-\nu)e_{\theta\theta} + \nu e_{rr}]$$

$$\sigma_{zz} = \frac{\nu E}{(1+\nu)(1-2\nu)}[e_{rr} + e_{\theta\theta}]$$

$$\sigma_{r\theta} = \frac{E}{1+\nu}e_{r\theta}$$

$$\sigma_{rz} = \sigma_{\theta z} = 0$$

10. Write out the independent components of Hooke's generalized law for strain in terms of stress, and stress in terms of strain, in Cartesian coordinates. Express your results using the parameters ν and E. (Assume a linear elastic, homogeneous, isotropic material.)

11. Write out the independent components of Hooke's generalized law for strain in terms of stress, and stress in terms of strain, in cylindrical coordinates. Express your results using the parameters ν and E. (Assume a linear elastic, homogeneous, isotropic material.)

12. Write out the independent components of Hooke's generalized law for strain in terms of stress, and stress in terms of strain in spherical coordinates. Express your results using the parameters ν and E. (Assume a linear elastic, homogeneous, isotropic material.)

13. For a linear elastic, homogeneous, isotropic material assume there exists a state of plane strain in Cartesian coordinates. Verify the equilibrium equations are

$$\frac{\partial \sigma_{xx}}{\partial x} + \frac{\partial \sigma_{xy}}{\partial y} + \varrho b_x = 0$$

$$\frac{\partial \sigma_{yx}}{\partial x} + \frac{\partial \sigma_{yy}}{\partial y} + \varrho b_y = 0$$

$$\frac{\partial \sigma_{zz}}{\partial z} + \varrho b_z = 0$$

Hint: See problem 14, Exercise 2.3.

▶ **14 .** For a linear elastic, homogeneous, isotropic material assume there exists a state of plane strain in polar coordinates. Verify the equilibrium equations are

$$\frac{\partial \sigma_{rr}}{\partial r} + \frac{1}{r}\frac{\partial \sigma_{r\theta}}{\partial \theta} + \frac{1}{r}(\sigma_{rr} - \sigma_{\theta\theta}) + \varrho b_r = 0$$

$$\frac{\partial \sigma_{r\theta}}{\partial r} + \frac{1}{r}\frac{\partial \sigma_{\theta\theta}}{\partial \theta} + \frac{2}{r}\sigma_{r\theta} + \varrho b_\theta = 0$$

$$\frac{\partial \sigma_{zz}}{\partial z} + \varrho b_z = 0$$

Hint: See problem 15, Exercise 2.3.

▶ **15.** For a linear elastic, homogeneous, isotropic material assume there exists a state of plane stress in Cartesian coordinates. Verify the equilibrium equations are

$$\frac{\partial \sigma_{xx}}{\partial x} + \frac{\partial \sigma_{xy}}{\partial y} + \varrho b_x = 0$$

$$\frac{\partial \sigma_{yx}}{\partial x} + \frac{\partial \sigma_{yy}}{\partial y} + \varrho b_y = 0$$

▶ **16.** Determine the compatibility equations in terms of the Airy stress function ϕ when there exists a state of plane stress. Assume the body forces are derivable from a potential function V.

▶ **17.** For a linear elastic, homogeneous, isotropic material assume there exists a state of plane stress in polar coordinates. Verify the equilibrium equations are

$$\frac{\partial \sigma_{rr}}{\partial r} + \frac{1}{r}\frac{\partial \sigma_{r\theta}}{\partial \theta} + \frac{1}{r}(\sigma_{rr} - \sigma_{\theta\theta}) + \varrho b_r = 0$$

$$\frac{\partial \sigma_{r\theta}}{\partial r} + \frac{1}{r}\frac{\partial \sigma_{\theta\theta}}{\partial \theta} + \frac{2}{r}\sigma_{r\theta} + \varrho b_\theta = 0$$

▶ **18.** Express each of the physical components of plane stress in polar coordinates, σ_{rr}, $\sigma_{\theta\theta}$, and $\sigma_{r\theta}$ in terms of the physical components of stress in Cartesian coordinates σ_{xx}, σ_{yy}, σ_{xy}. Hint: Consider the transformation law $\overline{\sigma}_{ij} = \sigma_{ab}\dfrac{\partial x^a}{\partial \overline{x}^i}\dfrac{\partial x^b}{\partial \overline{x}^j}$.

▶ **19.** Use the results from problem 19 and assume the stresses are derivable from the relations

$$\sigma_{xx} = V + \frac{\partial^2 \phi}{\partial y^2}, \qquad \sigma_{xy} = -\frac{\partial^2 \phi}{\partial x \partial y}, \qquad \sigma_{yy} = V + \frac{\partial^2 \phi}{\partial x^2}$$

where V is a potential function and ϕ is the Airy stress function. Show that upon changing to polar coordinates the Airy equations for stress become

$$\sigma_{rr} = V + \frac{1}{r}\frac{\partial \phi}{\partial r} + \frac{1}{r^2}\frac{\partial^2 \phi}{\partial \theta^2}, \qquad \sigma_{r\theta} = \frac{1}{r^2}\frac{\partial \phi}{\partial \theta} - \frac{1}{r}\frac{\partial^2 \phi}{\partial r \partial \theta}, \qquad \sigma_{\theta\theta} = V + \frac{\partial^2 \phi}{\partial r^2}.$$

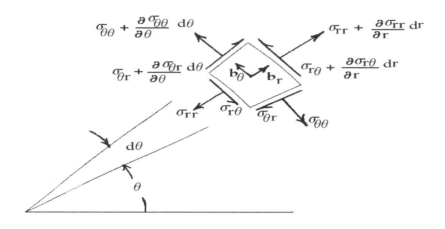

Figure 2.4-4. Polar element in equilibrium.

▶ **20.** Figure 2.4-4 illustrates the state of equilibrium on an element in polar coordinates assumed to be of unit length in the z-direction. Verify the stresses given in the figure and then sum the forces in the r and θ directions to derive the same equilibrium laws developed in the previous exercise.

Hint: Resolve the stresses into components in the r and θ directions. Use the results that $\sin \frac{d\theta}{2} \approx \frac{d\theta}{2}$ and $\cos \frac{d\theta}{2} \approx 1$ for small values of $d\theta$. Sum forces and then divide by $r\,dr\,d\theta$ and take the limit as $dr \to 0$ and $d\theta \to 0$.

▶ **21.** Verify that the Airy stress equations in polar coordinates, given in problem 20, satisfy the equilibrium equations in polar coordinates derived in problem 17.

22. In Cartesian coordinates show that the traction boundary conditions, equations (2.3.11), can be written in terms of the constants λ and μ as

$$T_1 = \lambda n_1 e_{kk} + \mu \left[2n_1 \frac{\partial u_1}{\partial x^1} + n_2 \left(\frac{\partial u_1}{\partial x^2} + \frac{\partial u_2}{\partial x^1} \right) + n_3 \left(\frac{\partial u_1}{\partial x^3} + \frac{\partial u_3}{\partial x^1} \right) \right]$$

$$T_2 = \lambda n_2 e_{kk} + \mu \left[n_1 \left(\frac{\partial u_2}{\partial x^1} + \frac{\partial u_1}{\partial x^2} \right) + 2n_2 \frac{\partial u_2}{\partial x^2} + n_3 \left(\frac{\partial u_2}{\partial x^3} + \frac{\partial u_3}{\partial x^2} \right) \right]$$

$$T_3 = \lambda n_3 e_{kk} + \mu \left[n_1 \left(\frac{\partial u_3}{\partial x^1} + \frac{\partial u_1}{\partial x^3} \right) + n_2 \left(\frac{\partial u_3}{\partial x^2} + \frac{\partial u_2}{\partial x^3} \right) + 2n_3 \frac{\partial u_3}{\partial x^3} \right]$$

where (n_1, n_2, n_3) are the direction cosines of the unit normal to the surface, u_1, u_2, u_3 are the components of the displacements and T_1, T_2, T_3 are the surface tractions.

▶ **23.** Consider an infinite plane subject to tension in the x−direction only. Assume a state of plane strain and let $\sigma_{xx} = T$ with $\sigma_{xy} = \sigma_{yy} = 0$. Find the strain components e_{xx}, e_{yy} and e_{xy}. Also find the displacement field $u = u(x, y)$ and $v = v(x, y)$.

276

▶ **24.** Consider an infinite plane subject to tension in the y-direction only. Assume a state of plane strain and let $\sigma_{yy} = T$ with $\sigma_{xx} = \sigma_{xy} = 0$. Find the strain components e_{xx}, e_{yy} and e_{xy}. Also find the displacement field $u = u(x, y)$ and $v = v(x, y)$.

▶ **25.** Consider an infinite plane subject to tension in both the x and y directions. Assume a state of plane strain and let $\sigma_{xx} = T$, $\sigma_{yy} = T$ and $\sigma_{xy} = 0$. Find the strain components e_{xx}, e_{yy} and e_{xy}. Also find the displacement field $u = u(x, y)$ and $v = v(x, y)$.

▶ **26.** An infinite cylindrical rod of radius R_0 has an external pressure P_0 as illustrated in figure 2.5-5. Find the stress and displacement fields.

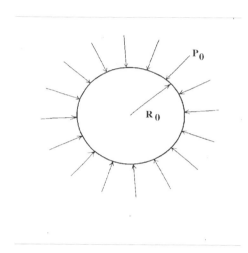

Figure 2.4-5. External pressure on a rod.

▶ **27.** An infinite plane has a circular hole of radius R_1 with an internal pressure P_1 as illustrated in the figure 2.4-6. Find the stress and displacement fields.

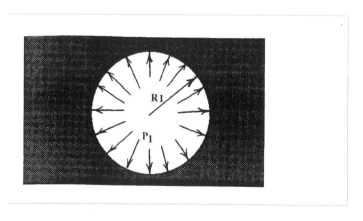

Figure 2.4-6. Internal pressure on circular hole.

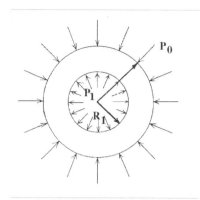

Figure 2.4-7. Tube with internal and external pressure.

28. A tube of inner radius R_1 and outer radius R_0 has an internal pressure of P_1 and an external pressure of P_0 as illustrated in the figure 2.4-7. Verify the stress and displacement fields derived in example 2.4-7.

29. Use Cartesian tensors and combine the equations of equilibrium $\sigma_{ij,j} + \varrho b_i = 0$, Hooke's law $\sigma_{ij} = \lambda e_{kk} \delta_{ij} + 2\mu e_{ij}$ and the strain tensor $e_{ij} = \dfrac{1}{2}(u_{i,j} + u_{j,i})$ and derive the Navier equations of equilibrium

$$\sigma_{ij,j} + \varrho b_i = (\lambda + \mu)\frac{\partial \Theta}{\partial x^i} + \mu \frac{\partial^2 u_i}{\partial x^k \partial x^k} + \varrho b_i = 0,$$

where $\Theta = e_{11} + e_{22} + e_{33}$ is the dilatation.

30. Show the Navier equations in problem 29 can be written in the tensor form

$$\mu u_{i,jj} + (\lambda + \mu)u_{j,ji} + \varrho b_i = 0$$

or the vector form

$$\mu \nabla^2 \vec{u} + (\lambda + \mu)\nabla (\nabla \cdot \vec{u}) + \varrho \vec{b} = \vec{0}.$$

31. Show that in an orthogonal coordinate system the components of $\nabla(\nabla \cdot \vec{u})$ can be expressed in terms of physical components by the relation

$$[\nabla (\nabla \cdot \vec{u})]_i = \frac{1}{h_i}\frac{\partial}{\partial x^i}\left\{ \frac{1}{h_1 h_2 h_3}\left[\frac{\partial (h_2 h_3 u(1))}{\partial x^1} + \frac{\partial (h_1 h_3 u(2))}{\partial x^2} + \frac{\partial (h_1 h_2 u(3))}{\partial x^3} \right] \right\}$$

32. Show that in orthogonal coordinates the components of $\nabla^2 \vec{u}$ can be written

$$\left[\nabla^2 \vec{u}\right]_i = g^{jk} u_{i,jk} = A_i$$

and in terms of physical components one can write

$$h_i A(i) = \sum_{j=1}^{3} \frac{1}{h_j^2}\left[\frac{\partial^2 (h_i u(i))}{\partial x^j \partial x^j} - 2\sum_{m=1}^{3}\begin{Bmatrix} m \\ i\,j \end{Bmatrix}\frac{\partial (h_m u(m))}{\partial x^j} - \sum_{m=1}^{3}\begin{Bmatrix} m \\ j\,j \end{Bmatrix}\frac{\partial (h_i u(i))}{\partial x^m} \right.$$

$$\left. - \sum_{m=1}^{3} h_m u(m)\left(\frac{\partial}{\partial x^j}\begin{Bmatrix} m \\ i\,j \end{Bmatrix} - \sum_{p=1}^{3}\begin{Bmatrix} m \\ i\,p \end{Bmatrix}\begin{Bmatrix} p \\ j\,j \end{Bmatrix} - \sum_{p=1}^{3}\begin{Bmatrix} m \\ j\,p \end{Bmatrix}\begin{Bmatrix} p \\ i\,j \end{Bmatrix} \right) \right]$$

278

▶ **33.** Use the results in problem 32 to show in Cartesian coordinates the physical components of $[\nabla^2 \vec{u}]_i = A_i$ can be represented

$$[\nabla^2 \vec{u}] \cdot \hat{e}_1 = A(1) = \frac{\partial^2 u}{\partial x^2} + \frac{\partial^2 u}{\partial y^2} + \frac{\partial^2 u}{\partial z^2}$$

$$[\nabla^2 \vec{u}] \cdot \hat{e}_2 = A(2) = \frac{\partial^2 v}{\partial x^2} + \frac{\partial^2 v}{\partial y^2} + \frac{\partial^2 v}{\partial z^2}$$

$$[\nabla^2 \vec{u}] \cdot \hat{e}_3 = A(3) = \frac{\partial^2 w}{\partial x^2} + \frac{\partial^2 w}{\partial y^2} + \frac{\partial^2 w}{\partial z^2}$$

where (u, v, w) are the components of the displacement vector $\vec{u}$.

▶ **34.** Use the results in problem 32 to show in cylindrical coordinates the physical components of $[\nabla^2 \vec{u}]_i = A_i$ can be represented

$$[\nabla^2 \vec{u}] \cdot \hat{e}_r = A(1) = \nabla^2 u_r - \frac{1}{r^2} u_r - \frac{2}{r^2} \frac{\partial u_\theta}{\partial \theta}$$

$$[\nabla^2 \vec{u}] \cdot \hat{e}_\theta = A(2) = \nabla^2 u_\theta + \frac{2}{r^2} \frac{\partial u_r}{\partial \theta} - \frac{1}{r^2} u_\theta$$

$$[\nabla^2 \vec{u}] \cdot \hat{e}_z = A(3) = \nabla^2 u_z$$

where u_r, u_θ, u_z are the physical components of $\vec{u}$ and $\nabla^2 \alpha = \frac{\partial^2 \alpha}{\partial r^2} + \frac{1}{r} \frac{\partial \alpha}{\partial r} + \frac{1}{r^2} \frac{\partial^2 \alpha}{\partial \theta^2} + \frac{\partial^2 \alpha}{\partial z^2}$

▶ **35.** Use the results in problem 32 to show in spherical coordinates the physical components of $[\nabla^2 \vec{u}]_i = A_i$ can be represented

$$[\nabla^2 \vec{u}] \cdot \hat{e}_\rho = A(1) = \nabla^2 u_\rho - \frac{2}{\rho^2} u_\rho - \frac{2}{\rho^2} \frac{\partial u_\theta}{\partial \theta} - \frac{2 \cot \theta}{\rho^2} u_\theta - \frac{2}{\rho^2 \sin \theta} \frac{\partial u_\phi}{\partial \phi}$$

$$[\nabla^2 \vec{u}] \cdot \hat{e}_\theta = A(2) = \nabla^2 u_\theta + \frac{2}{\rho^2} \frac{\partial u_\rho}{\partial \theta} - \frac{1}{\rho^2 \sin \theta} u_\theta - \frac{2 \cos \theta}{\rho^2 \sin^2 \theta} \frac{\partial u_\theta}{\partial \phi}$$

$$[\nabla^2 \vec{u}] \cdot \hat{e}_\phi = A(3) = \nabla^2 u_\phi - \frac{1}{\rho^2 \sin^2 \theta} u_\phi + \frac{2}{\rho^2 \sin \theta} \frac{\partial u_\rho}{\partial \phi} + \frac{2 \cos \theta}{\rho^2 \sin^2 \theta} \frac{\partial u_\theta}{\partial \phi}$$

where u_ρ, u_θ, u_ϕ are the physical components of $\vec{u}$ and where

$$\nabla^2 \alpha = \frac{\partial^2 \alpha}{\partial \rho^2} + \frac{2}{\rho} \frac{\partial \alpha}{\partial \rho} + \frac{1}{\rho^2} \frac{\partial^2 \alpha}{\partial \theta^2} + \frac{\cot \theta}{\rho^2} \frac{\partial \alpha}{\partial \theta} + \frac{1}{\rho^2 \sin^2 \theta} \frac{\partial^2 \alpha}{\partial \phi^2}$$

▶ **36.** Combine the results from problems 30,31,32 and 33 and write the Navier equations of equilibrium in Cartesian coordinates. Alternatively, write the stress-strain relations (2.4.29(b)) in terms of physical components and then use these results, together with the results from Exercise 2.3, problems 2 and 14, to derive the Navier equations.

▶ **37.** Combine the results from problems 30,31,32 and 34 and write the Navier equations of equilibrium in cylindrical coordinates. Alternatively, write the stress-strain relations (2.4.29(b)) in terms of physical components and then use these results, together with the results from Exercise 2.3, problems 3 and 15, to derive the Navier equations.

▶ **38.** Combine the results from problems 30,31,32 and 35 and write the Navier equations of equilibrium in spherical coordinates. Alternatively, write the stress-strain relations (2.4.29(b)) in terms of physical components and then use these results, together with the results from Exercise 2.3, problems 4 and 16, to derive the Navier equations.

39. Assume $\varrho\vec{b} = -\text{grad}\,V$ and let ϕ denote the Airy stress function defined by

$$\sigma_{xx} = V + \frac{\partial^2\phi}{\partial y^2}$$

$$\sigma_{yy} = V + \frac{\partial^2\phi}{\partial x^2}$$

$$\sigma_{xy} = -\frac{\partial^2\phi}{\partial x\partial y}$$

(a) Show that for conditions of plane strain the equilibrium equations in two dimensions are satisfied by the above definitions. (b) Express the compatibility equation

$$\frac{\partial^2 e_{xx}}{\partial y^2} + \frac{\partial^2 e_{yy}}{\partial x^2} = 2\frac{\partial^2 e_{xy}}{\partial x\partial y}$$

in terms of ϕ and V and show that

$$\nabla^4\phi + \frac{1-2\nu}{1-\nu}\nabla^2 V = 0.$$

40. Consider the case where the body forces are conservative and derivable from a scalar potential function such that $\varrho b_i = -V_{,i}$. Show that under conditions of plane strain in rectangular Cartesian coordinates the compatibility equation $e_{11,22} + e_{22,11} = 2e_{12,12}$ can be reduced to the form $\nabla^2\sigma_{ii} = \dfrac{1}{1-\nu}\nabla^2 V \qquad , i = 1, 2$ involving the stresses and the potential. Hint: Differentiate the equilibrium equations.

41. Use the relation $\sigma_j^i = 2\mu e_j^i + \lambda e_m^m \delta_j^i$ and solve for the strain in terms of the stress.

42. Derive the equation (2.4.26) from the equation (2.4.23).

43. In two dimensions assume that the body forces are derivable from a potential function V and $\varrho b^i = -g^{ij}V_{,j}$. Also assume that the stress is derivable from the Airy stress function and the potential function by employing the relations $\sigma^{ij} = \epsilon^{im}\epsilon^{jn}u_{m,n} + g^{ij}V \qquad i, j, m, n = 1, 2$ where $u_m = \phi_{,m}$ and ϵ^{pq} is the two dimensional epsilon permutation symbol and all indices have the range 1,2.

(a) Show that $\epsilon^{im}\epsilon^{jn}(\phi_m)_{,nj} = 0$.

(b) Show that $\sigma^{ij}_{,j} = -\varrho b^i$.

(c) Verify the stress laws for cylindrical and Cartesian coordinates given in problem 20 by using the above expression for σ^{ij}. Hint: Expand the contravariant derivative and convert all terms to physical components. Also recall that $\epsilon^{ij} = \frac{1}{\sqrt{g}}e^{ij}$.

44. Consider a material with body forces per unit volume $\rho F^i, i = 1, 2, 3$ and surface tractions denoted by $\sigma^r = \sigma^{rj}n_j$, where n_j is a unit surface normal. Further, let δu_i denote a small displacement vector associated with a small variation in the strain δe_{ij}.

(a) Show the work done during a small variation in strain is $\delta W = \delta W_B + \delta W_S$ where $\delta W_B = \displaystyle\int_V \rho F^i\delta u_i\,d\tau$ is a volume integral representing the work done by the body forces and $\delta W_S = \displaystyle\int_S \sigma^r\delta u_r\,dS$ is a surface integral representing the work done by the surface forces.

(b) Using the Gauss divergence theorem show that the work done can be represented as

$$\delta W = \frac{1}{2}\int_V c^{ijmn}\delta[e_{mn}e_{ij}]\,d\tau \quad\text{or}\quad W = \frac{1}{2}\int_V \sigma^{ij}e_{ij}\,d\tau.$$

The scalar quantity $\frac{1}{2}\sigma^{ij}e_{ij}$ is called the strain energy density or strain energy per unit volume.

Hint: Interchange subscripts, add terms and calculate $2W = \int_V \sigma^{ij}[\delta u_{i,j} + \delta u_{j,i}]\,d\tau$.

▶ **45.** Consider a spherical shell subjected to an internal pressure p_i and external pressure p_o. Let a denote the inner radius and b the outer radius of the spherical shell. Find the displacement and stress fields in spherical coordinates (ρ, θ, ϕ).

Hint: Assume symmetry in the θ and ϕ directions and let the physical components of displacements satisfy the relations $u_\rho = u_\rho(\rho), \quad u_\theta = u_\phi = 0$.

▶ **46.** (a) Verify the average normal stress is proportional to the dilatation, where the proportionality constant is the bulk modulus of elasticity. i.e. Show that $\frac{1}{3}\sigma_i^i = \frac{E}{1-2\nu}\frac{1}{3}e_i^i = ke_i^i$ where k is the bulk modulus of elasticity.

(b) Define the quantities of strain deviation and stress deviation in terms of the average normal stress $s = \frac{1}{3}\sigma_i^i$ and average cubic dilatation $e = \frac{1}{3}e_i^i$ as follows

$$\text{strain deviator} \qquad \varepsilon_j^i = e_j^i - e\delta_j^i$$
$$\text{stress deviator} \qquad s_j^i = \sigma_j^i - s\delta_j^i$$

Show that zero results when a contraction is performed on the stress and strain deviators. (The above definitions are used to split the strain tensor into two parts. One part represents pure dilatation and the other part represents pure distortion.)

(c) Show that $(1 - 2\nu)s = Ee \quad$ or $\quad s = (3\lambda + 2\mu)e$

(d) Express Hooke's law in terms of the strain and stress deviator and show

$$E(\varepsilon_j^i + e\delta_j^i) = (1 + \nu)s_j^i + (1 - 2\nu)s\delta_j^i$$

which simplifies to $s_j^i = 2\mu\varepsilon_j^i$.

▶ **47.** Show the strain energy density (problem 44) can be written in terms of the stress and strain deviators (problem 46) and

$$W = \frac{1}{2}\int_V \sigma^{ij}e_{ij}\,d\tau = \frac{1}{2}\int_V (3se + s^{ij}\varepsilon_{ij})\,d\tau$$

and from Hooke's law

$$W = \frac{3}{2}\int_V \left((3\lambda + 2\mu)e^2 + \frac{2\mu}{3}\varepsilon^{ij}\varepsilon_{ij}\right)d\tau.$$

▶ **48.** Find the stress $\sigma_{rr}, \sigma_{r\theta}$ and $\sigma_{\theta\theta}$ in an infinite plate with a small circular hole, which is traction free, when the plate is subjected to a pure shearing force F_{12}. Determine the maximum stress.

▶ **49.** Show that in terms of E and ν

$$C_{1111} = \frac{E(1 - \nu)}{(1 + \nu)(1 - 2\nu)} \qquad C_{1122} = \frac{E\nu}{(1 + \nu)(1 - 2\nu)} \qquad C_{1212} = \frac{E}{2(1 + \nu)}$$

▶ **50.** Show that in Cartesian coordinates the quantity

$$S = \sigma_{xx}\sigma_{yy} + \sigma_{yy}\sigma_{zz} + \sigma_{zz}\sigma_{xx} - (\sigma_{xy})^2 - (\sigma_{yz})^2 - (\sigma_{xz})^2$$

is a stress invariant. Hint: First verify that in tensor form $S = \frac{1}{2}(\sigma_{ii}\sigma_{jj} - \sigma_{ij}\sigma_{ij})$.

51. Show that in Cartesian coordinates for a state of plane strain where the displacements are given by $u = u(x, y), v = v(x, y)$ and $w = 0$, the stress components must satisfy the equations

$$\frac{\partial \sigma_{xx}}{\partial x} + \frac{\partial \sigma_{xy}}{\partial y} + \varrho b_x = 0$$

$$\frac{\partial \sigma_{yx}}{\partial x} + \frac{\partial \sigma_{yy}}{\partial y} + \varrho b_y = 0$$

$$\nabla^2(\sigma_{xx} + \sigma_{yy}) = \frac{-\varrho}{1-\nu}\left(\frac{\partial b_x}{\partial x} + \frac{\partial b_y}{\partial y}\right)$$

52. Show that in Cartesian coordinates for a state of plane stress where $\sigma_{xx} = \sigma_{xx}(x, y)$, $\sigma_{yy} = \sigma_{yy}(x, y)$, $\sigma_{xy} = \sigma_{xy}(x, y)$ and $\sigma_{xz} = \sigma_{yz} = \sigma_{zz} = 0$ the stress components must satisfy

$$\frac{\partial \sigma_{xx}}{\partial x} + \frac{\partial \sigma_{xy}}{\partial y} + \varrho b_x = 0$$

$$\frac{\partial \sigma_{yx}}{\partial x} + \frac{\partial \sigma_{yy}}{\partial y} + \varrho b_y = 0$$

$$\nabla^2(\sigma_{xx} + \sigma_{yy}) = -\varrho(\nu + 1)\left(\frac{\partial b_x}{\partial x} + \frac{\partial b_y}{\partial y}\right)$$

53. Consider a state of plane stress.

(a) Show that the equilibrium equations are

$$\frac{\partial \sigma_{xx}}{\partial x} + \frac{\partial \sigma_{xy}}{\partial y} + \varrho b_x = 0, \qquad \frac{\partial \sigma_{xy}}{\partial x} + \frac{\partial \sigma_{yy}}{\partial y} + \varrho b_y = 0$$

(b) Show the strain-stress relations become

$$e_{xx} = \frac{1}{E}(\sigma_{xx} - \nu\sigma_{yy}), \quad e_{xy} = \frac{1+\nu}{E}\sigma_{xy}, \quad e_{yy} = \frac{1}{E}(\sigma_{yy} - \nu\sigma_{xx})$$

(c) Show the strain-displacement relations simplify to

$$e_{xx} = \frac{\partial u}{\partial x}, \quad e_{xy} = \frac{1}{2}\left(\frac{\partial u}{\partial y} + \frac{\partial v}{\partial x}\right), \quad e_{yy} = \frac{\partial v}{\partial y}$$

(d) Using algebra, eliminate all strain components from the equations in parts (a)(b) and (c) above and show

$$\sigma_{xx} = \frac{E}{1-\nu^2}\left(\frac{\partial u}{\partial x} + \nu\frac{\partial v}{\partial y}\right), \quad \sigma_{xy} = \mu\left(\frac{\partial u}{\partial y} + \frac{\partial v}{\partial x}\right), \quad \sigma_{yy} = \frac{E}{1-\nu^2}\left(\frac{\partial v}{\partial y} + \nu\frac{\partial u}{\partial x}\right)$$

(e) Substitute equations from part (d) into equilibrium equations and derive the Navier equations for plane stress

$$\mu\nabla^2 u + \frac{E}{2(1-\nu)}\frac{\partial}{\partial x}\left(\frac{\partial u}{\partial x} + \frac{\partial v}{\partial y}\right) + \varrho b_x = 0$$

$$\mu\nabla^2 v + \frac{E}{2(1-\nu)}\frac{\partial}{\partial y}\left(\frac{\partial u}{\partial x} + \frac{\partial v}{\partial y}\right) + \varrho b_y = 0$$

282

§2.5 CONTINUUM MECHANICS (FLUIDS)

Let us consider a fluid medium and use Cartesian tensors to derive the mathematical equations that describe how a fluid behaves. A fluid continuum, like a solid continuum, is characterized by equations describing:

1. Conservation of linear momentum

$$\sigma_{ij,j} + \varrho b_i = \varrho \dot{v}_i \tag{2.5.1}$$

2. Conservation of angular momentum $\sigma_{ij} = \sigma_{ji}$.

3. Conservation of mass (continuity equation)

$$\frac{\partial \varrho}{\partial t} + \frac{\partial \varrho}{\partial x_i} v_i + \varrho \frac{\partial v_i}{\partial x_i} = 0 \quad \text{or} \quad \frac{D\varrho}{Dt} + \varrho \nabla \cdot \vec{V} = 0. \tag{2.5.2}$$

In the above equations $v_i, i = 1, 2, 3$ is a velocity field, ϱ is the density of the fluid, σ_{ij} is the stress tensor and b_j is an external force per unit mass. In the cgs system of units of measurement, the above quantities have dimensions

$$[\dot{v}_j] = \text{cm/sec}^2, \quad [b_j] = dynes/g, \quad [\sigma_{ij}] = dyne/cm^2, \quad [\varrho] = g/cm^3. \tag{2.5.3}$$

The displacement field $u_i, i = 1, 2, 3$ can be represented in terms of the velocity field $v_i, i = 1, 2, 3$, by the relation

$$u_i = \int_0^t v_i \, dt. \tag{2.5.4}$$

The strain tensor components of the medium can then be represented in terms of the velocity field as

$$e_{ij} = \frac{1}{2}(u_{i,j} + u_{j,i}) = \int_0^t \frac{1}{2}(v_{i,j} + v_{j,i}) \, dt = \int_0^t D_{ij} \, dt, \tag{2.5.5}$$

where

$$D_{ij} = \frac{1}{2}(v_{i,j} + v_{j,i}) \tag{2.5.6}$$

is called the *rate of deformation tensor*, *velocity strain tensor*, or *rate of strain tensor*.

Note the difference in the equations describing a solid continuum compared with those for a fluid continuum. In describing a solid continuum we were primarily interested in calculating the displacement field $u_i, i = 1, 2, 3$ when the continuum was subjected to external forces. In describing a fluid medium, we calculate the velocity field $v_i, i = 1, 2, 3$ when the continuum is subjected to external forces. We therefore replace the strain tensor relations by the velocity strain tensor relations in all future considerations concerning the study of fluid motion.

Constitutive Equations for Fluids

In addition to the above basic equations, we will need a set of constitutive equations which describe the material properties of the fluid. Toward this purpose consider an arbitrary point within the fluid medium and pass an imaginary plane through the point. The orientation of the plane is determined by a unit normal n_i, $i = 1, 2, 3$ to the planar surface. For a fluid at rest we wish to determine the stress vector $t_i^{(n)}$ acting on the plane element passing through the selected point P. We desire to express $t_i^{(n)}$ in terms of the stress tensor σ_{ij}. The superscript (n) on the stress vector is to remind you that the stress acting on the planar element depends upon the orientation of the plane through the point.

283

We make the assumption that $t_i^{(n)}$ is colinear with the normal vector to the surface passing through the selected point. It is also assumed that for fluid elements at rest, there are no shear forces acting on the planar element through an arbitrary point and therefore the stress tensor σ_{ij} should be independent of the orientation of the plane. That is, we desire for the stress vector σ_{ij} to be an isotropic tensor. This requires σ_{ij} to have a specific form. To find this specific form we let σ_{ij} denote the stress components in a general coordinate system x^i, $i=1,2,3$ and let $\overline{\sigma}_{ij}$ denote the components of stress in a barred coordinate system $\overline{x}^i, i=1,2,3$. Since σ_{ij} is a tensor, it must satisfy the transformation law

$$\overline{\sigma}_{mn} = \sigma_{ij}\frac{\partial x^i}{\partial \overline{x}^m}\frac{\partial x^j}{\partial \overline{x}^n}, \quad i,j,m,n = 1,2,3. \tag{2.5.7}$$

We desire for the stress tensor σ_{ij} to be an invariant under an arbitrary rotation of axes. Consider therefore the special coordinate transformations illustrated in the figures 2.5-1(a) and (b).

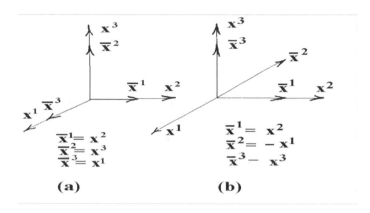

Figure 2.5-1. Coordinate transformations due to rotations

For the transformation equations given in figure 2.5-1(a), the stress tensor in the barred system of coordinates is

$$\begin{array}{lll}
\overline{\sigma}_{11} = \sigma_{22} & \overline{\sigma}_{21} = \sigma_{32} & \overline{\sigma}_{31} = \sigma_{12} \\
\overline{\sigma}_{12} = \sigma_{23} & \overline{\sigma}_{22} = \sigma_{33} & \overline{\sigma}_{32} = \sigma_{13} \\
\overline{\sigma}_{13} = \sigma_{21} & \overline{\sigma}_{23} = \sigma_{31} & \overline{\sigma}_{33} = \sigma_{11}.
\end{array} \tag{2.5.8}$$

If σ_{ij} is to be isotropic, we desire that $\overline{\sigma}_{11} = \sigma_{11}$, $\overline{\sigma}_{22} = \sigma_{22}$ and $\overline{\sigma}_{33} = \sigma_{33}$. If the equations (2.5.8) are to produce these results, we require that σ_{11}, σ_{22} and σ_{33} must be equal. We denote these common values by $(-p)$. In particular, the equations (2.5.8) show that if $\overline{\sigma}_{11} = \sigma_{11}$, $\overline{\sigma}_{22} = \sigma_{22}$ and $\overline{\sigma}_{33} = \sigma_{33}$, then we must require that $\sigma_{11} = \sigma_{22} = \sigma_{33} = -p$. If $\overline{\sigma}_{12} = \sigma_{12}$ and $\overline{\sigma}_{23} = \sigma_{23}$, then we also require that $\sigma_{12} = \sigma_{23} = \sigma_{31}$. We note that if $\overline{\sigma}_{13} = \sigma_{13}$ and $\overline{\sigma}_{32} = \sigma_{32}$, then we require that $\sigma_{21} = \sigma_{32} = \sigma_{13}$. If the equations (2.5.7) are expanded using the transformation given in figure 2.5-1(b), we obtain the additional requirements that

$$\begin{array}{lll}
\overline{\sigma}_{11} = \sigma_{22} & \overline{\sigma}_{21} = -\sigma_{12} & \overline{\sigma}_{31} = \sigma_{32} \\
\overline{\sigma}_{12} = -\sigma_{21} & \overline{\sigma}_{22} = \sigma_{11} & \overline{\sigma}_{32} = -\sigma_{31} \\
\overline{\sigma}_{13} = \sigma_{23} & \overline{\sigma}_{23} = -\sigma_{13} & \overline{\sigma}_{33} = \sigma_{33}.
\end{array} \tag{2.5.9}$$

Analysis of these equations implies that if σ_{ij} is to be isotropic, then $\overline{\sigma}_{21} = \sigma_{21} = -\sigma_{12} = -\sigma_{21}$

or $\sigma_{21} = 0$ which implies $\quad \sigma_{12} = \sigma_{23} = \sigma_{31} = \sigma_{21} = \sigma_{32} = \sigma_{13} = 0.$ \hfill (2.5.10)

The above analysis demonstrates that if the stress tensor σ_{ij} is to be isotropic, it must have the form

$$\sigma_{ij} = -p\delta_{ij}. \tag{2.5.11}$$

Use the traction condition (2.3.11), and express the stress vector as

$$t_j^{(n)} = \sigma_{ij}n_i = -pn_j. \tag{2.5.12}$$

This equation is interpreted as representing the stress vector at a point on a surface with outward unit normal n_i, where p is the pressure (hydrostatic pressure) stress magnitude assumed to be positive. The negative sign in equation (2.5.12) denotes a compressive stress.

Imagine a submerged object in a fluid medium. We further imagine the object to be covered with unit normal vectors emanating from each point on its surface. The equation (2.5.12) shows that the hydrostatic pressure always acts on the object in a compressive manner. A force results from the stress vector acting on the object. The direction of the force is opposite to the direction of the unit outward normal vectors. It is a compressive force at each point on the surface of the object.

The above considerations were for a fluid at rest (hydrostatics). For a fluid in motion (hydrodynamics) a different set of assumptions must be made. Hydrodynamical experiments show that the shear stress components are not zero and so we assume a stress tensor having the form

$$\sigma_{ij} = -p\delta_{ij} + \tau_{ij}, \quad i, j = 1, 2, 3, \tag{2.5.13}$$

where τ_{ij} is called the viscous stress tensor. Note that all real fluids are both viscous and compressible.

Definition: (Viscous/inviscid fluid) If the viscous stress tensor τ_{ij} is zero for all i, j, then the fluid is called an inviscid, non-viscous, ideal or perfect fluid. The fluid is called viscous when τ_{ij} is different from zero.

In these notes it is assumed that the equation (2.5.13) represents the basic form for constitutive equations describing fluid motion.

285

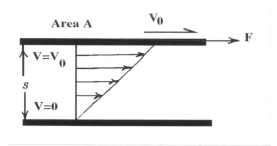

Figure 2.5-2. Viscosity experiment.

Viscosity

Most fluids are characterized by the fact that they cannot resist shearing stresses. That is, if you put a shearing stress on the fluid, the fluid gives way and flows. Consider the experiment illustrated in the figure 2.5-2 which illustrates a fluid moving between two parallel plane surfaces. Let S denote the distance between the two planes. Now keep the lower surface fixed or stationary and move the upper surface parallel to the lower surface with a constant velocity $\vec{V_0}$. If you measure the force F required to maintain the constant velocity of the upper surface, you discover that the force F varies directly as the area A of the surface and the ratio V_0/S. This is expressed in the form

$$\frac{F}{A} = \mu^* \frac{V_0}{S}. \tag{2.5.14}$$

The constant μ^* is a proportionality constant called the coefficient of viscosity. The viscosity usually depends upon temperature, but throughout our discussions we will assume the temperature is constant. A dimensional analysis of the equation (2.5.14) implies that the basic dimension of the viscosity is $[\mu^*] = ML^{-1}T^{-1}$. For example, $[\mu^*] = \text{gm}/(\text{cm sec})$ in the cgs system of units. The viscosity is usually measured in units of centipoise where one centipoise represents one-hundredth of a poise, where the unit of 1 poise= 1 gram per centimeter per second. The result of the above experiment shows that the stress is proportional to the change in velocity with change in distance or gradient of the velocity.

Linear Viscous Fluids

The above experiment with viscosity suggest that the viscous stress tensor τ_{ij} is dependent upon both the gradient of the fluid velocity and the density of the fluid.

In Cartesian coordinates, the simplest model suggested by the above experiment is that the viscous stress tensor τ_{ij} is proportional to the velocity gradient $v_{i,j}$ and so we write

$$\tau_{ik} = c_{ikmp}v_{m,p}, \tag{2.5.15}$$

where c_{ikmp} is a proportionality constant which is dependent upon the fluid density.

The viscous stress tensor must be independent of any reference frame, and hence we assume that the proportionality constants c_{ikmp} can be represented by an isotropic tensor. Recall that an isotropic tensor has the basic form

$$c_{ikmp} = \lambda^* \delta_{ik}\delta_{mp} + \mu^*(\delta_{im}\delta_{kp} + \delta_{ip}\delta_{km}) + \nu^*(\delta_{im}\delta_{kp} - \delta_{ip}\delta_{km}) \tag{2.5.16}$$

where λ^*, μ^* and ν^* are constants. Examining the results from equations (2.5.11) and (2.5.13) we find that if the viscous stress is symmetric, then $\tau_{ij} = \tau_{ji}$. This requires ν^* be chosen as zero and enables the viscous stress tensor to be represented in the form

$$\tau_{ik} = \lambda^* \delta_{ik} v_{p,p} + \mu^* (v_{k,i} + v_{i,k}). \tag{2.5.17}$$

The coefficient μ^* is called the first coefficient of viscosity and the coefficient λ^* is called the second coefficient of viscosity. Sometimes it is convenient to define

$$\zeta = \lambda^* + \frac{2}{3} \mu^* \tag{2.5.18}$$

as "another second coefficient of viscosity," or "bulk coefficient of viscosity." The condition of zero bulk viscosity is known as Stokes hypothesis. Many fluids problems assume the Stoke's hypothesis. This requires that the bulk coefficient be zero or very small. Under these circumstances the second coefficient of viscosity is related to the first coefficient of viscosity by the relation $\lambda^* = -\frac{2}{3} \mu^*$. In the study of shock waves and acoustic waves the Stoke's hypothesis is not applicable.

There are many tables and empirical formulas where the viscosity of different types of fluids or gases can be obtained. For example, in the study of the kinetic theory of gases the viscosity can be calculated from the Sutherland formula $\mu^* = \dfrac{C_1 g T^{3/2}}{T + C_2}$ where C_1, C_2 are constants for a specific gas. These constants can be found in certain tables. The quantity g is the gravitational constant and T is the temperature in degrees Rankine ($^oR = 460 + {}^oF$). Many other empirical formulas like the above exist. Also many graphs and tabular values of viscosity can be found. The table 5.1 lists the approximate values of the viscosity of some selected fluids and gases.

Table 5.1	Viscosity of selected fluids and gases in units of $\frac{\text{gram}}{\text{cm–sec}}$ = Poise at Atmospheric Pressure.			
Substance	$0°C$	$20°C$	$60°C$	$100°C$
Water	0.01798	0.01002	0.00469	0.00284
Alcohol	0.01773			
Ethyl Alcohol		0.012	0.00592	
Glycol		0.199	0.0495	0.0199
Mercury	0.017	0.0157	0.013	0.0100
Air	$1.708(10^{-4})$			$2.175(10^{-4})$
Helium	$1.86(10^{-4})$	$1.94(10^{-4})$		$2.28(10^{-4})$
Nitrogen	$1.658(10^{-4})$	$1.74(10^{-4})$	$1.92(10^{-4})$	$2.09(10^{-4})$

The viscous stress tensor given in equation (2.5.17) may also be expressed in terms of the rate of deformation tensor defined by equation (2.5.6). This representation is

$$\tau_{ij} = \lambda^* \delta_{ij} D_{kk} + 2\mu^* D_{ij}, \tag{2.5.19}$$

where $2D_{ij} = v_{i,j} + v_{j,i}$ and $D_{kk} = D_{11} + D_{22} + D_{33} = v_{1,1} + v_{2,2} + v_{3,3} = v_{i,i} = \Theta$ is the rate of change of the dilatation considered earlier. In Cartesian form, with velocity components u, v, w, the viscous stress tensor components are

$$\tau_{xx} = (\lambda^* + 2\mu^*)\frac{\partial u}{\partial x} + \lambda^*\left(\frac{\partial v}{\partial y} + \frac{\partial w}{\partial z}\right) \qquad \tau_{yx} = \tau_{xy} = \mu^*\left(\frac{\partial u}{\partial y} + \frac{\partial v}{\partial x}\right)$$

$$\tau_{yy} = (\lambda^* + 2\mu^*)\frac{\partial v}{\partial y} + \lambda^*\left(\frac{\partial u}{\partial x} + \frac{\partial w}{\partial z}\right) \qquad \tau_{zx} = \tau_{xz} = \mu^*\left(\frac{\partial w}{\partial x} + \frac{\partial u}{\partial z}\right)$$

$$\tau_{zz} = (\lambda^* + 2\mu^*)\frac{\partial w}{\partial z} + \lambda^*\left(\frac{\partial u}{\partial x} + \frac{\partial v}{\partial y}\right) \qquad \tau_{zy} = \tau_{yz} = \mu^*\left(\frac{\partial v}{\partial z} + \frac{\partial w}{\partial y}\right)$$

In cylindrical form, with velocity components v_r, v_θ, v_z, the viscous stress tensor components are

$$\tau_{rr} = 2\mu^*\frac{\partial v_r}{\partial r} + \lambda^*\nabla\cdot\vec{V}$$

$$\tau_{\theta\theta} = 2\mu^*\left(\frac{1}{r}\frac{\partial v_\theta}{\partial\theta} + \frac{v_r}{r}\right) + \lambda^*\nabla\cdot\vec{V}$$

$$\tau_{\theta r} = \tau_{r\theta} = \mu^*\left(\frac{1}{r}\frac{\partial v_r}{\partial\theta} + \frac{\partial v_\theta}{\partial r} - \frac{v_\theta}{r}\right)$$

$$\tau_{zz} = 2\mu^*\frac{\partial v_z}{\partial z} + \lambda^*\nabla\cdot\vec{V}$$

$$\tau_{rz} = \tau_{zr} = \mu^*\left(\frac{\partial v_r}{\partial z} + \frac{\partial v_z}{\partial r}\right)$$

$$\text{where}\quad \nabla\cdot\vec{V} = \frac{1}{r}\frac{\partial}{\partial r}(rv_r) + \frac{1}{r}\frac{\partial v_\theta}{\partial\theta} + \frac{\partial v_z}{\partial z}$$

$$\tau_{z\theta} = \tau_{\theta z} = \mu^*\left(\frac{1}{r}\frac{\partial v_z}{\partial\theta} + \frac{\partial v_\theta}{\partial z}\right)$$

In spherical coordinates, with velocity components v_ρ, v_θ, v_ϕ, the viscous stress tensor components have the form

$$\tau_{\rho\rho} = 2\mu^*\frac{\partial v_\rho}{\partial\rho} + \lambda^*\nabla\cdot\vec{V}$$

$$\tau_{\theta\theta} = 2\mu^*\left(\frac{1}{\rho}\frac{\partial v_\theta}{\partial\theta} + \frac{v_\rho}{\rho}\right) + \lambda^*\nabla\cdot\vec{V}$$

$$\tau_{\rho\theta} = \tau_{\theta\rho} = \mu^*\left(\rho\frac{\partial}{\partial\rho}\left(\frac{v_\theta}{\rho}\right) + \frac{1}{\rho}\frac{\partial v_\rho}{\partial\theta}\right)$$

$$\tau_{\phi\phi} = 2\mu^*\left(\frac{1}{\rho\sin\theta}\frac{\partial v_\phi}{\partial\phi} + \frac{v_\rho}{\rho} + \frac{v_\theta\cot\theta}{\rho}\right) + \lambda^*\nabla\cdot\vec{V}$$

$$\tau_{\phi\rho} = \tau_{\rho\phi} = \mu^*\left(\frac{1}{\rho\sin\theta}\frac{\partial v_r}{\partial\phi} + \rho\frac{\partial}{\partial\rho}\left(\frac{v_\theta}{\rho}\right)\right)$$

$$\text{where}\quad \nabla\cdot\vec{V} = \frac{1}{\rho^2}\frac{\partial}{\partial\rho}\left(\rho^2 v_\rho\right) + \frac{1}{\rho\sin\theta}\frac{\partial}{\partial\theta}(\sin\theta v_\theta) + \frac{1}{\rho\sin\theta}\frac{\partial v_\phi}{\partial\phi}$$

$$\tau_{\theta\phi} = \tau_{\phi\theta} = \mu^*\left(\frac{\sin\theta}{\rho}\frac{\partial}{\partial\theta}\left(\frac{v_\phi}{\sin\theta}\right) + \frac{1}{\rho\sin\theta}\frac{\partial v_\theta}{\partial\phi}\right)$$

Note that the viscous stress tensor is a linear function of the rate of deformation tensor D_{ij}. Such a fluid is called a *Newtonian fluid*. In cases where the viscous stress tensor is a nonlinear function of D_{ij} the fluid is called *non-Newtonian*.

> **Definition: (Newtonian Fluid)** If the viscous stress tensor τ_{ij} is expressible as a linear function of the rate of deformation tensor D_{ij}, the fluid is called a Newtonian fluid. Otherwise, the fluid is called a non-Newtonian fluid.

Important note: Do not assume an arbitrary form for the constitutive equations unless there is experimental evidence to support your assumption. A constitutive equation is a very important step in the modeling processes as it describes the material you are working with. One cannot arbitrarily assign a form to the viscous stress and expect the mathematical equations to describe the correct fluid behavior. The form of the viscous stress is an important part of the modeling process and by assigning different forms to the viscous stress tensor then various types of materials can be modeled. We restrict our study in these notes to Newtonian fluids.

288

In Cartesian coordinates the rate of deformation-stress constitutive equations for a Newtonian fluid can be written as

$$\sigma_{ij} = -p\delta_{ij} + \lambda^*\delta_{ij}D_{kk} + 2\mu^*D_{ij} \qquad (2.5.20)$$

which can also be written in the alternative form

$$\sigma_{ij} = -p\delta_{ij} + \lambda^*\delta_{ij}v_{k,k} + \mu^*(v_{i,j} + v_{j,i}) \qquad (2.5.21)$$

involving the gradient of the velocity.

Upon transforming from a Cartesian coordinate system $y^i, i = 1,2,3$ to a more general system of coordinates $\overline{x}^i, i = 1,2,3$, we write

$$\overline{\sigma}_{mn} = \sigma_{ij}\frac{\partial y^i}{\partial \overline{x}^m}\frac{\partial y^j}{\partial \overline{x}^n}. \qquad (2.5.22)$$

Now using the divergence from equation (2.1.3) and substituting equation (2.5.21) into equation (2.5.22) we obtain a more general expression for the constitutive equation. Performing the indicated substitutions there results

$$\overline{\sigma}_{mn} = \left[-p\delta_{ij} + \lambda^*\delta_{ij}v^k_{,k} + \mu^*(v_{i,j} + v_{j,i})\right]\frac{\partial y^i}{\partial \overline{x}^m}\frac{\partial y^j}{\partial \overline{x}^n}$$

$$\overline{\sigma}_{mn} = -p\overline{g}_{mn} + \lambda^*\overline{g}_{mn}\overline{v}^k_{,k} + \mu^*(\overline{v}_{m,n} + \overline{v}_{n,m}).$$

Dropping the bar notation, the stress-velocity strain relationships in the general coordinates $x^i, i = 1,2,3$, is

$$\sigma_{mn} = -pg_{mn} + \lambda^*g_{mn}g^{ik}v_{i,k} + \mu^*(v_{m,n} + v_{n,m}). \qquad (2.5.23)$$

Summary

The basic equations which describe the motion of a Newtonian fluid are :

Continuity equation (Conservation of mass)

$$\frac{\partial \varrho}{\partial t} + \left(\varrho v^i\right)_{,i} = 0, \quad \text{or} \quad \frac{D\varrho}{Dt} + \varrho\nabla \cdot \vec{V} = 0 \qquad 1 \text{ equation.} \qquad (2.5.24)$$

Conservation of linear momentum $\quad \sigma^{ij}_{,j} + \varrho b^i = \varrho\dot{v}^i, \qquad 3 \text{ equations}$

$$\text{or in vector form} \quad \varrho\frac{D\vec{V}}{Dt} = \varrho\vec{b} + \nabla \cdot \boldsymbol{\sigma} = \varrho\vec{b} - \nabla p + \nabla \cdot \boldsymbol{\tau} \qquad (2.5.25)$$

where $\boldsymbol{\sigma} = \sum_{i=1}^{3}\sum_{j=1}^{3}(-p\delta_{ij} + \tau_{ij})\,\hat{e}_i\,\hat{e}_j$ and $\boldsymbol{\tau} = \sum_{i=1}^{3}\sum_{j=1}^{3}\tau_{ij}\,\hat{e}_i\,\hat{e}_j$ are second order tensors. Conservation of angular momentum $\sigma^{ij} = \sigma^{ji}$, (Reduces the set of equations (2.5.23) to 6 equations.) Rate of deformation tensor (Velocity strain tensor)

$$D_{ij} = \frac{1}{2}\left(v_{i,j} + v_{j,i}\right), \qquad 6 \text{ equations.} \qquad (2.5.26)$$

Constitutive equations

$$\sigma_{mn} = -pg_{mn} + \lambda^*g_{mn}g^{ik}v_{i,k} + \mu^*(v_{m,n} + v_{n,m}), \qquad 6 \text{ equations.} \qquad (2.5.27)$$

In the cgs system of units the above quantities have the following units of measurements in Cartesian coordinates

$$v_i \quad \text{is the velocity field} , i = 1, 2, 3, \qquad [v_i] = \text{cm/sec}$$

$$\sigma_{ij} \quad \text{is the stress tensor}, i, j = 1, 2, 3, \qquad [\sigma_{ij}] = \text{dyne/cm}^2$$

$$\varrho \quad \text{is the fluid density} \qquad [\varrho] = \text{gm/cm}^3$$

$$b^i \quad \text{is the external body forces per unit mass} \qquad [b^i] = \text{dyne/gm}$$

$$D_{ij} \quad \text{is the rate of deformation tensor} \qquad [D_{ij}] = \text{sec}^{-1}$$

$$p \quad \text{is the pressure} \qquad [p] = \text{dyne/cm}^2$$

$$\lambda^*, \mu^* \quad \text{are coefficients of viscosity} \qquad [\lambda^*] = [\mu^*] = \text{Poise}$$

$$\text{where 1 Poise} = 1 \text{gm/cm sec}$$

If we assume the external body forces per unit mass are known, then the equations (2.5.24), (2.5.25), (2.5.26), and (2.5.27) represent 16 equations in the 16 unknowns

$$\varrho, v_1, v_2, v_3, \sigma_{11}, \sigma_{12}, \sigma_{13}, \sigma_{22}, \sigma_{23}, \sigma_{33}, D_{11}, D_{12}, D_{13}, D_{22}, D_{23}, D_{33}.$$

Navier-Stokes-Duhem Equations of Fluid Motion

Substituting the stress tensor from equation (2.5.27) into the linear momentum equation (2.5.25), and assuming that the viscosity coefficients are constants, we obtain the Navier-Stokes-Duhem equations for fluid motion. In Cartesian coordinates these equations can be represented in any of the equivalent forms

$$\varrho \dot{v}_i = \varrho b_i - p_{,j} \delta_{ij} + (\lambda^* + \mu^*) v_{k,ki} + \mu^* v_{i,jj}$$

$$\varrho \frac{\partial v_i}{\partial t} + \varrho v_j v_{i,j} = \varrho b_i + (-p \delta_{ij} + \tau_{ij})_{,j}$$

$$\frac{\partial \varrho v_i}{\partial t} + (\varrho v_i v_j + p \delta_{ij} - \tau_{ij})_{,j} = \varrho b_i \qquad (2.5.28)$$

$$\varrho \frac{D\vec{v}}{Dt} = \varrho \vec{b} - \nabla p + (\lambda^* + \mu^*) \nabla (\nabla \cdot \vec{v}) + \mu^* \nabla^2 \vec{v}$$

where $\dfrac{D\vec{v}}{Dt} = \dfrac{\partial \vec{v}}{\partial t} + (\vec{v} \cdot \nabla) \vec{v}$ is the material derivative, substantial derivative or convective derivative. This derivative is represented as

$$\dot{v}_i = \frac{\partial v_i}{\partial t} + \frac{\partial v_i}{\partial x^j} \frac{dx^j}{dt} = \frac{\partial v_i}{\partial t} + \frac{\partial v_i}{\partial x^j} v^j = \frac{\partial v_i}{\partial t} + v_{i,j} v^j. \qquad (2.5.29)$$

In the vector form of equations (2.5.28), the terms on the right-hand side of the equation represent force terms. The term $\varrho \vec{b}$ represents external body forces per unit volume. If these forces are derivable from a potential function ϕ, then the external forces are conservative and can be represented in the form $-\varrho \nabla \phi$. The term $-\nabla p$ is the gradient of the pressure and represents a force per unit volume due to hydrostatic pressure. The above statement is verified in the exercises that follow this section. The remaining terms can be written

$$\vec{f}_{viscous} = (\lambda^* + \mu^*) \nabla (\nabla \cdot \vec{v}) + \mu^* \nabla^2 \vec{v} \qquad (2.5.30)$$

and are given the physical interpretation of an internal force per unit volume. These internal forces arise from the shearing stresses in the moving fluid. If $\vec{f}_{viscous}$ is zero the vector equation in (2.5.28) is called Euler's equation.

If the viscosity coefficients are nonconstant, then the Navier-Stokes equations can be written in the Cartesian form

$$\varrho[\frac{\partial v_i}{\partial t} + v_j\frac{\partial v_i}{\partial x_j}] = \varrho b_i + \frac{\partial}{\partial x_j}\left[-p\delta_{ij} + \lambda^*\delta_{ij}\frac{\partial v_k}{\partial x_k} + \mu^*\left(\frac{\partial v_i}{\partial x_j} + \frac{\partial v_j}{\partial x_i}\right)\right]$$

$$= \varrho b_i - \frac{\partial p}{\partial x_i} + \frac{\partial}{\partial x_i}\left(\lambda^*\frac{\partial v_k}{\partial x_k}\right) + \frac{\partial}{\partial x^j}\left[\mu^*\left(\frac{\partial v_i}{\partial x_j} + \frac{\partial v_j}{\partial x_i}\right)\right]$$

which can also be written in terms of the bulk coefficient of viscosity $\zeta = \lambda^* + \frac{2}{3}\mu^*$ as

$$\varrho[\frac{\partial v_i}{\partial t} + v_j\frac{\partial v_i}{\partial x_j}] = \varrho b_i - \frac{\partial p}{\partial x_i} + \frac{\partial}{\partial x_i}\left((\zeta - \frac{2}{3}\mu^*)\frac{\partial v_k}{\partial x_k}\right) + \frac{\partial}{\partial x^j}\left[\mu^*\left(\frac{\partial v_i}{\partial x_j} + \frac{\partial v_j}{\partial x_i}\right)\right]$$

$$= \varrho b_i - \frac{\partial p}{\partial x_i} + \frac{\partial}{\partial x_i}\left(\zeta\frac{\partial v_k}{\partial x_k}\right) + \frac{\partial}{\partial x^j}\left[\mu^*\left(\frac{\partial v_i}{\partial x_j} + \frac{\partial v_j}{\partial x_i} - \frac{2}{3}\delta_{ij}\frac{\partial v_k}{\partial x_k}\right)\right]$$

These equations form the basics of viscous flow theory.

In the case of orthogonal coordinates, where $g_{(i)(i)} = h_i^2$ (no summation) and $g_{ij} = 0$ for $i \neq j$, general expressions for the Navier-Stokes equations in terms of the physical components $v(1), v(2), v(3)$ are:

Navier-Stokes-Duhem equations for compressible fluid in terms of physical components: $(i \neq j$

$$\varrho\left[\frac{\partial v(i)}{\partial t} + \frac{v(1)}{h_1}\frac{\partial v(i)}{\partial x_1} + \frac{v(2)}{h_2}\frac{\partial v(i)}{\partial x_2} + \frac{v(3)}{h_3}\frac{\partial v(i)}{\partial x_3}\right.$$

$$\left. - \frac{v(j)}{h_ih_j}\left(v(j)\frac{\partial h_j}{\partial x_i} - v(i)\frac{\partial h_i}{\partial x_j}\right) + \frac{v(k)}{h_ih_k}\left(v(i)\frac{\partial h_i}{\partial x_k} - v(k)\frac{\partial h_k}{\partial x_i}\right)\right] =$$

$$\varrho\frac{b(i)}{h_i} - \frac{1}{h_i}\frac{\partial p}{\partial x_i} + \frac{1}{h_i}\frac{\partial}{\partial x_i}\left(\lambda^*\nabla\cdot\vec{V}\right) + \frac{\mu^*}{h_ih_j}\left[\frac{h_j}{h_i}\frac{\partial}{\partial x_i}\left(\frac{v(j)}{h_j}\right) + \frac{h_i}{h_j}\frac{\partial}{\partial x_j}\left(\frac{v(i)}{h_i}\right)\right]\frac{\partial h_i}{\partial h_j}$$

$$+ \frac{\mu^*}{h_ih_k}\left[\frac{h_i}{h_k}\frac{\partial}{\partial x_k}\left(\frac{v(i)}{h_i}\right) + \frac{h_k}{h_i}\frac{\partial}{\partial x_i}\left(\frac{v(k)}{h_k}\right)\right]\frac{\partial h_i}{\partial x_k} - \frac{2\mu^*}{h_ih_j}\left[\frac{1}{h_j}\frac{\partial v(j)}{\partial x_j} + \frac{v(k)}{h_jh_k}\frac{\partial h_j}{\partial x_k} + \frac{v(i)}{h_ih_j}\frac{\partial h_j}{\partial x_i}\right]$$

$$- \frac{2\mu^*}{h_ih_k}\left[\frac{1}{h_k}\frac{\partial v(k)}{\partial x_k} + \frac{v(i)}{h_ih_k}\frac{\partial h_k}{\partial x_i} + \frac{v(k)}{h_kh_j}\frac{\partial h_k}{\partial x_i}\right]\frac{\partial h_k}{\partial x_i} + \frac{1}{h_ih_jh_k}\left[\frac{\partial}{\partial x_i}\left\{2\mu^*h_jh_k\left(\frac{1}{h_i}\frac{\partial v(i)}{\partial x_i} + \frac{v(j)}{h_ih_j}\frac{\partial h_i}{\partial h_j} + \frac{v(k)}{h_ih_k}\frac{\partial h_i}{\partial x_k}\right)\right\}\right.$$

$$\left. + \frac{\partial}{\partial x_j}\left\{\mu^*h_ih_k\left(\frac{h_j}{h_i}\frac{\partial}{\partial x_i}\left(\frac{v(j)}{h_j}\right) + \frac{h_i}{h_j}\frac{\partial}{\partial x_j}\left(\frac{v(i)}{h_i}\right)\right)\right\} + \frac{\partial}{\partial x_k}\left\{\mu^*h_ih_j\left(\frac{h_i}{h_k}\frac{\partial}{\partial x_k}\left(\frac{v(i)}{h_i}\right) + \frac{h_k}{h_i}\frac{\partial}{\partial x_i}\left(\frac{v(k)}{h_k}\right)\right)\right\}\right]$$

$$\tag{2.5.31}$$

where $\nabla\cdot\vec{v}$ is found in equation (2.1.4).

In the above equation, cyclic values are assigned to i, j and k. That is, for the x_1 components assign the values $i = 1, j = 2, k = 3$; for the x_2 components assign the values $i = 2, j = 3, k = 1$; and for the x_3 components assign the values $i = 3, j = 1, k = 2$.

The tables 5.2, 5.3 and 5.4 show the expanded form of the Navier-Stokes equations in Cartesian, cylindrical and spherical coordinates respectively.

$$\varrho\frac{DV_x}{Dt} = \varrho b_x - \frac{\partial p}{\partial x} + \frac{\partial}{\partial x}\left[2\mu^*\frac{\partial V_x}{\partial x} + \lambda^*\nabla\cdot\vec{V}\right] + \frac{\partial}{\partial y}\left[\mu^*\left(\frac{\partial V_x}{\partial y} + \frac{\partial V_y}{\partial x}\right)\right] + \frac{\partial}{\partial z}\left[\mu^*\left(\frac{\partial V_x}{\partial z} + \frac{\partial V_z}{\partial x}\right)\right]$$

$$\varrho\frac{DV_y}{Dt} = \varrho b_y - \frac{\partial p}{\partial y} + \frac{\partial}{\partial x}\left[\mu^*\left(\frac{\partial V_y}{\partial x} + \frac{\partial V_x}{\partial y}\right)\right] + \frac{\partial}{\partial y}\left[2\mu^*\frac{\partial V_y}{\partial y} + \lambda^*\nabla\cdot\vec{V}\right] + \frac{\partial}{\partial z}\left[\mu^*\left(\frac{\partial V_y}{\partial z} + \frac{\partial V_z}{\partial y}\right)\right]$$

$$\varrho\frac{DV_z}{Dt} = \varrho b_z - \frac{\partial p}{\partial z} + \frac{\partial}{\partial x}\left[\mu^*\left(\frac{\partial V_z}{\partial x} + \frac{\partial V_x}{\partial z}\right)\right] + \frac{\partial}{\partial y}\left[\mu^*\left(\frac{\partial V_z}{\partial y} + \frac{\partial V_y}{\partial z}\right)\right] + \frac{\partial}{\partial z}\left[2\mu^*\frac{\partial V_z}{\partial z} + \lambda^*\nabla\cdot\vec{V}\right]$$

where $\quad\dfrac{D}{Dt}( ) = \dfrac{\partial( )}{\partial t} + V_x\dfrac{\partial( )}{\partial x} + V_y\dfrac{\partial( )}{\partial y} + V_z\dfrac{\partial( )}{\partial z}$

and $\quad\nabla\cdot\vec{V} = \dfrac{\partial V_x}{\partial x} + \dfrac{\partial V_y}{\partial y} + \dfrac{\partial V_z}{\partial z}$

$$(2.5.31a)$$

Table 5.2 Navier-Stokes equations for compressible fluids in Cartesian coordinates.

$$\varrho\left[\frac{DV_r}{Dt} - \frac{V_\theta^2}{r}\right] = \varrho b_r - \frac{\partial p}{\partial r} + \frac{\partial}{\partial r}\left[2\mu^*\frac{\partial V_r}{\partial r} + \lambda^*\nabla\cdot\vec{V}\right] + \frac{1}{r}\frac{\partial}{\partial\theta}\left[\mu^*\left(\frac{1}{r}\frac{\partial V_r}{\partial\theta} + \frac{\partial V_\theta}{\partial r} - \frac{V_\theta}{r}\right)\right]$$
$$+ \frac{\partial}{\partial z}\left[\mu^*\left(\frac{\partial V_r}{\partial z} + \frac{\partial V_z}{\partial r}\right)\right] + \frac{2\mu^*}{r}\left(\frac{\partial V_r}{\partial r} - \frac{1}{r}\frac{\partial V_\theta}{\partial\theta} - \frac{V_r}{r}\right)$$

$$\varrho\left[\frac{DV_\theta}{Dt} + \frac{V_r V_\theta}{r}\right] = \varrho b_0 - \frac{1}{r}\frac{\partial p}{\partial\theta} + \frac{\partial}{\partial r}\left[\mu^*\left(\frac{1}{r}\frac{\partial V_r}{\partial\theta} + \frac{\partial V_\theta}{\partial r} - \frac{V_\theta}{r}\right)\right] + \frac{1}{r}\frac{\partial}{\partial\theta}\left[2\mu^*\left(\frac{1}{r}\frac{\partial V_\theta}{\partial\theta} + \frac{V_r}{r}\right) + \lambda^*\nabla\cdot\vec{V}\right]$$
$$+ \frac{\partial}{\partial z}\left[\mu^*\left(\frac{1}{r}\frac{\partial V_z}{\partial\theta} + \frac{\partial V_\theta}{\partial z}\right)\right] + \frac{2\mu^*}{r}\left[\frac{1}{r}\frac{\partial V_r}{\partial\theta} + \frac{\partial V_\theta}{\partial r} - \frac{V_\theta}{r}\right]$$

$$\varrho\frac{DV_z}{Dt} = \varrho b_z - \frac{\partial p}{\partial z} + \frac{1}{r}\frac{\partial}{\partial r}\left[\mu^* r\left(\frac{\partial V_r}{\partial z} + \frac{\partial V_z}{\partial r}\right)\right] + \frac{1}{r}\frac{\partial}{\partial\theta}\left[\mu^*\left(\frac{1}{r}\frac{\partial V_z}{\partial\theta} + \frac{\partial V_\theta}{\partial z}\right)\right] + \frac{\partial}{\partial z}\left[2\mu^*\frac{\partial V_z}{\partial z} + \lambda^*\nabla\cdot\vec{V}\right]$$

where $\quad\dfrac{D}{Dt}( ) = \dfrac{\partial( )}{\partial t} + V_r\dfrac{\partial( )}{\partial r} + \dfrac{V_\theta}{r}\dfrac{\partial( )}{\partial\theta} + V_z\dfrac{\partial( )}{\partial z}$

and $\quad\nabla\cdot\vec{V} = \dfrac{1}{r}\dfrac{\partial(rV_r)}{\partial r} + \dfrac{1}{r}\dfrac{\partial V_\theta}{\partial\theta} + \dfrac{\partial V_z}{\partial z}$

$$(2.5.31b)$$

Table 5.3 Navier-Stokes equations for compressible fluids in cylindrical coordinates.

Observe that for incompressible flow $\frac{D\varrho}{Dt} = 0$ which implies $\nabla \cdot \vec{V} = 0$. Therefore, the assumptions of constant viscosity and incompressibility of the flow will simplify the above equations. If on the other hand the viscosity is temperature dependent and the flow is compressible, then one should add to the above equations the continuity equation, an energy equation and an equation of state. The energy equation comes from the first law of thermodynamics applied to a control volume within the fluid and will be considered in the sections ahead. The equation of state is a relation between thermodynamic variables which is added so that the number of equations equals the number of unknowns. Such a system of equations is known as a closed system. An example of an equation of state is the ideal gas law where pressure p is related to gas density ϱ and temperature T by the relation $p = \varrho RT$ where R is the universal gas constant.

$$\varrho\left[\frac{DV_\rho}{Dt} - \frac{V_\theta^2 + V_\phi^2}{\rho}\right] = \varrho b_\rho - \frac{\partial p}{\partial \rho} + \frac{\partial}{\partial \rho}\left[2\mu^*\frac{\partial V_\rho}{\partial \rho} + \lambda^*\nabla \cdot \vec{V}\right] + \frac{1}{\rho}\frac{\partial}{\partial \theta}\left[\mu^*\rho\frac{\partial}{\partial \rho}\left(\frac{V_\theta}{\rho}\right) + \frac{\mu^*}{\rho}\frac{\partial V_\rho}{\partial \theta}\right]$$

$$+ \frac{1}{\rho\sin\theta}\frac{\partial}{\partial \phi}\left[\frac{\mu^*}{\rho\sin\theta}\frac{\partial V_\rho}{\partial \phi} + \mu^*\rho\frac{\partial}{\partial \rho}\left(\frac{V_\phi}{\rho}\right)\right]$$

$$+ \frac{\mu^*}{\rho}\left[4\frac{\partial V_\rho}{\partial \rho} - \frac{2}{\rho}\frac{\partial V_\theta}{\partial \theta} - \frac{4V_\rho}{\rho} - \frac{2}{\rho\sin\theta}\frac{\partial V_\phi}{\partial \phi} - \frac{2V_\theta\cot\theta}{\rho} + \rho\cot\theta\frac{\partial}{\partial \rho}\left(\frac{V_\theta}{\rho}\right) + \frac{\cot\theta}{\rho}\frac{\partial V_\rho}{\partial \theta}\right]$$

$$\varrho\left[\frac{DV_\theta}{Dt} + \frac{V_\rho V_\theta}{\rho} - \frac{V_\phi^2\cot\theta}{\rho}\right] = \varrho b_\theta - \frac{1}{\rho}\frac{\partial p}{\partial \theta} + \frac{\partial}{\partial \rho}\left[\mu^*\rho\frac{\partial}{\partial \rho}\left(\frac{V_\theta}{\rho}\right) + \frac{\mu^*}{\rho}\frac{\partial V_\rho}{\partial \theta}\right]$$

$$+ \frac{1}{\rho}\frac{\partial}{\partial \theta}\left[\frac{2\mu^*}{\rho}\left(\frac{\partial V_\theta}{\partial \theta} + V_\rho\right) + \lambda^*\nabla \cdot \vec{V}\right]$$

$$+ \frac{1}{\rho\sin\theta}\frac{\partial}{\partial \phi}\left[\frac{\mu^*\sin\theta}{\rho}\frac{\partial}{\partial \theta}\left(\frac{V_\phi}{\sin\theta}\right) + \frac{\mu^*}{\rho\sin\theta}\frac{\partial V_\theta}{\partial \phi}\right]$$

$$+ \frac{\mu^*}{\rho}\left[2\cot\theta\left(\frac{1}{\rho}\frac{\partial V_\theta}{\partial \theta} - \frac{1}{\rho\sin\theta}\frac{\partial V_\phi}{\partial \phi} - \frac{V_\theta\cot\theta}{\rho}\right) + 3\left(\rho\frac{\partial}{\partial \rho}\left(\frac{V_\theta}{\rho}\right) + \frac{1}{\rho}\frac{\partial V_\rho}{\partial \theta}\right)\right]$$

$$\varrho\left[\frac{DV_\phi}{Dt} + \frac{V_\theta V_\phi}{\rho} + \frac{V_\theta V_\phi\cot\theta}{\rho}\right] = \varrho b_\phi - \frac{1}{\rho\sin\theta}\frac{\partial p}{\partial \phi} + \frac{\partial}{\partial \rho}\left[\frac{\mu^*}{\rho\sin\theta}\frac{\partial V_\rho}{\partial \phi} + \mu^*\rho\frac{\partial}{\partial \rho}\left(\frac{V_\phi}{\rho}\right)\right]$$

$$+ \frac{1}{\rho}\frac{\partial}{\partial \theta}\left[\frac{\mu^*\sin\theta}{\rho}\frac{\partial}{\partial \theta}\left(\frac{V_\phi}{\sin\theta}\right) + \frac{\mu^*}{\rho\sin\theta}\frac{\partial V_\theta}{\partial \phi}\right]$$

$$+ \frac{1}{\rho\sin\theta}\frac{\partial}{\partial \phi}\left[\frac{2\mu^*}{\rho}\left(\frac{1}{\sin\theta}\frac{\partial V_\phi}{\partial \phi} + V_\rho + V_\theta\cot\theta\right) + \lambda^*\nabla \cdot \vec{V}\right]$$

$$+ \frac{\mu^*}{\rho}\left[\frac{3}{\rho\sin\theta}\frac{\partial V_\rho}{\partial \phi} + 3\rho\frac{\partial}{\partial \rho}\left(\frac{V_\phi}{\rho}\right) + 2\cot\theta\left(\frac{\sin\theta}{\rho}\frac{\partial}{\partial \theta}\left(\frac{V_\phi}{\sin\theta}\right) + \frac{1}{\rho\sin\theta}\frac{\partial V_\theta}{\partial \phi}\right)\right]$$

where $\quad \frac{D}{Dt}() = \frac{\partial()}{\partial t} + V_\rho\frac{\partial()}{\partial \rho} + \frac{V_\theta}{\rho}\frac{\partial()}{\partial \theta} + \frac{V_\phi}{\rho\sin\theta}\frac{\partial()}{\partial \phi}$

and $\quad \nabla \cdot \vec{V} = \frac{1}{\rho^2}\frac{\partial(\rho^2 V_\rho)}{\partial \rho} + \frac{1}{\rho\sin\theta}\frac{\partial V_\theta\sin\theta}{\partial \theta} + \frac{1}{\rho\sin\theta}\frac{\partial V_\phi}{\partial \phi}$

$$(2.5.31c)$$

Table 5.4 Navier-Stokes equations for compressible fluids in spherical coordinates.

We now consider various special cases of the Navier-Stokes-Duhem equations.

293

Special Case 1: Assume that $\vec{b}$ is a conservative force such that $\vec{b} = -\nabla\phi$. Also assume that the viscous force terms are zero. Consider steady flow ($\frac{\partial\vec{v}}{\partial t} = 0$) and show that equation (2.5.28) reduces to the equation

$$(\vec{v}\cdot\nabla)\,\vec{v} = \frac{-1}{\varrho}\nabla\,p - \nabla\,\phi \quad \varrho \text{ is constant.} \tag{2.5.32}$$

Employing the vector identity

$$(\vec{v}\cdot\nabla)\,\vec{v} = (\nabla\times\vec{v})\times\vec{v} + \frac{1}{2}\nabla(\vec{v}\cdot\vec{v}), \tag{2.5.33}$$

we take the dot product of equation (2.5.32) with the vector $\vec{v}$. Noting that $\vec{v}\cdot[(\nabla\times\vec{v})\times\vec{v}] = \vec{0}$ we obtain

$$\vec{v}\cdot\nabla\left[\frac{p}{\varrho} + \phi + \frac{1}{2}v^2\right] = 0. \tag{2.5.34}$$

This equation shows that for steady flow we will have

$$\frac{p}{\varrho} + \phi + \frac{1}{2}v^2 = \text{constant} \tag{2.5.35}$$

along a streamline. This result is known as Bernoulli's theorem. In the special case where $\phi = gh$ is a force due to gravity, the equation (2.5.35) reduces to $\frac{p}{\varrho} + \frac{v^2}{2} + gh = constant$. This equation is known as Bernoulli's equation. It is a conservation of energy statement which has many applications in fluids.

Special Case 2: Assume that $\vec{b} = -\nabla\phi$ is conservative and define the quantity $\vec{\Omega}$ by

$$\vec{\Omega} = \nabla\times\vec{v} = \text{curl}\,\vec{v} \qquad \vec{\omega} = \frac{1}{2}\vec{\Omega} \tag{2.5.36}$$

as the vorticity vector associated with the fluid flow and observe that its magnitude is equivalent to twice the angular velocity of a fluid particle. Then using the identity from equation (2.5.33) we can write the Navier-Stokes-Duhem equations in terms of the vorticity vector. We obtain the hydrodynamic equations

$$\frac{\partial\vec{v}}{\partial t} + \vec{\Omega}\times\vec{v} + \frac{1}{2}\nabla\,v^2 = -\frac{1}{\varrho}\nabla\,p - \nabla\,\phi + \frac{1}{\varrho}\vec{f}_{viscous}, \tag{2.5.37}$$

where $\vec{f}_{viscous}$ is defined by equation (2.5.30). In the special case of nonviscous flow this further reduces to the Euler equation

$$\frac{\partial\vec{v}}{\partial t} + \vec{\Omega}\times\vec{v} + \frac{1}{2}\nabla\,v^2 = -\frac{1}{\varrho}\nabla\,p - \nabla\,\phi.$$

If the density ϱ is a function of the pressure only it is customary to introduce the function

$$P = \int_c^p \frac{dp}{\varrho} \quad \text{so that} \quad \nabla P = \frac{dP}{dp}\nabla p = \frac{1}{\varrho}\nabla p$$

then the Euler equation becomes

$$\frac{\partial\vec{v}}{\partial t} + \vec{\Omega}\times\vec{v} = -\nabla(P + \phi + \frac{1}{2}v^2).$$

Some examples of vorticies are smoke rings, hurricanes, tornadoes, and some sun spots. You can create a vortex by letting water stand in a sink and then remove the plug. Watch the water and you will see that a rotation or vortex begins to occur. Vortices are associated with circulating motion.

Pick an arbitrary simple closed curve C and place it in the fluid flow and define the line integral $K = \oint_C \vec{v} \cdot \hat{e}_t \, ds$, where ds is an element of arc length along the curve C, $\vec{v}$ is the vector field defining the velocity, and $\hat{e}_t$ is a unit tangent vector to the curve C. The integral K is called the circulation of the fluid around the closed curve C. The circulation is the summation of the tangential components of the velocity field along the curve C. The local vorticity at a point is defined as the limit

$$\lim_{\text{Area} \to 0} \frac{\text{Circulation around } C}{\text{Area inside } C} = \text{circulation per unit area.}$$

By Stokes theorem, if $\text{curl}\,\vec{v} = \vec{0}$, then the fluid is called irrotational and the circulation is zero. Otherwise the fluid is rotational and possesses vorticity.

If we are only interested in the velocity field we can eliminate the pressure by taking the curl of both sides of the equation (2.5.37). If we further assume that the fluid is incompressible we obtain the special equations

$$\nabla \cdot \vec{v} = 0 \qquad \text{Incompressible fluid, } \varrho \text{ is constant.}$$

$$\vec{\Omega} = \text{curl}\,\vec{v} \qquad \text{Definition of vorticity vector.} \tag{2.5.38}$$

$$\frac{\partial \vec{\Omega}}{\partial t} + \nabla \times (\vec{\Omega} \times \vec{v}) = \frac{\mu^*}{\varrho} \nabla^2 \vec{\Omega} \qquad \text{Results because curl of gradient is zero.}$$

Note that when Ω is identically zero, we have irrotational motion and the above equations reduce to the Cauchy-Riemann equations. Note also that if the term $\nabla \times (\vec{\Omega} \times \vec{v})$ is neglected, then the last equation in equation (2.5.38) reduces to a diffusion equation. This suggests that the vorticity diffuses through the fluid once it is created.

Vorticity can be caused by a rigid rotation or by shear flow. For example, in cylindrical coordinates let $\vec{V} = r\omega\,\hat{e}_\theta$, with r, ω constants, denote a rotational motion, then $\text{curl}\,\vec{V} = \nabla \times \vec{V} = 2\omega\,\hat{e}_z$, which shows the vorticity is twice the rotation vector. Shear can also produce vorticity. For example, consider the velocity field $\vec{V} = y\,\hat{e}_1$ with $y \geq 0$. Observe that this type of flow produces shear because $|\vec{V}|$ increases as y increases. For this flow field we have $\text{curl}\,\vec{V} = \nabla \times \vec{V} = -\hat{e}_3$. The right-hand rule tells us that if an imaginary paddle wheel is placed in the flow it would rotate clockwise because of the shear effects.

Scaled Variables

In the Navier-Stokes-Duhem equations for fluid flow we make the assumption that the external body forces are derivable from a potential function ϕ and write $\vec{b} = -\nabla\phi\,[dyne/gm]$ We also want to write the Navier-Stokes equations in terms of scaled variables

$$\bar{\vec{v}} = \frac{\vec{v}}{v_0} \qquad \bar{\varrho} = \frac{\varrho}{\varrho_0} \qquad \bar{\phi} = \frac{\phi}{gL}, \qquad \bar{y} = \frac{y}{L}$$

$$\bar{p} = \frac{p}{p_0} \qquad \bar{t} = \frac{t}{\tau} \qquad \bar{x} = \frac{x}{L} \qquad \bar{z} = \frac{z}{L}$$

which can be referred to as the barred system of dimensionless variables. Dimensionless variables are introduced by scaling each variable associated with a set of equations by an appropriate constant term called a characteristic constant associated with that variable. Usually the characteristic constants are chosen from various parameters used in the formulation of the set of equations. The characteristic constants assigned to each variable are not unique and so problems can be scaled in a variety of ways. The characteristic constants assigned to each variable are scales, of the appropriate dimension, which act as reference quantities which reflect the order of magnitude changes expected of that variable over a certain range or area of interest associated with the problem. An inappropriate magnitude selected for a characteristic constant can result in a scaling where significant information concerning the problem can be lost. This is analogous to selecting an inappropriate mesh size in a numerical method. The numerical method might give you an answer but details of the answer might be lost.

In the above scaling of the variables occurring in the Navier-Stokes equations we let v_0 denote some characteristic speed, p_0 a characteristic pressure, ϱ_0 a characteristic density, L a characteristic length, g the acceleration of gravity and τ a characteristic time (for example $\tau = L/v_0$), then the barred variables $\overline{v}$, $\overline{p}$, $\overline{\varrho}, \overline{\phi}, \overline{t}, \overline{x}, \overline{y}$ and $\overline{z}$ are dimensionless. Define the barred gradient operator by

$$\overline{\nabla} = \frac{\partial}{\partial \overline{x}}\,\hat{e}_1 + \frac{\partial}{\partial \overline{y}}\,\hat{e}_2 + \frac{\partial}{\partial \overline{z}}\,\hat{e}_3$$

where all derivatives are with respect to the barred variables. The above change of variables reduces the Navier-Stokes-Duhem equations

$$\varrho\frac{\partial \vec{v}}{\partial t} + \varrho(\vec{v}\cdot\nabla)\,\vec{v} = -\varrho\nabla\phi - \nabla p + (\lambda^* + \mu^*)\nabla\,(\nabla\cdot\vec{v}) + \mu^*\nabla^2\,\vec{v}, \tag{2.5.39}$$

to the form
$$\left(\frac{\varrho_0 v_0}{\tau}\right)\overline{\varrho}\frac{\partial \overline{\vec{v}}}{\partial \overline{t}} + \left(\frac{\varrho_0 v_0^2}{L}\right)\overline{\varrho}\left(\overline{\vec{v}}\cdot\overline{\nabla}\right)\overline{\vec{v}} = -\varrho_0 g\overline{\varrho}\,\overline{\nabla}\,\overline{\phi} - \left(\frac{p_0}{L}\right)\overline{\nabla}\overline{p}$$
$$+ \frac{(\lambda^* + \mu^*)}{L^2}v_0\overline{\nabla}\left(\overline{\nabla}\cdot\overline{\vec{v}}\right) + \left(\frac{\mu^* v_0}{L^2}\right)\overline{\nabla}^2\overline{\vec{v}}. \tag{2.5.40}$$

Now if each term in the equation (2.5.40) is divided by the coefficient $\varrho_0 v_0^2/L$, we obtain the equation

$$S\overline{\varrho}\frac{\partial \overline{\vec{v}}}{\partial \overline{t}} + \overline{\varrho}\left(\overline{\vec{v}}\cdot\overline{\nabla}\right)\overline{\vec{v}} = \frac{-1}{F}\overline{\varrho}\overline{\nabla}\,\overline{\phi} - E\overline{\nabla}\overline{p} + \left(\frac{\lambda^*}{\mu^*}+1\right)\frac{1}{R}\overline{\nabla}\left(\overline{\nabla}\cdot\overline{\vec{v}}\right) + \frac{1}{R}\overline{\nabla}^2\overline{\vec{v}} \tag{2.5.41}$$

which has the dimensionless coefficients

$E = \dfrac{p_0}{\varrho_0 v_0^2} =$ Euler number $\qquad\qquad\qquad R = \dfrac{\varrho_0 V_0 L}{\mu^*} =$ Reynolds number

$F = \dfrac{v_0^2}{gL} =$ Froude number, g is acceleration of gravity $\qquad S = \dfrac{L}{\tau v_0} =$ Strouhal number.

Dropping the bars over the symbols, we write the dimensionless equation using the above coefficients. The scaled equation is found to have the form

$$S\varrho\frac{\partial \vec{v}}{\partial t} + \varrho(\vec{v}\cdot\nabla)\vec{v} = -\frac{1}{F}\varrho\nabla\phi - E\nabla p + \left(\frac{\lambda^*}{\mu^*}+1\right)\frac{1}{R}\nabla\,(\nabla\cdot\vec{v}) + \frac{1}{R}\nabla^2\vec{v} \tag{2.5.42}$$

Boundary Conditions

Fluids problems can be classified as internal flows or external flows. An example of an internal flow problem is that of fluid moving through a converging-diverging nozzle. An example of an external flow problem is fluid flow around the boundary of an aircraft. For both types of problems there is some sort of boundary which influences how the fluid behaves. In these types of problems the fluid is assumed to adhere to a boundary. Let $\vec{r}_b$ denote the position vector to a point on a boundary associated with a moving fluid, and let $\vec{r}$ denote the position vector to a general point in the fluid. Define $\vec{v}(\vec{r})$ as the velocity of the fluid at the point $\vec{r}$ and define $\vec{v}(\vec{r}_b)$ as the known velocity of the boundary. The boundary might be moving within the fluid or it could be fixed in which case the velocity at all points on the boundary is zero. We define the boundary condition associated with a moving fluid as an adherence boundary condition.

> **Definition: (Adherence Boundary Condition)**
> An adherence boundary condition associated with a fluid in motion
> is defined as the limit $\lim_{\vec{r}\to\vec{r}_b} \vec{v}(\vec{r}) = \vec{v}(\vec{r}_b)$ where $\vec{r}_b$ is the position
> vector to a point on the boundary.

Sometimes, when no finite boundaries are present, it is necessary to impose conditions on the components of the velocity far from the origin. Such conditions are referred to as boundary conditions at infinity.

Summary and Additional Considerations

Throughout the development of the basic equations of continuum mechanics we have neglected thermodynamical and electromagnetic effects. The inclusion of thermodynamics and electromagnetic fields adds additional terms to the basic equations of a continua. These basic equations describing a continuum are:

Conservation of mass

The conservation of mass is a statement that the total mass of a body is unchanged during its motion. This is represented by the continuity equation

$$\frac{\partial \varrho}{\partial t} + (\varrho v^k)_{,k} = 0 \quad \text{or} \quad \frac{D\varrho}{Dt} + \varrho \nabla \cdot \vec{V} = 0$$

where ϱ is the mass density and v^k is the velocity.

Conservation of linear momentum

The conservation of linear momentum requires that the time rate of change of linear momentum equal the resultant of all forces acting on the body. In symbols, we write

$$\frac{D}{Dt}\int_{\mathcal{V}} \varrho v^i \, d\tau = \int_{\mathcal{S}} F^i_{(s)} n_i \, dS + \int_{\mathcal{V}} \varrho F^i_{(b)} \, d\tau + \sum_{\alpha=1}^{n} F^i_{(\alpha)} \tag{2.5.43}$$

where $\frac{Dv^i}{Dt} = \frac{\partial v^i}{\partial t} + \frac{\partial v^i}{\partial x^k} v^k$ is the material derivative, $F^i_{(s)}$ are the surface forces per unit area, $F^i_{(b)}$ are the body forces per unit mass and $F^i_{(\alpha)}$ represents isolated external forces. Here $\mathcal{S}$ represents the surface and $\mathcal{V}$ represents the volume of the control volume. The right-hand side of this conservation law represents the resultant force coming from the consideration of all surface forces and body forces acting on a control volume.

Surface forces acting upon the control volume involve such things as pressures and viscous forces, while body forces are due to such things as gravitational, magnetic and electric fields.

Conservation of angular momentum

The conservation of angular momentum states that the time rate of change of angular momentum (moment of linear momentum) must equal the total moment of all forces and couples acting upon the body. In symbols,

$$\frac{D}{Dt}\int_V \varrho e_{ijk}x^j v^k \, d\tau = \int_S e_{ijk}x^j F^k_{(s)} \, dS + \int_V \varrho e_{ijk}x^j F^k_{(b)} \, d\tau + \sum_{\alpha=1}^n (e_{ijk}x^j_{(\alpha)}F^k_{(\alpha)} + M^i_{(\alpha)}) \tag{2.5.44}$$

where $M^i_{(\alpha)}$ represents concentrated couples and $F^k_{(\alpha)}$ represents isolated forces.

Conservation of energy

The conservation of energy law requires that the time rate of change of kinetic energy plus internal energies is equal to the sum of the rate of work from all forces and couples plus a summation of all external energies that enter or leave a control volume per unit of time. The energy equation results from the first law of thermodynamics and can be written

$$\frac{D}{Dt}(E + K) = \dot{W} + \dot{Q}_h \tag{2.5.45}$$

where E is the internal energy, K is the kinetic energy, $\dot{W}$ is the rate of work associated with surface and body forces, and $\dot{Q}_h$ is the input heat rate from surface and internal effects.

Let e denote the internal specific energy density within a control volume, then $E = \int_V \varrho e \, d\tau$ represents the total internal energy of the control volume. The kinetic energy of the control volume is expressed as $K = \frac{1}{2}\int_V \varrho g_{ij}v^i v^j \, d\tau$ where v^i is the velocity, ϱ is the density and $d\tau$ is a volume element. The energy (rate of work) associated with the body and surface forces is represented

$$\dot{W} = \int_S g_{ij}F^i_{(s)}v^j \, dS + \int_V \varrho g_{ij}F^i_{(b)}v^j \, d\tau + \sum_{\alpha=1}^n (g_{ij}F^i_{(\alpha)}v^j + g_{ij}M^i_{(\alpha)}\omega^j)$$

where ω^j is the angular velocity of the point $x^i_{(\alpha)}$, $F^i_{(\alpha)}$ are isolated forces, and $M^i_{(\alpha)}$ are isolated couples. Two external energy sources due to thermal sources are heat flow q^i and rate of internal heat production $\frac{\partial Q}{\partial t}$ per unit volume. The conservation of energy can thus be represented

$$\frac{D}{Dt}\int_V \varrho(e + \frac{1}{2}g_{ij}v^i v^j) \, d\tau = \int_S (g_{ij}F^i_{(s)}v^j - q_i n^i) \, dS + \int_V (\varrho g_{ij}F^i_{(b)}v^j + \frac{\partial Q}{\partial t}) \, d\tau$$
$$+ \sum_{\alpha=1}^n (g_{ij}F^i_{(\alpha)}v^j + g_{ij}M^i_{(\alpha)}\omega^j + U_{(\alpha)}) \tag{2.5.46}$$

where $U_{(\alpha)}$ represents all other energies resulting from thermal, mechanical, electric, magnetic or chemical sources which influx the control volume and D/Dt is the material derivative.

In equation (2.5.46) the left hand side is the material derivative of an integral of the total energy $e_t = \varrho(e + \frac{1}{2}g_{ij}v^i v^j)$ over the control volume. Material derivatives are not like ordinary derivatives and so

298

we cannot interchange the order of differentiation and integration in this term. Here we must use the result that

$$\frac{D}{Dt}\int_V e_t\, d\tau = \int_V \left(\frac{\partial e_t}{\partial t} + \nabla\cdot(e_t\vec{V})\right) d\tau.$$

To prove this result we consider a more general problem. Let $\mathcal{A}$ denote the amount of some quantity per unit mass. The quantity $\mathcal{A}$ can be a scalar, vector or tensor. The total amount of this quantity inside the control volume is $A = \int_V \varrho\mathcal{A}\, d\tau$ and therefore the rate of change of this quantity is

$$\frac{\partial A}{\partial t} = \int_V \frac{\partial(\varrho\mathcal{A})}{\partial t}\, d\tau = \frac{D}{Dt}\int_V \varrho\mathcal{A}\, d\tau - \int_S \varrho\mathcal{A}\vec{V}\cdot\hat{n}\, dS,$$

which represents the rate of change of material within the control volume plus the influx into the control volume. The minus sign is because $\hat{n}$ is always a unit outward normal. By converting the surface integral to a volume integral, by the Gauss divergence theorem, and rearranging terms we find that

$$\frac{D}{Dt}\int_V \varrho\mathcal{A}\, d\tau = \int_V \left[\frac{\partial(\varrho\mathcal{A})}{\partial t} + \nabla\cdot(\varrho\mathcal{A}\vec{V})\right] d\tau.$$

In equation (2.5.46) we neglect all isolated external forces and substitute $F^i_{(s)} = \sigma^{ij}n_j$, $F^i_{(b)} = b^i$ where $\sigma_{ij} = -p\delta_{ij} + \tau_{ij}$. We then replace all surface integrals by volume integrals and find that the conservation of energy can be represented in the form

$$\frac{\partial e_t}{\partial t} + \nabla\cdot(e_t\vec{V}) = \nabla(\boldsymbol{\sigma}\cdot\vec{V}) - \nabla\cdot\vec{q} + \varrho\vec{b}\cdot\vec{V} + \frac{\partial Q}{\partial t} \tag{2.5.47}$$

where $e_t = \varrho e + \varrho(v_1^2 + v_2^2 + v_3^2)/2$ is the total energy and $\boldsymbol{\sigma} = \sum_{i=1}^3\sum_{j=1}^3 \sigma_{ij}\hat{e}_i\hat{e}_j$ is the second order stress tensor. Here

$$\boldsymbol{\sigma}\cdot\vec{V} = -p\vec{V} + \sum_{j=1}^3 \tau_{1j}v_j\,\hat{e}_1 + \sum_{j=1}^3 \tau_{2j}v_j\,\hat{e}_2 + \sum_{j=1}^3 \tau_{3j}v_j\,\hat{e}_3 = -p\vec{V} + \boldsymbol{\tau}\cdot\vec{V}$$

and $\tau_{ij} = \mu^*(v_{i,j} + v_{j,i}) + \lambda^*\delta_{ij}v_{k,k}$ is the viscous stress tensor. Using the identities

$$\varrho\frac{D(e_t/\varrho)}{Dt} = \frac{\partial e_t}{\partial t} + \nabla\cdot(e_t\vec{V}) \quad\text{and}\quad \varrho\frac{D(e_t/\varrho)}{Dt} = \varrho\frac{De}{Dt} + \varrho\frac{D(V^2/2)}{Dt}$$

together with the momentum equation (2.5.25) dotted with $\vec{V}$ as

$$\varrho\frac{D\vec{V}}{Dt}\cdot\vec{V} = \varrho\vec{b}\cdot\vec{V} - \nabla p\cdot\vec{V} + (\nabla\cdot\boldsymbol{\tau})\cdot\vec{V}$$

the energy equation (2.5.47) can then be represented in the form

$$\varrho\frac{De}{Dt} + p(\nabla\cdot\vec{V}) = -\nabla\cdot\vec{q} + \frac{\partial Q}{\partial t} + \Phi \tag{2.5.48}$$

where Φ is the dissipation function and can be represented

$$\Phi = (\tau_{ij}v_i)_{,j} - v_i\tau_{ij,j} = \nabla\cdot(\boldsymbol{\tau}\cdot\vec{V}) - (\nabla\cdot\boldsymbol{\tau})\cdot\vec{V}.$$

As an exercise it can be shown that the dissipation function can also be represented as $\Phi = 2\mu^* D_{ij}D_{ij} + \lambda^*\Theta^2$ where Θ is the dilatation. The heat flow vector is determined from the Fourier law of heat conduction in terms of the temperature T as $\vec{q} = -\kappa \nabla T$, where κ is the thermal conductivity. Therefore, the energy equation can be written as

$$\varrho \frac{De}{Dt} + p(\nabla \cdot \vec{V}) = \frac{\partial Q}{\partial t} + \Phi + \nabla(k \nabla T). \tag{2.5.49}$$

In Cartesian coordinates (x, y, z) we use

$$\frac{D}{Dt} = \frac{\partial}{\partial t} + V_x \frac{\partial}{\partial x} + V_y \frac{\partial}{\partial y} + V_z \frac{\partial}{\partial z}$$

$$\nabla \cdot \vec{V} = \frac{\partial V_x}{\partial x} + \frac{\partial V_y}{\partial y} + \frac{\partial V_z}{\partial z}$$

$$\nabla \cdot (\kappa \nabla T) = \frac{\partial}{\partial x}\left(\kappa \frac{\partial T}{\partial x}\right) + \frac{\partial}{\partial y}\left(\kappa \frac{\partial T}{\partial y}\right) + \frac{\partial}{\partial z}\left(\kappa \frac{\partial T}{\partial z}\right)$$

In cylindrical coordinates (r, θ, z)

$$\frac{D}{Dt} = \frac{\partial}{\partial t} + V_r \frac{\partial}{\partial r} + \frac{V_\theta}{r} \frac{\partial}{\partial \theta} + V_z \frac{\partial}{\partial z}$$

$$\nabla \cdot \vec{V} = \frac{1}{r} \frac{\partial}{\partial r}(r V_r) + \frac{1}{r^2} \frac{\partial V_\theta}{\partial \theta} + \frac{\partial V_z}{\partial z}$$

$$\nabla \cdot (\kappa \nabla T) = \frac{1}{r} \frac{\partial}{\partial r}\left(r \kappa \frac{\partial T}{\partial r}\right) + \frac{1}{r^2} \frac{\partial}{\partial \theta}\left(\kappa \frac{\partial T}{\partial \theta}\right) + \frac{\partial}{\partial z}\left(\kappa \frac{\partial T}{\partial z}\right)$$

and in spherical coordinates (ρ, θ, ϕ)

$$\frac{D}{Dt} = \frac{\partial}{\partial t} + V_\rho \frac{\partial}{\partial \rho} + \frac{V_\theta}{\rho} \frac{\partial}{\partial \theta} \frac{V_\phi}{\rho \sin \theta} \frac{\partial}{\partial \phi}$$

$$\nabla \cdot \vec{V} = \frac{1}{\rho^2} \frac{\partial}{\partial \rho}(\rho V_\rho) + \frac{1}{\rho \sin \theta} \frac{\partial}{\partial \theta}(V_\theta \sin \theta) + \frac{1}{\rho \sin \theta} \frac{\partial V_\phi}{\partial \phi}$$

$$\nabla \cdot (\kappa \nabla T) = \frac{1}{\rho^2} \frac{\partial}{\partial \rho}\left(\rho^2 \kappa \frac{\partial T}{\partial \rho}\right) + \frac{1}{\rho^2 \sin \theta} \frac{\partial}{\partial \theta}\left(\kappa \sin \theta \frac{\partial T}{\partial \theta}\right) + \frac{1}{\rho^2 \sin^2 \theta} \frac{\partial}{\partial \phi}\left(\kappa \frac{\partial T}{\partial \phi}\right)$$

The combination of terms $h = e + p/\varrho$ is known as enthalpy and at times is used to express the energy equation in the form

$$\varrho \frac{Dh}{Dt} = \frac{Dp}{Dt} + \frac{\partial Q}{\partial t} - \nabla \cdot \vec{q} + \Phi.$$

The derivation of this equation is left as an exercise.

Conservative Systems

Let Q denote some physical quantity per unit volume. Here Q can be either a scalar, vector or tensor field. Place within this field an imaginary simple closed surface S which encloses a volume V. The total amount of Q within the surface is given by $\iiint_V Q \, d\tau$ and the rate of change of this amount with respect to time is $\frac{\partial}{\partial t} \iiint Q \, d\tau$. The total amount of Q within S changes due to sources (or sinks) within the volume and by transport processes. Transport processes introduce a quantity $\vec{J}$, called current, which represents a flow per unit area across the surface S. The inward flux of material into the volume is denoted $\iint_S -\vec{J} \cdot \hat{n} \, d\sigma$ ($\hat{n}$ is a unit outward normal.) The sources (or sinks) S_Q denotes a generation (or loss) of material per unit volume so that $\iiint_V S_Q \, d\tau$ denotes addition (or loss) of material to the volume. For a fixed volume we then have the material balance

$$\iiint_V \frac{\partial Q}{\partial t} \, d\tau = -\iint_S \vec{J} \cdot \hat{n} \, d\sigma + \iiint_V S_Q \, d\tau.$$

300

Using the divergence theorem of Gauss one can derive the general conservation law

$$\frac{\partial Q}{\partial t} + \nabla \cdot \vec{J} = S_Q \tag{2.5.50}$$

The continuity equation and energy equations are examples of a scalar conservation law in the special case where $S_Q = 0$. In Cartesian coordinates, we can represent the continuity equation by letting

$$Q = \varrho \quad \text{and} \quad \vec{J} = \varrho\vec{V} = \varrho(V_x\,\hat{e}_1 + V_y\,\hat{e}_2 + V_z\,\hat{e}_3) \tag{2.5.51}$$

The energy equation conservation law is represented by selecting $Q = e_t$ and neglecting the rate of internal heat energy we let

$$\vec{J} = \left[(e_t + p)v_1 - \sum_{i=1}^{3} v_i\tau_{xi} + q_x\right]\hat{e}_1 +$$
$$\left[(e_t + p)v_2 - \sum_{i=1}^{3} v_i\tau_{yi} + q_y\right]\hat{e}_2 + \tag{2.5.52}$$
$$\left[(e_t + p)v_3 - \sum_{i=1}^{3} v_i\tau_{zi} + q_z\right]\hat{e}_3.$$

In a general orthogonal system of coordinates (x_1, x_2, x_3) the equation (2.5.50) is written

$$\frac{\partial}{\partial t}((h_1h_2h_3Q)) + \frac{\partial}{\partial x_1}((h_2h_3J_1)) + \frac{\partial}{\partial x_2}((h_1h_3J_2)) + \frac{\partial}{\partial x_3}((h_1h_2J_3)) = 0,$$

where h_1, h_2, h_3 are scale factors obtained from the transformation equations to the general orthogonal coordinates.

The momentum equations are examples of a vector conservation law having the form

$$\frac{\partial \vec{a}}{\partial t} + \nabla \cdot (\boldsymbol{T}) = \varrho\vec{b} \tag{2.5.53}$$

where $\vec{a}$ is a vector and $\boldsymbol{T}$ is a second order symmetric tensor $\boldsymbol{T} = \sum_{k=1}^{3}\sum_{j=1}^{3} T_{jk}\,\hat{e}_j\,\hat{e}_k$. In Cartesian coordinates we let $\vec{a} = \varrho(V_x\,\hat{e}_1 + V_y\,\hat{e}_2 + V_z\,\hat{e}_3)$ and $T_{ij} = \varrho v_iv_j + p\delta_{ij} - \tau_{ij}$. In general coordinates (x_1, x_2, x_3) the momentum equations result by selecting $\vec{a} = \varrho\vec{V}$ and $T_{ij} = \varrho v_iv_j + p\delta_{ij} - \tau_{ij}$. In a general orthogonal system the conservation law (2.5.53) has the general form

$$\frac{\partial}{\partial t}((h_1h_2h_3\vec{a})) + \frac{\partial}{\partial x_1}((h_2h_3\boldsymbol{T}\cdot\hat{e}_1)) + \frac{\partial}{\partial x_2}((h_1h_3\boldsymbol{T}\cdot\hat{e}_2)) + \frac{\partial}{\partial x_3}((h_1h_2\boldsymbol{T}\cdot\hat{e}_3)) = \varrho\vec{b}. \tag{2.5.54}$$

Neglecting body forces and internal heat production, the continuity, momentum and energy equations can be expressed in the strong conservative form

$$\frac{\partial U}{\partial t} + \frac{\partial E}{\partial x} + \frac{\partial F}{\partial y} + \frac{\partial G}{\partial z} = 0 \tag{2.5.55}$$

where

$$U = \begin{bmatrix} \rho \\ \rho V_x \\ \rho V_y \\ \rho V_z \\ e_t \end{bmatrix} \tag{2.5.56}$$

$$E = \begin{bmatrix} \rho V_x \\ \rho V_x^2 + p - \tau_{xx} \\ \rho V_x V_y - \tau_{xy} \\ \rho V_x V_z - \tau_{xz} \\ (e_t + p)V_x - V_x\tau_{xx} - V_y\tau_{xy} - V_z\tau_{xz} + q_x \end{bmatrix} \tag{2.5.57}$$

$$F = \begin{bmatrix} \rho V_y \\ \rho V_x V_y - \tau_{xy} \\ \rho V_y^2 + p - \tau_{yy} \\ \rho V_y V_z - \tau_{yz} \\ (e_t + p)V_y - V_x\tau_{yx} - V_y\tau_{yy} - V_z\tau_{yz} + q_y \end{bmatrix} \tag{2.5.58}$$

$$G = \begin{bmatrix} \rho V_z \\ \rho V_x V_z - \tau_{xz} \\ \rho V_y V_z - \tau_{yz} \\ \rho V_z^2 + p - \tau_{zz} \\ (e_t + p)V_z - V_x\tau_{zx} - V_y\tau_{zy} - V_z\tau_{zz} + q_z \end{bmatrix} \tag{2.5.59}$$

where the shear stresses are $\tau_{ij} = \mu^*(V_{i,j} + V_{j,i}) + \delta_{ij}\lambda^* V_{k,k}$ for $i,j,k = 1,2,3$.

EXAMPLE 2.5-1. (One-dimensional fluid flow)

Construct an x-axis running along the center line of a long cylinder with cross sectional area A. Consider the motion of a gas driven by a piston and moving with velocity $v_1 = u$ in the x-direction. From an Eulerian point of view we imagine a control volume fixed within the cylinder and assume zero body forces. We require the following equations be satisfied.

Conservation of mass $\frac{\partial \varrho}{\partial t} + \text{div}(\varrho \vec{V}) = 0$ which in one-dimension reduces to $\frac{\partial \varrho}{\partial t} + \frac{\partial}{\partial x}(\varrho u) = 0$.

Conservation of momentum, equation (2.5.28) reduces to $\frac{\partial}{\partial t}(\varrho u) + \frac{\partial}{\partial x}(\varrho u^2) + \frac{\partial p}{\partial x} = 0$.

Conservation of energy, equation (2.5.48) in the absence of heat flow and internal heat production, becomes in one dimension $\varrho\left(\frac{\partial e}{\partial t} + u\frac{\partial e}{\partial x}\right) + p\frac{\partial u}{\partial x} = 0$. Using the conservation of mass relation this equation can be written in the form $\frac{\partial}{\partial t}(\varrho e) + \frac{\partial}{\partial x}(\varrho e u) + p\frac{\partial u}{\partial x} = 0$.

In contrast, from a Lagrangian point of view we let the control volume move with the flow and consider advection terms. This gives the following three equations which can then be compared with the above Eulerian equations of motion.

Conservation of mass $\frac{d}{dt}(\varrho J) = 0$ which in one-dimension is equivalent to $\frac{D\varrho}{Dt} + \varrho\frac{\partial u}{\partial x} = 0$.

Conservation of momentum, equation (2.5.25) in one-dimension $\varrho\frac{Du}{Dt} + \frac{\partial p}{\partial x} = 0$.

Conservation of energy, equation (2.5.48) in one-dimension $\varrho\frac{De}{Dt} + p\frac{\partial u}{\partial x} = 0$. In the above equations $\frac{D()}{Dt} = \frac{\partial}{\partial t}() + u\frac{\partial}{\partial x}()$. The Lagrangian viewpoint gives three equations in the three unknowns ρ, u, e.

In both the Eulerian and Lagrangian equations the pressure p represents the total pressure $p = p_g + p_v$ where p_g is the gas pressure and p_v is the viscous pressure which causes loss of kinetic energy. The gas pressure is a function of ϱ, e and is determined from the ideal gas law $p_g = \varrho RT = \varrho(c_p - c_v)T = \varrho(\frac{c_p}{c_v} - 1)c_v T$ or $p_g = \varrho(\gamma - 1)e$. Some kind of assumption is usually made to represent the viscous pressure p_v as a function of e, u. The above equations are then subjected to boundary and initial conditions and are usually solved numerically.

∎

EXAMPLE 2.5-2. The Navier-Stokes equations for two-dimensional flow of an incompressible fluid with constant viscosity under adiabatic, isothermal conditions are a special case of the equations (2.5.55). These equations reduce to

$$\frac{\partial}{\partial x}(\varrho u) + \frac{\partial}{\partial y}(\varrho v) = 0$$

$$\frac{\partial(\varrho u)}{\partial t} + \frac{\partial(\varrho u^2)}{\partial x} + \frac{\partial(\varrho uv)}{\partial y} + \frac{\partial p}{\partial x} - \mu^* \left(\frac{\partial^2 u}{\partial x^2} + \frac{\partial^2 u}{\partial y^2} \right) = 0$$

$$\frac{\partial(\varrho v)}{\partial t} + \frac{\partial(\varrho uv)}{\partial x} + \frac{\partial(\varrho v^2)}{\partial y} + \frac{\partial p}{\partial y} - \mu^* \left(\frac{\partial^2 v}{\partial x^2} + \frac{\partial^2 v}{\partial y^2} \right) = 0$$

where $u = V_x$ and $v = V_y$ are the x and y components of fluid velocity. Note that for an incompressible fluid we have $\varrho =$constant, which implies $\frac{\partial u}{\partial x} + \frac{\partial v}{\partial y} = 0$. It is this continuity equation which simplifies the second and third equations in the system (2.5.55). The additional assumption of an adiabatic isothermal system allows the energy equation to be neglected. An adiabatic system is one that can not absorb or generate heat, together with constant temperature, means heat generated by the viscous terms will be negligible. Boundary conditions must also be assigned and depend upon whether the flow is internal or external. This system is usually solved numerically. Unusual boundary conditions can sometimes produce difficult numerical problems.

Computational Coordinates

To transform the conservative system (2.5.55) from a physical (x, y, z) domain to a computational (ξ, η, ζ) domain requires that a general change of variables take place. Consider the following general transformation of the independent variables

$$\xi = \xi(x, y, z) \qquad \eta = \eta(x, y, z) \qquad \zeta = \zeta(x, y, z) \tag{2.5.60}$$

with Jacobian different from zero. The chain rule for changing variables in equation (2.5.55) requires the operators

$$\frac{\partial(\ )}{\partial x} = \frac{\partial(\ )}{\partial \xi}\xi_x + \frac{\partial(\ )}{\partial \eta}\eta_x + \frac{\partial(\ )}{\partial \zeta}\zeta_x$$

$$\frac{\partial(\ )}{\partial y} = \frac{\partial(\ )}{\partial \xi}\xi_y + \frac{\partial(\ )}{\partial \eta}\eta_y + \frac{\partial(\ )}{\partial \zeta}\zeta_y \tag{2.5.61}$$

$$\frac{\partial(\ )}{\partial z} = \frac{\partial(\ )}{\partial \xi}\xi_z + \frac{\partial(\ )}{\partial \eta}\eta_z + \frac{\partial(\ )}{\partial \zeta}\zeta_z$$

The partial derivatives in these equations occur in the differential expressions

$$\begin{aligned} d\xi &= \xi_x \, dx + \xi_y \, dy + \xi_z \, dz \\ d\eta &= \eta_x \, dx + \eta_y \, dy + \eta_z \, dz \quad \text{or} \\ d\zeta &= \zeta_x \, dx + \zeta_y \, dy + \zeta_z \, dz \end{aligned} \qquad \begin{bmatrix} d\xi \\ d\eta \\ d\zeta \end{bmatrix} = \begin{bmatrix} \xi_x & \xi_y & \xi_z \\ \eta_x & \eta_y & \eta_z \\ \zeta_x & \zeta_y & \zeta_z \end{bmatrix} \begin{bmatrix} dx \\ dy \\ dz \end{bmatrix} \tag{2.5.62}$$

In a similar manner from the inverse transformation equations

$$x = x(\xi, \eta, \zeta) \qquad y = y(\xi, \eta, \zeta) \qquad z = z(\xi, \eta, \zeta) \tag{2.5.63}$$

we can write the differentials

$$dx = x_\xi \, d\xi + x_\eta \, d\eta + x_\zeta \, d\zeta$$
$$dy = y_\xi \, d\xi + y_\eta \, d\eta + y_\zeta \, d\zeta \quad \text{or} \quad \begin{bmatrix} dx \\ dy \\ dz \end{bmatrix} = \begin{bmatrix} x_\xi & x_\eta & x_\zeta \\ y_\xi & y_\eta & y_\zeta \\ z_\xi & z_\eta & z_\zeta \end{bmatrix} \begin{bmatrix} d\xi \\ d\eta \\ d\zeta \end{bmatrix} \tag{2.5.64}$$
$$dz = z_\xi \, d\xi + z_\zeta \, d\zeta + z_\zeta \, d\zeta$$

The transformations (2.5.62) and (2.5.64) are inverses of each other and so we can write

$$\begin{bmatrix} \xi_x & \xi_y & \xi_z \\ \eta_x & \eta_y & \eta_z \\ \zeta_x & \zeta_y & \zeta_z \end{bmatrix} = \begin{bmatrix} x_\xi & x_\eta & x_\zeta \\ y_\xi & y_\eta & y_\zeta \\ z_\xi & z_\eta & z_\zeta \end{bmatrix}^{-1}$$

$$= J \begin{bmatrix} y_\eta z_\zeta - y_\zeta z_\eta & -(x_\eta z_\zeta - x_\zeta z_\eta) & x_\eta y_\zeta - x_\zeta y_\eta \\ -(y_\xi z_\zeta - y_\zeta z_\xi) & x_\xi z_\zeta - x_\zeta z_\xi & -(x_\xi y_\zeta - x_\zeta y_\xi) \\ y_\xi z_\eta - y_\eta z_\xi & -(x_\xi z_\eta - x_\eta z_\xi) & x_\xi y_\eta - x_\eta y_\xi \end{bmatrix} \tag{2.5.65}$$

By comparing like elements in equation (2.5.65) we obtain the relations

$$\xi_x = J(y_\eta z_\zeta - y_\zeta z_\eta) \qquad \eta_x = -J(y_\xi z_\zeta - y_\zeta z_\xi) \qquad \zeta_x = J(y_\xi z_\eta - y_\eta z_\xi)$$
$$\xi_y = -J(x_\eta z_\zeta - x_\zeta z_\eta) \qquad \eta_y = J(x_\xi z_\zeta - z_\zeta z_\xi) \qquad \zeta_y = -J(x_\xi z_\eta - x_\eta z_\xi) \tag{2.5.66}$$
$$\xi_z = J(x_\eta y_\zeta - x_\zeta y_\eta) \qquad \eta_z = -J(x_\xi y_\zeta - x_\zeta y_\xi) \qquad \zeta_z = J(x_\xi y_\eta - x_\eta y_\xi)$$

The equations (2.5.55) can now be written in terms of the new variables (ξ, η, ζ) as

$$\frac{\partial U}{\partial t} + \frac{\partial E}{\partial \xi}\xi_x + \frac{\partial E}{\partial \eta}\eta_x + \frac{\partial E}{\partial \zeta}\zeta_x + \frac{\partial F}{\partial \xi}\xi_y + \frac{\partial F}{\partial \eta}\eta_y + \frac{\partial F}{\partial \zeta}\zeta_y + \frac{\partial G}{\partial \xi}\xi_z + \frac{\partial G}{\partial \eta}\eta_z + \frac{\partial G}{\partial \zeta}\zeta_z = 0 \tag{2.5.67}$$

Now divide each term by the Jacobian J and write the equation (2.5.67) in the form

$$\frac{\partial}{\partial t}\left(\frac{U}{J}\right) + \frac{\partial}{\partial \xi}\left(\frac{E\zeta_x + F\xi_y + G\xi_z}{J}\right)$$
$$+ \frac{\partial}{\partial \eta}\left(\frac{E\eta_x + F\eta_y + G\eta_z}{J}\right)$$
$$+ \frac{\partial}{\partial \zeta}\left(\frac{E\zeta_x + F\zeta_y + G\zeta_z}{J}\right)$$
$$- E\left\{\frac{\partial}{\partial \xi}\left(\frac{\xi_x}{J}\right) + \frac{\partial}{\partial \eta}\left(\frac{\eta_x}{J}\right) + \frac{\partial}{\partial \zeta}\left(\frac{\zeta_x}{J}\right)\right\} \tag{2.5.68}$$
$$- F\left\{\frac{\partial}{\partial \xi}\left(\frac{\xi_y}{J}\right) + \frac{\partial}{\partial \eta}\left(\frac{\eta_y}{J}\right) + \frac{\partial}{\partial \zeta}\left(\frac{\zeta_y}{J}\right)\right\}$$
$$- G\left\{\frac{\partial}{\partial \xi}\left(\frac{\xi_z}{J}\right) + \frac{\partial}{\partial \eta}\left(\frac{\eta_z}{J}\right) + \frac{\partial}{\partial \zeta}\left(\frac{\zeta_z}{J}\right)\right\} = 0$$

Using the relations given in equation (2.5.66) one can show that the curly bracketed terms above are all zero and so the transformed equations (2.5.55) can also be written in the conservative form

$$\frac{\partial \widehat{U}}{\partial t} + \frac{\partial \widehat{E}}{\partial \xi} + \frac{\partial \widehat{F}}{\partial \eta} + \frac{\partial \widehat{G}}{\partial \zeta} = 0 \tag{2.5.69}$$

where

$$\widehat{U} = \frac{U}{J}$$
$$\widehat{E} = \frac{E\xi_x + F\xi_y + G\xi z}{J}$$
$$\widehat{F} = \frac{E\eta_x + F\eta_y + G\eta_z}{J} \tag{2.5.70}$$
$$\widehat{G} = \frac{E\zeta_x + F\zeta_y + G\zeta_z}{J}$$

Fourier law of heat conduction

The Fourier law of heat conduction can be written $q_i = -\kappa T_{,i}$ for isotropic material and $q_i = -\kappa_{ij}T_{,j}$ for anisotropic material. The Prandtl number is a nondimensional constant defined as $Pr = \frac{c_p \mu^*}{\kappa}$ so that the heat flow terms can be represented in Cartesian coordinates as

$$q_x = -\frac{c_p \mu^*}{Pr}\frac{\partial T}{\partial x} \qquad q_y = -\frac{c_p \mu^*}{Pr}\frac{\partial T}{\partial y} \qquad q_z = -\frac{c_p \mu^*}{Pr}\frac{\partial T}{\partial z}$$

Now one can employ the equation of state relations $P = \varrho e(\gamma - 1)$, $c_p = \frac{\gamma R}{\gamma - 1}$, $c_p T = \frac{\gamma RT}{\gamma - 1}$ and write the above equations in the alternate forms

$$q_x = -\frac{\mu^*}{Pr(\gamma - 1)}\frac{\partial}{\partial x}\left(\frac{\gamma P}{\varrho}\right) \qquad q_y = -\frac{\mu^*}{Pr(\gamma - 1)}\frac{\partial}{\partial y}\left(\frac{\gamma P}{\varrho}\right) \qquad q_z = -\frac{\mu^*}{Pr(\gamma - 1)}\frac{\partial}{\partial z}\left(\frac{\gamma P}{\varrho}\right)$$

The speed of sound is given by $a = \sqrt{\frac{\gamma P}{\varrho}} = \sqrt{\gamma RT}$ and so one can substitute a^2 in place of the ratio $\frac{\gamma P}{\varrho}$ in the above equations.

Equilibrium and Nonequilibrium Thermodynamics

High temperature gas flows require special considerations. In particular, the specific heat for monotonic and diatomic gases are different and are in general a function of temperature. The energy of a gas can be written as $e = e_t + e_r + e_v + e_e + e_n$ where e_t represents translational energy, e_r is rotational energy, e_v is vibrational energy, e_e is electronic energy, and e_n is nuclear energy. The gases follow a Boltzmann distribution for each degree of freedom and consequently at very high temperatures the rotational, translational and vibrational degrees of freedom can each have their own temperature. Under these conditions the gas is said to be in a state of nonequilibrium. In such a situation one needs additional energy equations. The energy equation developed in these notes is for equilibrium thermodynamics where the rotational, translational and vibrational temperatures are the same.

Equation of state

It is assumed that an equation of state such as the universal gas law or perfect gas law $pV = nRT$ holds which relates pressure $p\,[N/m^2]$, volume $V\,[m^3]$, amount of gas $n\,[mol]$, and temperature $T\,[K]$ where $R\,[J/mol-K]$ is the universal molar gas constant. If the ideal gas law is represented in the form $p = \varrho RT$ where $\varrho\,[Kg/m^3]$ is the gas density, then the universal gas constant must be expressed in units of $[J/Kg-K]$ (See Appendix A). Many gases deviate from this ideal behavior. In order to account for the intermolecular forces associated with high density gases, an empirical equation of state of the form

$$p = \rho RT + \sum_{n=1}^{M_1} \beta_n \rho^{n+r_1} + e^{-\gamma_1 \rho - \gamma_2 \rho^2} \sum_{n=1}^{M_2} c_n \rho^{n+r_2}$$

involving constants $M_1, M_2, \beta_n, c_n, r_1, r_2, \gamma_1, \gamma_2$ is often used. For a perfect gas the relations

$$e = c_v T \qquad \gamma = \frac{c_p}{c_v} \qquad c_v = \frac{R}{\gamma - 1} \qquad c_p = \frac{\gamma R}{\gamma - 1} \qquad h = c_p T$$

hold, where R is the universal gas constant, c_v is the specific heat at constant volume, c_p is the specific heat at constant pressure, γ is the ratio of specific heats and h is the enthalpy. For c_v and c_p constants the relations $p = (\gamma - 1)\varrho e$ and $RT = (\gamma - 1)e$ can be verified.

Entropy inequality

Energy transfer is not always reversible. Many energy transfer processes are irreversible. The second law of thermodynamics allows energy transfer to be reversible only in special circumstances. In general, the second law of thermodynamics can be written as an entropy inequality, known as the Clausius-Duhem inequality. This inequality states that the time rate of change of the total entropy is greater than or equal to the total entropy change occurring across the surface and within the body of a control volume. The Clausius-Duhem inequality places restrictions on the constitutive equations. This inequality can be expressed in the form

$$\underbrace{\frac{D}{Dt}\int_V \varrho s\, d\tau}_{\text{Rate of entropy increase}} \geq \underbrace{\int_S s^i n_i\, dS + \int_V \rho b\, d\tau + \sum_{\alpha=1}^n B_{(\alpha)}}_{\text{Entropy input rate into control volume}}$$

where s is the specific entropy density, s^i is an entropy flux, b is an entropy source and $B_{(\alpha)}$ are isolated entropy sources. Irreversible processes are characterized by the use of the inequality sign while for reversible processes the equality sign holds. The Clausius-Duhem inequality is assumed to hold for all independent thermodynamical processes.

If in addition there are electric and magnetic fields to consider, then these fields place additional forces upon the material continuum and we must add all forces and moments due to these effects. In particular we must add the following equations

Gauss's law for magnetism $\qquad \nabla \cdot \vec{B} = 0 \qquad\qquad \frac{1}{\sqrt{g}}\frac{\partial}{\partial x^i}(\sqrt{g}B^i) = 0.$

Gauss's law for electricity $\qquad \nabla \cdot \vec{D} = \varrho_e \qquad\qquad \frac{1}{\sqrt{g}}\frac{\partial}{\partial x^i}(\sqrt{g}D^i) = \varrho_e.$

Faraday's law $\qquad\qquad \nabla \times \vec{E} = -\frac{\partial \vec{B}}{\partial t} \qquad\qquad \epsilon^{ijk}E_{k,j} = -\frac{\partial B^i}{\partial t}.$

Ampere's law $\qquad\qquad \nabla \times \vec{H} = \vec{J} + \frac{\partial \vec{D}}{\partial t} \qquad\qquad \epsilon^{ijk}H_{k,j} = J^i + \frac{\partial D^i}{\partial t}.$

where ϱ_e is the charge density, J^i is the current density, $D_i = \epsilon_i^j E_j + P_i$ is the electric displacement vector, H_i is the magnetic field, $B_i = \mu_i^j H_j + M_i$ is the magnetic induction, E_i is the electric field, M_i is the magnetization vector and P_i is the polarization vector. Taking the divergence of Ampere's law produces the law of conservation of charge which requires that

$$\frac{\partial \varrho_e}{\partial t} + \nabla \cdot \vec{J} = 0 \qquad\qquad \frac{\partial \varrho_e}{\partial t} + \frac{1}{\sqrt{g}}\frac{\partial}{\partial x^i}(\sqrt{g}J^i) = 0.$$

The figure 2.5-3 is constructed to suggest some of the interactions that can occur between various variables which define the continuum. Pyroelectric effects occur when a change in temperature causes changes in the electrical properties of a material. Temperature changes can also change the mechanical properties of materials. Similarly, piezoelectric effects occur when a change in either stress or strain causes changes in the electrical properties of materials. Photoelectric effects are said to occur if changes in electric

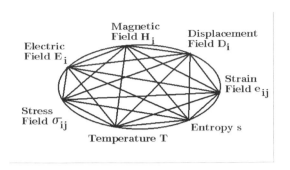

Figure 2.5-3. Interaction of various fields.

or mechanical properties effect the refractive index of a material. Such changes can be studied by modifying the constitutive equations to include the effects being considered.

From figure 2.5-3 we see that there can exist a relationship between the displacement field D_i and electric field E_i. When this relationship is linear we can write $D_i = \epsilon_{ji}E_j$ and $E_j = \beta_{jn}D_n$, where ϵ_{ji} are dielectric constants and β_{jn} are dielectric impermabilities. Similarly, when linear piezoelectric effects exist we can write linear relations between stress and electric fields such as $\sigma_{ij} = -g_{kij}E_k$ and $E_i = -e_{ijk}\sigma_{jk}$, where g_{kij} and e_{ijk} are called piezoelectric constants. If there is a linear relation between strain and an electric fields, this is another type of piezoelectric effect whereby $e_{ij} = d_{ijk}E_k$ and $E_k = -h_{ijk}e_{jk}$, where d_{ijk} and h_{ijk} are another set of piezoelectric constants. Similarly, entropy changes can cause pyroelectric effects. Piezooptical effects (photoelasticity) occurs when mechanical stresses change the optical properties of the material. Electrical and heat effects can also change the optical properties of materials. Piezoresistivity occurs when mechanical stresses change the electric resistivity of materials. Electric field changes can cause variations in temperature, another pyroelectric effect. When temperature effects the entropy of a material this is known as a heat capacity effect. When stresses effect the entropy in a material this is called a piezocaloric effect. Some examples of the representation of these additional effects are as follows. The piezoelectric effects are represented by equations of the form

$$\sigma_{ij} = -h_{mij}D_m \quad D_i = d_{ijk}\sigma_{jk} \quad e_{ij} = g_{kij}D_k \quad D_i = e_{ijk}e_{jk}$$

where h_{mij}, d_{ijk}, g_{kij} and e_{ijk} are piezoelectric constants.

Knowledge of the material or electric interaction can be used to help modify the constitutive equations. For example, the constitutive equations can be modified to included temperature effects by expressing the constitutive equations in the form

$$\sigma_{ij} = c_{ijkl}e_{kl} - \beta_{ij}\Delta T \quad \text{and} \quad e_{ij} = s_{ijkl}\sigma_{kl} + \alpha_{ij}\Delta T$$

where for isotropic materials the coefficients α_{ij} and β_{ij} are constants. As another example, if the strain is modified by both temperature and an electric field, then the constitutive equations would take on the form

$$e_{ij} = s_{ijkl}\sigma_{kl} + \alpha_{ij}\Delta T + d_{mij}E_m.$$

Note that these additional effects are additive under conditions of small changes. That is, we may use the principal of superposition to calculate these additive effects.

If the electric field and electric displacement are replaced by a magnetic field and magnetic flux, then piezomagnetic relations can be found to exist between the variables involved. One should consult a handbook to determine the order of magnitude of the various piezoelectric and piezomagnetic effects. For a large majority of materials these effects are small and can be neglected when the field strengths are weak.

The Boltzmann Transport Equation

The modeling of the transport of particle beams through matter, such as the motion of energetic protons or neutrons through bulk material, can be approached using ideas from the classical kinetic theory of gases. Kinetic theory is widely used to explain phenomena in such areas as: statistical mechanics, fluids, plasma physics, biological response to high-energy radiation, high-energy ion transport and various types of radiation shielding. The problem is basically one of describing the behavior of a system of interacting particles and their distribution in space, time and energy. The average particle behavior can be described by the Boltzmann equation which is essentially a continuity equation in a six-dimensional phase space (x, y, z, V_x, V_y, V_z). We will be interested in examining how the particles in a volume element of phase space change with time. We introduce the following notation:

(i) $\vec{r}$ the position vector of a typical particle of phase space and $d\tau = dxdydz$ the corresponding spatial volume element at this position.

(ii) $\vec{V}$ the velocity vector associated with a typical particle of phase space and $d\tau_v = dV_x dV_y dV_z$ the corresponding velocity volume element.

(iii) $\vec{\Omega}$ a unit vector in the direction of the velocity $\vec{V} = v\vec{\Omega}$.

(iv) $E = \frac{1}{2}mv^2$ kinetic energy of particle.

(v) $d\vec{\Omega}$ is a solid angle about the direction $\vec{\Omega}$ and $d\tau\, dE\, d\vec{\Omega}$ is a volume element of phase space involving the solid angle about the direction Ω.

(vi) $n = n(\vec{r}, E, \vec{\Omega}, t)$ the number of particles in phase space per unit volume at position $\vec{r}$ per unit velocity at position $\vec{V}$ per unit energy in the solid angle $d\vec{\Omega}$ at time t and $N = N(\vec{r}, E, \vec{\Omega}, t) = vn(\vec{r}, E, \vec{\Omega}, t)$ the number of particles per unit volume per unit energy in the solid angle $d\vec{\Omega}$ at time t. The quantity $N(\vec{r}, E, \vec{\Omega}, t)d\tau\, dE\, d\vec{\Omega}$ represents the number of particles in a volume element around the position $\vec{r}$ with energy between E and $E + dE$ having direction $\vec{\Omega}$ in the solid angle $d\vec{\Omega}$ at time t.

(vii) $\phi(\vec{r}, E, \vec{\Omega}, t) = vN(\vec{r}, E, \vec{\Omega}, t)$ is the particle flux (number of particles/cm^2 − Mev − sec).

(viii) $\Sigma(E' \to E, \vec{\Omega}' \to \vec{\Omega})$ a scattering cross-section which represents the fraction of particles with energy E' and direction $\vec{\Omega}'$ that scatter into the energy range between E and $E + dE$ having direction $\vec{\Omega}$ in the solid angle $d\vec{\Omega}$ per particle flux.

(ix) $\Sigma_s(E, \vec{r})$ fractional number of particles scattered out of volume element of phase space per unit volume per flux.

(x) $\Sigma_a(E, \vec{r})$ fractional number of particles absorbed in a unit volume of phase space per unit volume per flux.

Consider a particle at time t having a position $\vec{r}$ in phase space as illustrated in the figure 2.5-4. This particle has a velocity $\vec{V}$ in a direction $\vec{\Omega}$ and has an energy E. In terms of $d\tau = dx\, dy\, dz$, $\vec{\Omega}$ and E an

308

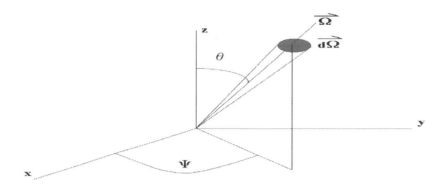

Figure 2.5-4. Volume element and solid angle about position $\vec{r}$.

element of volume of phase space can be denoted $d\tau dE d\vec{\Omega}$, where $d\vec{\Omega} = d\vec{\Omega}(\theta, \psi) = \sin\theta d\theta d\psi$ is a solid angle about the direction $\vec{\Omega}$.

The Boltzmann transport equation represents the rate of change of particle density in a volume element $d\tau\, dE\, d\vec{\Omega}$ of phase space and is written

$$\frac{d}{dt}N(\vec{r}, E, \vec{\Omega}, t)\, d\tau\, dE\, d\vec{\Omega} = D_C N(\vec{r}, E, \vec{\Omega}, t) \tag{2.5.71}$$

where D_C is a collision operator representing gains and losses of particles to the volume element of phase space due to scattering and absorption processes. The gains to the volume element are due to any sources $S(\vec{r}, E, \vec{\Omega}, t)$ per unit volume of phase space, with units of number of particles/sec per volume of phase space, together with any scattering of particles into the volume element of phase space. That is particles entering the volume element of phase space with energy E, which experience a collision, leave with some energy $E - \Delta E$ and thus will be lost from our volume element. Particles entering with energies $E' > E$ may, depending upon the cross-sections, exit with energy $E' - \Delta E = E$ and thus will contribute a gain to the volume element. In terms of the flux ϕ the gains due to scattering into the volume element are denoted by

$$\int d\vec{\Omega}' \int dE' \Sigma(E' \to E, \vec{\Omega}' \to \vec{\Omega})\phi(\vec{r}, E', \vec{\Omega}, t)\, d\tau\, dE\, d\vec{\Omega}$$

and represents the particles at position $\vec{r}$ experiencing a scattering collision with a particle of energy E' and direction $\vec{\Omega}'$ which causes the particle to end up with energy between E and $E + dE$ and direction $\vec{\Omega}$ in $d\vec{\Omega}$. The summations are over all possible initial energies.

In terms of ϕ the losses are due to those particles leaving the volume element because of scattering and are $\Sigma_s(E, \vec{r})\phi(\vec{r}, E, \vec{\Omega}, t)d\tau\, dE\, d\vec{\Omega}$. The particles which are lost due to absorption processes are $\Sigma_a(E, \vec{r})\phi(\vec{r}, E, \vec{\Omega}, t)\, d\tau\, dE\, d\vec{\Omega}$. The total change to the number of particles in an element of phase space per unit of time is obtained by summing all gains and losses. This total change is

$$\begin{aligned}\frac{dN}{dt}\, d\tau\, dE\, d\Omega = &\int d\vec{\Omega}' \int dE' \Sigma(E' \to E, \vec{\Omega}' \to \vec{\Omega})\phi(\vec{r}, E', \vec{\Omega}, t)\, d\tau\, dE\, d\vec{\Omega} \\ &- \Sigma_s(E, \vec{r})\phi(\vec{r}, E, \vec{\Omega}, t)d\tau\, dE\, d\Omega \\ &- \Sigma_a(E, \vec{r})\phi(\vec{r}, E, \vec{\Omega}, t)\, d\tau\, dE\, d\vec{\Omega} + S(\vec{r}, E, \vec{\Omega}, t)d\tau\, dE\, d\vec{\Omega}.\end{aligned} \tag{2.5.72}$$

The rate of change $\frac{dN}{dt}$ on the left-hand side of equation (2.5.72) expands to

$$\frac{dN}{dt} = \frac{\partial N}{\partial t} + \frac{\partial N}{\partial x}\frac{dx}{dt} + \frac{\partial N}{\partial y}\frac{dy}{dt} + \frac{\partial N}{\partial z}\frac{dz}{dt}$$
$$+ \frac{\partial N}{\partial V_x}\frac{dV_x}{dt} + \frac{\partial N}{\partial V_y}\frac{dV_y}{dt} + \frac{\partial N}{\partial V_z}\frac{dV_z}{dt}$$

which can be written as

$$\frac{dN}{dt} = \frac{\partial N}{\partial t} + \vec{V} \cdot \nabla_{\vec{r}} N + \frac{\vec{F}}{m} \cdot \nabla_{\vec{V}} N \tag{2.5.73}$$

where $\frac{d\vec{V}}{dt} = \frac{\vec{F}}{m}$ represents any forces acting upon the particles. The Boltzmann equation can then be expressed as

$$\frac{\partial N}{\partial t} + \vec{V} \cdot \nabla_{\vec{r}} N + \frac{\vec{F}}{m} \cdot \nabla_{\vec{V}} N = \text{Gains} - \text{Losses}. \tag{2.5.74}$$

If the right-hand side of the equation (2.5.74) is zero, the equation is known as the Liouville equation. In the special case where the velocities are constant and do not change with time the above equation (2.5.74) can be written in terms of the flux ϕ and has the form

$$\left[\frac{1}{v}\frac{\partial}{\partial t} + \vec{\Omega} \cdot \nabla_{\vec{r}} + \Sigma_s(E, \vec{r}) + \Sigma_a(E, \vec{r}) \right] \phi(\vec{r}, E, \vec{\Omega}, t) = D_C \phi \tag{2.5.75}$$

where

$$D_C \phi = \int d\vec{\Omega}' \int dE' \Sigma(E' \to E, \vec{\Omega}' \to \vec{\Omega})\phi(\vec{r}, E', \vec{\Omega}', t) + S(\vec{r}, E, \vec{\Omega}, t).$$

The above equation represents the Boltzmann transport equation in the case where all the particles are the same. In the case of atomic collisions of particles one must take into consideration the generation of secondary particles resulting from the collisions.

Let there be a number of particles of type j in a volume element of phase space. For example $j = p$ (protons) and $j = n$ (neutrons). We consider steady state conditions and define the quantities

(i) $\phi_j(\vec{r}, E, \vec{\Omega})$ as the flux of the particles of type j.

(ii) $\sigma_{jk}(\vec{\Omega}, \vec{\Omega}', E, E')$ the collision cross-section representing processes where particles of type k moving in direction $\vec{\Omega}'$ with energy E' produce a type j particle moving in the direction $\vec{\Omega}$ with energy E.

(iii) $\sigma_j(E) = \Sigma_s(E, \vec{r}) + \Sigma_a(E, \vec{r})$ the cross-section for type j particles.

The steady state form of the equation (2.5.75) can then be written as

$$\vec{\Omega} \cdot \nabla\phi_j(\vec{r}, E, \vec{\Omega}) + \sigma_j(E)\phi_j(\vec{r}, E, \vec{\Omega}) = \sum_k \int \sigma_{jk}(\vec{\Omega}, \vec{\Omega}', E, E')\phi_k(\vec{r}, E', \vec{\Omega}')d\vec{\Omega}'dE' + S(\vec{r}, E, \vec{\Omega}) \tag{2.5.76}$$

where the summation is over all particles $k \neq j$.

The Boltzmann transport equation can be represented in many different forms. These various forms are dependent upon the assumptions made during the derivation, the type of particles, coordinate systems and collision cross-sections. In general the collision cross-sections are dependent upon three components.

(1) *Elastic collisions.* Here the nucleus is not excited by the collision but energy is transferred by projectile recoil.

(2) *Inelastic collisions.* Here some particles are raised to a higher energy state but the excitation energy is not sufficient to produce any particle emissions due to the collision.

310

(3) *Non-elastic collisions.* Here the nucleus is left in an excited state due to the collision processes and some of its nucleons (protons or neutrons) are ejected. The remaining nucleons interact to form a stable structure and usually produce a distribution of low energy particles which is isotropic in character.

Various forms of the Boltzmann equation depend upon how the term $\vec{\Omega} \cdot \nabla \phi$ is represented. Various forms are:

(i) Rectangular coordinates $\phi = \phi(x, y, z; \mu, \chi)$:
$$\vec{\Omega} \cdot \nabla \phi = \mu \frac{\partial \phi}{\partial z} + \sqrt{1 - \mu^2} \left(\cos \chi \frac{\partial \phi}{\partial x} + \sin \chi \frac{\partial \phi}{\partial y} \right)$$
where γ, χ are angles associated with the solid angle. The flux is defined in terms of $\mu = \vec{\Omega} \cdot \hat{e}_z = \cos \gamma$ and χ is the angle between the planes formed by the unit vectors $\hat{e}_x, \hat{e}_z$ and $\vec{\Omega}, \hat{e}_z$. These angles vary over the range $0 \leq \gamma \leq \pi$ and $0 \leq \chi \leq 2\pi$.

(ii) Cylindrical coordinates $\phi = \phi(r, \theta, z; \mu, \chi)$:
$$\vec{\Omega} \cdot \nabla \phi = \mu \frac{\partial \phi}{\partial z} + \sqrt{1 - \mu^2} \cos \chi \frac{\partial \phi}{\partial r} + \frac{\sqrt{1 - \mu^2}}{r} \sin \chi \left(\frac{\partial \phi}{\partial \theta} - \frac{\partial \phi}{\partial \chi} \right)$$
where γ, χ are associated with the solid angle. The flux is defined in terms of $\mu = \vec{\Omega} \cdot \hat{e}_z = \cos \gamma$ and χ is the angle between the planes formed by the unit vectors $\Omega, \hat{e}_z$ and $\vec{r}, \hat{e}_z$.

(iii) Spherical coordinates $\phi = \phi(r, \theta, \varphi; \mu, \omega)$:
$$\vec{\Omega} \cdot \nabla \phi = \mu \frac{\partial \phi}{\partial r} + \frac{\sqrt{1 - \mu^2}}{r} \frac{\sin \omega}{\sin \theta} \frac{\partial \phi}{\partial \varphi} + \frac{\sqrt{1 - \mu^2}}{r} \cos \omega \frac{\partial \phi}{\partial \theta} + \frac{1 - \mu^2}{r} \frac{\partial \phi}{\partial \mu} - \frac{\sqrt{1 - \mu^2}}{r} \sin \omega \cot \theta \frac{\partial \phi}{\partial \omega}$$
with $\mu = \vec{\Omega} \cdot \vec{r}$ and ω the angle between the planes formed by the vectors $\vec{\Omega}, \vec{r}$ and $\vec{r}, \hat{e}_z$.

Various assumptions can be made concerning the particle flux. The resulting form of Boltzmann's equation must be modified to reflect these additional assumptions. As an example, we consider modifications to Boltzmann's equation in order to describe the motion of a massive ion moving into a region filled with a homogeneous material. Here it is assumed that the mean-free path for nuclear collisions is large in comparison with the mean-free path for ion interaction with electrons. In addition, the following assumptions are made

(i) All collision interactions are non-elastic.

(ii) The secondary particles produced have the same direction as the original particle. This is called the straight-ahead approximation.

(iii) Secondary particles never have kinetic energies greater than the original projectile that produced them.

(iv) A charged particle will eventually transfer all of its kinetic energy and stop in the media. This stopping distance is called the range of the projectile. The stopping power $S_j(E) = \frac{dE}{dx}$ represents the energy loss per unit length traveled in the media and determines the range by the relation $\frac{dR_j}{dE} = \frac{1}{S_j(E)}$ or $R_j(E) = \int_0^E \frac{dE'}{S_j(E')}$. Using the above assumptions Wilson, et.al.[1] show that the steady state linearized Boltzmann equation for homogeneous materials takes on the form

$$\vec{\Omega} \cdot \nabla \phi_j(\vec{r}, E, \vec{\Omega}) - \frac{1}{A_j} \frac{\partial}{\partial E} (S_j(E) \phi_j(\vec{r}, E, \vec{\Omega})) + \sigma_j(E) \phi_j(\vec{r}, E, \vec{\Omega})$$
$$= \sum_{k \neq j} \int dE' \, d\vec{\Omega}' \, \sigma_{jk}(\vec{\Omega}, \vec{\Omega}', E, E') \phi_k(\vec{r}, E', \vec{\Omega}') \tag{2.5.77}$$

---

[1] John W. Wilson, Lawrence W. Townsend, Walter Schimmerling, Govind S. Khandelwal, Ferdous Kahn, John E. Nealy, Francis A. Cucinotta, Lisa C. Simonsen, Judy L. Shinn, and John W. Norbury, *Transport Methods and Interactions for Space Radiations*, NASA Reference Publication 1257, December 1991.

where A_j is the atomic mass of the ion of type j and $\phi_j(\vec{r}, E, \vec{\Omega})$ is the flux of ions of type j moving in the direction $\vec{\Omega}$ with energy E.

Observe that in most cases the left-hand side of the Boltzmann equation represents the time rate of change of a distribution type function in a phase space while the right-hand side of the Boltzmann equation represents the time rate of change of this distribution function within a volume element of phase space due to scattering and absorption collision processes.

Boltzmann Equation for gases

Consider the Boltzmann equation in terms of a particle distribution function $f(\vec{r}, \vec{V}, t)$ which can be written as

$$\left(\frac{\partial}{\partial t} + \vec{V} \cdot \nabla_{\vec{r}} + \frac{\vec{F}}{m} \cdot \nabla_{\vec{V}} \right) f(\vec{r}, \vec{V}, t) = D_C f(\vec{r}, \vec{V}, t) \tag{2.5.78}$$

for a single species of gas particles where there is only scattering and no absorption of the particles. An element of volume in phase space (x, y, z, V_x, V_y, V_z) can be thought of as a volume element $d\tau = dx\,dy\,dz$ for the spatial elements together with a volume element $d\tau_v = dV_x dV_y dV_z$ for the velocity elements. These elements are centered at position $\vec{r}$ and velocity $\vec{V}$ at time t. In phase space a constant velocity V_1 can be thought of as a sphere since $V_1^2 = V_x^2 + V_y^2 + V_z^2$. The phase space volume element $d\tau d\tau_v$ changes with time since the position $\vec{r}$ and velocity $\vec{V}$ change with time. The position vector $\vec{r}$ changes because of velocity and the velocity vector changes because of the acceleration $\frac{\vec{F}}{m}$. Here $f(\vec{r}, \vec{V}, t)d\tau d\tau_v$ represents the expected number of particles in the phase space element $d\tau d\tau_v$ at time t.

Assume there are no collisions, then each of the gas particles in a volume element of phase space centered at position $\vec{r}$ and velocity $\vec{V}_1$ move during a time interval dt to a phase space element centered at position $\vec{r} + \vec{V}_1 dt$ and $\vec{V}_1 + \frac{\vec{F}}{m}dt$. If there were no loss or gains of particles, then the number of particles must be conserved and so these gas particles must move smoothly from one element of phase space to another without any gains or losses of particles. Because of scattering collisions in $d\tau$ many of the gas particles move into or out of the velocity range $\vec{V}_1$ to $\vec{V}_1 + d\vec{V}_1$. These collision scattering processes are denoted by the collision operator $D_C f(\vec{r}, \vec{V}, t)$ in the Boltzmann equation.

Consider two identical gas particles which experience a binary collision. Imagine that particle 1 with velocity $\vec{V}_1$ collides with particle 2 having velocity $\vec{V}_2$. Denote by $\sigma(\vec{V}_1 \to \vec{V}_1', \vec{V}_2 \to \vec{V}_2')\,d\tau_{V_1}d\tau_{V_2}$ the conditional probability that particle 1 is scattered from velocity $\vec{V}_1$ to between $\vec{V}_1'$ and $\vec{V}_1' + d\vec{V}_1'$ and the struck particle 2 is scattered from velocity $\vec{V}_2$ to between $\vec{V}_2'$ and $\vec{V}_2' + d\vec{V}_2'$. We will be interested in collisions of the type $(\vec{V}_1', \vec{V}_2') \to (\vec{V}_1, \vec{V}_2)$ for a fixed value of $\vec{V}_1$ as this would represent the number of particles scattered into $d\tau_{V_1}$. Also of interest are collisions of the type $(\vec{V}_1, \vec{V}_2) \to (\vec{V}_1', \vec{V}_2')$ for a fixed value $\vec{V}_1$ as this represents particles scattered out of $d\tau_{V_1}$. Imagine a gas particle in $d\tau$ with velocity $\vec{V}_1'$ subjected to a beam of particles with velocities $\vec{V}_2'$. The incident flux on the element $d\tau d\tau_{V_1'}$ is $|\vec{V}_1' - \vec{V}_2'|f(\vec{r}, \vec{V}_2', t)d\tau_{V_2'}$ and hence

$$\sigma(\vec{V}_1 \to \vec{V}_1', \vec{V}_2 \to \vec{V}_2')\,d\tau_{V_1}d\tau_{V_2}dt\,|\vec{V}_1' - \vec{V}_2'|f(\vec{r}, \vec{V}_2', t)\,d\tau_{V_2'} \tag{2.5.79}$$

represents the number of collisions, in the time interval dt, which scatter from $\vec{V}_1'$ to between $\vec{V}_1$ and $\vec{V}_1 + d\vec{V}_1$ as well as scattering $\vec{V}_2'$ to between $\vec{V}_2$ and $\vec{V}_2 + d\vec{V}_2$. Multiply equation (2.5.79) by the density of particles

312

in the element $d\tau d\tau_{V_1'}$ and integrate over all possible initial velocities $\vec{V}_1'$, $\vec{V}_2'$ and final velocities $\vec{V}_2$ not equal to $\vec{V}_1$. This gives the number of particles in $d\tau$ which are scattered into $d\tau_{V_1}dt$ as

$$Ns_{in} = d\tau d\tau_{V_1}dt \int d\tau_{V_2}d\tau_{V_2'} \int d\tau_{V_1'}\, \sigma(\vec{V}_1' \to \vec{V}_1, \vec{V}_2' \to \vec{V}_2)|\vec{V}_1' - \vec{V}_2'|f(\vec{r},\vec{V}_1',t)f(\vec{r},\vec{V}_2',t). \qquad (2.5.80)$$

In a similar manner the number of particles in $d\tau$ which are scattered out of $d\tau_{V_1}dt$ is

$$Ns_{out} = d\tau d\tau_{V_1}dt f(\vec{r},\vec{V}_1,t) \int d\tau_{V_2} \int d\tau_{V_2'} \int d\tau_{V_1'}\, \sigma(\vec{V}_1' \to \vec{V}_1, \vec{V}_2' \to \vec{V}_2)|\vec{V}_2 - \vec{V}_1|f(\vec{r},\vec{V}_2,t). \qquad (2.5.81)$$

Let

$$W(\vec{V}_1' \to \vec{V}_1, \vec{V}_2' \to \vec{V}_2) = |\vec{V}_1 - \vec{V}_2|\,\sigma(\vec{V}_1' \to \vec{V}_1, \vec{V}_2' \to \vec{V}_2) \qquad (2.5.82)$$

define a symmetric scattering kernel and use the relation $D_C f(\vec{r},\vec{V},t) = Ns_{in} - Ns_{out}$ to represent the Boltzmann equation for gas particles in the form

$$\left(\frac{\partial}{\partial t} + \vec{V}\cdot\nabla_{\vec{r}} + \frac{\vec{F}}{m}\cdot\nabla_{\vec{V}}\right)f(\vec{r},\vec{V}_1,t) =$$
$$\int d\tau_{V_1'} \int d\tau_{V_2'} \int d\tau_{V_2} W(\vec{V}_1 \to \vec{V}_1', \vec{V}_2 \to \vec{V}_2')\left[f(\vec{r},\vec{V}_1',t)f(\vec{r},\vec{V}_2',t) - f(\vec{r},\vec{V}_1,t)f(\vec{r},\vec{V}_2,t)\right]. \qquad (2.5.83)$$

Take the moment of the Boltzmann equation (2.5.83) with respect to an arbitrary function $\phi(\vec{V}_1)$. That is, multiply equation (2.5.83) by $\phi(\vec{V}_1)$ and then integrate over all elements of velocity space $d\tau_{V_1}$. Define the following averages and terminology:

• The particle density per unit volume

$$n = n(\vec{r},t) = \int d\tau_V f(\vec{r},\vec{V},t) = \int\!\!\!\int\!\!\!\int_{-\infty}^{+\infty} f(\vec{r},\vec{V},t)dV_xdV_ydV_z \qquad (2.5.84)$$

where $\rho = nm$ is the mass density.

• The mean velocity

$$\overline{\vec{V}_1} = \vec{V} = \frac{1}{n}\int\!\!\!\int\!\!\!\int_{-\infty}^{+\infty} \vec{V}_1 f(\vec{r},\vec{V}_1,t)dV_{1x}dV_{1y}dV_{1z}$$

For any quantity $Q = Q(\vec{V}_1)$ define the barred quantity

$$\overline{Q} = \overline{Q(\vec{r},t)} = \frac{1}{n(\vec{r},t)}\int Q(\vec{V})f(\vec{r},\vec{V},t)\,d\tau_V = \frac{1}{n}\int\!\!\!\int\!\!\!\int_{-\infty}^{+\infty} Q(\vec{V})f(\vec{r},\vec{V},t)dV_xdV_ydV_z. \qquad (2.5.85)$$

Further, assume that $\frac{\vec{F}}{m}$ is independent of $\vec{V}$, then the moment of equation (2.5.83) produces the result

$$\frac{\partial}{\partial t}\left(n\overline{\phi}\right) + \sum_{i=1}^{3}\frac{\partial}{\partial x^i}\left(n\overline{V_{1i}\phi}\right) - n\sum_{i=1}^{3}\frac{F_i}{m}\overline{\frac{\partial\phi}{\partial V_{1i}}} = 0 \qquad (2.5.86)$$

known as the Maxwell transfer equation. The first term in equation (2.5.86) follows from the integrals

$$\int \frac{\partial f(\vec{r},\vec{V}_1,t)}{\partial t}\phi(\vec{V}_1)d\tau_{V_1} = \frac{\partial}{\partial t}\int f(\vec{r},\vec{V}_1,t)\phi(\vec{V}_1)\,d\tau_{V_1} = \frac{\partial}{\partial t}(n\overline{\phi}) \qquad (2.5.87)$$

where differentiation and integration have been interchanged. The second term in equation (2.5.86) follows from the integral

$$\int \vec{V}_1 \nabla_{\vec{r}} f \, \phi(\vec{V}_1) d\tau_{V_1} = \int \sum_{i=1}^{3} V_{1i} \frac{\partial f}{\partial x^i} \phi \, d\tau_{V_1} = \sum_{i=1}^{3} \frac{\partial}{\partial x^i} \left(\int V_{1i} \phi f \, d\tau_{V_1} \right) = \sum_{i=1}^{3} \frac{\partial}{\partial x^i} \left(n \overline{V_{1i} \phi} \right). \qquad (2.5.88)$$

The third term in equation (2.5.86) is obtained from the following integral where integration by parts is employed

$$\begin{aligned}
\int \frac{\vec{F}}{m} \nabla_{\vec{V}_1} f \phi \, d\tau_{V_1} &= \int \sum_{i=1}^{3} \left(\frac{F_i}{m} \frac{\partial f}{\partial V_{1i}} \right) \phi \, d\tau_{V_1} \\
&= \int\!\!\!\int\!\!\!\int_{-\infty}^{+\infty} \sum_{i=1}^{3} \phi \left(\frac{F_i}{m} \frac{\partial f}{\partial V_{1i}} \right) dV_{1x} dV_{1y} dV_{1y} \\
&= -\int \frac{\partial}{\partial V_{1i}} \left(\frac{F_i}{m} \phi \right) f \, d\tau_{V_1} \\
&= -n \overline{\frac{\partial}{\partial V_{1i}} \left(\frac{F_i}{m} \phi \right)} = -\frac{F_i}{m} \overline{\frac{\partial \phi}{\partial V_{1i}}}
\end{aligned} \qquad (2.5.89)$$

since F_i does not depend upon $\vec{V}_1$ and $f(\vec{r}, \vec{V}, t)$ equals zero for V_i equal to $\pm\infty$. The right-hand side of equation (2.5.86) represents the integral of $(D_C f)\phi$ over velocity space. This integral is zero because of the symmetries associated with the right-hand side of equation (2.5.83). Physically, the integral of $(D_c f)\phi$ over velocity space must be zero since collisions with only scattering terms cannot increase or decrease the number of particles per cubic centimeter in any element of phase space.

In equation (2.5.86) we write the velocities V_{1i} in terms of the mean velocities (u, v, w) and random velocities (U_r, V_r, W_r) with

$$V_{11} = U_r + u, \qquad V_{12} = V_r + v, \qquad V_{13} = W_r + w \qquad (2.5.90)$$

or $\vec{V}_1 = \vec{V}_r + \vec{V}$ with $\overline{\vec{V}_1} = \overline{\vec{V}_r + \vec{V}} = \vec{V}$ since $\overline{\vec{V}_r} = 0$ (i.e. the average random velocity is zero.) For future reference we write equation (2.5.86) in terms of these random velocities and the material derivative. Substitution of the velocities from equation (2.5.90) in equation (2.5.86) gives

$$\frac{\partial (n\overline{\phi})}{\partial t} + \frac{\partial}{\partial x} \left(\overline{n(U_r + u)\phi} \right) + \frac{\partial}{\partial y} \left(\overline{n(V_r + v)\phi} \right) + \frac{\partial}{\partial z} \left(\overline{n(W_r + w)\phi} \right) - n \sum_{i=1}^{3} \frac{F_i}{m} \overline{\frac{\partial \phi}{\partial V_{1i}}} = 0 \qquad (2.5.91)$$

or

$$\begin{aligned}
\frac{\partial (n\overline{\phi})}{\partial t} &+ \frac{\partial}{\partial x} \left(n\overline{u\phi} \right) + \frac{\partial}{\partial y} \left(n\overline{v\phi} \right) + \frac{\partial}{\partial z} \left(n\overline{w\phi} \right) \\
&+ \frac{\partial}{\partial x} \left(n\overline{U_r \phi} \right) + \frac{\partial}{\partial y} \left(n\overline{V_r \phi} \right) + \frac{\partial}{\partial z} \left(n\overline{W_r \phi} \right) - n \sum_{i=1}^{3} \frac{F_i}{m} \overline{\frac{\partial \phi}{\partial V_{1i}}} = 0.
\end{aligned} \qquad (2.5.92)$$

Observe that

$$n\overline{u\phi} = \int\!\!\!\int\!\!\!\int_{-\infty}^{+\infty} u\phi f(\vec{r}, \vec{V}, t) dV_x dV_y dV_z = nu\overline{\phi} \qquad (2.5.93)$$

and similarly $\overline{nv\phi} = nv\overline{\phi}$, $\overline{nw\phi} = nw\overline{\phi}$. This enables the equation (2.5.92) to be written in the form

$$
\begin{aligned}
& n\frac{\partial\overline{\phi}}{\partial t} + nu\frac{\partial\overline{\phi}}{\partial x} + nv\frac{\partial\overline{\phi}}{\partial y} + nw\frac{\partial\overline{\phi}}{\partial z} \\
& + \overline{\phi}\left[\frac{\partial n}{\partial t} + \frac{\partial}{\partial x}(nu) + \frac{\partial}{\partial y}(nv) + \frac{\partial}{\partial z}(nw)\right] \\
& + \frac{\partial}{\partial x}\left(n\overline{U_r\phi}\right) + \frac{\partial}{\partial y}\left(n\overline{V_r\phi}\right) + \frac{\partial}{\partial z}\left(n\overline{W_r\phi}\right) - n\sum_{i=1}^{3}\frac{F_i}{m}\overline{\frac{\partial\phi}{\partial V_{1i}}} = 0.
\end{aligned}
\tag{2.5.94}
$$

The middle bracketed sum in equation (2.5.94) is recognized as the continuity equation when multiplied by m and hence is zero. The moment equation (2.5.86) now has the form

$$
n\frac{D\overline{\phi}}{Dt} + \frac{\partial}{\partial x}\left(n\overline{U_r\phi}\right) + \frac{\partial}{\partial y}\left(n\overline{V_r\phi}\right) + \frac{\partial}{\partial z}\left(n\overline{W_r\phi}\right) - n\sum_{i=1}^{3}\frac{F_i}{m}\overline{\frac{\partial\phi}{\partial V_{1i}}} = 0.
\tag{2.5.95}
$$

Note that from the equations (2.5.86) or (2.5.95) one can derive the basic equations of fluid flow from continuum mechanics developed earlier. We consider the following special cases of the Maxwell transfer equation.

(i) In the special case $\phi = m$ the equation (2.5.86) reduces to the continuity equation for fluids. That is, equation (2.5.86) becomes

$$
\frac{\partial}{\partial t}(nm) + \nabla\cdot(nm\overline{\vec{V}_1}) = 0
\tag{2.5.96}
$$

which is the continuity equation

$$
\frac{\partial\rho}{\partial t} + \nabla\cdot(\rho\vec{V}) = 0
\tag{2.5.97}
$$

where ρ is the mass density and $\vec{V}$ is the mean velocity defined earlier.

(ii) In the special case $\phi = m\vec{V}_1$ is momentum, the equation (2.5.86) reduces to the momentum equation for fluids. To show this, we write equation (2.5.86) in terms of the dyadic $\vec{V}_1\vec{V}_1$ in the form

$$
\frac{\partial}{\partial t}\left(nm\overline{\vec{V}_1}\right) + \nabla\cdot(nm\overline{\vec{V}_1\vec{V}_1}) - n\vec{F} = 0
\tag{2.5.98}
$$

or

$$
\frac{\partial}{\partial t}\left(\rho(\vec{V}_r + \vec{V})\right) + \nabla\cdot(\rho(\vec{V}_r + \vec{V})(\vec{V}_r + \vec{V})) - n\vec{F} = 0.
\tag{2.5.99}
$$

Let $\boldsymbol{\sigma} = -\rho\vec{V}_r\vec{V}_r$ denote a stress tensor which is due to the random motions of the gas particles and write equation (2.5.99) in the form

$$
\rho\frac{\partial\vec{V}}{\partial t} + \vec{V}\frac{\partial\rho}{\partial t} + \rho\vec{V}(\nabla\cdot\vec{V}) + \vec{V}(\nabla\cdot(\rho\vec{V})) - \nabla\cdot\boldsymbol{\sigma} - n\vec{F} = 0.
\tag{2.5.100}
$$

The term $\vec{V}\left(\frac{\partial\rho}{\partial t} + \nabla\cdot(\rho\vec{V})\right) = 0$ because of the continuity equation and so equation (2.5.100) reduces to the momentum equation

$$
\rho\left(\frac{\partial\vec{V}}{\partial t} + \vec{V}\nabla\cdot\vec{V}\right) = n\vec{F} + \nabla\cdot\boldsymbol{\sigma}.
\tag{2.5.101}
$$

For $\vec{F} = q\vec{E} + q\vec{V} \times \vec{B} + m\vec{b}$, where q is charge, $\vec{E}$ and $\vec{B}$ are electric and magnetic fields, and $\vec{b}$ is a body force per unit mass, together with

$$\sigma = \sum_{i=1}^{3}\sum_{j=1}^{3}(-p\delta_{ij} + \tau_{ij})\widehat{e}_i\widehat{e}_j \tag{2.5.102}$$

the equation (2.5.101) becomes the momentum equation

$$\rho\frac{D\vec{V}}{Dt} = \rho\vec{b} - \nabla p + \nabla \cdot \tau + nq(\vec{E} + \vec{V} \times \vec{B}). \tag{2.5.103}$$

In the special case were $\vec{E}$ and $\vec{B}$ vanish, the equation (2.5.103) reduces to the previous momentum equation (2.5.25) .

(iii) In the special case $\phi = \frac{m}{2}\vec{V}_1 \cdot \vec{V}_1 = \frac{m}{2}(V_{11}^2 + V_{12}^2 + V_{13}^2)$ is the particle kinetic energy, the equation (2.5.86) simplifies to the energy equation of fluid mechanics. To show this we substitute ϕ into equation (2.5.95) and simplify. Note that

$$\begin{aligned}\overline{\phi} &= \frac{m}{2}\left[\overline{(U_r + u)^2} + \overline{(V_r + v)^2} + \overline{(W_r + w)^2}\right]\\ \overline{\phi} &= \frac{m}{2}\left[\overline{U_r^2} + \overline{V_r^2} + \overline{W_r^2} + u^2 + v^2 + w^2\right]\end{aligned} \tag{2.5.104}$$

since $u\overline{U_r} = v\overline{V_r} = w\overline{W_r} = 0$. Let $V^2 = u^2 + v^2 + w^2$ and $\overline{C_r^2} = \overline{U_r^2} + \overline{V_r^2} + \overline{W_r^2}$ and write equation (2.5.104) in the form

$$\overline{\phi} = \frac{m}{2}\left(\overline{C_r^2} + V^2\right). \tag{2.5.105}$$

Also note that

$$\begin{aligned}n\overline{U_r\phi} &= \frac{nm}{2}\left[\overline{U_r(U_r + u)^2} + \overline{U_r(V_r + v)^2} + \overline{U_r(W_r + w)^2}\right]\\ &= \frac{nm}{2}\left[\overline{\frac{U_r C_r^2}{2}} + u\overline{U_r^2} + v\overline{U_r V_r} + w\overline{U_r W_r}\right]\end{aligned} \tag{2.5.106}$$

and that

$$n\overline{V_r\phi} = \frac{nm}{2}\left[\overline{V_r C_r^2} + u\overline{V_r U_r} + v\overline{V_r^2} + w\overline{V_r W_r}\right] \tag{2.5.107}$$

$$n\overline{W_r\phi} = \frac{nm}{2}\left[\overline{W_r C_r^2} + u\overline{W_r U_r} + v\overline{W_r V_r} + w\overline{W_r^2}\right] \tag{2.5.108}$$

are similar results.

We use $\frac{\partial}{\partial V_{1i}}(\phi) = mV_{1i}$ together with the previous results substituted into the equation (2.5.95), and find that the Maxwell transport equation can be expressed in the form

$$\begin{aligned}\rho\frac{D}{Dt}\left(\frac{\overline{C_r^2}}{2} + \frac{V^2}{2}\right) = &-\frac{\partial}{\partial x}\left(\rho[u\overline{U_r^2} + v\overline{U_r V_r} + w\overline{U_r W_r}]\right)\\ &-\frac{\partial}{\partial y}\left(\rho[u\overline{V_r U_r} + v\overline{V_r^2} + w\overline{V_r W_r}]\right)\\ &-\frac{\partial}{\partial z}\left(\rho[u\overline{W_r U_r} + v\overline{W_r V_r} + w\overline{W_r^2}]\right)\\ &-\frac{\partial}{\partial x}\left(\rho\frac{\overline{U_r C_r^2}}{2}\right) - \frac{\partial}{\partial y}\left(\rho\frac{\overline{V_r C_r^2}}{2}\right) - \frac{\partial}{\partial z}\left(\rho\frac{\overline{W_r C_r^2}}{2}\right) + n\vec{F} \cdot \vec{V}.\end{aligned} \tag{2.5.109}$$

Compare the equation (2.5.109) with the energy equation (2.5.48)

$$\rho \frac{De}{Dt} + \rho \frac{D}{Dt}\left(\frac{V^2}{2}\right) = \nabla(\boldsymbol{\sigma} \cdot \vec{V}) - \nabla \cdot \vec{q} + \rho \vec{b} \cdot \vec{V} \qquad (2.5.110)$$

where the internal heat energy has been set equal to zero. Let $e = \frac{\overline{C_r^2}}{2}$ denote the internal energy due to random motion of the gas particles, $\vec{F} = m\vec{b}$, and let

$$\begin{aligned}
\nabla \cdot \vec{q} &= -\frac{\partial}{\partial x}\left(\rho \frac{\overline{U_r C_r^2}}{2}\right) - \frac{\partial}{\partial y}\left(\rho \frac{\overline{V_r C_r^2}}{2}\right) - \frac{\partial}{\partial z}\left(\rho \frac{\overline{W_r C_r^2}}{2}\right) \\
&= -\frac{\partial}{\partial x}\left(k\frac{\partial T}{\partial x}\right) - \frac{\partial}{\partial y}\left(k\frac{\partial T}{\partial y}\right) - \frac{\partial}{\partial z}\left(k\frac{\partial T}{\partial z}\right)
\end{aligned} \qquad (2.5.111)$$

represent the heat conduction terms due to the transport of particle energy $\frac{mC_r^2}{2}$ by way of the random particle motion. The remaining terms are related to the rate of change of work and surface stresses giving

$$\begin{aligned}
-\frac{\partial}{\partial x}\left(\rho[u\overline{U_r^2} + v\overline{U_r V_r} + w\overline{U_r W_r}]\right) &= \frac{\partial}{\partial x}\left(u\sigma_{xx} + v\sigma_{xy} + w\sigma_{xz}\right) \\
-\frac{\partial}{\partial y}\left(\rho[u\overline{V_r U_r} + v\overline{V_r^2} + w\overline{V_r W_r}]\right) &= \frac{\partial}{\partial y}\left(u\sigma_{yx} + v\sigma_{yy} + w\sigma_{yz}\right) \\
-\frac{\partial}{\partial z}\left(\rho[u\overline{W_r U_r} + v\overline{W_r V_r} + w\overline{W_r^2}]\right) &= \frac{\partial}{\partial z}\left(u\sigma_{zx} + v\sigma_{zy} + w\sigma_{zz}\right).
\end{aligned} \qquad (2.5.112)$$

This gives the stress relations due to random particle motion

$$\begin{array}{lll}
\sigma_{xx} = -\rho\overline{U_r^2} & \sigma_{yx} = -\rho\overline{V_r U_r} & \sigma_{zx} = -\rho\overline{W_r U_r} \\
\sigma_{xy} = -\rho\overline{U_r V_r} & \sigma_{yy} = -\rho\overline{V_r^2} & \sigma_{zy} = -\rho\overline{W_r V_r} \\
\sigma_{xz} = -\rho\overline{U_r W_r} & \sigma_{yz} = -\rho\overline{V_r W_r} & \sigma_{zz} = -\rho\overline{W_r^2}.
\end{array} \qquad (2.5.113)$$

The Boltzmann equation is a basic macroscopic model used for the study of individual particle motion where one takes into account the distribution of particles in both space, time and energy. The Boltzmann equation for gases assumes only binary collisions as three-body or multi-body collisions are assumed to rarely occur. Another assumption used in the development of the Boltzmann equation is that the actual time of collision is thought to be small in comparison with the time between collisions. The basic problem associated with the Boltzmann equation is to find a velocity distribution, subject to either boundary and/or initial conditions, which describes a given gas flow.

The continuum equations involve trying to obtain the macroscopic variables of density, mean velocity, stress, temperature and pressure which occur in the basic equations of continuum mechanics considered earlier. Note that the moments of the Boltzmann equation, derived for gases, also produced these same continuum equations and so they are valid for gases as well as liquids.

In certain situations one can assume that the gases approximate a Maxwellian distribution

$$f(\vec{r}, \vec{V}, t) \approx n(\vec{r}, t)\left(\frac{m}{2\pi kT}\right)^{3/2}\exp\left(-\frac{m}{2kT}\vec{V}\cdot\vec{V}\right) \qquad (2.5.114)$$

thereby enabling the calculation of the pressure tensor and temperature from statistical considerations.

In general, one can say that the Boltzmann integral-differential equation and the Maxwell transfer equation are two important formulations in the kinetic theory of gases. The Maxwell transfer equation depends upon some gas-particle property ϕ which is assumed to be a function of the gas-particle velocity. The Boltzmann equation depends upon a gas-particle velocity distribution function f which depends upon position $\vec{r}$, velocity $\vec{V}$ and time t. These formulations represent two distinct and important viewpoints considered in the kinetic theory of gases.

EXERCISE 2.5

1. Let $p = p(x, y, z)$, [dyne/cm^2] denote the pressure at a point (x, y, z) in a fluid medium at rest (hydrostatics), and let ΔV denote an element of fluid volume situated at this point as illustrated in the figure 2.5-5.

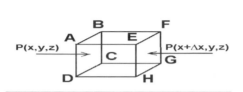

Figure 2.5-5. Pressure acting on a volume element.

(a) Show that the force acting on the face $ABCD$ is $p(x, y, z)\Delta y \Delta z\, \hat{e}_1$.

(b) Show that the force acting on the face $EFGH$ is

$$-p(x + \Delta x, y, z)\Delta y \Delta z\, \hat{e}_1 = -\left[p(x, y, z) + \frac{\partial p}{\partial x}\Delta x + \frac{\partial^2 p}{\partial x^2}\frac{(\Delta x)^2}{2!} + \cdots \right]\Delta y \Delta z\, \hat{e}_1.$$

(c) In part (b) neglect terms with powers of Δx greater than or equal to 2 and show that the resultant force in the x-direction is $-\dfrac{\partial p}{\partial x}\Delta x \Delta y \Delta z\, \hat{e}_1$.

(d) What has been done in the x-direction can also be done in the y and z-directions. Show that the resultant forces in these directions are $-\dfrac{\partial p}{\partial y}\Delta x \Delta y \Delta z\, \hat{e}_2$ and $-\dfrac{\partial p}{\partial z}\Delta x \Delta y \Delta z\, \hat{e}_3$. (e) Show that $-\nabla p = -\left(\dfrac{\partial p}{\partial x}\,\hat{e}_1 + \dfrac{\partial p}{\partial y}\,\hat{e}_2 + \dfrac{\partial p}{\partial z}\,\hat{e}_3 \right)$ is the force per unit volume acting at the point (x, y, z) of the fluid medium.

2. Follow the example of exercise 1 above but use cylindrical coordinates and find the force per unit volume at a point (r, θ, z). Hint: An element of volume in cylindrical coordinates is given by $\Delta V = r\Delta r \Delta \theta \Delta z$.

3. Follow the example of exercise 1 above but use spherical coordinates and find the force per unit volume at a point (ρ, θ, ϕ). Hint: An element of volume in spherical coordinates is $\Delta V = \rho^2 \sin\theta \Delta\rho\Delta\theta\Delta\phi$.

4. Show that if the density $\varrho = \varrho(x, y, z, t)$ is a constant, then $v^r_{,r} = 0$.

5. Assume that λ^* and μ^* are zero. Such a fluid is called a nonviscous or perfect fluid. (a) Show the Cartesian equations describing conservation of linear momentum are

$$\frac{\partial u}{\partial t} + u\frac{\partial u}{\partial x} + v\frac{\partial u}{\partial y} + w\frac{\partial u}{\partial z} = b_x - \frac{1}{\varrho}\frac{\partial p}{\partial x}$$

$$\frac{\partial v}{\partial t} + u\frac{\partial v}{\partial x} + v\frac{\partial v}{\partial y} + w\frac{\partial v}{\partial z} = b_y - \frac{1}{\varrho}\frac{\partial p}{\partial y}$$

$$\frac{\partial w}{\partial t} + u\frac{\partial w}{\partial x} + v\frac{\partial w}{\partial y} + w\frac{\partial w}{\partial z} = b_z - \frac{1}{\varrho}\frac{\partial p}{\partial z}$$

where (u, v, w) are the physical components of the fluid velocity. (b) Show that the continuity equation can be written

$$\frac{\partial \varrho}{\partial t} + \frac{\partial}{\partial x}(\varrho u) + \frac{\partial}{\partial y}(\varrho v) + \frac{\partial}{\partial z}(\varrho w) = 0$$

318

▶ **6.** Assume $\lambda^* = \mu^* = 0$ so that the fluid is ideal or nonviscous. Use the results given in problem 5 and make the following additional assumptions:

- The density is constant and so the fluid is incompressible.
- The body forces are zero.
- Steady state flow exists.
- Only two dimensional flow in the x-y plane is considered such that $u = u(x, y)$, $v = v(x, y)$ and $w = 0$. (a) Employ the above assumptions and simplify the equations in problem 5 and verify the results

$$u\frac{\partial u}{\partial x} + v\frac{\partial u}{\partial y} + \frac{1}{\varrho}\frac{\partial p}{\partial x} = 0$$

$$u\frac{\partial v}{\partial x} + v\frac{\partial v}{\partial y} + \frac{1}{\varrho}\frac{\partial p}{\partial y} = 0$$

$$\frac{\partial u}{\partial x} + \frac{\partial v}{\partial y} = 0$$

(b) Make the additional assumption that the flow is irrotational and show that this assumption produces the results

$$\frac{\partial v}{\partial x} - \frac{\partial u}{\partial y} = 0 \quad \text{and} \quad \frac{1}{2}\left(u^2 + v^2\right) + \frac{1}{\varrho}p = constant.$$

(c) Point out the Cauchy-Riemann equations and Bernoulli's equation in the above set of equations.

▶ **7.** Assume the body forces are derivable from a potential function ϕ such that $b_i = -\phi_{,i}$. Show that for an ideal fluid with constant density the equations of fluid motion can be written in either of the forms

$$\frac{\partial v^r}{\partial t} + v^r_{,s}v^s = -\frac{1}{\varrho}g^{rm}p_{,m} - g^{rm}\phi_{,m} \quad \text{or} \quad \frac{\partial v_r}{\partial t} + v_{r,s}v^s = -\frac{1}{\varrho}p_{,r} - \phi_{,r}$$

▶ **8.** The vector identities $\nabla^2\vec{v} = \nabla(\nabla \cdot \vec{v}) - \nabla \times (\nabla \times \vec{v})$ and $(\vec{v} \cdot \nabla)\vec{v} = \frac{1}{2}\nabla(\vec{v} \cdot \vec{v}) - \vec{v} \times (\nabla \times \vec{v})$ are used to express the Navier-Stokes-Duhem equations in alternate forms involving the vorticity $\Omega = \nabla \times \vec{v}$. (a) Use Cartesian tensor notation and derive the above identities. (b) Show the second identity can be written in generalized coordinates as $v^j v^m_{,j} = g^{mj}v^k v_{k,j} - \epsilon^{mnp}\epsilon^{ijk}g_{pi}v_n v_{k,j}$. Hint: Show that $\frac{\partial v^2}{\partial x^j} = 2v^k v_{k,j}$.

▶ **9.** Use problem 8 and show that the results in problem 7 can be written

$$\frac{\partial v^r}{\partial t} - \epsilon^{rnp}\Omega_p v_n = -g^{rm}\frac{\partial}{\partial x^m}\left(\frac{p}{\varrho} + \phi + \frac{v^2}{2}\right)$$

or

$$\frac{\partial v_i}{\partial t} - \epsilon_{ijk}v^j\Omega^k = -\frac{\partial}{\partial x^i}\left(\frac{p}{\varrho} + \phi + \frac{v^2}{2}\right)$$

▶ **10.** In terms of physical components, show that in generalized orthogonal coordinates, for $i \neq j$, the rate of deformation tensor D_{ij} can be written $D(ij) = \frac{1}{2}\left[\frac{h_i}{h_j}\frac{\partial}{\partial x^j}\left(\frac{v(i)}{h_i}\right) + \frac{h_j}{h_i}\frac{\partial}{\partial x^i}\left(\frac{v(j)}{h_j}\right)\right]$, no summations and for $i = j$ there results $D(ii) = \frac{1}{h_i}\frac{\partial v(i)}{\partial x^i} - \frac{v(i)}{h_i^2}\frac{\partial h_i}{\partial x^i} + \sum_{k=1}^{3}\frac{1}{h_i h_k}v(k)\frac{\partial h_i}{\partial x^k}$, no summations. (Hint: See Problem 17 Exercise 2.1.)

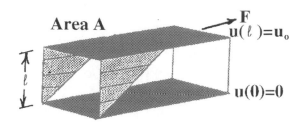

Area A

$\mathbf{F}$

$\mathbf{u}(\ell) = \mathbf{u_o}$

ℓ

$\mathbf{u(0)=0}$

Figure 2.5-6. Plane Couette flow

▶ **11.** Find the physical components of the rate of deformation tensor D_{ij} in Cartesian coordinates. (Hint: See problem 10.)

12. Find the physical components of the rate of deformation tensor in cylindrical coordinates. (Hint: See problem 10.)

13. (Plane Couette flow) Assume a viscous fluid with constant density is between two plates as illustrated in the figure 2.5-6.

(a) Define $\nu = \frac{\mu^*}{\varrho}$ as the kinematic viscosity and show the equations of fluid motion can be written

$$\frac{\partial v^i}{\partial t} + v^i_{,s} v^s = -\frac{1}{\varrho} g^{im} p_{,m} + \nu g^{jm} v^i_{,mj} + g^{ij} b_j, \quad i = 1, 2, 3$$

(b) Let $\vec{v} = (u, v, w)$ denote the physical components of the fluid flow and make the following assumptions

- $u = u(y)$, $v = w = 0$
- Steady state flow exists
- The top plate, with area A, is a distance ℓ above the bottom plate. The bottom plate is fixed and a constant force F is applied to the top plate to keep it moving with a velocity $u_0 = u(\ell)$.
- p and ϱ are constants
- The body force components are zero.

Find the velocity $u = u(y)$

(c) Show the tangential stress exerted by the moving fluid is $\dfrac{F}{A} = \sigma_{21} = \sigma_{xy} = \sigma_{yx} = \mu^* \dfrac{u_0}{\ell}$. This example illustrates that the stress is proportional to u_0 and inversely proportional to ℓ.

▶ **14.** In the continuity equation make the change of variables

$$\bar{t} = \frac{t}{\tau}, \quad \bar{\varrho} = \frac{\varrho}{\varrho_0}, \quad \vec{\bar{v}} = \frac{\vec{v}}{v_0}, \quad \bar{x} = \frac{x}{L}, \quad \bar{y} = \frac{y}{L}, \quad \bar{z} = \frac{z}{L}$$

and write the continuity equation in terms of the barred variables and the Strouhal parameter.

▶ **15.** Simplify the Navier-Stokes-Duhem equations using the assumption that there is incompressible and irrotational flow.

▶ **16.** Let $\zeta = \lambda^* + \frac{2}{3}\mu^*$ and show the constitutive equations (2.5.21) for fluid motion can be written in the form

$$\sigma_{ij} = -p\delta_{ij} + \mu^* \left[v_{i,j} + v_{j,i} - \frac{2}{3}\delta_{ij} v_{k,k} \right] + \zeta \delta_{ij} v_{k,k}.$$

320

▶ **17.** (Plane Poiseuille flow)   Consider two flat plates parallel to one another as illustrated in the figure 2.5-7. One plate is at $y = 0$ and the other plate is at $y = 2\ell$. Let $\vec{v} = (u, v, w)$ denote the physical components of the fluid velocity and make the following assumptions concerning the flow The body forces are zero. The derivative $\dfrac{\partial p}{\partial x} = -p_0$ is a constant and $\dfrac{\partial p}{\partial y} = \dfrac{\partial p}{\partial z} = 0$. The velocity in the x-direction is a function of y only with $u = u(y)$ and $v = w = 0$ with boundary values $u(0) = u(2\ell) = 0$. The density is constant and $\nu = \mu^*/\varrho$ is the kinematic viscosity.

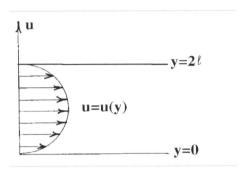

Figure 2.5-7. Plane Poiseuille flow

(a) Show the equation of fluid motion is $\nu\dfrac{d^2 u}{dy^2} + \dfrac{p_0}{\varrho} = 0,$ $u(0) = u(2\ell) = 0$

(b) Find the velocity $u = u(y)$ and find the maximum velocity in the x-direction. (c) Let M denote the mass flow rate across the plane $x = x_0 = constant$, , where $0 \le y \le 2\ell$, and $0 \le z \le 1$. Show that $M = \dfrac{2}{3\mu^*}\varrho p_0 \ell^3$. Note that as μ^* increases, M decreases.

▶ **18.** The heat equation (or diffusion equation) can be expressed div (k grad u) $+ H = \dfrac{\partial(\delta c u)}{\partial t}$, where c is the specific heat [cal/gm C], δ is the volume density [gm/cm^3], H is the rate of heat generation [cal/sec cm^3], u is the temperature [C], k is the thermal conductivity [cal/sec cm C]. Assume constant thermal conductivity, volume density and specific heat and express the boundary value problem

$$k\frac{\partial^2 u}{\partial x^2} = \delta c \frac{\partial u}{\partial t}, \quad 0 < x < L$$
$$u(0, t) = 0, \quad u(L, t) = u_1, \quad u(x, 0) = f(x)$$

in a form where all the variables are dimensionless. Assume u_1 is constant.

▶ **19.** Simplify the Navier-Stokes-Duhem equations using the assumption that there is incompressible flow.

▶ **20.** (Rayleigh impulsive flow)   The figure 2.5-8 illustrates fluid motion in the plane where $y > 0$ above a plate located along the axis where $y = 0$. The plate along $y = 0$ has zero velocity for all negative time and at time $t = 0$ the plate is given an instantaneous velocity u_0 in the positive x-direction. Assume the physical components of the velocity are $\vec{v} = (u, v, w)$ which satisfy $u = u(y, t)$, $v = w = 0$. Assume that the density of the fluid is constant, the gradient of the pressure is zero, and the body forces are zero. (a) Show that the velocity in the x-direction is governed by the differential equation

$$\frac{\partial u}{\partial t} = \nu \frac{\partial^2 u}{\partial y^2}, \quad \text{with} \quad \nu = \frac{\mu^*}{\varrho}.$$

Assume u satisfies the initial condition $u(0,t) = u_0 H(t)$ where H is the Heaviside step function. Also assume there exist a condition at infinity $\lim_{y \to \infty} u(y,t)$. This latter condition requires a bounded velocity at infinity. (b) Use any method to show the velocity is

$$u(y,t) = u_0 - u_0 \operatorname{erf}\left(\frac{y}{2\sqrt{\nu t}}\right) = u_0 \operatorname{erfc}\left(\frac{y}{2\sqrt{\nu t}}\right)$$

where erf and erfc are the error function and complimentary error function respectively. Pick a point on the line $y = y_0 = 2\sqrt{\nu}$ and plot the velocity as a function of time. How does the viscosity effect the velocity of the fluid along the line $y = y_0$?

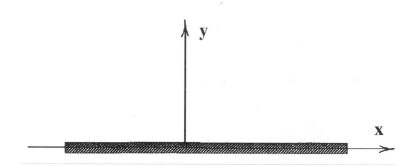

Figure 2.5-8. Rayleigh impulsive flow

21. (a) Write out the Navier-Stokes-Duhem equation for two dimensional flow in the x-y direction under the assumptions that

- $\lambda^* + \frac{2}{3}\mu^* = 0$ (This condition is referred to as Stoke's flow.)
- The fluid is incompressible
- There is a gravitational force $\vec{b} = -g\nabla h$ Hint: Express your answer as two scalar equations involving the variables $v_1, v_2, h, g, \varrho, p, t, \mu^*$ plus the continuity equation. (b) In part (a) eliminate the pressure and body force terms by cross differentiation and subtraction. (i.e. take the derivative of one equation with respect to x and take the derivative of the other equation with respect to y and then eliminate any common terms.) (c) Assume that $\vec{\omega} = \omega\,\hat{e}_3$ where $\omega = \frac{1}{2}\left(\dfrac{\partial v_2}{\partial x} - \dfrac{\partial v_1}{\partial y}\right)$ and derive the vorticity-transport equation

$$\frac{d\omega}{dt} = \nu\nabla^2\omega \qquad \text{where} \qquad \frac{d\omega}{dt} = \frac{\partial\omega}{\partial t} + v_1\frac{\partial\omega}{\partial x} + v_2\frac{\partial\omega}{\partial y}.$$

Hint: The continuity equation makes certain terms zero. (d) Define a stream function $\psi = \psi(x,y)$ satisfying $v_1 = \dfrac{\partial\psi}{\partial y}$ and $v_2 = -\dfrac{\partial\psi}{\partial x}$ and show the continuity equation is identically satisfied. Show also that $\omega = -\frac{1}{2}\nabla^2\psi$ and that

$$\nabla^4\psi = \frac{1}{\nu}\left[\frac{\partial\nabla^2\psi}{\partial t} + \frac{\partial\psi}{\partial y}\frac{\partial\nabla^2\psi}{\partial x} - \frac{\partial\psi}{\partial x}\frac{\partial\nabla^2\psi}{\partial y}\right].$$

If ν is very large, show that $\nabla^4\psi \approx 0$.

▶ **22.** In generalized orthogonal coordinates, show that the physical components of the rate of deformation stress can be written, for $i \neq j$

$$\sigma(ij) = \mu^* \left[\frac{h_i}{h_j} \frac{\partial}{\partial x^j} \left(\frac{v(i)}{h_i} \right) + \frac{h_j}{h_i} \frac{\partial}{\partial x^i} \left(\frac{v(j)}{h_j} \right) \right], \quad \text{no summation,}$$

and for $i \neq j \neq k$

$$\sigma(ii) = -p + 2\mu^* \left[\frac{1}{h_i} \frac{\partial v(i)}{\partial x^i} + \frac{1}{h_i h_j} v(j) \frac{\partial h_i}{\partial x^j} + \frac{1}{h_i h_k} v(k) \frac{\partial h_i}{\partial x^k} \right]$$
$$+ \frac{\lambda^*}{h_1 h_2 h_3} \left[\frac{\partial}{\partial x^1} \{ h_2 h_3 v(1) \} + \frac{\partial}{\partial x^2} \{ h_1 h_3 v(2) \} + \frac{\partial}{\partial x^3} \{ h_1 h_2 v(3) \} \right], \quad \text{no summation}$$

▶ **23.** Find the physical components for the rate of deformation stress in Cartesian coordinates. Hint: See problem 22.

▶ **24.** Find the physical components for the rate of deformation stress in cylindrical coordinates. Hint: See problem 22.

▶ **25.** Verify the Navier-Stokes equations for an incompressible fluid can be written $\dot{v}_i = -\frac{1}{\varrho} p_{,i} + \nu v_{i,mm} + b_i$ where $\nu = \frac{\mu^*}{\varrho}$ is called the kinematic viscosity.

▶ **26.** Verify the Navier-Stokes equations for a compressible fluid with zero bulk viscosity can be written $\dot{v}_i = -\frac{1}{\varrho} p_{,i} + \frac{\nu}{3} v_{m,mi} + \nu v_{i,mm} + b_i$ with $\nu = \frac{\mu^*}{\varrho}$ the kinematic viscosity.

▶ **27.** The constitutive equation for a certain non-Newtonian Stokesian fluid is $\sigma_{ij} = -p\delta_{ij} + \beta D_{ij} + \gamma D_{ik} D_{kj}$. Assume that β and γ are constants (a) Verify that $\sigma_{ij,j} = -p_{,i} + \beta D_{ij,j} + \gamma(D_{ik} D_{kj,j} + D_{ik,j} D_{kj})$
(b) Write out the Cauchy equations of motion in Cartesian coordinates. (See page 237).

▶ **28.** Let the constitutive equations relating stress and strain for a solid material take into account thermal stresses due to a temperature T. The constitutive equations have the form $e_{ij} = \frac{1+\nu}{E} \sigma_{ij} - \frac{\nu}{E} \sigma_{kk} \delta_{ij} + \alpha T \delta_{ij}$ where α is a coefficient of linear expansion for the material and T is the absolute temperature. Solve for the stress in terms of strains.

▶ **29.** Derive equation (2.5.53) and then show that when the bulk coefficient of viscosity is zero, the Navier-Stokes equations, in Cartesian coordinates, can be written in the conservation form

$$\frac{\partial(\varrho u)}{\partial t} + \frac{\partial(\varrho u^2 + p - \tau_{xx})}{\partial x} + \frac{\partial(\varrho uv - \tau_{xy})}{\partial y} + \frac{\partial(\varrho uw - \tau_{xz})}{\partial z} = \varrho b_x$$

$$\frac{\partial(\varrho v)}{\partial t} + \frac{\partial(\varrho uv - \tau_{xy})}{\partial x} + \frac{\partial(\varrho v^2 + p - \tau_{yy})}{\partial y} + \frac{\partial(\varrho vw - \tau_{yz})}{\partial z} = \varrho b_y$$

$$\frac{\partial(\varrho w)}{\partial t} + \frac{\partial(\varrho uw - \tau_{xz})}{\partial x} + \frac{\partial(\varrho vw - \tau_{yz})}{\partial y} + \frac{\partial(\varrho w^2 + p - \tau_{zz})}{\partial z} = \varrho b_z$$

where $v_1 = u, v_2 = v, v_3 = w$ and $\tau_{ij} = \mu^*(v_{i,j} + v_{j,i} - \frac{2}{3} \delta_{ij} v_{k,k})$. Hint: Alternatively, consider 2.5.29 and use the continuity equation.

30. Show that for a perfect gas, where $\lambda^* = -\frac{2}{3}\mu^*$ and $\eta = \mu^*$ is a function of position, the vector form of equation (2.5.25) is

$$\varrho\frac{D\vec{v}}{Dt} = \varrho\vec{b} - \nabla p + \frac{4}{3}\nabla(\eta\nabla\cdot\vec{v}) + \nabla(\vec{v}\cdot\nabla\eta) - \vec{v}\nabla^2\eta + (\nabla\eta)\times(\nabla\times\vec{v}) - (\nabla\cdot\vec{v})\nabla\eta - \nabla\times(\nabla\times(\eta\vec{v}))$$

31. Derive the energy equation $\varrho\dfrac{D\,h}{Dt} = \dfrac{D\,p}{Dt} + \dfrac{\partial Q}{\partial t} - \nabla\cdot\vec{q} + \Phi$. Hint: Use the continuity equation.

32. Show that in Cartesian coordinates the Navier-Stokes equations of motion for a compressible fluid can be written

$$\rho\frac{Du}{Dt} = \rho b_x - \frac{\partial p}{\partial x} + \frac{\partial}{\partial x}\left(2\mu^*\frac{\partial u}{\partial x} + \lambda^*\nabla\cdot\vec{V}\right) + \frac{\partial}{\partial y}\left(\mu^*(\frac{\partial u}{\partial y} + \frac{\partial v}{\partial x})\right) + \frac{\partial}{\partial z}\left(\mu^*(\frac{\partial w}{\partial x} + \frac{\partial u}{\partial z})\right)$$

$$\rho\frac{Dv}{Dt} = \rho b_y - \frac{\partial p}{\partial y} + \frac{\partial}{\partial y}\left(2\mu^*\frac{\partial v}{\partial y} + \lambda^*\nabla\cdot\vec{V}\right) + \frac{\partial}{\partial z}\left(\mu^*(\frac{\partial v}{\partial z} + \frac{\partial w}{\partial y})\right) + \frac{\partial}{\partial x}\left(\mu^*(\frac{\partial w}{\partial y} + \frac{\partial w}{\partial x})\right)$$

$$\rho\frac{Dv}{Dt} = \rho b_z - \frac{\partial p}{\partial z} + \frac{\partial}{\partial z}\left(2\mu^*\frac{\partial w}{\partial z} + \lambda^*\nabla\cdot\vec{V}\right) + \frac{\partial}{\partial x}\left(\mu^*(\frac{\partial w}{\partial x} + \frac{\partial u}{\partial z})\right) + \frac{\partial}{\partial y}\left(\mu^*(\frac{\partial v}{\partial z} + \frac{\partial w}{\partial y})\right)$$

where $(V_x, V_y, V_z) = (u, v, w)$.

33. Show that in cylindrical coordinates the Navier-Stokes equations of motion for a compressible fluid can be written

$$\varrho\left(\frac{DV_r}{Dt} - \frac{V_\theta^2}{r}\right) = \varrho b_r - \frac{\partial p}{\partial r} + \frac{\partial}{\partial r}\left(2\mu^*\frac{\partial V_r}{\partial r} + \lambda^*\nabla\cdot\vec{V}\right) + \frac{1}{r}\frac{\partial}{\partial\theta}\left(\mu^*(\frac{1}{r}\frac{\partial V_r}{\partial\theta} + \frac{\partial V_\theta}{\partial r} - \frac{V_\theta}{r})\right)$$

$$+ \frac{\partial}{\partial z}\left(\mu^*(\frac{\partial V_r}{\partial z} + \frac{\partial V_z}{\partial r})\right) + \frac{2\mu^*}{r}(\frac{\partial V_r}{\partial r} - \frac{1}{r}\frac{\partial V_\theta}{\partial\theta} - \frac{V_r}{r})$$

$$\varrho\left(\frac{DV_\theta}{Dt} + \frac{V_r V_\theta}{r}\right) = \varrho b_\theta - \frac{1}{r}\frac{\partial p}{\partial\theta} + \frac{1}{r}\frac{\partial}{\partial\theta}\left(2\mu^*(\frac{1}{r}\frac{\partial V_\theta}{\partial\theta} + \frac{V_r}{r}) + \lambda^*\nabla\cdot\vec{V}\right) + \frac{\partial}{\partial z}\left(\mu^*(\frac{1}{r}\frac{\partial V_z}{\partial\theta} + \frac{\partial V_\theta}{\partial z})\right)$$

$$+ \frac{\partial}{\partial r}\left(\mu^*(\frac{1}{r}\frac{\partial V_r}{\partial\theta} + \frac{\partial V_\theta}{\partial r} - \frac{V_\theta}{r})\right) + \frac{2\mu^*}{r}(\frac{1}{r}\frac{\partial V_r}{\partial\theta} + \frac{\partial V_\theta}{\partial r} - \frac{V_\theta}{r})$$

$$\varrho\frac{DV_z}{Dt} = \varrho b_z - \frac{\partial p}{\partial z} + \frac{\partial}{\partial z}\left(2\mu^*\frac{\partial V_z}{\partial z} + \lambda^*\nabla\cdot\vec{V}\right) + \frac{1}{r}\frac{\partial}{\partial r}\left(\mu^* r(\frac{\partial V_r}{\partial z} + \frac{\partial V_z}{\partial r})\right)$$

$$+ \frac{1}{r}\frac{\partial}{\partial\theta}\left(\mu^*(\frac{1}{r}\frac{\partial V_z}{\partial\theta} + \frac{\partial V_\theta}{\partial z})\right)$$

34. Show that the dissipation function Φ can be written as $\Phi = 2\mu^* D_{ij}D_{ij} + \lambda^*\Theta^2$.

35. Verify the identities:

$$(a)\quad \varrho\frac{D}{Dt}(e_t/\varrho) = \frac{\partial e_t}{\partial t} + \nabla\cdot(e_t\vec{V}) \qquad (b)\quad \varrho\frac{D}{Dt}(e_t/\varrho) = \varrho\frac{De}{Dt} + \varrho\frac{D}{Dt}(V^2/2).$$

36. Show that the conservation law for heat flow is given by $\dfrac{\partial T}{\partial t} + \nabla\cdot(T\vec{v} - \kappa\nabla T) = S_Q$ where κ is the thermal conductivity of the material, T is the temperature, $\vec{J}_{advection} = T\vec{v}$, $\vec{J}_{conduction} = -\kappa\nabla T$ and S_Q is a source term. Note that in a solid material there is no flow and so $\vec{v} = 0$ and the above equation reduces to the heat equation. Assign units of measurements to each term in the above equation and make sure the equation is dimensionally homogeneous.

324

▶ **37.** Show that in spherical coordinates the Navier-Stokes equations of motion for a compressible fluid can be written

$$\varrho(\frac{DV_\rho}{Dt} - \frac{V_\theta^2 + V_\phi^2}{\rho}) = \varrho b_\rho - \frac{\partial p}{\partial \rho} + \frac{\partial}{\partial \rho}\left(2\mu^*\frac{\partial V_\rho}{\partial \rho} + \lambda^*\nabla\cdot\vec{V}\right) + \frac{1}{\rho}\frac{\partial}{\partial\theta}\left(\mu^*(\rho\frac{\partial}{\partial\rho}(V_\theta/\rho) + \frac{1}{\rho}\frac{\partial V_\rho}{\partial\theta})\right)$$

$$+ \frac{1}{\rho\sin\theta}\frac{\partial}{\partial\phi}\left(\mu^*(\frac{1}{\rho\sin\theta}\frac{\partial V_\rho}{\partial\phi} + \rho\frac{\partial}{\partial\rho}(V_\phi/\rho))\right)$$

$$+ \frac{\mu^*}{\rho}(4\frac{\partial V_\rho}{\partial\rho} - \frac{2}{\rho}\frac{\partial V_\theta}{\partial\theta} - \frac{4V_\phi}{\rho} - \frac{2}{\rho\sin\theta}\frac{\partial V_\phi}{\partial\phi} - \frac{2V_\theta\cot\theta}{\rho} + \rho\cot\theta\frac{\partial}{\partial\rho}(V_\theta/\rho) + \frac{\cot\theta}{\rho}\frac{\partial V_\rho}{\partial\theta})$$

$$\varrho(\frac{DV_\theta}{Dt} + \frac{V_\rho V_\theta}{\rho} - \frac{V_\phi^2\cot\theta}{\rho}) = \varrho b_\theta - \frac{1}{\rho}\frac{\partial p}{\partial\theta} + \frac{1}{\rho}\frac{\partial}{\partial\theta}\left(\frac{2\mu^*}{\rho}(\frac{\partial V_\theta}{\partial\theta} + V_\rho) + \lambda^*\nabla\cdot\vec{V}\right)$$

$$+ \frac{1}{\rho\sin\theta}\frac{\partial}{\partial\phi}\left(\mu^*(\frac{\sin\theta}{\rho}\frac{\partial}{\partial\theta}(V_\phi/\sin\theta) + \frac{1}{\rho\sin\theta}\frac{\partial V_\theta}{\partial\phi})\right) + \frac{\partial}{\partial\rho}\left(\mu^*(\rho\frac{\partial}{\partial\rho}(V_\theta/\rho) + \frac{1}{\rho}\frac{\partial V_\rho}{\partial\theta})\right)$$

$$+ \frac{\mu^*}{\rho}\left[2\left(\frac{1}{\rho}\frac{\partial V_\theta}{\partial\theta} - \frac{1}{\rho\sin\theta}\frac{\partial V_\phi}{\partial\phi} - \frac{V_\theta\cot\theta}{\rho}\right)\cot\theta + 3\left(\rho\frac{\partial}{\partial\rho}(V_\theta/\rho) + \frac{1}{\rho}\frac{\partial V_\rho}{\partial\theta}\right)\right]$$

$$\varrho\left(\frac{DV_\phi}{Dt} + \frac{V_\phi V_\rho}{\rho} + \frac{V_\theta V_\phi\cot\theta}{\rho}\right) = \varrho b_\phi - \frac{1}{\rho\sin\theta}\frac{\partial p}{\partial\phi} + \frac{\partial}{\partial\rho}\left(\mu^*\left(\frac{1}{\rho\sin\theta}\frac{\partial V_\rho}{\partial\phi} + \rho\frac{\partial}{\partial\rho}(V_\phi/\rho)\right)\right)$$

$$+ \frac{1}{\rho\sin\theta}\frac{\partial}{\partial\phi}\left(\frac{2\mu^*}{\rho}\left(\frac{1}{\sin\theta}\frac{\partial V_\phi}{\partial\phi} + V_\rho + V_\theta\cot\theta\right) + \lambda^*\nabla\cdot\vec{V}\right)$$

$$+ \frac{1}{\rho}\frac{\partial}{\partial\theta}\left(\mu^*\left(\frac{\sin\theta}{\rho}\frac{\partial}{\partial\theta}(V_\phi/\sin\theta) + \frac{1}{\rho\sin\theta}\frac{\partial V_\theta}{\partial\phi}\right)\right)$$

$$+ \frac{\mu^*}{\rho}\left[3\left(\frac{1}{\rho\sin\theta}\frac{\partial V_\rho}{\partial\phi} + \rho\frac{\partial}{\partial\rho}(V_\phi/\rho)\right) + 2\cot\theta\left(\frac{\sin\theta}{\rho}\frac{\partial}{\partial\theta}(V_\phi/\sin\theta) + \frac{1}{\rho\sin\theta}\frac{\partial V_\theta}{\partial\phi}\right)\right]$$

▶ **38.** Verify all the equations (2.5.28).

▶ **39.** Use the conservation of energy equation (2.5.47) together with the momentum equation (2.5.25) to derive the equation (2.5.48).

▶ **40.** Verify the equation (2.5.55).

▶ **41.** Consider nonviscous flow and write the 3 linear momentum equations and the continuity equation and make the following assumptions: (i) The density ϱ is constant. (ii) Body forces are zero. (iii) Steady state flow only. (iv) Consider only two dimensional flow with non-zero velocity components $u = u(x, y)$ and $v = v(x, y)$. Show that there results the system of equations

$$u\frac{\partial u}{\partial x} + v\frac{\partial u}{\partial y} + \frac{1}{\varrho}\frac{\partial P}{\partial x} = 0, \qquad u\frac{\partial v}{\partial x} + v\frac{\partial v}{\partial y} + \frac{1}{\varrho}\frac{\partial P}{\partial y} = 0, \qquad \frac{\partial u}{\partial x} + \frac{\partial v}{\partial y} = 0.$$

Recognize that the last equation in the above set as one of the Cauchy-Riemann equations that $f(z) = u - iv$ be an analytic function of a complex variable. Further assume that the fluid flow is irrotational so that $\frac{\partial v}{\partial x} - \frac{\partial u}{\partial y} = 0$. Show that this implies that $\frac{1}{2}(u^2 + v^2) + \frac{P}{\varrho} = $ Constant. If in addition u and v are derivable from a potential function $\phi(x, y)$, such that $u = \frac{\partial\phi}{\partial x}$ and $v = \frac{\partial\phi}{\partial y}$, then show that ϕ is a harmonic function. By constructing the conjugate harmonic function $\psi(x, y)$ the complex potential $F(z) = \phi(x, y) + i\psi(x, y)$ is such that $F'(z) = u(x, y) - iv(x, y)$ and $\overline{F'(z)}$ gives the velocity. The family of curves $\phi(x, y) =$constant are called equipotential curves and the family of curves $\psi(x, y) = $ constant are called streamlines. Show that these families are an orthogonal family of curves.

§2.6 ELECTRIC AND MAGNETIC FIELDS

Introduction

In electromagnetic theory the mks system of units and the Gaussian system of units are the ones most often encountered. In this section the equations will be given in the mks system of units. If you want the equations in the Gaussian system of units make the replacements given in the column 3 of Table 1.

Table 1. MKS AND GAUSSIAN UNITS

MKS symbol	MKS units	Replacement symbol	GAUSSIAN units
$\vec{E}$ (Electric field)	volt/m	$\vec{E}$	statvolt/cm
$\vec{B}$ (Magnetic field)	weber/m^2	$\frac{\vec{B}}{c}$	gauss
$\vec{D}$ (Displacement field)	coulomb/m^2	$\frac{\vec{D}}{4\pi}$	statcoulomb/cm^2
$\vec{H}$ (Auxiliary Magnetic field)	ampere/m	$\frac{c\vec{H}}{4\pi}$	oersted
$\vec{J}$ (Current density)	ampere/m^2	$\vec{J}$	statampere/cm^2
$\vec{A}$ (Vector potential)	weber/m	$\frac{\vec{A}}{c}$	gauss-cm
$\mathcal{V}$ (Electric potential)	volt	$\mathcal{V}$	statvolt
ϵ (Dielectric constant)		$\frac{\epsilon}{4\pi}$	
μ (Magnetic permeability)		$\frac{4\pi\mu}{c^2}$	

Electrostatics

A basic problem in electrostatic theory is to determine the force $\vec{F}$ on a charge Q placed a distance r from another charge q. The solution to this problem is Coulomb's law

$$\vec{F} = \frac{1}{4\pi\epsilon_0}\frac{qQ}{r^2}\,\hat{\mathbf{e}}_r \qquad (2.6.1)$$

where q, Q are measured in coulombs, $\epsilon_0 = 8.85 \times 10^{-12}$ coulomb^2/N · m^2 is called the permittivity in a vacuum, r is in meters, $[\vec{F}]$ has units of Newtons and $\hat{\mathbf{e}}_r$ is a unit vector pointing from q to Q if q, Q have the same sign or pointing from Q to q if q, Q are of opposite sign. The quantity $\vec{E} = \vec{F}/Q$ is called the electric field produced by the charges. In the special case $Q = 1$, we have $\vec{E} = \vec{F}$ and so $Q = 1$ is called a test charge. This tells us that the electric field at a point P can be viewed as the force per unit charge exerted on a test charge Q placed at the point P. The test charge Q is always positive and so is repulsed if q is positive and attracted if q is negative.

The electric field associated with many charges is obtained by the principal of superposition. For example, let $q_1, q_2, \ldots, q_n$ denote n-charges having respectively the distances $r_1, r_2, \ldots, r_n$ from a test charge Q placed at a point P. The force exerted on Q is

$$\vec{F} = \vec{F}_1 + \vec{F}_2 + \cdots + \vec{F}_n$$
$$\vec{F} = \frac{1}{4\pi\epsilon_0}\left(\frac{q_1 Q}{r_1^2}\,\hat{\mathbf{e}}_{r_1} + \frac{q_2 Q}{r_2^2}\,\hat{\mathbf{e}}_{r_2} + \cdots + \frac{q_n Q}{r_n^2}\,\hat{\mathbf{e}}_{r_n}\right)$$
$$\text{or}\quad \vec{E} = \vec{E}(P) = \frac{\vec{F}}{Q} = \frac{1}{4\pi\epsilon_0}\sum_{i=1}^{n}\frac{q_i}{r_i^2}\,\hat{\mathbf{e}}_{r_i} \qquad (2.6.2)$$

where $\vec{E} = \vec{E}(P)$ is the electric field associated with the system of charges. The equation (2.6.2) can be generalized to other situations by defining other types of charge distributions. We introduce a line charge density λ^*, (coulomb/m), a surface charge density μ^*, (coulomb/m^2), a volume charge density ρ^*, (coulomb/m^3), then we can calculate the electric field associated with these other types of charge distributions. For example, if there is a charge distribution $\lambda^* = \lambda^*(s)$ along a curve C, where s is an arc length parameter, then we would have

$$\vec{E}(P) = \frac{1}{4\pi\epsilon_0} \int_C \frac{\widehat{\mathbf{e}}_r}{r^2} \lambda^* ds \tag{2.6.3}$$

as the electric field at a point P due to this charge distribution. The integral in equation (2.6.3) being a line integral along the curve C and where ds is an element of arc length. Here equation (2.6.3) represents a continuous summation of the charges along the curve C. For a continuous charge distribution over a surface S, the electric field at a point P is

$$\vec{E}(P) = \frac{1}{4\pi\epsilon_0} \int\int_S \frac{\widehat{\mathbf{e}}_r}{r^2} \mu^* d\sigma \tag{2.6.4}$$

where $d\sigma$ represents an element of surface area on S. Similarly, if ρ^* represents a continuous charge distribution throughout a volume V, then the electric field is represented

$$\vec{E}(P) = \frac{1}{4\pi\epsilon_0} \int\int\int_V \frac{\widehat{\mathbf{e}}_r}{r^2} \rho^* d\tau \tag{2.6.5}$$

where $d\tau$ is an element of volume. In the equations (2.6.3), (2.6.4), (2.6.5) we let (x, y, z) denote the position of the test charge and let (x', y', z') denote a point on the line, on the surface or within the volume, then

$$\vec{r} = (x - x')\widehat{\mathbf{e}}_1 + (y - y')\widehat{\mathbf{e}}_2 + (z - z')\widehat{\mathbf{e}}_3 \tag{2.6.6}$$

represents the distance from the point P to an element of charge $\lambda^* ds$, $\mu^* d\sigma$ or $\rho^* d\tau$ with $r = |\vec{r}|$ and $\widehat{\mathbf{e}}_r = \frac{\vec{r}}{r}$.

If the electric field is conservative, then $\nabla \times \vec{E} = 0$, and so it is derivable from a potential function $\mathcal{V}$ by taking the negative of the gradient of $\mathcal{V}$ and

$$\vec{E} = -\nabla\mathcal{V}. \tag{2.6.7}$$

For these conditions note that $\nabla\mathcal{V} \cdot d\vec{r} = -\vec{E} \cdot d\vec{r}$ is an exact differential so that the potential function can be represented by the line integral

$$\mathcal{V} = \mathcal{V}(P) = -\int_\alpha^P \vec{E} \cdot d\vec{r} \tag{2.6.8}$$

where α is some reference point (usually infinity, where $\mathcal{V}(\infty) = 0$). For a conservative electric field the line integral will be independent of the path connecting any two points a and b so that

$$\mathcal{V}(b) - \mathcal{V}(a) = -\int_\alpha^b \vec{E} \cdot d\vec{r} - \left(-\int_\alpha^a \vec{E} \cdot d\vec{r}\right) = -\int_a^b \vec{E} \cdot d\vec{r} = \int_a^b \nabla\mathcal{V} \cdot d\vec{r}. \tag{2.6.9}$$

Let $\alpha = \infty$ in equation (2.6.8), then the potential function associated with a point charge moving in the radial direction $\widehat{\mathbf{e}}_r$ is

$$\mathcal{V}(r) = -\int_\infty^r \vec{E} \cdot d\vec{r} = \frac{-q}{4\pi\epsilon_0} \int_\infty^r \frac{1}{r^2} dr = \frac{q}{4\pi\epsilon_0} \frac{1}{r}\Big|_\infty^r = \frac{q}{4\pi\epsilon_0 r}.$$

By superposition, the potential at a point P for a continuous volume distribution of charges is given by $\mathcal{V}(P) = \dfrac{1}{4\pi\epsilon_0}\iiint_V \dfrac{\rho^*}{r}\,d\tau$ and for a surface distribution of charges $\mathcal{V}(P) = \dfrac{1}{4\pi\epsilon_0}\iint_S \dfrac{\mu^*}{r}\,d\sigma$ and for a line distribution of charges $\mathcal{V}(P) = \dfrac{1}{4\pi\epsilon_0}\int_C \dfrac{\lambda^*}{r}\,ds$; and for a discrete distribution of point charges $\mathcal{V}(P) = \dfrac{1}{4\pi\epsilon_0}\sum_{i=1}^{N} \dfrac{q_i}{r_i}$. When the potential functions are defined from a common reference point, then the principal of superposition applies.

The potential function $\mathcal{V}$ is related to the work done W in moving a charge within the electric field. The work done in moving a test charge Q from point a to point b is an integral of the force times distance moved. The electric force on a test charge Q is $\vec{F} = Q\vec{E}$ and so the force $\vec{F} = -Q\vec{E}$ is in opposition to this force as you move the test charge. The work done is

$$W = \int_a^b \vec{F}\cdot d\vec{r} = \int_a^b -Q\vec{E}\cdot d\vec{r} = Q\int_a^b \nabla V \cdot d\vec{r} = Q[\mathcal{V}(b) - \mathcal{V}(a)]. \qquad (2.6.10)$$

The work done is independent of the path joining the two points and depends only on the end points and the change in the potential. If one moves Q from infinity to point b, then the above becomes $W = Q V(b)$.

An electric field $\vec{E} = \vec{E}(P)$ is a vector field which can be represented graphically by constructing vectors at various selected points in the space. Such a plot is called a vector field plot. A field line associated with a vector field is a curve such that the tangent vector to a point on the curve has the same direction as the vector field at that point. Field lines are used as an aid for visualization of an electric field and vector fields in general. The tangent to a field line at a point has the same direction as the vector field $\vec{E}$ at that point. For example, in two dimensions let $\vec{r} = x\,\hat{\mathbf{e}}_1 + y\,\hat{\mathbf{e}}_2$ denote the position vector to a point on a field line. The tangent vector to this point has the direction $d\vec{r} = dx\,\hat{\mathbf{e}}_1 + dy\,\hat{\mathbf{e}}_2$. If $\vec{E} = \vec{E}(x, y) = -N(x, y)\,\hat{\mathbf{e}}_1 + M(x, y)\,\hat{\mathbf{e}}_2$ is the vector field constructed at the same point, then $\vec{E}$ and $d\vec{r}$ must be colinear. Thus, for each point (x, y) on a field line we require that $d\vec{r} = K\vec{E}$ for some constant K. Equating like components we find that the field lines must satisfy the differential relation.

$$\frac{dx}{-N(x, y)} = \frac{dy}{M(x, y)} = K$$
$$\text{or} \qquad M(x, y)\,dx + N(x, y)\,dy = 0. \qquad (2.6.11)$$

In two dimensions, the family of equipotential curves $\mathcal{V}(x, y) = C_1 =$ constant, are orthogonal to the family of field lines and are described by solutions of the differential equation

$$N(x, y)\,dx - M(x, y)\,dy = 0$$

obtained from equation (2.6.11) by taking the negative reciprocal of the slope. The field lines are perpendicular to the equipotential curves because at each point on the curve $\mathcal{V} = C_1$ we have $\nabla\mathcal{V}$ being perpendicular to the curve $\mathcal{V} = C_1$ and so it is colinear with $\vec{E}$ at this same point. Field lines associated with electric fields are called electric lines of force. The density of the field lines drawn per unit cross sectional area are proportional to the magnitude of the vector field through that area.

328

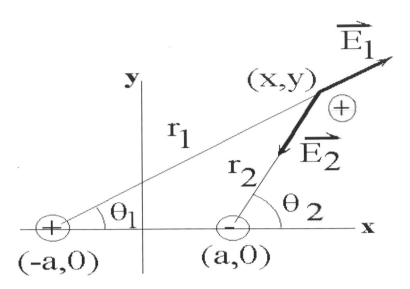

Figure 2.6-1. Electric forces due to a positive charge at $(-a, 0)$ and negative charge at $(a, 0)$.

EXAMPLE 2.6-1.

Find the field lines and equipotential curves associated with a positive charge q located at the point $(-a, 0)$ and a negative charge $-q$ located at the point $(a, 0)$.

Solution: With reference to the figure 2.6-1, the total electric force $\vec{E}$ on a test charge $Q = 1$ place at a general point (x, y) is, by superposition, the sum of the forces from each of the isolated charges and is $\vec{E} = \vec{E}_1 + \vec{E}_2$. The electric force vectors due to each individual charge are

$$\vec{E}_1 = \frac{kq(x + a)\,\widehat{\mathbf{e}}_1 + kqy\,\widehat{\mathbf{e}}_2}{r_1^3} \quad \text{with} \quad r_1^2 = (x + a)^2 + y^2$$

$$\vec{E}_2 = \frac{-kq(x - a)\,\widehat{\mathbf{e}}_1 - kqy\,\widehat{\mathbf{e}}_2}{r_2^3} \quad \text{with} \quad r_2^2 = (x - a)^2 + y^2$$

(2.6.12)

where $k = \dfrac{1}{4\pi\epsilon_0}$ is a constant. This gives

$$\vec{E} = \vec{E}_1 + \vec{E}_2 = \left[\frac{kq(x + a)}{r_1^3} - \frac{kq(x - a)}{r_2^3}\right]\widehat{\mathbf{e}}_1 + \left[\frac{kqy}{r_1^3} - \frac{kqy}{r_2^3}\right]\widehat{\mathbf{e}}_2.$$

This determines the differential equation of the field lines

$$\frac{dx}{\frac{kq(x+a)}{r_1^3} - \frac{kq(x-a)}{r_2^3}} = \frac{dy}{\frac{kqy}{r_1^3} - \frac{kqy}{r_2^3}}.$$

(2.6.13)

To solve this differential equation we make the substitutions

$$\cos\theta_1 = \frac{x + a}{r_1} \quad \text{and} \quad \cos\theta_2 = \frac{x - a}{r_2}$$

(2.6.14)

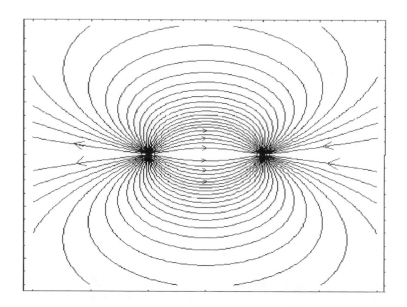

Figure 2.6-2. Lines of electric force between two opposite sign charges.

as suggested by the geometry from figure 2.6-1. From the equations (2.6.12) and (2.6.14) we obtain the relations

$$-\sin\theta_1\,d\theta_1 = \frac{r_1 dx - (x+a)\,dr_1}{r_1^2}$$

$$2r_1 dr_1 = 2(x+a)\,dx + 2y\,dy$$

$$-\sin\theta_2\,d\theta_2 = \frac{r_2\,dx - (x-a)dr_2}{r_2^2}$$

$$2r_2\,dr_2 = 2(x-a)\,dx + 2y\,dy$$

which implies that

$$-\sin\theta_1\,d\theta_1 = -\frac{(x+a)y\,dy}{r_1^3} + \frac{y^2\,dx}{r_1^3}$$

$$-\sin\theta_2\,d\theta_2 = -\frac{(x-a)y\,dy}{r_2^3} + \frac{y^2\,dx}{r_2^3}$$

(2.6.15)

Now compare the results from equation (2.6.15) with the differential equation (2.6.13) and determine that y is an integrating factor of equation (2.6.13) . This shows that the differential equation (2.6.13) can be written in the much simpler form of the exact differential equation

$$-\sin\theta_1\,d\theta_1 + \sin\theta_2\,d\theta_2 = 0 \qquad (2.6.16)$$

in terms of the variables θ_1 and θ_2. The equation (2.6.16) is easily integrated to obtain

$$\cos\theta_1 - \cos\theta_2 = C \qquad (2.6.17)$$

where C is a constant of integration. In terms of x, y the solution can be written

$$\frac{x+a}{\sqrt{(x+a)^2+y^2}} - \frac{x-a}{\sqrt{(x-a)^2+y^2}} = C. \qquad (2.6.18)$$

These field lines are illustrated in the figure 2.6-2.

The differential equation for the equipotential curves is obtained by taking the negative reciprocal of the slope of the field lines. This gives

$$\frac{dy}{dx} = \frac{\frac{kq(x-a)}{r_2^3} - \frac{kq(x+a)}{r_1^3}}{\frac{kqy}{r_1^3} - \frac{kqy}{r_2^3}}.$$

This result can be written in the form

$$-\left[\frac{(x+a)dx + ydy}{r_1^3}\right] + \left[\frac{(x-a)dx + ydy}{r_2^3}\right] = 0$$

which simplifies to the easily integrable form

$$-\frac{dr_1}{r_1^2} + \frac{dr_2}{r_2^2} = 0$$

in terms of the new variables r_1 and r_2. An integration produces the equipotential curves

$$\frac{1}{r_1} - \frac{1}{r_2} = C_2$$

or

$$\frac{1}{\sqrt{(x+a)^2 + y^2}} - \frac{1}{\sqrt{(x-a)^2 + y^2}} = C_2.$$

The potential function for this problem can be interpreted as a superposition of the potential functions $V_1 = -\frac{kq}{r_1}$ and $V_2 = \frac{kq}{r_2}$ associated with the isolated point charges at the points $(-a, 0)$ and $(a, 0)$. ■

Observe that the electric lines of force move from positive charges to negative charges and they do not cross one another. Where field lines are close together the field is strong and where the lines are far apart the field is weak. If the field lines are almost parallel and equidistant from one another the field is said to be uniform. The arrows on the field lines show the direction of the electric field $\vec{E}$. If one moves along a field line in the direction of the arrows the electric potential is decreasing and they cross the equipotential curves at right angles. Also, when the electric field is conservative we will have $\nabla \times \vec{E} = 0$.

In three dimensions the situation is analogous to what has been done in two dimensions. If the electric field is $\vec{E} = \vec{E}(x, y, z) = P(x, y, z)\,\hat{\mathbf{e}}_1 + Q(x, y, z)\,\hat{\mathbf{e}}_2 + R(x, y, z)\,\hat{\mathbf{e}}_3$ and $\vec{r} = x\,\hat{\mathbf{e}}_1 + y\,\hat{\mathbf{e}}_2 + z\,\hat{\mathbf{e}}_3$ is the position vector to a variable point (x, y, z) on a field line, then at this point $d\vec{r}$ and $\vec{E}$ must be colinear so that $d\vec{r} = K\vec{E}$ for some constant K. Equating like coefficients gives the system of equations

$$\frac{dx}{P(x, y, z)} = \frac{dy}{Q(x, y, z)} = \frac{dz}{R(x, y, z)} = K. \tag{2.6.19}$$

From this system of equations one must try to obtain two independent integrals, call them $u_1(x, y, z) = c_1$ and $u_2(x, y, z) = c_2$. These integrals represent one-parameter families of surfaces. When any two of these surfaces intersect, the result is a curve which represents a field line associated with the vector field $\vec{E}$. These type of field lines in three dimensions are more difficult to illustrate.

The electric flux ϕ_E of an electric field $\vec{E}$ over a surface S is defined as the summation of the normal component of $\vec{E}$ over the surface and is represented

$$\phi_E = \iint_S \vec{E} \cdot \hat{\mathbf{n}}\, d\sigma \quad \text{with units of } \frac{\text{N m}^2}{C} \tag{2.6.20}$$

where $\hat{\mathbf{n}}$ is a unit normal to the surface. The flux ϕ_E can be thought of as being proportional to the number of electric field lines passing through an element of surface area. If the surface is a closed surface we have by the divergence theorem of Gauss

$$\phi_E = \iiint_V \nabla \cdot \vec{E}\, d\tau = \iint_S \vec{E} \cdot \hat{\mathbf{n}}\, d\sigma$$

where V is the volume enclosed by S.

Gauss Law

Let $d\sigma$ denote an element of surface area on a surface S. A cone is formed if all points on the boundary of $d\sigma$ are connected by straight lines to the origin. The cone need not be a right circular cone. The situation is illustrated in the figure 2.6-3.

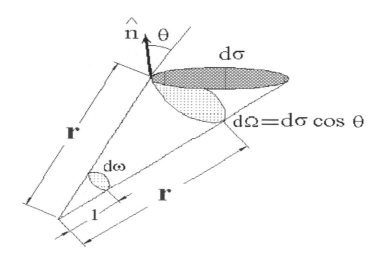

Figure 2.6-3. Solid angle subtended by element of area.

We let $\vec{r}$ denote a position vector from the origin to a point on the boundary of $d\sigma$ and let $\hat{\mathbf{n}}$ denote a unit outward normal to the surface at this point. We then have $\hat{\mathbf{n}} \cdot \vec{r} = r \cos\theta$ where $r = |\vec{r}|$ and θ is the angle between the vectors $\hat{\mathbf{n}}$ and $\vec{r}$. Construct a sphere, centered at the origin, having radius r. This sphere intersects the cone in an element of area $d\Omega$. The solid angle subtended by $d\sigma$ is defined as $d\omega = \dfrac{d\Omega}{r^2}$. Note that this is equivalent to constructing a unit sphere at the origin which intersect the cone in an element of area $d\omega$. Solid angles are measured in steradians. The total solid angle about a point equals the area of the sphere divided by its radius squared or 4π steradians. The element of area $d\Omega$ is the projection of $d\sigma$ on the constructed sphere and $d\Omega = d\sigma \cos\theta = \dfrac{\hat{\mathbf{n}} \cdot \vec{r}}{r} d\sigma$ so that $d\omega = \dfrac{\hat{\mathbf{n}} \cdot \vec{r}}{r^3} d\sigma = \dfrac{d\Omega}{r^2}$. Observe that sometimes the dot product $\hat{\mathbf{n}} \cdot \vec{r}$ is negative, the sign depending upon which of the normals to the surface is constructed. (i.e. the inner or outer normal.)

The Gauss law for electrostatics in a vacuum states that the flux through any surface enclosing many charges is the total charge enclosed by the surface divided by ϵ_0. The Gauss law is written

$$\iint_S \vec{E} \cdot \hat{\mathbf{n}}\, d\sigma = \begin{cases} \frac{Q_e}{\epsilon_0} & \text{for charges inside } S \\ 0 & \text{for charges outside } S \end{cases} \tag{2.6.21}$$

where Q_e represents the total charge enclosed by the surface S with $\hat{\mathbf{n}}$ the unit outward normal to the surface. The proof of Gauss's theorem follows. Consider a single charge q within the closed surface S. The electric field at a point on the surface S due to the charge q within S is represented $\vec{E} = \dfrac{1}{4\pi\epsilon_0}\dfrac{q}{r^2}\,\hat{\mathbf{e}}_r$ and so the flux integral is

$$\phi_E = \iint_S \vec{E}\cdot\hat{\mathbf{n}}\,d\sigma = \iint_S \frac{q}{4\pi\epsilon_0}\frac{\hat{\mathbf{e}}_r\cdot\hat{\mathbf{n}}}{r^2}\,d\sigma = \frac{q}{4\pi\epsilon_0}\iint_S \frac{d\Omega}{r^2} = \frac{q}{\epsilon_0} \tag{2.6.22}$$

since $\dfrac{\hat{\mathbf{e}}_r\cdot\hat{\mathbf{n}}}{r^2} = \dfrac{\cos\theta\,d\sigma}{r^2} = \dfrac{d\Omega}{r^2} = d\omega$ and $\iint_S d\omega = 4\pi$. By superposition of the charges, we obtain a similar result for each of the charges within the surface. Adding these results gives $Q_e = \sum\limits_{i=1}^{n} q_i$. For a continuous distribution of charge inside the volume we can write $Q_e = \iiint_V \rho^*\,d\tau$, where ρ^* is the charge distribution per unit volume. Note that charges outside of the closed surface do not contribute to the total flux across the surface. This is because the field lines go in one side of the surface and go out the other side. In this case $\iint_S \vec{E}\cdot\hat{\mathbf{n}}\,d\sigma = 0$ for charges outside the surface. Also the position of the charge or charges within the volume does not effect the Gauss law.

The equation (2.6.21) is the Gauss law in integral form. We can put this law in differential form as follows. Using the Gauss divergence theorem we can write for an arbitrary volume that

$$\iint_S \vec{E}\cdot\hat{\mathbf{n}}\,d\sigma = \iiint_V \nabla\cdot\vec{E}\,d\tau = \iiint_V \frac{\rho^*}{\epsilon_0}\,d\tau = \frac{Q_e}{\epsilon_0} = \frac{1}{\epsilon_0}\iiint_V \rho^*\,d\tau$$

which for an arbitrary volume implies

$$\nabla\cdot\vec{E} = \frac{\rho^*}{\epsilon_0}. \tag{2.6.23}$$

The equations (2.6.23) and (2.6.7) can be combined so that the Gauss law can also be written in the form $\nabla^2\mathcal{V} = -\dfrac{\rho^*}{\epsilon_0}$ which is called Poisson's equation.

EXAMPLE 2.6-2

Find the electric field associated with an infinite plane sheet of positive charge.

Solution: Assume there exists a uniform surface charge μ^* and draw a circle at some point on the plane surface. Now move the circle perpendicular to the surface to form a small cylinder which extends equal distances above and below the plane surface. We calculate the electric flux over this small cylinder in the limit as the height of the cylinder goes to zero. The charge inside the cylinder is $\mu^* A$ where A is the area of the circle. We find that the Gauss law requires that

$$\iint_S \vec{E}\cdot\hat{\mathbf{n}}\,d\sigma = \frac{Q_e}{\epsilon_0} = \frac{\mu^* A}{\epsilon_0} \tag{2.6.24}$$

where $\hat{\mathbf{n}}$ is the outward normal to the cylinder as we move over the surface S. By the symmetry of the situation the electric force vector is uniform and must point away from both sides to the plane surface in the direction of the normals to both sides of the surface. Denote the plane surface normals by $\hat{\mathbf{e}}_n$ and $-\hat{\mathbf{e}}_n$ and assume that $\vec{E} = \beta\hat{\mathbf{e}}_n$ on one side of the surface and $\vec{E} = -\beta\hat{\mathbf{e}}_n$ on the other side of the surface for some constant β. Substituting this result into the equation (2.6.24) produces

$$\iint_S \vec{E}\cdot\hat{\mathbf{n}}\,d\sigma = 2\beta A \tag{2.6.25}$$

since only the ends of the cylinder contribute to the above surface integral. On the sides of the cylinder we will have $\hat{n} \cdot \pm \hat{e}_n = 0$ and so the surface integral over the sides of the cylinder is zero. By equating the results from equations (2.6.24) and (2.6.25) we obtain the result that $\beta = \dfrac{\mu^*}{2\epsilon_0}$ and so one can write $\vec{E} = \dfrac{\mu^*}{2\epsilon_0} \hat{e}_n$ where $\hat{e}_n$ represents one of the normals to the surface. ∎

Note an electric field will always undergo a jump discontinuity when crossing a surface charge μ^*. As in the above example we have $\vec{E}_{up} = \dfrac{\mu^*}{2\epsilon_0} \hat{e}_n$ and $\vec{E}_{down} = -\dfrac{\mu^*}{2\epsilon} \hat{e}_n$ so that the difference is

$$\vec{E}_{up} - \vec{E}_{down} = \frac{\mu^*}{\epsilon_0} \hat{e}_n \qquad \text{or} \qquad E^i n_i^{(1)} + E^i n_i^{(2)} + \frac{\mu^*}{\epsilon_0} = 0. \tag{2.6.26}$$

It is this difference which causes the jump discontinuity.

EXAMPLE 2.6-3.

Calculate the electric field associated with a uniformly charged sphere of radius a.

Solution: We proceed as in the previous example. Let μ^* denote the uniform charge distribution over the surface of the sphere and let $\hat{e}_r$ denote the unit normal to the sphere. The total charge then is written as $q = \displaystyle\iint_{S_a} \mu^* d\sigma = 4\pi a^2 \mu^*$. If we construct a sphere of radius $r > a$ around the charged sphere, then we have by the Gauss theorem

$$\iiint_{S_r} \vec{E} \cdot \hat{e}_r \, d\sigma = \frac{Q_e}{\epsilon_0} = \frac{q}{\epsilon_0}. \tag{2.6.27}$$

Again, we can assume symmetry for $\vec{E}$ and assume that it points radially outward in the direction of the surface normal $\hat{e}_r$ and has the form $\vec{E} = \beta \hat{e}_r$ for some constant β. Substituting this value for $\vec{E}$ into the equation (2.6.27) we find that

$$\iint_{S_r} \vec{E} \cdot \hat{e}_r \, d\sigma = \beta \iint_{S_r} d\sigma = 4\pi \beta r^2 = \frac{q}{\epsilon_0}. \tag{2.6.28}$$

This gives $\vec{E} = \dfrac{1}{4\pi\epsilon_0} \dfrac{q}{r^2} \hat{e}_r$ where $\hat{e}_r$ is the outward normal to the sphere. This shows that the electric field outside the sphere is the same as if all the charge were situated at the origin. ∎

For S a piecewise closed surface enclosing a volume V and $F^i = F^i(x^1, x^2, x^3)$ $i = 1, 2, 3$, a continuous vector field with continuous derivatives the Gauss divergence theorem enables us to replace a flux integral of F^i over S by a volume integral of the divergence of F^i over the volume V such that

$$\iint_S F^i n_i \, d\sigma = \iiint_V F^i_{,i} \, d\tau \quad \text{or} \quad \iint_S \vec{F} \cdot \hat{n} \, d\sigma = \iiint_V \operatorname{div}\vec{F} \, d\tau. \tag{2.6.29}$$

If V contains a simple closed surface Σ where F^i is discontinuous we must modify the above Gauss divergence theorem.

EXAMPLE 2.6-4.

We examine the modification of the Gauss divergence theorem for spheres in order to illustrate the concepts. Let V have surface area S which encloses a surface Σ. Consider the figure 2.6-4 where the volume V enclosed by S and containing Σ has been cut in half.

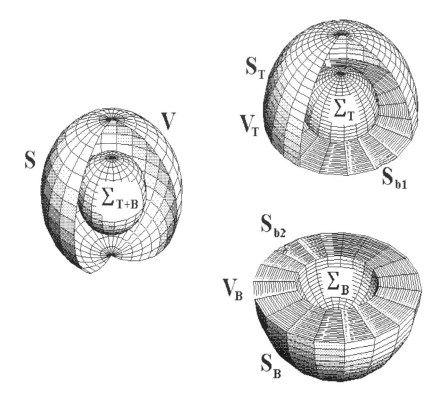

Figure 2.6-4. Sphere S containing sphere Σ.

Applying the Gauss divergence theorem to the top half of figure 2.6-4 gives

$$\iint_{S_T} F^i n_i^T \, d\sigma + \iint_{S_{b1}} F^i n_i^{b_T} \, d\sigma + \iint_{\Sigma_T} F^i n_i^{\Sigma_T} \, d\sigma = \iiint_{V_T} F^i_{,i} \, d\tau \qquad (2.6.30)$$

where the n_i are the unit outward normals to the respective surfaces S_T, S_{b1} and Σ_T. Applying the Gauss divergence theorem to the bottom half of the sphere in figure 2.6-4 gives

$$\iint_{S_B} F^i n_i^B \, d\sigma + \iint_{S_{b2}} F^i n_i^{b_B} \, d\sigma + \iint_{\Sigma_B} F^i n_i^{\Sigma_B} \, d\sigma = \iiint_{V_B} F^i_{,i} \, d\tau \qquad (2.6.31)$$

Observe that the unit normals to the surfaces S_{b1} and S_{b2} are equal and opposite in sign so that adding the equations (2.6.30) and (2.6.31) we obtain

$$\iint_S F^i n_i \, d\sigma + \iint_\Sigma F^i n_i^{(1)} \, d\sigma = \iiint_{V_T + V_B} F^i_{,i} \, d\tau \qquad (2.6.32)$$

where $S = S_T + S_B$ is the total surface area of the outside sphere and $\Sigma = \Sigma_T + \Sigma_B$ is the total surface area of the inside sphere, and $n_i^{(1)}$ is the inward normal to the sphere Σ when the top and bottom volumes are combined. Applying the Gauss divergence theorem to just the isolated small sphere Σ we find

$$\iint_\Sigma F^i n_i^{(2)}\, d\sigma = \iiint_{V_\Sigma} F^i_{,i}\, d\tau \tag{2.6.33}$$

where $n_i^{(2)}$ is the outward normal to Σ. By adding the equations (2.6.33) and (2.6.32) we find that

$$\iint_S F^i n_i\, d\sigma + \iint_\Sigma \left(F^i n_i^{(1)} + F^i n_i^{(2)}\right) d\sigma = \iiint_V F^i_{,i}\, d\tau \tag{2.6.34}$$

where $V = V_T + V_B + V_\Sigma$. The equation (2.6.34) can also be written as

$$\iint_S F^i n_i\, d\sigma = \iiint_V F^i_{,i}\, d\tau - \iint_\Sigma \left(F^i n_i^{(1)} + F^i n_i^{(2)}\right) d\sigma. \tag{2.6.35}$$

In the case that V contains a surface Σ the total electric charge inside S is

$$Q_e = \iiint_V \rho^*\, d\tau + \iint_\Sigma \mu^*\, d\sigma \tag{2.6.36}$$

where μ^* is the surface charge density on Σ and ρ^* is the volume charge density throughout V. The Gauss theorem requires that

$$\iint_S E^i n_i\, d\sigma = \frac{Q_e}{\epsilon_0} = \frac{1}{\epsilon_0}\iiint_V \rho^*\, d\tau + \frac{1}{\epsilon_0}\iint_\Sigma \mu^*\, d\sigma. \tag{2.6.37}$$

In the case of a jump discontinuity across the surface Σ we use the results of equation (2.6.34) and write

$$\iint_S E^i n_i\, d\sigma = \iiint_V E^i_{,i}\, d\tau - \iint_\Sigma \left(E^i n_i^{(1)} + E^i n_i^{(2)}\right) d\sigma. \tag{2.6.38}$$

Subtracting the equation (2.6.37) from the equation (2.6.38) gives

$$\iiint_V \left(E^i_{,i} - \frac{\rho^*}{\epsilon_0}\right) d\tau - \iint_\Sigma \left(E^i n_i^{(1)} + E^i n_i^{(2)} + \frac{\mu^*}{\epsilon_0}\right) d\sigma = 0. \tag{2.6.39}$$

For arbitrary surfaces S and Σ, this equation implies the differential form of the Gauss law

$$E^i_{,i} = \frac{\rho^*}{\epsilon_0}. \tag{2.6.40}$$

Further, on the surface Σ, where there is a surface charge distribution we have

$$E^i n_i^{(1)} + E^i n_i^{(2)} + \frac{\mu^*}{\epsilon_0} = 0 \tag{2.6.41}$$

which shows the electric field undergoes a discontinuity when you cross a surface charge μ^*. ∎

Electrostatic Fields in Materials

When charges are introduced into materials it spreads itself throughout the material. Materials in which the spreading occurs quickly are called conductors, while materials in which the spreading takes a long time are called nonconductors or dielectrics. Another electrical property of materials is the ability to hold local charges which do not come into contact with other charges. This property is called induction. For example, consider a single atom within the material. It has a positively charged nucleus and negatively charged electron cloud surrounding it. When this atom experiences an electric field $\vec{E}$ the negative cloud moves opposite to $\vec{E}$ while the positively charged nucleus moves in the direction of $\vec{E}$. If $\vec{E}$ is large enough it can ionize the atom by pulling the electrons away from the nucleus. For moderately sized electric fields the atom achieves an equilibrium position where the positive and negative charges are offset. In this situation the atom is said to be polarized and have a dipole moment $\vec{p}$.

Definition: When a pair of charges $+q$ and $-q$ are separated by a distance $2\vec{d}$ the electric dipole moment is defined by $\vec{p} = 2\vec{d}q$, where $\vec{p}$ has dimensions of [C m].

In the special case where $\vec{d}$ has the same direction as $\vec{E}$ and the material is symmetric we say that $\vec{p}$ is proportional to $\vec{E}$ and write $\vec{p} = \alpha \vec{E}$, where α is called the atomic polarizability. If in a material subject to an electric field their results many such dipoles throughout the material then the dielectric is said to be polarized. The vector quantity $\vec{P}$ is introduced to represent this effect. The vector $\vec{P}$ is called the polarization vector having units of [C/m^2], and represents an average dipole moment per unit volume of material. The vectors P_i and E_i are related through the displacement vector D_i such that

$$P_i = D_i - \epsilon_0 E_i. \tag{2.6.42}$$

For an anisotropic material (crystal)

$$D_i = \epsilon_i^j E_j \qquad \text{and} \qquad P_i = \alpha_i^j E_j \tag{2.6.43}$$

where ϵ_i^j is called the dielectric tensor and α_i^j is called the electric susceptibility tensor. Consequently, the polarization can be represented

$$P_i = \alpha_i^j E_j = \epsilon_i^j E_j - \epsilon_0 E_i = (\epsilon_i^j - \epsilon_0 \delta_i^j) E_j \quad \text{so that} \quad \alpha_i^j = \epsilon_i^j - \epsilon_0 \delta_i^j. \tag{2.6.44}$$

A dielectric material is called homogeneous if the electric force and displacement vector are the same for any two points within the medium. This requires that the electric force and displacement vectors be constant parallel vector fields. It is left as an exercise to show that the condition for homogeneity is that $\epsilon_{i,k}^j = 0$. A dielectric material is called isotropic if the electric force vector and displacement vector have the same direction. This requires that $\epsilon_i^j = \epsilon \delta_j^i$ where δ_j^i is the Kronecker delta. The term $\epsilon = \epsilon_0 K_e$ is called the dielectric constant of the medium. The constant $\epsilon_0 = 8.85(10)^{-12}\,\text{coul}^2/\text{N} \cdot \text{m}^2$ is the permittivity of free space and the quantity $k_e = \frac{\epsilon}{\epsilon_0}$ is called the relative dielectric constant (relative to ϵ_0). For free space $k_e = 1$. Similarly for an isotropic material we have $\alpha_i^j = \epsilon_0 \alpha_e \delta_i^j$ where α_e is called the electric susceptibility. For a linear medium the vectors $\vec{P}$, $\vec{D}$ and $\vec{E}$ are related by

$$D_i = \epsilon_0 E_i + P_i = \epsilon_0 E_i + \epsilon_0 \alpha_e E_i = \epsilon_0(1 + \alpha_e)E_i = \epsilon_0 K_e E_i = \epsilon E_i \tag{2.6.45}$$

where $K_e = 1 + \alpha_e$ is the relative dielectric constant. The equation (2.6.45) are constitutive equations for dielectric materials.

The effect of polarization is to produce regions of bound charges ρ_b within the material and bound surface charges μ_b together with free charges ρ_f which are not a result of the polarization. Within dielectrics we have $\nabla \cdot \vec{P} = \rho_b$ for bound volume charges and $\vec{P} \cdot \hat{\mathbf{e}}_n = \mu_b$ for bound surface charges, where $\hat{\mathbf{e}}_n$ is a unit normal to the bounding surface of the volume. In these circumstances the expression for the potential function is written

$$V = \frac{1}{4\pi\epsilon_0} \iiint_V \frac{\rho_b}{r}\, d\tau + \frac{1}{4\pi\epsilon_0} \iint_S \frac{\mu_b}{r}\, d\sigma \tag{2.6.46}$$

and the Gauss law becomes

$$\epsilon_0 \nabla \cdot \vec{E} = \rho^* = \rho_b + \rho_f = -\nabla \cdot \vec{P} + \rho_f \qquad \text{or} \qquad \nabla(\epsilon_0 \vec{E} + \vec{P}) = \rho_f. \tag{2.6.47}$$

Since $\vec{D} = \epsilon_0 \vec{E} + \vec{P}$ the Gauss law can also be written in the form

$$\nabla \cdot \vec{D} = \rho_f \qquad \text{or} \qquad D^i_{,i} = \rho_f. \tag{2.6.48}$$

When no confusion arises we replace ρ_f by ρ. In integral form the Gauss law for dielectrics is written

$$\iint_S \vec{D} \cdot \hat{\mathbf{n}}\, d\sigma = Q_{fe} \tag{2.6.49}$$

where Q_{fc} is the total free charge density within the enclosing surface.

Magnetostatics

A stationary charge generates an electric field $\vec{E}$ while a moving charge generates a magnetic field $\vec{B}$. Magnetic field lines associated with a steady current moving in a wire form closed loops as illustrated in the figure 2.6-5.

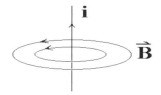

Figure 2.6-5. Magnetic field lines.

The direction of the magnetic force is determined by the right hand rule where the thumb of the right hand points in the direction of the current flow and the fingers of the right hand curl around in the direction of the magnetic field $\vec{B}$. The force on a test charge Q moving with velocity $\vec{V}$ in a magnetic field is

$$\vec{F}_m = Q(\vec{V} \times \vec{B}). \tag{2.6.50}$$

The total electromagnetic force acting on Q is the electric force plus the magnetic force and is

$$\vec{F} = Q\left[\vec{E} + (\vec{V} \times \vec{B})\right] \tag{2.6.51}$$

which is known as the Lorentz force law. The magnetic force due to a line charge density λ^* moving along a curve C is the line integral

$$\vec{F}_{mag} = \int_C \lambda^* ds(\vec{V} \times \vec{B}) = \int_C \vec{I} \times \vec{B} ds. \qquad (2.6.52)$$

Similarly, for a moving surface charge density moving on a surface

$$\vec{F}_{mag} = \iint_S \mu^* d\sigma(\vec{V} \times \vec{B}) = \iint_S \vec{K} \times \vec{B} \, d\sigma \qquad (2.6.53)$$

and for a moving volume charge density

$$\vec{F}_{mag} = \iiint_V \rho^* d\tau(\vec{V} \times \vec{B}) = \iint_V \vec{J} \times \vec{B} \, d\tau \qquad (2.6.54)$$

where the quantities $\vec{I} = \lambda^* \vec{V}$, $\vec{K} = \mu^* \vec{V}$ and $\vec{J} = \rho^* \vec{V}$ are respectively the current, the current per unit length, and current per unit area.

A conductor is any material where the charge is free to move. The flow of charge is governed by Ohm's law. Ohm's law states that the current density vector J_i is a linear function of the electric intensity or $J_i = \sigma_{im} E_m$, where σ_{im} is the conductivity tensor of the material. For homogeneous, isotropic conductors $\sigma_{im} = \sigma \delta_{im}$ so that $J_i = \sigma E_i$ where σ is the conductivity and $1/\sigma$ is called the resistivity.

Surround a charge density ρ^* with an arbitrary simple closed surface S having volume V and calculate the flux of the current density across the surface. We find by the divergence theorem

$$\iint_S \vec{J} \cdot \hat{\mathbf{n}} \, d\sigma = \iiint_V \nabla \cdot \vec{J} \, d\tau. \qquad (2.6.55)$$

If charge is to be conserved, the current flow out of the volume through the surface must equal the loss due to the time rate of change of charge within the surface which implies

$$\iint_S \vec{J} \cdot \hat{\mathbf{n}} \, d\sigma = \iiint_V \nabla \cdot \vec{J} \, d\tau = -\frac{d}{dt} \iiint_V \rho^* \, d\tau = -\iiint_V \frac{\partial \rho^*}{\partial t} \, d\tau \qquad (2.6.56)$$

or

$$\iiint_V \left[\nabla \cdot \vec{J} + \frac{\partial \rho^*}{\partial t} \right] d\tau = 0. \qquad (2.6.57)$$

This implies that for an arbitrary volume we must have

$$\nabla \cdot \vec{J} = -\frac{\partial \rho^*}{\partial t}. \qquad (2.6.58)$$

Note that equation (2.6.58) has the same form as the continuity equation (2.3.73) for mass conservation and so it is also called a continuity equation for charge conservation. For magnetostatics there exists steady line currents or stationary current so $\frac{\partial \rho^*}{\partial t} = 0$. This requires that $\nabla \cdot \vec{J} = 0$.

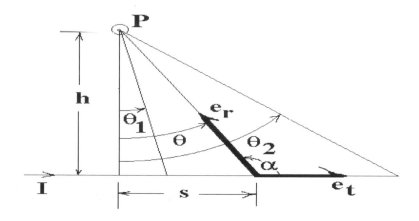

Figure 2.6-6. Magnetic field around wire.

Biot-Savart Law

The Biot-Savart law for magnetostatics describes the magnetic field at a point P due to a steady line current moving along a curve C and is

$$\vec{B}(P) = \frac{\mu_0}{4\pi} \int_C \frac{\vec{I} \times \hat{\mathbf{e}}_r}{r^2}\, ds \qquad (2.6.59)$$

with units [N/amp·m] and where the integration is in the direction of the current flow. In the Biot-Savart law we have the constant $\mu_0 = 4\pi \times 10^{-7}\,\mathrm{N/amp^2}$ which is called the permeability of free space, $\vec{I} = I\,\hat{\mathbf{e}}_t$ is the current flowing in the direction of the unit tangent vector $\hat{\mathbf{e}}_t$ to the curve C, $\hat{\mathbf{e}}_r$ is a unit vector directed from a point on the curve C toward the point P and r is the distance from a point on the curve to the general point P. Note that for a steady current to exist along the curve the magnitude of $\vec{I}$ must be the same everywhere along the curve. Hence, this term can be brought out in front of the integral. For surface currents $\vec{K}$ and volume currents $\vec{J}$ the Biot-Savart law is written

$$\vec{B}(P) = \frac{\mu_0}{4\pi} \iint_S \frac{\vec{K} \times \hat{\mathbf{e}}_r}{r^2}\, d\sigma$$

and $$\vec{B}(P) = \frac{\mu_0}{4\pi} \iiint_V \frac{\vec{J} \times \hat{\mathbf{e}}_r}{r^2}\, d\tau.$$

EXAMPLE 2.6-5.

Calculate the magnetic field $\vec{B}$ a distance h perpendicular to a wire carrying a constant current $\vec{I}$.
Solution: The magnetic field circles around the wire. For the geometry of the figure 2.6-6, the magnetic field points out of the page. We can write

$$\vec{I} \times \hat{\mathbf{e}}_r = I\,\hat{\mathbf{e}}_t \times \hat{\mathbf{e}}_r = I\hat{\mathbf{e}} \sin\alpha$$

where $\hat{\mathbf{e}}$ is a unit vector tangent to the circle of radius h which encircles the wire and cuts the wire perpendicularly.

For this problem the Biot-Savart law is

$$\vec{B}(P) = \frac{\mu_0 I}{4\pi} \int \frac{\hat{\mathbf{e}}}{r^2} \, ds.$$

In terms of θ we find from the geometry of figure 2.6-6

$$\tan\theta = \frac{s}{h} \quad \text{with} \quad ds = h \sec^2\theta \, d\theta \quad \text{and} \quad \cos\theta = \frac{h}{r}.$$

Therefore,

$$\vec{B}(P) = \frac{\mu_0}{\pi} \int_{\theta_1}^{\theta_2} \frac{I\hat{\mathbf{e}} \sin\alpha \, h \sec^2\theta}{h^2/\cos^2\theta} \, d\theta.$$

But, $\alpha = \pi/2 + \theta$ so that $\sin\alpha = \cos\theta$ and consequently

$$\vec{B}(P) = \frac{\mu_0 I \hat{\mathbf{e}}}{4\pi h} \int_{\theta_1}^{\theta_2} \cos\theta \, d\theta = \frac{\mu_0 I \hat{\mathbf{e}}}{4\pi h} (\sin\theta_2 - \sin\theta_1).$$

For a long straight wire $\theta_1 \to -\pi/2$ and $\theta_2 \to \pi/2$ to give the magnetic field $\vec{B}(P) = \frac{\mu_0 I \hat{\mathbf{e}}}{2\pi h}$. ■

For volume currents the Biot-Savart law is

$$\vec{B}(P) = \frac{\mu_0}{4\pi} \iiint_V \frac{\vec{J} \times \hat{\mathbf{e}}_r}{r^2} \, d\tau \tag{2.6.60}$$

and therefore (see exercises)

$$\nabla \cdot \vec{B} = 0. \tag{2.6.61}$$

Recall the divergence of an electric field is $\nabla \cdot \vec{E} = \frac{\rho^*}{\epsilon_0}$ is known as the Gauss's law for electric fields and so in analogy the divergence $\nabla \cdot \vec{B} = 0$ is sometimes referred to as Gauss's law for magnetic fields. If $\nabla \cdot \vec{B} = 0$, then there exists a vector field $\vec{A}$ such that $\vec{B} = \nabla \times \vec{A}$. The vector field $\vec{A}$ is called the vector potential of $\vec{B}$. Note that $\nabla \cdot \vec{B} = \nabla \cdot (\nabla \times \vec{A}) = 0$. Also the vector potential $\vec{A}$ is not unique since $\vec{B}$ is also derivable from the vector potential $\vec{A} + \nabla\phi$ where ϕ is an arbitrary continuous and differentiable scalar.

Ampere's Law

Ampere's law is associated with the work done in moving around a simple closed path. For example, consider the previous example 2.6-5. In this example the integral of $\vec{B}$ around a circular path of radius h which is centered at some point on the wire can be associated with the work done in moving around this path. The summation of force times distance is

$$\oint_C \vec{B} \cdot d\vec{r} = \oint_C \vec{B} \cdot \hat{\mathbf{e}} \, ds = \frac{\mu_0 I}{2\pi h} \oint_C ds = \mu_0 I \tag{2.6.62}$$

where now $d\vec{r} = \hat{\mathbf{e}} \, ds$ is a tangent vector to the circle encircling the wire and $\oint_C ds = 2\pi h$ is the distance around this circle. The equation (2.6.62) holds not only for circles, but for any simple closed curve around the wire. Using the Stoke's theorem we have

$$\oint_C \vec{B} \cdot d\vec{r} = \iint_S (\nabla \times \vec{B}) \cdot \hat{\mathbf{e}}_n \, d\sigma = \mu_0 I = \iint_S \mu_0 \vec{J} \cdot \hat{\mathbf{e}}_n \, d\sigma \tag{2.6.63}$$

where $\iint_S \vec{J} \cdot \hat{e}_n \, d\sigma$ is the total flux (current) passing through the surface which is created by encircling some curve about the wire. Equating like terms in equation (2.6.63) gives the differential form of Ampere's law

$$\nabla \times \vec{B} = \mu_0 \vec{J}. \tag{2.6.64}$$

Magnetostatics in Materials

Similar to what happens when charges are introduced into materials we have magnetic fields whenever there are moving charges within materials. For example, when electrons move around an atom tiny current loops are formed. These current loops create what are called magnetic dipole moments $\vec{m}$ throughout the material. When a magnetic field $\vec{B}$ is applied to a material medium there is a net alignment of the magnetic dipoles. The quantity $\vec{M}$, called the magnetization vector is introduced. Here $\vec{M}$ is associated with a dielectric medium and has the units [amp/m] and represents an average magnetic dipole moment per unit volume and is analogous to the polarization vector $\vec{P}$ used in electrostatics. The magnetization vector $\vec{M}$ acts a lot like the previous polarization vector in that it produces bound volume currents $\vec{J}_b$ and surface currents $\vec{K}_b$ where $\nabla \times \vec{M} = \vec{J}_b$ is a volume current density throughout some volume and $\vec{M} \times \hat{e}_n = \vec{K}_b$ is a surface current on the boundary of this volume.

From electrostatics note that the time derivative of $\epsilon_0 \frac{\partial \vec{E}}{\partial t}$ has the same units as current density. The total current in a magnetized material is then $\vec{J}_t = \vec{J}_b + \vec{J}_f + \epsilon_0 \frac{\partial \vec{E}}{\partial t}$ where $\vec{J}_b$ is the bound current, $\vec{J}_f$ is the free current and $\epsilon_0 \frac{\partial \vec{E}}{\partial t}$ is the induced current. Ampere's law, equation (2.6.64), in magnetized materials then becomes

$$\nabla \times \vec{B} = \mu_0 \vec{J}_t = \mu_0 (\vec{J}_b + \vec{J}_f + \epsilon_0 \frac{\partial \vec{E}}{\partial t}) = \mu_0 \vec{J} + \mu_0 \epsilon_0 \frac{\partial \vec{E}}{\partial t} \tag{2.6.65}$$

where $\vec{J} = \vec{J}_b + \vec{J}_f$. The term $\epsilon_0 \frac{\partial \vec{E}}{\partial t}$ is referred to as a displacement current or as a Maxwell correction to the field equation. This term implies that a changing electric field induces a magnetic field.

An auxiliary magnet field $\vec{H}$ defined by

$$H_i = \frac{1}{\mu_0} B_i - M_i \tag{2.6.66}$$

is introduced which relates the magnetic force vector $\vec{B}$ and magnetization vector $\vec{M}$. This is another constitutive equation which describes material properties. For an anisotropic material (crystal)

$$B_i = \mu_i^j H_j \qquad \text{and} \qquad M_i = \chi_i^j H_j \tag{2.6.67}$$

where μ_i^j is called the magnetic permeability tensor and χ_i^j is called the magnetic permeability tensor. Both of these quantities are dimensionless. For an isotropic material

$$\mu_i^j = \mu \delta_i^j \qquad \text{where} \qquad \mu = \mu_0 k_m. \tag{2.6.68}$$

Here $\mu_0 = 4\pi \times 10^{-7}\,\mathrm{N/amp^2}$ is the permeability of free space and $k_m = \frac{\mu}{\mu_0}$ is the relative permeability coefficient. Similarly, for an isotropic material we have $\chi_i^j = \chi_m \delta_i^j$ where χ_m is called the magnetic susceptibility coefficient and is dimensionless. The magnetic susceptibility coefficient has positive values for materials called paramagnets and negative values for materials called diamagnets. For a linear medium the quantities $\vec{B}$, $\vec{M}$ and $\vec{H}$ are related by

$$B_i = \mu_0(H_i + M_i) = \mu_0 H_i + \mu_0 \chi_m H_i = \mu_0(1 + \chi_m)H_i = \mu_0 k_m H_i = \mu H_i \tag{2.6.69}$$

where $\mu = \mu_0 k_m = \mu_0(1 + \chi_m)$ is called the permeability of the material.

Note: The auxiliary magnetic vector $\vec{H}$ for magnetostatics in materials plays a role similar to the displacement vector $\vec{D}$ for electrostatics in materials. Be careful in using electromagnetic equations from different texts as many authors interchange the roles of $\vec{B}$ and $\vec{H}$. Some authors call $\vec{H}$ the magnetic field. However, the quantity $\vec{B}$ should be the fundamental quantity.[1]

Electrodynamics

In the nonstatic case of electrodynamics there is an additional quantity $\vec{J}_p = \frac{\partial \vec{P}}{\partial t}$ called the polarization current which satisfies

$$\nabla \cdot \vec{J}_p = \nabla \cdot \frac{\partial \vec{P}}{\partial t} = \frac{\partial}{\partial t} \nabla \cdot \vec{P} = -\frac{\partial \rho_b}{\partial t} \tag{2.6.70}$$

and the current density has three parts

$$\vec{J} = \vec{J}_b + \vec{J}_f + \vec{J}_p = \nabla \times \vec{M} + \vec{J}_f + \frac{\partial \vec{P}}{\partial t} \tag{2.6.71}$$

consisting of bound, free and polarization currents.

Faraday's law states that a changing magnetic field creates an electric field. In particular, the electromagnetic force induced in a closed loop circuit C is proportional to the rate of change of flux of the magnetic field associated with any surface S connected with C. Faraday's law states

$$\oint_C \vec{E} \cdot d\vec{r} = -\frac{\partial}{\partial t} \iint_S \vec{B} \cdot \hat{\mathbf{e}}_n \, d\sigma.$$

Using the Stoke's theorem, we find

$$\iint_S (\nabla \times \vec{E}) \cdot \hat{\mathbf{e}}_n \, d\sigma = -\iint_S \frac{\partial \vec{B}}{\partial t} \cdot \hat{\mathbf{e}}_n \, d\sigma.$$

The above equation must hold for an arbitrary surface and loop. Equating like terms we obtain the differential form of Faraday's law

$$\nabla \times \vec{E} = -\frac{\partial \vec{B}}{\partial t}. \tag{2.6.72}$$

This is the first electromagnetic field equation of Maxwell.

Ampere's law, equation (2.6.65), written in terms of the total current from equation (2.6.71) , becomes

$$\nabla \times \vec{B} = \mu_0(\nabla \times \vec{M} + \vec{J}_f + \frac{\partial \vec{P}}{\partial t}) + \mu_0 \epsilon_0 \frac{\partial \vec{E}}{\partial t} \tag{2.6.73}$$

which can also be written as

$$\nabla \times (\frac{1}{\mu_0}\vec{B} - \vec{M}) = \vec{J}_f + \frac{\partial}{\partial t}(\vec{P} + \epsilon_0 \vec{E})$$

[1]D.J. Griffiths, Introduction to Electrodynamics, Prentice Hall, 1981. P.232.

or

$$\nabla \times \vec{H} = \vec{J}_f + \frac{\partial \vec{D}}{\partial t}. \qquad (2.6.74)$$

This is Maxwell's second electromagnetic field equation.

To the equations (2.6.74) and (2.6.73) we add the Gauss's law for magnetization, equation (2.6.61) and Gauss's law for electrostatics, equation (2.6.48). These four equations produce the Maxwell's equations of electrodynamics and are now summarized. The general form of Maxwell's equations involve the quantities

$E_i,$ Electric force vector, $[E_i] = $ Newton/coulomb

$B_i,$ Magnetic force vector, $[B_i] = $ Weber/m^2

$H_i,$ Auxilary magnetic force vector, $[H_i] = $ ampere/m

$D_i,$ Displacement vector, $[D_i] = $ coulomb/m^2

$J_i,$ Free current density, $[J_i] = $ ampere/m^2

$P_i,$ Polarization vector, $[P_i] = $ coulomb/m^2

$M_i,$ Magnetization vector, $[M_i] = $ ampere/m

for $i = 1, 2, 3$. There are also the quantities

$\varrho,$ representing the free charge density, with units $[\varrho] = $ coulomb/m^3

$\epsilon_0,$ Permittivity of free space, $[\epsilon_0] = $ farads/m   or   coulomb^2/Newton $\cdot$ m^2 .

$\mu_0,$ Permeability of free space, $[\mu_0] = $ henrys/m   or   kg $\cdot$ m/coulomb^2

In addition, there arises the material parameters:

$\mu_j^i,$ magnetic permeability tensor, which is dimensionless

$\epsilon_j^i,$ dielectric tensor, which is dimensionless

$\alpha_j^i,$ electric susceptibility tensor, which is dimensionless

$\chi_j^i,$ magnetic susceptibility tensor, which is dimensionless

These parameters are used to express variations in the electric field E_i and magnetic field B_i when acting in a material medium. In particular, P_i, D_i, M_i and H_i are defined from the equations

$$D_i = \epsilon_i^j E_j = \epsilon_0 E_i + P_i \qquad \epsilon_j^i = \epsilon_0 \delta_j^i + \alpha_i^j$$
$$B_i = \mu_i^j H_j = \mu_0 H_i + \mu_0 M_i, \qquad \mu_j^i = \mu_0(\delta_j^i + \chi_j^i)$$
$$P_i = \alpha_i^j E_j, \qquad \text{and} \qquad M_i = \chi_i^j H_j \quad \text{for } i = 1, 2, 3.$$

The above quantities obey the following laws:

Faraday's Law   This law states the line integral of the electromagnetic force around a loop is proportional to the rate of flux of magnetic induction through the loop. This gives rise to the first electromagnetic field equation:

$$\nabla \times \vec{E} = -\frac{\partial \vec{B}}{\partial t} \qquad \text{or} \qquad \epsilon^{ijk} E_{k,j} = -\frac{\partial B^i}{\partial t}. \qquad (2.6.75)$$

Ampere's Law This law states the line integral of the magnetic force vector around a closed loop is proportional to the sum of the current through the loop and the rate of flux of the displacement vector through the loop. This produces the second electromagnetic field equation:

$$\nabla \times \vec{H} = \vec{J}_f + \frac{\partial \vec{D}}{\partial t} \qquad \text{or} \qquad \epsilon^{ijk} H_{k,j} = J_f^i + \frac{\partial D^i}{\partial t}. \tag{2.6.76}$$

Gauss's Law for Electricity This law states that the flux of the electric force vector through a closed surface is proportional to the total charge enclosed by the surface. This results in the third electromagnetic field equation:

$$\nabla \cdot \vec{D} = \rho_f \qquad \text{or} \qquad D^i_{,i} = \rho_f \qquad \text{or} \qquad \frac{1}{\sqrt{g}} \frac{\partial}{\partial x^i} \left(\sqrt{g} D^i \right) = \rho_f. \tag{2.6.77}$$

Gauss's Law for Magnetism This law states the magnetic flux through any closed volume is zero. This produces the fourth electromagnetic field equation:

$$\nabla \cdot \vec{B} = 0 \qquad \text{or} \qquad B^i_{,i} = 0 \qquad \text{or} \qquad \frac{1}{\sqrt{g}} \frac{\partial}{\partial x^i} \left(\sqrt{g} B^i \right) = 0. \tag{2.6.78}$$

When no confusion arises it is convenient to drop the subscript f from the above Maxwell equations. Special expanded forms of the above Maxwell equations are given on the pages 176 to 179.

Electromagnetic Stress and Energy

Let V denote the volume of some simple closed surface S. Let us calculate the rate at which electromagnetic energy is lost from this volume. This represents the energy flow per unit volume. Begin with the first two Maxwell's equations in Cartesian form

$$\epsilon_{ijk} E_{k,j} = - \frac{\partial B_i}{\partial t} \tag{2.6.79}$$

$$\epsilon_{ijk} H_{k,j} = J_i + \frac{\partial D_i}{\partial t}. \tag{2.6.80}$$

Now multiply equation (2.6.79) by H_i and equation (2.6.80) by E_i. This gives two terms with dimensions of energy per unit volume per unit of time which we write

$$\epsilon_{ijk} E_{k,j} H_i = - \frac{\partial B_i}{\partial t} H_i \tag{2.6.81}$$

$$\epsilon_{ijk} H_{k,j} E_i = J_i E_i + \frac{\partial D_i}{\partial t} E_i. \tag{2.6.82}$$

Subtracting equation (2.6.82) from equation (2.6.81) we find

$$\epsilon_{ijk}(E_{k,j} H_i - H_{k,j} E_i) = - J_i E_i - \frac{\partial D_i}{\partial t} E_i - \frac{\partial B_i}{\partial t} H_i$$

$$\epsilon_{ijk} \left[(E_k H_i)_{,j} - E_k H_{i,j} + H_{i,j} E_k \right] = - J_i E_i - \frac{\partial D_i}{\partial t} E_i - \frac{\partial B_i}{\partial t} H_i$$

Observe that $\epsilon_{jki}(E_k H_i)_{,j}$ is the same as $\epsilon_{ijk}(E_j H_k)_{,i}$ so that the above simplifies to

$$\epsilon_{ijk}(E_j H_k)_{,i} + J_i E_i = - \frac{\partial D_i}{\partial t} E_i - \frac{\partial B_i}{\partial t} H_i. \tag{2.6.83}$$

Now integrate equation (2.6.83) over a volume and apply Gauss's divergence theorem to obtain

$$\iint_S \epsilon_{ijk}E_jH_kn_i\,d\sigma + \iiint_V J_iE_i\,d\tau = -\iiint_V \left(\frac{\partial D_i}{\partial t}E_i + \frac{\partial B_i}{\partial t}H_i\right)d\tau. \tag{2.6.84}$$

The first term in equation (2.6.84) represents the outward flow of energy across the surface enclosing the volume. The second term in equation (2.6.84) represents the loss by Joule heating and the right-hand side is the rate of decrease of stored electric and magnetic energy. The equation (2.6.84) is known as Poynting's theorem and can be written in the vector form

$$\iint_S (\vec{E}\times\vec{H})\cdot\hat{\mathbf{n}}\,d\sigma = \iiint_V \left(-\vec{E}\cdot\frac{\partial\vec{D}}{\partial t} - \vec{H}\cdot\frac{\partial\vec{B}}{\partial t} - \vec{E}\cdot\vec{J}\right)d\tau. \tag{2.6.85}$$

For later use we define the quantity

$$S_i = \epsilon_{ijk}E_jH_k \quad\text{or}\quad \vec{S} = \vec{E}\times\vec{H} \quad [\text{Watts/m}^2] \tag{2.6.86}$$

as Poynting's energy flux vector and note that S_i is perpendicular to both E_i and H_i and represents units of energy density per unit time which crosses a unit surface area within the electromagnetic field.

Electromagnetic Stress Tensor

Instead of calculating energy flow per unit volume, let us calculate force per unit volume. Consider a region containing charges and currents but is free from dielectrics and magnetic materials. To obtain terms with units of force per unit volume we take the cross product of equation (2.6.79) with D_i and the cross product of equation (2.6.80) with B_i and subtract to obtain

$$-\epsilon_{irs}\epsilon_{ijk}(E_{k,j}D_s + H_{k,j}B_s) = \epsilon_{ris}J_iB_s + \epsilon_{ris}\left(\frac{\partial D_i}{\partial t}B_s + \frac{\partial B_s}{\partial t}D_i\right)$$

which simplifies using the $e-\delta$ identity to

$$-(\delta_{rj}\delta_{sk} - \delta_{rk}\delta_{sj})(E_{k,j}D_s + H_{k,j}B_s) = \epsilon_{ris}J_iB_s + \epsilon_{ris}\frac{\partial}{\partial t}(D_iB_s)$$

which further simplifies to

$$-E_{s,r}D_s + E_{r,s}D_s - H_{s,r}B_s + H_{r,s}B_s = \epsilon_{ris}J_iB_s + \frac{\partial}{\partial t}(\epsilon_{ris}D_iB_s). \tag{2.6.87}$$

Observe that the first two terms in the equation (2.6.87) can be written

$$\begin{aligned}E_{r,s}D_s - E_{s,r}D_s &= E_{r,s}D_s - \epsilon_0 E_{s,r}E_s\\ &= (E_rD_s)_{,s} - E_rD_{s,s} - \epsilon_0(\tfrac{1}{2}E_sE_s)_{,r}\\ &= (E_rD_s)_{,s} - \rho E_r - \tfrac{1}{2}(E_jD_j\delta_{sr})_{,s}\\ &= (E_rD_s - \tfrac{1}{2}E_jD_j\delta_{rs})_{,s} - \rho E_r\end{aligned}$$

which can be expressed in the form

$$E_{r,s}D_s - E_{s,r}D_s = T^E_{rs,s} - \rho E_r$$

where

$$T^E_{rs} = E_r D_s - \frac{1}{2} E_j D_j \delta_{rs} \qquad (2.6.88)$$

is called the electric stress tensor. In matrix form the stress tensor is written

$$T^E_{rs} = \begin{bmatrix} E_1 D_1 - \frac{1}{2} E_j D_j & E_1 D_2 & E_1 D_3 \\ E_2 D_1 & E_2 D_2 - \frac{1}{2} E_j D_j & E_2 D_3 \\ E_3 D_1 & E_3 D_2 & E_3 D_3 - \frac{1}{2} E_j D_j \end{bmatrix}. \qquad (2.6.89)$$

By performing similar calculations we can transform the third and fourth terms in the equation (2.6.87) and obtain

$$H_{r,s} B_s - H_{s,r} B_s = T^M_{rs,s} \qquad (2.6.90)$$

where

$$T^M_{rs} = H_r B_S - \frac{1}{2} H_j B_j \delta_{rs} \qquad (2.6.91)$$

is the magnetic stress tensor. In matrix form the magnetic stress tensor is written

$$T^M_{rs} = \begin{bmatrix} B_1 H_1 - \frac{1}{2} B_j H_j & B_1 H_2 & B_1 H_3 \\ B_2 H_1 & B_2 H_2 - \frac{1}{2} B_j H_j & B_2 H_3 \\ B_3 H_1 & B_3 H_2 & B_3 H_3 - \frac{1}{2} B_j H_j \end{bmatrix}. \qquad (2.6.92)$$

The total electromagnetic stress tensor is

$$T_{rs} = T^E_{rs} + T^M_{rs}. \qquad (2.6.93)$$

Then the equation (2.6.87) can be written in the form

$$T_{rs,s} - \rho E_r = \epsilon_{ris} J_i B_s + \frac{\partial}{\partial t}(\epsilon_{ris} D_i B_s)$$

or

$$\rho E_r + \epsilon_{ris} J_i B_S = T_{rs,s} - \frac{\partial}{\partial t}(\epsilon_{ris} D_i B_s). \qquad (2.6.94)$$

For free space $D_i = \epsilon_0 E_i$ and $B_i = \mu_0 H_i$ so that the last term of equation (2.6.94) can be written in terms of the Poynting vector as

$$\mu_0 \epsilon_0 \frac{\partial S_r}{\partial t} = \frac{\partial}{\partial t}(\epsilon_{ris} D_i B_s). \qquad (2.6.95)$$

Now integrate the equation (2.6.94) over the volume to obtain the total electromagnetic force

$$\iiint_V \rho E_r \, d\tau + \iiint_V \epsilon_{ris} J_i B_s \, d\tau = \iiint_V T_{rs,s} \, d\tau - \mu_0 \epsilon_0 \iiint_V \frac{\partial S_r}{\partial t} \, d\tau.$$

Applying the divergence theorem of Gauss gives

$$\iiint_V \rho E_r \, d\tau + \iiint_V \epsilon_{ris} J_i B_s \, d\tau = \iint_S T_{rs} n_s \, d\sigma - \mu_0 \epsilon_0 \iiint_V \frac{\partial S_r}{\partial t} \, d\tau. \qquad (2.6.96)$$

The left side of the equation (2.6.96) represents the forces acting on charges and currents contained within the volume element. If the electric and magnetic fields do not vary with time, then the last term on the right is zero. In this case the forces can be expressed as an integral of the electromagnetic stress tensor.

EXERCISE 2.6

1. Find the field lines and equipotential curves associated with a positive charge q located at $(-a,0)$ and a positive charge q located at $(a,0)$. The field lines are illustrated in the figure 2.6-7.

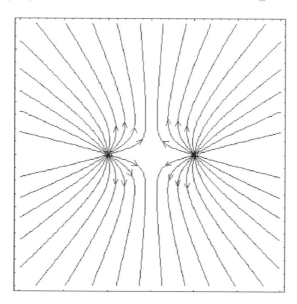

Figure 2.6-7. Lines of electric force between two charges of the same sign.

2. Calculate the lines of force and equipotential curves associated with the electric field $\vec{E} = \vec{E}(x,y) = 2y\,\widehat{\mathbf{e}}_1 + 2x\,\widehat{\mathbf{e}}_2$. Sketch the lines of force and equipotential curves. Put arrows on the lines of force to show direction of the field lines.

3. A right circular cone is defined by

$$x = u\sin\theta_0\cos\phi, \qquad y = u\sin\theta_0\sin\phi, \qquad z = u\cos\theta_0$$

with $0 \le \phi \le 2\pi$ and $u \ge 0$. Show the solid angle subtended by this cone is $\Omega = \frac{A}{r^2} = 2\pi(1-\cos\theta_0)$.

▶ **4.** A charge $+q$ is located at the point $(0,a)$ and a charge $-q$ is located at the point $(0,-a)$. Show that the electric force $\vec{E}$ at the position $(x,0)$, where $x > a$ is $\vec{E} = \frac{1}{4\pi\epsilon_0}\frac{-2aq}{(a^2+x^2)^{3/2}}\,\widehat{\mathbf{e}}_2$.

▶ **5.** Let the circle $x^2 + y^2 = a^2$ carry a line charge λ^*. Show the electric field at the point $(0,0,z)$ is $\vec{E} = \frac{1}{4\pi\epsilon_0}\frac{\lambda^* az(2\pi)\widehat{\mathbf{e}}_3}{(a^2+z^2)^{3/2}}$.

▶ **6.** Use superposition to find the electric field associated with two infinite parallel plane sheets each carrying an equal but opposite sign surface charge density μ^*. Find the field between the planes and outside of each plane. Hint: Fields are of magnitude $\pm\frac{\mu^*}{2\epsilon_0}$ and perpendicular to plates.

▶ **7.** For a volume current $\vec{J}$ the Biot-Savart law gives $\vec{B} = \frac{\mu_0}{4\pi}\iiint_V \frac{\vec{J}\times\widehat{\mathbf{e}}_r}{r^2}\,d\tau$. Show that $\nabla\cdot\vec{B} = 0$.

Hint: Let $\widehat{\mathbf{e}}_r = \frac{\vec{r}}{r}$ and consider $\nabla\cdot(\vec{J}\times\frac{\vec{r}}{r^3})$. Then use numbers 13 and 10 of the appendix C. Also note that $\nabla\times\vec{J} = 0$ because $\vec{J}$ does not depend upon position.

▶ **8.** A homogeneous dielectric is defined by D_i and E_i having parallel vector fields. Show that for a homogeneous dielectric $e^j_{i,k} = 0$.

▶ **9.** Show that for a homogeneous, isotropic dielectric medium that ϵ is a constant.

▶ **10.** Show that for a homogeneous, isotropic linear dielectric in Cartesian coordinates

$$P_{i,i} = \frac{\alpha_e}{1 + \alpha_e} \rho_f.$$

▶ **11.** Verify the Maxwell's equations in Gaussian units for a charge free isotropic homogeneous dielectric.

$$\nabla \cdot \vec{E} = \frac{1}{\epsilon} \nabla \cdot \vec{D} = 0 \qquad \nabla \times \vec{E} = -\frac{1}{c} \frac{\partial \vec{B}}{\partial t} = -\frac{\mu}{c} \frac{\partial \vec{H}}{\partial t}$$

$$\nabla \cdot \vec{B} = \mu \nabla \vec{H} = 0 \qquad \nabla \times \vec{H} = \frac{1}{c} \frac{\partial \vec{D}}{\partial t} + \frac{4\pi}{c} \vec{J} = \frac{\epsilon}{c} \frac{\partial \vec{E}}{\partial t} + \frac{4\pi}{c} \sigma \vec{E}$$

▶ **12.** Verify the Maxwell's equations in Gaussian units for an isotropic homogeneous dielectric with a charge.

$$\nabla \cdot \vec{D} = 4\pi\rho \qquad \nabla \times \vec{E} = -\frac{1}{c} \frac{\partial \vec{B}}{\partial t}$$

$$\nabla \cdot \vec{B} = 0 \qquad \nabla \times \vec{H} = \frac{4\pi}{c} \vec{J} + \frac{1}{c} \frac{\partial \vec{D}}{\partial t}$$

▶ **13.** For a volume charge ρ in an element of volume $d\tau$ located at a point (ξ, η, ζ) Coulombs law is

$$\vec{E}(x, y, z) = \frac{1}{4\pi\epsilon_0} \iiint_V \frac{\rho}{r^2} \widehat{\mathbf{e}}_r d\tau$$

(a) Show that $r^2 = (x - \xi)^2 + (y - \eta)^2 + (z - \zeta)^2$.

(b) Show that $\widehat{\mathbf{e}}_r = \frac{1}{r} ((x - \xi) \widehat{\mathbf{e}}_1 + (y - \eta) \widehat{\mathbf{e}}_2 + (z - \zeta) \widehat{\mathbf{e}}_3)$.

(c) Show that

$$\vec{E}(x, y, z) = \frac{1}{4\pi\epsilon_0} \iiint_V \frac{(x - \xi) \widehat{\mathbf{e}}_1 + (y - \eta) \widehat{\mathbf{e}}_2 + (z - \zeta) \widehat{\mathbf{e}}_3}{[(x - \xi)^2 + (y - \eta)^2 + (z - \zeta)^2]^{3/2}} \rho \, d\xi d\eta d\zeta = \frac{1}{4\pi\epsilon_0} \iiint_V \nabla \left(\frac{\widehat{\mathbf{e}}_r}{r^2} \right) \rho \, d\xi d\eta d\zeta$$

(d) Show that the potential function for $\vec{E}$ is $V = \frac{1}{4\pi\epsilon_0} \iiint_V \frac{\rho(\xi, \eta, \zeta)}{[(x - \xi)^2 + (y - \eta)^2 + (z - \zeta)^2]^{1/2}} d\xi d\eta d\zeta$

(e) Show that $\vec{E} = -\nabla V$.

(f) Show that $\nabla^2 V = -\frac{\rho}{\epsilon}$ Hint: Note that the integrand is zero everywhere except at the point where $(\xi, \eta, \zeta) = (x, y, z)$. Consider the integral split into two regions. One region being a small sphere about the point (x, y, z) in the limit as the radius of this sphere approaches zero. Observe the identity $\nabla_{(x,y,z)} \left(\frac{\widehat{\mathbf{e}}_r}{r^2} \right) = -\nabla(\xi, \eta, \zeta) \left(\frac{\widehat{\mathbf{e}}_r}{r^2} \right)$ enables one to employ the Gauss divergence theorem to obtain a surface integral. Use a mean value theorem to show $-\frac{\rho}{4\pi\epsilon_0} \iint_S \frac{\widehat{\mathbf{e}}_r}{r^2} \cdot \hat{n} dS = \frac{\rho}{4\pi\epsilon_0} 4\pi$ since $\hat{n} = -\widehat{\mathbf{e}}_r$.

▶ **14.** Show that for a point charge in space $\rho^* = q\delta(x - x_0)\delta(y - y_0)\delta(z - z_0)$, where δ is the Dirac delta function, the equation (2.6.5) can be reduced to the equation (2.6.1).

▶ **15.**

(a) Show the electric field $\vec{E} = \frac{1}{r^2} \widehat{\mathbf{e}}_r$ is irrotational. Here $\widehat{\mathbf{e}}_r = \frac{\vec{r}}{r}$ is a unit vector in the direction of r.

(b) Find the potential function V such that $\vec{E} = -\nabla V$ which satisfies $V(r_0) = 0$ for $r_0 > 0$.

16.

(a) If $\vec{E}$ is a conservative electric field such that $\vec{E} = -\nabla V$, then show that $\vec{E}$ is irrotational and satisfies $\nabla \times \vec{E} = \operatorname{curl} \vec{E} = 0$.

(b) If $\nabla \times \vec{E} = \operatorname{curl} \vec{E} = 0$, show that $\vec{E}$ is conservative. (i.e. Show $\vec{E} = -\nabla V$.)

Hint: The work done on a test charge $Q = 1$ along the straight line segments from (x_0, y_0, z_0) to (x, y_0, z_0) and then from (x, y_0, z_0) to (x, y, z_0) and finally from (x, y, z_0) to (x, y, z) can be written

$$V = V(x, y, z) = -\int_{x_0}^{x} E_1(x, y_0, z_0)\, dx - \int_{y_0}^{y} E_2(x, y, z_0)\, dy - \int_{z_0}^{z} E_3(x, y, z)\, dz.$$

Now note that

$$\frac{\partial V}{\partial y} = -E_2(x, y, z_0) - \int_{z_0}^{z} \frac{\partial E_3(x, y, z)}{\partial y}\, dz$$

and from $\nabla \times \vec{E} = 0$ we find $\dfrac{\partial E_3}{\partial y} = \dfrac{\partial E_2}{\partial z}$, which implies $\dfrac{\partial V}{\partial y} = -E_2(x, y, z)$. Similar results are obtained for $\dfrac{\partial V}{\partial x}$ and $\dfrac{\partial V}{\partial z}$ Hence show $-\nabla V = \vec{E}$.

17.

(a) Show that if $\nabla \cdot \vec{B} = 0$, then there exists some vector field $\vec{A}$ such that $\vec{B} = \nabla \times \vec{A}$. The vector field $\vec{A}$ is called the vector potential of $\vec{B}$.

Hint: Let $\vec{A}(x, y, z) = \displaystyle\int_0^1 s\vec{B}(sx, sy, sz) \times \vec{r}\, ds$ where $\vec{r} = x\,\widehat{\mathbf{e}}_1 + y\,\widehat{\mathbf{e}}_2 + z\,\widehat{\mathbf{e}}_3$

and integrate $\displaystyle\int_0^1 \frac{dB_i}{ds} s^2\, ds$ by parts.

(b) Show that $\nabla \cdot (\nabla \times \vec{A}) = 0$.

18. Use Faraday's law and Ampere's law to show

$$g^{im}(E^j_{,j})_{,m} - g^{jm} E^i_{,mj} = -\mu_0 \frac{\partial}{\partial t}\left[J^i + \epsilon_0 \frac{\partial E^i}{\partial t}\right]$$

19. Assume that $\vec{J} = \sigma \vec{E}$ where σ is the conductivity. Show that for $\rho = 0$ Maxwell's equations produce

$$\mu_0 \sigma \frac{\partial \vec{E}}{\partial t} + \mu_0 \epsilon_0 \frac{\partial^2 \vec{E}}{\partial t^2} = \nabla^2 \vec{E}$$

$$\text{and} \qquad \mu_0 \sigma \frac{\partial \vec{B}}{\partial t} + \mu_0 \epsilon_0 \frac{\partial^2 \vec{B}}{\partial t^2} = \nabla^2 \vec{B}.$$

Here both $\vec{E}$ and $\vec{B}$ satisfy the same equation which is known as the telegrapher's equation.

20. Show that Maxwell's equations (2.6.75) through (2.6.78) for the electric field under electrostatic conditions reduce to

$$\nabla \times \vec{E} = 0$$
$$\nabla \cdot \vec{D} = \rho_f$$

Now $\vec{E}$ is irrotational so that $\vec{E} = -\nabla V$. Show that $\nabla^2 V = -\dfrac{\rho_f}{\epsilon}$.

350

▶ **21.** Show that Maxwell's equations (2.6.75) through (2.6.78) for the magnetic field under magnetostatic conditions reduce to $\nabla \times \vec{H} = \vec{J}$ and $\nabla \cdot \vec{B} = 0$. The divergence of $\vec{B}$ being zero implies $\vec{B}$ can be derived from a vector potential function $\vec{A}$ such that $\vec{B} = \nabla \times \vec{A}$. Here $\vec{A}$ is not unique, see problem 24. If we select $\vec{A}$ such that $\nabla \cdot \vec{A} = 0$ then show for a homogeneous, isotropic material, free of any permanent magnets, that $\nabla^2 \vec{A} = -\mu \vec{J}$.

▶ **22.** Show that under nonsteady state conditions of electrodynamics the Faraday law from Maxwell's equations (2.6.75) through (2.6.78) does not allow one to set $\vec{E} = -\nabla \mathcal{V}$. Why is this? Observe that $\nabla \cdot \vec{B} = 0$ so we can write $\vec{B} = \nabla \times \vec{A}$ for some vector potential $\vec{A}$. Using this vector potential show that Faraday's law can be written $\nabla \times \left(\vec{E} + \dfrac{\partial \vec{A}}{\partial t} \right) = 0$. This shows that the quantity inside the parenthesis is conservative and so we can write $\vec{E} + \dfrac{\partial \vec{A}}{\partial t} = -\nabla \mathcal{V}$ for some scalar potential $\mathcal{V}$. The representation

$$\vec{E} = -\nabla \mathcal{V} - \frac{\partial \vec{A}}{\partial t}$$

is a more general representation of the electric potential. Observe that for steady state conditions $\dfrac{\partial \vec{A}}{\partial t} = 0$ so that this potential representation reduces to the previous one for electrostatics.

▶ **23.** Using the potential formulation $\vec{E} = -\nabla \mathcal{V} - \dfrac{\partial \vec{A}}{\partial t}$ derived in problem 22, show that in a vacuum

(a) Gauss law can be written $\nabla^2 \mathcal{V} + \dfrac{\partial \nabla \cdot \vec{A}}{\partial t} = -\dfrac{\rho}{\epsilon_0}$

(b) Ampere's law can be written

$$\nabla \times \left(\nabla \times \vec{A} \right) = \mu_0 \vec{J} - \mu_0 \epsilon_0 \nabla \left(\frac{\partial \mathcal{V}}{\partial t} \right) - \mu_0 \epsilon_0 \frac{\partial^2 \vec{A}}{\partial t^2}$$

(c) Show the result in part (b) can also be expressed in the form

$$\left(\nabla^2 \vec{A} - \mu_0 \epsilon_0 \frac{\partial \vec{A}}{\partial t} \right) - \nabla \left(\nabla \cdot \vec{A} + \mu_0 \epsilon_0 \frac{\partial \mathcal{V}}{\partial t} \right) = -\mu_0 \vec{J}$$

▶ **24.** The Maxwell equations in a vacuum have the form

$$\nabla \times \vec{E} = -\frac{\partial \vec{B}}{\partial t} \qquad \nabla \times \vec{H} = \frac{\partial \vec{D}}{\partial t} + \rho \vec{V} \qquad \nabla \cdot \vec{D} = \rho \qquad \nabla \cdot \vec{B} = 0$$

where $\vec{D} = \epsilon_0 \vec{E}$, $\vec{B} = \mu_0 \vec{H}$ with ϵ_0 and μ_0 constants satisfying $\epsilon_0 \mu_0 = 1/c^2$ where c is the speed of light. Introduce the vector potential $\vec{A}$ and scalar potential $\mathcal{V}$ defined by $\vec{B} = \nabla \times \vec{A}$ and $\vec{E} = -\dfrac{\partial \vec{A}}{\partial t} - \nabla \mathcal{V}$. Note that the vector potential is not unique. For example, given ψ as a scalar potential we can write $\vec{B} = \nabla \times \vec{A} = \nabla \times (\vec{A} + \nabla \psi)$, since the curl of a gradient is zero. Therefore, it is customary to impose some kind of additional requirement on the potentials. These additional conditions are such that $\vec{E}$ and $\vec{B}$ are not changed. One such condition is that $\vec{A}$ and $\mathcal{V}$ satisfy $\nabla \cdot \vec{A} + \dfrac{1}{c^2} \dfrac{\partial \mathcal{V}}{\partial t} = 0$. This relation is known as the Lorentz relation or Lorentz gauge. Find the Maxwell's equations in a vacuum in terms of $\vec{A}$ and $\mathcal{V}$ and show that

$$\left[\nabla^2 - \frac{1}{c^2} \frac{\partial^2}{\partial t^2} \right] \mathcal{V} = -\frac{\rho}{\epsilon_0} \qquad \text{and} \qquad \left[\nabla^2 - \frac{1}{c^2} \frac{\partial^2}{\partial t^2} \right] \vec{A} = -\mu_0 \rho \vec{V}.$$

25. In a vacuum show that $\vec{E}$ and $\vec{B}$ satisfy

$$\nabla^2 \vec{E} = \frac{1}{c^2}\frac{\partial^2 \vec{E}}{\partial t^2} \qquad \nabla^2 \vec{B} = \frac{1}{c^2}\frac{\partial^2 \vec{B}}{\partial t^2} \qquad \nabla \cdot \vec{E} = 0 \qquad \nabla \vec{B} = 0$$

26.

(a) Show that the wave equations in problem 25 have solutions in the form of waves traveling in the x- direction given by

$$\vec{E} = \vec{E}(x,t) = \vec{E}_0 e^{i(kx \pm \omega t)} \qquad \text{and} \qquad \vec{B} = \vec{B}(x,t) = \vec{B}_0 e^{i(kx \pm \omega t)}$$

where $\vec{E}_0$ and $\vec{B}_0$ are constants. Note that wave functions of the form $u = Ae^{i(kx \pm \omega t)}$ are called plane harmonic waves. Sometimes they are called monochromatic waves. Here $i^2 = -1$ is an imaginary unit. Euler's identity shows that the real and imaginary parts of these type wave functions have the form

$$A\cos(kx \pm \omega t) \qquad \text{and} \qquad A\sin(kx \pm \omega t).$$

These represent plane waves. The constant A is the amplitude of the wave , ω is the angular frequency, and $k/2\pi$ is called the wave number. The motion is a simple harmonic motion both in time and space. That is, at a fixed point x the motion is simple harmonic in time and at a fixed time t, the motion is harmonic in space. By examining each term in the sine and cosine terms we find that x has dimensions of length, k has dimension of reciprocal length, t has dimensions of time and ω has dimensions of reciprocal time or angular velocity. The quantity $c = \omega/k$ is the wave velocity. The value $\lambda = 2\pi/k$ has dimension of length and is called the wavelength and $1/\lambda$ is called the wave number. The wave number represents the number of waves per unit of distance along the x-axis. The period of the wave is $T = \lambda/c = 2\pi/\omega$ and the frequency is $f = 1/T$. The frequency represents the number of waves which pass a fixed point in a unit of time.

(b) Show that $\omega = 2\pi f$

(c) Show that $c = f\lambda$

(d) Is the wave motion $u = \sin(kx - \omega t) + \sin(kx + \omega t)$ a traveling wave? Explain.

(e) Show that in general the wave equation $\nabla^2 \phi = \frac{1}{c^2}\frac{\partial^2 \phi}{\partial t^2}$ have solutions in the form of waves traveling in either the $+x$ or $-x$ direction given by

$$\phi = \phi(x,t) = f(x + ct) + g(x - ct)$$

where f and g are arbitrary twice differentiable functions.

(f) Assume a plane electromagnetic wave is moving in the $+x$ direction. Show that the electric field is in the xy−plane and the magnetic field is in the xz−plane.
Hint: Assume solutions $E_x = g_1(x - ct)$, $E_y = g_2(x - ct)$, $E_z = g_3(x - ct)$, $B_x = g_4(x - ct)$, $B_y = g_5(x - ct)$, $B_z = g_6(x - ct)$ where $g_i, i = 1,...,6$ are arbitrary functions. Then show that E_x does not satisfy $\nabla \cdot \vec{E} = 0$ which implies g_1 must be independent of x and so not a wave function. Do the same for the components of $\vec{B}$. Since both $\nabla \cdot \vec{E} = \nabla \cdot \vec{B} = 0$ then $E_x = B_x = 0$. Such waves are called transverse waves because the electric and magnetic fields are perpendicular to the direction of propagation. Faraday's law implies that the $\vec{E}$ and $\vec{B}$ waves must be in phase and be mutually perpendicular to each other.

§2.7 GENERALIZATIONS

Consider the concept that — "Everything is a special case of something more general." In many instances the pursuit of a more general concept or idea can lead to some interesting results. For example, the set of complex numbers $z = x + iy$ is a generalization of the real numbers $z = x$. In order to make this generalization it was necessary to introduce the concept of an imaginary component, denoted by the symbol i, where $i^2 = -1$. In the case of complex numbers we are "adding" two entirely different quantities, a real quantity and an imaginary quantity. Unless you are prepared to approach generalizations with the idea that new concepts might arise, accepting the symbol i and mixing different quantities by addition will be a hard concept to grasp. This idea of "adding different types of quantities" will be investigated in more detail.

Quaternions

Sir William R. Hamilton (1805-1865) invented quaternions by using imaginary quantities. Quaternions can be represented in different ways. One form is

$$q = s + v_1\iota_1 + v_2\iota_2 + v_3\iota_3 \tag{2.7.1}$$

where $\iota_1, \iota_2, \iota_3$ are distinct imaginary quantities satisfying the multiplication properties

$$
\begin{array}{lll}
\iota_1\iota_1 = -1 & \iota_1\iota_2 = \iota_3 & \iota_1\iota_3 = -\iota_2 \\
\iota_2\iota_1 = -\iota_3 & \iota_2\iota_2 = -1 & \iota_2\iota_3 = \iota_1 \\
\iota_3\iota_1 = \iota_2 & \iota_3\iota_2 = -\iota_1 & \iota_3\iota_3 = -1
\end{array}
\tag{2.7.2}
$$

which can be written as the multiplication table

$$
\begin{array}{|c|c|c|c|c|}
\hline
 & 1 & \iota_1 & \iota_2 & \iota_3 \\
\hline
1 & 1 & \iota_1 & \iota_2 & \iota_3 \\
\hline
\iota_1 & \iota_1 & -1 & \iota_3 & -\iota_2 \\
\hline
\iota_2 & \iota_2 & -\iota_3 & -1 & \iota_1 \\
\hline
\iota_3 & \iota_3 & \iota_2 & -\iota_1 & -1 \\
\hline
\end{array}
\tag{2.7.3}
$$

The quantities $\iota_1, \iota_2, \iota_3$ are treated like unit vectors but they represent orthogonal imaginary directions. The first part of q is the quantity s which is treated as a scalar. The second quantity $v_1\iota_1 + v_2\iota_2 + v_3\iota_3$ is to be treated as a vector.

Another form is to write the quaternion as a 4-tuple of numbers

$$q = (s, v_1, v_2, v_3) \tag{2.7.4}$$

where the basis elements $(1, i_1, i_2, i_3)$ are to be understood. The imaginary units $(\iota_1, \iota_2, \iota_3)$ act like orthogonal unit vectors and so quaternions can also be represented in the form of a real number plus a 3-vector

$$q = (s, \vec{v}). \tag{2.7.5}$$

Here s is called the real part of q and $\vec{v} = v_1\iota_1 + v_2\iota_2 + v_3\iota_3$ is called the pure imaginary part. Let us examine some properties of quaternions using the above notations.

Properties of quaternions

1. Addition and subtraction involve using like components. For example, if

$$q_1 = s_1 + \alpha_1 \imath_1 + \alpha_2 \imath_2 + \alpha_3 \imath_3$$
$$q_2 = s_2 + \beta_1 \imath_1 + \beta_2 \imath_2 + \beta_3 \imath_3$$

$$\text{form 1} \qquad (2.7.6)$$

or

$$q_1 = (s_1, \alpha_1, \alpha_2, \alpha_3)$$
$$q_2 = (s_2, \beta_1, \beta_2, \beta_3)$$

$$\text{form 2} \qquad (2.7.7)$$

or

$$q_1 = (s_1, \vec{\alpha})$$
$$q_2 = (s_2, \vec{\beta})$$

$$\text{form 3} \qquad (2.7.8)$$

then addition requires adding like components and can be represented in any of the following ways.

$$q_1 + q_2 = (s_1 + s_2) + (\alpha_1 + \beta_1)\imath_1 + (\alpha_2 + \beta_2)\imath_2 + (\alpha_3 + \beta_3)\imath_3$$
$$q_1 + q_2 = (s_1 + s_2, \alpha_1 + \beta_1, \alpha_2 + \beta_2, \alpha_3 + \beta_3) \qquad (2.7.9)$$
$$q_1 + a_2 = (s_1 + s_2, \vec{\alpha} + \vec{\beta}).$$

Similarly, subtraction requires subtracting like components and can be represented in one of the forms

$$q_1 - q_2 = (s_1 - s_2) + (\alpha_1 - \beta_1)\imath_1 + (\alpha_2 - \beta_2)\imath_2 + (\alpha_3 - \beta_3)\imath_3$$
$$q_1 - q_2 = (s_1 - s_2, \alpha_1 - \beta_1, \alpha_2 - \beta_2, \alpha_3 - \beta_3) \qquad (2.7.10)$$
$$q_1 - q_2 = (s_1 - s_2, \vec{\alpha} - \vec{\beta})$$

2. Scalar multiplication changes the magnitude of each component. For example,

$$\gamma q_1 = \gamma s_1 + \gamma \alpha_1 i_1 + \gamma \alpha_2 i_2 + \gamma \alpha_3 i_3$$
$$\gamma q_1 = (\gamma s_1, \gamma \alpha_1, \gamma \alpha_2, \gamma \alpha_3) \qquad (2.7.11)$$
$$\gamma q_1 = (\gamma s_1, \gamma \vec{\alpha})$$

where γ represents a real number.

3. Multiplication follows the regular rules of algebra. For example

$$q_1 q_2 = (s_1 + \alpha_1 \imath_1 + \alpha_2 \imath_2 + \alpha_3 \imath_3)(s_2 + \beta_1 \imath_1 + \beta_2 \imath_2 + \beta_3 \imath_3)$$
$$q_1 q_2 = s_1(s_2 + \beta_1 \imath_1 + \beta_2 \imath_2 + \beta_3 \imath_3)$$
$$+ \alpha_1 \imath_1 (s_2 + \beta_1 \imath_1 + \beta_2 \imath_2 + \beta_3 \imath_3)$$
$$+ \alpha_2 \imath_2 (s_2 + \beta_1 \imath_1 + \beta_2 \imath_2 + \beta_3 \imath_3)$$
$$+ \alpha_3 \imath_3 (s_2 + \beta_1 \imath_1 + \beta_2 \imath_2 + \beta_3 \imath_3).$$

This simplifies when we use the multiplication properties given in equations (2.7.2) and

$$q_1 q_2 = s_1 s_2 + s_1 \beta_1 \imath_1 + s_1 \beta_2 \imath_2 + s_1 \beta_3 \imath_3$$
$$+ \alpha_1 s_2 \imath_1 - \alpha_1 \beta_1 + \alpha_1 \beta_2 \imath_3 - \alpha_1 \beta_3 \imath_2$$
$$+ \alpha_2 s_2 \imath_2 - \alpha_2 \beta_1 \imath_3 - \alpha_2 \beta_2 + \alpha_2 \beta_3 \imath_1$$
$$+ \alpha_3 s_2 \imath_3 + \alpha_3 \beta_1 \imath_2 - \alpha_3 \beta_2 \imath_1 - \alpha_3 \beta_3$$

$$(2.7.12)$$

or after rearranging terms one can express the product in the form

$$q_1 q_2 = s_1 s_2 - \alpha_1 \beta_1 - \alpha_2 \beta_2 - \alpha_3 \beta_3$$
$$+ s_1(\beta_1 \imath_1 + \beta_2 \imath_2 + \beta_3 \imath_3) + s_2(\alpha_1 \imath_1 + \alpha_2 \imath_2 + \alpha_3 \imath_3) \qquad (2.7.13)$$
$$+ \imath_1(\alpha_2 \beta_3 - \alpha_3 \beta_2) + \imath_2(\alpha_3 \beta_1 - \alpha_1 \beta_3) + \imath_3(\alpha_1 \beta_2 - \alpha_2 \beta_1).$$

An easier form to remember multiplication of quaternions is to use the scalar 3-vector form for multiplication

$$q_1 q_2 = (s_1, \vec{\alpha})(s_2, \vec{\beta}) = (s_1 s_2 - \vec{\alpha} \cdot \vec{\beta}, s_1 \vec{\beta} + s_2 \vec{\alpha} + \vec{\alpha} \times \vec{\beta}) \qquad (2.7.14)$$

where $\cdot$ is the usual dot product for vectors and $\times$ is the usual cross product for vectors. Verifying that equation (2.7.14) reduces to equation (2.7.13) is left as an exercise. Still another form for obtaining the product of two quaternions is a matrix form. Observe that the equation (2.7.12) can be written in the form

$$q_1 q_2 = q_3 = s_3 + \gamma_1 \imath_1 + \gamma_2 \imath_2 + \gamma_3 \imath_3$$
$$\text{where} \qquad s_3 = s_1 s_2 - \alpha_1 \beta_1 - \alpha_2 \beta_2 - \alpha_3 \beta_3$$
$$\gamma_1 = \alpha_1 s_2 + s_1 \beta_1 - \alpha_3 \beta_2 + \alpha_2 \beta_3 \qquad (2.7.15)$$
$$\gamma_2 = \alpha_2 s_2 + \alpha_3 \beta_1 + s_1 \beta_2 - \alpha_1 \beta_3$$
$$\gamma_3 = \alpha_3 s_2 - \alpha_2 \beta_1 + \alpha_1 \beta_2 + s_1 \beta_3$$

which we observe has the matrix form

$$\begin{bmatrix} s_3 \\ \gamma_1 \\ \gamma_2 \\ \gamma_3 \end{bmatrix} = \begin{bmatrix} s_1 & -\alpha_1 & -\alpha_2 & -\alpha_3 \\ \alpha_1 & s_1 & -\alpha_3 & \alpha_2 \\ \alpha_2 & \alpha_3 & s_1 & -\alpha_1 \\ \alpha_3 & -\alpha_2 & \alpha_1 & s_1 \end{bmatrix} \begin{bmatrix} s_2 \\ \beta_1 \\ \beta_2 \\ \beta_3 \end{bmatrix}. \qquad (2.7.16)$$

An alternative matrix form for the equations (2.7.15) is given by

$$\begin{bmatrix} s_3 \\ \gamma_1 \\ \gamma_2 \\ \gamma_3 \end{bmatrix} = \begin{bmatrix} s_2 & -\beta_1 & -\beta_2 & -\beta_3 \\ \beta_1 & s_2 & \beta_3 & -\beta_2 \\ \beta_2 & -\beta_3 & s_2 & \beta_1 \\ \beta_3 & \beta_2 & -\beta_1 & s_2 \end{bmatrix} \begin{bmatrix} s_1 \\ \alpha_1 \\ \alpha_2 \\ \alpha_3 \end{bmatrix}. \qquad (2.7.17)$$

Observe that both forms have a skew-symmetric coefficient matrix. These are alternative forms for calculating the product of two quaternions.

4. Conjugation is analogous to what is done with complex numbers. The conjugate of the quaternion q_1 is given by

$$q_1^* = s_1 - \alpha_1 \imath_1 - \alpha_2 \imath_2 - \alpha_3 \imath_3 \qquad \text{form 1}$$
$$\text{or} \qquad q_1^* = (s_1, -\alpha_1, -\alpha_2, -\alpha_3) \qquad \text{form 2} \qquad (2.7.18)$$
$$\text{or} \qquad q_1^* = (s_1, -\vec{\alpha}) \qquad \text{form 3}$$

The conjugate of a product is the product of conjugates in reverse order $(q_1 q_2)^* = q_2^* q_1^*$. The sum of a quaternion with its conjugate gives a scalar quantity $q_1 + q_1^* = 2s_1$.

5. Absolute value, length or norm $\| \ \|$ of the quaternion q_1 is defined as the square root of q_1 times its own conjugate. This gives

$$\|q_1\| = \sqrt{q_1 q_1^*} = \sqrt{s_1^2 + \alpha_1^2 + \alpha_2^2 + \alpha_3^2} \qquad (2.7.19)$$

Observe that the norm of a product equals the product of norms and

$$||q_1 q_2|| = ||q_1|| \, ||q_2||. \tag{2.7.20}$$

A quaternion of length unity is called a unit quaternion. The quaternion $q = q_0 + q_1 i_1 + q_2 i_2 + q_3 i_3$, with $q_1^2 + q_2^2 + q_3^2 \neq 0$ can be written in the form

$$q = ||q||(\cos\theta + \widehat{e}\sin\theta) \qquad 0 \le \theta < 2\pi \tag{2.7.21}$$

where

$$||q|| = \sqrt{q_0^2 + q_1^2 + q_2^2 + q_3^2} \qquad \cos\theta = \frac{q_0}{||q||} \qquad \sin\theta = \frac{\pm\sqrt{q_1^2 + q_2^2 + q_3^2}}{||q||}$$

where the angle θ is called the angle of the quaternion q. The quantity $\widehat{e}$ is a unit vector given by

$$\widehat{e} = \pm \frac{q_1 i_1 + q_2 i_2 + q_3 i_3}{\sqrt{q_1^2 + q_2^2 + q_3^2}} \qquad \text{and satisfies} \quad \widehat{e}\widehat{e} = -1.$$

In the special case $\sin\theta = 0$, then q is real so that $\widehat{e}$ can be arbitrary. The special case of a unit quaternion can be written in terms of an angle and a unit vector $\widehat{e}$ in the form

$$q = \cos\theta + \widehat{e}\sin\theta. \tag{2.7.22}$$

6. Inverse and division property. Using the relation from equation (2.7.19) we can write

$$q_1^* q_1 = q_1 q_1^* = ||q_1||^2 \qquad \text{or} \qquad q_1\left(\frac{q_1^*}{||q_1||^2}\right) = 1 \tag{2.7.23}$$

This gives a left and right inverse of q_1 which is defined

$$q_1^{-1} = \frac{q_1^*}{||q_1||^2} \tag{2.7.24}$$

In the special case where q_1 is a unit quaternion, then $q_1^{-1} = q_1^*$.

7. Special quaternions.

$q = (s,0,0,0)$ are real numbers (scalars)

$q = (x_1, y_1, 0, 0)$ are complex numbers $q = z_1 = x_1 + i_1 y_1$

$q = (x_2, 0, y_2, 0)$ another complex number $q = z_2 = x_2 + i_2 y_2$

$q = (x_3, 0, 0, y_3)$ another complex number $q = z_3 = x_3 + i_3 y_3$

$q = (0, y_1, y_2, y_3)$ a pure quaternion which is treated as a vector

$q = (s_1, y_1, y_2, y_3)$ a general quaternion

One can treat quaternions as an extension of the complex number system. It is easily verified from the quaternion multiplication properties that quaternion multiplication is associative $q_1(q_2 q_3) = (q_1 q_2)q_3$, but not commutative, since in general $q_1 q_2 \neq q_2 q_1$.

Isomorphism

There is an isomorphism between the basis elements $\{1, i_1, i_2, i_3\}$ and the matrices

$$I = \begin{bmatrix} 1 & 0 \\ 0 & 1 \end{bmatrix} \quad J_1 = \begin{bmatrix} 0 & -i \\ -i & 0 \end{bmatrix} \quad J_2 = \begin{bmatrix} 0 & -1 \\ 1 & 0 \end{bmatrix} \quad J_3 = \begin{bmatrix} -i & 0 \\ 0 & i \end{bmatrix} \tag{2.7.25}$$

where i denotes an imaginary unit, with $i^2 = -1$. The isomorphism can be seen from the multiplication table

	I	J_1	J_2	J_3
I	I	J_1	J_2	J_3
J_1	J_1	-I	J_3	$-J_2$
J_2	J_2	$-J_3$	-I	J_1
J_3	J_3	J_2	$-J_1$	-I

$$\tag{2.7.26}$$

Quaternions can also be represented in the form

$$Q = sI + v_1 J_1 + v_2 J_2 + v_3 J_3 \tag{2.7.27}$$

which simplifies to

$$Q = \begin{bmatrix} s + iv_3 & -v_2 - iv_1 \\ v_2 - iv_1 & s + iv_3 \end{bmatrix}. \tag{2.7.28}$$

For this form, the sum or product of two quaternions is obtained from the matrix sum or product. The conjugate of Q is denoted

$$Q^* = sI - v_1 J_1 - v_2 J_2 - v_3 J_3 = \begin{bmatrix} s + iv_3 & v_2 + iv_1 \\ -v_2 + iv_1 & s - iv_3 \end{bmatrix}$$

and $QQ^* = I|Q| = I(s^2 + v_1^2 + v_2^2 + v_3^2)$. The matrix Q is called a unitary matrix if the column vectors (row vectors) are mutually orthogonal unit vectors. The matrix Q is unitary if and only if $\overline{Q}^T Q = I$. Observe $Q^* = \overline{Q}^T$ where $\overline{Q}^T$ is the transpose of the complex conjugate of Q. It is readily verified from (2.7.28) that the column vectors (row vectors) are already orthogonal. In order that Q be unitary we require that

$$|Q| = s^2 + v_1^2 + v_2^2 + v_3^2 = 1.$$

Physical interpretations and rotations

A unit quaternion $q = q_0 + q_1 i_1 + q_2 i_2 + q_3 i_3$ satisfies $q_0^2 + q_1^2 + q_2^2 + q_3^2 = 1$ and can be written in the form

$$q = q_0 + \vec{q} = \cos\frac{\theta}{2} + \hat{e}\sin\frac{\theta}{2}. \tag{2.7.29}$$

This unit quaternion is used to rotate axes through an angle θ, which is twice the angle associated with the above quaternion. The operator

$$q^{-1}\vec{X}q = \vec{Y} \tag{2.7.30}$$

is used to convert the vector $\vec{X}$ to a vector $\vec{Y}$ by a matrix transformation which can be written in the matrix form $Y = AX$ where X and Y are column vectors. Observe that the operator $q\vec{X}q^{-1} = \vec{Y}$ produces the matrix transformation $Y = A^T X$ where A^T is the transpose of the matrix A. We will show that $q\vec{X}q^{-1}$

represents a vector rotation, while $q^{-1}\vec{X}q$ represents a frame rotation. To develop this matrix transformation we make use of the fact that q is a unit quaternion and so we can write

$$\vec{Y} = [q_0, -\vec{q}]\,[0, \vec{X}]\,[q_0, \vec{q}]$$

$$\vec{Y} = [q_0, -\vec{q}]\,[-\vec{X}\cdot\vec{q}, q_0\vec{X} + \vec{X}\times\vec{q}]$$

$$\vec{Y} = [q_0(-\vec{X}\cdot\vec{q}) + \vec{q}\cdot(q_0\vec{X} + \vec{X}\times\vec{q}), q_0(q_0\vec{X} + \vec{X}\times\vec{q}) + (\vec{X}\cdot\vec{q})\vec{q} - \vec{q}\times(q_0\vec{X} + \vec{X}\times\vec{q})]$$

Using the identity $\vec{q}\times(\vec{X}\times\vec{q}) = \vec{X}(\vec{q}\cdot\vec{q}) - \vec{q}(\vec{q}\cdot\vec{X})$ together with the fact that q is a unit quaternion so that $q_0^2 + \vec{q}\cdot\vec{q} = 1$ one can simplify this last equation to read

$$\vec{Y} = [0, (2q_0^2 - 1)\vec{X} + 2q_0(\vec{X}\times\vec{q}) + 2(\vec{q}\cdot\vec{X})\vec{q}] \tag{2.7.31}$$

In equation (2.7.31) $\vec{Y}$ is a vector quantity and each term in equation (2.7.31) can be written in matrix form. For example

$$(2q_0^2 - 1)\vec{X} = \begin{bmatrix} 2q_0^2 - 1 & 0 & 0 \\ 0 & 2q_0^2 - 1 & 0 \\ 0 & 0 & 2q_0^2 - 1 \end{bmatrix}\begin{bmatrix} x_1 \\ x_2 \\ x_3 \end{bmatrix} \tag{2.7.32}$$

$$2q_0(\vec{X}\times\vec{q}) = \begin{bmatrix} 0 & +2q_0q_3 & -2q_0q_2 \\ -2q_0q_3 & 0 & +2q_0q_1 \\ +2q_0q_2 & -2q_0q_1 & 0 \end{bmatrix}\begin{bmatrix} x_1 \\ x_2 \\ x_3 \end{bmatrix} \tag{2.7.33}$$

$$2(\vec{q}\cdot\vec{X})\vec{q} - \begin{bmatrix} 2q_1^2 & 2q_1q_2 & 2q_1q_3 \\ 2q_1q_2 & 2q_2^2 & 2q_2q_3 \\ 2q_1q_3 & 2q_2q_3 & 2q_3^2 \end{bmatrix}\begin{bmatrix} x_1 \\ x_2 \\ x_3 \end{bmatrix} \tag{2.7.34}$$

Adding these matrices we can write equation (2.7.31) in the matrix form

$$\begin{bmatrix} y_1 \\ y_2 \\ y_3 \end{bmatrix} = \begin{bmatrix} 2q_0^2 - 1 + 2q_1^2 & 2(q_1q_2 + q_0q_3) & 2(q_1q_3 - q_0q_2) \\ 2(q_1q_2 - q_0q_3) & 2q_0^2 - 1 + 2q_2^2 & 2(q_2q_3 + q_0q_1) \\ 2(q_1q_3 + q_0q_2) & 2(q_2q_3 - q_0q_1) & 2q_0^2 - 1 + 2q_3^2 \end{bmatrix}\begin{bmatrix} x_1 \\ x_2 \\ x_3 \end{bmatrix} \tag{2.7.35}$$

or using the identity $q_0^2 + q_1^2 + q_2^2 + q_3^2 = 1$ (q must be a unit quaternion) the equation (2.7.35) can be written in the alternative form

$$\begin{bmatrix} y_1 \\ y_2 \\ y_3 \end{bmatrix} = \begin{bmatrix} q_0^2 + q_1^2 - q_2^2 - q_3^2 & 2(q_1q_2 + q_0q_3) & 2(q_1q_3 - q_0q_2) \\ 2(q_1q_2 - q_0q_3) & q_0^2 - q_1^2 + q_2^2 - q_3^2 & 2(q_2q_3 + q_0q_1) \\ 2(q_1q_3 + q_0q_2) & 2(q_2q_3 - q_0q_1) & q_0^2 - q_1^2 - q_2^2 + q_3^2 \end{bmatrix}\begin{bmatrix} x_1 \\ x_2 \\ x_3 \end{bmatrix} \tag{2.7.36}$$

The transformation $\vec{Y} = q\vec{X}q^{-1}$ produces the transformation equation

$$\begin{bmatrix} y_1 \\ y_2 \\ y_3 \end{bmatrix} = \begin{bmatrix} q_0^2 + q_1^2 - q_2^2 - q_3^2 & 2(q_1q_2 - q_0q_3) & 2(q_1q_3 + q_0q_2) \\ 2(q_1q_2 + q_0q_3) & q_0^2 - q_1^2 + q_2^2 - q_3^2 & 2(q_2q_3 - q_0q_1) \\ 2(q_1q_3 - q_0q_2) & 2(q_2q_3 + q_0q_1) & q_0^2 - q_1^2 - q_2^2 + q_3^2 \end{bmatrix}\begin{bmatrix} x_1 \\ x_2 \\ x_3 \end{bmatrix} \tag{2.7.37}$$

used for rotation of vectors.

358

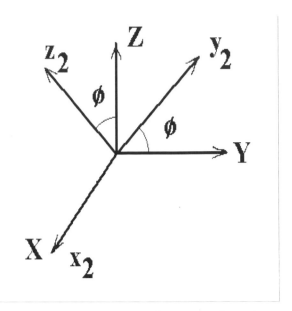

Figure 2.7-1 Rotation of yz plane.

EXAMPLE 2.7-1.

The rotation about an axis can be represented by matrix multiplication. Consider the roll, pitch and yaw of an aircraft. The roll can be represented by a rotation of the $yz-$plane about the $x-$axis through an angle ϕ as illustrated in the figure 2.7-1.

This rotation can be represented in the matrix form

$$
\begin{pmatrix} x_2 \\ y_2 \\ z_2 \end{pmatrix} = \begin{pmatrix} 1 & 0 & 0 \\ 0 & \cos\phi & \sin\phi \\ 0 & -\sin\phi & \cos\phi \end{pmatrix} \begin{pmatrix} X \\ Y \\ Z \end{pmatrix} \tag{2.7.38}
$$

This rotation can also be represented by the unit quaternion $q_1 = s_1 + \alpha_1 i_1$ where $s_1 = \cos\frac{\phi}{2}$ and $\alpha_1 = \sin\frac{\phi}{2}$ are called Euler parameters. The transformation (2.7.38) can then be written in terms of the Euler parameters as

$$
\begin{pmatrix} x_2 \\ y_2 \\ z_2 \end{pmatrix} = \begin{pmatrix} s_1^2 + \alpha_1^2 & 0 & 0 \\ 0 & s_1^2 - \alpha_1^2 & 2s_1\alpha_1 \\ 0 & -2s_1\alpha_1 & s_1^2 - \alpha_1^2 \end{pmatrix} \begin{pmatrix} X \\ Y \\ Z \end{pmatrix} \tag{2.7.39}
$$

as one can readily verify. Next consider the pitch illustrated in the figure 2.7-2 which we denoted by a rotation of the $xz-$plane through an angle θ.

This transformation is denoted by the matrix product

$$
\begin{pmatrix} x_3 \\ y_3 \\ z_3 \end{pmatrix} = \begin{pmatrix} \cos\theta & 0 & -\sin\theta \\ 0 & 1 & 0 \\ \sin\theta & 0 & \cos\theta \end{pmatrix} \begin{pmatrix} x_2 \\ y_2 \\ z_2 \end{pmatrix} \tag{2.7.40}
$$

and can be represented by the unit quaternion $q_2 = s_2 + \alpha_2 i_2$ with Euler parameters $s_2 = \cos\frac{\theta}{2}$ and $\alpha_2 = \sin\frac{\theta}{2}$. The transformation (2.7.40) can now be written in the matrix form

$$
\begin{pmatrix} x_3 \\ y_3 \\ z_3 \end{pmatrix} = \begin{pmatrix} s_2^2 - \alpha_2^2 & 0 & -2s_2\alpha_2 \\ 0 & s_2^2 + \alpha_2^2 & 0 \\ 2s_2\alpha_2 & 0 & s_2^2 - \alpha_2^2 \end{pmatrix} \begin{pmatrix} x_2 \\ y_2 \\ z_2 \end{pmatrix} \tag{2.7.41}
$$

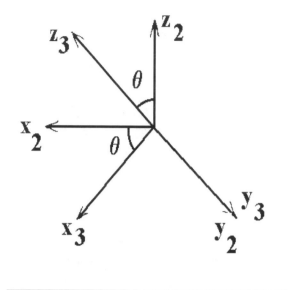

Figure 2.7-2 Rotation of the $xz-$plane.

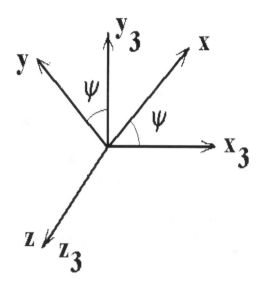

Figure 2.7-3 Rotation of the $xy-$plane.

Finally, the yaw motion, illustrated in the figure 2.7-3, can be represented by a rotation of the $xy-$plane through an angle ψ.

This yaw transformation can be represented in the matrix form

$$\begin{pmatrix} x \\ y \\ z \end{pmatrix} = \begin{pmatrix} \cos\psi & \sin\psi & 0 \\ -\sin\psi & \cos\psi & 0 \\ 0 & 0 & 1 \end{pmatrix} \begin{pmatrix} x_3 \\ y_3 \\ z_3 \end{pmatrix} \qquad (2.7.42)$$

This transformation can also be represented by the unit quaternion $q_3 = s_3 + \alpha_3 \imath_3$ with Euler parameters $s_3 = \cos\frac{\psi}{2}$ and $\alpha_3 = \sin\frac{\psi}{2}$ and written in the matrix form as

$$\begin{pmatrix} x \\ y \\ z \end{pmatrix} = \begin{pmatrix} s_3^2 - \alpha_3^2 & 2s_3\alpha_3 & 0 \\ -2s_3\alpha_3 & s_3^2 - \alpha_3^2 & 0 \\ 0 & 0 & s_3^2 + \alpha_3^2 \end{pmatrix} \begin{pmatrix} x_3 \\ y_3 \\ z_3 \end{pmatrix} \qquad (2.7.43)$$

Leonard Euler (1707-1783) showed that any two orthonormal coordinate systems can be related by a sequence of axis rotations, where no two successive rotations are about the same axis. The angles of rotation in the above examples are called Euler angles and the sequence of roll, pitch, yaw rotations, all in the positive sense, is called an Euler angle-axis sequence of rotations. The above roll, pitch, yaw sequence of rotations is denoted by the notation XYZ which is read– first a rotation about the x-axis, followed by a rotation about the y-axis, followed by a rotation about the z-axis. The restriction that successive axes of rotation be distinct permits twelve possible Euler angle sequences. These twelve sequences are denoted

$$\begin{array}{cccc} XYZ & XZY & XYX & XZX \\ YZX & YZX & YZY & YXY \\ ZXY & ZYX & ZXZ & ZYZ \end{array} \qquad (2.7.44)$$

That is, we can use any one of the above twelve Euler sequences to relate two arbitrary and different coordinate systems. Be careful in jumping from one textbook to another as physics, engineering and mathematics textbooks might not use the same sequence of rotations. The ideas associated with each sequence of rotations are the same as in our roll, pitch and yaw example with slight notation changes. Let us return now to our roll, pitch and yaw example. We use matrix multiplication to relate the axes (x_3, y_3, z_3) and (X, Y, Z) by the matrix product

$$\begin{pmatrix} x_3 \\ y_3 \\ z_3 \end{pmatrix} = \begin{pmatrix} s_2^2 - \alpha_2^2 & 0 & -2s_2\alpha_2 \\ 0 & s_2^2 + \alpha_2^2 & 0 \\ 2s_2\alpha_2 & 0 & s_2^2 - \alpha_2^2 \end{pmatrix} \begin{pmatrix} s_1^2 + \alpha_1^2 & 0 & 0 \\ 0 & s_1^2 - \alpha_1^2 & 2s_1\alpha_1 \\ 0 & -2s_1\alpha_1 & s_1^2 - \alpha_1^2 \end{pmatrix} \begin{pmatrix} X \\ Y \\ Z \end{pmatrix} \qquad (2.7.45)$$

which can be expanded to the form

$$\begin{pmatrix} x_3 \\ y_3 \\ z_3 \end{pmatrix} = \begin{pmatrix} (s_2^2 - \alpha_2^2)(s_1^2 + \alpha_1^2) & 4s_1s_2\alpha_1\alpha_2 & -2s_2\alpha_2(s_1^2 - \alpha_1)^2 \\ 0 & (s_2^2 + \alpha_2^2)(s_1^2 - \alpha_1)^2 & 2s_1\alpha_1(s_2^2 + \alpha_2^2) \\ 2s_2\alpha_2(s_1^2 + \alpha_1^2) & -2s_1\alpha_1(s_2^2 - \alpha_2^2) & (s_2^2 - \alpha_2^2)(s_1^2 + \alpha_1^2) \end{pmatrix} \begin{pmatrix} X \\ Y \\ Z \end{pmatrix} \qquad (2.7.46)$$

This matrix product corresponds to the quaternion product

$$\begin{aligned} \beta = q_1 q_2 &= (s_1 + \alpha_1 \imath_1)(s_2 + \alpha_2 \imath_2) = (s_1 s_2 + s_2 \alpha_1 \imath_1 + s_1 \alpha_2 \imath_2 + \alpha_1 \alpha_2 \imath_3) \\ &= \beta_0 + \beta_1 \imath_1 + \beta_2 \imath_2 + \beta_3 \imath_3 \end{aligned} \qquad (2.7.47)$$

with coefficients

$$\begin{aligned} \beta_0 &= \cos\frac{\phi}{2}\cos\frac{\theta}{2} = s_1 s_2 & \beta_2 &= \cos\frac{\phi}{2}\sin\frac{\theta}{2} = s_1 \alpha_2 \\ \beta_1 &= \sin\frac{\phi}{2}\cos\frac{\theta}{2} = \alpha_1 s_2 & \beta_3 &= \sin\frac{\phi}{2}\sin\frac{\theta}{2} = \alpha_1 \alpha_2 \end{aligned} \qquad (2.7.48)$$

The elements in the transformation matrix (2.7.46) can now be written in terms of the quaternion elements $\beta_0, \beta_1, \beta_2, \beta_3$. Observe that

$$(s_2^2 - \alpha_2^2)(s_1^2 + \alpha_1^2) = s_2^2 s_1^2 - \alpha_2^2 s_1^2 + s_2^2 \alpha_1^2 - \alpha_1^2 \alpha_2^2 = \beta_0^2 + \beta_1^2 - \beta_2^2 - \beta_3^2$$

$$(s_2^2 + \alpha_2^2)(s_1^2 - \alpha_1^2) = s_2^2 s_1^2 - \alpha_1^2 s_2^2 + \alpha_2^2 s_1^2 - \alpha_1^2 \alpha_2^2 = \beta_0^2 - \beta_1^2 + \beta_2^2 - \beta_3^2$$

$$(s_2^2 - \alpha_2^2)(s_1^2 - \alpha_1^2) = s_2^2 s_1^2 - \alpha_1^2 s_2^2 - \alpha_2^2 s_1^2 + \alpha_1^2 \alpha_2^2 = \beta_0^2 - \beta_1^2 - \beta_2^2 + \beta_3^2$$

$$4 s_1 s_2 \alpha_1 \alpha_2 = 2(s_1 \alpha_2)(s_2 \alpha_1) + 2(s_1 s_2)(\alpha_1 \alpha_2) = 2(\beta_1 \beta_2 + \beta_3 \beta_0)$$

$$-2 s_2 \alpha_2 s_1^2 + 2 s_2 \alpha_2 \alpha_1^2 = 2(\alpha_1 s_2)(\alpha_1 \alpha_2) - 2(s_1 s_2)(s_1 \alpha_2) = 2(\beta_4 \beta_3 - \beta_0 \beta_2)$$

with similar results for the other terms in equation (2.7.46). This leads to writing the equation (2.7.46) in the form of equation (2.7.36) to obtain

$$\begin{pmatrix} x_3 \\ y_3 \\ z_3 \end{pmatrix} = \begin{pmatrix} \beta_0^2 + \beta_1^2 - \beta_2^2 - \beta_3^2 & 2(\beta_1\beta_2 + \beta_0\beta_3) & 2(\beta_1\beta_3 - \beta_0\beta_2) \\ 2(\beta_1\beta_2 - \beta_0\beta_3) & \beta_0^2 - \beta_1^2 + \beta_2^2 - \beta_3^2 & 2(\beta_2\beta_3 + \beta_0\beta_1) \\ 2(\beta_1\beta_3 + \beta_0\beta_2) & 2(\beta_2\beta_3 - \beta_0\beta_1) & \beta_0^2 - \beta_1^2 - \beta_2^2 + \beta_3^2 \end{pmatrix} \begin{pmatrix} X \\ Y \\ Z \end{pmatrix} \qquad (2.7.49)$$

Similarly, the sequence of transformations for roll ϕ, pitch θ, and yaw ψ can be represented by multiplication of the respective transformation matrices. The results in terms of Euler parameters can be written

$$\begin{pmatrix} x \\ y \\ z \end{pmatrix} = \begin{pmatrix} \gamma_0^2 + \gamma_1^2 - \gamma_2^2 - \gamma_3^2 & 2(\gamma_1\gamma_2 + \gamma_0\gamma_3) & 2(\gamma_1\gamma_3 - \gamma_0\gamma_2) \\ 2(\gamma_1\gamma_2 - \gamma_0\gamma_3) & \gamma_0^2 - \gamma_1^2 + \gamma_2^2 - \gamma_3^2 & 2(\gamma_2\gamma_3 + \gamma_0\gamma_1) \\ 2(\gamma_1\gamma_3 + \gamma_0\gamma_2) & 2(\gamma_2\gamma_3 - \gamma_0\gamma_1) & \gamma_0^2 - \gamma_1^2 - \gamma_2^2 + \gamma_3^2 \end{pmatrix} \begin{pmatrix} X \\ Y \\ Z \end{pmatrix} \qquad (2.7.50)$$

where γ is obtained from the quaternion product $\gamma = q_1 q_2 q_3 = \gamma_0 + \gamma_1 \imath_1 + \gamma_2 \imath_2 + \gamma_3 \imath_3$ which gives

$$\gamma_0 = s_1 s_2 s_3 - \alpha_1 \alpha_2 \alpha_3 = \cos\frac{\phi}{2}\cos\frac{\theta}{2}\cos\frac{\psi}{2} - \sin\frac{\phi}{2}\sin\frac{\theta}{2}\sin\frac{\psi}{2}$$

$$\gamma_1 = s_3 s_2 \alpha_1 + s_1 \alpha_2 \alpha_3 = \sin\frac{\phi}{2}\cos\frac{\theta}{2}\cos\frac{\psi}{2} + \cos\frac{\phi}{2}\sin\frac{\theta}{2}\sin\frac{\psi}{2}$$

$$\gamma_2 = s_3 s_1 \alpha_2 - s_2 \alpha_1 \alpha_3 = -\sin\frac{\phi}{2}\cos\frac{\theta}{2}\sin\frac{\psi}{2} + \cos\frac{\phi}{2}\sin\frac{\theta}{2}\cos\frac{\psi}{2}$$

$$\gamma_3 = s_3 \alpha_1 \alpha_2 + s_1 s_2 \alpha_3 = \sin\frac{\phi}{2}\sin\frac{\theta}{2}\cos\frac{\psi}{2} + \cos\frac{\phi}{2}\cos\frac{\theta}{2}\sin\frac{\psi}{2}$$

$$(2.7.51)$$

This transformation can also be represented in terms of the Euler angles ϕ, θ, ψ as

$$\begin{pmatrix} x \\ y \\ z \end{pmatrix} = \begin{pmatrix} \cos\theta\cos\psi & \cos\psi\sin\theta\sin\phi + \cos\phi\sin\psi & -\cos\phi\cos\psi\sin\theta + \sin\theta\sin\psi \\ -\cos\theta\sin\psi & \cos\phi\cos\psi - \sin\theta\sin\phi\sin\psi & \cos\psi\sin\phi + \cos\phi\sin\theta\sin\psi \\ \sin\theta & -\cos\theta\sin\phi & \cos\theta\cos\phi \end{pmatrix} \begin{pmatrix} X \\ Y \\ Z \end{pmatrix} \qquad (2.7.52)$$

∎

EXAMPLE 2.7-2.

Using the quaternions from the previous example the sequence of transformations ZYX can be represented in terms of the Euler parameters r_0, r_1, r_2, r_3 where

$$r = q_3 q_2 q_1 = r_0 + r_1 \imath_1 + r_2 \imath_2 + r_3 \imath_3$$

with values

$$r_0 = s_1 s_2 s_3 + \alpha_1 \alpha_2 \alpha_3 = \cos\frac{\phi}{2}\cos\frac{\theta}{2}\cos\frac{\psi}{2} + \sin\frac{\phi}{2}\sin\frac{\theta}{2}\sin\frac{\psi}{2}$$

$$r_1 = s_3 s_2 \alpha_1 - \alpha_3 s_1 \alpha_2 = \cos\frac{\psi}{2}\cos\frac{\theta}{2}\sin\frac{\phi}{2} - \sin\frac{\psi}{2}\cos\frac{\phi}{2}\sin\frac{\theta}{2}$$

$$r_2 = s_3 s_1 \alpha_2 + s_2 \alpha_1 \alpha_3 = \cos\frac{\psi}{2}\cos\frac{\phi}{2}\sin\frac{\theta}{2} + \cos\frac{\theta}{2}\sin\frac{\phi}{2}\sin\frac{\psi}{2}$$

$$r_3 = s_1 s_2 \alpha_3 - s_3 \alpha_1 \alpha_2 = \cos\frac{\phi}{2}\cos\frac{\theta}{2}\sin\frac{\psi}{2} - \cos\frac{\psi}{2}\sin\frac{\phi}{2}\sin\frac{\theta}{2}$$

The resulting transformation can be expressed in the matrix form

$$\begin{pmatrix} x \\ y \\ z \end{pmatrix} = \begin{pmatrix} r_0^2 + r_1^2 - r_2^2 - r_3^2 & 2(r_1 r_2 + r_0 r_3) & 2(r_1 r_3 - r_0 r_2) \\ 2(r_1 r_2 - r_0 r_3) & r_0^2 - r_1^2 + r_2^2 - r_3^2 & 2(r_2 r_3 + r_0 r_1) \\ 2(r_1 r_3 + r_0 r_2) & 2(r_2 r_3 - r_0 r_1) & r_0^2 - r_1^2 - r_2^2 + r_3^2 \end{pmatrix} \begin{pmatrix} X \\ Y \\ Z \end{pmatrix} \qquad (2.7.53)$$

involving Euler parameters or

$$\begin{pmatrix} x \\ y \\ z \end{pmatrix} = \begin{pmatrix} \cos\theta\cos\psi & \cos\theta\sin\psi & -\sin\theta \\ \sin\theta\sin\phi\cos\psi - \cos\phi\sin\psi & \sin\theta\sin\phi\sin\psi + \cos\phi\cos\psi & \cos\theta\sin\phi \\ \sin\theta\cos\phi\cos\psi + \sin\phi\sin\psi & \sin\theta\cos\phi\sin\psi - \sin\phi\cos\psi & \cos\theta\cos\phi \end{pmatrix} \begin{pmatrix} X \\ Y \\ Z \end{pmatrix} \qquad (2.7.54)$$

involving Euler angles.

∎

Let $q = (q_0, \vec{Q})$ be a unit quaternion and let $b = (b_0, \vec{B})$ denote a non-scalar quaternion, then the mapping $c = q^{-1}bq$ produces a quaternion $c = (b_0, \vec{C})$ with the following properties:

1. The norm of quaternion b equals the norm of quaternion c. This follows from the properties of the norm.

$$||c|| = ||qbq^{-1}|| = ||q||\,||b||\,||q^{-1}|| = ||b||$$

2. Both the quaternions c and b have the same scalar values. This follows from the multiplication properties of quaternions. Recall that for a unit quaternion $q^{-1} = q^* = (q_0, -\vec{Q})$ so that

$$bq^{-1} = (b_0, \vec{B})(q_0, -\vec{Q}) = (b_0 q_0 + \vec{B}\cdot\vec{Q}, -b_0\vec{Q} + q_0\vec{B} - \vec{B}\times\vec{Q})$$

and so the scalar part of qbq^{-1} is

$$q_0 b_0 q_0 + b_0 \vec{Q}\cdot\vec{Q} = b_0(q_0^2 + \vec{Q}\cdot\vec{Q}) = b_0$$

since q is a unit quaternion we have $||q||^2 = q_0^2 + \vec{Q}\cdot\vec{Q} = 1$.

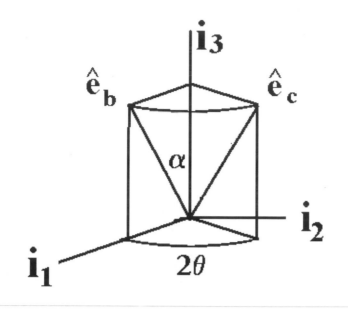

Figure 2.7-4 Conical rotation of unit vector $\hat{e}_b$ to vector $\hat{e}_c$.

3. The vector part of the quaternion c is obtained by revolving the vector part of the quaternion b in a conical fashion about the vector part of q. The rotation is through twice the angle associate with q. Recall that the quaternion q can be written in the form $q = \cos\theta + e\sin\theta$ where e is a unit vector associated with the vector part of q.

For simplification assume that e can be associated with the imaginary i_3 direction and we will use the representation

$$q = \cos\theta + i_3\sin\theta$$

where it is understood i_1, i_2, i_3 are treated as a set of orthonormal directions. We now show that the mapping qbq^{-1} produces a quaternion $c = (b_0, \vec{C})$ where the vector $\vec{C}$ is obtained by a conical revolution of the vector $\vec{B}$ about the axis i_3 through an angle 2θ as illustrated in the figure 2.7-4. Using ordinary algebra we find

$$qbq^{-1} = q(b_0, \vec{B})q^{-1} = (qb_0q^{-1}, q\vec{B}q^{-1}) = (b_0, q\vec{B}q^{-1}).$$

By writing the quaternion b in the form $b = ||b||(\cos\phi + \hat{e}_b\sin\phi)$ we can write the vector part of b in the form $\vec{B} = ||b||\sin\phi\,\hat{e}_b$, so that

$$q\vec{B}q^{-1} = ||b||\sin\phi\,q\hat{e}_bq^{-1}.$$

It can now be demonstrated that the transformation $q\hat{e}_bq^{-1}$ represents a conical revolution of $\hat{e}_b$ about i_3 illustrated in the figure 2.7-4. Assume the i_1, i_2, i_3 axes are oriented such that i_1 lies in the plane of $\hat{e}_b$ and i_3, then the unit vector $\hat{e}_b$ can be written in the component form

$$\hat{e}_b = \cos\alpha\,i_1 + \sin\alpha\,i_3$$

for some angle α. We can now write

$$q\hat{e}_bq^{-1} = \cos\alpha\,qi_1q^{-1} + \sin\alpha\,qi_3q^{-1}.$$

Here the vector part of q is colinear with $\imath_3$ so that $q\imath_3 q^{-1} = \imath_3$. Therefore,

$$
\begin{aligned}
q\imath_1 q^{-1} &= (\cos\theta + \imath_3 \sin\theta)\imath_1(\cos\theta - \imath_3 \sin\theta) \\
&= (\imath_1 \cos\theta + \imath_2 \sin\theta)(\cos\theta - \imath_3 \sin\theta) \\
&= \imath_1 \cos^2\theta + \imath_2 \sin\theta\cos\theta + \imath_2 \sin\theta\cos\theta - \imath_1 \sin^2\theta \\
&= \imath_1(\cos^2\theta - \sin^2\theta) + \imath_2(2\sin\theta\cos\theta) \\
&= \imath_1 \cos 2\theta + \imath_2 \sin 2\theta.
\end{aligned}
$$

This shows that the vector in the $\imath_1$ direction rotates to the vector $\hat{e}_1 = \imath_1 \cos 2\theta + \imath_2 \sin 2\theta$ and so $\hat{e}_b$ rotates to

$$
\hat{e}_c = \cos\alpha(\imath_1 \cos 2\theta + \imath_2 \sin 2\theta) + \sin\alpha\imath_3
$$

with components

$$
\hat{e}_c = \imath_1 \cos\alpha\cos 2\theta + \imath_2 \cos\alpha\sin 2\theta + \imath_3 \sin\alpha
$$

which represents a conical rotation through the angle 2θ as illustrated in the figure 2.7-4.

Special transformation of unit quaternion

In the special case $q = (0, e_A)$, where e_A is a unit vector, the angle θ associated with q is $\pi/2$. In this case the transformation $q\hat{e}q^{-1} = -e_A\hat{e}e_A$ rotates a unit vector $\hat{e}$ in a conical fashion through an angle of π radians about the vector e_A, as illustrated in the figure 2.7-5. If we write $e_A\hat{e}e_A$ as $-(-e_A\hat{e}e_A)$ we can interpret the transformation $e_A\hat{e}e_A$ as a reflection of $\hat{e}$ in the plane normal to the vector e_A. Thus, a rotation can be viewed as a reflection plus a reversal of the reflection.

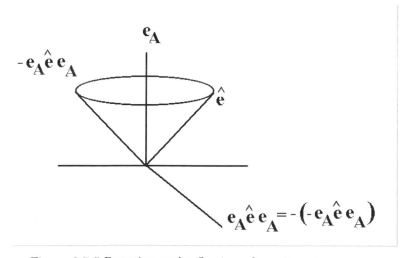

Figure 2.7-5 Rotation and reflection of a unit vector.

A sequence of rotations is denoted by $q_2(q_1 \hat{e} q_1^{-1}) q_2^{-1}$ where q_1, q_2 are unit quaternions. The product of rotations signifies q_1 followed by q_2 and is denoted $q_2 q_1$. Such a sequence is equivalent to a rotation about the unit vector associated with the quaternion product $q_2 q_1$ through an angle twice the angle associated with the quaternion product.

Clifford Algebra

William Kingdom Clifford (1845-1879) was professor of applied mathematics at University College London. He invented an algebra which he called "geometric algebra". The following is a brief introduction into Clifford's ideas.

In addition to the dot product $A \cdot B$ and cross product $A \times B$ of vectors, we define the wedge product $A \wedge B$ which creates a new quantity different from scalars and vectors. The object created is called a bivector and represents the parallelogram illustrated in the figure 2.7-6(a).

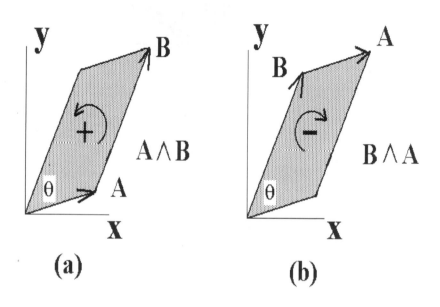

Figure 2.7-6 The bivectors $A \wedge B$ and $B \wedge A$.

To form a wedge product the origins of the vectors A and B are made to coincide, then the first vector A is moved parallel to itself along the second vector B. This creates a parallelogram element called the directed area created by moving A along B. The magnitude of the wedge product is defined to be the area of the parallelogram and so

$$|A \wedge B| = |A||B| \sin \theta. \tag{2.7.55}$$

The bivector has an orientation associated with its creation. If the parallelogram created is traversed around the boundary starting with the first vector A, then moving along the second direction B, the wedge product $A \wedge B$ is said to be positively oriented. In contrast, the wedge product $B \wedge A$ is negatively oriented and so the wedge product is anticommutative with the property

$$A \wedge B = -B \wedge A \tag{2.7.56}$$

366

The terminology of 'grades' is used to describe the type of object being discussed. For example, scalars (points) are of grade 0. Points moving in a line create a vector which is of grade 1. Vectors move along other vectors to create parallelograms called bivectors of grade 2. Moving a bivector $A \wedge B$ along another vector C produces a volume element $(A \wedge B) \wedge C$ called a trivector which has grade 3. A trivector is illustrated in the figure 2.7-7.

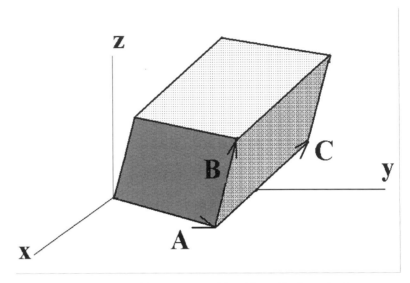

Figure 2.7-7 The trivector $(A \wedge B) \wedge C$ of grade 3.

The wedge product is associative $\quad (A \wedge B) \wedge C = A \wedge (B \wedge C)$ $\hfill (2.7.57)$

and distributive $\quad A \wedge (B + C) = A \wedge B + A \wedge C$ $\hfill (2.7.58)$

The wedge product of two parallel vectors is zero. Thus, the wedge product of a vector with itself gives zero, $A \wedge A = 0$.

A vector space which exhibits closures under some type of product operation is called an algebra. The Clifford geometric product of two vectors A and B is defined as a combination of a scalar and a bivector and written

$$AB = A \cdot B + A \wedge B \hfill (2.7.59)$$

The left side of equation (2.7.59) is the geometric product of A and B. The right side of equation (2.7.59) has two distinctly different quantities– the scalar dot product $A \cdot B$ and the bivector wedge product $A \wedge B$. This should not seem strange as complex number also added two distinctly different quantities. We use quite often the special geometric product $AA = ||A||^2 = A \cdot A$, which is a scalar. Note also that the geometric product is not commutative since

$$BA = B \cdot A + B \wedge A = A \cdot B - A \wedge B \hfill (2.7.60)$$

By adding and subtracting the equations (2.7.59) and (2.7.60) we find that in terms of geometric products we can write

$$A \cdot B = \frac{1}{2}(AB + BA) \qquad A \wedge B = \frac{1}{2}(AB - BA) \hfill (2.7.61)$$

Euclidean 2-dimensional space E_2

In E_2 the orthonormal vectors e_1, e_2 satisfy $e_1 \cdot e_1 = 1$, $e_2 \cdot e_2 = 1$ and $e_1 \cdot e_2 = 0$. The geometric product of the orthonormal vectors $e_1 e_2$ is a bivector and given the special symbol I and called a grade-2 pseudoscalar. Using the geometric product we find

$$
\begin{aligned}
I &= e_1 e_2 = e_1 \cdot e_2 + e_1 \wedge e_2 = e_1 \wedge e_2 \\
-I &= e_2 e_1 = e_2 \cdot e_1 + e_2 \wedge e_1 = -e_1 \wedge e_2
\end{aligned}
\tag{2.7.62}
$$

In addition we find the geometric products

$$
\begin{aligned}
e_1 e_1 &= e_1 \cdot e_1 + e_1 \wedge e_1 = 1 \\
e_2 e_2 &= e_2 \cdot e_2 + e_2 \wedge e_2 = 1
\end{aligned}
\tag{2.7.63}
$$

$$\text{and} \qquad I^2 = (e_1 e_2)(e_1 e_2) = (-e_2 e_1)(e_1 e_2) = -1.$$

Hence the geometric product of the unit vectors satisfies $e_1 e_2 = -e_2 e_1$ and the bivector $I = e_1 e_2$ behaves like an imaginary unit. Note also the bivector products with unit vectors with left and right multiplication gives

$$
\begin{aligned}
(e_1 \wedge e_2) e_1 &= (e_1 e_2) e_1 = (-e_2 e_1) e_1 = -e_2 \\
(e_1 \wedge e_2) e_2 &= (e_1 e_2) e_2 = e_1 \\
e_1 (e_1 \wedge e_2) &= e_1 (e_1 e_2) = e_2 \\
e_2 (e_1 \wedge e_2) &= e_2 (e_1 e_2) = e_2 (-e_2 e_1) = -e_1.
\end{aligned}
\tag{2.7.64}
$$

These products have important physical interpretations.

1. Left multiplication of unit vectors by the bivector $e_1 \wedge e_2$ produces a rotation of $\pi/2$ radians clockwise (negative sense).
2. Right multiplication of unit vector by the bivector $e_1 \wedge e_2$ produces a rotation of $\pi/2$ radians counterclockwise (positive sense). We will later discover that bivectors are closely associated with rotations.

EXAMPLE 2.7-3

It is well known that complex numbers $a + ib$ can be represented as matrices $\begin{pmatrix} a & b \\ -b & a \end{pmatrix}$. There is also a correspondence between complex numbers $a + ib$ and vectors $Z = a e_1 + b e_2$. Observe that if we left multiply Z by e_1 we obtain $z = e_1 Z = a + b e_1 e_2 = a + bI$ where I is the grade-2 pseudoscalar which is an imaginary component. Thus, there is a one-to-one mapping between complex points and two-dimensional vectors.

∎

All elements in E_2 are then linear combinations of the basis elements

$$
\underbrace{1}_{1-scalar} \qquad \underbrace{e_1, e_2}_{2-vectors} \qquad \underbrace{e_1 e_2}_{1-bivector}
\tag{2.7.65}
$$

which have grades 0,1 and 2 respectively. The resulting quantities are called multivectors. For example,

$$
\begin{aligned}
A &= a_0 + a_1 e_1 + a_2 e_2 + a_3 e_1 e_2 \\
B &= b_0 + b_1 e_2 + b_2 e_2 + b_3 e_1 e_2
\end{aligned}
\tag{2.7.66}
$$

with real coefficients, are called multivectors. Their sum is denoted

$$A + B = (a_0 + b_0) + (a_1 + b_1)e_1 + (a_2 + b_2)e_2 + (a_3 + b_3)e_1e_2 \qquad (2.7.67)$$

and their difference is denoted

$$A - B = (a_0 - b_0) + (a_1 - b_1)e_1 + (a_2 - b_2)e_2 + (a_3 - b_3)e_1e_2 \qquad (2.7.68)$$

The product is understood to be a geometric product which produces closure. We find

$$AB = \quad (a_0 + a_1e_1 + a_2e_2 + a_3e_1e_2)(b_0 + b_1e_1 + b_2e_2 + b_3e_1e_2)$$
$$AB = \quad a_0(b_0 + b_1e_1 + b_2e_2 + b_3e_1e_2)$$
$$+a_1e_1(b_0 + b_1e_1 + b_2e_2 + b_3e_1e_2)$$
$$+a_2e_2(b_0 + b_1e_1 + b_2e_2 + b_3e_1e_2)$$
$$+a_3e_1e_2(b_0 + b_1e_1 + b_2e_2 + b_3e_1e_2).$$

Expanding using the distributive law we find

$$AB = a_0b_0 + a_0b_1e_1 + a_0b_2e_2 + a_0b_3e_1e_2$$
$$+a_1b_0e_1 + a_1b_1e_1e_1 + a_1b_2e_1e_2 + a_1b_3e_1(e_1e_2)$$
$$+a_2b_0e_2 + a_2b_1e_2e_1 + a_2b_2e_2e_2 + a_2b_3e_2(e_1e_2)$$
$$+a_3b_0e_1e_2 + a_3b_1(e_1e_2)e_1 + a_3b_2(e_1e_2)e_2 + a_3b_3(e_1e_2)(e_1e_2)$$

which simplifies to

$$AB = \quad (a_0 + a_1e_1 + a_2e_2 + a_3e_1e_2)(b_0 + b_1e_1 + b_2e_2 + b_3e_1e_2)$$
$$AB = \quad (a_0b_0 + a_1b_1 + a_2b_2 - a_3b_3)$$
$$+(a_0b_1 + a_1b_0 - a_2b_3 + a_3b_2)e_1 \qquad (2.7.69)$$
$$+(a_0b_2 + a_1b_3 + a_2b_0 - a_3b_1)e_2$$
$$+(a_0b_3 + a_1b_2 - a_2b_1 + a_3b_0)e_1e_2.$$

Multivectors satisfy

The associative law $A(BC) = (AB)C = ABC$ for multiplication

The distributive law $A(B + C) = AB + BC$

The geometric product is not commutative $AB \neq BA$

The geometric product of two vectors gives a scalar plus a bivector

$$AB = (a_1e_1 + a_2e_2)(b_1e_1 + b_2e_2)$$
$$AB = a_1b_1e_1e_1 + a_2b_2e_2e_2 + a_1b_2e_1e_2 + a_2b_1e_2e_1$$
$$AB = (a_1b_1 + a_2b_2) + (a_1b_2 - a_2b_1)e_1e_2$$

We have found that the pseudoscalar I satisfies $I^2 = -1$, so I is treated as an imaginary unit. The special multivector $\cos\theta + \sin\theta I$ is written in the exponential form $e^{I\theta}$.

Euclidean 3-dimensional space E_3

In E_3 the orthonormal vectors e_1, e_2, e_3 satisfy $e_i \cdot e_j = \delta_{ij}$ $\quad i, j = 1, 2, 3$ and can be used to create bivectors $e_1 e_2, e_3 e_1, e_2 e_3$ of grade-2 and a trivector $I = e_1 e_2 e_3$. The trivector I is called a grade-3 pseudoscalar. All elements in E_3 are linear combinations of the basis elements

$$\underbrace{1}_{1-scalar} \qquad \underbrace{e_1, e_2, e_3}_{3-vectors} \qquad \underbrace{e_1 e_2, e_3 e_1, e_2 e_3}_{3-bivectors} \qquad \underbrace{e_1 e_2 e_3}_{1-trivector} \tag{2.7.70}$$

of grades 0,1,2 and 3 respectively. Algebraic quantities are denoted by the multivector

$$A = a_0 + a_1 e_1 + a_2 e_2 + a_3 e_3 + a_4 e_2 e_3 + a_5 e_3 e_1 + a_6 e_1 e_2 + a_7 e_1 e_2 e_3$$

which has eight components. The product of two such multivectors would involve geometric products of

scalars with scalars

scalars with vectors.

scalars with bivectors.

scalars with a trivector.

vectors with vectors.

vectors with bivectors.

vectors with trivectors.

bivectors with bivectors.

bivectors with a trivector.

trivector with trivector.

In order to calculate these various products we examine fundamental multiplication properties of the basis elements. Define the bivectors

$$i_1 = e_2 e_3 = -e_3 e_2 = e_2 \wedge e_3$$
$$i_2 = e_3 e_1 = -e_1 e_3 = e_3 \wedge e_1 \tag{2.7.71}$$
$$i_3 = e_1 e_2 = -e_2 e_1 = e_1 \wedge e_2$$

and note the cyclical property of the indices. Also observe

$$i_1^2 = (e_2 e_3)(e_2 e_3) = -(e_2 e_3)(e_3 e_2) = -1$$
$$i_2^2 = (e_3 e_1)(e_3 e_1) = -(e_3 e_1)(e_1 e_3) = -1 \tag{2.7.72}$$
$$i_3^2 = (e_1 e_2)(e_1 e_2) = -(e_2 e_1)(e_1 e_2) = -1$$

so that i_1, i_2, i_3 behave like imaginary quantities. Observe that Hamilton's $\imath_1, \imath_2, \imath_3$, used in quaternions are the negatives of i_1, i_2, i_3 listed above. This can be seen by checking the multiplication properties (2.7.2). Attention is directed toward the special multivector given by $q = q_0 + q_1 i_1 + q_2 i_2 + q_3 i_3$ which is called a quaternion. This form of the quaternion changes the multiplication properties. For example, using i_1, i_2, i_3 instead of $\imath_1, \imath_2, \imath_3$, the equation (2.7.14) becomes

$$q_1 q_2 = (s_1, \vec{\alpha})(s_2, \vec{\beta}) = (s_1 s_2 - \vec{\alpha} \cdot \vec{\beta}, s_1 \vec{\beta} + s_2 \vec{\alpha} - \vec{\alpha} \times \vec{\beta}). \tag{2.7.73}$$

The proof is left as an exercise. We discover that quaternions are elements from a Clifford algebra and are really multivectors involving scalars and bivectors.

370

The grade-3 pseudoscalar $I = e_1e_2e_3$ satisfies

$$I^2 = (e_1e_2e_3)(e_1e_2e_3) = -(e_1e_2e_3)(e_1e_3e_2) = (e_1e_2e_3)(e_3e_1e_2) = -(e_1e_2e_3)(e_3e_2e_1) = -1$$

so the trivector I also acts like an imaginary quantity. Again the special multivector $\cos\theta + \sin\theta I$ is written in the exponential form $e^{I\theta}$.

The product of trivectors with unit vectors gives

$$\begin{aligned}
e_1I = Ie_1 &= (e_1e_2e_3)e_1 = -e_1e_2e_1e_3 = e_1e_1e_2e_3 = e_2e_3 = i_1 \\
e_2I = Ie_2 &= (e_1e_2e_3)e_2 = -e_1e_2e_2e_3 = -e_1e_3 = e_3e_1 = i_2 \\
e_3I = Ie_3 &= (e_1e_2e_3)e_3 = e_1e_2 = i_3.
\end{aligned}$$
(2.7.74)

The product of the trivector with bivectors gives

$$\begin{aligned}
i_1I = Ii_1 &= (e_1e_2e_3)(e_2e_3) = -(e_1e_2e_3)(e_3e_2) = -e_1 \\
i_2I = Ii_2 &= (e_1e_2e_3)(e_3e_1) = e_1e_2e_1 = -e_2 \\
i_3I = Ii_3 &= (e_1e_2e_3)(e_1e_2) = (e_1e_3e_2)(e_2e_1) = e_1e_3e_1 = -e_3.
\end{aligned}$$
(2.7.75)

The operation of multiplying a vector by the grade-3 pseudoscalar I is called a duality transformation. It has the effect of interchanging planes and normals to the plane. For example, if $n = n_je_j$ is a vector, then $nI = n_je_jI = n_ji_j = N$ is a bivector, which represents a plane area associated with the given vector. Conversely, $-NI = -n_ji_jI = n_je_j$ gives back the original vector.

EXAMPLE 2.7-4

A vector V with origin on a plane can be resolved into components perpendicular and parallel to the plane. Let the component $V_\perp$ be normal to the plane and the component $V_\parallel$ be parallel to the plane as illustrated in the figure 2.7-8.

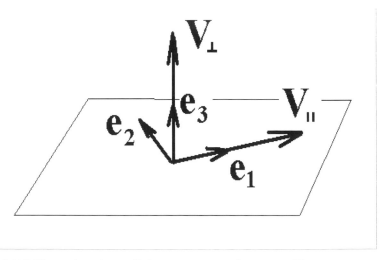

Figure 2.7-8 Normal and parallel components of a vector V.

Without loss of generality, we can assume the e_1 axis aligns with the parallel component, then $v_\perp = \alpha e_3$ and $V_\| = \beta e_1$ for some scalars α and β. We then have

$$V_\perp(e_1 \wedge e_2) = \alpha e_2(e_1 e_2) = \alpha I$$
$$(e_1 \wedge e_2)V_\perp = (e_1 e_2)(\alpha e_3) = \alpha I$$
$$V_\|(e_1 \wedge e_2) = (\beta e_1)(e_1 e_2) = \beta e_2$$
$$(e_1 \wedge e_2)V_\| = (e_1 e_2)\beta e_1 = -\beta e_2.$$

This shows the bivector operating on the parallel component $V_\|$ rotates the vector in the bivector plane. The bivector operating on the perpendicular component $V_\perp$ produces a trivector with both left and right multiplication.

■

EXAMPLE 2.7-5

The bivectors i_1, i_2, i_3 form an orthonormal frame of reference. Arbitrary bivectors $C = a \wedge b$ can be represented in terms of the components of the orthonormal frame.

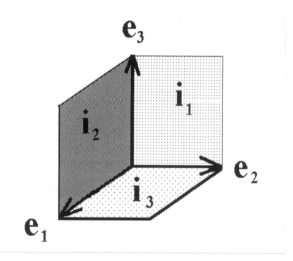

Figure 2.7-9 Orthonormal frame of reference

The cross product of two vectors produces another vector, for example

$$c = a \times b = (a_i e_i) \times (b_j e_j) = e_{ijk} a_j b_k e_i$$
$$= (a_2 b_3 - b_2 a_3)e_1 + (a_3 b_1 - a_1 b_3)e_2 + (a_1 b_2 - a_2 b_1)e_3$$

The wedge product of two vectors gives a parallelogram which can be expressed in terms of the orthonormal references i_1, i_2, i_3 since

$$C = a \wedge b = (a_1 e_1 + a_2 e_2 + a_3 e_3) \wedge (b_1 e_1 + b_2 e_2 + b_3 e_3)$$
$$C = a \wedge b = a_1 b_2\, e_1 \wedge e_2 + a_1 b_3\, e_1 \wedge e_3 + a_2 b_1\, e_2 \wedge e_1 + a_2 b_3\, e_2 \wedge e_3 + a_3 b_1\, e_3 \wedge e_1 + a_3 b_2\, e_3 \wedge e_2$$
$$C = a \wedge b = (a_2 b_3 - b_2 a_3)\, e_2 \wedge e_3 + (a_3 b_1 - a_1 b_3)\, e_3 \wedge e_1 + (a_1 b_2 - a_2 b_1)\, e_1 \wedge e_2$$
$$C = a \wedge b = (a_2 b_3 - b_2 a_3)\, i_1 + (a_3 b_1 - a_1 b_3)\, i_2 + (a_1 b_2 - a_2 b_1)\, i_3$$

To summarize, the wedge product behaves much like a cross product. In index notation

$$c = a \times b \qquad\qquad C = a \wedge b$$

$$c_i = e_{ijk} a_j b_k \qquad\qquad C_i = e_{ijk} a_j b_k$$

$$\text{or} \quad c = c_m e_m \qquad\qquad \text{or} \quad C = C_m i_m$$

Observe that by using the grade-3 pseudoscalar I we have

$$CI = c_1 i_1 I + c_2 i_2 I + c_3 i_3 I = -c_1 e_1 - c_2 e_2 - c_3 e_3.$$

This shows that the cross product can be written in terms of a wedge product by writing

$$a \times b = -I(a \wedge b) = -(a \wedge b)I \qquad \text{or} \qquad a \wedge b = (a \times b)I. \tag{2.7.76}$$

∎

Analogous to representing quaternions by a 4-tuple $(q_0, \vec{q})$, we denote multivectors by the 8-tuple

$$\mathcal{A} = (a_0, \vec{a}, \vec{A}, \alpha)$$

$$\begin{aligned}
\text{where} \quad a_0 &= \text{scalar component} \\
\vec{a} &= \text{vector components } (a_1, a_2, a_3) \\
\vec{A} &= \text{bivector components } (A_1, A_2, A_3) \\
\alpha &= \text{trivector component}
\end{aligned} \tag{2.7.77}$$

with the basis elements given by equation (2.7.70) understood. The geometric product of two multivectors can, with some effort, be written using the above notation. We find that

$$\mathcal{C} = \mathcal{AB} = (a_0, \vec{a}, \vec{A}, \alpha)(b_0, \vec{b}, \vec{B}, \beta) = (c_0, \vec{c}, \vec{C}, \gamma)$$

$$\begin{aligned}
\text{where} \quad c_0 &= a_0 b_0 + \vec{a} \cdot \vec{b} - \vec{A} \cdot \vec{B} - \alpha\beta \\
\vec{c} &= a_0 \vec{b} + b_a \vec{a} - \beta \vec{A} - \alpha \vec{B} - \vec{a} \times \vec{B} + \vec{b} \times \vec{A} \\
\vec{C} &= a_0 \vec{B} + b_0 \vec{A} + \alpha \vec{b} + \beta \vec{a} + \vec{a} \times \vec{b} - \vec{A} \times \vec{B} \\
\gamma &= a_0 \beta + b_0 \alpha + \vec{a} \cdot \vec{B} + \vec{A} \cdot \vec{b}
\end{aligned} \tag{2.7.78}$$

where dot products and cross products are the same as in ordinary vector algebra. These results can also be expressed using the index notation with $i, j, k = 1, 2, 3$

$$\begin{aligned}
c_0 &= a_0 b_0 + a_j b_j - A_j B_j - \alpha\beta \\
c_i &= a_0 b_i + b_0 a_i - \beta A_i - \alpha B_i - e_{ijk} a_j B_k + e_{ijk} b_j A_k \\
C_i &= a_0 B_i + b_0 A_i + \alpha b_i + \beta a_i + e_{ijk} a_j b_k - e_{ijk} A_j B_k \\
\gamma &= a_0 \beta + b_0 \alpha + a_j B_j + A_j b_j.
\end{aligned} \tag{2.7.79}$$

EXAMPLE 2.7-6

Using the results from equation (2.7.78) we can check our quaternion multiplication, since quaternions are special multivectors. We find

$$\mathcal{C} = \mathcal{AB} = (a_0, 0, \vec{A}, 0)(b_0, 0, \vec{B}, 0) = (a_0 b_0 - \vec{A} \cdot \vec{B}, 0, a_0 \vec{B} + b_0 \vec{A} - \vec{A} \times \vec{B}, 0)$$

which agrees with our previous concepts of quaternion multiplication. (See (2.7.73)).

∎

EXAMPLE 2.7-7

Using the results from equation (2.7.78) we can check our results for multiplication in E_2, since the E_2 multivectors are a subset of the E_3 multivectors. We find that in E_3 with $\vec{a} = (a_1, a_2, 0)$, $\vec{b} = (b_1, b_2, 0)$, $\vec{A} = (0, 0, a_3)$ and $\vec{B} = (0, 0, b_3)$ that

$$\mathcal{AB} = (a_0, \vec{a}, \vec{A}, 0)(b_0, \vec{b}, \vec{B}, 0) = (c_0, \vec{c}, \vec{C}, \gamma)$$

with
$$c_0 = a_0 b_0 + \vec{a} \cdot \vec{b} - \vec{A} \cdot \vec{B}$$
$$c_0 = a_0 b_0 + a_1 b_1 + a_2 b_2 - a_3 b_3$$
$$\vec{c} = a_0 \vec{b} + b_0 \vec{a} - \vec{a} \times \vec{B} + \vec{b} \times \vec{A}$$
$$\vec{c} = (a_0 b_1 + a_1 b_0 - a_2 b_3 + a_3 b_2) e_1$$
$$+ (a_0 b_2 + a_1 b_3 + a_2 b_0 - a_3 b_1) e_2$$
$$\vec{C} = a_0 \vec{B} + b_0 \vec{A} + \vec{a} \times \vec{b} - \vec{A} \times \vec{B}$$
$$\vec{C} = (a_0 b_3 + a_1 b_2 - a_2 b_1 + a_3 b_0) i_3$$
$$\gamma = \vec{a} \cdot \vec{B} + \vec{A} \cdot \vec{b} = 0$$

which agrees with our previous result.

∎

The inverse of a multivector element A is denoted A^{-1}. If $AA^{-1} = 1$, then A^{-1} is called a right inverse. If $A^{-1}A - 1$, then A^{-1} is called a left inverse. Not all multivectors have inverses. A simple multivector, called a blade, is defined as a quantity which can be written as the geometric product of anticommuting vectors. For A a product of m vectors, $A = A_1 A_2 A_3 \cdots A_m$, with $A_j A_k = -A_k A_j$ for $j \neq k$, for j, k over the range $j, k = 1, \ldots, m$, then A is a simple multivector. The reverse of A is denoted $\tilde{A}$ and defined as

$$\tilde{A} = A_m \cdots A_3 A_2 A_1. \tag{2.7.80}$$

For A a simple nonzero multivector, the product $\tilde{A}A$ is the scalar product of the norms squared of each vector. In this special case one can construct an inverse element since $(\tilde{A}A)A^{-1} = \tilde{A}(AA^{-1}) = \tilde{A}$, which gives $A^{-1} = \frac{\tilde{A}}{\tilde{A}A}$. For a single vector A_1, its inverse is given by $A_1^{-1} = \frac{A_1}{||A_1||^2}$. In the special case $A = e$ is a single unit vector, then one can write $e^{-1} = e$. The reversing the order of vectors, bivectors and trivectors gives the result that

$$\text{if} \quad \mathcal{A} = \underbrace{a_0}_{scalar} + \underbrace{\vec{a}}_{vector} + \underbrace{\vec{A}}_{bivector} + \underbrace{\alpha I}_{trivector}$$
$$\text{then} \quad \tilde{A} = a_0 + \vec{a} - \vec{A} - \alpha I \tag{2.7.81}$$

Rotations and Reflections

Consider the reflection of a vector V in a plane perpendicular to a given unit vector α which creates the reflected image $\bar{V}$ as illustrated in the figure 2.7-10.

The unit vector α and vector V determine a plane. In this plane we let $\alpha_\perp$ denote a unit vector perpendicular to α to form a right-hand system. The vector V can now be written in the form

$$V = v_1 \alpha + v_2 \alpha_\perp$$

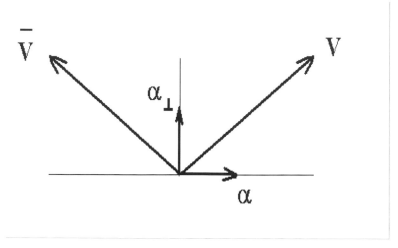

Figure 2.7-10 Reflection of vector V

where V_1 is the projection of V on α and V_2 is the projection of V on $\alpha_\perp$. The reflected image $\bar{V}$ is given by

$$\bar{V} = -V_1\,\alpha + V_2\,\alpha_\perp$$

where $V_1 = v \cdot \alpha$ is the projection of V on α and by vector addition $V_2\,\alpha_\perp = V - (V \cdot \alpha)\alpha$. In terms of geometric products, we can write,

$$V_2\alpha_\perp = V - (V \cdot \alpha)\alpha = (V\alpha - V \cdot \alpha)\alpha = (V \wedge \alpha)\,\alpha$$

The reflected image $\bar{V}$ can now be represented as a geometric product since

$$\begin{aligned}
\bar{V} &= -v_2\,\alpha + V_2\,\alpha_\perp = -(V \cdot \alpha)\,\alpha + (V \wedge \alpha)\,\alpha \\
\bar{V} &= -(\alpha \cdot V + \alpha \wedge V)\,\alpha = -\alpha V \alpha
\end{aligned} \tag{2.7.82}$$

This represents a reflection in a plane perpendicular to the unit vector α. This result has the same form as our quaternion rotation developed earlier. Consider a second reflection being performed on $\bar{V}$, in a plane orthogonal to a given unit vector β, to create a reflected image $\bar{\bar{V}}$. This second reflection is denoted

$$\bar{\bar{V}} = -\beta\bar{V}\beta$$

so that the original vector V gets transformed by two successive reflections represented by

$$\bar{\bar{V}} = -\beta(-\alpha V \alpha)\beta = \beta\alpha V \alpha\beta. \tag{2.7.83}$$

The quantity $R = \beta\alpha$ is the geometric product of two unit vectors and is called a rotor. By definition,

$$R = \beta\alpha = \beta \cdot \alpha + \beta \wedge \alpha = \cos\theta + \beta \wedge \alpha \tag{2.7.84}$$

where θ is the angle between the unit vectors when their origins are made to coincide. The bivector $\beta \wedge \alpha$ has magnitude $|\beta|\,|\alpha|\,\sin\theta$. Define the unit right-handed bivector

$$\hat{B} = \frac{\alpha \wedge \beta}{\sin\theta}, \qquad \sin\theta \neq 0 \tag{2.7.85}$$

which satisfies

$$\hat{B}^2 = \hat{B}\hat{B} = \frac{(\alpha \wedge \beta)(\alpha \wedge \beta)}{\sin^2\theta} = \frac{(\alpha \times \beta)I(\alpha \times \beta)I}{\sin^2\theta} = -\frac{\sin^2\theta}{\sin^2\theta} = -1 \qquad (2.7.86)$$

Hence, the rotor can be written in the form

$$R = \beta\alpha = \cos\theta - \sin\theta\hat{B} = e^{-\hat{B}\theta} \qquad \text{with} \qquad \tilde{R} = \alpha\beta = e^{\hat{B}\theta} \qquad (2.7.87)$$

The transformation equation (2.7.83) can then be expressed as

$$\bar{\bar{V}} = RV\tilde{R} = e^{-\hat{B}\theta}Ve^{\hat{B}\theta} \qquad (2.7.88)$$

where $\hat{B}$ is a unit bivector. This transformation represents a positive rotation through an angle 2θ in the plane $\alpha \wedge \beta$ of the bivector $\hat{B}$. To show this we can write the vector V in the form $V = V_{\parallel} + V_{\perp}$, where $V_{\parallel}$ lies in the $\hat{B}$ plane and $V_{\perp}$ is perpendicular to the $\hat{B}$ plane. We examine the transformation equation (2.7.88) and find

$$\bar{\bar{V}} = (\cos\theta - \sin\theta\hat{B})(V_{\parallel} + V_{\perp})(\cos\theta + \sin\theta\hat{B})$$
$$\bar{\bar{V}} = (\cos\theta - \sin\theta\hat{B})(\cos\theta V_{\parallel} + \sin\theta V_{\parallel}\hat{B} + \cos\theta V_{\perp} + \sin\theta V_{\perp}\hat{B})$$
$$\bar{\bar{V}} = (\cos^2\theta + \sin^2\theta)V_{\perp} + (\cos^2\theta + 2\sin\theta\cos\theta\hat{B} + \cos^2\theta\hat{B}\hat{B}) \qquad (2.7.89)$$
$$\bar{\bar{V}} = V_{\perp} + (\cos\theta + \sin\theta\hat{B})^2 V_{\parallel}$$
$$\bar{\bar{V}} = V_{\perp} + V_{\parallel}e^{2\hat{B}\theta}$$

This shows the perpendicular component is unchanged while the projected component gets rotated through an angle of 2θ.

Conclusions

Clifford algebra fits into tensors by way of transformations. Multivectors are related to tensors by way of multilinear forms which we have previously discussed. Clifford algebras are currently being used for many applied purposes. Its use has created new incites into the structure of physics. The following is a short list of applied areas where Clifford algebras have been used.

1. Geometric representations.

2. Geometric calculus of differentiation and integration.

3. New representations of the basic laws of physics.

   (a) Classical mechanics.

   (b) Electricity and magnetism.

   (c) Quantum mechanics.

   (d) Relativity.

4. Computer graphics and simulations.

5. String theory- Theory of everything.

This ends our brief introduction into Clifford algebra. For more information see the D. Hestenes and G. Sobczyk reference in the bibliography listing.

EXERCISE 2.7

▶ **1.** For quaternions $q_1 = 2 + 4i_3$ and $q_2 = 3 + 5i_2$, find the product $q = q_1 q_2$.

▶ **2.** For the quaternion $q_1 = 3 - 4i_1 + 5i_2$ find q_1^{-1} and verify that $q_1 q_1^{-1} = 1$.

▶ **3.** Show that for unit quaternions $q_1 = \cos\alpha + \hat{e}\sin\alpha$ and $q_2 = \cos\beta + \hat{e}\sin\beta$ the product is given by $q_3 = q_1 q_2 = \cos\gamma + \hat{e}\sin\gamma$ where $\gamma = \alpha + \beta$.

▶ **4.** Assume a, b, c are known quaternions, with nonzero norms such that $||a||^2 \neq ||b||^2$.

    (a) Show that the quaternion equation $aq + qb = c$ has the solution $q = \dfrac{ac - cb}{||b||^2 - ||a||^2}$.

    (b) Show that the quaternion equation $qa + bq = c$ has the solution $q = \dfrac{ca - bc}{||a||^2 - ||b||^2}$.

▶ **5.**

    (a) Show that the quaternion $q = (q_0, \vec{q})$ with $\vec{q} = q_1 i_1 + q_2 i_2 + q_3 i_3$ must satisfy the quadratic equation $q^2 - 2q_0 q + ||q||^2 = 0$ which is called the principal equation of q.

    (b) Show q^* also satisfies this equation.

▶ **6.** For $u = \alpha_0 + \alpha_1 i_1 + \alpha_2 i_2$ and $v = \beta_0 + \beta_3 i_3$ solve the quaternion equation $uq = v$

▶ **7.** Prove that the norm of a quaternion product equals the product of the norms.

▶ **8.**

    (a) Solve $q^2 + 1 = 0$ in the complex domain and show the complex roots are $q = \pm i$, $-\pi \leq arg(q) < \pi$.

    (b) To solve $q^2 + 1 = 0$ in the quaternion domain, we set $q = (q_0, \vec{q})$ where $\vec{q} = q_1 i_1 + q_2 i_2 + q_3 i_3$, then if

$$q^2 + 1 = (q_0^2 - q_1^2 - q_2^2 - q_3^2 + 1, 2q_0\vec{q}) = 0$$

we require that $q_0^2 - q_1^2 - q_2^2 - q_3^2 + 1 = 0$ and $2q_0\vec{q} = 0$. Examine the following cases and find the roots.

        (i) $q_0 = 0$, $q_1 = \pm 1$ (iii) $q_0 = 0$, $q_3 = \pm 1$

        (ii) $q_0 = 0$, $q_2 = \pm 1$ (iv) $q_0 = e^{i(\pi/2 + 2n\pi)}$, $n = 0, 1, 2, \ldots$, $q_1 = q_2 = q_3 = 0$

This shows the quaternion equation $q^2 + 1 = 0$ can have an infinite number of solutions.

▶ **9.** For α and β given quaternions, find the solutions to the quaternion equations $\alpha q = \beta$ and $q\alpha = \beta$. Do these equations have the same solution? Under what conditions will the solutions be the same?

▶ **10.** Express the sequence of transforms ZXZ, involving the Euler angles ϕ, θ, ψ, in terms of quaternion products and Euler parameters. Express the Euler parameters in terms of the Euler angles. (Hint: See EXAMPLE 2.2-5)

▶ **11.**

    (a) For geometric algebra in Euclidean space E_4 write out the basis elements and note the relation to Pascal's triangle.

$$
\begin{array}{ccccccccccc}
 & & & & & 1 & & & & & \\
 & & & & 1 & & 1 & & & & \\
 & & & 1 & & 2 & & 1 & & & \\
 & & 1 & & 3 & & 3 & & 1 & & \\
 & 1 & & 4 & & 6 & & 4 & & 1 & \\
1 & & 5 & & 10 & & 10 & & 5 & & 1 \\
\end{array}
$$

    (b) Write out the basis elements for the 5-dimensional Euclidean space E_5.

12. The Cayley-Klein parameters $(\alpha, \beta, \gamma, \delta)$ involve only the symbol i from complex variable theory, where $i^2 = -1$, instead of the quaternions i_1, i_2, i_3. Show that the substitutions

$$q_0 = \frac{\alpha + \delta}{2}, \quad q_1 = \frac{\beta - \gamma}{2}, \quad q_2 = \frac{\beta + \gamma}{2i}, \quad q_3 = \frac{\alpha - \delta}{2i}$$

where

$$\alpha = \cos\frac{\theta}{2}e^{\frac{i}{2}(\phi+\psi)} \qquad\qquad \beta = i\sin\frac{\theta}{2}e^{\frac{i}{2}(\phi-\psi)}$$

$$\gamma = i\sin\frac{\theta}{2}e^{-\frac{i}{2}(\phi-\psi)} \qquad\qquad \delta = \cos\frac{\theta}{2}e^{-\frac{i}{2}(\phi+\psi)}$$

replaces the transformation equation (2.7.50) by

$$\begin{pmatrix} x \\ y \\ z \end{pmatrix} = \begin{pmatrix} \frac{1}{2}(\alpha^2 + \beta^2 + \gamma^2 + \delta^2) & \frac{i}{2}(-\alpha^2 - \beta^2 + \gamma^2 + \delta^2) & i(\alpha\gamma + \beta\delta) \\ \frac{i}{2}(\alpha^2 - \beta^2 + \gamma^2 - \delta^2) & \frac{1}{2}(\alpha^2 - \beta^2 - \gamma^2 + \delta^2) & \beta\delta - \alpha\gamma \\ -i(\alpha\beta + \gamma\delta) & \gamma\delta - \alpha\beta & \alpha\delta + \beta\gamma \end{pmatrix} \begin{pmatrix} X \\ Y \\ Z \end{pmatrix}$$

13. In E_2 with $x = x_1 e_1 + x_2 e_2$ and $y = y_1 e_1 + y_2 e_2$

(a) Show the bivector $x \wedge y$ is represented $x \wedge y = (x_1 y_2 - x_2 y_1)e_1 e_2$.

(b) Calculate the area of the parallelogram with sides x and y in terms of x_1, x_2, y_1, y_2.

14. For $A = e_1 + e_2 + e_3$, $B = e_1 + 2e_2 + 3e_3$ and $C = 3e_1 + 3e_2 + e_3$ find the volume associated with the trivector $A \wedge B \wedge C$.

15. Verify that the bivectors satisfy the geometric product $i_m i_n = e_{nm(p)} i_{(p)}$ (no summation on p index) and $m \neq n \neq p$.

16. For $I = e_1 e_2$, interpret $e^{I\phi} = \cos\phi + \sin\phi\, I$ as a multivector analogous to a complex variable.

(a) Show that the transformation $w = e^{I\phi} z$ represents a rotation of the vector $z = x + y\, I$ through an angle ϕ.

(b) In terms of vectors, show that the transformation $R = Xe_1 + Ye_2 = e^{I\phi}(xe_1 + ye_2)$ also represents a rotation through an angle ϕ.

17. Verify the geometric product $e_i e_j = e_i \cdot e_j + e_i \wedge e_j = \delta_{ij} + e_{ijk} e_k I$

18.

(a) Show the geometric product of three vectors $V_1 = (0, \vec{v}_1, 0, 0)$, $V_2 = (0, \vec{v}_2, 0, 0)$ and $V_3 = (0, \vec{v}_3, 0, 0)$ gives a vector plus a trivector given by

$$V_1 V_2 V_3 = (0, (\vec{v}_1 \cdot \vec{v}_2)\vec{v}_3 + \vec{v}_3 \times (\vec{v}_1 \times \vec{v}_2), 0, \vec{v}_3 \cdot (\vec{v}_1 \times \vec{v}_2))$$

(b) Show the vector component can be written as

$$(\vec{v}_1 \cdot \vec{v}_2)\vec{v}_3 - (\vec{v}_1 \cdot \vec{v}_3)\vec{v}_2 + (\vec{v}_3 \cdot \vec{v}_2)\vec{v}_1.$$

(c) Show that in terms of geometric products

$$V_1 V_2 V_3 + V_2 V_3 V_1 + V_3 V_1 V_2 - V_1 V_3 V_2 - V_3 V_2 V_1 - V_2 V_1 V_3 = (0, 0, 0, 6\vec{v}_1 \cdot (\vec{v}_2 \times \vec{v}_3))$$

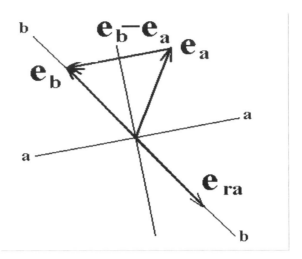

Figure 2.7-11 Rotation in terms of reflections.

▶ **19.** N-dimensional rotation. Consider the rotation of a unit vector e_a into another unit vector e_b. The rotation is achieved by a series of reflections. The first reflection is with respect to a plane a-a which is normal to the line which bisects the unit vectors e_a and e_b as illustrated in the figure 2.7-11. The second reflection is about plane perpendicular to line b-b.

(a) Show the vector to the midpoint of vector $e_b - e_a$ is given by $N = e_a + \frac{1}{2}(e_b - e_a)$.

(b) Show a unit vector in the direction of N is $n = \dfrac{e_b + e_a}{|e_b + e_a|}$.

(c) Show reflection of e_a in plane perpendicular to n gives the reflected vector $e_{ra} = -n e_a n$.

(d) Show a reflection of e_{ra} about plane perpendicular to e_b is represented $e_b = R e_a \tilde{R}$ where $R = \dfrac{1 + e_b e_a}{\sqrt{2(1 + e_b \cdot e_a)}}$ This transformation law works in N-dimensions and can be applied to objects of different grades.

▶ **20.**

(a) Let $\vec{a}$ denote a unit vector perpendicular to a plane passing through the origin and construct an arbitrary nonzero vector $\vec{x}$ from the origin. Let $\vec{x}$ be reflected to $\vec{\tilde{x}}$ with respect to the plane through the origin. Show that $\vec{\tilde{x}} = \vec{x} - 2(\vec{a} \cdot \vec{x})\vec{a}$.

(b) Show that in terms of geometric products

$$\vec{a}\vec{x} + \vec{x}\vec{a} = 2(\vec{a} \cdot \vec{x})$$

so that
$$\vec{\tilde{x}} = x - (\vec{a}\vec{x} + \vec{x}\vec{a})\vec{a}$$

$$\vec{\tilde{x}} = \vec{x} - \vec{a}\vec{x}\vec{a} - \vec{x}\vec{a}\vec{a}$$

$$\vec{\tilde{x}} = -\vec{a}\vec{x}\vec{a}$$

A transformation law having the same form as the quaternion transformation for a reflection. Thus, a rotation can be treated as two successive reflections.

21. The Dirac matrices

$$\hat{\gamma}_1 = \begin{bmatrix} 0 & 0 & 0 & 1 \\ 0 & 0 & 1 & 0 \\ 0 & 1 & 0 & 0 \\ 1 & 0 & 0 & 0 \end{bmatrix} \qquad \hat{\gamma}_2 = \begin{bmatrix} 0 & 0 & 1 & 0 \\ 0 & 0 & 0 & -1 \\ 1 & 0 & 0 & 0 \\ 0 & -1 & 0 & 0 \end{bmatrix} \qquad \hat{\gamma}_3 = \begin{bmatrix} 1 & 0 & 0 & 0 \\ 0 & 1 & 0 & 0 \\ 0 & 0 & -1 & 0 \\ 0 & 0 & 0 & -1 \end{bmatrix}$$

form the set $I, \hat{\gamma}_1, \hat{\gamma}_2, \hat{\gamma}_3, \hat{\gamma}_2\hat{\gamma}_3, \hat{\gamma}_3\hat{\gamma}_1, \hat{\gamma}_1\hat{\gamma}_2, \hat{\gamma}_1\hat{\gamma}_2\hat{\gamma}_3$ such that any linear combination of elements from this set forms an algebra under ordinary matrix multiplication. The algebra is called a Clifford algebra. Some multiple of the identity matrix I is called a grade 0-vector (scalar). A linear combination of $\hat{\gamma}_1, \hat{\gamma}_2, \hat{\gamma}_3$ is called a 1-vector. A linear combination of $\hat{\gamma}_2\hat{\gamma}_3, \hat{\gamma}_3\hat{\gamma}_1, \hat{\gamma}_1\hat{\gamma}_2$ is called a 2-vector (bivector) and a multiple of $\hat{\gamma}_1\hat{\gamma}_2\hat{\gamma}_3$ is called a 3-vector (trivector).

(a)   Show that

$$\hat{\gamma}_2\hat{\gamma}_3 = \begin{bmatrix} 0 & 0 & -1 & 0 \\ 0 & 0 & 0 & 1 \\ 1 & 0 & 0 & 0 \\ 0 & -1 & 0 & 0 \end{bmatrix} \qquad \hat{\gamma}_1\hat{\gamma}_2 = \begin{bmatrix} 0 & -1 & 0 & 0 \\ 1 & 0 & 0 & 0 \\ 0 & 0 & 0 & -1 \\ 0 & 0 & 1 & 0 \end{bmatrix}$$

$$\hat{\gamma}_3\hat{\gamma}_1 = \begin{bmatrix} 0 & 0 & 0 & 1 \\ 0 & 0 & 1 & 0 \\ 0 & -1 & 0 & 0 \\ -1 & 0 & 0 & 0 \end{bmatrix} \qquad \hat{\gamma}_1\hat{\gamma}_2\hat{\gamma}_3 = \begin{bmatrix} 0 & -1 & 0 & 0 \\ 1 & 0 & 0 & 0 \\ 0 & 0 & 0 & 1 \\ 0 & 0 & -1 & 0 \end{bmatrix}$$

(b)   Show that $\hat{\gamma}_i\hat{\gamma}_j = -\hat{\gamma}_j\hat{\gamma}_i$ for $i \neq j$.

(c)   Show the product ab of the two vectors $a = a_1\hat{\gamma}_1 + a_2\hat{\gamma}_2 + a_3\hat{\gamma}_3$ and $b = b_1\hat{\gamma}_1 + b_2\hat{\gamma}_2 + b_3\hat{\gamma}_3$ gives a scalar plus a bivector

(d)   Show the product $ba \neq ab$.

22.   The Pauli matrices

$$\hat{\gamma}_1 = \begin{bmatrix} 0 & 1 \\ 1 & 0 \end{bmatrix} \qquad \hat{\gamma}_2 = \begin{bmatrix} 0 & -i \\ i & 0 \end{bmatrix} \qquad \hat{\gamma}_3 = \begin{bmatrix} 1 & 0 \\ 0 & -1 \end{bmatrix}$$

form the set $I, \hat{\gamma}_1, \hat{\gamma}_2, \hat{\gamma}_3, \hat{\gamma}_2\hat{\gamma}_3, \hat{\gamma}_3\hat{\gamma}_1, \hat{\gamma}_1\hat{\gamma}_2, \hat{\gamma}_1\hat{\gamma}_2\hat{\gamma}_3$ such that any linear combination of elements from this set forms an algebra under ordinary matrix multiplication. The algebra is also called a Clifford algebra.

(a)   Calculate the matrices $\hat{\gamma}_2\hat{\gamma}_3, \hat{\gamma}_3\hat{\gamma}_1, \hat{\gamma}_1\hat{\gamma}_2, \hat{\gamma}_1\hat{\gamma}_2\hat{\gamma}_3$ and show that $\hat{\gamma}_i\hat{\gamma}_j = -\hat{\gamma}_j\hat{\gamma}_i$ for $i \neq j$.

(b)   Show the product ab of the two vectors $a = a_1\hat{\gamma}_1 + a_2\hat{\gamma}_2 + a_3\hat{\gamma}_3$ and $b = b_1\hat{\gamma}_1 + b_2\hat{\gamma}_2 + b_3\hat{\gamma}_3$ gives a scalar plus a bivector

(c)   Show the product $ba \neq ab$.

23.   The matrices

$$\hat{\gamma}_0 = \begin{bmatrix} 1 & 0 & 0 & 0 \\ 0 & 1 & 0 & 0 \\ 0 & 0 & -1 & 0 \\ 0 & 0 & 0 & -1 \end{bmatrix} \qquad \hat{\gamma}_2 = \begin{bmatrix} 0 & 0 & 0 & i \\ 0 & 0 & -i & 0 \\ 0 & -i & 0 & 0 \\ i & 0 & 0 & 0 \end{bmatrix}$$

$$\hat{\gamma}_1 = \begin{bmatrix} 0 & 0 & 0 & -1 \\ 0 & 0 & -1 & 0 \\ 0 & 1 & 0 & 0 \\ 1 & 0 & 0 & 0 \end{bmatrix} \qquad \hat{\gamma}_3 = \begin{bmatrix} 0 & 0 & -1 & 0 \\ 0 & 0 & 0 & 1 \\ 1 & 0 & 0 & 0 \\ 0 & -1 & 0 & 0 \end{bmatrix}$$

together with the identity matrix I are used in Minkowski 4-space. Vectors are represented $\vec{x} = ct\hat{\gamma}_0 + x^1\hat{\gamma}_1 + x^2\hat{\gamma}_2 + x^3\hat{\gamma}_3$. These matrices satisfy $\hat{\gamma}_i\hat{\gamma}_j = -\hat{\gamma}_j\hat{\gamma}_i$ for $i \neq j$. Show that $\vec{x}\vec{x} = I[(ct)^2 - (x^1)^2 - (x^2)^2 - (x^3)^2]$.

BIBLIOGRAPHY

- Abramowitz, M. and Stegun, I.A., *Handbook of Mathematical Functions*, 10th ed, New York:Dover, 1972.

- Akivis, M.A., Goldberg, V.V., *An Introduction to Linear Algebra and Tensors*, New York:Dover, 1972.

- Aris, R., *Vectors, Tensors, and the Basic Equations of Fluid Mechanics*, Englewood Cliffs, N.J.:Prentice-Hall, 1962.

- Atkin, R.J., Fox, N., *An Introduction to the Theory of Elasticity*, London:Longman Group Limited, 1980.

- Bishop, R.L., Goldberg, S.I., *Tensor Analysis on Manifolds*, New York:Dover, 1968.

- Borisenko, A.I., Tarapov, I.E., *Vector and Tensor Analysis with Applications*, New York:Dover, 1968.

- Brand, L., *Vector and Tensor Analysis*, John Wiley and Sons, 1947.

- Chorlton, F., *Vector and Tensor Methods*, Chichester,England:Ellis Horwood Ltd, 1976.

- Dodson, C.T.J., Poston, T., *Tensor Geometry*, London:Pittman Publishing Co., 1979.

- Eisenhart, L.P., *Riemannian Geometry*, Princeton, N.J.:Univ. Princeton Press, 1960.

- Eringen, A.C., *Mechanics of Continua*, Huntington, N.Y.:Robert E. Krieger, 1980.

- Flügge, W., *Tensor Analysis and Continuum Mechanics*, New York:Springer-Verlag, 1972.

- Fung, Y.C., *A First Course in Continuum Mechanics*, Englewood Cliffs,N.J.:Prentice-Hall, 1969.

- Goodbody, A.M., *Cartesian Tensors*, Chichester, England:Ellis Horwood Ltd, 1982.

- Griffiths, D.J., *Introduction to Electrodynamics*, Prentice Hall, 1981.

- Hay, G.E., *Vector and Tensor Analysis*, New York:Dover, 1953.

- Hestenes, D., Sobczyk, G., Clifford Algebra to Geometric Calculus, D.Reidel Publishing Co., Dordrecht, Holland, 1984.

- Hughes, W.F., Gaylord, E.W., *Basic Equations of Engineering Science*, New York:McGraw-Hill, 1964.

- Jeffreys, H., *Cartesian Tensors*, Cambridge, England:Cambridge Univ. Press, 1974.

- Lass, H., *Vector and Tensor Analysis*, New York:McGraw-Hill, 1950.

- Levi-Civita, T., *The Absolute Differential Calculus*, London:Blackie and Son Limited, 1954.

- Lovelock, D., Rund, H. , *Tensors, Differential Forms, and Variational Principles*, New York:Dover, 1989.

- Malvern, L.E., *Introduction to the Mechanics of a Continuous Media*, Englewood Cliffs, N.J.:Prentice-Hall, 1969.

- McConnell, A.J., *Application of Tensor Analysis*, New York:Dover, 1947.

- Newell, H.E., *Vector Analysis*, New York:McGraw Hill, 1955.

- Schouten, J.A., *Tensor Analysis for Physicists*, New York:Dover, 1989.

- Scipio, L.A., *Principles of Continua with Applications*, New York:John Wiley and Sons, 1967.

- Sokolnikoff, I.S., *Tensor Analysis*, New York:John Wiley and Sons, 1958.

- Spiegel, M.R., *Vector Analysis*, New York:Schaum Outline Series, 1959.

- Synge, J.L., Schild, A., *Tensor Calculus*, Toronto:Univ. Toronto Press, 1956.

APPENDIX A
UNITS OF MEASUREMENT

The following units, abbreviations and prefixes are from the
Système International d'Unitès   (designated SI in all Languages.)

Prefixes.

Abbreviations		
Prefix	Multiplication factor	Symbol
tera	10^{12}	T
giga	10^9	G
mega	10^6	M
kilo	10^3	K
hecto	10^2	h
deka	10	da
deci	10^{-1}	d
centi	10^{-2}	c
milli	10^{-3}	m
micro	10^{-6}	μ
nano	10^{-9}	n
pico	10^{-12}	p

Basic Units.

Basic units of measurement		
Unit	Name	Symbol
Length	meter	m
Mass	kilogram	kg
Time	second	s
Electric current	ampere	A
Temperature	degree Kelvin	$^\circ$K
Luminous intensity	candela	cd

Supplementary units		
Unit	Name	Symbol
Plane angle	radian	rad
Solid angle	steradian	sr

DERIVED UNITS		
Name	Units	Symbol
Area	square meter	m^2
Volume	cubic meter	m^3
Frequency	hertz	Hz (s^{-1})
Density	kilogram per cubic meter	kg/m^3
Velocity	meter per second	m/s
Angular velocity	radian per second	rad/s
Acceleration	meter per second squared	m/s^2
Angular acceleration	radian per second squared	rad/s^2
Force	newton	N $(kg \cdot m/s^2)$
Pressure	newton per square meter	N/m^2
Kinematic viscosity	square meter per second	m^2/s
Dynamic viscosity	newton second per square meter	$N \cdot s/m^2$
Work, energy, quantity of heat	joule	J $(N \cdot m)$
Power	watt	W (J/s)
Electric charge	coulomb	C $(A \cdot s)$
Voltage, Potential difference	volt	V (W/A)
Electromotive force	volt	V (W/A)
Electric force field	volt per meter	V/m
Electric resistance	ohm	Ω (V/A)
Electric capacitance	farad	F $(A \cdot s/V)$
Magnetic flux	weber	Wb $(V \cdot s)$
Inductance	henry	H $(V \cdot s/A)$
Magnetic flux density	tesla	T (Wb/m^2)
Magnetic field strength	ampere per meter	A/m
Magnetomotive force	ampere	A

Physical constants.

$$4 \arctan 1 = \pi = 3.14159\,26535\,89793\,23846\,2643\ldots$$

$$\lim_{n \to \infty} \left(1 + \frac{1}{n}\right)^n = e = 2.71828\,18284\,59045\,23536\,0287\ldots$$

Euler's constant $\quad \gamma = 0.57721\,56649\,01532\,86060\,6512\ldots$

$$\gamma = \lim_{n \to \infty} \left(1 + \frac{1}{2} + \frac{1}{3} + \cdots + \frac{1}{n} - \log n\right)$$

speed of light in vacuum $= 2.997925(10)^8\, m\ s^{-1}$

electron charge $= 1.60210(10)^{-19}\, C$

Avogadro's constant $= 6.02252(10)^{23}\, mol^{-1}$

Plank's constant $= 6.6256(10)^{-34}\, J\,s$

Universal gas constant $= 8.3143\, J\,K^{-1}\,mol^{-1} = 8314.3\, J\,Kg^{-1}\,K^{-1}$

Boltzmann constant $= 1.38054(10)^{-23}\, J\,K^{-1}$

Stefan–Boltzmann constant $= 5.6697(10)^{-8}\, W\,m^{-2}\,K^{-4}$

Gravitational constant $= 6.67(10)^{-11}\, N\,m^2 kg^{-2}$

APPENDIX B
CHRISTOFFEL SYMBOLS OF SECOND KIND

1. <u>**Cylindrical coordinates**</u> $(r, \theta, z) = (x^1, x^2, x^3)$

$$x = r \cos \theta \qquad r \geq 0 \qquad\qquad h_1 = 1$$
$$y = r \sin \theta \qquad 0 \leq \theta \leq 2\pi \qquad\qquad h_2 = r$$
$$z = z \qquad\qquad -\infty < z < \infty \qquad\qquad h_3 = 1$$

The coordinate curves are formed by the intersection of the coordinate surfaces

$$x^2 + y^2 = r^2, \qquad\qquad \text{Cylinders}$$
$$y/x = \tan \theta \qquad\qquad \text{Planes}$$
$$z = Constant \qquad\qquad \text{Planes.}$$

$$\left\{ \begin{matrix} 1 \\ 2\,2 \end{matrix} \right\} = -r \qquad\qquad \left\{ \begin{matrix} 2 \\ 1\,2 \end{matrix} \right\} = \left\{ \begin{matrix} 2 \\ 2\,1 \end{matrix} \right\} = \frac{1}{r}$$

2. <u>**Spherical coordinates**</u> $(\rho, \theta, \phi) = (x^1, x^2, x^3)$

$$x = \rho \sin \theta \cos \phi \qquad \rho \geq 0 \qquad\qquad h_1 = 1$$
$$y = \rho \sin \theta \sin \phi \qquad 0 \leq \theta \leq \pi \qquad\qquad h_2 = \rho$$
$$z = \rho \cos \theta \qquad\qquad 0 \leq \phi \leq 2\pi \qquad\qquad h_3 = \rho \sin \theta$$

The coordinate curves are formed by the intersection of the coordinate surfaces

$$x^2 + y^2 + z^2 = \rho^2 \qquad\qquad \text{Spheres}$$
$$x^2 + y^2 = \tan^2 \theta \, z \qquad\qquad \text{Cones}$$
$$y = x \tan \phi \qquad \text{Planes.}$$

$$\left\{ \begin{matrix} 1 \\ 2\,2 \end{matrix} \right\} = -\rho \qquad\qquad \left\{ \begin{matrix} 2 \\ 1\,2 \end{matrix} \right\} = \left\{ \begin{matrix} 2 \\ 2\,1 \end{matrix} \right\} = \frac{1}{\rho}$$

$$\left\{ \begin{matrix} 1 \\ 3\,3 \end{matrix} \right\} = -\rho \sin^2 \theta \qquad\qquad \left\{ \begin{matrix} 3 \\ 1\,3 \end{matrix} \right\} = \left\{ \begin{matrix} 3 \\ 3\,1 \end{matrix} \right\} = \frac{1}{\rho}$$

$$\left\{ \begin{matrix} 2 \\ 3\,3 \end{matrix} \right\} = -\sin \theta \cos \theta \qquad\qquad \left\{ \begin{matrix} 3 \\ 3\,2 \end{matrix} \right\} = \left\{ \begin{matrix} 3 \\ 2\,3 \end{matrix} \right\} = \cot \theta$$

3. <u>**Parabolic cylindrical coordinates**</u> $(\xi, \eta, z) = (x^1, x^2, x^3)$

$$x = \xi\eta \qquad\qquad -\infty < \xi < \infty \qquad\qquad h_1 = \sqrt{\xi^2 + \eta^2}$$

$$y = \frac{1}{2}(\xi^2 - \eta^2) \qquad -\infty < z < \infty \qquad\qquad h_2 = \sqrt{\xi^2 + \eta^2}$$

$$z = z \qquad\qquad\qquad \eta \geq 0 \qquad\qquad\qquad h_3 = 1$$

The coordinate curves are formed by the intersection of the coordinate surfaces

$$x^2 = -2\xi^2(y - \frac{\xi^2}{2}) \qquad\qquad \text{Parabolic cylinders}$$

$$x^2 = 2\eta^2(y + \frac{\eta^2}{2}) \qquad\qquad \text{Parabolic cylinders}$$

$$z = Constant \qquad\qquad \text{Planes.}$$

$$\left\{ \begin{matrix} 1 \\ 1\,1 \end{matrix} \right\} = \frac{\xi}{\xi^2 + \eta^2} \qquad\qquad \left\{ \begin{matrix} 1 \\ 2\,2 \end{matrix} \right\} = \frac{-\xi}{\xi^2 + \eta^2}$$

$$\left\{ \begin{matrix} 2 \\ 2\,2 \end{matrix} \right\} = \frac{\eta}{\xi^2 + \eta^2} \qquad\qquad \left\{ \begin{matrix} 1 \\ 1\,2 \end{matrix} \right\} = \left\{ \begin{matrix} 1 \\ 2\,1 \end{matrix} \right\} = \frac{\eta}{\xi^2 + \eta^2}$$

$$\left\{ \begin{matrix} 2 \\ 1\,1 \end{matrix} \right\} = \frac{-\eta}{\xi^2 + \eta^2} \qquad\qquad \left\{ \begin{matrix} 2 \\ 2\,1 \end{matrix} \right\} = \left\{ \begin{matrix} 2 \\ 1\,2 \end{matrix} \right\} = \frac{\xi}{\xi^2 + \eta^2}$$

4. <u>**Parabolic coordinates**</u> $(\xi, \eta, \phi) = (x^1, x^2, x^3)$

$$x = \xi\eta\cos\phi \qquad\quad \xi \geq 0 \qquad\qquad h_1 = \sqrt{\xi^2 + \eta^2}$$

$$y = \xi\eta\sin\phi \qquad\quad \eta \geq 0 \qquad\qquad h_2 = \sqrt{\xi^2 + \eta^2}$$

$$z = \frac{1}{2}(\xi^2 - \eta^2) \quad 0 < \phi < 2\pi \qquad\qquad h_3 = \xi\eta$$

The coordinate curves are formed by the intersection of the coordinate surfaces

$$x^2 + y^2 = -2\xi^2(z - \frac{\xi^2}{2}) \qquad\qquad \text{Paraboloids}$$

$$x^2 + y^2 = 2\eta^2(z + \frac{\eta^2}{2}) \qquad\qquad \text{Paraboloids}$$

$$y = x\tan\phi \qquad\qquad \text{Planes.}$$

$$\left\{ \begin{matrix} 1 \\ 1\,1 \end{matrix} \right\} = \frac{\xi}{\xi^2 + \eta^2} \qquad\qquad \left\{ \begin{matrix} 1 \\ 3\,3 \end{matrix} \right\} = \frac{-\xi\eta^2}{\xi^2 + \eta^2}$$

$$\left\{ \begin{matrix} 2 \\ 2\,2 \end{matrix} \right\} = \frac{\eta}{\xi^2 + \eta^2} \qquad\qquad \left\{ \begin{matrix} 1 \\ 2\,1 \end{matrix} \right\} = \left\{ \begin{matrix} 1 \\ 2\,1 \end{matrix} \right\} = \frac{\eta}{\xi^2 + \eta^2}$$

$$\left\{ \begin{matrix} 1 \\ 2\,2 \end{matrix} \right\} = \frac{-\xi}{\xi^2 + \eta^2} \qquad\qquad \left\{ \begin{matrix} 2 \\ 2\,1 \end{matrix} \right\} = \left\{ \begin{matrix} 2 \\ 1\,2 \end{matrix} \right\} = \frac{\xi}{\xi^2 + \eta^2}$$

$$\left\{ \begin{matrix} 2 \\ 1\,1 \end{matrix} \right\} = \frac{-\eta}{\xi^2 + \eta^2} \qquad\qquad \left\{ \begin{matrix} 3 \\ 3\,2 \end{matrix} \right\} = \left\{ \begin{matrix} 3 \\ 2\,3 \end{matrix} \right\} = \frac{1}{\eta}$$

$$\left\{ \begin{matrix} 2 \\ 3\,3 \end{matrix} \right\} = \frac{-\eta\xi^2}{\xi^2 + \eta^2} \qquad\qquad \left\{ \begin{matrix} 3 \\ 1\,3 \end{matrix} \right\} = \left\{ \begin{matrix} 3 \\ 3\,1 \end{matrix} \right\} = \frac{1}{\xi}$$

5. Elliptic cylindrical coordinates $(\xi, \eta, z) = (x^1, x^2, x^3)$

$$x = \cosh \xi \cos \eta \quad \xi \geq 0 \qquad\qquad h_1 = \sqrt{\sinh^2 \xi + \sin^2 \eta}$$

$$y = \sinh \xi \sin \eta \quad 0 \leq \eta \leq 2\pi \qquad h_2 = \sqrt{\sinh^2 \xi + \sin^2 \eta}$$

$$z = z \qquad\qquad -\infty < z < \infty \qquad h_3 = 1$$

The coordinate curves are formed by the intersection of the coordinate surfaces

$$\frac{x^2}{\cosh^2 \xi} + \frac{y^2}{\sinh^2 \xi} = 1 \qquad \text{Elliptic cylinders}$$

$$\frac{x^2}{\cos^2 \eta} - \frac{y^2}{\sin^2 \eta} = 1 \qquad \text{Hyperbolic cylinders}$$

$$z = Constant \qquad \text{Planes.}$$

$$\left\{ \begin{matrix} 1 \\ 1\,1 \end{matrix} \right\} = \frac{\sinh \xi \cosh \xi}{\sinh^2 \xi + \sin^2 \eta} \qquad\qquad \left\{ \begin{matrix} 2 \\ 2\,2 \end{matrix} \right\} = \frac{\sin \eta \cos \eta}{\sinh^2 \xi + \sin^2 \eta}$$

$$\left\{ \begin{matrix} 1 \\ 2\,2 \end{matrix} \right\} = \frac{-\sinh \xi \cosh \xi}{\sinh^2 \xi + \sin^2 \eta} \qquad\qquad \left\{ \begin{matrix} 2 \\ 1\,1 \end{matrix} \right\} = \frac{-\sin \eta \cos \eta}{\sinh^2 \xi + \sin^2 \eta}$$

$$\left\{ \begin{matrix} 1 \\ 1\,2 \end{matrix} \right\} = \left\{ \begin{matrix} 1 \\ 2\,1 \end{matrix} \right\} = \frac{\sin \eta \cos \eta}{\sinh^2 \xi + \sin^2 \eta} \qquad \left\{ \begin{matrix} 2 \\ 1\,2 \end{matrix} \right\} = \left\{ \begin{matrix} 2 \\ 2\,1 \end{matrix} \right\} = \frac{\sinh \xi \cosh \xi}{\sinh^2 \xi + \sin^2 \eta}$$

6. Elliptic coordinates $(\xi, \eta, \phi) = (x^1, x^2, x^3)$

$$x = \sqrt{(1-\eta^2)(\xi^2-1)} \cos \phi \quad 1 \leq \xi < \infty \qquad h_1 = \sqrt{\frac{\xi^2 - \eta^2}{\xi^2 - 1}}$$

$$y = \sqrt{(1-\eta^2)(\xi^2-1)} \sin \phi \quad -1 \leq \eta \leq 1 \qquad h_2 = \sqrt{\frac{\xi^2 - \eta^2}{1 - \eta^2}}$$

$$z = \xi \eta \qquad\qquad 0 \leq \phi < 2\pi \qquad h_3 = \sqrt{(1-\eta^2)(\xi^2-1)}$$

The coordinate curves are formed by the intersection of the coordinate surfaces

$$\frac{x^2}{\xi^2 - 1} + \frac{y^2}{\xi^2 - 1} + \frac{z^2}{\xi^2} = 1 \qquad \text{Prolate ellipsoid}$$

$$\frac{z^2}{\eta^2} - \frac{x^2}{1 - \eta^2} - \frac{y^2}{1 - \eta^2} = 1 \qquad \text{Two-sheeted hyperboloid}$$

$$y = x \tan \phi \quad \text{Planes}$$

$$\left\{ \begin{matrix} 1 \\ 1\,1 \end{matrix} \right\} = -\frac{\xi}{-1 + \xi^2} + \frac{\xi}{\xi^2 - \eta^2} \qquad \left\{ \begin{matrix} 2 \\ 3\,3 \end{matrix} \right\} = \frac{(-1 + \xi^2)\,\eta\,(1 - \eta^2)}{\xi^2 - \eta^2}$$

$$\left\{ \begin{matrix} 2 \\ 2\,2 \end{matrix} \right\} = \frac{\eta}{1 - \eta^2} - \frac{\eta}{\xi^2 - \eta^2} \qquad \left\{ \begin{matrix} 1 \\ 1\,2 \end{matrix} \right\} = -\frac{\eta}{\xi^2 - \eta^2}$$

$$\left\{ \begin{matrix} 1 \\ 2\,2 \end{matrix} \right\} = -\frac{\xi\,(-1 + \xi^2)}{(1 - \eta^2)\,(\xi^2 - \eta^2)} \qquad \left\{ \begin{matrix} 2 \\ 2\,1 \end{matrix} \right\} = \frac{\xi}{\xi^2 - \eta^2}$$

$$\left\{ \begin{matrix} 1 \\ 3\,3 \end{matrix} \right\} = -\frac{\xi\,(-1 + \xi^2)\,(1 - \eta^2)}{\xi^2 - \eta^2} \qquad \left\{ \begin{matrix} 3 \\ 3\,1 \end{matrix} \right\} = \frac{\xi}{-1 + \xi^2}$$

$$\left\{ \begin{matrix} 2 \\ 1\,1 \end{matrix} \right\} = \frac{\eta\,(1 - \eta^2)}{(-1 + \xi^2)\,(\xi^2 - \eta^2)} \qquad \left\{ \begin{matrix} 3 \\ 3\,2 \end{matrix} \right\} = -\frac{\eta}{1 - \eta^2}$$

7. **Bipolar coordinates** $(u, v, z) = (x^1, x^2, x^3)$

$$x = \frac{a \sinh v}{\cosh v - \cos u}, \qquad 0 \le u < 2\pi \qquad\qquad h_1^2 = h_2^2$$

$$y = \frac{a \sin u}{\cosh v - \cos u}, \qquad -\infty < v < \infty \qquad\qquad h_2^2 = \frac{a^2}{(\cosh v - \cos u)^2}$$

$$z = z \qquad\qquad -\infty < z < \infty \qquad\qquad h_3^2 = 1$$

The coordinate curves are formed by the intersection of the coordinate surfaces

$$(x - a \coth v)^2 + y^2 = \frac{a^2}{\sinh^2 v} \qquad\qquad \text{Cylinders}$$

$$x^2 + (y - a \cot u)^2 = \frac{a^2}{\sin^2 u} \qquad\qquad \text{Cylinders}$$

$$z = Constant \qquad\qquad \text{Planes.}$$

$$\begin{Bmatrix} 1 \\ 1\,1 \end{Bmatrix} = \frac{\sin u}{\cos u - \cosh v} \qquad\qquad \begin{Bmatrix} 2 \\ 1\,1 \end{Bmatrix} = \frac{\sinh v}{-\cos u + \cosh v}$$

$$\begin{Bmatrix} 2 \\ 2\,2 \end{Bmatrix} = \frac{\sinh v}{\cos u - \cosh v} \qquad\qquad \begin{Bmatrix} 1 \\ 1\,2 \end{Bmatrix} = \frac{\sinh v}{\cos u - \cosh v}$$

$$\begin{Bmatrix} 1 \\ 2\,2 \end{Bmatrix} = \frac{\sin u}{-\cos u + \cosh v} \qquad\qquad \begin{Bmatrix} 2 \\ 2\,1 \end{Bmatrix} = \frac{\sin u}{\cos u - \cosh v}$$

8. **Conical coordinates** $(u, v, w) = (x^1, x^2, x^3)$

$$x = \frac{uvw}{ab}, \qquad b^2 > v^2 > a^2 > w^2, \quad u \ge 0 \qquad\qquad h_1^2 = 1$$

$$y = \frac{u}{a}\sqrt{\frac{(v^2 - a^2)(w^2 - a^2)}{a^2 - b^2}} \qquad\qquad h_2^2 = \frac{u^2(v^2 - w^2)}{(v^2 - a^2)(b^2 - v^2)}$$

$$z = \frac{v}{b}\sqrt{\frac{(v^2 - b^2)(w^2 - b^2)}{b^2 - a^2}} \qquad\qquad h_3^2 = \frac{u^2(v^2 - w^2)}{(w^2 - a^2)(w^2 - b^2)}$$

The coordinate curves are formed by the intersection of the coordinate surfaces

$$x^2 + y^2 + z^2 = u^2 \qquad\qquad \text{Spheres}$$

$$\frac{x^2}{v^2} + \frac{y^2}{v^2 - a^2} + \frac{z^2}{v^2 - b^2} = 0, \qquad\qquad \text{Cones}$$

$$\frac{x^2}{w^2} + \frac{y^2}{w^2 - a^2} + \frac{z^2}{w^2 - b^2} = 0, \qquad\qquad \text{Cones.}$$

$$\begin{Bmatrix} 2 \\ 2\,2 \end{Bmatrix} = \frac{v}{b^2 - v^2} - \frac{v}{-a^2 + v^2} + \frac{v}{v^2 - w^2} \qquad \begin{Bmatrix} 3 \\ 2\,2 \end{Bmatrix} = \frac{w\left(-a^2 + w^2\right)\left(-b^2 + w^2\right)}{\left(b^2 - v^2\right)\left(-a^2 + v^2\right)\left(v^2 - w^2\right)}$$

$$\begin{Bmatrix} 3 \\ 3\,3 \end{Bmatrix} = -\frac{w}{v^2 - w^2} - \frac{w}{-a^2 + w^2} - \frac{w}{-b^2 + w^2} \qquad \begin{Bmatrix} 2 \\ 2\,1 \end{Bmatrix} = \frac{1}{u}$$

$$\begin{Bmatrix} 1 \\ 2\,2 \end{Bmatrix} = -\frac{u\left(v^2 - w^2\right)}{\left(b^2 - v^2\right)\left(-a^2 + v^2\right)} \qquad \begin{Bmatrix} 2 \\ 2\,3 \end{Bmatrix} = -\frac{w}{v^2 - w^2}$$

$$\begin{Bmatrix} 1 \\ 3\,3 \end{Bmatrix} = -\frac{u\left(v^2 - w^2\right)}{\left(-a^2 + w^2\right)\left(-b^2 + w^2\right)} \qquad \begin{Bmatrix} 3 \\ 3\,1 \end{Bmatrix} = \frac{1}{u}$$

$$\begin{Bmatrix} 2 \\ 3\,3 \end{Bmatrix} = -\frac{v\left(b^2 - v^2\right)\left(-a^2 + v^2\right)}{\left(v^2 - w^2\right)\left(-a^2 + w^2\right)\left(-b^2 + w^2\right)} \qquad \begin{Bmatrix} 3 \\ 3\,2 \end{Bmatrix} = \frac{v}{v^2 - w^2}$$

9. **Prolate spheroidal coordinates** $(u, v, \phi) = (x^1, x^2, x^3)$

$$x = a \sinh u \sin v \cos \phi, \quad u \geq 0 \qquad\qquad h_1^2 = h_2^2$$

$$y = a \sinh u \sin v \sin \phi, \quad 0 \leq v \leq \pi \qquad h_2^2 = a^2(\sinh^2 u + \sin^2 v)$$

$$z = a \cosh u \cos v, \quad 0 \leq \phi < 2\pi \qquad h_3^2 = a^2 \sinh^2 u \sin^2 v$$

The coordinate curves are formed by the intersection of the coordinate surfaces

$$\frac{x^2}{(a \sinh u)^2} + \frac{y^2}{(a \sinh u)^2} + \frac{z^2}{(a \cosh u)^2} = 1, \qquad \text{Prolate ellipsoids}$$

$$\frac{x^2}{(a \cos v)^2} - \frac{y^2}{(a \sin v)^2} - \frac{z^2}{(a \cos v)^2} = 1, \qquad \text{Two-sheeted hyperpoloid}$$

$$y = x \tan \phi, \qquad\qquad \text{Planes.}$$

$$\left\{ \begin{matrix} 1 \\ 1\,1 \end{matrix} \right\} = \frac{\cosh u \sinh u}{\sin^2 v + \sinh^2 u} \qquad\qquad \left\{ \begin{matrix} 2 \\ 3\,3 \end{matrix} \right\} = -\frac{\cos v \sin v \sinh^2 u}{\sin^2 v + \sinh^2 u}$$

$$\left\{ \begin{matrix} 2 \\ 2\,2 \end{matrix} \right\} = \frac{\cos v \sin v}{\sin^2 v + \sinh^2 u} \qquad\qquad \left\{ \begin{matrix} 1 \\ 1\,2 \end{matrix} \right\} = \frac{\cos v \sin v}{\sin^2 v + \sinh^2 u}$$

$$\left\{ \begin{matrix} 1 \\ 2\,2 \end{matrix} \right\} = -\frac{\cosh u \sinh u}{\sin^2 v + \sinh^2 u} \qquad\qquad \left\{ \begin{matrix} 2 \\ 2\,1 \end{matrix} \right\} = \frac{\cosh u \sinh u}{\sin^2 v + \sinh^2 u}$$

$$\left\{ \begin{matrix} 1 \\ 3\,3 \end{matrix} \right\} = -\frac{\sin^2 v \cosh u \sinh u}{\sin^2 v + \sinh^2 u} \qquad\qquad \left\{ \begin{matrix} 3 \\ 3\,1 \end{matrix} \right\} = \frac{\cosh u}{\sinh u}$$

$$\left\{ \begin{matrix} 2 \\ 1\,1 \end{matrix} \right\} = -\frac{\cos v \sin v}{\sin^2 v + \sinh^2 u} \qquad\qquad \left\{ \begin{matrix} 3 \\ 3\,2 \end{matrix} \right\} = \frac{\cos v}{\sin v}$$

10. **Oblate spheroidal coordinates** $(\xi, \eta, \phi) = (x^1, x^2, x^3)$

$$x = a \cosh \xi \cos \eta \cos \phi, \qquad \xi \geq 0 \qquad\qquad h_1^2 = h_2^2$$

$$y = a \cosh \xi \cos \eta \sin \phi, \qquad -\frac{\pi}{2} \leq \eta \leq \frac{\pi}{2} \qquad h_2^2 = a^2(\sinh^2 \xi + \sin^2 \eta)$$

$$z = a \sinh \xi \sin \eta, \qquad 0 \leq \phi \leq 2\pi \qquad h_3^2 = a^2 \cosh^2 \xi \cos^2 \eta$$

The coordinate curves are formed by the intersection of the coordinate surfaces

$$\frac{x^2}{(a \cosh \xi)^2} + \frac{y^2}{(a \cosh \xi)^2} + \frac{z^2}{(a \sinh \xi)^2} = 1, \qquad \text{Oblate ellipsoids}$$

$$\frac{x^2}{(a \cos \eta)^2} + \frac{y^2}{(a \cos \eta)^2} - \frac{z^2}{(a \sin \eta)^2} = 1, \qquad \text{One-sheet hyperboloids}$$

$$y = x \tan \phi, \qquad\qquad \text{Planes.}$$

$$\left\{ \begin{matrix} 1 \\ 1\,1 \end{matrix} \right\} = \frac{\cosh \xi \sinh \xi}{\sin^2 \eta + \sinh^2 \xi} \qquad\qquad \left\{ \begin{matrix} 2 \\ 3\,3 \end{matrix} \right\} = \frac{\cos \eta \sin \eta \cosh^2 \xi}{\sin^2 \eta + \sinh^2 \xi}$$

$$\left\{ \begin{matrix} 2 \\ 2\,2 \end{matrix} \right\} = \frac{\cos \eta \sin \eta}{\sin^2 \eta + \sinh^2 \xi} \qquad\qquad \left\{ \begin{matrix} 1 \\ 1\,2 \end{matrix} \right\} = \frac{\cos \eta \sin \eta}{\sin^2 \eta + \sinh^2 \xi}$$

$$\left\{ \begin{matrix} 1 \\ 2\,2 \end{matrix} \right\} = -\frac{\cosh \xi \sinh \xi}{\sin^2 \eta + \sinh^2 \xi} \qquad\qquad \left\{ \begin{matrix} 2 \\ 2\,1 \end{matrix} \right\} = \frac{\cosh \xi \sinh \xi}{\sin^2 \eta + \sinh^2 \xi}$$

$$\left\{ \begin{matrix} 1 \\ 3\,3 \end{matrix} \right\} = -\frac{\cos^2 \eta \cosh \xi \sinh \xi}{\sin^2 \eta + \sinh^2 \xi} \qquad\qquad \left\{ \begin{matrix} 3 \\ 3\,1 \end{matrix} \right\} = \frac{\sinh \xi}{\cosh \xi}$$

$$\left\{ \begin{matrix} 2 \\ 1\,1 \end{matrix} \right\} = -\frac{\cos \eta \sin \eta}{\sin^2 \eta + \sinh^2 \xi} \qquad\qquad \left\{ \begin{matrix} 3 \\ 3\,2 \end{matrix} \right\} = -\frac{\sin \eta}{\cos \eta}$$

11. **Toroidal coordinates** $(u, v, \phi) = (x^1, x^2, x^3)$

$$x = \frac{a \sinh v \cos \phi}{\cosh v - \cos u}, \quad 0 \le u < 2\pi$$

$$y = \frac{a \sinh v \sin \phi}{\cosh v - \cos u}, \quad -\infty < v < \infty$$

$$z = \frac{a \sin u}{\cosh v - \cos u}, \quad 0 \le \phi < 2\pi$$

$$h_1^2 = h_2^2$$

$$h_2^2 = \frac{a^2}{(\cosh v - \cos u)^2}$$

$$h_3^2 = \frac{a^2 \sinh^2 v}{(\cosh v - \cos u)^2}$$

The coordinate curves are formed by the intersection of the coordinate surfaces

$$x^2 + y^2 + \left(z - \frac{a \cos u}{\sin u}\right)^2 = \frac{a^2}{\sin^2 u}, \qquad \text{Spheres}$$

$$\left(\sqrt{x^2 + y^2} - a\frac{\cosh v}{\sinh v}\right)^2 + z^2 = \frac{a^2}{\sinh^2 v}, \qquad \text{Tores}$$

$$y = x \tan \phi, \qquad \text{planes}$$

$$\left\{\begin{matrix} 1 \\ 1\,1 \end{matrix}\right\} = \frac{\sin u}{\cos u - \cosh v}$$

$$\left\{\begin{matrix} 2 \\ 2\,2 \end{matrix}\right\} = \frac{\sinh v}{\cos u - \cosh v}$$

$$\left\{\begin{matrix} 1 \\ 2\,2 \end{matrix}\right\} = \frac{\sin u}{-\cos u + \cosh v}$$

$$\left\{\begin{matrix} 1 \\ 3\,3 \end{matrix}\right\} = \frac{\sin u \sinh v^2}{-\cos u + \cosh v}$$

$$\left\{\begin{matrix} 2 \\ 1\,1 \end{matrix}\right\} = \frac{\sinh v}{-\cos u + \cosh v}$$

$$\left\{\begin{matrix} 2 \\ 3\,3 \end{matrix}\right\} = -\frac{\sinh v (\cos u \cosh v - 1)}{\cos u - \cosh v}$$

$$\left\{\begin{matrix} 1 \\ 1\,2 \end{matrix}\right\} = \frac{\sinh v}{\cos u - \cosh v}$$

$$\left\{\begin{matrix} 2 \\ 2\,1 \end{matrix}\right\} = \frac{\sin u}{\cos u - \cosh v}$$

$$\left\{\begin{matrix} 3 \\ 3\,1 \end{matrix}\right\} = \frac{\sin u}{\cos u - \cosh v}$$

$$\left\{\begin{matrix} 3 \\ 3\,2 \end{matrix}\right\} = \frac{\cos u \cosh v - 1}{\cos u \sinh v - \cosh v \sinh v}$$

12. **Confocal ellipsoidal coordinates** $(u, v, w) = (x^1, x^2, x^3)$

$$x^2 = \frac{(a^2 - u)(a^2 - v)(a^2 - w)}{(a^2 - b^2)(a^2 - c^2)}, \qquad u < c^2 < b^2 < a^2$$

$$y^2 = \frac{(b^2 - u)(b^2 - v)(b^2 - w)}{(b^2 - a^2)(b^2 - c^2)}, \qquad c^2 < v < b^2 < a^2$$

$$z^2 = \frac{(c^2 - u)(c^2 - v)(c^2 - w)}{(c^2 - a^2)(c^2 - b^2)}, \qquad c^2 < b^2 < v < a^2$$

$$h_1^2 = \frac{(u - v)(u - w)}{4(a^2 - u)(b^2 - u)(c^2 - u)}$$

$$h_2^2 = \frac{(v - u)(v - w)}{4(a^2 - v)(b^2 - v)(c^2 - v)}$$

$$h_3^2 = \frac{(w - u)(w - v)}{4(a^2 - w)(b^2 - w)(c^2 - w)}$$

$$\left\{ \begin{matrix} 1 \\ 1\,1 \end{matrix} \right\} = \frac{1}{2\,(a^2 - u)} + \frac{1}{2\,(b^2 - u)} + \frac{1}{2\,(c^2 - u)} + \frac{1}{2\,(u - v)} + \frac{1}{2\,(u - w)}$$

$$\left\{ \begin{matrix} 2 \\ 2\,2 \end{matrix} \right\} = \frac{1}{2\,(a^2 - v)} + \frac{1}{2\,(b^2 - v)} + \frac{1}{2\,(c^2 - v)} + \frac{1}{2\,(-u + v)} + \frac{1}{2\,(v - w)}$$

$$\left\{ \begin{matrix} 3 \\ 3\,3 \end{matrix} \right\} = \frac{1}{2\,(a^2 - w)} + \frac{1}{2\,(b^2 - w)} + \frac{1}{2\,(c^2 - w)} + \frac{1}{2\,(-u + w)} + \frac{1}{2\,(-v + w)}$$

$$\left\{ \begin{matrix} 1 \\ 2\,2 \end{matrix} \right\} = \frac{\left(a^2 - u\right)\left(b^2 - u\right)\left(c^2 - u\right)(v - w)}{2\left(a^2 - v\right)\left(b^2 - v\right)\left(c^2 - v\right)(u - v)(u - w)} \qquad\qquad \left\{ \begin{matrix} 1 \\ 1\,2 \end{matrix} \right\} = \frac{-1}{2\,(u - v)}$$

$$\left\{ \begin{matrix} 1 \\ 3\,3 \end{matrix} \right\} = \frac{\left(a^2 - u\right)\left(b^2 - u\right)\left(c^2 - u\right)(-v + w)}{2\,(u - v)\left(a^2 - w\right)\left(b^2 - w\right)\left(c^2 - w\right)(u - w)} \qquad\qquad \left\{ \begin{matrix} 1 \\ 1\,3 \end{matrix} \right\} = \frac{-1}{2\,(u - w)}$$

$$\left\{ \begin{matrix} 2 \\ 1\,1 \end{matrix} \right\} = \frac{\left(a^2 - v\right)\left(b^2 - v\right)\left(c^2 - v\right)(u - w)}{2\left(a^2 - u\right)\left(b^2 - u\right)\left(c^2 - u\right)(-u + v)(v - w)} \qquad\qquad \left\{ \begin{matrix} 2 \\ 2\,1 \end{matrix} \right\} = \frac{-1}{2\,(-u + v)}$$

$$\left\{ \begin{matrix} 2 \\ 3\,3 \end{matrix} \right\} = \frac{\left(a^2 - v\right)\left(b^2 - v\right)\left(c^2 - v\right)(-u + w)}{2\,(-u + v)\left(a^2 - w\right)\left(b^2 - w\right)\left(c^2 - w\right)(v - w)} \qquad\qquad \left\{ \begin{matrix} 2 \\ 2\,3 \end{matrix} \right\} = \frac{-1}{2\,(v - w)}$$

$$\left\{ \begin{matrix} 3 \\ 1\,1 \end{matrix} \right\} = \frac{(u - v)\left(a^2 - w\right)\left(b^2 - w\right)\left(c^2 - w\right)}{2\left(a^2 - u\right)\left(b^2 - u\right)\left(c^2 - u\right)(-u + w)(-v + w)} \qquad\qquad \left\{ \begin{matrix} 3 \\ 3\,1 \end{matrix} \right\} = \frac{-1}{2\,(-u + w)}$$

$$\left\{ \begin{matrix} 3 \\ 2\,2 \end{matrix} \right\} = \frac{(-u + v)\left(a^2 - w\right)\left(b^2 - w\right)\left(c^2 - w\right)}{2\left(a^2 - v\right)\left(b^2 - v\right)\left(c^2 - v\right)(-u + w)(-v + w)} \qquad\qquad \left\{ \begin{matrix} 3 \\ 3\,2 \end{matrix} \right\} = \frac{-1}{2\,(-v + w)}$$

APPENDIX C
VECTOR IDENTITIES

The following identities assume that $\vec{A}, \vec{B}, \vec{C}, \vec{D}$ are differentiable vector functions of position while f, f_1, f_2 are differentiable scalar functions of position.

1.	$\vec{A} \cdot (\vec{B} \times \vec{C}) = \vec{B} \cdot (\vec{C} \times \vec{A}) = \vec{C} \cdot (\vec{A} \times \vec{B})$
2.	$\vec{A} \times (\vec{B} \times \vec{C}) = \vec{B}(\vec{A} \cdot \vec{C}) - \vec{C}(\vec{A} \cdot \vec{B})$
3.	$(\vec{A} \times \vec{B}) \cdot (\vec{C} \times \vec{D}) = (\vec{A} \cdot \vec{C})(\vec{B} \cdot \vec{D}) - (\vec{A} \cdot \vec{D})(\vec{B} \cdot \vec{C})$
4.	$\vec{A} \times (\vec{B} \times \vec{C}) + \vec{B} \times (\vec{C} \times \vec{A}) + \vec{C} \times (\vec{A} \times \vec{B}) = \vec{0}$
5.	$(\vec{A} \times \vec{B}) \times (\vec{C} \times \vec{D}) = \vec{B}(\vec{A} \cdot \vec{C} \times \vec{D}) - \vec{A}(\vec{B} \cdot \vec{C} \times \vec{D})$ $= \vec{C}(\vec{A} \cdot \vec{B} \times \vec{C}) - \vec{D}(\vec{A} \cdot \vec{B} \times \vec{C})$
6.	$(\vec{A} \times \vec{B}) \cdot (\vec{B} \times \vec{C}) \times (\vec{C} \times \vec{A}) = (\vec{A} \cdot \vec{B} \times \vec{C})^2$
7.	$\nabla(f_1 + f_2) = \nabla f_1 + \nabla f_2$
8.	$\nabla \cdot (\vec{A} + \vec{B}) = \nabla \cdot \vec{A} + \nabla \cdot \vec{B}$
9.	$\nabla \times (\vec{A} + \vec{B}) = \nabla \times \vec{A} + \nabla \times \vec{B}$
10.	$\nabla(f\vec{A}) = (\nabla f) \cdot \vec{A} + f\nabla \cdot \vec{A}$
11.	$\nabla(f_1 f_2) = f_1 \nabla f_2 + f_2 \nabla f_1$
12.	$\nabla \times (f\vec{A}) =)\nabla f) \times \vec{A} + f(\nabla \times \vec{A})$
13.	$\nabla \cdot (\vec{A} \times \vec{B}) = \vec{B} \cdot (\nabla \times \vec{A}) - \vec{A} \cdot (\nabla \times \vec{B})$
14.	$(\vec{A} \cdot \nabla)\vec{A} = \nabla\left(\dfrac{|\vec{A}|^2}{2}\right) - \vec{A} \times (\nabla \times \vec{A})$
15.	$\nabla(\vec{A} \cdot \vec{B}) = (\vec{B} \cdot \nabla)\vec{A} + (\vec{A} \cdot \nabla)\vec{B} + \vec{B} \times (\nabla \times \vec{A}) + \vec{A} \times (\nabla \times \vec{B})$
16.	$\nabla \times (\vec{A} \times \vec{B}) = (\vec{B} \cdot \nabla)\vec{A} - \vec{B}(\nabla \cdot \vec{A}) - (\vec{A} \cdot \nabla)\vec{B} + \vec{A}(\nabla \cdot \vec{B})$
17.	$\nabla \cdot (\nabla f) = \nabla^2 f$
18.	$\nabla \times (\nabla f) = \vec{0}$
19.	$\nabla \cdot (\nabla \times \vec{A}) = 0$
20.	$\nabla \times (\nabla \times \vec{A}) = \nabla(\nabla \cdot \vec{A}) - \nabla^2\vec{A}$

APPENDIX D
SOLUTIONS TO SELECTED EXERCISES

SELECTED SOLUTIONS EXERCISE 1.1

► 1.

(a) e_{ijn} (c) e_{smn} (e) δ_{in}

2.

(a) 3 (c) 0 (e) 0

► 3.

(a) $\vec{A} \cdot (\vec{B} \times \vec{C})$ has components $A_i e_{ijk} B_j C_k$

(c) $\vec{B}(\vec{A} \cdot \vec{C})$ has components $B_i(A_j C_j)$

► 4. Each transposition causes a sign change so that $e_{ijk} = -e_{jik} = e_{jki}$ etc.

► 5. Let $\vec{D} = \vec{B} \times \vec{C}$ and $\vec{F} = \vec{A} \times \vec{D} = \vec{A} \times (\vec{B} \times \vec{C})$ then $D_i = e_{ijk} B_j C_k$ and

$$F_n = e_{mnp} A_n D_p = e_{mnp} A_n e_{pjk} B_j C_k$$

$$F_n = e_{pjk} e_{pmn} A_n B_j C_k = (\delta_{jm}\delta_{kn} - \delta_{jn}\delta_{km}) A_n B_j C_k$$

$$F_n = B_m A_k C_k - C_m A_n B_n \quad \text{or} \quad \vec{F} = \vec{B}(\vec{A} \cdot \vec{C}) - \vec{C}(\vec{A} \cdot \vec{B})$$

► 6.

(a) $y_i = a_{ij} x_j$ and $x_j = a_{nj} z_n$ so that $y_i = a_{ij} a_{nj} z_n$

(c)

$$\begin{pmatrix} y_1 \\ y_2 \end{pmatrix} = \begin{pmatrix} a_{11} & a_{12} \\ a_{21} & a_{22} \end{pmatrix} \begin{pmatrix} a_{11} & a_{21} \\ a_{21} & a_{22} \end{pmatrix} \begin{pmatrix} z_1 \\ z_2 \end{pmatrix} = \begin{pmatrix} a_{11}a_{11} + a_{11}a_{12} & a_{11}a_{21} + a_{12}a_{22} \\ a_{21}a_{11} + a_{22}a_{12} & a_{21}a_{21} + a_{22}a_{22} \end{pmatrix} \begin{pmatrix} z_1 \\ z_2 \end{pmatrix}$$

► 7. (a) $e_{ijk}e_{jik} = -e_{ijk}e_{ijk} = -[\delta_{jj}\delta_{kk} - \delta_{jk}\delta_{jk}] = -[3 \cdot 3 - 3] = -6$

► 8.

(a)

$$A_i e_{ijk} B_j C_k = \vec{A} \cdot \vec{B} \times \vec{C}$$

$$B_j e_{jki} C_k A_i = \vec{B} \cdot \vec{C} \times \vec{A}$$

$$C_k e_{kij} A_i B_j = \vec{C} \cdot \vec{A} \times \vec{B}$$

► 9.

(b) Let $\vec{D} = \vec{B} \times \vec{C}$, $\vec{E} = \vec{C} \times \vec{A}$, $\vec{F} = \vec{A} \times \vec{B}$. Now let $\vec{H} = \vec{A} \times (\vec{B} \times \vec{C}) + \vec{B} \times (\vec{C} \times \vec{A}) + \vec{C} \times (\vec{A} \times \vec{B})$ then

$$H_i = e_{ijk} A_j D_k + e_{ijk} B_j E_k + e_{ijk} C_j F_k$$

$$H_i = e_{ijk} A_j e_{kmn} B_m C_n + e_{ijk} B_j e_{kmn} C_m A_n + e_{ijk} C_j e_{kmn} A_m B_n$$

$$H_i = e_{kij} e_{kmn} A_j B_m C_n + e_{kij} e_{kmn} B_j C_m A_n + e_{kij} e_{kmn} C_j A_m B_n$$

$$H_i = (\delta_{im}\delta_{jn} - \delta_{jm}\delta_{in}) A_j B_m C_n + (\delta_{im}\delta_{jn} - \delta_{jm}\delta_{in}) A_n B_j C_m + (\delta_{im}\delta_{jn} - \delta_{jm}\delta_{in}) A_m B_n C_j$$

$$H_i = A_j B_i C_j - A_m B_m C_i + C_i B_n A_n - A_i C_j B_j + A_i C_n B_n - B_i C_m A_m = 0$$

EXERCISE 1.1

▶ **10.** (a) $\vec{C} = (-2, -2, 1)$ (b) 7 (c) cross and dot product of vectors

▶ **11.** $\dfrac{dy_i}{dt} = a_{ij}y_j$

▶ **12.** $\nabla^2\Phi = \dfrac{\partial^2\Phi}{\partial r^2} + \dfrac{1}{r}\dfrac{\partial\Phi}{\partial r} + \dfrac{1}{r^2}\dfrac{\partial^2\Phi}{\partial\theta^2}$

▶ **13.**   86

▶ **15.** Using $\widehat{\mathbf{e}}_m = (\delta_{1m}, \delta_{2m}, \delta_{3m})$ we have $\widehat{\mathbf{e}}_i \cdot (\widehat{\mathbf{e}}_j \times \widehat{\mathbf{e}}_k) = \begin{vmatrix} \delta_{1i} & \delta_{2i} & \delta_{3i} \\ \delta_{1j} & \delta_{2j} & \delta_{3j} \\ \delta_{1k} & \delta_{2k} & \delta_{3k} \end{vmatrix} = e_{ijk}$

▶ **16.** (a) $B^{i\ell} = 0$ (b) $A_{i\ell} = 0$

▶ **17.** Sum on m, then sum each term on n. If x^i and y^i are arbitrary then let

$$x^1 = 1, \quad x^2 = 0, \quad x^3 = 0, \quad y^1 = 1, \quad y^2 = 0, \quad y^3 = 0 \quad \text{to show} \quad A_{11} = 0$$
$$x^1 = 1, \quad x^2 = 0, \quad x^3 = 0, \quad y^1 = 0, \quad y^2 = 1, \quad y^3 = 0 \quad \text{to show} \quad A_{12} = 0$$
$$x^1 = 1, \quad x^2 = 0, \quad x^3 = 0, \quad y^1 = 0, \quad y^2 = 0, \quad y^3 = 1 \quad \text{to show} \quad A_{13} = 0$$
$$x^1 = 0, \quad x^2 = 1, \quad x^3 = 0, \quad y^1 = 1, \quad y^2 = 0, \quad y^3 = 0 \quad \text{to show} \quad A_{21} = 0$$
$$x^1 = 0, \quad x^2 = 1, \quad x^3 = 0, \quad y^1 = 0, \quad y^2 = 1, \quad y^3 = 0 \quad \text{to show} \quad A_{22} = 0$$
$$x^1 = 0, \quad x^2 = 1, \quad x^3 = 0, \quad y^1 = 0, \quad y^2 = 0, \quad y^3 = 1 \quad \text{to show} \quad A_{23} = 0$$
$$x^1 = 0, \quad x^2 = 0, \quad x^3 = 1, \quad y^1 = 1, \quad y^2 = 0, \quad y^3 = 0 \quad \text{to show} \quad A_{31} = 0$$
$$x^1 = 0, \quad x^2 = 0, \quad x^3 = 1, \quad y^1 = 0, \quad y^2 = 1, \quad y^3 = 0 \quad \text{to show} \quad A_{32} = 0$$
$$x^1 = 0, \quad x^2 = 0, \quad x^3 = 1, \quad y^1 = 0, \quad y^2 = 0, \quad y^3 = 1 \quad \text{to show} \quad A_{33} = 0$$

▶ **18.**

(a) If a_{mn} is skew-symmetric, then $a_{mn} = -a_{nm}$ and $a_{mn}x^m x^n = -a_{nm}x^m x^n$ or $2a_{mn}x^m x^n = 0$ or $a_{mn}x^m x^n = 0$.

(b) If $a_{mn}x^m x^n = 0$ for all x^i, then interchange m and n to obtain

$$a_{mn}x^m x^n = 0$$
$$a_{nm}x^m x^n = 0$$

Adding these results we find $(a_{mn} + a_{nm})x^m x^n = 0$ for all x^i, hence $a_{mn} = -a_{nm}$ (See problem 17.)

▶ **19.** Let $c_{ij} = a_{im}b_{mj}$ then

$$\begin{aligned} \det C = |C| &= e_{ijk}c_{i1}c_{i2}c_{i3} \\ &= e_{ijk}(a_{im}b_{m1})(a_{jn}b_{n2})(a_{kp}b_{p3}) \\ &= e_{ijk}a_{im}a_{jn}a_{kp}b_{m1}b_{n2}b_{p3} \\ &= |A|e_{mnp}b_{m1}b_{n2}b_{p3} \\ &= |A||B| \end{aligned}$$

EXERCISE 1.1

20.

(a) Let $a^i_j = \frac{\partial u^i}{\partial s^j}$ and $b^j_m = \frac{\partial s^j}{\partial x^m}$, then $c^i_m = \frac{\partial u^i}{\partial s^j}\frac{\partial s^j}{\partial x^m} = a^i_j b^j_m$ so that

$$
\begin{aligned}
\det C = |C| = |c^i_m| &= e_{ijk}c^i_1 c^j_2 c^k_3\\
&= e_{ijk}(a^i_p b^p_1)(a^j_q b^q_2)(a^k_r b^r_3)\\
&= e_{ijk}a^i_p a^j_q a^k_r b^p_1 b^q_2 b^r_3\\
&= |A|e_{pqr}b^p_1 b^q_2 b^r_3\\
&= |A||B| \qquad \text{or}
\end{aligned}
$$

$$
\left|\frac{\partial u^i}{\partial x^m}\right| = \left|\frac{\partial u^i}{\partial s^j}\frac{\partial s^j}{\partial x^m}\right| = \left|\frac{\partial u^i}{\partial j^j}\right|\left|\frac{\partial s^j}{\partial x^m}\right|
$$

(b) In the special case $\frac{\partial x^i}{\partial \bar{x}^j}\frac{\partial \bar{x}^j}{\partial x^m} = \frac{\partial x^i}{\partial x^m} = \delta^i_m$ we obtain $J\left(\frac{\bar{x}}{x}\right)J\left(\frac{x}{\bar{x}}\right) = 1$.

21. If $a_{\ell mn}$ is completely symmetric, then $a_{\ell mn} = a_{m\ell n} = a_{\ell nm} = a_{mn\ell} = a_{nm\ell} = a_{n\ell m}$.

(i) There are $3^3 = 27$ elements.

(ii)

Case	Index relation	Elements	# terms	Independent components
1	$\ell = m = n$	$a_{111}, a_{222}, a_{333}$	3	3
2	$\ell = m \neq n$	$a_{112} = a_{121} = a_{211}$ $a_{221} = a_{212} = a_{122}$ $a_{331} = a_{213} = a_{133}$ $a_{113} = a_{131} = a_{311}$ $a_{223} = a_{232} = a_{322}$ $a_{332} = a_{323} = a_{233}$	18	6
3	$\ell \neq m \neq n$	$a_{123} = a_{132} = a_{312}$ $= a_{321} = a_{231} = a_{213}$	6	1
		Totals	27	10

22. If $b_{\ell mn}$ is completely skew-symmetric, then $b_{\ell mn} = -b_{m\ell n} = b_{mn\ell} = -b_{nm\ell} = b_{n\ell m} = -b_{\ell nm}$.

(i) There are $3^3 = 27$ elements.

(ii)

Case	Index relation	Elements	# terms	Independent components
1	$\ell = m = n$	$b_{111} = b_{222} = b_{333} = 0$	3	0
2	$\ell = m \neq n$	$b_{112} = b_{121} = b_{211} = 0$ $b_{221} = b_{212} = b_{122} = 0$ $b_{331} = b_{213} = b_{133} = 0$ $b_{113} = b_{131} = b_{311} = 0$ $b_{223} = b_{232} = b_{322} = 0$ $b_{332} = b_{323} = b_{233} = 0$	18	0
3	$\ell \neq m \neq n$	$b_{123} = -b_{132} = b_{312}$ $= -b_{321} = b_{231} = -b_{213}$	6	1
		Totals	27	1

Selecting b_{123} as the independent component we can write $b_{\ell mn} = e_{\ell mn}b_{123}$.

23. Sum on j, then sum each term on k to show $\sigma_{ij} = \sigma_{ji}$. That is σ_{ij} is symmetric.

EXERCISE 1.1

▶ **24.** See exercise 1.1, number 17. Write $C_{mn} = A_{mn} - B_{mn}$ and show $C_{mn} = 0$.

▶ **25.** If $B_{mn} = B_{nm}$ and $B_{mn}x^m x^n = 0$ then $B_{nm}x^n x^m = 0$. Adding these results and use number 17 above to show $2B_{mn} = 0$ or $B_{mn} = 0$.

▶ **26.**

(a) $\delta^{ijk}_{mnp} = e^{ijk}e_{mnp}$ hence $\delta^{123}_{mnp} = e_{mnp}$

(c) $\delta^{ij}_{mn} = \begin{vmatrix} \delta^i_m & \delta^i_n \\ \delta^j_m & \delta^j_n \end{vmatrix} = \delta^i_m \delta^j_n - \delta^i_n \delta^j_m = e^{ij}e_{mn}$ since permutation of rows and columns

(e) $\delta^{rst}_{pst} = e^{rst}e_{tps} = \delta^r_p \delta^s_s - \delta^r_s \delta^s_p = 3\delta^r_p - \delta^r_p = 2\delta^r_p$

▶ **27.**

(a) Cofactor of a^r_i is $A^i_r = \frac{1}{2!}e^{ijk}e_{rst}a^s_j a^t_k$, therefore

$$
\begin{aligned}
e^{rst}A^i_r &= e^{rst}\frac{1}{2}e^{ijk}e_{rmn}a^m_j a^t_k \\
&= \frac{1}{2}e^{ijk}\left(\delta^s_m \delta^t_n - \delta^s_n \delta^t_m\right)a^m_j a^n_k \\
&= \frac{1}{2}e^{ijk}(a^s_j a^t_k - a^t_j a^s_k) \\
&= \frac{1}{2}(e^{ijk}a^s_j a^t_k - e^{ikj}a^t_k a^s_j) = e^{ijk}a^s_j a^t_k
\end{aligned}
$$

▶ **28.**

(b) There are N^3 elements. Consider the cases given below.

Case	Index relation	Elements	# terms	Independent components
1	$i = j = k$	$A_{(i)(i)(i)} = A_{(i)(i)(i)}$	N	N
2	$i = j \neq k$	$A_{(i)(i)(k)} = A_{(i)(i)(k)}$	$N(N-1)$	$N(N-1)$
3	$i = k \neq j$	$A_{(i)j(i)} = A_{(i)(i)j}$	$2N(N-1)$	$N(N-1)$
4	$i \neq j \neq k$	$A_{ijk} = A_{jik}$	$N(N-1)(N-2)$	$N(N-1)(N-2)/2$
		Totals	N^3	$\frac{1}{2}N^2(N+1)$

▶ **29.**   0

▶ **32.**

$$
\begin{aligned}
A_{i1} &= e_{ijk}a_{j2}a_{k3} \\
&= \frac{1}{2}[e_{123}e_{ijk}a_{j2}a_{k3} + e_{132}e_{ikj}a_{j2}a_{k3}] \\
&= \frac{1}{2}[e_{123}e_{ijk}a_{j2}a_{k3} + e_{132}e_{ijk}a_{k2}a_{j3}] \\
&= \frac{1}{2}[e_{12m}e_{ijk}a_{j2}a_{km} + e_{13m}e_{ijk}a_{j3}a_{km}] \\
&= \frac{1}{2!}e_{1nm}e_{ijk}a_{jn}a_{km}
\end{aligned}
$$

▶ **33.**

$$
\begin{aligned}
a_{pm}A_{im} &= \frac{1}{2}e_{mst}e_{ijk}a_{js}a_{kt}a_{pm} \\
&= \frac{1}{2}e_{ijk}(e_{mst}a_{pm}a_{js}a_{kt}) = \frac{1}{2}e_{ijk}e_{pjk}|A| \\
&= \frac{1}{2}e_{kij}e_{kpj}|A| = \frac{1}{2}(\delta_{ip}\delta_{jj} - \delta_{jp}\delta_{ij})|A| \\
&= \frac{1}{2}(3\delta_{ip} - \delta_{ip})|A| = \delta_{ip}|A|
\end{aligned}
$$

EXERCISE 1.1

34. 33

35. 64

36. 83

37.

$$(a) \quad 4A_{ijk\ell}x_ix_jx_kx_\ell \qquad\qquad (b) \quad 3P_{ijk}x^ix^jx^k$$

$$(c) \quad \frac{\partial x^i}{\partial x^j} = \delta^i_j \qquad\qquad (d) \quad 0$$

38. (a) Permutation of columns (b) Permutation of rows and columns

39. Consider permutation of rows and columns of $\begin{vmatrix} \delta^1_1 & \delta^1_2 & \delta^1_3 \\ \delta^2_1 & \delta^2_2 & \delta^2_3 \\ \delta^3_1 & \delta^3_2 & \delta^3_3 \end{vmatrix}$

40. From number 39

$$e^{ijk}e_{mnp}A^{mnp} = \begin{vmatrix} \delta^1_1 & \delta^1_2 & \delta^1_3 \\ \delta^2_1 & \delta^2_2 & \delta^2_3 \\ \delta^3_1 & \delta^3_2 & \delta^3_3 \end{vmatrix} A^{mnp}$$

$$= [\delta^i_m(\delta^j_n\delta^k_p - \delta^k_n\delta^j_p) - \delta^i_n(\delta^j_m\delta^k_p - \delta^k_m\delta^j_p) + \delta^i_p(\delta^j_m\delta^k_n - \delta^k_m\delta^j_n)]A^{mnp}$$

$$= A^{ijk} - A^{ikj} + A^{kij} - A^{jik} + A^{jki} - A^{kji}$$

41. (d) n!

42. True

43. $|A| = e_{ijk}a_{1i}a_{2j}a_{3k}$ and

$$\frac{d|A|}{dt} = e_{ijk}[\frac{da_{1i}}{dt}a_{2j}a_{3k} + a_{1i}\frac{da_{2j}}{dt}a_{3k} + a_{1i}a_{2j}\frac{da_{3k}}{dt}]$$

$$\frac{d|A|}{dt} = \begin{vmatrix} \frac{da_{11}}{dt} & \frac{da_{12}}{dt} & \frac{da_{13}}{dt} \\ a_{21} & a_{22} & a_{23} \\ a_{31} & a_{32} & a_{33} \end{vmatrix} + \begin{vmatrix} a_{11} & a_{12} & a_{13} \\ \frac{da_{21}}{dt} & \frac{da_{22}}{dt} & \frac{da_{23}}{dt} \\ a_{31} & a_{32} & a_{33} \end{vmatrix} + \begin{vmatrix} a_{11} & a_{12} & a_{13} \\ a_{21} & a_{22} & a_{23} \\ \frac{da_{31}}{dt} & \frac{da_{32}}{dt} & \frac{da_{33}}{dt} \end{vmatrix}$$

44. $\operatorname{grad}\phi = \phi_{,i} = \frac{d\phi}{df}f_{,i} = \frac{d\phi}{df}\frac{\partial f}{\partial x^i}$

45.

(b) Let $\vec{B} = \nabla \times \vec{A}$ with components $B_i = e_{ijk}A_{k,j}$, then

$$\nabla \cdot \vec{B} = \nabla \cdot \nabla \times \vec{A} = (e_{ijk}A_{k,j})_{,i} = e_{ijk}A_{k,ji}$$

$$\text{sum on k} \quad \nabla \cdot \vec{B} = e_{ij1}A_{1,ji} + e_{ij2}A_{2,ji} + e_{ij3}A_{3,ji}$$

$$\text{sum on j} \quad \nabla \cdot \vec{B} = e_{i21}A_{1,2i} + e_{i31}A_{1,3i} + e_{i12}A_{2,1i}$$
$$+ e_{i32}a_{2,3i} + e_{i13}A_{3,1i} + e_{i23}A_{3,2i}$$

$$\text{sum on i} \quad \nabla \cdot \vec{B} = e_{321}A_{1,23} + e_{231}A_{1,32} + e_{312}A_{2,13}$$
$$+ e_{132}A_{2,31} + e_{213}A_{3,12} + e_{123}A_{3,21} = 0$$

46. $C = 0$

47.

$$\frac{\partial \bar{A}_{mn}}{\partial \bar{x}^k} = A_{ij}\frac{\partial x^i}{\partial \bar{x}^m}\frac{\partial^2 x^j}{\partial \bar{x}^m \partial \bar{x}^k} + A_{ij}\frac{\partial^2 x^i}{\partial \bar{x}^m \partial \bar{x}^k}\frac{\partial x^j}{\partial \bar{x}^n} + \left(\frac{\partial A_{ij}}{\partial x^q}\frac{\partial x^q}{\partial \bar{x}^k}\right)\frac{\partial x^i}{\partial \bar{x}^m}\frac{\partial x^j}{\partial \bar{x}^n}$$

EXERCISE 1.1

▶ **48.** Let $|A| = e_{ijk}a_{i1}a_{j2}a_{k3}$ Now interchange rows i and j, to obtain

$$e_{ijk}a_{j1}a_{i2}a_{k3} = -e_{jik}a_{j1}a_{i2}a_{k3} = -e_{ijk}a_{i1}a_ja_{k3} = -|A|$$

▶ **51.** $e_{ijk}A_{i\ell}A_{jm}A_{kn} = |A|e_{\ell mn}$ therefore $\phi = |A|e_{\ell mn}e_{\ell mn} = 6|A|$.

▶ **52.** (a) Eight components. (b) Four zero components and 2 independent components.

▶ **53.** Twenty-seven components all are zero.

▶ **54.** If $T_{ij} = A_{ij} + B_{ij}$ where $A_{ji} = A_{ij}$ and $B_{ji} = -B_{ij}$, then one can write

$$T_{ji} = A_{ji} + B_{ji} = A_{ij} - B_{ij}$$

Therefore by addition $\quad A_{ij} = \dfrac{1}{2}(T_{ij} + T_{ji})$

and by subtraction $\quad B_{ij} = \dfrac{1}{2}(T_{ij} - T_{ji}).$

Hence one can always write $T_{ij} = \dfrac{1}{2}(T_{ij} + T_{ji}) + \dfrac{1}{2}(T_{ij} - T_{ji}).$

▶ **55.** (a) There are $4^3 = 64$ components. (b) If $A_{ijk} = -A_{ikj}$, then consider the cases given below.

Case	Index relation	Elements	# terms	Independent components
1	$i = j = k$	$A_{(i)(i)(i)} = 0$	4	0
2	$i = j \neq k$	$A_{(i)(i)(k)} = -A_{(i)(k)(i)}$	24	12
3	$i \neq j = k$	$A_{(i)(j)(j)} = -A_{(i)(j)(j)}$	12	0
4	$i \neq j \neq k$	$A_{ijk} = -A_{jik}$	24	12
		Totals	64	24

(c) If in addition $A_{ijk} + A_{jki} + A_{kij} = 0$ then there are 20 independent A_{ijk} terms. i.e. case 4 above reduces from 12 to 8 independent terms.

▶ **56.** (a) $x_i = x_i^0 + tA_i$ (b) $n_i(x_i - x_i^0) = 0$

(c) $\dfrac{x_1 - x_1^0}{A_1} = \dfrac{x_2 - x_2^0}{A_2} = \dfrac{x_3 - x_3^0}{A_3}$ (d) $n_1(x_1 - x_1^0) + n_2(x_2 - x_2^0) + n_3(x_3 - x_3^0) = 0$

There are also the parametric representation (a) $x = x_0 + tA_1$, $y = y_0 + tA_2$, $z = z_0 + tA_3$ where $x_0 = x_1^0, y_0 = x_2^0, z_0 = x_3^0$

(d) $x = x(u,v) = \alpha_1 u + \beta_1 v + \gamma_1$, $y = y(u,v) = \alpha_2 u + \beta_2 v + \gamma_2$, $z = z(u,v) = \alpha_3 u + \beta_3 v + \gamma_3$

Parametric representations are not unique. (e) One representation of the solution is

$$x = 6 - 15t, \quad y = -12 + 34t, \quad z = 6 - 12t$$

(f) One representation of the solution is $2(x-5) + 10(y-3) + 8(z-2) = 0$. Normal to plane has components $(2, 10, 8)$. Note if $\vec{N}$ is normal to surface at a point, then $-\vec{N}$ is also normal to surface at same point.

▶ **58.** Show $|A| = (-1)^N|A|$

▶ **59.** (a) $2A_{im}x_i$, (b) $2A_{km}$

▶ **60.** a_{ij} is skew-symmetric and $U_n = \frac{1}{2}a_{ij}e_{ijn}$

▶ **61.** $\dfrac{A_1}{B_1} = \dfrac{A_2}{B_2} = \cdots = \dfrac{A_N}{B_N} = \lambda$

▶ **62.** zero

▶ **63.** (a) 256 (b) 36 (c) 21

▶ **64.** $\dfrac{\partial \bar{A}_i}{\partial \bar{x}_m} = a_{ij}a_{pm}\dfrac{\partial A_j}{\partial x_p}$

EXERCISE 1.1

SELECTED SOLUTIONS EXERCISE 1.2

1. (i) Identity $\alpha = 0$, (ii) Inverse of $T_\theta = T_{-\theta}$, (iii) $T_{\theta_1}T_{\theta_2} = T_{\theta_1+\theta_2}$

2. (i) Identity $\alpha = 1$, Inverse of $T_\alpha = T_{1/\alpha}$, $T_\alpha T_\beta = T_{\alpha\beta}$, $T_{(\alpha\beta)\gamma} = T_{\alpha(\beta\gamma)}$

3. Identity $\alpha = 0$, Inverse of $T_\alpha = T_{-\alpha}$, $T_\alpha T_\beta = T_{\alpha+\beta}$

4. Identity $v = 0$,

$$T_{v_1} : \left\{ \bar{x}^1 = b_1(x^1 - v_1 x^4), \quad \bar{x}^2 = x^2, \quad \bar{x}^3 = x^3, \quad \bar{x}^4 = b_1(x^4 - \frac{v_1}{c^2}x^1) \right\} \quad \text{where} \quad b_1 = (1 - \frac{v_1^2}{c^2})^{-1/2}$$

$$T_{v_2} : \left\{ \bar{\bar{x}}^1 = b_2(\bar{x}^1 - v_2\bar{x}^4), \quad \bar{\bar{x}}^2 = \bar{x}^2, \quad \bar{\bar{x}}^3 = \bar{x}^3, \quad \bar{\bar{x}}^4 = b_2(\bar{x}^4 - \frac{v_2}{c^2}\bar{x}^1) \right\} \quad \text{where} \quad b_2 = (1 - \frac{v_2^2}{c^2})^{-1/2}$$

Show $\bar{\bar{x}}^1 = b_3(x^1 - v_3 x^4)$ and $\bar{\bar{x}}^4 = b_3(x^4 - \frac{v_3}{c^2}x^1)$

Hint:

$$b_3 = b_1 b_2(1 + \frac{v_1 v_2}{c^2}) = \left(1 - \frac{v_1^2}{c^2}\right)^{-1/2}\left(1 - \frac{v_2^2}{c^2}\right)^{-1/2}\left[\left(1 + \frac{v_1 v_2}{c^2}\right)^2\right]^{1/2}$$

$$= \left[\frac{\left(1 - \frac{v_1^2}{c^2}\right)\left(1 - \frac{v_2^2}{c^2}\right)}{\left(1 + \frac{v_1 v_2}{c^2}\right)^2}\right]^{-1/2} = \left[\frac{1 - \frac{v_1^2}{c^2} - \frac{v_2^2}{c^2} + \frac{v_1^2 v_2^2}{c^4}}{\left(1 + \frac{v_1 v_2}{c^2}\right)^2}\right]^{-1/2}$$

$$= \left[\frac{1 + \frac{2v_1 v_2}{c^2} + \frac{v_1^2 v_2^2}{c^4} - \frac{1}{c^2}(v_1^2 + 2v_1 v_2 + v_2^2)}{\left(1 + \frac{v_1 v_2}{c^2}\right)^2}\right]^{-1/2} = \left(1 - \frac{v_3^2}{c^2}\right)^{-1/2}$$

5. (a) Show $\vec{E}^i \cdot \vec{E}_j = \delta^i_j$. That is

$$\vec{E}^1 \cdot \vec{E}_1 = \frac{1}{V}\vec{E}_1 \cdot (\vec{E}_2 \times \vec{E}_3) = 1 \quad \text{for} \quad V = \vec{E}_1 \cdot (\vec{E}_1 \times \vec{E}_3)$$

$$\vec{E}^1 \cdot \vec{E}_2 = \frac{1}{V}\vec{E}_2 \cdot (\vec{E}_2 \times \vec{E}_3) = 0$$

$$etc.$$

(b) If $\vec{E}^j = \alpha\vec{E}_1 + \beta\vec{E}_2 + \gamma\vec{E}_3$ then $\alpha = g^{j1}$, $\beta = g^{j2}$, $\gamma = g^{j3}$ so that $\vec{E}^j = g^{ji}\vec{E}_i = g^{ij}\vec{E}_i$.

6. (a)

Transformation	with inverse transformation
$x = r\cos\beta$	$r = \sqrt{x^2 + y^2}$
$y = r\sin\beta$	$\beta = \arctan(y/x)$
$z = z$	$z = z$

EXERCISE 1.2

6(b)

$$\vec{E}_1 = \frac{\partial \vec{r}}{\partial r} = \cos\beta \,\widehat{\mathbf{e}}_1 + \sin\beta \,\widehat{\mathbf{e}}_2$$

$$\vec{E}_2 = \frac{\partial \vec{r}}{\partial \beta} = -r\sin\beta \,\widehat{\mathbf{e}}_1 + r\cos\beta \,\widehat{\mathbf{e}}_2$$

$$\vec{E}_3 = \frac{\partial \vec{r}}{\partial z} = \widehat{\mathbf{e}}_3$$

$$g_{ij} = \begin{pmatrix} 1 & 0 & 0 \\ 0 & r^2 & 0 \\ 0 & 0 & 1 \end{pmatrix}$$

$$\vec{E}^1 = \cos\beta \,\widehat{\mathbf{e}}_1 + \sin\beta \,\widehat{\mathbf{e}}_2$$

$$\vec{E}^2 = \frac{1}{r}(-\sin\beta)\,\widehat{\mathbf{e}}_1 + \frac{1}{r}\cos\beta \,\widehat{\mathbf{e}}_2$$

$$g^{ij} = \begin{pmatrix} 1 & 0 & 0 \\ 0 & \frac{1}{r^2} & 0 \\ 0 & 0 & 1 \end{pmatrix}$$

$$\vec{E}^3 = \widehat{\mathbf{e}}_3$$

$$\widehat{\mathbf{e}}_r = \cos\beta \,\widehat{\mathbf{e}}_1 + \sin\beta \,\widehat{\mathbf{e}}_2$$

$$\widehat{\mathbf{e}}_\beta = -\sin\beta \,\widehat{\mathbf{e}}_1 + \cos\beta \,\widehat{\mathbf{e}}_2$$

$$\widehat{\mathbf{e}}_z = \widehat{\mathbf{e}}_3$$

6(c)

$$A^1 = \vec{A} \cdot \vec{E}^1 = A_x \cos\beta + A_y \sin\beta$$

$$A^2 = \vec{A} \cdot \vec{E}^2 = -A_x \frac{\sin\beta}{r} + A_y \frac{\cos\beta}{r}$$

$$A^3 = \vec{A} \cdot \vec{E}^3 = A_z$$

$$A_1 = \vec{A} \cdot \vec{E}_1 = A_x \cos\beta + A_y \sin\beta$$

$$A_2 = \vec{A} \cdot \vec{E}_2 = -A_x r \cos\beta + A_y r \sin\beta$$

$$A_3 = \vec{A} \cdot \vec{E}_3 = A_z$$

$$A_r = \vec{A} \cdot \widehat{\mathbf{e}}_r = A_x \cos\beta + A_y \sin\beta$$

$$A_\beta = \vec{A} \cdot \widehat{\mathbf{e}}_\beta = -A_x \sin\beta + A_y \cos\beta$$

$$A_z = \vec{A} \cdot \widehat{\mathbf{e}}_z = A_z$$

▶ 7. (a)

Transformation   with inverse transformation

$$x = \rho \sin\alpha \cos\beta \qquad \rho = \sqrt{x^2 + y^2 + z^2}$$

$$y = \rho \sin\alpha \sin\beta \qquad \beta = \arctan(y/x)$$

$$z = \rho \cos\alpha \qquad \alpha = \arctan(\sqrt{x^2 + y^2}/z)$$

EXERCISE 1.2

7(b)

$$\vec{E}_1 = \sin\alpha\cos\beta\,\widehat{\mathbf{e}}_1 + \sin\alpha\sin\beta\,\widehat{\mathbf{e}}_2 + \cos\alpha\,\widehat{\mathbf{e}}_3$$

$$\vec{E}_2 = \rho\cos\alpha\cos\beta\,\widehat{\mathbf{e}}_1 + \rho\cos\alpha\sin\beta\,\widehat{\mathbf{e}}_2 - \rho\sin\alpha\,\widehat{\mathbf{e}}_3$$

$$\vec{E}_3 = -\rho\sin\alpha\sin\beta\,\widehat{\mathbf{e}}_1 + \rho\sin\alpha\cos\beta\,\widehat{\mathbf{e}}_3$$

$$\vec{E}^1 = \sin\alpha\cos\beta\,\widehat{\mathbf{e}}_1 + \sin\alpha\sin\beta\,\widehat{\mathbf{e}}_2 + \cos\alpha\,\widehat{\mathbf{e}}_3$$

$$\vec{E}^2 = \frac{1}{\rho}\left(\cos\alpha\cos\beta\,\widehat{\mathbf{e}}_1 + \cos\alpha\sin\beta\,\widehat{\mathbf{e}}_2 - \sin\alpha\,\widehat{\mathbf{e}}_3\right)$$

$$\vec{E}^3 = \frac{1}{\rho\sin\alpha}\left(-\sin\beta\,\widehat{\mathbf{e}}_1 + \cos\beta\,\widehat{\mathbf{e}}_2\right)$$

$$\widehat{\mathbf{e}}_\rho = \sin\alpha\cos\beta\,\widehat{\mathbf{e}}_1 + \sin\alpha\sin\beta\,\widehat{\mathbf{e}}_2 + \cos\alpha\,\widehat{\mathbf{e}}_3$$

$$\widehat{\mathbf{e}}_\alpha = \cos\alpha\cos\beta\,\widehat{\mathbf{e}}_1 + \cos\alpha\sin\beta\,\widehat{\mathbf{e}}_2 - \sin\alpha\,\widehat{\mathbf{e}}_3$$

$$\widehat{\mathbf{e}}_\beta = -\sin\beta\,\widehat{\mathbf{e}}_1 + \cos\beta\,\widehat{\mathbf{e}}_2$$

7(c)

$$A^1 = A_x\sin\alpha\cos\beta + A_y\sin\alpha\sin\beta + A_z\cos\alpha$$

$$A^2 = \frac{1}{\rho}\left(A_x\cos\alpha\cos\beta + A_y\cos\alpha\sin\beta - A_z\sin\alpha\right)$$

$$A^3 = \frac{1}{\rho\sin\alpha}\left(-A_x\sin\beta + A_y\cos\beta\right)$$

$$A_1 = A_x\sin\alpha\cos\beta + A_y\sin\alpha\sin\beta + A_z\cos\alpha$$

$$A_2 = A_x\rho\cos\alpha\cos\beta + A_y\rho\cos\alpha\sin\beta - A_z\rho\sin\alpha$$

$$A_3 = -A_x\rho\sin\alpha\sin\beta + A_y\rho\sin\alpha\cos\beta$$

$$A_\rho = A_x\sin\alpha\cos\beta + A_y\sin\alpha\sin\beta + A_z\cos\alpha$$

$$A_\alpha = A_x\cos\alpha\cos\beta + A_y\cos\alpha\sin\beta - A_z\sin\alpha$$

$$A_\beta = -A_x\sin\beta + A_y\cos\beta$$

► 9.

$$(a) \quad \vec{V}^1 = \frac{1}{24}(5\widehat{\mathbf{e}}_1 - \widehat{\mathbf{e}}_2)$$

$$\vec{V}^2 = \frac{1}{24}(-\widehat{\mathbf{e}}_1 + 5\widehat{\mathbf{e}}_2)$$

$$\vec{V}^3 = \widehat{\mathbf{e}}_3$$

(e) $g_{ij} = \begin{pmatrix} 26 & 10 & 0 \\ 10 & 26 & 0 \\ 0 & 0 & 1 \end{pmatrix}$ Note that $\vec{V}_i \cdot \vec{V}_j = g_{ij}$ and $\vec{V}^i \cdot \vec{V}^j = g^{ij}$

► 11. (a) Sketch family of curves $x - 2y = u_0$ and $x - y = -v_0$

(d) $A_1 = -\alpha_1 - \alpha_2$, $A_2 = -2\alpha_1 - \alpha_2$, $A_3 = \alpha_3$ (e) $g_{ij} = \begin{pmatrix} 2 & 3 & 0 \\ 3 & 5 & 0 \\ 0 & 0 & 1 \end{pmatrix}$ (i) $\left(\frac{-\alpha_1 - \alpha_2}{\sqrt{2}}, \frac{-2\alpha_1 - \alpha_2}{\sqrt{5}}, \alpha_3\right)$

► 14. Representation is not unique. $x^1 = 1 + 13t$, $x^2 = 2 + 5t$, $x^3 = 3 - 6t$

► 16. $S = \pi b^2 \sin\alpha$

EXERCISE 1.2

▶ **17.**   Representation is not unique. Sketch the following vectors: $\vec{r}_1 = (1,2,3)$, $\vec{r}_3 = (5,5,5)$ and $\vec{r}_2 = (14,7,-3)$, with

$$\frac{\partial \vec{r}}{\partial u} = \alpha_1 \,\widehat{\mathbf{e}}_1 + \alpha_2 \,\widehat{\mathbf{e}}_2 + \alpha_3 \,\widehat{\mathbf{e}}_3 = \vec{r}_3 - \vec{r}_1$$

$$\frac{\partial \vec{r}}{\partial v} = \beta_1 \,\widehat{\mathbf{e}}_1 + \beta_2 \,\widehat{\mathbf{e}}_2 + \beta_3 \,\widehat{\mathbf{e}}_3 = \vec{r}_2 - \vec{r}_1$$

$$\vec{r}(0,0) = \gamma_1 \,\widehat{\mathbf{e}}_1 + \gamma_2 \,\widehat{\mathbf{e}}_2 + \gamma_3 \,\widehat{\mathbf{e}}_3$$

Then one can show $x^1 = 4u + 13v + 1$, $x^2 = 3u + 5v + 2$, $x^3 = 2u - 6v + 3$ Another representation is when $\frac{\partial \vec{r}}{\partial v} = \vec{r}_3 - \vec{r}_2$. In this case the plane has the representation $x^1 = 4u + 9v + 1$, $x^2 = 3u + 2v + 2$, $x^3 = 2u - 8v + 3$

▶ **18.** $(1,1,1)$ $(2,4,8)$

▶ **21.**   Hypothesis: $\bar{A}^i = A^j \dfrac{\partial \bar{x}^i}{\partial x^j}$ and $\bar{B}^k = B^t \dfrac{\partial \bar{x}^k}{\partial x^t}$ Therefore: $\bar{C}^{ik} = \bar{A}^i \bar{B}^k = A^j B^t \dfrac{\partial \bar{x}^i}{\partial x^j} \dfrac{\partial \bar{x}^k}{\partial x^t} = C^{jt} \dfrac{\partial \bar{x}^i}{\partial x^j} \dfrac{\partial \bar{x}^k}{\partial x^t}$

▶ **23.**   By hypothesis $\bar{A}^m_{np} = A^i_{jk} \dfrac{\partial \bar{x}^m}{\partial x^i} \dfrac{\partial x^j}{\partial \bar{x}^n} \dfrac{\partial x^k}{\partial \bar{x}^p}$, $\bar{B}^r_s = B^k_n \dfrac{\partial \bar{x}^r}{\partial x^k} \dfrac{\partial x^n}{\partial \bar{x}^s}$, $\bar{C}^q_{tu} = C^i_{jn} \dfrac{\partial \bar{x}^q}{\partial x^i} \dfrac{\partial x^j}{\partial \bar{x}^t} \dfrac{\partial x^n}{\partial \bar{x}^u}$,

Therefore, $\bar{A}^m_{np} \bar{B}^r_s = A^i_{jk} B^t_u \dfrac{\partial \bar{x}^m}{\partial x^i} \dfrac{\partial x^j}{\partial \bar{x}^n} \dfrac{\partial x^k}{\partial \bar{x}^p} \dfrac{\partial \bar{x}^r}{\partial x^t} \dfrac{\partial x^u}{\partial \bar{x}^s}$

Now contract on the indices p and r

$$\bar{A}^m_{np} \bar{B}^p_s = A^i_{jk} B^t_u \frac{\partial \bar{x}^m}{\partial x^i} \frac{\partial x^j}{\partial \bar{x}^n} \frac{\partial x^k}{\partial \bar{x}^p} \frac{\partial \bar{x}^p}{\partial x^t} \frac{\partial x^u}{\partial \bar{x}^s}$$

$$\bar{A}^m_{np} \bar{B}^p_s = A^i_{jk} B^t_u \delta^k_t \frac{\partial \bar{x}^m}{\partial x^i} \frac{\partial x^j}{\partial \bar{x}^n} \frac{\partial x^u}{\partial \bar{x}^s}$$

$$\bar{A}^m_{np} \bar{B}^p_s = A^i_{jk} B^k_u \delta^k_t \frac{\partial \bar{x}^m}{\partial x^i} \frac{\partial x^j}{\partial \bar{x}^n} \frac{\partial x^u}{\partial \bar{x}^s}$$

$$\bar{A}^m_{np} \bar{B}^p_s = C^i_{ju} \frac{\partial \bar{x}^m}{\partial x^i} \frac{\partial x^j}{\partial \bar{x}^n} \frac{\partial x^u}{\partial \bar{x}^s} = \bar{C}^m_{ns}$$

▶ **26.** (a) and (b) (i) $(1,0,0)$ (ii) $(0,1,0)$ (iii) $(0,0,1)$

▶ **27.**   Let $\bar{A}_{ij} = A_{mn} \dfrac{\partial x^m}{\partial \bar{x}^i} \dfrac{\partial x^n}{\partial \bar{x}^j}$ and $\bar{A}^{ij} = A^{pq} \dfrac{\partial \bar{x}^i}{\partial x^p} \dfrac{\partial \bar{x}^j}{\partial x^q}$, then we have

$$\lambda = \bar{A}_{ij} \bar{A}^{ij} = A_{mn} A^{pq} \frac{\partial x^m}{\partial \bar{x}^i} \frac{\partial \bar{x}^i}{\partial x^p} \frac{\partial x^n}{\partial \bar{x}^j} \frac{\partial \bar{x}^j}{\partial x^q} = A_{mn} A^{pq} \delta^m_p \delta^n_q = A_{pq} A^{pq}$$

▶ **28.**   Write out the derivatives for $\bar{a}_{mn}$, $\bar{a}_{nk}$ and $\bar{a}_{km}$ with respect to the barred variables and then add the results to obtain

$$\frac{\partial \bar{a}_{mn}}{\partial \bar{x}^k} + \frac{\partial \bar{a}_{nk}}{\partial \bar{x}^m} + \frac{\partial \bar{a}_{km}}{\partial \bar{x}^n} = (a_{ij} + a_{ji}) \frac{\partial x^i}{\partial \bar{x}^m} \frac{\partial^2 x^i}{\partial \bar{x}^k \partial \bar{x}^n}$$

$$+ (a_{ij} + a_{ji}) \frac{\partial x^j}{\partial \bar{x}^n} \frac{\partial^2 x^j}{\partial \bar{x}^k \partial \bar{x}^m}$$

$$+ (a_{ij} + a_{ji}) \frac{\partial x^i}{\partial \bar{x}^k} \frac{\partial^2 x^j}{\partial \bar{x}^m \partial \bar{x}^n}$$

$$+ \left(\frac{\partial a_{ij}}{\partial x^\ell} + \frac{\partial a_{j\ell}}{\partial x^i} + \frac{\partial a_{\ell i}}{\partial x^j} \right) \frac{\partial x^\ell}{\partial \bar{x}^k} \frac{\partial x^i}{\partial \bar{x}^m} \frac{\partial x^j}{\partial \bar{x}^n}$$

The first three terms on the right-hand side add to zero, since a_{ij} is skew-symmetric. See also problem No. 21, Exercise 1.1.

EXERCISE 1.2

29. Show

$$A_1 x + B_1 y + C_1 = \bar{A}_1 \bar{x} + \bar{B}_1 \bar{y} + \bar{C}_1$$

$$A_2 x + B_2 y + C_2 = \bar{A}_2 \bar{x} + \bar{B}_2 \bar{y} + \bar{C}_2$$

where

$$\bar{A}_1 = A_1 \cos\theta + B_1 \sin\theta$$

$$\bar{B}_1 = B_1 \cos\theta - A_1 \sin\theta$$

$$\bar{C}_1 = A_1 h + B_1 k + C_1$$

$$\bar{A}_2 = A_2 \cos\theta + B_2 \sin\theta$$

$$\bar{B}_2 = B_2 \cos\theta - A_2 \sin\theta$$

$$\bar{C}_2 = A_2 h + B_2 k + C_2$$

Now show using algebra that
$$\begin{aligned} \bar{A}_1 \bar{B}_2 - \bar{A}_2 \bar{B}_1 &= A_1 B_2 - A_2 B_1 \\ \bar{A}_1 \bar{A}_2 + \bar{B}_1 \bar{B}_2 &= A_1 A_2 + B_1 B_2 \end{aligned}$$
are invariants.

30. Show

$$y' = \frac{dy}{dx} = \frac{dy}{d\bar{x}} \frac{d\bar{x}}{dx} = \frac{\sin\theta + \frac{d\bar{y}}{d\bar{x}}\cos\theta}{\cos\theta - \frac{d\bar{y}}{d\bar{x}}\sin\theta}$$

$$y'' = \frac{d^2 y}{dx^2} = \frac{d}{d\bar{x}}\left(\frac{dy}{dx}\right)\frac{d\bar{x}}{dx} = \frac{\left(\cos\theta - \frac{d\bar{y}}{d\bar{x}}\sin\theta\right)\frac{d^2\bar{y}}{d\bar{x}^2}\cos\theta - \left(\sin\theta + \frac{d\bar{y}}{d\bar{x}}\cos\theta\right)\left(-\frac{d^2\bar{y}}{d\bar{x}^2}\sin\theta\right)}{\left(\cos\theta - \frac{d\bar{y}}{d\bar{x}}\sin\theta\right)^3}$$

Substitute y' and y'' into κ to show that $\kappa = \dfrac{\frac{d^2\bar{y}}{d\bar{x}^2}}{\left(1 + \left(\frac{d\bar{y}}{d\bar{x}}\right)^2\right)^{3/2}}$. Hence, κ is an invariant.

34. $\bar{a}_{11} = 10/4, \bar{a}_{12} = -1/2, \bar{a}_{21} = -1, \bar{a}_{22} = 0$

37. Weight 7.

42. Vector x^i is perpendicular to both vectors A_i and B_i and so must be proportional to their cross product.

43. (a) $x = 3u + 4v, y = 4u + 7v, z = w$

(b) $\bar{E}^1 = \dfrac{7}{5}\hat{e}_1 - \dfrac{4}{5}\hat{e}_2, \quad \bar{E}^2 = -\dfrac{4}{5}\hat{e}_1 + \dfrac{3}{5}\hat{e}_2, \quad \bar{E}^3 = \hat{e}_3$

(c) $(3y + 4z, 4y + 7z, x)$ \qquad (d) $\left(\dfrac{7}{5}y - \dfrac{4}{5}z, -\dfrac{4}{5}y + \dfrac{3}{5}z, x\right)$

46.

$$\bar{a}_{ij} = a_{mn}\frac{\partial x^m}{\partial \bar{x}^i}\frac{\partial x^n}{\partial \bar{x}^j}$$

$$\bar{a}_{ji} = a_{mn}\frac{\partial x^m}{\partial \bar{x}^j}\frac{\partial x^n}{\partial \bar{x}^i} = a_{nm}\frac{\partial x^n}{\partial \bar{x}^i}\frac{\partial x^m}{\partial \bar{x}^j} = \bar{a}_{ij}$$

hence symmetric

48. If $\bar{A}_i = A_j \dfrac{\partial x^j}{\partial \bar{x}^i}$, then $\bar{A}_i \dfrac{\partial \bar{x}^i}{\partial x^m} = A_j \dfrac{\partial x^j}{\partial \bar{x}^i}\dfrac{\partial \bar{x}^i}{\partial x^m} = A_j \delta_m^j = A_m$

EXERCISE 1.2

SELECTED SOLUTIONS EXERCISE 1.3

▶ **1.**

(a)

$$\bar{g}_{ij} = g_{ab}\frac{\partial x^a}{\partial \bar{x}^i}\frac{\partial x^b}{\partial \bar{x}^j}$$

$$\bar{g}_{ij}\frac{\partial \bar{x}^i}{\partial x^m}\frac{\partial \bar{x}^j}{\partial x^n} = g_{ab}\frac{\partial x^a}{\partial \bar{x}^i}\frac{\partial \bar{x}^i}{\partial x^m}\frac{\partial x^b}{\partial \bar{x}^j}\frac{\partial \bar{x}^j}{\partial x^n}$$

$$= g_{ab}\delta^a_m \delta^b_n = g_{mn}$$

(b) Assume $g_{ab} = g_{ba}$, For $\quad \bar{g}_{ji} = g_{ab}\frac{\partial x^a}{\partial \bar{x}^j}\frac{\partial x^b}{\partial \bar{x}^i}$, then

$\bar{g}_{ij} = g_{ab}\frac{\partial x^a}{\partial \bar{x}^i}\frac{\partial x^b}{\partial \bar{x}^j} = g_{ab}\frac{\partial x^a}{\partial \bar{x}^i}\frac{\partial x^b}{\partial \bar{x}^j}$ Now subtract to obtain

$\bar{g}_{ji} - \bar{g}_{ij} = (g_{ba} - g_{ab})\frac{\partial x^a}{\partial \bar{x}^i}\frac{\partial x^b}{\partial \bar{x}^j} = 0$ which implies $\bar{g}_{ij} = \bar{g}_{ji}$ hence symmetric in all coordinate systems.

▶ **2.**

$$g_{ij} = \frac{\partial y^m}{\partial x^i}\frac{\partial y^m}{\partial x^j} \quad \text{multiply by } \frac{\partial x^i}{\partial y^s}$$

$$\frac{\partial x^i}{\partial y^s}g_{ij} = \frac{\partial x^i}{\partial y^s}\frac{\partial y^m}{\partial x^i}\frac{\partial y^m}{\partial x^j} = \delta^m_s\frac{\partial y^m}{\partial x^j} = \frac{\partial y^s}{\partial x^j} \quad \text{multiply by } g^{jr}$$

$$\frac{\partial x^i}{\partial y^s}g_{ij}g^{jr} = \frac{\partial y^s}{\partial x^j}g^{jr}$$

$$\frac{\partial x^i}{\partial y^s}\delta^r_i = \frac{\partial x^r}{\partial y^s} = \frac{\partial y^s}{\partial x^j}g^{jr} \quad \text{multiply by } \frac{\partial x^m}{\partial y^s}$$

$$\frac{\partial x^m}{\partial y^s}\frac{\partial x^r}{\partial y^s} = \frac{\partial x^m}{\partial y^s}\frac{\partial y^s}{\partial x^j}g^{jr} = \delta^m_j g^{jr} = g^{mr}$$

▶ **6.**

$$g^{ij} = \frac{1}{g}\begin{pmatrix} r^2(1 - \cos^2\phi\cos^2\alpha) & r\sin\phi\cos\phi\cos^2\alpha & -r^2\sin\phi\cos\alpha \\ r\sin\phi\cos\phi\cos^2\alpha & 1 - \sin^2\phi\cos^2\alpha & -r\cos\phi\cos\alpha \\ -r^2\sin\phi\cos\alpha & -r\cos\phi\cos\alpha & r^2 \end{pmatrix}$$

where $g = r^2\sin^2\alpha$.

▶ **7.**

$$\bar{A}_1 = \frac{1}{3}x_1^2 - \frac{2}{3}x_2^2 + \frac{2}{3}x_3^2$$

$$\bar{A}_2 = -\frac{14}{15}x_1^2 - \frac{1}{3}x_2^2 + \frac{2}{15}x_3^2$$

$$\bar{A}_3 = \frac{2}{15}x_1^2 - \frac{2}{3}x_2^2 - \frac{11}{15}x_3^2$$

where

$$x_1 = \frac{5}{15}\bar{x}_1 - \frac{2}{3}\bar{x}_2 + \frac{10}{15}\bar{x}_3$$

$$x_2 = -\frac{14}{15}\bar{x}_1 - \frac{1}{3}\bar{x}_2 + \frac{2}{15}\bar{x}_3$$

$$x_3 = \frac{2}{15}\bar{x}_1 - \frac{2}{3}\bar{x}_2 - \frac{11}{15}\bar{x}_3$$

Remember $\bar{A}_i$ must be a function of the barred coordinates.

12.

(a)

$$dS_1 = |\vec{E}_2 \times \vec{E}_3| dvdw$$

$$dS_1 = \sqrt{(\vec{E}_2 \times \vec{E}_3) \cdot (\vec{E}_2 \times \vec{E}_3)} dvdw$$

$$dS_1 = \sqrt{(\vec{E}_2 \cdot \vec{E}_2)(\vec{E}_3 \cdot \vec{E}_3) - (\vec{E}_2 \cdot \vec{E}_3)(\vec{E}_2 \cdot \vec{E}_3)} dvdw$$

$$dS_1 = \sqrt{g_{22}g_{33} - (g_{23})^2} dvdw$$

16.

(b)

$$\bar{A}_1 = A_x \sin\theta\cos\phi + A_y \sin\theta\sin\phi + A_z \cos\theta$$

$$\bar{A}_2 = A_x \rho\cos\theta\cos\phi + A_y \rho\cos\theta\sin\phi - A_z \rho\sin\theta$$

$$\bar{A}_3 = -A_x \rho\sin\theta\sin\phi + A_y \rho\sin\theta\cos\phi$$

$$\bar{A}^1 = A_x \sin\theta\cos\phi + A_y \sin\theta\sin\phi + A_z \cos\theta$$

$$\bar{A}^2 = A_x \frac{\cos\theta\cos\phi}{\rho} + A_y \frac{\cos\theta\sin\phi}{\rho} - A_z \frac{\sin\theta}{\rho}$$

$$\bar{A}^3 = -A_x \frac{\sin\phi}{\rho\sin\theta} + A_y \frac{\cos\phi}{\rho\sin\theta}$$

18. $\bar{x}_i = \ell_{ij}x_j$ and $\ell_{im}\bar{x}_i = \ell_{im}\ell_{ij}x_j = \delta_{mj}x_j = x_m$ so that $\dfrac{\partial x_m}{\partial \bar{x}_j} = \ell_{im}\delta_{ij} = \ell_{jm}$ Hence

$$\bar{C}_{rstn} = [\lambda\delta_{ik}\delta_{mp} + \mu(\delta_{im}\delta_{kp} + \delta_{ip}\delta_{km}) + \nu(\delta_{im}\delta_{kp} - \delta_{ip}\delta_{km})]\ell_{ri}\ell_{sk}\ell_{tm}\ell_{np}$$

$$\bar{C}_{rstn} = \lambda\ell_{rk}\ell_{sk}\ell_{tp}\ell_{np} + \mu(\ell_{rm}\ell_{tm}\ell_{sp}\ell_{np} + \ell_{rp}\ell_{np}\ell_{sm}\ell_{tm}) + \nu(\ell_{rm}\ell_{tm}\ell_{sp}\ell_{np} - \ell_{rp}\ell_{np}\ell_{sm}\ell_{tm})$$

$$\bar{C}_{rstn} = \lambda\delta_{rs}\delta_{tn} + \mu(\delta_{rt}\delta_{sn} + \delta_{rn}\delta_{st}) + \nu(\delta_{rt}\delta_{sn} - \delta_{rn}\delta_{st}) = C_{rstn}$$

20.

$$\bar{T}_{mnpq} = T_{ijk\ell} \frac{\partial x^i}{\partial \bar{x}^m} \frac{\partial x^j}{\partial \bar{x}^n} \frac{\partial x^k}{\partial \bar{x}^p} \frac{\partial x^\ell}{\partial \bar{x}^q}$$

$$\bar{T}_{mnqp} = T_{ijk\ell} \frac{\partial x^i}{\partial \bar{x}^m} \frac{\partial x^j}{\partial \bar{x}^n} \frac{\partial x^k}{\partial \bar{x}^q} \frac{\partial x^\ell}{\partial \bar{x}^p} \qquad \text{Now interchange } k \text{ and } \ell$$

$$\bar{T}_{mnqp} = T_{ij\ell k} \frac{\partial x^i}{\partial \bar{x}^m} \frac{\partial x^j}{\partial \bar{x}^n} \frac{\partial x^\ell}{\partial \bar{x}^q} \frac{\partial x^k}{\partial \bar{x}^p} \qquad \text{Now add the first and third equations}$$

$$\bar{T}_{mnpq} + \bar{T}_{mnqp} = (T_{ijk\ell} + T_{ij\ell k}) \frac{\partial x^i}{\partial \bar{x}^m} \frac{\partial x^j}{\partial \bar{x}^n} \frac{\partial x^\ell}{\partial \bar{x}^q} \frac{\partial x^k}{\partial \bar{x}^p}$$

Hence if $T_{ijk\ell} + T_{ij\ell k} = 0$, then $\bar{T}_{mnpq} + \bar{T}_{mnqp} = 0$.

22. True

24. Hypothesis $A^{rs} = -A^{sr}$, $c_r = \frac{1}{2}\epsilon_{rmn}A^{mn}$ and

$$\bar{A}^{mn} = A^{pq}\frac{\partial \bar{x}^m}{\partial x^p}\frac{\partial \bar{x}^n}{\partial x^q} \qquad \bar{\epsilon}_{rmn} = \epsilon_{ijk}\frac{\partial x^i}{\partial \bar{x}^r}\frac{\partial x^j}{\partial \bar{x}^m}\frac{\partial x^k}{\partial \bar{x}^n}$$

then

$$\bar{c}_r = \frac{1}{2}\bar{A}^{mn}\bar{\epsilon}_{rmn} = \frac{1}{2}A^{pq}\epsilon_{ijk}\frac{\partial \bar{x}^m}{\partial x^p}\frac{\partial \bar{x}^n}{\partial x^q}\frac{\partial x^i}{\partial \bar{x}^r}\frac{\partial x^j}{\partial \bar{x}^m}\frac{\partial x^k}{\partial \bar{x}^n}$$

$$= \frac{1}{2}A^{pq}\epsilon_{ijk}\delta_p^j\delta_q^k\frac{\partial x^i}{\partial \bar{x}^r} = \frac{1}{2}A^{pq}\epsilon_{ipq}\frac{\partial x^i}{\partial \bar{x}^r} = c_i\frac{\partial x^i}{\partial \bar{x}^r}$$

Hence c_i transforms as a first order oovariant tensor.

$$c_1 = \sqrt{g}A^{23} \quad c_2 = \sqrt{g}A^{31} \quad c_3 = \sqrt{g}A^{12}$$

EXERCISE 1.3

▶ **31.** $\quad \bar{g}^{nj} = \dfrac{1}{\lambda} g^{jn}$

▶ **32.** Show $A_{ijk} = \epsilon_{rjk} A_i^r + \epsilon_{irk} A_j^r + \epsilon_{ijr} A_k^r$ is completely skew-symmetric so that $A_{ijk} = e_{ijk} A_{123}$. Then show $A_{123} = A_r^r \sqrt{g}$

▶ **34.** Using properties of determinants with $g = |g_{ij}|$ we have $e^{ijk} g_{im} g_{jn} g_{kp} = g e_{mnp}$ or $\dfrac{1}{\sqrt{g}} e^{ijk} g_{im} g_{jn} g_{kp} = \sqrt{g} e_{mnp}$, hence $\epsilon^{ijk} g_{im} g_{jn} g_{kp} = \epsilon_{mnp}$

▶ **36.** $\quad (i) \quad$ Calculate $\bar{T}_n^n$ from transformation law for $\bar{T}_n^m$.

$$(ii) \quad \frac{1}{2}\left[\left(\bar{T}_i^i\right)^2 - \bar{T}_m^i \bar{T}_i^m\right] = \frac{1}{2}\left[\left(T_i^i\right)^2 - \left(T_\beta^\alpha \frac{\partial \bar{x}^i}{\partial x^\alpha} \frac{\partial x^\beta}{\partial \bar{x}^m}\right)\left(T_s^r \frac{\partial \bar{x}^m}{\partial x^r} \frac{\partial x^s}{\partial \bar{x}^i}\right)\right]$$

$$= \frac{1}{2}\left[\left(T_i^i\right)^2 - \left(T_\beta^\alpha \delta_\alpha^s T_s^r \delta_r^\beta\right)\right] = \frac{1}{2}\left[\left(T_i^i\right)^2 - T_r^\alpha T_\alpha^r\right]$$

$$(iii) \quad det[\bar{T}_j^i] = \bar{e}^{ijk} \bar{T}_i^1 \bar{T}_j^2 \bar{T}_k^3$$

$$= J e^{rst} \frac{\partial \bar{x}^i}{\partial x^r} \frac{\partial \bar{x}^j}{\partial x^s} \frac{\partial \bar{x}^k}{\partial x^t} T_\beta^\alpha \frac{\partial \bar{x}^1}{\partial x^\alpha} \frac{\partial x^\beta}{\partial \bar{x}^i} T_b^c \frac{\partial \bar{x}^2}{\partial x^i} \frac{\partial x^b}{\partial \bar{x}^j} T_p^s \frac{\partial \bar{x}^3}{\partial x^s} \frac{\partial x^p}{\partial \bar{x}^k}$$

$$= J e^{rst} T_r^\alpha T_b^c T_p^s \delta_r^\beta \delta_s^b \delta_t^p \frac{\partial \bar{x}^1}{\partial x^\alpha} \frac{\partial \bar{x}^2}{\partial x^c} \frac{\partial \bar{x}^3}{\partial x^s} = J e^{rst} T_r^\alpha T_s^c T_t^s \frac{\partial \bar{x}^1}{\partial x^\alpha} \frac{\partial \bar{x}^2}{\partial x^c} \frac{\partial \bar{x}^3}{\partial x^s}$$

$$= J |T_j^i| e^{\alpha cs} \frac{\partial \bar{x}^1}{\partial x^\alpha} \frac{\partial \bar{x}^2}{\partial x^c} \frac{\partial \bar{x}^3}{\partial x^s} = J |T_j^i| J^{-1} = |T_j^i| = det[T_j^i]$$

▶ **41.**

$$\frac{1}{\sqrt{g}}\left[\frac{\partial}{\partial x^1}\left(\sqrt{g} \varrho V^1\right) + \frac{\partial}{\partial x^2}\left(\sqrt{g} \varrho V^2\right) + \frac{\partial}{\partial x^3}\left(\sqrt{g} \varrho V^3\right)\right] + \frac{\partial \varrho}{\partial t} = 0$$

(b) Let $\quad V_r = h_1 V^1 = V^1, \quad V_\theta = h_2 V^2 = r V^2, \quad V_z = h_3 V^3 = V^3, \quad g = r^2$

$$\frac{1}{r}\frac{\partial}{\partial r}\left(r \varrho V_r\right) + \frac{1}{r}\frac{\partial}{\partial \theta}\left(\varrho V_\theta\right) + \frac{\partial}{\partial z}\left(\varrho V_z\right) + \frac{\partial \varrho}{\partial t} = 0$$

▶ **43.**

$$(V^{ij}) = \begin{pmatrix} 0 & f & -e & d \\ -f & 0 & c & -b \\ e & -c & 0 & a \\ -d & b & -a & 0 \end{pmatrix}$$

▶ **48.**

$$g = e_{ijk} g_{1i} g_{2j} g_{3k}$$

$$\frac{\partial g}{\partial x^m} = e_{ijk} g_{1i} g_{2j} \frac{\partial g_{3k}}{\partial x^m} + e_{ijk} g_{1i} \frac{\partial g_{2j}}{\partial x^m} g_{3k} + e_{ijk} \frac{\partial g_{1i}}{\partial x^m} g_{2j} g_{3k}$$

$$\frac{\partial g}{\partial x^m} = e_{ijk} g_{1i} g_{2j} \delta_k^s \frac{\partial g_{3s}}{\partial x^m} + e_{ijk} \delta_j^s g_{1i} \frac{\partial g_{2s}}{\partial x^m} g_{3k} + e_{ijk} \delta_i^s \frac{\partial g_{1s}}{\partial x^m} g_{2j} g_{3k}$$

$$= e_{ijk} g_{1i} g_{2j} g_{nk} g^{ns} \frac{\partial g_{3s}}{\partial x^m} + e_{ijk} g_{nj} g^{ns} g_{1i} \frac{\partial g_{2s}}{\partial x^m} g_{3k} + e_{ijk} g_{ni} g^{ns} \frac{\partial g_{1s}}{\partial x^m} g_{2j} g_{3k}$$

$$= e_{ijk} g_{1i} g_{2j} g_{3k}\left[g^{3s} \frac{\partial g_{3s}}{\partial x^m} + g^{2s} \frac{\partial g_{2s}}{\partial x^m} + g^{1s} \frac{\partial g_{1s}}{\partial x^m}\right]$$

$$\frac{\partial g}{\partial x^m} = g g^{is} \frac{\partial g_{is}}{\partial x^m}$$

SELECTED SOLUTIONS EXERCISE 1.4

1-5. See Appendix B

6.

$$(g_{ij}) - \begin{pmatrix} 1 & 0 & \sin\phi\cos\alpha \\ 0 & r^2 & r\cos\phi\cos\alpha \\ \sin\phi\cos\alpha & r\cos\phi\cos\alpha & 1 \end{pmatrix}$$

Nonzero Christoffel symbols are:

$$[12,2] = r \qquad\qquad [22,1] = -r \qquad\qquad \begin{Bmatrix} 1 \\ 22 \end{Bmatrix} = -r$$

$$[12,3] = \cos\phi\cos\alpha \qquad [21,3] = \cos\phi\cos\alpha$$

$$[21,2] = r \qquad\qquad [22,3] = -r\sin\phi\cos\alpha \qquad \begin{Bmatrix} 2 \\ 21 \end{Bmatrix} = \begin{Bmatrix} 2 \\ 12 \end{Bmatrix} = \frac{1}{r}$$

9. (a) $\bar{\epsilon}_{abc,d} = \epsilon_{ijk,m}\dfrac{\partial x^i}{\partial \bar{x}^a}\dfrac{\partial x^j}{\partial \bar{x}^b}\dfrac{\partial x^k}{\partial \bar{x}^c}\dfrac{\partial x^m}{\partial \bar{x}^d}$ is a fourth order covariant tensor.

(b) If $\epsilon_{ijk,m} = 0$ in one coordinate system, then $\bar{\epsilon}_{abc,d} = 0$ in all coordinate systems. In Cartesian coordinates

$$e_{ijk} = \begin{cases} +1 \\ -1 \\ 0 \end{cases}$$ is a numerical tensor with covariant derivative zero. i.e. $e_{rst,p} = 0$

(c) $(\sqrt{g}e_{rst})_{,p} = 0$ gives $\sqrt{g}e_{rst,p} + (\sqrt{g})_{,p}e_{rst} = 0$ or $(\sqrt{g})_{,p} = 0$

11.

$$A_{i,kj} = \frac{\partial^2 A_i}{\partial x^k \partial x^j} - \frac{\partial A_\sigma}{\partial x^j}\begin{Bmatrix} \sigma \\ ik \end{Bmatrix} - \frac{\partial A_m}{\partial x^k}\begin{Bmatrix} m \\ ij \end{Bmatrix} - \frac{\partial A_i}{\partial x^m}\begin{Bmatrix} m \\ jk \end{Bmatrix}$$

$$- A_\sigma\left[\frac{\partial}{\partial x^j}\begin{Bmatrix} \sigma \\ ij \end{Bmatrix} - \begin{Bmatrix} \sigma \\ im \end{Bmatrix}\begin{Bmatrix} m \\ kj \end{Bmatrix} - \begin{Bmatrix} \sigma \\ mk \end{Bmatrix}\begin{Bmatrix} m \\ ij \end{Bmatrix}\right]$$

13. (b) Using the results from problem 44, for $\phi = \sqrt{g}$ and $W = 1$, we have $(\sqrt{g})_{,k} = \dfrac{\partial\sqrt{g}}{\partial x^k} - W\begin{Bmatrix} r \\ kr \end{Bmatrix}\sqrt{g}$.

Since $(\sqrt{g})_{,k} = 0$, then $\dfrac{\partial\sqrt{g}}{\partial x^k} = \begin{Bmatrix} r \\ kr \end{Bmatrix}\sqrt{g}$ or $\begin{Bmatrix} m \\ km \end{Bmatrix} = \dfrac{1}{\sqrt{g}}\dfrac{\partial\sqrt{g}}{\partial x^k} = \dfrac{\partial}{\partial x^k}\ln\sqrt{g}$

18.

$$C^i_{j,k} = \frac{\partial C^i_j}{\partial x^k} + C^\sigma_j\begin{Bmatrix} i \\ \sigma k \end{Bmatrix} - C^i_\sigma\begin{Bmatrix} \sigma \\ jk \end{Bmatrix} \qquad \text{With } C^i_j = A^i B_j \text{ this becomes}$$

$$(A^i B_j)_{,k} = \frac{\partial A^i B_j}{\partial x^k} + A^\sigma B_j\begin{Bmatrix} i \\ \sigma k \end{Bmatrix} - A^i B_\sigma\begin{Bmatrix} \sigma \\ jk \end{Bmatrix}$$

$$= A^i\frac{\partial B_j}{\partial x^k} + \frac{\partial A^i}{\partial x^k}B_j + A^\sigma B_j\begin{Bmatrix} i \\ \sigma k \end{Bmatrix} - A^i B_\sigma\begin{Bmatrix} \sigma \\ jk \end{Bmatrix}$$

$$= A^i\left[\frac{\partial B_j}{\partial x^k} - B_\sigma\begin{Bmatrix} \sigma \\ jk \end{Bmatrix}\right] + B_j\left[\frac{\partial A^i}{\partial x^k} + A^\sigma\begin{Bmatrix} i \\ \sigma k \end{Bmatrix}\right]$$

$$(A^i B_j)_{,k} = A^i B_{j,k} + B_j A^i_{,k}$$

EXERCISE 1.4

406

▶ **23.** (a) $\vec{A} = A^j \vec{E}_j$, with

$$d\vec{A} = \frac{\partial \vec{A}}{\partial x^k} dx^k = \left(\frac{\partial A^j}{\partial x^k} \vec{E}_j + A^j \frac{\partial \vec{E}_j}{\partial x^k} \right) dx^k$$

$$d\vec{A} = \left(\frac{\partial A^j}{\partial x^k} \vec{E}_j + A^j \left\{ \begin{matrix} i \\ j\,k \end{matrix} \right\} \vec{E}_i \right) dx^k$$

$$d\vec{A} = \left(\frac{\partial A^j}{\partial x^k} + \left\{ \begin{matrix} j \\ m\,k \end{matrix} \right\} A^m \right) \vec{E}_j\, dx^k$$

$$d\vec{A} = A^j_{,k}\, dx^k\, \vec{E}_j$$

▶ **27.** (a) $A_r = r, \quad A_\theta = r\cos\theta \quad A_z = z\sin\theta$

(b)

$$\begin{matrix} A_{rr} = 1 & A_{r\theta} = -\cos\theta & A_{rz} = 0 \\ A_{\theta r} = \cos\theta & A_{\theta\theta} = 1-\sin\theta & A_{\theta z} = 0 \\ A_{zr} = 0 & A_{z\theta} = \frac{z}{r}\cos\theta & A_{zz} = \sin\theta \end{matrix}$$

▶ **28.** $C_m = g^{pj}\epsilon_{mp\ell}A^\ell_{,j}$

▶ **30.**

$$[11,1] = \frac{1}{2}\frac{\partial g_{11}}{\partial u^1} \quad [11,2] = -\frac{1}{2}\frac{\partial g_{11}}{\partial u^2} \quad [21,2]=[12,2]=\frac{1}{2}\frac{\partial g_{22}}{\partial u^1}$$

$$[22,2] = \frac{1}{2}\frac{\partial g_{22}}{\partial u^2} \quad [22,1] = -\frac{1}{2}\frac{\partial g_{22}}{\partial u^1} \quad [12,1]=[21,1]=\frac{1}{2}\frac{\partial g_{11}}{\partial u^2}$$

$$\left\{ \begin{matrix} 1 \\ 1\,1 \end{matrix} \right\} = \frac{1}{2g_{11}}\frac{\partial g_{11}}{\partial u^1} \qquad \left\{ \begin{matrix} 2 \\ 1\,1 \end{matrix} \right\} = \frac{-1}{2g_{22}}\frac{\partial g_{11}}{\partial u^2}$$

$$\left\{ \begin{matrix} 1 \\ 2\,2 \end{matrix} \right\} = \frac{-1}{2g_{11}}\frac{\partial g_{22}}{\partial u^1} \qquad \left\{ \begin{matrix} 2 \\ 2\,2 \end{matrix} \right\} = \frac{1}{2g_{22}}\frac{\partial g_{22}}{\partial u^2}$$

$$\left\{ \begin{matrix} 1 \\ 1\,2 \end{matrix} \right\} = \left\{ \begin{matrix} 1 \\ 2\,1 \end{matrix} \right\} = \frac{1}{2g_{11}}\frac{\partial g_{11}}{\partial u^2} \qquad \left\{ \begin{matrix} 2 \\ 1\,2 \end{matrix} \right\} = \left\{ \begin{matrix} 2 \\ 2\,1 \end{matrix} \right\} = \frac{1}{2g_{22}}\frac{\partial g_{22}}{\partial u^1}$$

▶ **32.** Nonzero Christoffel symbols are:

$$[22,1] = -a^2\sin u^1\cos u^1 \qquad [21,2]=[12,2]=a^2\sin u^1\cos u^1$$

$$\left\{ \begin{matrix} 1 \\ 2\,2 \end{matrix} \right\} = -\sin u^1\cos u^1 \qquad \left\{ \begin{matrix} 2 \\ 2\,1 \end{matrix} \right\} = \left\{ \begin{matrix} 2 \\ 1\,2 \end{matrix} \right\} = \cot u^1$$

▶ **37.** (a) Cartesian (x,y,z)

$$\begin{matrix} V^1 = \frac{dx}{dt} & V^2 = \frac{dy}{dt} & V^3 = \frac{dz}{dt} \\ f^1 = \frac{d^2x}{dt^2} & f^2 = \frac{d^2y}{dt^2} & f^3 = \frac{d^2z}{dt^2} \\ V_1 = \frac{dx}{dt} & V_2 = \frac{dy}{dt} & V_3 = \frac{dz}{dt} \\ f_1 = \frac{d^2x}{dt^2} & f_2 = \frac{d^2y}{dt^2} & f_3 = \frac{d^2z}{dt^2} \\ V_x = \frac{dx}{dt} & V_y = \frac{dy}{dt} & V_z = \frac{dz}{dt} \\ f_x = \frac{d^2x}{dt^2} & f_y = \frac{d^2y}{dt^2} & f_z = \frac{d^2z}{dt^2} \end{matrix}$$

(b) Cylindrical (r,θ,z)

$$\begin{matrix} V^1 = \frac{dr}{dt} & V^2 = \frac{d\theta}{dt} & V^3 = \frac{dz}{dt} \\ f^1 = \frac{d^2r}{dt^2} - r\left(\frac{d\theta}{dt}\right)^2 & f^2 = \frac{d^2\theta}{dt^2} + \frac{2}{r}\frac{dr}{dt}\frac{d\theta}{dt} & f^3 = \frac{d^2z}{dt^2} \\ V_1 = \frac{dr}{dt} & V_2 = r^2\frac{d\theta}{dt} & V_3 = \frac{dz}{dt} \\ f_1 = \frac{d^2r}{dt^2} - r\left(\frac{d\theta}{dt}\right)^2 & f_2 = r^2\frac{d^2\theta}{dt^2} + 2r\frac{dr}{dt}\frac{d\theta}{dt} & f_3 = \frac{d^2z}{dt^2} \\ V_r = \frac{dr}{dt} & V_\theta = r\frac{d\theta}{dt} & V_z = \frac{dz}{dt} \\ f_r = \frac{d^2r}{dt^2} - r\left(\frac{d\theta}{dt}\right)^2 & f_\theta = r\frac{d^2\theta}{dt^2} + 2\frac{dr}{dt}\frac{d\theta}{dt} & f_z = \frac{d^2z}{dt^2} \end{matrix}$$

EXERCISE 1.4

41. By hypothesis S is a scalar of weight one so that $\overline{S} = JS$ and A^i_{jk} is a third order relative tensor of weight W so that $\overline{A}^m_{np} = J^W A^i_{jk} \frac{\partial \bar{x}^m}{\partial x^i} \frac{\partial x^j}{\partial \bar{x}^n} \frac{\partial x^k}{\partial \bar{x}^p}$. Therefore

$$\overline{S}^{-W} \overline{A}^m_{np} = J^{-W} S^{-W} J^W A^i_{jk} \frac{\partial \bar{x}^m}{\partial x^i} \frac{\partial x^j}{\partial \bar{x}^n} \frac{\partial x^k}{\partial \bar{x}^p}$$

$$\left(\overline{S}^{-W} \overline{A}^m_{np} \right) = \left(S^{-W} A^i_{jk} \right) \frac{\partial \bar{x}^m}{\partial x^i} \frac{\partial x^j}{\partial \bar{x}^n} \frac{\partial x^k}{\partial \bar{x}^p}$$

Hence $S^{-W} A^i_{jk}$ is an absolute third order tensor.

42. We have $X^m = \overline{Y}^i \frac{\partial x^m}{\partial \bar{x}^i}$ and $\frac{dx^i}{dt} = \frac{\partial x^i}{\partial \bar{y}^j} \frac{d\bar{y}^j}{dt}$.

$$g_{mn} x^m \frac{dx^n}{dt} = \left(\bar{g}_{pq} \frac{\partial \bar{y}^p}{\partial x^m} \frac{\partial \bar{y}^q}{\partial x^n} \right) \left(\overline{Y}^i \frac{\partial x^m}{\partial \bar{y}^i} \right) \left(\frac{\partial x^n}{\partial \bar{y}^j} \frac{d\bar{y}^j}{dt} \right)$$

$$= \bar{g}_{pq} \delta^p_i \delta^q_j \overline{Y}^i \frac{d\bar{y}^j}{dt} = \bar{g}_{ij} \overline{Y}^i \frac{d\bar{y}^j}{dt} = \cos \theta = \text{constant}$$

43. $J = J\left(\frac{x}{\bar{x}}\right) = \left| \frac{\partial x}{\partial \bar{x}} \right| = e_{ijk} \frac{\partial x^i}{\partial \bar{x}^1} \frac{\partial x^j}{\partial \bar{x}^2} \frac{\partial x^k}{\partial \bar{x}^3}$

$$\frac{\partial J}{\partial x^m} = e_{ijk} \frac{\partial x^i}{\partial \bar{x}^1} \frac{\partial x^j}{\partial \bar{x}^2} \frac{\partial^2 x^k}{\partial \bar{x}^3 \partial \bar{x}^p} \frac{\partial \bar{x}^p}{\partial x^m} + e_{ijk} \frac{\partial x^i}{\partial \bar{x}^1} \frac{\partial^2 x^j}{\partial \bar{x}^2 \partial \bar{x}^p} \frac{\partial \bar{x}^p}{\partial x^m} + e_{ijk} \frac{\partial^2 x^i}{\partial \bar{x}^1 \partial \bar{x}^p} \frac{\partial x^j}{\partial \bar{x}^2} \frac{\partial x^k}{\partial \bar{x}^3} \frac{\partial \bar{x}^p}{\partial x^m}$$

Now use the result (1.4.7) and write

$$\frac{\partial J}{\partial x^m} = e_{ijk} \frac{\partial x^i}{\partial \bar{x}^1} \frac{\partial x^j}{\partial \bar{x}^2} \left[\left\{ \begin{matrix} \mu \\ 3\,p \end{matrix} \right\} \frac{\partial x^k}{\partial \bar{x}^\mu} - \left\{ \begin{matrix} k \\ a\,c \end{matrix} \right\} \frac{\partial x^a}{\partial \bar{x}^3} \frac{\partial x^c}{\partial \bar{x}^p} \right] \frac{\partial \bar{x}^p}{\partial x^m}$$

$$+ e_{ijk} \frac{\partial x^i}{\partial \bar{x}^1} \left[\left\{ \begin{matrix} \mu \\ 2\,p \end{matrix} \right\} - \left\{ \begin{matrix} j \\ a\,c \end{matrix} \right\} \frac{\partial x^u}{\partial \bar{x}^2} \frac{\partial x^c}{\partial \bar{x}^p} \right] \frac{\partial x^k}{\partial \bar{x}^3} \frac{\partial x^p}{\partial x^m}$$

$$+ e_{ijk} \left[\left\{ \begin{matrix} \mu \\ 1\,p \end{matrix} \right\} \frac{\partial x^i}{\partial \bar{x}^\mu} - \left\{ \begin{matrix} i \\ a\,c \end{matrix} \right\} \frac{\partial x^a}{\partial \bar{x}^1} \frac{\partial x^c}{\partial \bar{x}^p} \right] \frac{\partial x^j}{\partial \bar{x}^2} \frac{\partial x^k}{\partial \bar{x}^3} \frac{\partial \bar{x}^p}{\partial x^m}$$

which simplifies to

$$\frac{\partial J}{\partial x^m} = e_{ijk} \frac{\partial x^i}{\partial \bar{x}^1} \frac{\partial x^j}{\partial \bar{x}^2} \frac{\partial x^k}{\partial \bar{x}^3} \overline{\left\{ \begin{matrix} \alpha \\ \alpha\,p \end{matrix} \right\}} \frac{\partial \bar{x}^p}{\partial x^m} - e_{ijk} \frac{\partial x^i}{\partial \bar{x}^1} \frac{\partial x^j}{\partial \bar{x}^2} \left\{ \begin{matrix} k \\ a\,m \end{matrix} \right\} \frac{\partial x^a}{\partial \bar{x}^3}$$

$$- e_{ijk} \frac{\partial x^i}{\partial \bar{x}^1} \frac{\partial x^a}{\partial \bar{x}^2} \left\{ \begin{matrix} j \\ a\,m \end{matrix} \right\} \frac{\partial x^k}{\partial \bar{x}^3} - e_{ijk} \frac{\partial x^a}{\partial \bar{x}^1} \left\{ \begin{matrix} i \\ a\,m \end{matrix} \right\} \frac{\partial x^j}{\partial \bar{x}^2} \frac{\partial x^k}{\partial \bar{x}^3} \qquad (A)$$

Now let $a^i_r = \frac{\partial x^i}{\partial \bar{x}^r}$ with A^r_i the cofactor of a^i_r in Jacobian determinant. One can then write (see number 27 Exercise 1.1)

$$e_{ijk} \frac{\partial x^i}{\partial \bar{x}^1} \frac{\partial x^j}{\partial \bar{x}^2} = e_{kij} \frac{\partial x^i}{\partial \bar{x}^1} \frac{\partial x^j}{\partial \bar{x}^2} = e_{r12} A^r_k$$

$$\text{similarly} \quad e_{ijk} \frac{\partial x^i}{\partial \bar{x}^1} \frac{\partial x^k}{\partial \bar{x}^3} = e_{jki} \frac{\partial x^k}{\partial \bar{x}^3} \frac{\partial x^i}{\partial \bar{x}^1} = e_{r31} A^r_j$$

$$\text{and} \quad e_{ijk} \frac{\partial x^j}{\partial \bar{x}^2} \frac{\partial x^k}{\partial \bar{x}^3} = e_{r23} A^r_i$$

Then one can show that the right-hand side of equation (A) above simplifies to the result

$$\frac{\partial J}{\partial x^m} = J \overline{\left\{ \begin{matrix} \alpha \\ \alpha\,p \end{matrix} \right\}} \frac{\partial \bar{x}^p}{\partial x^m} - J \left\{ \begin{matrix} k \\ k\,m \end{matrix} \right\}$$

47. The given quantity T^{jk} is not a tensor.

EXERCISE 1.4

SELECTED SOLUTIONS EXERCISE 1.5

▶ **2.**

(a) This is the plane $z = 0$ with tangents to the coordinate curves $\frac{\partial \vec{r}}{\partial u} = \hat{\mathbf{e}}_1$ and $\frac{\partial \vec{r}}{\partial v} = \hat{\mathbf{e}}_2$. Coordinate curves are $x = u_0 = $ constant and $y = v_0 = $ constant

(b) This is again the plane $z = 0$ with coordinate curves $x^2 + y^2 = u_0^2$ which are circles and $y/x = \tan v_0$ which are straight lines. (i.e. polar coordinates). The tangents to the coordinate curves are
$\frac{\partial \vec{r}}{\partial u} = \cos v \, \hat{\mathbf{e}}_1 + \sin v \, \hat{\mathbf{e}}_2$ and $\frac{\partial \vec{r}}{\partial v} = -u \sin v \, \hat{\mathbf{e}}_1 + u \cos v \, \hat{\mathbf{e}}_2$.

(c) This is again the plane $z = 0$ with coordinate curves $(x - u_0)^2 + y^2 = u_0^2$ and $x^2 + (y - v_0)^2 = v_0^2$ which are circles, all passing through the orgin. See figure

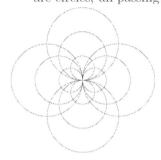

▶ **3.** (a)   (b)

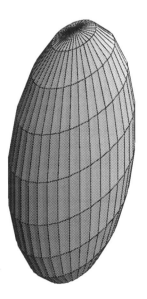
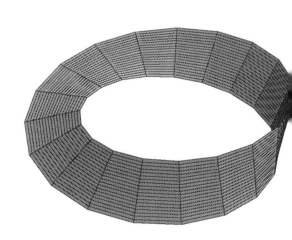

EXERCISE 1.5

3. (c)

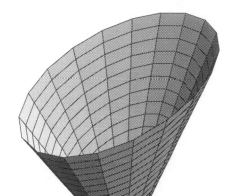

(d)

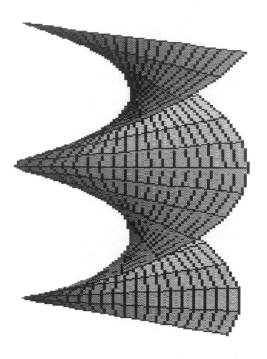

(e)

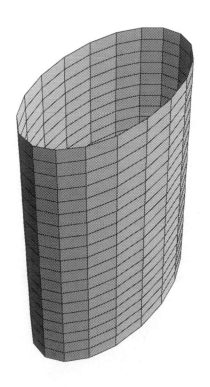

(f)

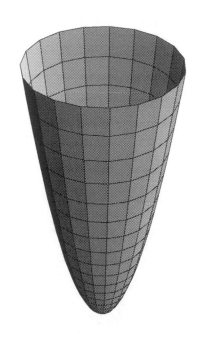

EXERCISE 1.5

410

▶ **5.** Use $[\alpha\beta, \gamma] = \dfrac{1}{2}\left[\dfrac{\partial a_{\beta\gamma}}{\partial x^{\alpha}} + \dfrac{\partial a_{\alpha\gamma}}{\partial u^{\beta}} - \dfrac{\partial a_{\alpha\beta}}{\partial u^{\gamma}}\right]$ with $\alpha, \beta = 1, 2$.

▶ **7.** Use $\begin{Bmatrix} i \\ j\,k \end{Bmatrix} = a^{im}[jk, m] = a^{i1}[jk, 1] + a^{i2}[jk, 2]$ with

$$a_{ij} = \begin{pmatrix} E & F \\ F & G \end{pmatrix} \text{ and } a^{ij} = \tfrac{1}{EG-F^2}\begin{pmatrix} G & -F \\ -F & E \end{pmatrix}$$

▶ **10.**

$$(\vec{r}_u \times \vec{r}_{uu}) \cdot \hat{n} = \frac{(\vec{r}_u \times \vec{r}_{uu}) \cdot (\vec{r}_u \times \vec{r}_v)}{\sqrt{EG - F^2}} = \frac{(\vec{r}_u \cdot \vec{r}_u)(\vec{r}_{uu} \cdot \vec{r}_v) - (\vec{r}_u \cdot \vec{r}_v)(\vec{r}_{uu} \cdot \vec{r}_u)}{\sqrt{EG - F^2}}$$

$$(\vec{r}_u \times \vec{r}_{uu}) \cdot \hat{n} = \frac{E[11, 2] - F[11, 1]}{\sqrt{EG - F^2}}$$

$$(\vec{r}_u \times \vec{r}_{uu}) \cdot \hat{n} = \left(g^{21}[11, 1] + g^{22}[11, 2]\right) = \begin{Bmatrix} 2 \\ 11 \end{Bmatrix}\sqrt{EG - F^2}$$

with similar results for the other relations.

▶ **11.** Use the results $\lambda_1\lambda_2 = \dfrac{Ef - Fe}{Fg - fG}$ and $\lambda_1 + \lambda_2 = \dfrac{-(Eg - Ge)}{Fg - fG}$ to obtain desired result.

▶ **12.** Use $\delta_{mn}^{rs} = \delta_m^r\delta_n^s - \delta_n^r\delta_m^s$, expand and make use of R_{ijkl} symmetries.

▶ **15.** Use $\dfrac{\partial}{\partial x^k}\left(g_{im}\begin{Bmatrix} m \\ j\,l \end{Bmatrix}\right) = g_{im}\dfrac{\partial}{\partial x^k}\begin{Bmatrix} m \\ j\,l \end{Bmatrix} + \dfrac{\partial g_{im}}{\partial x^k}\begin{Bmatrix} m \\ j\,l \end{Bmatrix}$ and

$\dfrac{\partial}{\partial x^k}[jl, i] = g_{im}\dfrac{\partial}{\partial x^k}\begin{Bmatrix} m \\ j\,l \end{Bmatrix} + ([ik, m] + [mk, i])\begin{Bmatrix} m \\ j\,l \end{Bmatrix}$ to show that

$$R_{irst} = \frac{\partial}{\partial x^s}[rt, i] - ([is, m] + [ms, i])\begin{Bmatrix} m \\ r\,t \end{Bmatrix}$$

$$- \frac{\partial}{\partial x^t}[rs, i] + ([it, m] + [mt, i])\begin{Bmatrix} m \\ r\,s \end{Bmatrix} + \begin{Bmatrix} p \\ r\,t \end{Bmatrix}[ps, i] - \begin{Bmatrix} p \\ r\,s \end{Bmatrix}[tp, i]$$

which simplifies to the desired result.

▶ **22.** Use $R_{2112} = -R_{1212}$, $R_{2121} = R_{1212}$ and $R_{1221} = -R_{1212}$ and show from the transformation law for R_{ijkl} that

$$\overline{R}_{1212} = R_{1212}\left(\frac{\partial x^1}{\partial \bar{x}^1}\frac{\partial x^2}{\partial \bar{x}^2}\frac{\partial x^1}{\partial \bar{x}^1}\frac{\partial x^2}{\partial \bar{x}^2} - 2\frac{\partial x^1}{\partial \bar{x}^1}\frac{\partial x^2}{\partial \bar{x}^2}\frac{\partial x^1}{\partial \bar{x}^2}\frac{\partial x^2}{\partial \bar{x}^1} + \frac{\partial x^1}{\partial \bar{x}^2}\frac{\partial x^2}{\partial \bar{x}^1}\frac{\partial x^1}{\partial \bar{x}^2}\frac{\partial x^2}{\partial \bar{x}^1}\right)$$

$$= R_{1212}\left(\frac{\partial x^1}{\partial \bar{x}^1}\frac{\partial x^2}{\partial \bar{x}^2} - \frac{\partial x^1}{\partial \bar{x}^2}\frac{\partial x^2}{\partial \bar{x}^1}\right)^2$$

which simplifies to the desired result.

27. Show that

$$R_{1212} = \frac{\partial}{\partial u^1}[22,1] - \frac{\partial}{\partial u^2}[21,1] + g^{\sigma m}[12,\sigma][21,m] - g^{\sigma m}[11,\sigma][22,m]$$

Expand and verify that

$$R_{1212} = -\frac{1}{2}\left\{\frac{\partial}{\partial u^1}\left[-\frac{\partial a_{22}}{\partial u^1}\right] + \frac{\partial}{\partial u^2}\left[\frac{\partial a_{11}}{\partial u^2}\right]\right\} + g^{11}[12,1][21,1] + g^{22}[12,2][21,2]$$
$$- g^{11}[11,1][22,1] - g^{22}[11,2][22,2]$$

from which

$$K = \frac{R_{1212}}{a} = -\frac{1}{2\sqrt{a}}\left[\frac{1}{\sqrt{a}}\frac{\partial}{\partial u^1}\left(\frac{\partial a_{22}}{\partial u^1}\right) + \frac{1}{\sqrt{a}}\frac{\partial}{\partial u^2}\left(\frac{\partial a_{11}}{\partial u^2}\right) - \frac{1}{2\sqrt{a}\,a_{11}}\left(\frac{\partial a_{11}}{\partial u^2}\right)^2 - \frac{1}{2\sqrt{a}\,a_{22}}\left(\frac{\partial a_{22}}{\partial u^1}\right)^2 \right.$$
$$\left. - \frac{1}{2\sqrt{a}\,a_{11}}\frac{\partial a_{11}}{\partial u^1}\frac{\partial a_{22}}{\partial u^1} - \frac{1}{2\sqrt{a}\,a_{22}}\frac{\partial a_{11}}{\partial u^2}\frac{\partial a_{22}}{\partial u^2}\right]$$

Note that for $a = a_{11}a_{22}$

$$\frac{\partial}{\partial u^i}\left(\frac{1}{\sqrt{a}}\right) = -\frac{1}{2\sqrt{a}}\left(\frac{1}{a_{22}}\frac{\partial a_{22}}{\partial u^i} + \frac{1}{a_{11}}\frac{\partial a_{11}}{\partial u^i}\right) \qquad i=1,2$$

so that the above can be expressed

$$K = -\frac{1}{2\sqrt{a}}\left[\frac{1}{\sqrt{a}}\frac{\partial}{\partial u^1}\left(\frac{\partial a_{22}}{\partial u^1}\right) + \frac{1}{\sqrt{a}}\frac{\partial}{\partial u^2}\left(\frac{\partial a_{11}}{\partial u^2}\right) + \frac{\partial a_{22}}{\partial u^1}\frac{\partial}{\partial u^1}\left(\frac{1}{\sqrt{a}}\right) + \frac{\partial a_{11}}{\partial u^2}\frac{\partial}{\partial u^2}\left(\frac{1}{\sqrt{a}}\right)\right]$$

which simplifies to the desired result.

29.

$$\frac{\delta T^i}{\delta s} = T^i_{,j}\frac{dx^j}{dt} = \left[\frac{\partial T^i}{\partial x^j} + T^k\left\{\begin{matrix}i\\j\,k\end{matrix}\right\}\right]\frac{dx^j}{dt}\kappa N^i$$

with similar results for the other equations.

32.

$$c_{\alpha\beta} - 2Hb_{\alpha\beta} + Ka_{\alpha\beta} = 0$$
$$a^{\alpha\beta}c_{\alpha\beta} = 2Ha^{\alpha\beta}b_{\alpha\beta} - Ka^{\alpha\beta}a_{\alpha\beta}$$
$$= 2H(2H) - K\delta^\alpha_\alpha = 4H^2 - 2K$$

36.

$$\frac{d^2\xi}{ds^2} + \frac{\sin\xi}{b}(a+b\cos\xi)\left(\frac{d\eta}{ds}\right)^2 = 0$$
$$\frac{d^2\eta}{ds^2} - \frac{2b\sin\xi}{a+b\cos\xi}\frac{d\eta}{ds}\frac{d\xi}{ds} = 0$$

42.

$$(a) \qquad \hat{\mathbf{n}} = \frac{f_x\hat{\mathbf{e}}_1 + f_y\hat{\mathbf{e}}_2 + f_z\hat{\mathbf{e}}_3}{\sqrt{f_x^2 + f_y^2 + 1}}$$

$$(b) \qquad \hat{\mathbf{n}} = \frac{F_x\hat{\mathbf{e}}_1 + F_y\hat{\mathbf{e}}_2 + F_z\hat{\mathbf{e}}_3}{\sqrt{F_x^2 + F_y^2 + F_z^2}}$$

$$(c) \qquad \hat{\mathbf{n}} = \frac{\vec{r}_u \times \vec{r}_v}{\sqrt{EG - F^2}}$$

EXERCISE 1.5

▶ **1.** $\operatorname{grad} f = \dfrac{\partial f}{\partial r}\hat{\mathbf{e}}_r + \dfrac{1}{r}\dfrac{\partial f}{\partial \theta}\hat{\mathbf{e}}_\theta + \dfrac{\partial f}{\partial z}\hat{\mathbf{e}}_z$

▶ **3.** $\operatorname{curl}\vec{A} = \dfrac{1}{r}\left(\dfrac{\partial A_z}{\partial \theta} - \dfrac{\partial (rA_\theta)}{\partial z}\right)\hat{\mathbf{e}}_r + \left(\dfrac{\partial A_r}{\partial z} - \dfrac{\partial A_z}{\partial r}\right)\hat{\mathbf{e}}_\theta + \dfrac{1}{r}\left(\dfrac{\partial (rA_\theta)}{\partial r} - \dfrac{\partial A_r}{\partial \theta}\right)\hat{\mathbf{e}}_z$

▶ **5.** $\operatorname{grad} f = \dfrac{\partial f}{\partial \rho}\hat{\mathbf{e}}_\rho + \dfrac{1}{\rho}\dfrac{\partial f}{\partial \theta}\hat{\mathbf{e}}_\theta + \dfrac{1}{\rho \sin\theta}\dfrac{\partial f}{\partial \phi}\hat{\mathbf{e}}_\phi$

▶ **7.** $\operatorname{curl}\vec{A} = \dfrac{1}{\rho^2 \sin\theta}\left[\dfrac{\partial}{\partial \theta}(\rho \sin\theta A_\phi) - \dfrac{\partial}{\partial \phi}(\rho A_\theta)\right]\hat{\mathbf{e}}_\rho + \dfrac{1}{\rho \sin\theta}\left[\dfrac{\partial A_\rho}{\partial \phi} - \dfrac{\partial}{\partial \rho}(\rho \sin\theta A_\phi)\right]\hat{\mathbf{e}}_\theta + \dfrac{1}{\rho}\left[\dfrac{\partial}{\partial \rho}(\rho A_\theta) - \dfrac{\partial A_\rho}{\partial \theta}\right]\hat{\mathbf{e}}_\phi$

▶ **9.**

$$(a)\quad \frac{\vec{r}}{r} \quad (b)\quad mr^{m-2}\vec{r} \quad (c)\quad -\frac{\vec{r}}{r^3} \quad (d)\quad \frac{\vec{r}}{r^2} \quad (e)\quad \phi'(r)\frac{\vec{r}}{r}$$

▶ **11.** (a) 0   (b) 0

▶ **17.**

$$(a)\quad D_{ij} = \frac{1}{2}\left[\frac{\partial V_i}{\partial x^j} - V_m \left\{\begin{matrix} m \\ i\,j \end{matrix}\right\} + \frac{\partial V_j}{\partial x^i} - V_m \left\{\begin{matrix} m \\ j\,i \end{matrix}\right\}\right]$$

when $j = i$ $\quad D_{(i)(i)} = \dfrac{1}{2}\left[2\dfrac{\partial V_{(i)}}{\partial x^i} - 2\left(V_1\left\{\begin{matrix} 1 \\ i\,i \end{matrix}\right\} + V_2\left\{\begin{matrix} 2 \\ i\,i \end{matrix}\right\} + V_3\left\{\begin{matrix} 3 \\ i\,i \end{matrix}\right\}\right)\right]$

In terms of physical components

$$h_i h_i D(ii) = \frac{\partial}{\partial x^i}\left(h_i V(i)\right) - \sum_{j=1}^{3} h_j V(j)\left\{\begin{matrix} j \\ i\,i \end{matrix}\right\} \qquad \text{no sum on i}$$

When summed $j = 1, 2, 3$, one of these values must equal i, hence

$$h_i^2 D(ii) = h_i \frac{\partial V(i)}{\partial x^i} + V(i)\frac{\partial h_i}{\partial x^i} - \sum_{j \neq i} h_j V(j)\left\{\begin{matrix} j \\ i\,i \end{matrix}\right\} - h_i V(i)\left\{\begin{matrix} i \\ i\,i \end{matrix}\right\} \qquad \text{no sum on i}$$

Now $\left\{\begin{matrix} j \\ i\,i \end{matrix}\right\} = -\dfrac{h_i}{h_j^2}\dfrac{\partial h_i}{\partial x^j}$ and $\left\{\begin{matrix} i \\ i\,i \end{matrix}\right\} = \dfrac{1}{h_i}\dfrac{\partial h_i}{\partial x^i}$ no sum on i

These substitutions simplify the above to the result given.

▶ **19.**

$$D_{rr} = \frac{\partial V_r}{\partial r} \qquad\qquad D_{r\theta} = D_{\theta r} = \frac{1}{2}\left[\frac{1}{r}\frac{\partial V_r}{\partial \theta} + r\frac{\partial}{\partial r}\left(\frac{V_\theta}{r}\right)\right]$$

$$D_{\theta\theta} = \frac{1}{r}\frac{\partial V_\theta}{\partial \theta} + \frac{V_r}{r} \qquad D_{rz} = D_{zr} = \frac{1}{2}\left[\frac{\partial V_r}{\partial z} + \frac{\partial V_z}{\partial r}\right]$$

$$D_{zz} = \frac{\partial V_z}{\partial z} \qquad\qquad D_{\theta z} = D_{z\theta} = \frac{1}{2}\left[r\frac{\partial V_\theta}{\partial z} + \frac{1}{r}\frac{\partial V_z}{\partial \theta}\right]$$

▶ **21.**

$$(\lambda + 2\mu)\nabla\phi - 2\mu\nabla \times \vec{\omega} + \vec{F} = \vec{0}$$

$$(\lambda + 2\mu)g^{im}\phi_{,m} - 2\mu\epsilon^{ijk}\omega_{k,j} + F^i = 0$$

23.

$$(\lambda + 2\mu)\frac{\partial \phi}{\partial r} - \frac{2\mu}{r}\left[\frac{\partial \omega_z}{\partial \theta} - \frac{\partial (r\omega_\theta)}{\partial z}\right] + F_r = 0$$

$$(\lambda + 2\mu)\frac{1}{r^2}\frac{\partial \phi}{\partial \theta} - \frac{2\mu}{r}\left[\frac{\partial \omega_r}{\partial z} - \frac{\partial \omega_z}{\partial r}\right] + \frac{1}{r}F_\theta = 0$$

$$(\lambda + 2\mu)\frac{\partial \phi}{\partial z} - \frac{2\mu}{r}\left[\frac{\partial (r\omega_\theta)}{\partial r} - \frac{\partial \omega_r}{\partial \theta}\right] + F_z = 0$$

25.

$$\frac{1}{\xi^2 + \eta^2}\left[\frac{\partial}{\partial \xi}(\sqrt{\xi^2 + \eta^2}A_\xi) + \frac{\partial}{\partial \eta}(\sqrt{\xi^2 + \eta^2}A_\eta) + \frac{\partial}{\partial z}((\xi^2 + \eta^2)A_z)\right]$$

27.

$$\frac{1}{\sinh^2 \xi + \sin^2 \eta}\left[\frac{\partial}{\partial \xi}(\sqrt{\sinh^2 \xi + \sin^2 \eta}A_\xi) + \frac{\partial}{\partial \eta}(\sqrt{\sinh^2 \xi + \sin^2 \eta}A_\eta) + \frac{\partial}{\partial z}((\sinh^2 \xi + \sin^2 \eta)A_z)\right]$$

29.

$$\frac{d}{dt}\left(\epsilon_{ijk}A^i B^j C^k\right) = \epsilon_{ijk}\frac{dA^i}{dt}B^j C^k + \epsilon_{ijk}A^i\frac{dB^j}{dt}C^k + \epsilon_{ijk}A^i B^j\frac{dC^k}{dt}$$

31. $\frac{1}{c}\frac{\partial H^i}{\partial t} = -\epsilon^{ijk}E_{k,j}$

33. $\operatorname{curl}\vec{B} + \vec{F} = \vec{0}$

35. $\frac{\partial \rho}{\partial t} + \nabla(\rho\vec{V}) = 0$

37. $I_{11} = \int_{y_1=0}^{b}\int_{y_2=0}^{\frac{h}{b}y_1+h} y_2^2 dy_2 dy_1 = \frac{bh^3}{12}$

38. $\bar{I}_{ij} = I_{ab}\frac{\partial y^a}{\partial \bar{y}^i}\frac{\partial y^b}{\partial \bar{y}^j}$ where $\begin{array}{l} y_1 = \bar{y}_1\cos\theta - \bar{y}_2\sin\theta \\ y_2 = \bar{y}_1\sin\theta + \bar{y}_2\cos\theta \end{array}$ $\bar{I}_{11}$ is a maximum when $\frac{d\bar{I}_{11}}{d\theta} = 0$

47. Eigenvalues $-1, 1, 4$, Eigenvectors $\begin{pmatrix} -1 \\ 0 \\ 1 \end{pmatrix}, \begin{pmatrix} 1 \\ -2 \\ 1 \end{pmatrix}, \begin{pmatrix} 1 \\ 1 \\ 1 \end{pmatrix}$

48. Eigenvalues $1, 1-\sqrt{5}, 1+\sqrt{5}$, Eigenvectors $\begin{pmatrix} 0 \\ -1 \\ 2 \end{pmatrix}, \begin{pmatrix} -\sqrt{5} \\ 2 \\ 1 \end{pmatrix}, \begin{pmatrix} \sqrt{5} \\ 2 \\ 1 \end{pmatrix}$

49. Eigenvalues $1, 1-\sqrt{2}, 1+\sqrt{2}$, Eigenvectors $\begin{pmatrix} -1 \\ 0 \\ 1 \end{pmatrix}, \begin{pmatrix} 1 \\ -\sqrt{2} \\ 1 \end{pmatrix}, \begin{pmatrix} 1 \\ \sqrt{2} \\ 1 \end{pmatrix}$

50. (b) $\nabla^4\phi = \nabla^2\nabla^2\phi = \left(\frac{\partial^2}{\partial r^2} + \frac{1}{r}\frac{\partial}{\partial r} + \frac{1}{r^2}\frac{\partial^2}{\partial \theta^2}\right)\left(\frac{\partial^2\phi}{\partial r^2} + \frac{1}{r}\frac{\partial\phi}{\partial r} + \frac{1}{r^2}\frac{\partial^2\phi}{\partial \theta^2}\right)$

EXERCISE 2.1

SELECTED SOLUTIONS EXERCISE 2.2

▶ **1.** Solution not unique $x = 1 + t \quad y = 1 + 2t \quad z = 1 + 3t \quad \vec{T} = \dfrac{d\vec{r}}{ds} = \dfrac{\hat{\mathbf{e}}_1 + 2\,\hat{\mathbf{e}}_2 + 3\,\hat{\mathbf{e}}_3}{\sqrt{14}}$

▶ **3.** For $t = 1, (1, 1, 1)$ For $t = 2, (2, 4, 8)$ For $t = 3, (3, 9, 27)$

▶ **5.** Tangent plane $x + y = \sqrt{2}$. Tangent line at $\vec{r}_0$ is $\vec{r} = \vec{r}_0 + \lambda \vec{T}_1$ where $\vec{r}_0 = \dfrac{1}{\sqrt{2}}\,\hat{\mathbf{e}}_1 + \dfrac{1}{\sqrt{2}}\,\hat{\mathbf{e}}_2 + \dfrac{1}{2}\,\hat{\mathbf{e}}_3$ and

$\vec{T}_1 = \dfrac{\frac{1}{\sqrt{2}}\,\hat{\mathbf{e}}_1 - \frac{1}{\sqrt{2}}\,\hat{\mathbf{e}}_2 + \frac{2}{\pi}\,\hat{\mathbf{e}}_3}{\sqrt{1 + \frac{4}{\pi^2}}}$ and the scalar curvature is $\kappa = \dfrac{1}{1 + \frac{4}{\pi^2}}$.

▶ **13.** mass $= M = \displaystyle\iiint_V \rho\, d\tau$ Assume element of mass is located at general position x^i, then moment

of mass about the origin is given by $M_0^i = \displaystyle\iiint_V x^i \rho\, d\tau$. For ξ^i the center of mass, we want $M\xi^i = M_0^i$,

therefore $\xi^i = \dfrac{M_0^i}{M} = \dfrac{\iiint_V x^i \rho\, d\tau}{\iiint_V \rho\, d\tau}$

▶ **17.**
$$(m_1 + m_2 + m_3)\frac{d^2 y_1}{dt^2} + (m_3 - m_2)\frac{d^2 y_2}{dt^2} - (m_1 - m_2 - m_3)g = 0$$
$$(m_3 - m_2)\frac{d^2 y_1}{dt^2} + (m_2 + m_3)\frac{d^2 y_2}{dt^2} + (m_2 - m_3)g = 0$$

▶ **19.** $\dfrac{\delta T}{\delta t} = m f_i \dot{x}^i$ and for $m f_i = Q_i = -\dfrac{\partial V}{\partial x^i}$ we have $\dfrac{\delta T}{\delta t} = m f_i \dot{x}^i = Q_i v^i = -\dfrac{\partial V}{\partial x^i} v^i = -\dfrac{\delta V}{\delta t}$ or

$\dfrac{\delta}{\delta t}(T + V) = 0$ which shows $T + V = $ constant.

▶ **21.**
$$B^i = \epsilon^{ijk} T_j N_k = \epsilon^{ijk} g_{jm} T^m g_{kn} N^n \qquad \frac{\delta B^i}{\delta s} = \epsilon^{ijk} g_{jm} g_{kn} \left[T^m \frac{\delta N^n}{\delta s} + N^n \frac{\delta T^m}{\delta s} \right]$$
$$\frac{\delta B^i}{\delta s} = \epsilon^{ijk} g_{jm} g_{kn} \left[T^m(\tau B^n - \kappa T^n) + N^n(\kappa N^m) \right] = \epsilon^{ijk} \left[\tau T_j B_k - \kappa T_j T_k + \kappa N_k N_j \right]$$
$$= -\epsilon^{ijk} \tau B_k T_j = -\tau N^i$$

▶ **24.**
$$(m_1 + m_2 + m_3)L_1^2 \frac{d^2\theta_1}{dt^2} + (m_2 + m_3)L_1 L_2 \frac{d^2\theta_2}{dt^2}\cos(\theta_1 - \theta_2) + m_3 L_1 L_3 \frac{d^2\theta_3}{dt^2}\cos(\theta_1 - \theta_3)$$
$$+ (m_2 + m_3)L_1 L_2 \left(\frac{d\theta_2}{dt}\right)^2 \sin(\theta_1 - \theta_2) + m_3 L_1 L_3 \left(\frac{d\theta_3}{dt}\right)^2 \sin(\theta_1 - \theta_3) + (m_1 + m_2 + m_3)g L_1 \sin\theta_1 = 0$$

$$(m_2 + m_3)L_2^2 \frac{d^2\theta_2}{dt^2} + (m_2 + m_3)L_1 L_2 \frac{d^2\theta_1}{dt^2}\cos(\theta_1 - \theta_2) + m_3 L_2 L_3 \frac{d^2\theta_3}{dt^2}\cos(\theta_2 - \theta_3)$$
$$- (m_2 + m_3)L_1 L_2 \left(\frac{d\theta_1}{dt}\right)^2 \sin(\theta_1 - \theta_2) + m_3 L_2 L_3 \left(\frac{d\theta_3}{dt}\right)^2 \sin(\theta_2 - \theta_3) + (m_2 + m_3)g L_2 \sin\theta_2 = 0$$

$$m_3 L_3^2 \frac{d^2\theta_3}{dt^2} + m_3 L_1 L_3 \frac{d^2\theta_1}{dt^2}\cos(\theta_1 - \theta_3) + m_3 L_2 L_3 \frac{d^2\theta_2}{dt^2}\cos(\theta_2 - \theta_3) - m_3 L_1 L_3 \left(\frac{d\theta_1}{dt}\right)^2 \sin(\theta_1 - \theta_3)$$
$$- m_3 L_2 L_3 \left(\frac{d\theta_2}{dt}\right)^2 \sin(\theta_2 - \theta_3) + m_3 g L_3 \sin\theta_3 = 0$$

▶ **26.**
$$\begin{pmatrix} x \\ y \\ z \end{pmatrix} = \begin{pmatrix} \cos\psi & -\sin\psi & 0 \\ \sin\psi & \cos\psi & 0 \\ 0 & 0 & 1 \end{pmatrix} \begin{pmatrix} 1 & 0 & 0 \\ 0 & \cos\phi & -\sin\phi \\ 0 & \sin\phi & \cos\phi \end{pmatrix} \begin{pmatrix} \cos\theta & 0 & -\sin\theta \\ 0 & 1 & 0 \\ \sin\theta & 0 & \cos\theta \end{pmatrix} \begin{pmatrix} X \\ Y \\ Z \end{pmatrix}$$

EXERCISE 2.2

SELECTED SOLUTIONS EXERCISE 2.3

19.

$$|a|^2 = g_{ij}a^i a^j \qquad\qquad 2|a|\delta a = g_{ij}a^i\,\delta a^j + g_{ij}\delta a^i\,a^j$$

$$2|a|\delta a = g_{ij}a^i(u^j_{,s}a^s) + g_{ij}(u^i_{,p}a^p)a^j = a^i a^s(u_{i,s} + u_{s,i})$$

$$2|a|\delta a = 2e_{is}a^i a^s$$

Hence by definition

$$e = \frac{\delta a}{|a|} = \frac{e_{rs}a^r a^s}{|a|\,|a|} = e_{rs}\lambda^r\lambda^s \qquad \lambda^r = \frac{a^r}{|a|}$$

Now consider the cases

$$\lambda^1 = 1,\ \lambda^2 = 0,\ \lambda^3 = 0 \qquad \delta a = e = e_{11}$$

$$\lambda^1 = 0,\ \lambda^2 = 1,\ \lambda^3 = 0 \qquad \delta a = e = e_{22}$$

$$\lambda^1 = 0,\ \lambda^2 = 0,\ \lambda^3 = 1 \qquad \delta a = e = e_{33}$$

20. $A^i = a^i + u^i_{,j}a^j$ gives

$$A^1 = \frac{2}{3}\epsilon + \frac{2}{3}\epsilon(y + x) \times (10)^{-2}$$

$$A^2 = \frac{2}{3}\epsilon + \frac{1}{3}\epsilon(2z + y) \times (10)^{-2}$$

$$A^3 = \frac{1}{3}\epsilon + \frac{1}{3}\epsilon(2z + x) \times (10)^{-2}$$

21.

$$(e_{ij}) = \begin{pmatrix} 2 & 5/2 & 4 \\ 5/2 & 1 & 9/2 \\ 4 & 9/2 & 2 \end{pmatrix} \qquad (\omega_{ij}) = \begin{pmatrix} 0 & 1/2 & -2 \\ -1/2 & 0 & 3/2 \\ 2 & -3/2 & 0 \end{pmatrix}$$

22. Convective operator $\vec{C} = (\vec{V}\cdot\nabla)\vec{A}$ or $C^i = V^m A^i_{,m} = V^m\left(\dfrac{\partial A^i}{\partial x^m} + \left\{\begin{matrix} i \\ m\,k \end{matrix}\right\}A^k\right)$ Use the results

$$\left\{\begin{matrix} j \\ i\,i \end{matrix}\right\} = -\frac{h_i}{h_j^2}\frac{\partial h_i}{\partial x^j}, \qquad \left\{\begin{matrix} i \\ i\,j \end{matrix}\right\} = \frac{1}{h_i}\frac{\partial h_i}{\partial x^j}, \qquad \left\{\begin{matrix} i \\ i\,i \end{matrix}\right\} = \frac{1}{h_i}\frac{\partial h_i}{\partial x^i} \quad \text{no summations}$$

and $\left\{\begin{matrix} i \\ j\,k \end{matrix}\right\} = 0$ for $i \neq j \neq k$. Expand and use physical components $V^r = \frac{V(r)}{h_r}$, etc.

23. In the ℓ−direction let ℓ_1, ℓ_2, ℓ_3 denote the changes due to the forces. We then have

$$\text{In the } \quad \ell - \text{direction} \quad -P = E\frac{\Delta\ell_1}{\ell}$$

$$\text{In the } \quad \omega - \text{direction} \quad \frac{\Delta\ell_2}{\ell} = \nu\frac{P}{E}$$

$$\text{In the } \quad \text{h-direction} \quad \frac{\Delta\ell_3}{\ell} = \frac{\nu P}{E}$$

$$\text{Hence} \quad \frac{\Delta\ell}{\ell} = \frac{\Delta\ell_1}{\ell} + \frac{\Delta\ell_2}{\ell} + \frac{\Delta\ell_3}{\ell} = \frac{-P}{E}(1 - 2\nu)$$

With similar results in the other directions.

27. (a) Hooke's law in y−direction $\sigma_{yy} = Ee_{yy}$ and in the x−direction we have $e_{xx} = -\nu e_{yy} = \dfrac{-\nu}{E}\sigma_{yy}$.

28. Another way is to use $\dfrac{\partial V_i}{\partial X^m} = \left(\dfrac{\partial V_i}{\partial x_1}\dfrac{\partial x_1}{\partial X_m} + \dfrac{\partial V_i}{\partial x_2}\dfrac{\partial x_2}{\partial X_m} + \dfrac{\partial V_i}{\partial x_3}\dfrac{\partial x_3}{\partial X_m}\right)$, then show terms like

$$e^{mnp}\frac{\partial V_1}{\partial x_2}\frac{\partial x_2}{\partial X_m}\frac{\partial x_2}{\partial X_n}\frac{\partial x_3}{\partial X_p} = 0, \text{ etc.}$$

EXERCISE 2.3

416

▶ **3.** From $\lambda = \dfrac{\nu E}{(1+\nu)(1-2\nu)}$ we have by algebra $2\nu^2 + \nu\left(1+\dfrac{E}{\lambda}\right) - 1 = 0$ so that

$$\nu = -\frac{E+\lambda}{4\lambda} \pm \sqrt{\left(\frac{E+\lambda}{4\lambda}\right)^2 - \frac{4(2)(-1)}{16}} = \frac{\sqrt{(E+\lambda)^2 + 8\lambda^2} - (E+\lambda)}{4\lambda}$$

Note ν must be positive. All of the results in problems 2,3,4,5,6 result using substitutions and basic algebra.

▶ **10.**

$$e_{xx} = \frac{\sigma_{xx}}{E} - \frac{\nu}{E}(\sigma_{yy} + \sigma_{zz}) \qquad e_{xy} = \frac{1+\nu}{E}\sigma_{xy}$$

$$e_{yy} = \frac{\sigma_{yy}}{E} - \frac{\nu}{E}(\sigma_{xx} + \sigma_{zz}) \qquad e_{xz} = \frac{1+\nu}{E}\sigma_{xz}$$

$$e_{zz} = \frac{\sigma_{zz}}{E} - \frac{\nu}{E}(\sigma_{xx} + \sigma_{yy}) \qquad e_{yz} = \frac{1+\nu}{E}\sigma_{yz}$$

$$\sigma_{xx} = \frac{E}{(1+\nu)(1-2\nu)}[(1-\nu)e_{xx} + \nu(e_{yy}+e_{zz})] \qquad \sigma_{xy} = \frac{E}{1+\nu}e_{xy}$$

$$\sigma_{yy} = \frac{E}{(1+\nu)(1-2\nu)}[(1-\nu)e_{yy} + \nu(e_{xx}+e_{zz})] \qquad \sigma_{xz} = \frac{E}{1+\nu}e_{xz}$$

$$\sigma_{zz} = \frac{E}{(1+\nu)(1-2\nu)}[(1-\nu)e_{zz} + \nu(e_{yy}+e_{xx})] \qquad \sigma_{yz} = \frac{E}{1+\nu}e_{yz}$$

▶ **11.**

$$e_{rr} = \frac{1}{E}[\sigma_{rr} - \nu(\sigma_{\theta\theta} + \sigma_{zz})] \qquad e_{r\theta} = \frac{1+\nu}{E}\sigma_{r\theta}$$

$$e_{\theta\theta} = \frac{1}{E}[\sigma_{\theta\theta} - \nu(\sigma_{rr} + \sigma_{zz})] \qquad e_{rz} = \frac{1+\nu}{E}\sigma_{rz}$$

$$e_{zz} = \frac{1}{E}[\sigma_{zz} - \nu(\sigma_{\theta\theta} + \sigma_{rr})] \qquad e_{\theta z} = \frac{1+\nu}{E}\sigma_{\theta z}$$

$$\sigma_{rr} = \frac{E}{(1+\nu)(1-2\nu)}[(1-\nu)e_{rr} + \nu(e_{\theta\theta}+e_{zz})] \qquad \sigma_{r\theta} = \frac{E}{1+\nu}e_{r\theta}$$

$$\sigma_{\theta\theta} = \frac{E}{(1+\nu)(1-2\nu)}[(1-\nu)e_{\theta\theta} + \nu(e_{rr}+e_{zz})] \qquad \sigma_{rz} = \frac{E}{1+\nu}e_{rz}$$

$$\sigma_{zz} = \frac{E}{(1+\nu)(1-2\nu)}[(1-\nu)e_{zz} + \nu(e_{\theta\theta}+e_{rr})] \qquad \sigma_{\theta z} = \frac{E}{1+\nu}e_{\theta z}$$

▶ **12.**

$$e_{\rho\rho} = \frac{\sigma_{\rho\rho}}{E} - \frac{\nu}{E}(\sigma_{\theta\theta} + \sigma_{\phi\phi}) \qquad e_{\rho\theta} = \frac{1+\nu}{E}\sigma_{\rho\theta}$$

$$e_{\theta\theta} = \frac{\sigma_{\theta\theta}}{E} - \frac{\nu}{E}(\sigma_{\rho\rho} + \sigma_{\phi\phi}) \qquad e_{\rho\phi} = \frac{1+\nu}{E}\sigma_{\rho\phi}$$

$$e_{\phi\phi} = \frac{\sigma_{\phi\phi}}{E} - \frac{\nu}{E}(\sigma_{\theta\theta} + \sigma_{\rho\rho}) \qquad e_{\theta\phi} = \frac{1+\nu}{E}\sigma_{\theta\phi}$$

▶ **16.** For state of plane stress

$$\frac{\partial^2 e_{xx}}{\partial y^2} + \frac{\partial^2 e_{yy}}{\partial x^2} = 2\frac{\partial^2 e_{xy}}{\partial x \partial y}$$

is the compatibility equation. Substitute into this equation

$$e_{xx} = \frac{1}{E}(\sigma_{xx} - \nu\sigma_{yy}) \qquad e_{yy} = \frac{1}{E}(\sigma_{yy} - \nu\sigma_{xx}) \qquad e_{xy} = \frac{1+\nu}{E}\sigma_{xy}$$

EXERCISE 2.4

Then differentiate the plane stress equations $\frac{\partial \sigma_{xx}}{\partial x} + \frac{\partial \sigma_{xy}}{\partial y} \varrho b_x = 0$, $\frac{\partial \sigma_{yx}}{\partial x} + \frac{\partial \sigma_{yy}}{\partial y} + \varrho b_y = 0$ after substituting $\varrho \vec{b} = -\operatorname{grad} V$. Simplify and express the results in terms of the Airy stress function to obtain $\nabla^4 \phi + (1 - \nu)\nabla^2 V = 0$

19.

$$\bar{\sigma}_{ij} = \sigma_{ab}\frac{\partial x^a}{\partial \bar{x}^i}\frac{\partial x^b}{\partial \bar{x}^j} \qquad \begin{aligned} \sigma_{rr} &= \sigma_{xx}\cos^2\theta + 2\sin\theta\cos\theta\sigma_{xy} + \sin^2\theta\sigma_{yy} \\ \sigma_{r\theta} &= -\sin\theta\cos\theta\sigma_{xx} + (\cos^2\theta - \sin^2\theta)\sigma_{xy} + \sin\theta\cos\theta\sigma_{yy} \\ \sigma_{\theta\theta} &= \sin^2\theta\sigma_{xx} - 2\sin\theta\cos\theta\sigma_{xy} + \cos^2\theta\sigma_{yy} \end{aligned}$$

► **20.** Use chain rule differentiation $\frac{\partial \phi}{\partial r} = \frac{\partial \phi}{\partial x}\frac{\partial x}{\partial r} + \frac{\partial \phi}{\partial y}\frac{\partial y}{\partial r}$

$$\frac{\partial^2 \phi}{\partial r^2} = \frac{\partial \phi}{\partial x}\frac{\partial^2 x}{\partial r^2} + \frac{\partial x}{\partial r}\left[\frac{\partial^2 \phi}{\partial x^2}\frac{\partial x}{\partial r} + \frac{\partial^2 \phi}{\partial x\partial y}\frac{\partial y}{\partial r}\right] + \frac{\partial \phi}{\partial y}\frac{\partial^2 y}{\partial r^2} + \frac{\partial y}{\partial r}\left[\frac{\partial^2 \phi}{\partial y\partial x}\frac{\partial x}{\partial r} + \frac{\partial^2 \phi}{\partial y^2}\frac{\partial y}{\partial r}\right]$$

with similar results for the other derivatives.

23. $e_{xx} = \frac{(1 - \nu^2)T}{E}$ $\qquad e_{yy} = -\frac{\nu(1 + \nu)T}{E}$ $\qquad e_{xy} = 0$

$\qquad u = u(x, y) = (1 - \nu^2)\frac{Tx}{E} + c_0 y + c_1$ $\qquad v = v(x, y) = -\nu(1 + \nu)\frac{Ty}{E} - c_0 x + c_2$.

► **26.** See EXAMPLE 2.4-7, let $R_1 \to 0$ and $P_1 \to 0$ to obtain special case.

36. u–equation is

$$\mu\left[\frac{\partial^2 u}{\partial x^2} + \frac{\partial^2 u}{\partial y^2} + \frac{\partial^2 u}{\partial z^2}\right] + (\lambda + \mu)\left[\frac{\partial^2 u}{\partial x^2} + \frac{\partial^2 v}{\partial y\partial x} + \frac{\partial^2 w}{\partial z\partial x}\right] + \varrho b_x = 0$$

► **37.** u_r–equation is
$$\mu\left(\nabla^2 u_r - \frac{1}{r^2}u_r - \frac{2}{r^2}\frac{\partial u_\theta}{\partial \theta}\right) + (\lambda + \mu)\left[\frac{\partial}{\partial r}\left(\frac{1}{r}\left[\frac{\partial}{\partial r}(ru_r) + \frac{\partial}{\partial \theta}(u_\theta) + \frac{\partial}{\partial z}(ru_z)\right]\right)\right] + \varrho b_r = 0$$

► **38.**

u_ρ–equation is $\quad \mu\left(\nabla^2 u_\rho - \frac{2}{\rho^2}u_\rho - \frac{2}{\rho^2}\frac{\partial u_\theta}{\partial \theta} - \frac{2\cot\theta}{\rho^2}u_\theta - \frac{2}{\rho^2\sin\theta}\frac{\partial u_\phi}{\partial \phi}\right)$

$+ (\lambda + \mu)\frac{\partial}{\partial \rho}\left(\frac{1}{\rho^2\sin\theta}\left[\frac{\partial}{\partial \rho}(\rho^2\sin\theta u_\rho) + \frac{\partial}{\partial \theta}(\rho\sin\theta u_\theta) + \frac{\partial}{\partial \phi}(\rho u_\phi)\right]\right) + \varrho b_\rho = 0$

► **43.**

$$\sigma^{ij} = \frac{1}{g}e^{im}e^{jn}\left[\frac{\partial u_m}{\partial x^n} - u_1\begin{Bmatrix}1\\mn\end{Bmatrix} - u_2\begin{Bmatrix}2\\mn\end{Bmatrix}\right] + g^{ij}V$$

Sum on m, then sum each term on n. In polar coordinates

$$g_{ij} = \begin{pmatrix}1 & 0\\0 & r^2\end{pmatrix} \qquad g^{ij} = \begin{pmatrix}1 & 0\\0 & \frac{1}{r^2}\end{pmatrix} \qquad g = r^2 \qquad \sigma^{ij} = \frac{\sigma(ij)}{h_i h_j}$$

$$\sigma^{11} = \frac{1}{r^2}\left[\frac{\partial u_2}{\partial x_2} - u_1(-r)\right] + V \quad \text{becomes} \quad \sigma_{rr} = V + \frac{1}{r}\frac{\partial \phi}{\partial r} + \frac{1}{r^2}\frac{\partial^2 \phi}{\partial \theta^2}$$

► **45.** $\sigma_{\rho\theta} = \sigma_{\rho\phi} = \sigma_{\theta\phi} = 0$

$$u_\rho = \frac{c_1}{3}\rho + \frac{c_2}{\rho^2} \qquad \sigma_{\rho\rho} = \lambda\left(\frac{du_\rho}{d\rho} + \frac{2}{\rho}u_\rho\right) + 2\mu\frac{du_\rho}{d\rho} \qquad \sigma_{\theta\theta} = \sigma_{\phi\phi} = \lambda\left(\frac{du_\rho}{d\rho} + \frac{2}{\rho}u_\rho\right) + 2\mu\frac{u_\rho}{\rho}$$

Boundary conditions $\sigma_{\rho\rho}|_{\rho=a} = P_i$, $\sigma_{\rho\rho}|_{\rho=b} = -P_o$ gives $c_2 = \frac{(P_i - P_o)a^3 b^3}{4\mu(b^3 - a^3)}$, $\frac{c_1}{3} = \frac{(a^3 P_i - b^3 P_o)}{(b^3 - a^3)(2\mu + 3\lambda)}$

EXERCISE 2.4

SELECTED SOLUTIONS EXERCISE 2.5

▶ **1.** (b) Use Taylor series expansion. $p(x + \Delta x, y, z) = p(x, y, z) + \dfrac{\partial p}{\partial x} \Delta x + \cdots$

▶ **2.** In $r-$direction use Taylor series

$$[p(r, \theta, z)\,\widehat{\mathbf{e}}_r - p(r + \Delta r, \theta, z)\,\widehat{\mathbf{e}}_r] r \Delta \theta \Delta z = -\dfrac{\partial p}{\partial r} r \Delta r \Delta \theta \Delta z \,\widehat{\mathbf{e}}_r$$

▶ **3.** In $\rho-$direction use Taylor series

$$[p(\rho, \theta, \phi)\,\widehat{\mathbf{e}}_\rho - p(\rho + \Delta \rho, \theta, \phi)\,\widehat{\mathbf{e}}_\rho] \rho^2 \sin \theta \Delta \theta \Delta \phi = -\dfrac{\partial p}{\partial \rho} \rho^2 \sin \theta \Delta \rho \Delta \theta \Delta \phi \,\widehat{\mathbf{e}}_\rho$$

▶ **8.** Let $c^i = \nabla \times (\nabla \times \vec{v})\,\widehat{\mathbf{e}}_i$ then

$$c^r = g^{ir} \epsilon_{ijk} g^{jm} (\epsilon^{kst} v_{t,s})_{,m}$$

$$\text{and} \quad g^{ir} g^{jm} g^{ka} \epsilon_{ijk} g_{ab} = \epsilon^{rma} g_{ab}$$

$$\text{hence} \quad c^i = \epsilon^{ist} g_{tj} \epsilon^{jkm} v_{m,ks}$$

$$c^i = \epsilon^{ist} g_{tj} g^{pj} g^{qk} g^{rm} v_{m,ks} \epsilon_{pqr}$$

$$c^i = \epsilon^{ist} \delta_t^p \epsilon_{pqr} g^{qk} g^{rm} v_{m,ks} = \epsilon^{ist} \epsilon_{tqr} g^{qk} v^r_{,ks}$$

$$c^i = (\delta_q^i \delta_r^s - \delta_r^i \delta_q^s) g^{qk} v^r_{,ks}$$

$$c^i = g^{ik} v^s_{,ks} - g^{sk} v^i_{,ks}$$

$$c^i = g^{ik} (v^s_{,s})_{,k} - g^{sk} v^i_{,ks} \quad \text{This is contravariant form of}$$

$$\nabla \times (\nabla \times \vec{v}) = \nabla (\nabla \cdot \vec{v}) - \nabla^2 \vec{v}$$

▶ **10.** See No. 17 Exercise 2.1

▶ **12.** See No. 19 Exercise 2.1

▶ **13.** (b) $u = u(y) = \dfrac{u_0}{\ell} y \quad v = w = 0$

▶ **14.** $S\dfrac{\partial \bar{\varrho}}{\partial t} + \bar{\nabla}(\bar{\varrho}\vec{v}) = 0$

▶ **15.** $\dfrac{\partial \vec{v}}{\partial t} + \dfrac{1}{2} \nabla (v^2) = \dfrac{-1}{\varrho} \nabla p - \nabla \phi$

▶ **17.** (b) $u = u(y) = -\dfrac{p_0}{2\nu \varrho} y^2 + \dfrac{p_0 \ell}{\nu \varrho} y \quad u_{max} = \dfrac{p_0 \ell^2}{2\nu \varrho}$

▶ **18.** $U = \dfrac{u}{A}, \quad \tau = \dfrac{t}{B}, \quad \xi = \dfrac{x}{L}$ then $\dfrac{\partial^2 U}{\partial \xi^2} = \dfrac{\partial U}{\partial \tau}$ with $B = \frac{\delta c L^2}{k}$ with boundary conditions $U(1, \tau) = 1$ if we let $A = u_1$, also $U(\xi, 0) = \frac{f(\xi L)}{u_1}$ and $U(0, \tau) = 0$

▶ **20.** Use Laplace transforms or separation of variables.

▶ **22.** See Exercise 2.1 No. 17.

▶ **25.** See equation (2.5.29).

▶ **28.** In terms of λ and μ we find $\sigma_{ij} = 2\mu e_{ij} + \lambda e_{mm} \delta_{ij} - (2\mu + 3\lambda)\alpha T \delta_{ij}$

EXERCISE 2.5

30.

$$\varrho\frac{\partial v_i}{\partial t} + \varrho v_j v_{i,j} = \varrho b_i - p_{,i} + \tau_{ij,j}$$

where $\quad \tau_{ij} = \lambda^* D_{kk}\delta_{ij} + 2\mu^* D_{ij} \quad D_{ij} = \frac{1}{2}(v_{i,j} + v_{j,i}) \quad D_{kk} = v_{k,k}$

$$\varrho\frac{\partial v_i}{\partial t} + \varrho v_j v_{i,j} = \varrho b_i - p_{,i} + [\lambda^* v_{j,j}]_{,i} + \eta v_{i,jj} + \eta_{,j}v_{i,j} + \eta v_{j,ij} + \eta_{,j}v_{j,i}$$

$$= \varrho b_i - p_{,i} + [\lambda^* v_{j,j}]_{,i} + 2\eta v_{j,ji} + \eta_{,i}v_{j,j} + \eta_{,n}v_{n,i} - v_{j,j}\eta_{,i}$$

$$- \eta v_{j,ij} + \eta v_{i,jj} + \eta_{,j}v_{i,j}$$

We add and subtract certain terms to change the form of the equation

$$\varrho\frac{\partial v_i}{\partial t} + \varrho v_j v_{i,j} = \varrho b_i - p_{,i} + [(\lambda^* + 2\eta)v_{j,j}]_{,i} + v_j\eta_{,ji} + v_{j,i}\eta_{,j} - v_i\eta_{,jj}$$

$$+ \eta_{,n}v_{n,i} - \eta_{,m}v_{i,m} - v_{j,j}\eta_{,i}$$

$$- \eta v_{j,ij} - \eta_{,j}v_{j,i} - \eta_{,i}v_{j,j} - \eta_{,ij}v_j + \eta v_{i,jj} + \eta_{,j}v_{i,j} + \eta_{,j}v_{i,j} + \eta_{,jj}v_i$$

$$= \varrho b_i - p_{,i} + \frac{4}{3}[\eta v_{j,j}]_{,i} + (v_j\eta_{,j})_{,i} - v_i\nabla^2\eta + \eta_{,n}v_{n,i} - \eta_{,m}v_{i,m}$$

$$- v_{j,j}\eta_{,j} - (\eta v_j)_{,ij} + (\eta v_i)_{,jj}$$

$$= \varrho b_i - p_{,i} + \frac{4}{3}[\eta v_{j,j}]_{,i} + (v_j\eta_{,j})_{,i} - v_i\nabla^2\eta$$

$$+ [\delta_{im}\delta_{jn} - \delta_{in}\delta_{jnm}]\eta_{,j}v_{n,m} - v_{j,j}\eta_{,i} - [\delta_{is}\delta_{jt} - \delta_{js}\delta_{it}](\eta v_t)_{,sj}$$

Use $\epsilon - \delta$ identity

$$\varrho\frac{\partial v_i}{\partial t} + \varrho v_j v_{i,j} = \varrho b_i - p_{,i} + \frac{4}{3}(\eta v_{j,j})_{,i} + (v_j\eta_{,j})_{,i}$$

$$- v_i\nabla^2\eta + \epsilon_{kij}\eta_{,j}\epsilon_{kmn}v_{n,m} - v_{j,j}\eta_{,i} - \epsilon_{ijk}(\epsilon_{kst}(\eta v_t)_{,s})_{,j}$$

$$\varrho\frac{D\vec{v}}{Dt} = \varrho\vec{b} - \nabla p + \frac{4}{3}\nabla(\eta\nabla\cdot\vec{v}) + \nabla(\vec{v}\cdot\nabla\eta) - \vec{v}\nabla^2\eta + (\nabla\eta)\times(\nabla\times\vec{v}) - (\nabla\cdot\vec{v})\nabla\eta - \nabla\times(\nabla\times(\eta\vec{v}))$$

34.

$$\Phi = (\tau_{ij}v_i)_{,j} - (\tau_{ij,j}v_i) = \tau_{ij}v_{i,j}$$

Interchange i and j

$$\Phi = \tau_{ji}v_{j,i}$$

add to obtain $\quad 2\Phi = \tau_{ij}(v_{i,j} + v_{j,i})$

or $\quad \Phi = \tau_{ij}D_{ij} = (\lambda^*\delta_{ij}D_{kk} + 2\mu^* D_{ij})D_{ij} = 2\mu^* D_{ij}D_{ij} + \lambda^*(D_{kk})^2$

36. Use conservation law with $\vec{J} = \vec{J}_{advection} + \vec{J}_{conduction}$ or $\vec{J} = T\vec{v} - k\nabla T$

SELECTED SOLUTIONS EXERCISE 2.6

► **1.** Field lines $\dfrac{x+a}{\sqrt{(x+a)^2+y^2}} + \dfrac{x-a}{\sqrt{(x-a)^2+y^2}} = c_1$

Equipotential curves $\dfrac{1}{\sqrt{(x+a)^2+y^2}} + \dfrac{1}{\sqrt{(x-a)^2+y^2}} = c_2$

► **8.** If $D_i = \epsilon_i^j E_j$, then $D_{i,k} = \epsilon_i^j E_{j,k} + \epsilon_{i,k}^j E_j$ For parallel vector field $E_{j,k} = 0$ and $D_{i,k} = 0$, hence $\epsilon_{i,k}^j E_j = 0$ which for arbitrary E_j requires that $\epsilon_{i,k}^j = 0$.

► **12.**

$$\nabla \cdot \vec{D} = \rho \qquad\qquad\qquad \nabla \cdot \vec{D} = 4\pi\rho$$

$$\nabla \cdot \vec{B} = 0 \qquad\qquad\qquad \nabla \cdot \vec{B} = 0$$

$$\nabla \times \vec{E} = -\frac{\partial \vec{B}}{\partial t} \quad \text{becomes} \quad \nabla \times \vec{E} = -\frac{1}{c}\frac{\partial \vec{B}}{\partial t}$$

$$\nabla \times \vec{H} = \vec{J} + \frac{\partial \vec{D}}{\partial t} \qquad\qquad \nabla \times \vec{H} = \frac{4\pi}{c}\vec{J} + \frac{1}{c}\frac{\partial \vec{D}}{\partial t}$$

► **15.**

(a) $\vec{E} = \dfrac{\vec{r}}{r^3}$ where $r = |\vec{r}| = \sqrt{x^2+y^2+z^2}$ Show $\nabla \times \vec{E} = \begin{vmatrix} \widehat{\mathbf{e}}_1 & \widehat{\mathbf{e}}_2 & \widehat{\mathbf{e}}_3 \\ \frac{\partial}{\partial x} & \frac{\partial}{\partial y} & \frac{\partial}{\partial z} \\ \frac{x}{r^3} & \frac{y}{r^3} & \frac{z}{r^3} \end{vmatrix} = 0$ hence $\vec{E}$ is irrotational.

(b) Solve the system of equations $\dfrac{\partial \mathcal{V}}{\partial x} = -\dfrac{x}{r^2}, \quad \dfrac{\partial \mathcal{V}}{\partial y} = -\dfrac{y}{r^2}, \quad \dfrac{\partial \mathcal{V}}{\partial z} = -\dfrac{z}{r^2}$ to obtain solution $\mathcal{V} = \ln\left(\dfrac{r_0}{r}\right)$.

► **17.** Show $\nabla \times \vec{A} = \int_0^1 s\nabla \times (\vec{B} \times \vec{r})\, ds = \int_0^1 s[\vec{B}\nabla \cdot \vec{r} - \vec{r}\nabla \cdot \vec{B} + (\vec{r} \cdot \nabla)\vec{B} - (\vec{B} \cdot \nabla)\vec{r}]\, ds$ Use the facts that $\nabla \vec{r} = 3, \nabla \cdot \vec{B} = 0$ and

$$(\vec{r} \cdot \nabla)\vec{B} = \left(x\frac{\partial}{\partial x} + y\frac{\partial}{\partial y} + z\frac{\partial}{\partial z} \right)\vec{B}$$

$$(\vec{B} \cdot \nabla)\vec{r} = \left(B_1\frac{\partial}{\partial x} + B_2\frac{\partial}{\partial y} + B_3\frac{\partial}{\partial z} \right)\vec{r}$$

and show $\nabla \times \vec{A} = \int_0^1 \left[s(2\vec{B}) + s^2\frac{d\vec{B}}{ds} \right] ds$ Then $\nabla \times \vec{A} = \int_0^1 2s\vec{B}\, ds + s^2\vec{B} \big|_0^1 - \int_0^1 2s\vec{B}\, ds = \vec{B}$

► **20.** Under electrostatic conditions $\nabla \times \vec{E} = 0, \nabla \cdot \vec{D} = \rho_f$ and $\vec{D} = \epsilon\vec{E}$. Hence

$$\vec{E} = -\nabla \mathcal{V} \quad \text{and} \quad \nabla \cdot \vec{E} = -\nabla^2 \mathcal{V} = \frac{1}{\epsilon}\nabla \cdot \vec{D} = \frac{\rho_f}{\epsilon}$$

► **22.** If $\vec{E} = -\nabla \mathcal{V}$, then $\nabla \times (\nabla \mathcal{V}) = 0$ so that Maxwell's equations fail. If $\vec{B} = \nabla \times \vec{A}$ we can write Faraday's law as $\nabla \times \vec{E} + \frac{\partial}{\partial t}\nabla \times \vec{A} = \nabla \times (\vec{E} + \frac{\partial \vec{A}}{\partial t}) = 0$ so that $\vec{E} + \dfrac{\partial \vec{A}}{\partial t}$ is conservative. Therefore $\vec{E} + \frac{\partial \vec{A}}{\partial t} = -\nabla \mathcal{V}$ for some potential $\mathcal{V}$.

SELECTED SOLUTIONS EXERCISE 2.7

▸ **1.** $q = 6 - 20i_1 + 10i_2 + 12i_3$

▸ **6.** $q = \dfrac{\alpha_0\beta_0 - (\alpha_1\beta_0 + \alpha_2\beta_3)i_1 + (\alpha_1\beta_3 - \alpha_2\beta_0)i_2 + \alpha_0\beta_3 i_3}{\alpha_0^2 + \alpha_1^2 + \alpha_2^2}$

▸ **10.** Let $q_1 = \cos\phi/2 + i_3\sin\phi/2$, $q_2 = \cos\theta/2 + i_1\sin\theta/2$ and $q_3 = \cos\psi/2 + i_3\sin\psi/2$ and let $\gamma = q_1 q_2 q_3 = \gamma_0 + \gamma_1 i_1 + \gamma_2 i_3 + \gamma_3 i_3$. Then solution is given by (2.7.50) with

$$\gamma_0 = \cos\psi/2\cos\theta/2\cos\phi/2 - \sin\phi/2\cos\theta/2\sin\psi/2$$

$$\gamma_1 = \cos\phi/2\sin\theta/2\cos\psi/2 + \sin\phi/2\sin\theta/2\sin\psi/2$$

$$\gamma_2 = \sin\phi/2\sin\theta/2\cos\psi/2 - \cos\phi/2\sin\theta/2\sin\psi/2$$

$$\gamma_3 = \sin\phi/2\cos\theta/2\cos\psi/2 + \sin\psi/2\cos\theta/2\cos\phi/2$$

This gives the transformation equations given in EXAMPLE 2.2-5.

▸ **14.** Volume= 2

EXERCISE 2.7

Printed in the United States
by Baker & Taylor Publisher Services